THYRISTORISED POWER CONTROLLERS

THYRISTORISED POWER CONTROLLERS

G.K. Dubey
S.R. Doradla
A. Joshi
R.M.K. Sinha
Department of Electrical Engineering
Indian Institute of Technology
Kanpur, India

JOHN WILEY & SONS
New York Chichester Brisbane Toronto Singapore

First published in 1986
Reprinted 1987
WILEY EASTERN LIMITED
4835/24 Ansari Road, Daryaganj
New Delhi 110 002, India.

Distributors:

Australia and New Zealand:
Jacaranda-Wiley Ltd., Jacaranda Press,
JOHN WILEY & SONS, INC.
GPO Box 859, Brisbane, Queensland 4001, Australia

Canada:
JOHN WILEY & SONS CANADA LIMITED
22 Worcester Road, Rexdale, Ontario, Canada

Europe and Africa:
JOHN WILEY & SONS LIMITED
Baffins Lane, Chichester, West Sussex, England

South East Asia:
JOHN WILEY & SONS, INC.
05–05 Block B, Union Industrial Building
37 Jalan Pemimpin, Singapore 2057

Africa and South Asia:
WILEY EASTERN LIMITED
4835/24 Ansari Road, Daryaganj
New Delhi 110002, India

North and South America and rest of the world:
JOHN WILEY & SONS, INC.
605, Third Avenue, New York, N.Y. 10158, USA.

Library of Congress Cataloging in Publication Data

Dubey, G. K.
Thyristorised power controllers.

1. Power electronics. 2. Thyristor control.
I. Title.
TK7881.15.D83 1985 621.31'7 85–9444

ISBN 0–470–20220–3 John Wiley & Sons, Inc.
ISBN 0–85226–190–X Wiley Eastern Limited

Printed in India at Gopsons Papers Pvt. Ltd., Noida, India.

Preface

The subject of Power Conversion and Control has grown continuously since the invention of the Thyristor in 1957. Thyristor Controllers are now standard modules in different industrial plants and electric traction as well. The rolling mills, paper mills, printing presses, machine tools, fertiliser plants, mining and excavation machinery, cranes, elevators, satellite tracking stations, computer and air craft power supplies incorporate Thyristor Controllers in different forms. Along with this growth in applications, a large body of knowledge relating to Thyristor Controllers appeared in the form of research papers, monographs, reports, reference books and text books.

The authors conducted a variety of courses in the general area of Thyristor Power Controllers and Drives. Some of these were short term courses for practising engineers from the Indian industries, mostly in the Steel sector. The present book was written initially as a set of notes for these engineers. This was later modified to provide a comprehensive coverage of basics and to incorporate some advanced topics. The authors are of the view that the study of Thyristor Controllers is not complete without adequate coverage of the firing and control circuits. Therefore, the chapters on the Linear and Digital ICs and Microprocessors have been incorporated in the book.

The contents of this book have been extensively tested by the authors in the undergraduate and graduate level electives offered at IIT Kanpur. The topics presented in Chapters 2, 3, 5 and 7 can be used for a first course in 'Power Electronics'. In order to fit these topics in a one semester course, ($\simeq$ 40 hours of lectures) sections 3.2.10, 3.3.6, 5.2 and 7.3 can be dropped without loss of continuity. A second course, probably at graduate level, may include chapters 4, 8 and some of the sections dropped earlier. A parallel course on Logic and Control Circuits can be offered, based on chapter 9, 10 and 11. A set of problems are included at the end of each chapter (except 1 and 9) to test the understanding of the reader. Extensive bibliography is given at the end of each chapter for readers requiring in-depth study of a topic. The authors hope that the present book will be useful to the students, the teachers at various engineering colleges and institutes and the practising engineers in the industries.

The authors gratefully acknowledge the financial support given by the curriculum improvement cell of the Quality Improvement Programme at Indian Institute of Technology, Kanpur in the preparation of the manuscript including the drawings.

The authors greatly appreciate the assistance of Mr. Omprakash Arora and Mr. D.N. Joshi in coordinating the activities related to the preparation of the manuscript.

Thanks are also due to the persons involved in typing (Mr. C.M. Abraham, Mr. J.S. Rawat, Mr. Yogendra Chandra, Mr. H.V.C. Srivastava and Mrs. Kamla Devi), drawing (Mr. J.C. Verma and Mr. O.P. Sharma), xeroxing (Mr. R.S. Shukla, Mr. R.K. Chaurasia, Mr. M.L. Gupta), cyclostyling and collating (Mr. Ram Autar, Mr. Triveni Tiwari, Mr. Ganga Ram) and electronic stencils cutting (Mr. G.D. Sonkar).

The authors thank Mrs. Shakti Dubey, Mrs. Subhadra Devi Doradla, and Mrs. Lily Sinha for their patience and support during the preparation of this book. Cooperation of other family members is gratefully acknowledged.

Many other individuals and institutions have helped us in many different ways during the preparation of this book and their help is gratefully acknowledged. Special thanks are due to our students who used the early versions of the type written manuscript in their courses and provided valuable feedback. Special thanks are also due to Mr. Y.V.Y.S. Murty for experimental work concerning μP controlled dc drive.

GOPAL KRISHNA DUBEY
SESHAGIRI RAO DORADLA
AVINASH JOSHI
RAI MAHESH KUMAR SINHA

April 2, 1986
IIT Kanpur

Contents

1. INTRODUCTION

Thyristorised power controllers are now widely used in the industry. Conventional controllers involving magnetic amplifiers, mercury arc rectifiers, rotating amplifiers, resistance controllers etc. have been replaced by thyristorised power controllers in almost all applications. The prominent applications of such controllers are as follows:

1. Control of a.c. and d.c. drives in rolling mills, paper and textile mills, traction vehicles, mine winders, cranes, exacavators, rotary kilns, ventilation fans, gas compressors etc.
2. Uninterruptible and standby power supplies for critical loads. Special power supplies for aircraft and space applications.
3. Power control in metallurgical and chemical processes using arc melting, induction heating and melting, resistance heating, arc welding, electrolysis etc.
4. Static power compensators, transformer tap changers and static contactors for industrial power systems.
5. Power conversion at the terminals of a HVDC transmission system.
6. HV supplies for electrostatic precipitators, and X-ray generators.
7. Illumination control for lighting in home, stages and theaters.
8. Solid state controllers for home appliances.

Some of the important advantages associated with thyristorised power controllers are as follows:

1. High efficiency due to low loss in the thyristor.
2. Long life and reduced maintenance due to absence of mechanical wear.
3. Compact size and absence of special foundations.
4. Flexibility in operation due to digital controls.
5. Faster dynamic response compared to electromechanical converters.
6. Lower acoustic noise compared to electromagnetic controllers, e.g. relays and contactors.

The thyristorised power controllers also suffer from some disadvantages. In order to use these controllers effectively one must be aware of these disadvantages. The main disadvantages are as follows.

1. All types of thyristorised power controllers invariably generate harmonics. These harmonics adversely affect the performance of the loads connected to them. For example, when supplying motor loads there are additional losses, torque harmonics and increase in acoustic noise.

The harmonics are also injected into the supply lines and thus adversely affect other loads connected to the line. In some applications, e.g. traction there is interference with the communication circuits due to supply line harmonics and electromagnetic radiation.

2. The thyristorised controllers for a.c. to d.c. or a.c. to a.c. power conversion can operate at low power factors under some conditions. Special measures are then required for correcting the line power factor.

3. The thyristorised controllers have no short time overloading capacity and therefore must be rated for maximum loading conditions. This may lead to an increase in the cost in some applications.

Since the advantages largely outweigh the disadvantages, thyristorised controllers are very widely used. They have replaced the conventional controllers in all prominent applications.

Figure 1.1 shows the block diagram of the system employing thyristorised power controller. The power flow between the source and the load, shown by solid line, is regulated by the controller.

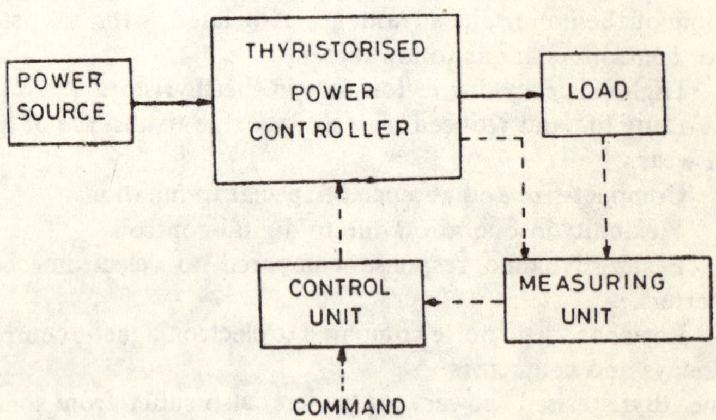

Fig. 1.1. Block diagram showing power flow (—) and control signal flow (- - - -) through a thyristorised power controller.

The controller also simultaneously conditions the power to suit the requirements of the load. For example, in a d.c. drive the controller receives the power from the a.c. mains. It converts the a.c. voltage to a d.c. voltage whose magnitude can be adjusted to control the speed of the d.c. motor driving a mechanical load. Another example is the a.c. drive using an induction motor. The controller receives power from a d.c. source and converts it into an a.c. voltage of variable frequency. The induction motor and the associated mechanical load can now be operated at different speeds by varying the frequency of the supply.

The thyristorised controllers can be classified into five broad categories.

1. Line Commutated Converters

These include diode rectifiers and single/three-phase controlled bridges. As the name suggests these circuits use the line voltage for the commutation of thyristors. These are also sometimes called naturally commutated converters. The power conversion in these circuits is from a fixed a.c. mains voltage to a variable d.c. voltage. Converters are widely used in d.c. drives, UPS, and HVDC systems.

2. Cycloconverters

These circuits convert power from a fixed a.c. mains voltage to a variable frequency and variable voltage. The cycloconverters are most commonly used for obtaining low frequency a.c. voltage. Two prominent applications of cycloconverters are traction vehicles and gearless rotary kilns.

3. A.C. Regulators

These circuits convert the constant frequency a.c. voltage to variable a.c. voltage at the same frequency using line commutation. A.C. regulators are mainly used for speed control of large fans and pumps.

4. Choppers

The choppers obtain power from a fixed voltage d.c. bus and convert it to a variable d.c. voltage. For this reason they are also called 'd.c. to d.c. converters'. In these circuits forced commutation is used to turn off thyristors. The choppers are further classified into several categories depending upon the direction of power flow and the type of commutation. The choppers are widely used in d.c. drives for suburban traction and switching power supplies.

5. Inverters

The inverters are used for power conversion from a fixed d.c. voltage to an a.c. voltage of variable frequency and fixed or variable magnitude. The inverters like choppers use forced commutation to turn off thyristors. The main application of inverters is in industrial a.c. drives using induction and synchronous motors.

The thyristor controllers mentioned above are discussed in detail in Chapters 3 to 7 respectively. The electrical characteristics of the thyristor itself are explained in Chapter 2. The efficient operation of an electrical system requires high power factor and low harmonic content. Chapter 8 describes in detail the methods for improvement of power factor, reduction of harmonics and control of reactive power.

The presence of harmonics in thyristorised systems gives rise to various measurement problems. Chapter 9 describes the instrumentation for thyristor systems. Furthermore, for monitoring the system or for closed loop operation (Fig. 1.1), different variables such as current, voltage, speed etc. are measured. The output from the measuring unit is processed by the control unit so as to realise the required control action. The various methods of sensing the above mentioned variables are also discussed in Chapter 9.

The controls for thyristorised controllers are built in the control unit which operates at a fairly low power level. The control unit consists of logic and firing circuits realised using discrete components like transistors or/and linear and digital integrated circuits. Chapter 10 deals in reasonable detail with linear and digital integrated circuits. Some basic firing circuits are given in Chapter 2. Specific firing circuits are given in Chapters 3 to 7.

Because of the flexibility available with microprocessors, recently they are being used in the control units particularly when sophisticated control strategies are required. Chapter 11 deals with the microprocessor and its application in the control of thyristorised controllers.

Thus all the major aspects related to the thyristorised converters and their control and firing circuits have been covered in this book.

2. INTRODUCTION TO THYRISTOR AND ITS CONTROL CIRCUITS

The term thyristor was earlier employed to denote a family of semiconductor devices consisting of silicon controlled rectifier (SCR), triac, diac etc. Out of these devices in thyristor family, SCR found much wider applications than any other device. Over the years the words thyristor and SCR have become synonymous. Now a days the term thyristor is more widely used for SCR rather than SCR itself. Here also term thyristor will be employed.

2.1 THYRISTOR AND ITS TERMINAL CHARACTERISTICS

Thyristor is a three terminal, four layer (*p–n–p–n*) semiconductor device. Details of thyristor construction are given in Fig. 2.1. Its diagrammatical representation and circuit symbol are shown in

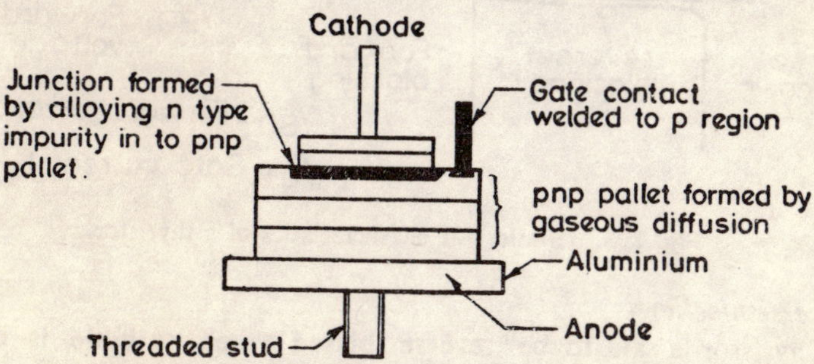

Fig. 2.1 Details of thyristor construction.

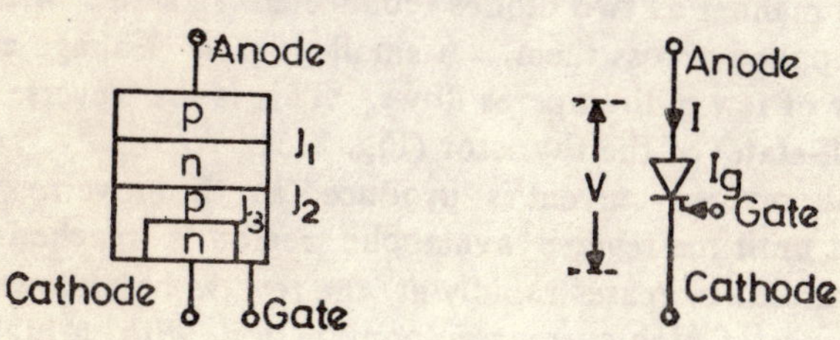

Fig. 2.2. The structure and electric circuit symbol of a thyristor.

Fig. 2.2. The terminal connected to end *p* region is called anode, the terminal connected to end *n* region is called cathode and the terminal connected to *p* region in the centre is called gate. In power thyristors, the anode (Fig. 2.1) is usually mounted onto a heat sink stud or is threaded to permit direct mounting onto an external heat-sink.

2.1.1 Modes of Operation

Thyristor has three basic modes of operation: reverse blocking, forward blocking and forward conduction. The static V–I characteristic of a thyristor is shown in Fig. 2.3.

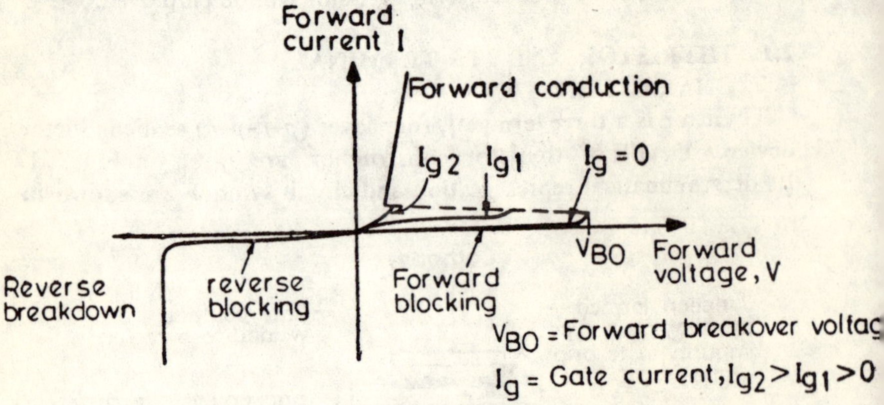

Fig. 2.3. Static V–I characteristics of a thyristor.

Reverse blocking

Thyristor is said to be reverse biased when cathode is made positive with respect to anode. Then junctions J_1 and J_3 are reverse biased, while junction J_2 is forward biased. The device behaves in the same manner as two diodes connected in series with reverse voltage applied across them. A small reverse leakage current of the order of few milliamperes flows. This is the reverse blocking mode (off-state) of the thyristor (Fig. 2.3).

A small reverse current is produced as the reverse voltage is increased until the reverse avalanche region is reached when the reverse current increases rapidly at the reverse breakdown voltage. The presence of high current in conjunction with a high voltage gives rise to large dissipation. This can damage the thyristor due to junction temperature crossing the permissible value. Therefore, when used in any circuit care is taken to ensure that the maximum reverse voltage across a thyristor does not exceed its breakdown voltage.

When a thyristor is reverse biased, it is advisable not to apply a positive signal between gate and cathode because of the following reasons. A thyristor is designed such that most of the reverse voltage drops across junction J_1. With a positive signal between gate and cathode junction J_3 becomes forward biased. The forward biased junction J_2 injects holes into reverse biased junction J_1. One effect of this is to increase the reverse leakage current. Increase of reverse current causes increased junction heating which releases more carriers to increase the reverse current still further. Therefore application of a positive gate voltage to a reverse biased thyristor should be strictly limited in order to prevent failure of a junction by thermal runaway. The maximum gate voltage during conditions of reverse bias is often specified by the manufacturers so that the heating effect is limited.

Forward blocking

With the anode potential maintained positive with respect to the cathode, junctions J_1 and J_3 are forward biased and junction J_2 is reverse biased. The reverse bias on junction J_2 limits forward current to the value of forward leakage current, of order up to several milliamps (Fig. 2.3). The thyristor is then said to be in the forward blocking state (off state). Application of a small positive gate current causes slight increase of forward leakage current.

Forward conduction

With forward bias voltage, a thyristor can be made to conduct (on state) by any of the four techniques

(a) Exceeding forward breakover voltage
(b) Gate triggering
(c) *dv/dt* turn on
(d) Irradiation of gate-cathode junction.

(a) *Exceeding forward breakover voltage*: Any increase of the anode voltage in the absence of gate signal causes avalanche breakdown of junction J_2 at a voltage called forward breakover voltage. At this voltage the thyristor changes its characteristic of having a high voltage across it with a low leakage current (off state) to a low voltage across it with a large forward current whose value is limited mainly by the load impedance. This is the on state of a thyristor. The forward voltage drop during the on state is of the order of 1 to 1.5 volts and increases slightly with the current. The forward breakover voltage is usually greater than the reverse breakover voltage.

If the anode circuit voltage is now reduced, since the reverse biased junction J_2 no longer exists due to the movement of carriers, thyristor will continue to stay on. Only when the forward current falls below a certain specific low value thyristor will go to blocking state. The value of current below which thyristor goes to the blocking state is called *holding* or sustaining current.

(b) *Gate triggering*; This is the most common method of triggering a thyristor. A thyristor having breakover voltage higher than the normal forward voltage that may appear across it is used. The thyristor, therefore, remains in the forward blocking state in the absence of a gate signal. When the thyristor turn on is desired a suitable positive voltage is applied at the gate to cathode terminals. The layer at the gate is flooded by electrons from the cathode and it looses its identity as a layer. Accordingly the thyristor becomes equivalent to a conducting diode. As shown in Fig. 2.3, a positive gate to cathode current reduces the forward breakover voltage. For a given forward voltage there is a minimum gate current at which a thyristor turns on. This gate current may vary from few milliamps to several milliamps.

Once the thyristor is on, the gate current is no longer required for the device to remain in conduction, so a gate pulse is enough. However, if the gate current is reduced to zero before the rising anode current reaches a value, called the *latching current*, the thyristor will turn off again. The gate pulse width should be chosen to ensure that the anode current exceeds latching current before it is removed.

Once the latching current is exceeded gate terminal loses control. Thyristor cannot be turned off (or the forward blocking state cannot be regained) unless the gate signal is removed and the forward current falls below the holding current value. The latching current is higher than the holding current and can be assumed to be three times the holding current when not specified in thyristor data sheet [1].

(c) *dv/dt turn on*: A thyristor may be triggered on by a rapid rate of increase of forward anode to cathode voltage, in the presence or absence of a gate signal. This, so called *dv/dt* effect, can be explained in terms of anode to gate and gate to cathode capacitances. A rapidly rising anode voltage produces transient gate current sufficient to turn on the thyristor.

This type of turn-on is not used. An unwanted turn-on may however occur due to spurious spikes of voltage emanating from other equipment. It is avoided by ensuring that *dv/dt* across the thyristor is less than the value at which turn on may occur. Typical *dv/dt* ratings to avoid spurious turn on are in the range of *10 to 200 volts per microseconds. dv/dt* value for turn on also depends on the immediate past state of the thyristor and, in particularly, on how recently it was conducting forward current. Its value can be increased by using a low external gate to cathode resistor.

(d) *Irradiation of gate-cathode junction*: A beam of light directed at the gate to cathode junction can turn on thyristor.

2.1.2 Dynamic Characteristics

2.1.2.1 *Dynamic Characteristics During Turn On*

During the initial part of turn-on only a small area near the gate terminal conducts anode current. If the anode current is allowed to increase at a fast rate a large current will flow through a comparatively narrow channel resulting in local heating which may damage the thyristor. Hence a maximum *di/dt* (rate of change of current) is specified for a given thyristor typically in the range of 20–300A/μs. Fast rising gate currents can be used to increase *di/dt* withstand capability of a thyristor. Rate of current rise can also be restricted by using an inductance in series. This inductance also reduces the turn on and turn off losses. Presence of inductance, however will also reduce the rate of rise of current even after conduction has spread across the junction and may give rise to large reverse voltage transients which may damage the thyristor. Former drawback can be overcome by using a saturable reactor which is made to saturate once the thyristor is fully conducting.

The nature of the dynamic characteristics of a thyristor during turn on is shown in Fig. 2.4. Turn on time is defined as the duration between the initiation of triggering, when the thyristor is in the forward blocking state, and the instant when steady state forward voltage drop is reached. Total turn on time is divided into three intervals, delay time (t_d), rise time (t_r) and spread time (t_p).

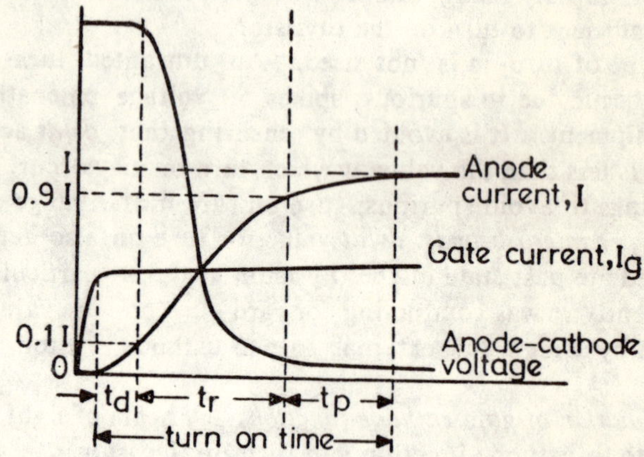

Fig. 2.4. Variation of current and voltage during turn-on.

The delay time is the duration between the front of the gate pulse and the beginning of rapid rate of increase of anode current. During this period current flows only in a narrow region near the gate. During the rise time t_r, thyristor power dissipation is the highest due to high voltage and current occurring together. Furthermore, this dissipation occurs only over a comparatively narrow region of the junction. The spread time t_p is the time taken for conduction to spread over the entire cross section of the thyristor. When the above waveforms are for resistive loads sum of t_d, t_r and t_p gives thyristor turn on time.

Time required for the device to turn on is of the order of *1 to 3 microseconds*, provided the load inductance does not limit the build up of anode current. This time, called the *turn-on time*, can be increased or decreased somewhat by modifying the gate current and voltage waveshapes. On a short time basis, thyristors may be generally considered to be charge controlled [1]. Thyristor conduction begins after a certain charge is injected into the gate region. Higher the magnitude of gate current, the less time it takes to inject this charge. The turn on time can therefore, be reduced by using higher values of gate currents. The best shape of a gate pulse is one with a sharp leading edge, and shorter its duration greater must be its magnitude for reliable turn on.

2.1.2.2 *Dynamic Characteristic During Turn Off*

Thyristor turn off means that the thyristor has regained forward voltage blocking capability after forward conduction. Process of turn off is often called commutation. Various methods used for turning off thyristors will be discussed in Sec. 2.4. It can however be noted at this stage that a thyristor can be turned off by reducing the forward current to zero and maintaining a reverse bias at least for a specified period.

Dynamic characteristics of thyristors during turn off by forced commutation are shown in Fig. 2.5. The turn-off time t_q is defined as the time required for thyristor to regain its forward voltage

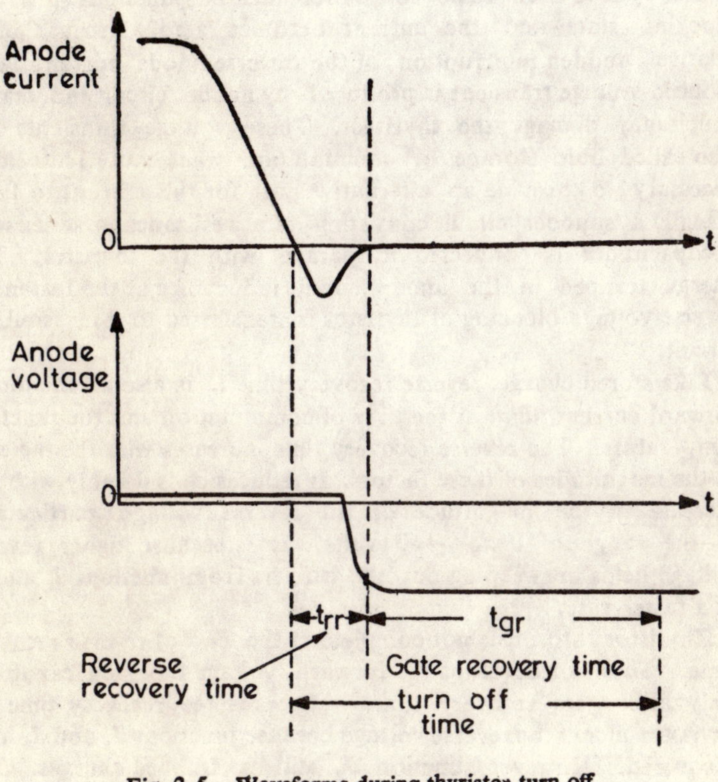

Fig. 2.5. Waveforms during thyristor turn off.

blocking capability after reduction of anode current to zero. The turn-off time consists of two parts: reverse recovery time t_{rr} and gate recovery time t_{gr}.

Because of the stored charge the thyristor continues to conduct and acts as a short circuit, with a small positive voltage across it, even after its current has reversed. The stored charge depends upon the peak forward current, *di/dt* at the time of commutation and the temperature of the junction, its value increases with the increase of these factors. Till such a time when about sixty per cent of stored charge gets recovered, the thyristor acts as a short circuit and the reverse current reaches a peak value with *di/dt* determined by external circuit. Beyond this point the thyristor develops reverse blocking capability. The reverse voltage depends on the thyristor decay current and external circuit impedance [2]. The reverse current must flow until most of the carriers of junctions J_1 and J_3 have been removed. After this the junctions revert to blocking state and the current becomes zero. Because of a relatively sudden interruption of the reverse anode current, large induced voltage transient is produced by anode circuit inductance which may damage the thyristor. These voltage transients are also called hole storage or commutation transients. Thus it is necessary to provide an alternative path for this current to flow. Usually a 'snubber' circuit consisting of a resistance in series with a capacitance is connected in parallel with the thyristor. The energy trapped in the anode circuit inductance at the instant of reverse voltage blocking of thyristor is transferred to the snubber circuit.

Like stored charge, reverse recovery time t_{rr} is also a function of forward current, *di/dt* at the time of commutation and the junction temperature. The reverse recovery time increases with the increase in the magnitudes of these factors. It reduces considerably with the increase in the magnitude of the reverse voltage, particularly in the range of 0 to —60 volts. It is because higher reverse voltage helps in sweeping out the carriers from junctions J_1 and J_3 at a faster rate.

Thyristor turn off is not complete at the end of reverse recovery time. Thyristor develops the forward voltage blocking capability only after gate recovery time. After reverse recovery time the thyristor blocks the reverse voltage because junctions J_1 and J_3 have recovered. However junction J_2 still has trapped charges. The thyristor blocks the forward voltage only when the excess carriers at junction J_2 have recombined. Therefore, reverse voltage must be maintained across the thyristor during the gate recovery time also, though the magnitude of this voltage is not too critical.

The duration for which a commutation circuit is able to provide reverse voltage across a thyristor after forward current has dropped to zero is called *circuit turn off time* t_c and t_c is chosen to be greater than thyristor turn off time t_q for a reliable turn off. If immediately after turn-off, forward voltage is likely to appear across the thyristor it is advisable to ensure that the forward dv/dt is less than the specified limit in order to avoid spurious turn on.

Turn off times of thyristors vary from 5 to 200 microseconds. Thyristors with fast turn off time ($t_q \leqslant 25$ microseconds) are called inverter grade thyristors and the thyristors with slower turn off are called converter grade thyristors. Inverter grade thyristors are employed in inverters, choppers, forced commutated converters and in other applications requiring fast turn off. Converter grade thyristors are cheaper and are used in applications where slower turn off in acceptable such as rectifiers a.c. regulators, cycloconverters etc.

2.1.3 Thyristor Gate Characteristics

The dc gate trigger characteristics of a thyristor are presented in the form of a graph similar to one shown in Fig. 2.6. The graph shows gate to cathode voltage V_g as a function of positive gate current I_g. The preferred gate drive area is abcdfgna.

The V_g and I_g characteristic varies not only for different types of thyristors but also for the thyristors of the same type. Curves (1) and (2) correspond to possible spread of the characteristics of thyristors of the same type. There are minimum values of gate voltage and current for reliable firing. The odd shape near the origin is to take into account the higher gate power needed at low junction temperatures. There are also maximum limits for gate voltage (V_{gm}), gate current (I_{gm}) and rated (average) gate power dissipation (P_g). A signal in excess of these limits will cause damage. Usually a non-triggering gate voltage is also prescribed for worst case, i.e. *at peak allowable junction temperature and peak forward anode voltage*. If firing circuit generates positive gate bias prior to triggering, care should be taken to ensure that it is less than non triggering gate voltage. Similarly all unwanted

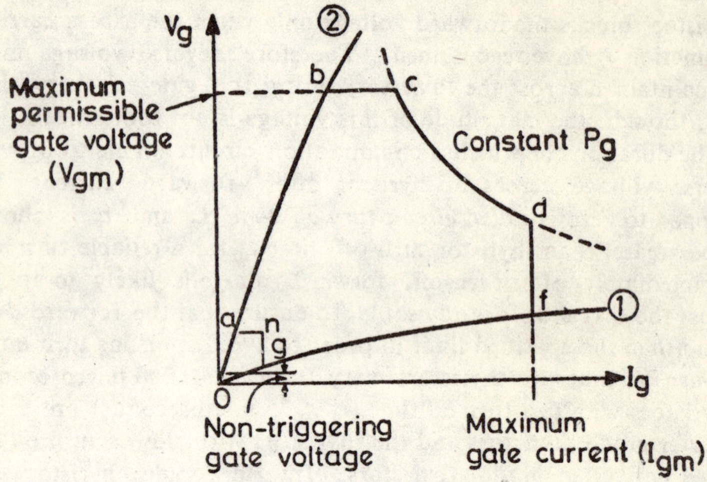

Fig. 2.6. Gate trigger characteristics.

and spurious gate signals (noise) should be below this value.

The load line of the gate supply must pass above and to the right of the firing point of the least sensitive device at the lowest temperature, as shown by the chain dotted line in Fig. 2.7. The load line

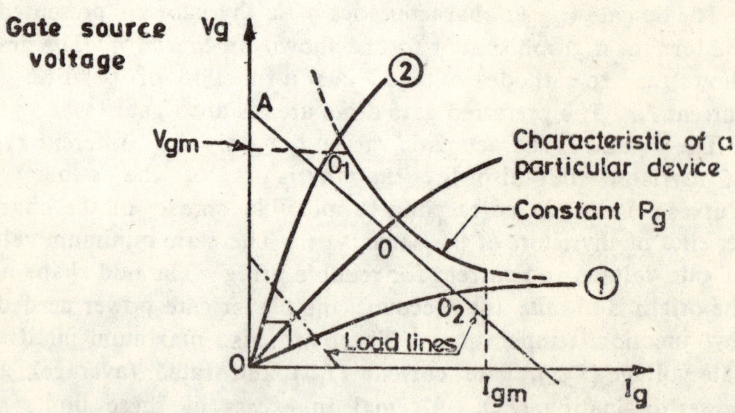

Fig. 2.7. Choice of gate trigger voltage and current.

AO_1O_2, which may give the operating point any where on the line O_1O_2, depending on the characteristics of a particular device, will be the most preferred one if the operating point lies as close as possible to P_g curve without exceeding the maximum allowable voltage and current.

Graph of Fig. 2.7 provides the gate drive requirements in terms of the continuous dc gate voltage and current. But usually a pulse or a train of pulses is used to trigger a thyristor. For pulse widths down to 100 microseconds, the dc data apply [1]. For short pulse widths gate voltage and current should be increased.

As stated earlier, on a short time basis thyristors may be considered to be charge controlled. Higher the magnitude of the gate current pulse, the less time it takes to inject the charge necessary for conduction to start. Turn on time can therefore be reduced by using a higher magnitude for the gate pulse. Gate current rise time should be of the order of several amperes per microsecond to minimise the turn on time, particularly when switching into high currents. This in turn results in minimum turn-on anode switching dissipation and minimum jitter [1]. The duration of the gate pulse should be long enough for anode current to reach the latching value. Trigger pulse strength is limited by a maximum allowable instantaneous gate power (P_{gm}) which is also prescribed. When a train of pulses is used, the peak allowable gate drive power is calculated from the following equation [1]:

$$(\text{Peak gate drive power} \times \text{pulse width}) \times (\text{pulse repetition rate}) \leqslant (\text{allowable average gate power})$$

There is also a maximum permissible value of negative voltage that could be applied across gate to cathode terminals. If the firing circuit is such that there is a possibility of gate becoming more negative than this limit, a diode is connected either in series with the gate, or from cathode to gate to limit the reverse gate voltage. Once the device is conducting, the gate cathode junction presents a voltage source to the gate circuit, with its magnitude approaching anode-cathode voltage. If due to some reason gate source voltage reverses or becomes zero, the internal gate-cathode voltage may supply a reverse gate current. This reverse gate current causes a section close to gate terminal to stop conducting. The anode current now flows through a small cross sectional area remote from the gate. This results in higher current density and excessive power dissipation in the section which is conducting and it may be damaged. The gate circuit must be designed to account for this.

During the forward blocking interval, negative gate bias has certain advantages. It increases the forward breakover voltage and *dv/dt* withstand capability.

2.2 RATINGS AND PROTECTION

The reliable operation of a thyristor can be ensured only when the device is operated within its ratings. The thyristor like any other semiconductor device has a limited capability with regard to on-state current or off-state voltage. The manufacturer after thoroughly testing the device prepares the specifications showing a comprehensive list of ratings under specified temperature conditions that the device can withstand during on or off state without causing any damage to the device. The objective of this section is to introduce the terminology and the importance of different ratings of a thyristor. For satisfactory operation and continuity in service, the thyristor must be safeguarded against unusual conditions. Therefore, some measures for protecting the device under abnormal operating conditions are also outlined in this section.

2.2.1 Current Ratings

The thyristor being made up of semiconductor material having a small thermal capacity is extremely sensitive when the current flowing through it larger than the rated value. The device junction temperature rises rapidly even for short overloads of current. Heat is produced within the device owing to the power dissipated by the device. The power generated in the junction region of a thyristor in normal operation consists of the following six components of dissipation.

1. Forward coduction
2. Turn-on switching
3. Turn-off or commutation
4. Forward blocking
5. Reverse blocking
6. Gate pulse triggering

In the line commutated converters and AC regulator circuits, the forward conduction loss is the major source of junction heating. In high frequency inverters, and in converters where the thyristors are switched on and off several times in each supply cycle, the turn on and turn off losses may also have to be taken into consideration while selecting the device ratings since the switching loss may constitute a significant portion of the total loss. For satisfactory operation of a thyristor the junction temperature under steady state conditions should not exceed the safe value which is usually 125°C. This is achieved by transmitting the heat produced at the junction to the ambient via the body and cooling medium. The device mountings and different methods of cooling are presented in Section 2.3. In what follows the device current ratings and other ratings that are related to the device current are explained. These ratings are established with both repetitive and non-repetitive waveforms of current.

2.2.1.1 *Average Current Rating (I_T)*

The power loss in a thyristor depends upon the average current flowing through the device. Fig. 2.8 illustrates the variation of the

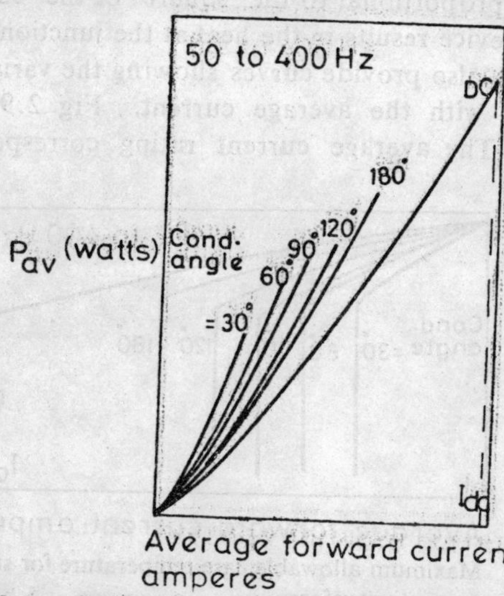

Fig. 2.8. Average forward power dissipation for sinusoidal current waveform.

average power loss with the average current of a thyristor for different conduction angles. This data is normally supplied by the manufacturer of thyristors. These characteristics are valid for the frequency range 50 Hz to 400 Hz. The curve marked DC is applicable for continuous direct current. The maximum continuous current rating without exceeding the maximum specified junction temperature is obtained corresponding to the terminal point of the curve marked DC. The curves also show the average current rating of a phase controlled thyristor resulting in different conduction angles. If the same average current is permitted for different conduction angles, the instantaneous current increases as the conduction angle decreases. This may result in an increased voltage drop across the device, and the average power dissipation may increase raising the junction temperature beyond the safe allowable limit. The permissible average current should therefore decrease with a decrease in the conduction angle. It may be noted in Fig. 2.8 that the curves for different conduction angles have been terminated at different values of the average current. At these terminal points, the maximum rms current rating of the device is reached. Since the device has resistive joints comprising of leads and metal contacts the device is also specified by the rms current rating as the heating in these resistive joints is proportional to the square of the current. The power loss in a device results in the heat at the junction. Therefore, the manufacturers also provide curves showing the variation of the case temperature with the average current. Fig. 2.9 illustrates such a relation. The average current rating corresponding to a

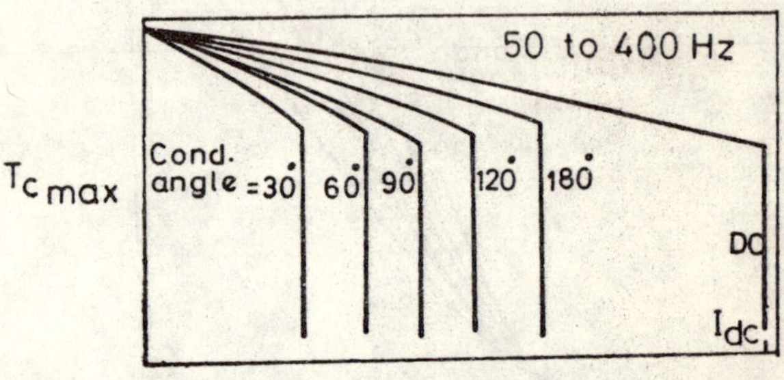

Fig. 2.9. Maximum allowable case temperature for sinusoidal current waveform.

particular case temperature is assured provided the case temperature is held constant. The heat produced in the body of the device is transmitted to the ambient, through an appropriate cooling method. By reducing the temperature of the body, the average current rating can be improved.

In these curves, it is assumed that the thyristors are switched into a purely resistive load. If there is an inductance in the load circuit, these curves need to be modified [1]. As the form factor of the current waveform decreases, a higher average current is permitted.

The switching loss dues to turn on and turn off may be significant for thyristors used in high frequency inverter and chopper circuits. In such cases, the switching loss must be included in the curves showing the relation between the average power loss and the average current. These curves are generally supplied by the manufacturers.

2.2.1.2 *RMS Current Rating* (I_{TR})

In order to prevent excessive heating in the resistive elements of a thyristor such as metallic joints, leads and interfaces, the rms current is given in the specification sheet of a thyristor by each manufacturer. Obviously this is the same as DC with continuous conduction as may be noted by the terminal point of the DC curve, say I_{dc} in Fig. 2.8. Similarly, the rms current for different conduction angles remains the same. However, the average current is limited depending upon the value of the form factor of the current waveshape. With 180° conduction, the ratio of the rms current to the average current (form factor of half-sine wave) is $\pi/2$. The average current is therefore limited to $I_{dc} \times 2/\pi$. In a similar manner, the average currents for other conduction angles can be calculated. It is assumed that the forward conduction loss is the major loss. The other components of power dissipation are neglected.

The anode currents of square wave shape with different duty cycles are also common in the inverter and chopper circuits. The specification sheets often take switching losses into account and provide curves relating power dissipation versus average current. These curves must be referred while selecting thyristors for such applications.

The current ratings I_T and I_{TR} are of repetitive type. They have been specified at the maximum junction temperature and the ratings vary with the case temperature. In addition to these ratings. there are non-recurrent or non-repetitive ratings. With these ratings, the junction temperature is permitted to exceed the maximum allowable limit provided they last for a short interval. The surge current which is non-repetitive is explained in the following section.

2.2.1.3 *Surge Current Rating* (I_{TSM})

This rating is specified to accommodate unusual conditions such as faults and short circuits resulting in a high current. These fault conditions will not occur more than a limited number of times during the life cycle of the equipment in which the thyristor is employed. In case the fault is repeated, the thyristor is guaranteed to withstand all other ratings provided the fault occurs after the junction temperature has decreased to the maximum specified limit or less.

The surge current is assumed to be sinusoidal with the supply frequency of 50 or 60 Hz. The one-cycle surge current rating is the peak amplitude of an allowable non-recurrent half-sine wave of 10 msec (50 Hz) duration. The specification sheet gives the number of surge cycles versus surge current rating which is the peak amplitude of a half-sine wave. The surge current rating decreases with an increase in the number of cycles. A sub-cycle surge current rating for the thyristor is also specified with a view to protect the device by the fast acting fuse. The sub-cycle surge current rating which is the peak of a sine wave at 50 Hz is obtained by equating the energy of one cycle surge and the sub-cycle surge as given by the following expression

$$I_{\text{subcycle}} \simeq I\sqrt{\frac{1}{100t}}, \tag{2.1}$$

where I = one cycle surge current rating and t = duration of sub-cycle surge.

2.2.1.4 *I^2t Rating*

This rating is required to coordinate with the fast acting fuse for protecting the device during an overload or fault conditions. The rating is a measure of the thermal energy that the device can absorb for a short time before the fault is cleared by the fuse. It is usually specified for fault currents lasting for less than or equal to one half-period at the supply frequency of 50 or 60 Hz.

2.2.1.5 *di/dt rating*

If the rate of rise of anode current is very rapid compared to the spreading velocity of carriers across the junctions during the turn-on period, the local 'hot-spot' heating occurs due to high current density in the junction regions. This increases the junction temperature beyond the safe limit and the device may be damaged. Therefore, the limit of the *di/dt* at turn on is specified for all thyristors usually in Amps/μsec. Typical values lie in the range 50 to 800 A/μsec.

2.2.2 Voltage Ratings

The structure of the thyristor showing the three junctions is explained in Section 2.1. The thyristor blocks the forward voltage and the reverse voltage by the middle and end semiconductor *p-n* junctions respectively. The device voltage rating is a measure of the maximum voltage that can be applied across the device without causing a breakdown in the junction area. The different ratings related to the voltage are discussed in this section.

2.2.2.1 *Peak Repetitive Forward Blocking Voltage* (V_{DRM})

This is the peak voltage which the thyristor can block in the forward direction. It is specified at the maximum allowable junction temperature with gate circuit open or with a specified biasing resistance between gate and cathode terminals.

2.2.2.2 *Peak Repetitive Reverse Voltage* (V_{RRM})

This is the peak reverse voltage which the thyristor can withstand without breakdown at the maximum allowable junction temperature. If this rating is substantially exceeded, the device may be destroyed by junction breakdown.

2.2.2.3 *Non-repetitive Peak Reverse Voltage* (V_{RSM})

This is the maximum transient reverse voltage which can be safely blocked by the thyristor. The transient reverse voltage rating can be increased by inserting a diode of equal current rating in series with the thyristor.

2.2.2.4 *Forward dv/dt Rating*

If the rate of rise of forward voltage is higher than the specified maximum value, it may cause switching from the off-state to the on-state. Since gate pulse triggering is commonly used for the turn on of a thyristor, the *dv/dt* causes unscheduled turn on of a thyristor. This method of switching is to be avoided as this may lead to the destruction of the thyristor through high local current density. For this reason the *dv/dt* rating is specified either by a linear or exponential waveform. In the latter case, the *dv/dt* is the slope of the line through the origin intersecting the curve, thyristor voltage versus time with a single time constant.

2.2.2.5 *Other Ratings*

In addition to the ratings discussed in Sections 2.2.1 and 2.2.2 there are some more ratings as given below:

(a) Turn-on time
(b) Turn-off time
(c) Gate circuit voltage and current
(d) Latching and holding currents
(e) Reverse recovered charge characteristic.

The ratings (a) to (d) are explained in Section 2.1. The reverse recovery charge characteristic is required while designing the *dv/dt* protection (snubber) circuit for the thyristor. During the process of turning off, the current flows through the thyristor in the reverse direction for a very short interval due to the stored charge in the junctions. The amount of recovered charge during the negative current flow depends upon the magnitude of the forward current and the rate at which it is reduced.

Detailed specifications during on state and off state can be obtained from the manufacturer by quoting the appropriate specification sheet number of the desired thyristor.

2.2.3 Thyristor Protection

The different ratings of a thyristor are discussed in Sections 2.2.1 and 2.2.2. If the ratings are exceeded due to overload or other abnormal operating conditions, the device may be destroyed permanently. For satisfactory and reliable operation of a thyristor circuit and equipment, the devices should be protected against abnormal conditions. Usually some safety margin is provided in the equipment by selecting devices with ratings larger than those adequate for normal operation. It is not economical to select the device rating considering the abnormal operating conditions on a steady state basis. Under such unusual transient conditions, the devices are to be protected. Some of the protection techniques against overcurrent, overvoltage etc. are discussed in this section.

2.2.3.1 *Transient Overvoltage Protection*

Overvoltage is one of the single greatest causes of failure of thyristors, since many electrical systems produce transient voltage peak well in excess of the supply peak voltage. In the absence of any protective network, transient peaks that are five to ten times the recurrent peak voltage are not uncommon in conventional circuits. The transient voltages often lead to either malfunction of circuit by the unscheduled turn on of a thyristor or permanent destruction of a device due to reverse breakdown. The overvoltages may be of internal type or external type. The former arise from the semiconductor device operation while the latter come from the supply mains or the load source.

(a) INTERNAL OVERVOLTAGES

During the commutation of a thyristor, the device continues to conduct a reverse current after the forward current has decreased to zero due to the earlier stored charge. At the end of the reverse recovery interval, the device interrupts. Since the interruption takes place in a very short time, the current rise is therefore very short. Because of the inductance of the commutation circuit, a high voltage is produced. In the absence of any protection, this internal voltage may be several times the break over voltage rating of a thyristor.

(b) EXTERNAL OVERVOLTAGES

(i) If a thyristor converter supplied by a line transformer is in the blocking mode, only magnetizing current flows through the primary of the supply transformer. Whenever the magnetizing current is interrupted by opening the switch in the primary, a voltage transient is generated in the secondary circuit. This voltage will be maximum if the opening of the primary switch occurs at a current peak and it may be several times the thyristor breakover voltage.

(ii) High voltage peaks arrive as travelling waves caused by switching on of long lines on the input of a transformer. Because of leakage reactance and winding capacitance, overvoltages are produced.

(iii) Lightning surges cause voltage peaks on the a.c. lines to which thyristor converters are connected.

(iv) If a converter feeds a highly inductive load, the interruption of the current produces a large voltage in the primary which is in turn transferred to the secondary side. This may cause intolerable overvoltage across the thyristors connected to the secondary side.

(v) Fuses produce arc voltages whenever they are interrupted. As very rapid interrupting fuses have inherently high arc voltages, overvoltages well in excess of breakover voltages of thyristors may be produced.

(vi) Often switches are provided on the DC side. These switches are opened whenever the current on the DC side is intolerably high. The sudden interruption of these switches produce arc voltages opposing the supply voltage. This may also give rise to overvoltages across the thyristors.

To protect against transient voltages, each thyristor and diode in a converter circuit is provided with a parallel $R-C$ network. The series connection of $R-C$ is also called a snubber circuit. The snubber network shown in Fig. 2.10 protects the device against internal voltage transients due to reverse recovery. The equivalent circuit of the converter during the commutation of the diogonally connected thyristors is shown in Fig. 2.11. The source inductance is shown as L_s and the source resistance is neglected. The equivalent snubber resistance R' is $R/2$, and the equivalent snubber capacitor C' is $2C$. The general equivalent circuit for the design of snubber elements is shown in Fig. 2.12. After the thyristor is commutated,

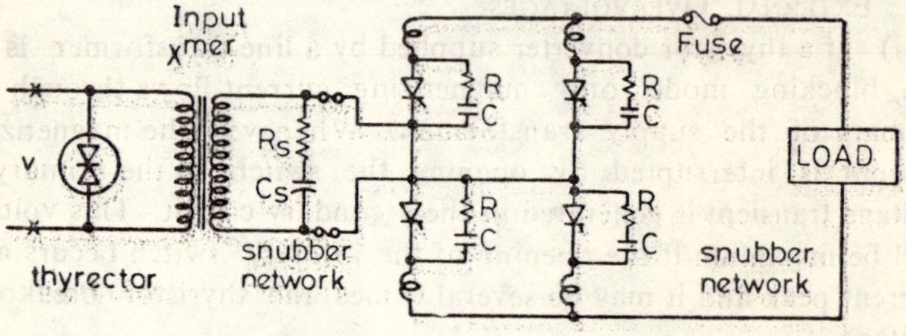

Fig. 2.10. Protective networks against voltage transients and overcurrents.

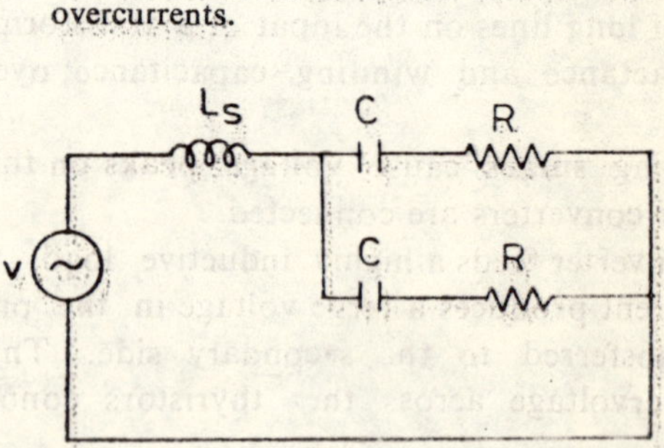

Fig. 2.11. Equivalent circuit during commutation in a single phase bridge converter.

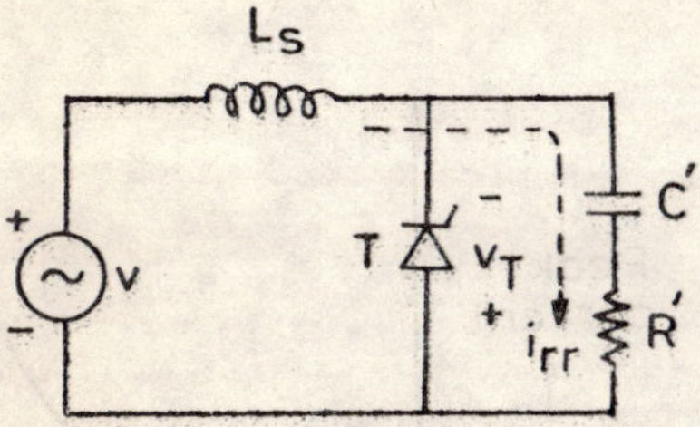

Fig. 2.12. Equivalent circuit showing the diversion of reverse recovery current (i_{rr}) through the snubber circuit.

the reverse recovery current is diverted into the snubber circuit consisting of the elements R' and C' connected in series as shown in Fig. 2.12. If the elements R', C' and L' ($=L_S$) are suitably chosen, then the voltage across the thyristor can be limited to a suitable value. If the damping is too small, there will be an appreciable voltage across the capacitor. On the other hand, if the damping is too strong, the voltage across the resistance R' will be too high at the instant of commutation. In both cases, the thyristor voltage will be appreciable. The optimum solution is obtained by taking the approximate relation between the elements R', C', and L' as $R' \simeq \sqrt{L'/C'}$ [3]. The equivalent circuit of Fig. 2.12 can be analysed and the peak reverse voltage across each thyristor can be limited to about 85 per cent of the peak reverse voltage rating of the device. This provides enough safety margin. Also, the capacitor must be properly dimensioned so that the maximum dv/dt appearing across the device is within the specification of the device.

The R_S-C_S network connected across the input of the converter (Fig. 2.10) suppresses the transient voltage during switching on or off of the input transformer.

The lightning and switching surges along the input line may damage the transformer and/or the converter. Therefore a nonlinear resistor called 'thyrector' (Fig. 2.10) is connected across the input lines. The thyrector provides a low resistance path to the transient high voltage and thus bypasses the transformer and the converter. The typical $V-I$ characteristic of a non-polarized thyrector is shown in Fig. 2.13.

The thyrector limits the forward and reverse voltage to a definite value. It does not produce any ringing oscillations. Also it does not draw any charging current. However, this device cannot

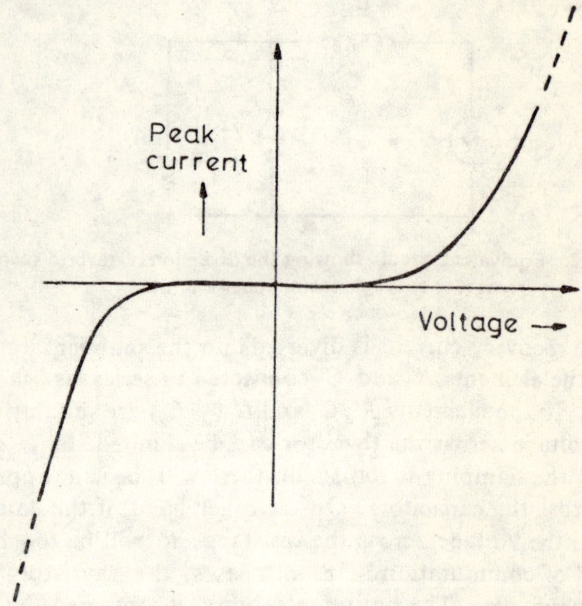

Fig. 2.13. Nonpolarized *V—I* characteristic of a thyrector.

influence *dv/dt* rating. These are the relative merits of a thyrector over a *RC* protective network.

2.2.3.2 *Overcurrent Protection*

The overcurrent in the thyristor circuits can occur due to internal or external short circuits. The causes of internal short circuits are:

(a) Misalignment of firing pulses.

(b) Failure of thyristors to block forward or reverse voltages.

(c) Short circuit of the output terminals of the converter due to faults in the load or connecting cables.

The causes of external short circuits are:

(a) Short circuit of the load circuit.

(b) Sustained overload, e.g., a stalled motor in a drive system.

Thyristor systems can be protected against sustained overloads by conventional overcurrent protective devices such as ordinary fuses, relays, contactors and circuit breakers. Another approach is the electronic monitoring of the output current. The detection of the abnormal current results in the suppression of the gating signals or lowering of the output voltage by the control circuit. The electronic protection is easily provided since most converters have a regulating loop for the output current.

The type of protection that should be used against short circuit and fault currents depends on whether the system has inherent capability to restrict either the magnitude of the overcurrent or the rate of rise of current or not. If sufficient impedance remains in the circuit during a fault, the current can be limited below the multi-cycle surge rating of the thyristor being protected. This type of situation will arise when the thyristor circuit is fed by a weak supply system. Large internal impedance of the source will limit the current below the multi cycle surge rating of the thyristor. Another example of this type occurs in the machine tool drives and excavator drives when the motor is stalled due to overloads. The current is limited by the source and the motor impedance. Similarly if sufficient inductance remains in the circuit during a fault, the rate of rise of the current will be limited and therefore the current will remain less than the multi cycle surge rating of the thyristor. The filter inductance commonly used in dc and some ac drives can play this role. Protection against this type of fault can also be achieved by employing conventional overcurrent protective elements described above. Proper coordination is essential to ensure that the fault current is interrupted before the thyristor surge current rating is exceeded, and that only the faulty branch of the circuit is isolated.

In those circuits where there is negligible impedance to limit the magnitude and the rate of rise of fault current, one half-cycle surge current rating may be exceeded. Hence, the conventional overcurrent protective elements and the thyristor gate blocking schemes will be too slow to have satisfactory protection. Special high speed current-limiting fuses capable of interrupting the fault current within few milliseconds before it reaches its first peak, are required.

The action of the fuse is illustrated in Fig. 2.14. If a fault occurs at A in the absence of the fuse, the fault current would have increased as shown by the dotted line. A properly selected fuse melts at B and introduces an arc in series with the fault. The current may continue to increase for a brief interval reaching a maximum value called 'peak let through current' which is considerably less than the peak available current. As the arc resistance increases the fault current decreases. At point C, the arcing stops and the fault current is interrupted.

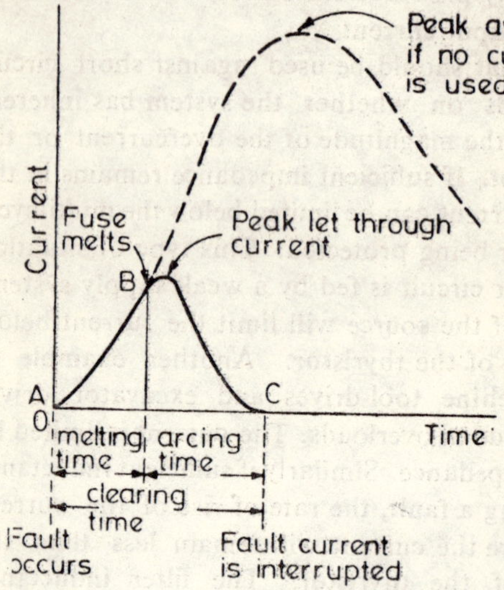

Fig. 2.14. Limiting action of the current limiting fuse in an a.c. circuit.

Proper co-ordination is required between the thyristor and the fuse [1]. The objective of the co-ordination is to ensure that the fuse let through energy under fault conditions remains below the tolerable energy of the thyristor under all service conditions for which the protection is desired. The fuse is selected such that its current rating is greater than the maximum expected load current under normal operating conditions taking the small marginal overload also into account. But the maximum let through current should be limited to a value below the subcycle surge current rating of the thyristor. It should also be insured that the arc voltage produced when the fuse interrupts is not more than twice the peak supply voltage, otherwise other thyristors in the off state may be damaged due to overvoltage transients.

2.3 MOUNTING, TEMPERATURE RISE AND COOLING OF THYRISTORS

Thyristors are presently available from low power to very high power ratings. As the power rating increases, the power dissipated by the device also increases. The power loss transforms itself into

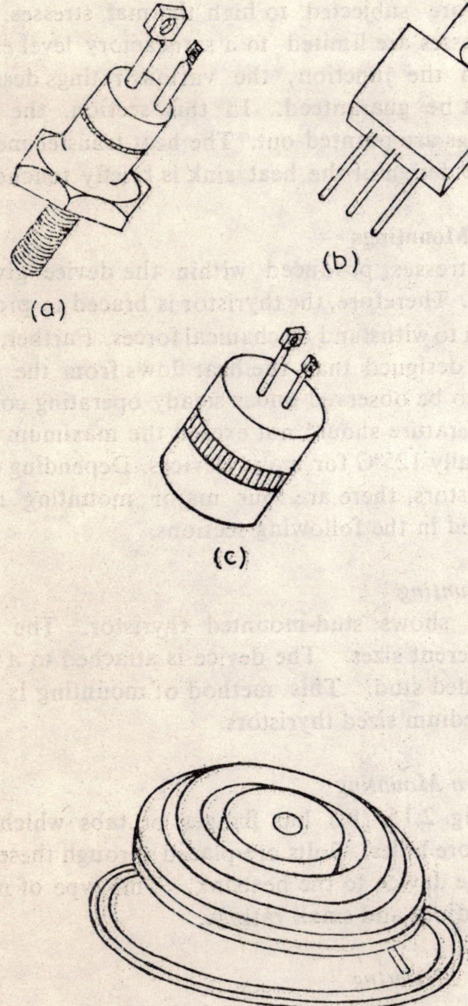

Fig. 2.15. Different Thyristor Mountings

(a) Stud mounting.

(b) Bolt down mounting.

(c) Press fit mounting.

(d) Pressure mounting.

heat at the junction within a very small volume of the device. The thyristor is therefore subjected to high thermal stresses. Unless these thermal stresses are limited to a satisfactory level extracting the heat out from the junction, the various ratings described in Section 2.2 cannot be guaranteed. In this section, the different thyristor mountings are pointed out. The heat transfer mechanism is explained. The design of the heat sink is briefly touched upon.

2.3.1. Thyristor Mountings

The thermal stresses produced within the device give rise to mechanical forces. Therefore, the thyristor is braced to provide the necessary strength to withstand mechanical forces. Further, the case or mounting is so designed that the heat flows from the junction to the case. It is to be observed under steady operating conditions, the junction temperature should not exceed the maximum specified limit which is usually 125°C for many devices. Depending upon the rating of the thyristors, there are four major mounting methods. These are described in the following sections.

2.3.1.1. *Stud Mounting*

Figure 2.15 (a) shows stud-mounted thyristor. The threaded stud can have different sizes. The device is attached to a heatsink through the threaded stud. This method of mounting is common with small and medium sized thyristors.

2.3.1.2 *Bolt Down Mounting*

The device (Fig. 2.15 (b)) has flanges or tabs which usually contain one or more holes. Bolts are placed through these holes in order to attach the device to the heatsink. This type of mounting is common in medium and small ratings.

2.3.1.3 *Press Fit Mounting*

This type of mounting is employed for thyristors of a large rating. The device is force fitted into a hole of the appropriate size with a large force (Fig. 2.15 (c)).

2.3.1.4 *Pressure Mounting*

This type of mounting is used for thyristors of very high current ratings. The device is clamped under a very large external force (Fig. 2.15 (d)). Such devices are also called Hockey-puk SCRs because of their shape. The clamping force is applied smoothly, evenly and perpendicularly to Hockey-puk to insure that there is no deformation to either the pole faces or the heat exchangers during mounting. It is possible to have double side cooling using air-cooled or liquid cooled heat exchangers.

2.3.2 Heat Transfer Process in Thyristors

Before discussing the heat transfer process in a thyristor, let us define the thermal resistance. The thermal system can be described as analogous to the electric circuit. The different thermal parameters and their analogues in the electric circuit are:

Heat produced by the device due to power dissipation (P_{av}):

electric current, I

Temperatures at different points (T): Voltages (V)

The thermal resistance between two points having temperatures T_1 and T_2 is given by

$$\theta_{12}(0°\text{C/watt}) = \frac{T_2 - T_1}{P_{av}} \tag{2.2}$$

The heat generated at the junction flows from the junction to the case, from the case to the heat exchanger or heat sink, and from the heat sink to the surrounding atmosphere. Heat transfer from the source to the surrounding atmosphere takes place by conduction, convection and radiation. Although all these modes of heat transfer are present, some are more predominant depending upon the type of cooling method adopted. The schematic diagram showing thermal resistances and temperatures at junction, case and ambient is shown in Fig. 2.16. The thermal resistances offered by the

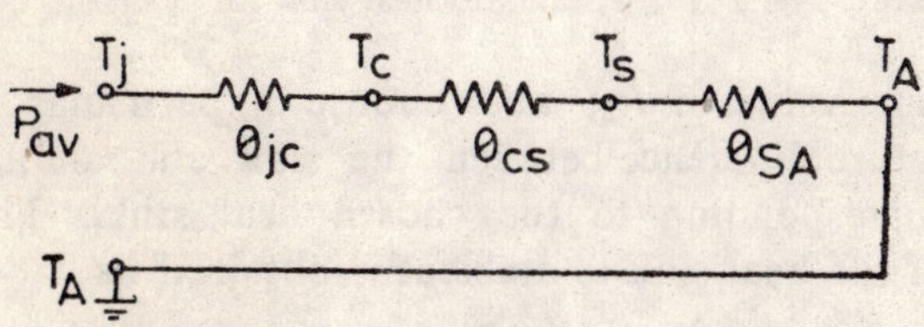

Fig. 2.16. Thermal Equivalent Circuit.

junction to case, θ_{jc} and case to the sink, θ_{cs} are generally supplied by the manufacturer. These depend upon the device encapsulation and the mounting. The thermal impedance θ_{SA} depends upon the heat sink. The parameters influencing this resistance are size, shape, surface finish and material of the heat sink. From Fig. 2.16, the equation relating power dissipation and temperature drop between the junction and the ambient under steady state conditions is

$$T_J - T_A = P_{av}(\theta_{JC} + \theta_{CS} + \theta_{SA})°\text{C} \tag{2.3}$$

The thermal impedance θ_{SA} can be determined from the data of the heat sink. The manufacturer of heat sinks provides the data in the form of curves relating temperature difference between heat sink and ambient versus power dissipation. Typical curves of standard heat sinks of aluminium extrusions are shown in Fig. 2.17.

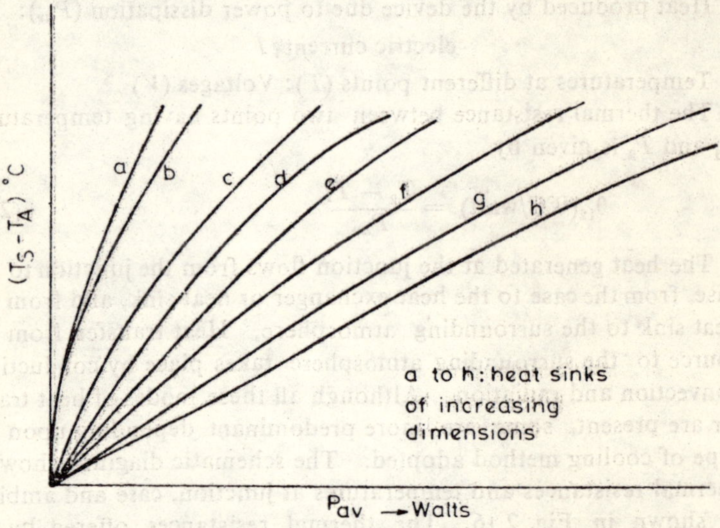

Fig. 2.17. Standard heat sink ratings.

Consider some value of P_{av} and choose a particular heat sink. The temperature difference between the sink and ambient can be read off corresponding to the chosen heat sink. The thermal resistance of the heat sink to ambient is obtained as

$$\theta_{SA} = \frac{T_S - T_A}{P_{av}} \tag{2.4}$$

Substituting for θ_{SA} in eqn. (2.3), the permissible value of P_{av} is determined for the maximum junction temperature and a known ambient temperature. This value is compared with the original value taken for P_{av}. If there is a difference, the data corresponding to another heat sink or another value of P_{av} is tried out. This procedure is continued until eqns. (2.3) and (2.4) are satisfied. After determining the permissible value of P_{av}, the average current rating for a given conduction angle is obtained from the appropriate curve in Fig. 2.8.

Alternatively, if we know the waveform of current, the average current can be determined. The average power dissipation can be read off from Fig. 2.8 corresponding to the average current of a known conduction angle. The thermal resistance can be determined from eqn. (2.3) knowing ambient temperature and assuming maximum junction temperature. The temperature difference between the heat sink and ambient is calculated from eqn. (2.4). An appropriate heat sink is then selected from Fig. 2.17, the details of which are usually supplied by the manufacturer of heat sinks.

The thermal resistance of a heat sink is also a function of the temperature difference between the heat sink and the ambient. As the temperature increases, the radiation from the sink increases thereby reducing the thermal resistance. This can be noted from the curves of Fig. 2.17 which are bending at high temperatures. Black surfaces radiate much more heat energy than shiny or light coloured surfaces. Often the black coated heat sinks are used.

2.4 THYRISTOR FIRING CIRCUITS

2.4.1 Methods of Turn On of a Thyristor

A thyristor can be switched from a non-conducting state to a conducting state by several ways as described below.

(a) Forward Breakover Voltage

If the voltage across the thyristor exceeds the rated forward breakover voltage the thyristor will start conducting due to avalanche breakdown. This process of turn on has been explained in Section 2.1.

(b) Rate of Change of Voltage

With a positive anode to cathode voltage across the device, the central junction of the four-layer device is reverse biased. The barrier developed at the central junction possesses a capacitance. If a forward voltage is suddenly applied, a charging current will flow tending to turn the device on. If the charge is denoted by Q and the capacitance by C_j and the voltage impressed across the device by V then

$$i_c = \frac{dQ}{dt} = \frac{d}{dt}(C_jV) = C_j\frac{dV}{dt} + V\frac{dC_j}{dt} \tag{2.5}$$

The rate of change of junction capacitance may be negligible as the junction capacitance is almost constant. The contribution to the charging current by the latter term is negligible. Hence, eqn. (2.5) reduces to

$$i_c = C_j\frac{dV}{dt} \tag{2.6}$$

Therefore, if the rate of change of voltage across the device is large, the device may turn on even though the voltage appearing across the device is small.

(c) Temperature

With the increase in temperature, the leakage current through a reverse-biased $p-n$ junction increases. At high temperature, the device may turn on.

(d) Light Triggering

Light sensitive turn-on action is obtained by irradiating the device. This may be used in conjunction with the gate triggering. Such triggering may be used in applications which require response to light or electrical isolation between the trigger signal and the load. It often offers light sensitive operation at power levels much above the capability of conventional photoresistive or photovoltaic devices.

(e) Gate Triggering

This is the most common method of turn-on of a thyristor. The gate triggering provides a large turn on gain which is the ratio of the anode current to the gate current. Also, the control so obtained consumes low power. This method of turn-on has been briefly explained in Section 2.1 with the help of Fig. 2.4. As the gate current injected increases, the forward breakover voltage V_{BO} shifts to the left, allowing the *SCR* to trigger at lower voltages.

The equivalent circuit with an excitation in the anode and the gate circuit is shown in Fig. 2.18. The relevant equations are:

$$e_s = v_G + i_G R_s \tag{2.7}$$

$$i_G = \frac{e_s - v_G}{R_s} \tag{2.8}$$

$$i_G = f(v_G, T_j, V_A) \tag{2.9}$$

where T_j is the junction temperature.

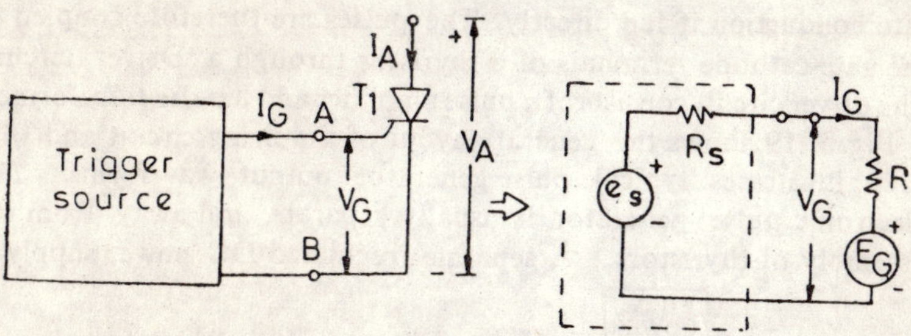

Fig. 2.18. Equivalent gate cathode circuit.

The equation (2.9) shows that the gate current characteristic is non-linear while the gate load line given by eqn. (2.8) is linear. The intersection of these two characteristics gives the required operating point. The voltage and current corresponding to the operating point are obtained from eqns. (2.8) and (2.9). As explained in Section 2.1 it is always desirable for reliable triggering to select the operating point as close to the maximum dissipation curve as possible without exceeding the gate current and voltage.

2.4.2 Principal Features of Firing Circuits

Of the different methods of turn on discussed in Section 2.4.1, the turn on of a thyristor by tne gate pulse control is by far the most commonly employed method. The gate control circuits are called firing or triggering circuits. These are generally low power electronic circuits. Any practical power circuit configuration, e.g., chopper, converter, inverter, and cycloconverter consists of several thyristors. The gate control circuit must fulfil two general functions.

(i) It is required to produce voltage pulses for each thyristor at the appropriate instant of time in a periodic manner and with a particular sequence depending upon the type of the power circuit. The trigger pulses with the desired sequence are generated using electronic circuits consisting of logic gates, flip-flops, counters etc. The availability of all these components in the integrated circuit (IC) configuration has simplified the control circuitry and brought about great changes in the field of power control.

(ii) The pulses produced by the control circuit are usually at low power level. They may not be able to trigger the thyristors into conduction if fed directly. The pulses are therefore coupled to the gate-cathode terminals of a thyristor through a 'Driver' circuit. The driver circuit consists of a pulse amplifier and a pulse transformer.

Fig. 2.19 shows the general layout of the firing circuit and Fig. 2.20 illustrates typical pulse generator output waveforms. The electronic pulse generator is usually separate and away from the assembly of thyristors. A separate regulated DC power supply is

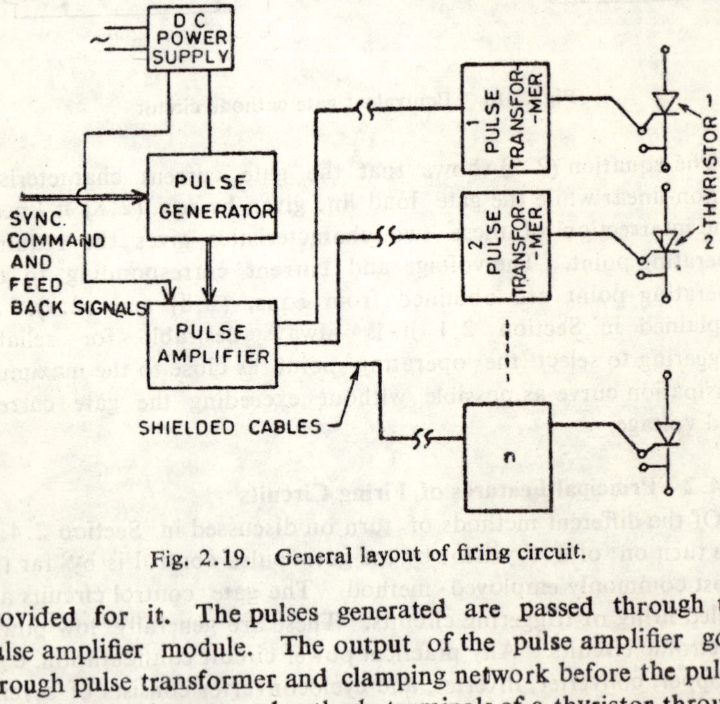

Fig. 2.19. General layout of firing circuit.

provided for it. The pulses generated are passed through the pulse amplifier module. The output of the pulse amplifier goes through pulse transformer and clamping network before the pulses are coupled to the gate and cathode terminals of a thyristor through shielded cables. The function of the pulse transformer is to isolate the low voltage gate control circuit from the high voltage

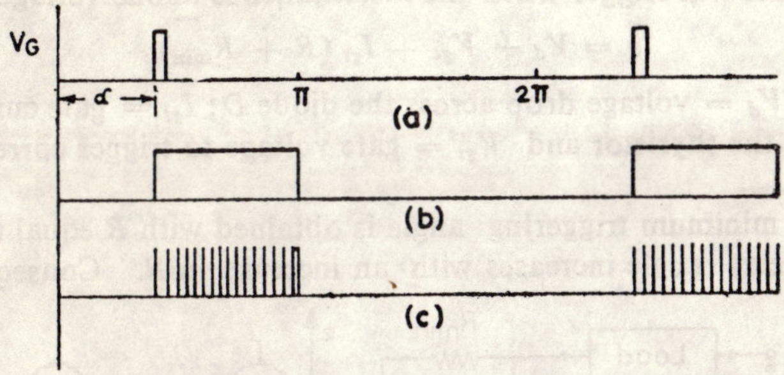

Fig. 2.20. Typical pulse generator outputs; (a) Single pulse, (b) D.C., (c) H.F. Pulse train.

anode circuit. The clamping network basically consists of a diode in series with the gate, another diode across the gate and cathode terminals and a suitable biasing resistance. The function of all these components is explained later while discussing a typical firing circuit.

Some basic firing circuits are reviewed in this section. The firing circuits for three-phase converters, cycloconverters and regulator circuits are discussed in Chapters 3, 4 and 5 respectively.

2.4.3 Simple *R* and *RC* Triggering Circuits

R and *RC* firing circuits are simple and economical [4]. However, their main drawback is that they do not have repeatability over a temperature range.

(a) *R*-Firing Circuit

Figure 2.21 (a) illustrates the arrangement. In order that the peak gate current of the thyristor I_{gfm} is not exceeded, a limiting resistor $R_{\min}$ is placed between anode and gate. For the worst case, that is, when the supply voltage has reached its peak, E

$$R_{min} \geqslant \frac{E}{I_{gfm}} \tag{2.10}$$

The stabilising resistor R_b is so chosen that the voltage across it does not exceed the forward gate voltage V_{gfm}. From the voltage distribution

$$R_b \leqslant \frac{(R+R_{\min})\, V_{gfm}}{E-V_{gfm}} \tag{2.11}$$

The SCR will trigger when the instantaneous anode voltage e is

$$e = V_d + V_{gt} + I_{gt}(R + R_{min}) \tag{2.12}$$

where V_d = voltage drop across the diode D; I_{gt} = gate current to trigger the thyristor and V_{gt} = gate voltage to trigger corresponding to I_{gt}.

The minimum triggering angle is obtained with R equal to zero. The delay angle increases with an increase in R. Consequently,

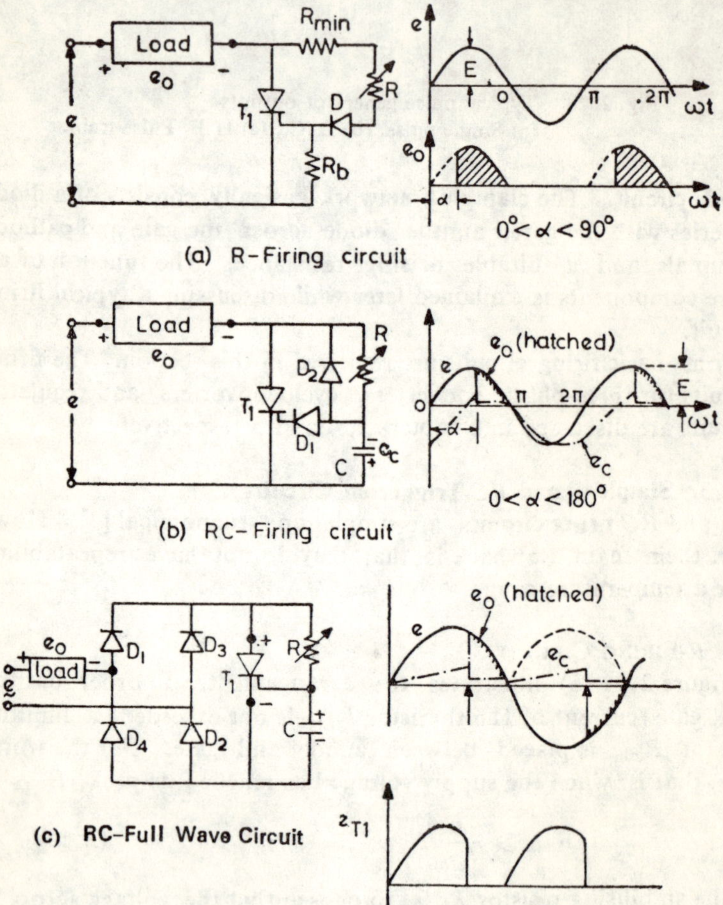

Fig. 2.21. R and RC triggering circuits.

the power delivered to the load decreases. Since the thyristor T_1 latches into conduction the first time I_{gt} is reached and, because the anode voltage and the gate current are in phase, the conduction of the thyristor T_1 cannot be delayed beyond 90° with this simple triggering circuit. The circuit provides control from 100 per cent of half-wave output to 50 per cent of half-wave output.

(b) *RC* FIRING CIRCUIT

Fig. 2.21 (b) illustrates the RC firing circuit. The role of the capacitor is to shift the phase of the anode voltage so that a positive gate current can be supplied even after the peak of the anode voltage. By varying the resistor R the firing angle can be controlled from 0° to 180°. The capacitor charges to the negative peak of the A C. voltage in every negative half-cycle through the diode D_2. During the positive half-cycle it begins to charge through the resistance R. When the voltage across the capacitor reaches the required positive value, the thyristor T_1 is triggered and the capacitor voltage remains almost constant. The diode D_1 prevents the breakdown of the gate to cathode junction during the negative half-cycle.

For a typical thyristor in the range of power frequencies it may be empirically shown [4] that RC for zero output must be chosen as follows:

$$RC \geqslant \frac{1.3T}{2} \simeq 4/\omega, \ \omega = 2\pi f \tag{2.13}$$

where $T = 1/f$ = period of A.C. line frequency in secs. When the capacitor voltage $e_c = v_{GT} + v_d$, where v_d is the voltage drop of the diode D_1, the thyristor will turn on provided the gate current I_{gt} is available. Hence, the maximum value of R is given by

$$e \geqslant I_{gt}\, R + e_c = I_{gt}\, R + v_{gt} + v_d \tag{2.14}$$

or
$$R \leqslant \frac{e - v_{gt} - v_d}{I_{gt}}$$

where e is the instantaneous supply voltage at which the thyristor will turn on. Suitable values of R and C can be obtained from eqns. (2.13) and (2.14).

(c) *RC* FULL WAVE CIRCUIT

There are many possible variations of the RC trigger circuit, all operating on the same basic principle. A simple circuit giving full wave output is shown in Fig. 2.21 (c). In this circuit the initial voltage from which the capacitor C charges is essentially zero. The capacitor C is reset to this voltage by the clamping action of the thyristor gate. For this reason, the charging time constant RC must be chosen longer than for the half-wave RC circuit in order to delay the triggering an equal amount for a given AC half-cycle of charging voltage. The RC value is chosen using the following empirical rule [4]

$$RC \geqslant \frac{50T}{2} \simeq \frac{157}{\omega} \tag{2.15}$$

Similar to eqn. (2.14), the value of R is selected from eqn. (2.16).

$$R \leqslant \frac{e - V_{gt}}{I_{gt}} \tag{2.16}$$

2.4.4 Unijunction Transistor (UJT) Pulse Trigger Circuit

The simple resistor and capacitor circuits that have been described depend heavily on the trigger characteristics of the thyristor used. The power dissipation in the gate circuit is quite high due to prolonged pulse. Also, they do not readily lend themselves to automatic, self programmed or feedback control systems.

On the other hand, pulse triggering can accommodate wide tolerances intriggering characteristics by instantaneously by overdriving the gate. The power level in such circuits is lower as the triggering energy can be stored slowly and discharged rapidly when the triggering is required.

The UJT is used as a relaxation oscillator to obtain sharp, repetitive pulses with good rise time. Also, it has good frequency stability against variation in the supply voltage and temperature. Fig. 2.22 shows UJT firing circuit along with some pertinent waveforms. The pulse transformer is used to isolate the control circuit from the main power circuit.

The unijunction breaks down when the voltage between the emitter E and the base B_1 reaches the peak voltage V_p given by

$$V_p = \eta V_Z + V_D, \tag{2.17}$$

η being intrinsic stand off ratio.

Then the capacitor discharges through the emitter and a pulse occurs to trigger the thyristor. The zener is used to clip the rectified voltage to a standard level to prevent erratic firing. Since V_Z goes to zero at the beginning of each half-cycle, the synchronisation of the trigger pulse with the supply voltage is achieved. The two

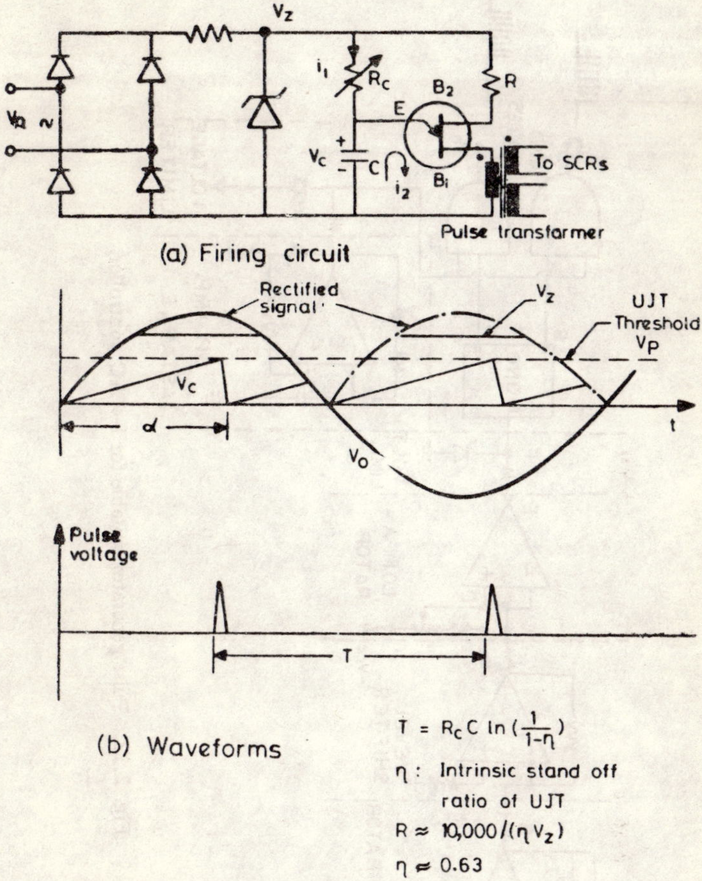

Fig. 2.22. Pulse firing using UJT.

windings of the pulse transformer secondary feed pulses to two thyristors in a single phase half-controlled converter or a single-phase AC regulator. Since pulses appear both in positive and negative half-cycles, the thyristor whichever is forward biased turns on. Since it is not desirable to apply gate pulses when the thyristor is reverse biased, with some additional logic circuitry (not shown in Fig. 2.22 (a)), the gate pulses appearing during the reverse bias can be eliminated. The firing angle is controlled by varying the resistor R_C. The total range of control is around 150°.

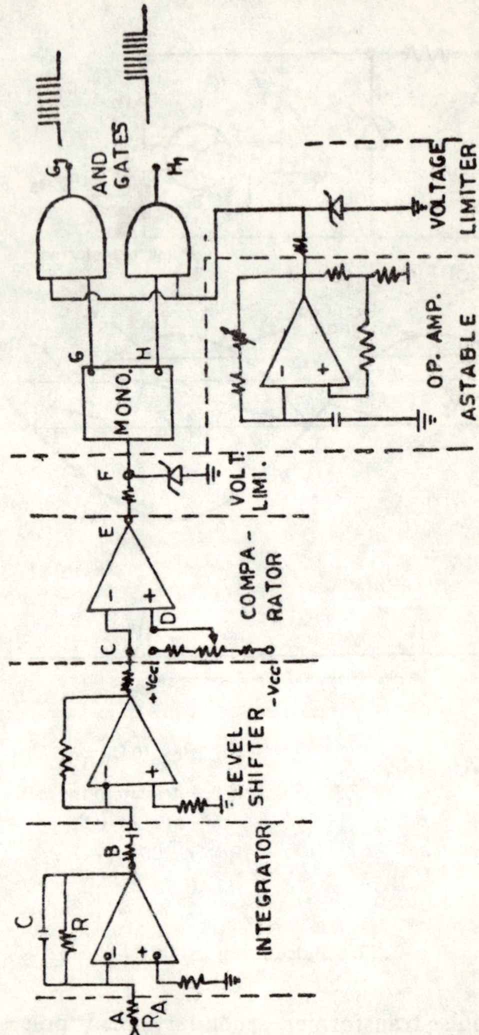

Fig. 2.23. Pulse generator suitable for 1-ϕ AC controller.

2.4.5 Triggering Circuit Suitable for Single-Phase Converters and AC Controllers

Figure 2.23 shows a single-phase firing circuit suitable for triggering thyristors in a single-phase AC voltage controller or converter circuit.

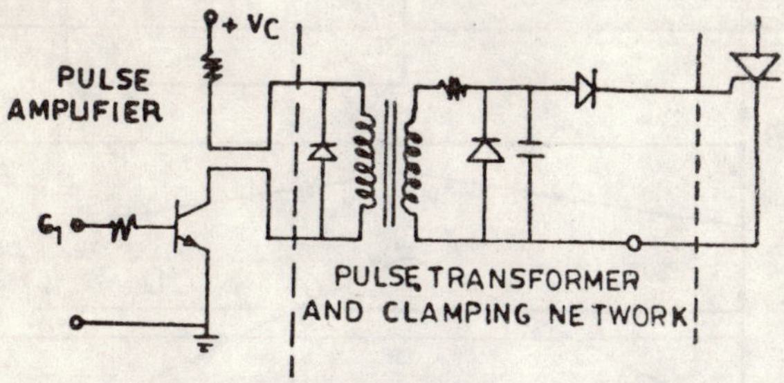

Fig. 2.24. Driver circuit shown for the pulse-G_1.

The waveforms at different points are illustrated in Fig. 2.25. The synchronising signal is derived from the same supply to which the thyristor power circuit is fed. This signal is given to the zero crossing detector. The square wave signal *A* so derived is integrated and level shifted to obtain the waveform *C*. *C* is compared with a variable DC voltage *D* to obtain *E* which is limited by a zener so that it is TTL compatible. The pulses at *F* are fed to a monostable which gives complementary outputs *G* and *H*. The pulse widths of *G* and *H* are adjusted to be 10 msec each. These are ANDed with a high frequency (10 KHz) pulse train I to obtain G_1 and H_1.

G_1 and H_1 are fed to transistor switches which act as pulse amplifiers. This is shown for the pulse G_1 in Fig. 2.24. The diode across the primary of the pulse transformer serves to protect the transistor. The gate of the thyristor should be protected from large reverse voltages. Hence, a diode is placed across the secondary of the transformer. The diode in series with the gate prevents a reverse current flowing from the gate during the conduction of the thyristor. The capacitor at the secondary prevents any spurious high frequency from triggering the thyristor.

2.4.6 Triggering Circuit suitable for a Single-Phase Inverter Circuit

It is always desirable to use long trigger pulses as shown in Fig. 2.20 (c) in inverter circuits. Fig. 2.26(a) shows the block diagram of a firing circuit using astable and logic gates. This circuit is

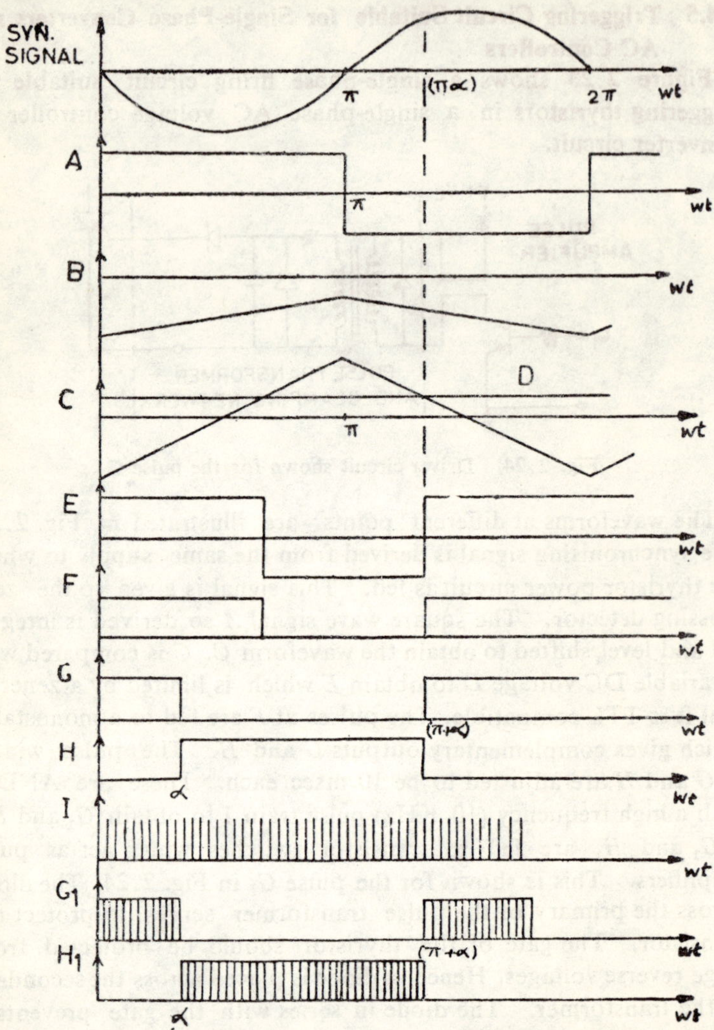

Fig. 2.25. Waveforms at Different points of Fig. 2.23.

quite suitable for triggering thyristors in a single-phase inverter circuit. The pertinent waveforms of this circuit are shown in Fig. 2.26 (b). The frequency of operation of the inverter is controlled by the astable block. The transistors are used to amplify the pulses so as to trigger the thyristors reliably. The pulse transformers provide an isolation between the low power firing circuit and the main power circuit.

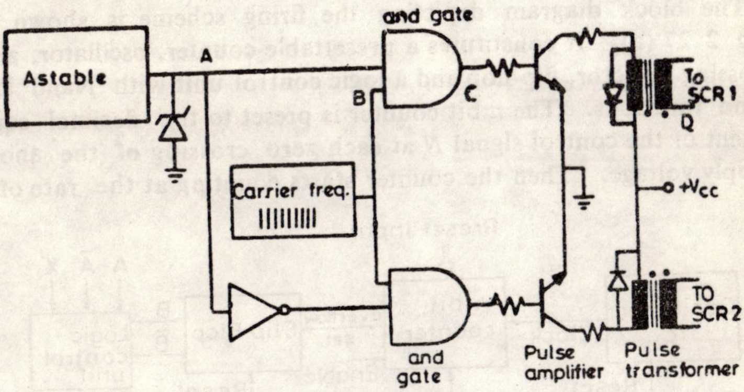

(a) Firing circuit.

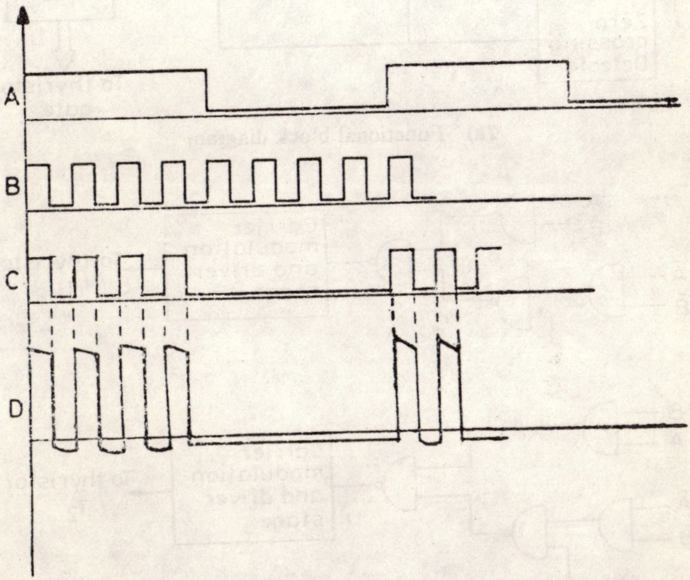

(b) Waveforms at points A, B, C and D.

Fig. 2.26. Maintained pulse firing with high frequency (10 KHz) modulation.

2.4.7 Digital Firing Scheme

In this section, a simple digital firing circuit is described. Although this is illustrated for a single-phase converter or an AC regulator, the same principle can be extended to cover firing circuits for inverters and choppers.

The block diagram depicting the firing scheme is shown in Fig. 2 27 (a). It constitutes a presettable counter, oscillator, zero crossing detector, flip-flop and a logic control unit with 'Nand' and 'And' functions. The n bit counter is preset to the decimal equivalent of the control signal N at each zero crossing of the anode supply voltage. Then the counter starts counting at the rate of f_c

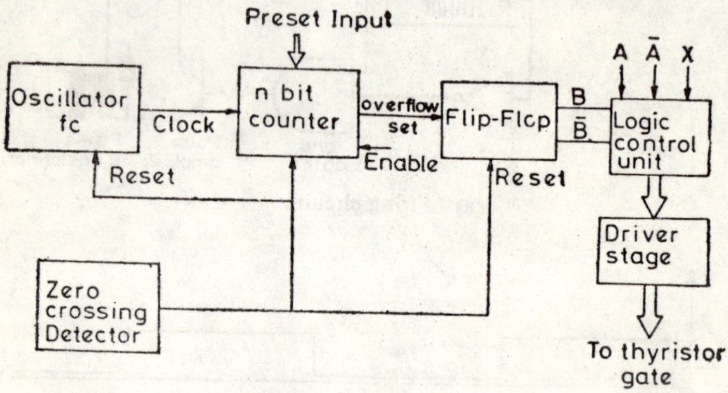

(a) Functional block diagram

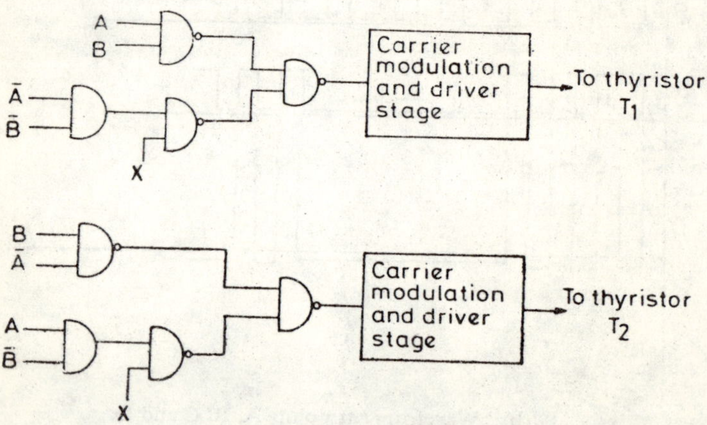

(b) Logic control driver blocks.

Fig. 2.27. Block diagram of a digital firing circuit.

counts per second. The counter overflow signal max/min is processed to trigger the thyristor. With the help of a flip-flop and a logic control unit and driver stage, the trigger pulses are produced which can be fed to the thyristors T_1 and T_2 (Fig. 2.27(b)). Thus by varying the preset input, it is possible to control the firing angle. The logic variable X equal to 0 or 1 enables to select the pulse duration either from α to π or from α to $(\pi + \alpha)$ respectively. Some waveforms corresponding to Fig. 2.27 with $X = 0$ are illustrated in Fig. 2.28.

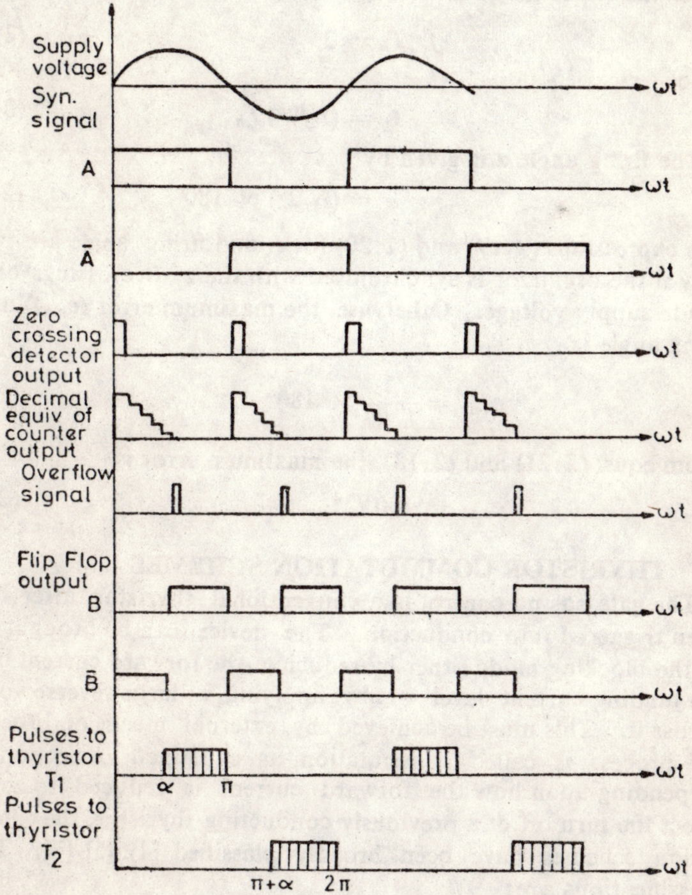

Fig. 2.28. Waveforms at different points of digital firing circuit (Fig. 2.27).

Let the counter be in the down counting mode with the variables defined below.

t_1 = overflow instant

T_h = duration of half cycle of AC supply

n = number of bits in the binary counter

N = decimal equivalent of control signal

f_c = frequency of oscillator signal

The total number of states of the counter is 2^n. In order to cover the entire range of α from 0° to 180° it is required, that

$$f_c \cdot T_h = 2^n \tag{2.18}$$

Also

$$t_1 = (N/2^n)\, T_h \tag{2.19}$$

The firing angle α is given by

$$\alpha = (N/2^n) \times 180° \tag{2.20}$$

The expressions (2.19) and (2.20) for t_1 and firing angle are valid only if the oscillator is synchronised with the zero crossings of the anode supply voltage. Otherwise, the maximum error (e_{max}) in the firing angle is

$$e_{max} = \frac{1}{f_c\, T_h} \times 180° \tag{2.21}$$

From eqns. (2.21) and (2.18), the maximum error is

$$e_{max} = 180°/2^n \tag{2.22}$$

2.5 THYRISTOR COMMUTATION SCHEMES

The gate has no control in a conventional thyristor after it has been triggered into conduction. The device can be brought back to the blocking mode either by reducing the forward current below the holding current level or by applying a large reverse voltage across it. This must be achieved by external means and the turn off process is called commutation as explained in Section 2.1. Depending upon how the forward current is reduced to zero to effect the turn off of a previously conducting thyristor, the commutation schemes have been broadly classified [1], [5]–[7]. These classifications are:

I. Line commutation

II. Load commutation

III. Forced commutation

IV. External pulse commutation

In the sections that follow, each commutation scheme is explained with the help of a thyristor circuit.

2.5.1 Line Commutation

When the thyristor circuit is excited by an AC supply source, the natural reversal of the line voltage is utilized to commutate a thyristor. Fig. 2.29(a) shows the circuit of a single-phase controlled

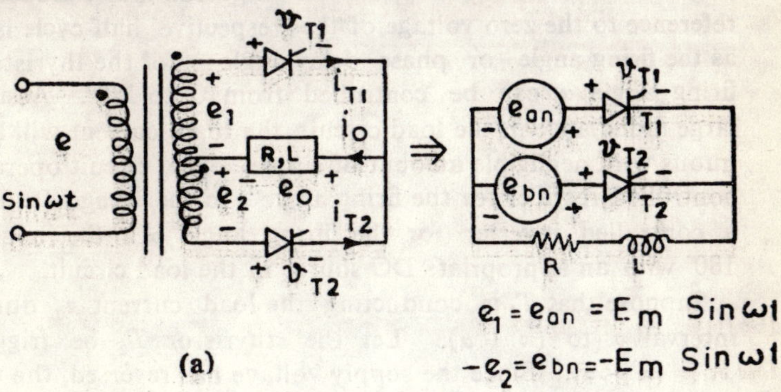

(a)

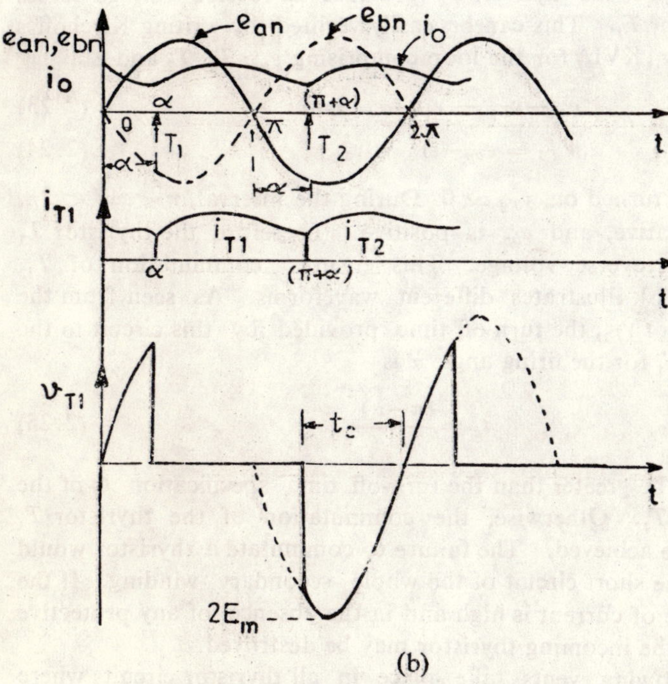

(b)

Fig. 2.29. (a) Single-phase line commutated converter.
(b) Waveforms illustrating AC line commutation.

rectifier. The operation of this circuit is described in Chapter 3. However, a brief description is given here in order to explain the turn off process. Thyristors T_1 and T_2 are triggered in the positive and negative half-cycles of the supply voltage, e respectively. The angle at which each thyristor is triggered into conduction with reference to the zero voltage of the respective half-cycle is defined as the firing angle or phase delay angle α of the thyristor. The firing angle α can be controlled from 0 to 180°. Assuming a large inductance in the load circuit, the load current will be continuous with negligible amount of ripple. The circuit operates as a controlled rectifier for the firing angle α in the range 0 to 90° and a controlled inverter for the firing angle α in the range 90° to 180° with an appropriate DC source in the load circuit.

Suppose that T_1 is conducting the load current i_0 during the interval α to $(\pi + \alpha)$. Let the thyristor T_2 be triggered at $\omega t = (\pi + \alpha)$. Since the supply voltage has reversed, the thyristor T_2 gets a forward voltage and hence it turns on at $\omega t = (\pi + \alpha)$. The turn on of the thyristor T_2 provides a reverse voltage across the thyristor T_1. This can be easily verified by writing Kirchoff's voltage law (KVL) for the loop comprising e_{an}, T_1, T_2 and e_{bn}.

$$e_{an} - v_{T_1} + v_{T2} - e_{bn} = 0 \tag{2.23}$$

$$V_{T1} = e_{an} - e_{bn} + v_{T2} \tag{2.24}$$

Since T_2 is turned on, $v_{T2} \simeq 0$. During the interval, $\pi < \omega t < 2\pi$, e_{an} is negative, and e_{bn} is positive. Therefore the thyristor T_1 receives a reverse voltage. This ensures commutation of T_1. Fig. 2.29(b) illustrates different waveforms. As seen from the waveform of v_{T_1}, the turn off time provided by this circuit to the thyristor T_1 for the firing angle α is

$$t_c = \frac{(\pi - \alpha)}{\omega} \tag{2.25}$$

This must be greater than the turn-off time specification t_q of the thyristor T_1. Otherwise, the commutation of the thyristor T_1 may not be achieved. The failure to commutate a thyristor would result in the short circuit of the whole secondary winding. If the rate of rise of current is high and in the absence of any protective elements, the incoming thyristor may be destroyed.

The following events take place in all thyristor circuits where line commutation is successfully employed.

1. The incoming thyristor must be forward biased when the commutation of the outgoing thyristor is desired.

2. When the incoming thyristor is triggered into conduction, the load current, if present is transferred from the outgoing thyristor to the incoming thyristor.

3. The outgoing thyristor must be reverse biased for an interval greater than its turn off time specification. The AC line voltage provides the necessary reverse bias in all line commutated converters. No additional commutating elements are required.

The line commutation is simple, reliable and is employed in AC to DC converters, AC regulators and cycloconverters where AC supply source is normally available.

2.5.2 Load Commutation

Load commutation is employed in thyristor circuits supplied from a DC source. Since AC source is not used in these circuits, the line commutation is obviously not possible. The load circuit parameters are selected in such a way that there is a natural tendency for the load current which passes through the thyristor to become zero. The thyristor is thus commutated. The principle of this method [1] is demonstrated by the circuit shown in Fig. 2.30. The elements L, C and R are chosen such that the circuit is under-damped. When the load resistance R is very low, the elements are connected in series as shown in Fig. 2.30. On the other hand, if the load resistance R is high, then the elements are connected as shown in Fig. 2.31(a). In this case also, the commutating elements L and C are selected such that the circuit is under-damped.

Let us consider the series resonant circuit of Fig. 2.30. Let the thyristor T_1 be turned on at $t = 0$ with the initial capacitor voltage zero. The circuit equation is

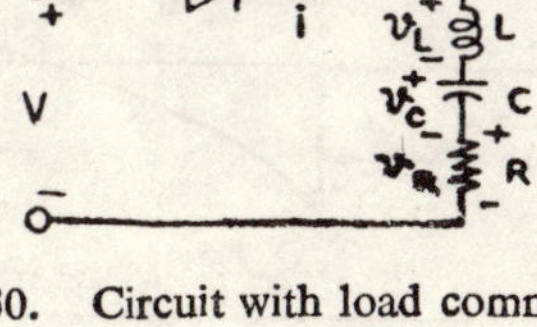

Fig. 2.30. Circuit with load commutation.

$$V = iR + L\frac{di}{dt} + \frac{1}{C}\int i\,dt \tag{2.26}$$

The initial conditions are:

$$i(0) = 0$$

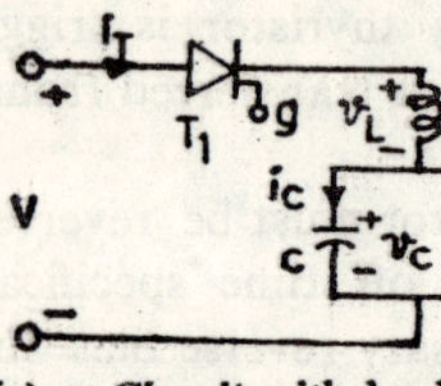

Fig. 2.31(a). Circuit with load commutation.

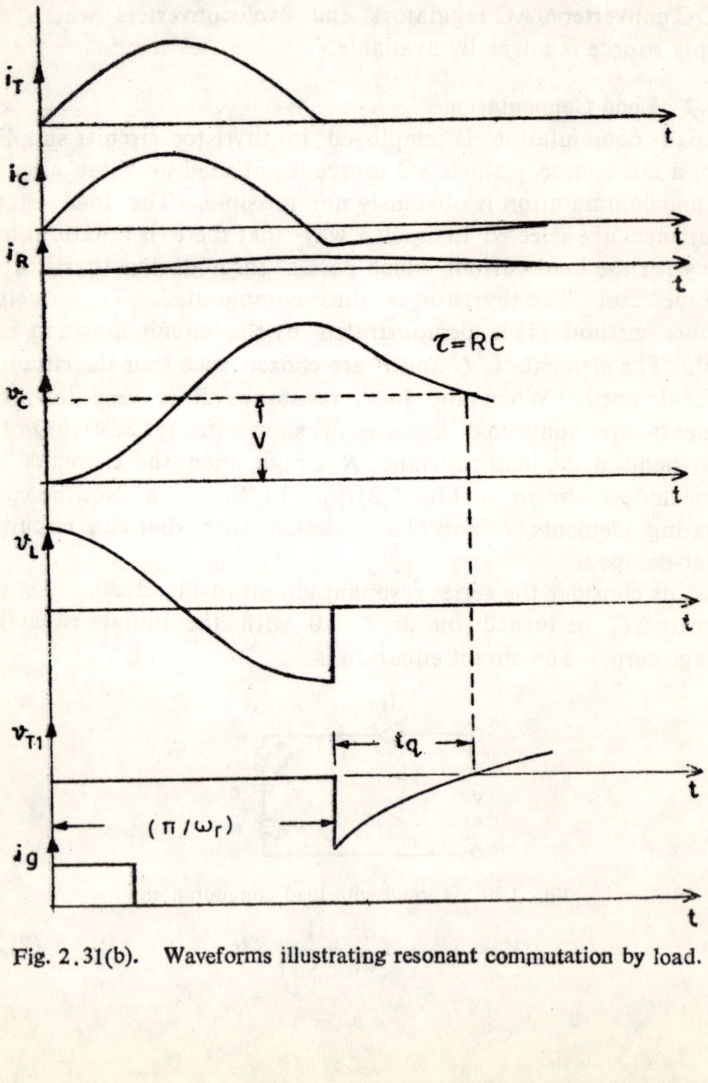

Fig. 2.31(b). Waveforms illustrating resonant commutation by load.

$$\frac{di}{dt}(0) = \frac{V}{L}$$

$$v_C(0) = 0$$

Solving eqn. (2.26) with the above initial conditions

$$i(t) = \frac{V}{\omega_r L} \exp(-\xi t) \sin \omega_r t \tag{2.27}$$

$$v_L = L\frac{di}{dt} = V.\frac{\omega_0}{\omega_r} \exp(-\xi t) \cos(\omega_r t + \psi) \tag{2.28}$$

$$v_R = iR = \frac{VR}{\omega_r L} \exp(-\xi t) \sin \omega_r t \tag{2.29}$$

$$v_C = V - v_L - v_R = V\left[1 - \frac{\omega_0}{\omega_r} \exp(-\xi t) \cos(\omega_r t - \psi)\right] \tag{2.30}$$

where $\xi = \frac{R}{2L}$, $R^2 < \frac{4L}{C}$, $\omega_0 = \frac{1}{\sqrt{LC}}$ and $\omega_r = \sqrt{\frac{1}{LC} - \frac{R^2}{4L^2}}$.

Equation (2.27) shows that the thyristor current i goes to zero at $\omega_r t = \pi$ or $t = \pi \Big/ \left(\frac{1}{LC} - \frac{R^2}{4L^2}\right)^{1/2}$. Since the capacitor voltage is higher than the supply voltage at the current zero, the thyristor gets a reverse bias after the current has gone to zero and it is therefore commutated. In order to turn on the thyristor again, the capacitor must be discharged through an alternate path which is not shown in the figure.

In a similar way, the circuit of Fig. 2.31(a) can be analysed and the following expressions can be easily derived after the thyristor T_1 is turned on with the initial conditions

$$v_C(0) = 0$$

$$\frac{di_T(0)}{dt} = \frac{V}{L}$$

$$i_C(t) = \frac{V}{\omega_r L} e^{-\xi t} \sin \omega_r t \tag{2.31}$$

$$v_C(t) = \frac{1}{C}\int_0^t i_C \, dt = V\left[1 - e^{-\xi t} \cos \omega_r t - \frac{\xi}{\omega_r} e^{-\xi t} \sin \omega_r t\right] \tag{2.32}$$

$$v_R(t) = \frac{V}{R}\left[1 - e^{-\xi t}\cos\omega_r t - \frac{\xi}{\omega_r}e^{-\xi t}\sin\omega_r t\right] \tag{2.33}$$

$$i_T(t) = i_C + i_R = \frac{V}{R}\left[1 - e^{-\xi t}\cos\omega_r t + \left(\frac{R}{\omega_r L} - \frac{\xi}{\omega_r}\right)e^{-\xi t}\sin\omega_r t\right] \tag{2.34}$$

$$v_L(t) = V - v_C = V\left[e^{-\xi t}\cos\omega_r t + \frac{\xi}{\omega_r}e^{-\xi t}\sin\omega_r t\right] \tag{2.35}$$

After the thyristor current goes to zero, the capacitor discharges into the load resistance, R. The voltage across the inductance becomes zero. The waveforms of different circuit variables given in Fig. 2.31(a) are illustrated in Fig. 2.31(b). It is important to select a proper value for the resistance R in order to commutate the thyristor T_1. For this circuit, $R \gg \sqrt{L/4C}$.

Since the commutating elements carry the load current on a continuous basis, their ratings are generally high. For low frequency operation the size of the commutating elements becomes large. Thus large values of L and C coupled with high current ratings make the cost of the commutation circuit prohibitive. This type of commutation becomes quite attractive when the frequency of operation is very high. Hence the load commutation is commonly employed in series inverters operating at frequencies of the order of a few KHz.

2.5.3 Forced Commutation

If a thyristor is commutated by using external elements L and C which do not carry the load current on a continuous basis, the turn off process is called 'Forced Commutation'. This is employed in thyristor circuits fed by DC or AC source. Since the commutation transient lasts for a short period, the turn off time in forced commutation is much smaller than in the line commutation. However, the thyristor to be commutated must be reverse biased for a duration greater than the turn off time specification. The thyristor which has been conducting the load current previously can be commutated by applying a pulse of large reverse voltage or by passing an external pulse of current in the reverse direction greater than the load current. Thus the forced commutation can be classified into two broad categories.

I. Voltage commutation

II. Current commutation

2.5.3.1 *Voltage Commutation*

Figure 2.32 shows different methods of introducing a reverse voltage in parallel to the thyristor while Fig. 2.33 shows the different ways of connecting a large reverse voltage in series with the thyristor to be commutated [6]. The reverse voltage can be generated through an auxiliary DC supply or a previously charged capacitor. This voltage can either be directly coupled or coupled through a transformer. The voltage pulse is introduced into the main thyristor circuit only when the commutation of the main thyristor is desired. The switch *S* shown in Figs. 2.32 and 2.33 serves this purpose remaining open most of the time except during the period of commutation. In most cases, another thyristor serves the function of the switch *S* shown in Figs. 2.32 and 2.33. If this thyristor is active during the commutation interval only and is dormant for the rest of the time period, this is called 'Auxiliary voltage commutation'. On the other hand, if the thyristor conducts the load current for some part of the total period besides commutating another load carrying thyristor, this is called 'Complementary Voltage Commutation'. Let us study thyristor circuits which illustrate these commutation schemes.

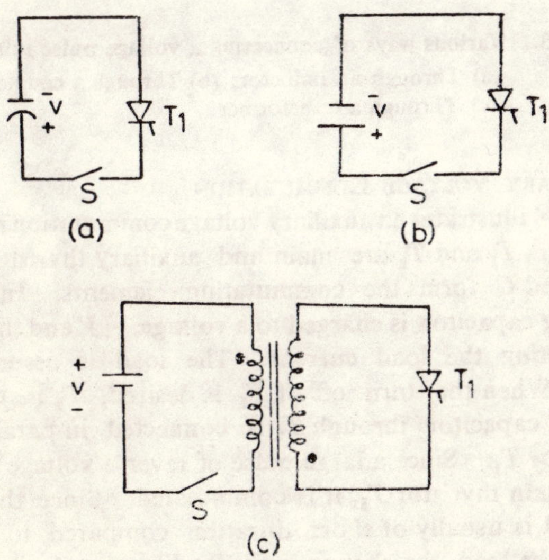

Fig. 2.32. Various ways of applying voltage pulse in parallel:

(a) Through a charged capacitor; (b) Auxiliary DC supply: (c) Auxiliary supply through a transformer.

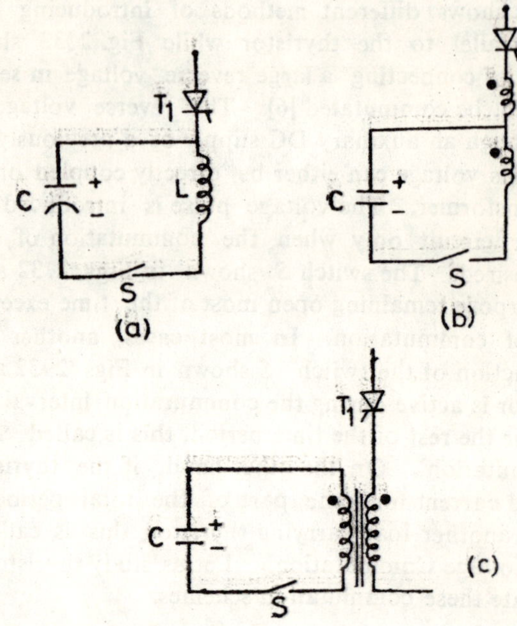

Fig. 2.33. Various ways of connecting a voltage pulse in series: (a) Through an inductor; (b) Through a coupled inductor; (c) Through a transformer.

(a) AUXILIARY VOLTAGE COMMUTATION

Figure 2.34 illustrates an auxiliary voltage commutation method [1]. The thyristors T_1 and T_2 are main and auxiliary thyristors respectively. L and C form the commutating elements. Initially the commutating capacitor is charged to a voltage, $-V$ and the thyristor T_1 is conducting the load current. The load is assumed to be inductive. When the turn off of T_1 is desired, T_2 is turned on. The charged capacitor through T_2 is connected in parallel to the main thyristor T_1. Since a large pulse of reverse voltage is applied across the main thyristor T_1, it is commutated. Since the commutation period is usually of short duration compared to the time constant of the load, the change in the load current during the commutation interval is very small and can be neglected. It is justified to assume a constant current during the commutation period. For the sake of clarity, the commutation interval is expanded in Fig. 2.34.

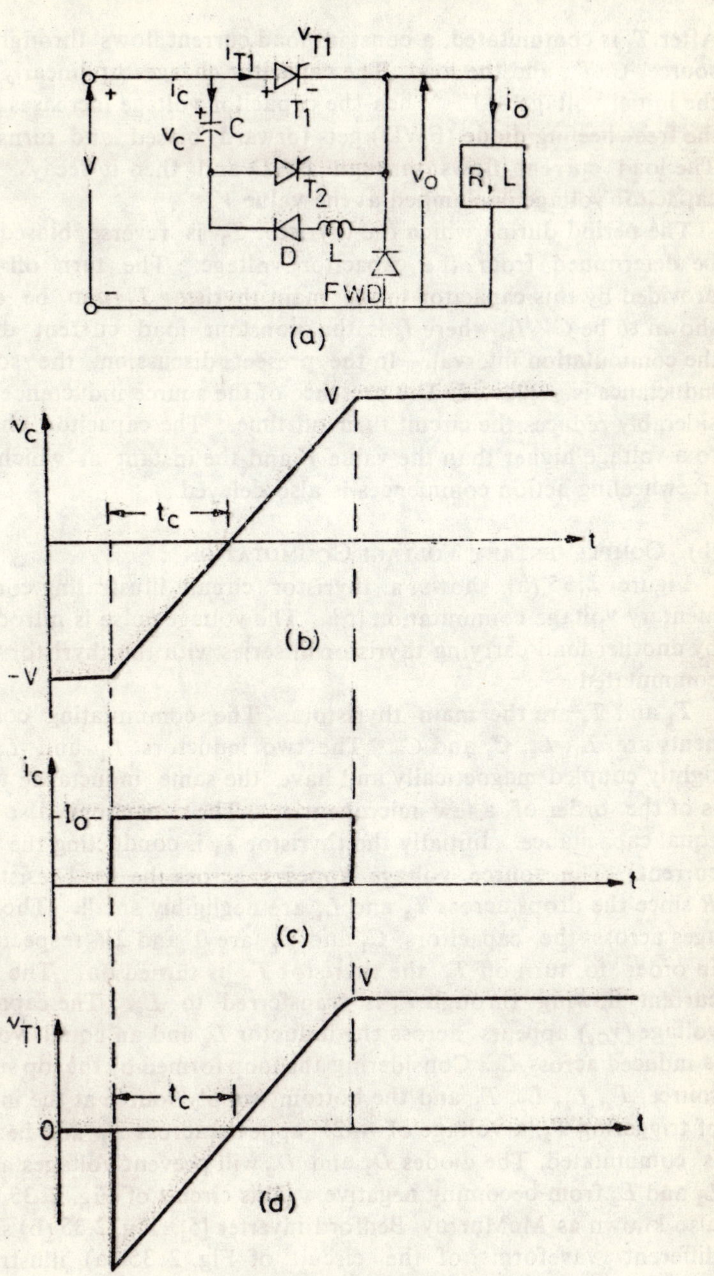

Fig. 2.34. Auxiliary voltage commutation:
(a) Circuit; (b) Waveforms v_C, i_C and v_{T1}.

After T_1 is commutated, a constant load current flows through the source, C, T_2, and the load. The capacitor charges up linearly with the initial voltage $-V$. When the capacitor voltage increases to V, the freewheeling diode (FWD) gets forward biased and turns on. The load current flows through FWD and then it decays. The capacitor voltage is clamped at the value V.

The period during which the thyristor T_1 is reverse biased can be determined from the capacitor voltage. The turn off time provided by this capacitor to the main thyristor T_1 can be easily shown to be CV/I_0, where I_0 is the constant load current during the commutation interval. In the present discussion, the source inductance is neglected. The presence of the source inductance considerably reduces the circuit turn off time. The capacitor charges to a voltage higher than the value V and the instant at which the freewheeling action commences is also delayed.

(b) Complementary Voltage Commutation

Figure 2.35 (a) shows a thyristor circuit illustrating complementary voltage commutation [5]. The voltage pulse is introduced by another load carrying thyristor in series with the thyristor to be commutated.

T_1 and T_2 are the main thyristors. The commutating components are L_1, L_2, C_1 and C_2. The two inductors L_1 and L_2 are tightly coupled magnetically and have the same inductance which is of the order of a few microhenries. The capacitors also have equal capacitance. Initially the thyristor T_1 is conducting the load current. The source voltage appears across the load resistance, R since the drops across T_1 and L_1 are negligibly small. The voltages across the capacitors C_1 and C_2 are 0 and $2V$ respectively. In order to turn off T_1 the thyristor T_2 is turned on. The load current flowing through T_1 is transferred to L_2. The capacitor voltage (v_{C2}) appears across the inductor L_2 and an equal voltage is induced across L_1. Considering the loop formed by the top supply source, T_1, L_1, L_2, T_2 and the bottom supply source at the instant of triggering T_2, a voltage of $-2V$ appears across T_1 and hence it is commutated. The diodes D_1 and D_2 will prevent voltages across L_1 and L_2 from becoming negative. This circuit of Fig. 2.35 (a) is also known as McMurray-Bedford inverter [5]. Fig. 2.35 (b) shows different waveforms of the circuit of Fig. 2.35 (a) illustrating complementary commutation. Especially with resistive load,

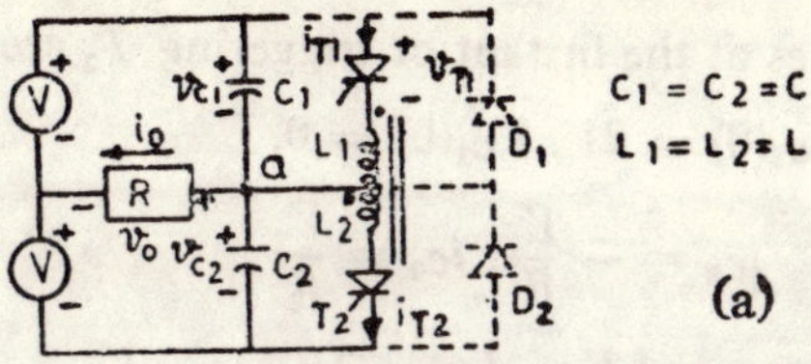

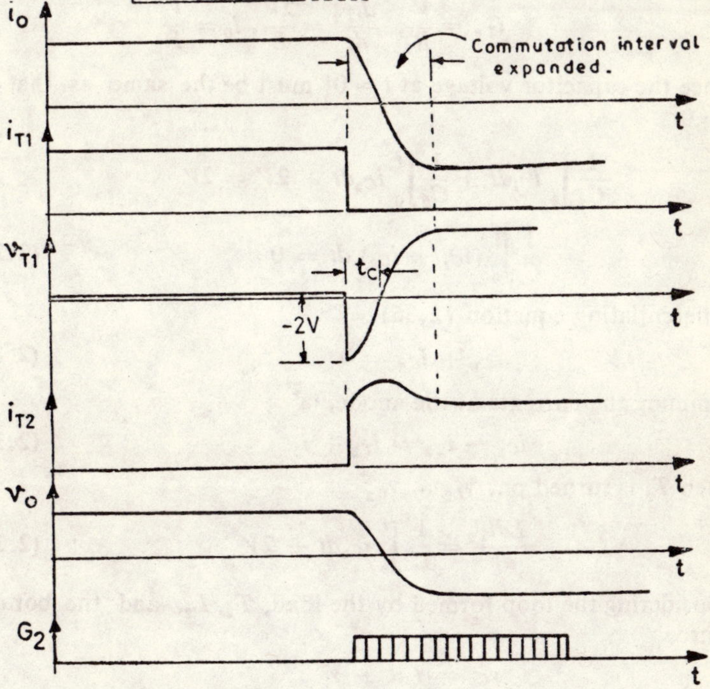

(b)

Fig. 2.35. (a) Circuit Operating on complementary Commutation.
(b) Waveforms of circuit illustrating complementary commutation with resistive load.

proper values are to be taken for L and C in relation to the load resistance R for the operation as a square wave inverter. The following equations can be written from the circuit of Fig. 2.35 (a).

$$i_0 = \frac{V}{R}$$

$$V_0 = V$$

$$V_{C_2} = 2V$$

$$V_{C_1} = 0$$

The circuit variables at the instant of triggering T_2 are ($t = 0^+$):

$$V_{C_2}(0) = 2V,\ V_{C_1}(0) = 0,$$

$$i_{C_2} = -\frac{V}{R},\ i_{C_1} = \frac{V}{R},$$

$$i_{T_2} = \frac{V}{R},\quad \frac{di_{T_2}}{dt} = \frac{2V}{L},\ i_0 = \frac{V}{R}$$

Since the capacitor voltage at $t = 0^+$ must be the same as that at $t = 0^-$

$$\frac{1}{C_1}\int_0^t i_{C_1} dt + \frac{1}{C_2}\int_0^t i_{C_2} dt + 2V = 2V$$

or
$$\frac{1}{C}\int_0^t (i_{C_1} + i_{C_2})\, dt = 0 \tag{2.36}$$

Differentiating equation (2.36)

$$i_{C_1} + i_{C_2} = 0 \tag{2.37}$$

Summing the currents at the anode, 'a'

$$i_{C_1} - i_{C_2} = i_{T_2} + i_0 \tag{2.38}$$

when T_2 is turned on, $v_{L_2} = v_{C_2}$

$$\frac{L di_{T_2}}{dt} = \frac{1}{C}\int_0^t i_{C_2} dt + 2V \tag{2.39}$$

Considering the loop formed by the load, T_2, L_2, and the bottom source

$$v_{C_2} = i_0 R + V$$

$$C\frac{dv_{C_2}}{dt} = RC\frac{di_0}{dt} = i_{C_2} \tag{2.40}$$

From equations (2.37) to (2.40), the following differential equation can be derived

$$\frac{d^2 i_{T_2}}{dt^2} + \frac{1}{2RC}\cdot\frac{di_{T_2}}{dt} + \frac{i_{T_2}}{2LC} = 0 \tag{2.41}$$

The differential equation (2.41) can be solved, and three possible solutions exist for i_{T_2} depending upon the values of L, C, and R.

The load resistance R must be less than $\sqrt{\frac{L}{8C}}$ so that the current (i_{T_2}) response is over-damped. Otherwise the response may become oscillatory. The waveforms illustrated in Fig. 2.35 (b) correspond to an over-damped condition. The turn off time is also shown in this figure.

2.5.3.2 *Current Commutation*

Figures 2.36 and 2.37 illustrate different ways of introducing a current pulse [6] in parallel and series with the thyristor to be

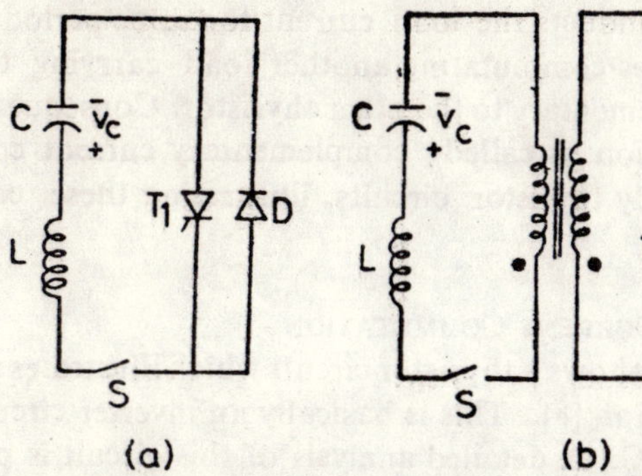

Fig. 2.36. Different ways of connecting current pulse in parallel to the thyristor:
(a) LC resonant commutation directly across T_1.
(b) LC resonant commutation through a transformer.

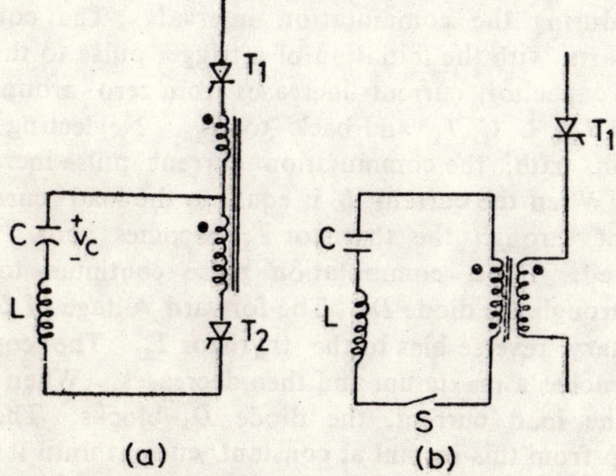

Fig. 2.37. Different ways of connecting a current pulse in series with the thyristor:
(a) LC resonant circuit through a coupled inductor.
(b) LC resonant circuit through a transformer.

commutated respectively. This current pulse usually generated by an L–C resonant circuit is coupled either directly or through a transformer. As in the case of voltage commutation, the switch S usually represents another thyristor switch. If the thyristor switch remains active only during the commutation of the main thyristor and dormant during the rest of period, this method of commutation is recognized as 'Auxiliary current commutation'. If the thyristor conducts the load current for some period of the output cycle besides commutating another load carrying thyristor, it becomes complementary to the other thyristor. Consequently, this type of commutation is called 'complementary current commutation'. Let us study thyristor circuits illustrating these commutation schemes.

(a) AUXILIARY CURRENT COMMUTATION

Figure 2.38 (a) shows a thyristor circuit which illustrates auxiliary current commutation [8]. This is basically an inverter circuit which converts DC to AC. A detailed analysis of this circuit is presented in Chapter 7. T_1 and T_2 are the main thyristors, and T_{1a} and T_{2a} are the auxiliary thyristors. Let us consider the commutation of T_1. Earlier to the commutation, T_1 is conducting the load current. The capacitor is charged to a voltage $-V$ by an earlier commutation. All other devices are in the blocking mode. The load is assumed to be inductive, and the load current is assumed to remain constant during the commutation interval. The commutation interval starts with the initiation of a trigger pulse to the thyristor T_{1a}. The capacitor current increases from zero around the loop formed by T_{1a}, L, C, T_1 and back to T_{1a}. Neglecting the resistance in this path, the commutation current pulse increases sinusoidally. When the current i_C is equal to the load current I_0, the net current through the thyristor T_1 becomes zero. T_1 is thus commutated. The commutation pulse continues to build up flowing through the diode D_1. The forward voltage of D_1 provides the necessary reverse bias to the thyristor T_1. The commutation current reaches a maximum and then decreases. When it is again equal to the load current, the diode D_1 blocks. The capacitor charges up from this instant at constant current until it is equal to the supply voltage. Diode D_2 turns on as it is now forward biased. The capacitor current decreases and the current through the diode D_2 increases gradually. The capacitor overcharges beyond the

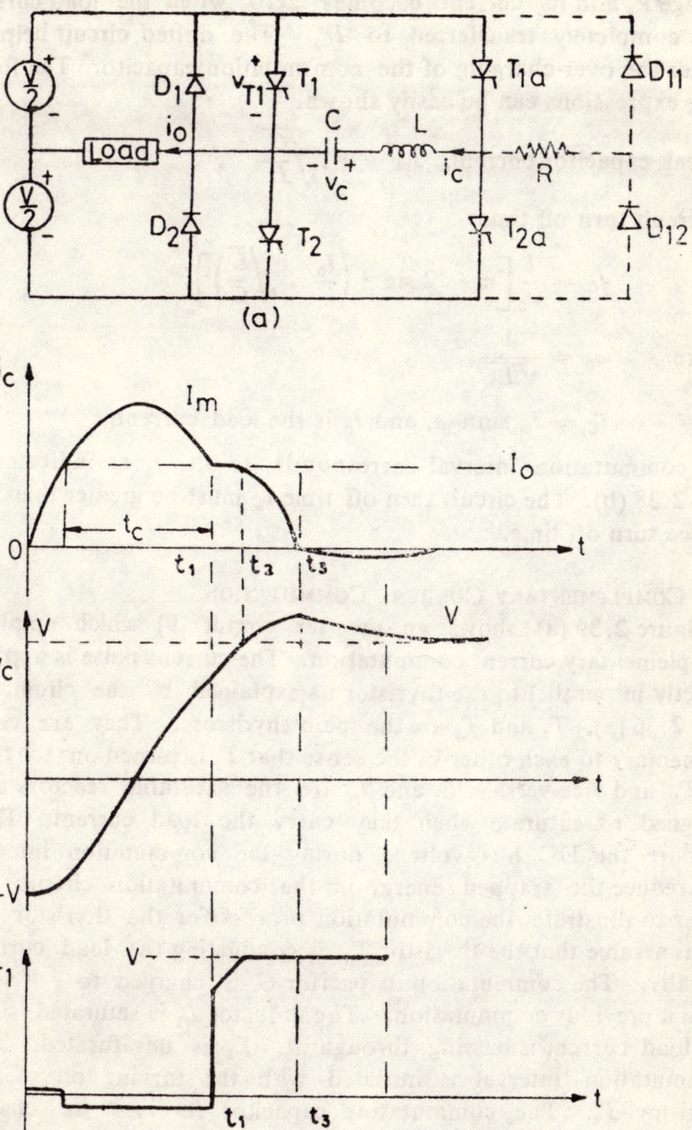

Fig. 2.38. Auxiliary current commutation:

(a) Circuit.

(b) Waveforms of some circuit variables.

voltage V, and its current becomes zero, when the load current gets completely transferred to D_2. The dotted circuit helps to reduce the over-charging of the commutation capacitor. The following expressions can be easily shown.

Peak capacitor current, $I_m = V.\sqrt{\frac{C}{L}}$

Circuit turn off time,

$$t_C = \frac{1}{\omega_0}\left[\pi - 2\sin^{-1}\left(\frac{I_0}{V} \cdot \sqrt{\frac{L}{C}}\right)\right]$$

where $$\omega_0 = \frac{1}{\sqrt{LC}}$$

$i_C = I_m \sin \omega_0 t$, and I_0 is the load current.

The commutation interval corresponds to 0 to t_3 as indicated in Fig. 2.38 (b). The circuit turn off time t_C must be greater than the device turn off time.

(b) Complementary Current Commutation

Figure 2.39 (a) shows an inverter circuit [9] which employs complementary current commutation. The current pulse is applied directly in parallel to the thyristor as explained by the circuit of Fig. 2.36 (a). T_1 and T_2 are the main thyristors. They are complementary to each other in the sense that T_2 is turned on to turn off T_1 and vice-versa. L_1 and L_2 are the saturable reactors and designed to saturate when they carry the load current. They support the DC bus voltage during the commutation interval and reduce the trapped energy in the commutation circuit. In order to illustrate the commutation process for the thyristor T_1, let us assume that the thyristor T_1 is conducting the load current initially. The commutation capacitor C is charged to $+V$ volts from a previous commutation. The inductor L_1 is saturated since the load current is passing through it. L_2 is unsaturated. The commutation interval is initiated with the turning on of the thyristor T_2. The commutating capacitor reverses its charge through T_1 and T_2. The inductor L_2 supports the entire DC voltage, V. If the loss in the LC resonant commutation circuit is neglected, the commutation current pulse is sinusoidal in response. After one half-cycle, the current becomes zero and hence the thyristor T_2 is commutated. But T_1 continues to conduct since the load current is passing through it. As in the previous commutation circuits, it is assumed that the load current is constant

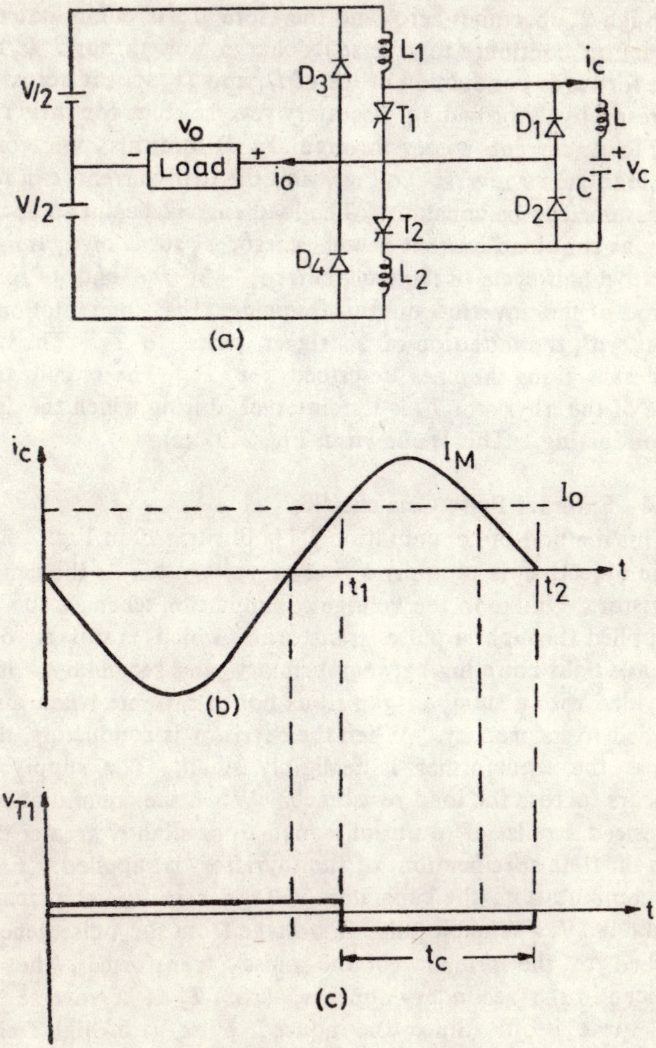

Fig. 2.39. Complementary current commutation:

(a) Thyristor circuit;

(b) Commutating capacitor current;

(c) Voltage across the thyristor T_1.

during the commutation period which is usually of the order of a few microseconds. The commutating capacitor reverses its charge again through D_2 and T_1 giving another pulse of current. When the capacitor current is equal to the load current, the net current through T_1 becomes zero and therefore T_1 is commutated. The capacitor continues to reverse its charge now through D_2 and D_1. The forward conduction drops of D_1 and D_2 appear across T_1 and T_2 respectively providing necessary reverse bias for their turn off. The load current passes through D_4. From this point onwards, the load gets a reverse voltage and the load current can no more be assumed to be constant. The load current begins to decay. As soon as the load current becomes zero, T_2 turns on again giving a negative half-cycle of the load current. At the end of one half period of the inverter output frequency, the commutation of T_2 starts with the initiation of a trigger pulse to T_1. The turn off process is along the lines described for T_1. The circuit turn off time of the thyristor T_1 is the interval during which the diode D_1 is conducting. This is shown in Fig. 2.39 (c).

2.5.4 External Pulse Commutation

This method of commutation [1] is illustrated in Fig. 2.40. The main objective is to apply a reverse voltage across the conducting thyristor. Unlike in the voltage commutation schemes, this voltage is applied through a pulse transformer which is suitably designed to have tight coupling between primary and secondary. It is also provided with a small air gap so as not to saturate when a pulse is applied to its primary. When the thyristor is conducting, the drop across the transformer is negligibly small. The supply voltage appears across the load resistance. When the commutation of T_1 is desired a pulse of duration equal to or slightly greater than the turn off time specification of the thyristor is applied. Earlier to the commutation, the capacitor voltage remains at a small value of about 1V. When a pulse of voltage from the pulse generator is applied to the primary of the pulse transformer, the voltage induced in the secondary appears across T_1 as a reverse voltage and turns it off. Since the induced pulse is of high frequency, the capacitor offers almost zero impedance. After T_1 is turned off, the load current decays to zero.

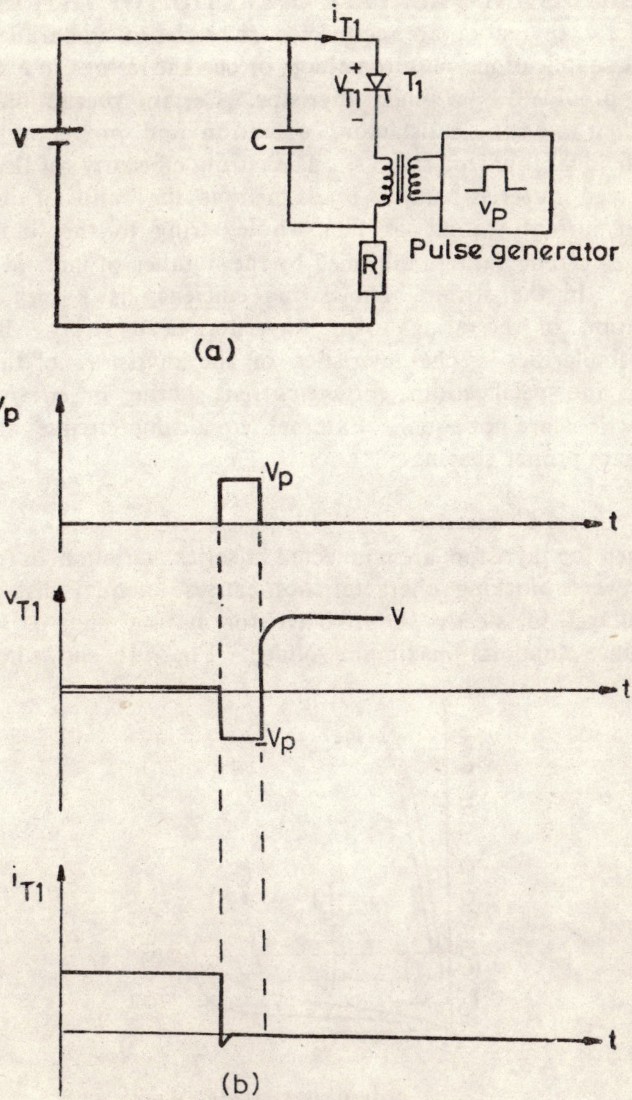

Fig. 2.40. External Pulse Commutation:
(a) Circuit;
(b) Typical waveforms.

2.6 SERIES AND PARALLEL OPERATION OF THYRISTORS

Several thyristors are connected in series or in parallel when circuit applications require voltage or current ratings in excess of those provided by a single thyristor. Certain precautions must be taken to ensure satisfactory operation and proper utilisation of individual thyristor ratings. The string efficiency of thyristors connected in series/parallel is defined as the ratio of the total voltage/current rating of the whole string to the individual voltage/current rating multiplied by the number of units in series/parallel in the string. The string efficiency is a measure of utilisation of the ratings of the thyristors in the string. Because of dissimilarities in characteristics of the thyristors of the same ratings and specifications, voltage/current sharing in series/parallel connections are not equal. External equalising circuits are used to ensure proper sharing.

2.6.1 Series Connection

When the thyristors are connected in series, variation in forward and reverse blocking characteristics causes unequal distribution of voltage in steady state. Thyristor having highest leakage resistance supports maximum voltage. Fig. 2.41 shows how two

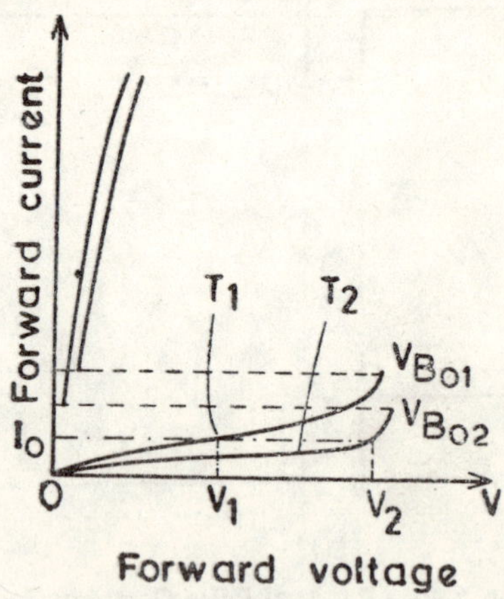

Fig. 2.41. Sharing of voltage between series connected thyristors T_1 and T_2.

thyristors, each having forward blocking capability of V_2 volts, share the voltage. While, thyristor T_2 is on the point of breakdown, thyristor T_1 is supporting only a small voltage. The maximum voltage that the string can block is only (V_1+V_2) and not $2V_2$. Full forward blocking voltage capabilities of each thyristor can only be realised by equal distribution of forward blocking voltages.

A nearly equal distribution of voltages in steady state can be realised by connecting a suitable resistance in parallel with each thyristor such that each parallel combination has the same resistance. For this, resistances of different values will be required. A simple arrangement, permitting reasonably uniform distribution of voltages, will be to use the same resistance in parallel with each thyristor (Fig. 2.42). This is called the static equalising circuit. Value of parallel resistances can be obtained as follows [1].

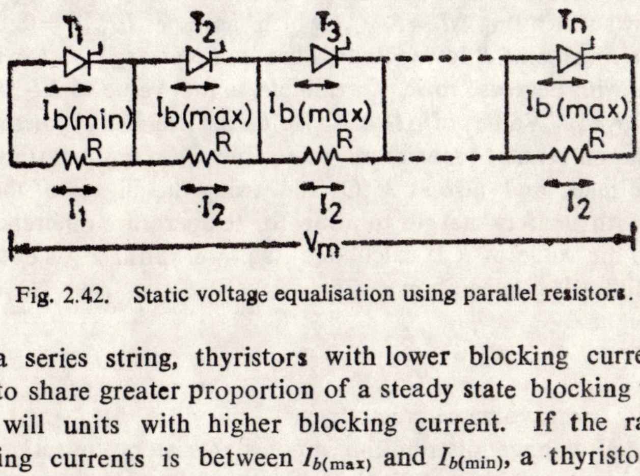

Fig. 2.42. Static voltage equalisation using parallel resistors.

In a series string, thyristors with lower blocking current will have to share greater proportion of a steady state blocking voltage than will units with higher blocking current. If the range of blocking currents is between $I_{b(max)}$ and $I_{b(min)}$, a thyristor in a series string consisting of n-thyristors will be required to share the maximum voltage when it has a blocking current of $I_{b(min)}$ and the rest $(n-1)$ thyristors have the blocking current $I_{b(max)}$. This worst situation is shown in Fig. 2.42.

Here $I_{b(min)} + I_1 = I_{b(max)} + I_2$

or
$$I_{b(max)} - I_{b(min)} = I_1 - I_2 = \Delta I_b \text{ (say)} \tag{2.42}$$

If V_p is the maximum permissible blocking voltage then

$$V_p = I_1 R \tag{2.43}$$

String voltage V_m is given by

$$V_m = V_p + (n-1)\, RI_2 \tag{2.44}$$

Substituting from eqn. (2.42) in eqn. (2.44)

$$V_m = V_p + (n-1)\,R\,(I_1 - \Delta I_b)$$
$$= V_p + (n-1)I_1R - (n-1)R\,\Delta I_b$$

Substituting from eqn. (2.53) gives

$$V_m = nV_p - (n-1)R\,\Delta I_b$$

Now,

$$R < \frac{nV_p - V_m}{(n-1)\,\Delta I_b} \tag{2.45}$$

In general thyristor data sheet provides only the maximum blocking current (which is applicable to both forward and reverse blocking) and rarely ΔI_b. If only maximum blocking current is known, a conservate estimate of R is obtained from equation (2.43) by substituting $\Delta I_b = I_{b(max)}$ which assumes $I_{b(min)} = 0$. This will give a value of R lower than what is necessary. Since losses increase with decrease in R, if available, actual value of ΔI_b should be employed. Value of ΔI_b can also be determined by measuring blocking currents at maximum rated junction temperature and rated voltage, and also at 25°C; and using the higher of the two values with a safety margin to allow for temperature differences.

Once the value of R is calculated its power rating P_R is obtained by the following expression

$$P_R = \frac{V_r^2}{R}$$

where V_r = rms voltage across R.

Unequal voltage distribution among thyristors in series also occur during transient conditions of turn off, turn on and high frequency operation. If one thyristor in the string has short reverse recovery time, it will alone support the string reverse voltage. A thyristor which has highest turn on time in the string will have to momentarily support the full string voltage. One method that can be used to minimise turn on time differences is to use gate pulses with fast rise time and magnitude as large as the rating allows.

Shunt capacitors are more effective than shunt resistances in maintaining equal voltage distribution during turn off, turn on and high frequency operation. During turn off when a thyristor with fast reverse recovery stops passing reverse current, the capacitor carries the reverse recovery current to the rest of the string, until other thyristors also cease passing reverse current.

The shunt capacitance also forces equal voltage sharing during sudden changes in applied voltage since it swamps out differences between thyristor self-capacitances and inequalities between distributed circuit capacitances.

Since the shunt capacitor discharges through the thyristor during turn on, resistance R_c is provided to limit the magnitude and rate of change of current. It also damps out the high frequency oscillations due to ringing of shunt capacitor with circuit inductance. Although the value of R_c should be sufficient to limit the magnitude and rate of rise of current during turn on, it should not be so large that it either nullifies the effect of shunt capacitors or develops an excessively high voltage during flow of reverse recovery current through it [1]. This C and R_c combination is called dynamic equalising network and is shown in Fig. 2.43.

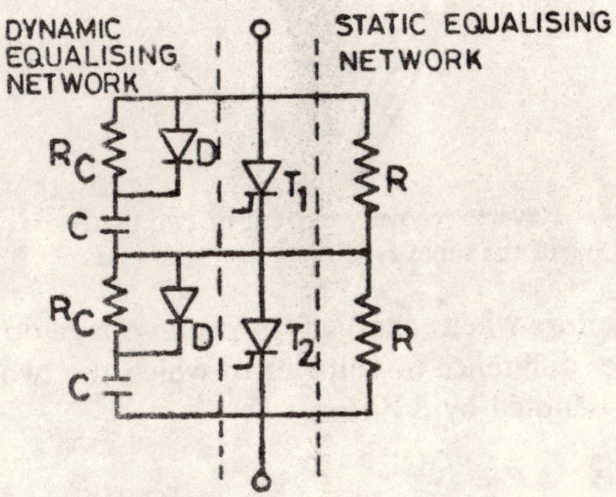

Fig. 2.43. Voltage equalising arrangement for thyristors in series.

Diode D can be placed across damping resistor R_c with a view to bypass it for forward voltages. This increases the effectiveness of capacitor C in controlling dv/dt across the thyristor. R_c still remains in the circuit to limit current and di/dt due to discharge of capacitor through the thyristor during turn on. Some small amount of damping resistance should still be used in series with the diode to suppress ringing. The diodes should have soft recovery characteristics to avoid high voltage spikes which may be generated due to the abrupt recovery action of diodes.

Value of capacitor C in Fig. 2.43 is calculated as follows:

Since the maximum unbalance of voltage under transient

conditions occur during turn off due to dissimilarities in reverse recovery characteristics, the capacitor C is chosen based on the reverse recovery characteristics. Reverse recovery characteristics of two thyristors in a series string is shown in Fig. 2.44. Shaded area ΔQ represents the difference in reverse recovery charges of the two thyristors. Due to the difference in recovery characteristics,

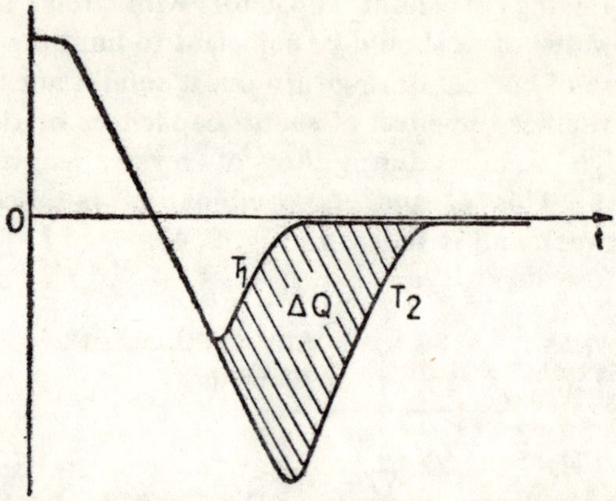

Fig. 2.44. Reverse recovery current of two thyristors (T_1, T_2) of the same type.

the shunt capacitors when charged to peak voltage do not charge equally. If the difference in voltages to which the two capacitors are charged is denoted by ΔV then

$$\Delta V = \frac{\Delta Q}{C} \tag{2.46}$$

The maximum unbalance in voltages will occur when a series string with n thyristors has one thyristor with the fastest recovery and the remaining $(n - 1)$ thyristors with the slowest recovery among that type of thyristors.

Let ΔQ_{max} be the maximum difference in recovery charge and ΔV_{max} the maximum difference in voltage under this situation. If V_p is the maximum permissible blocking voltage then,

$$\text{voltage across slow thyristors} = (V_p - \Delta V_{max})$$

The string voltage V_m will now be

$$V_m = V_p + (n - 1)(V_p - \Delta V_{max})$$

or

$$(n - 1)\Delta V_{max} = nV_p - V_m \tag{2.47}$$

since

$$\Delta V_{max} = \Delta Q_{max}/C$$

$$C \geqslant \frac{(n-1)\,\Delta Q_{\max}}{nV_p - V_m} \tag{2.48}$$

Since the cathodes of thyristors in the series string operate at different voltages there should be isolation between the firing circuit of each thyristor. This is done by using either a single pulse transformer with properly insulated multiple secondary windings, as shown in Fig. 2.45 or by using multiple pulse transformers with

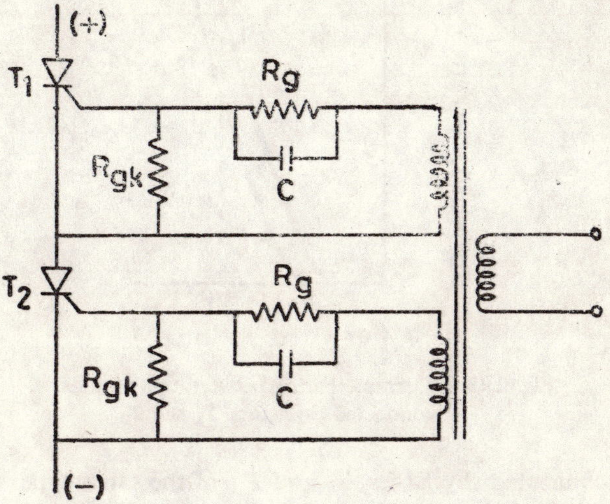

Fig. 2.45. Simultaneous triggering of series connected thyristors through a pulse transformer.

their primaries connected in parallel and secondaries connected to the gates of each individual thyristor. The interwinding insulation should be able to support at least the peak supply voltage. Resistance R_{gk} is added to the gate circuit to prevent unequal loading of the trigger signal due to difference in the gate to cathode impedance of the individual units. When the total energy available for triggering is limited, resistors R_{gk} are replaced by capacitors C. Capacitors tend to equalise charge supplied to each thyristor gate without additional energy dissipation. Resistance R_g is connected from gate to cathode of each thyristor to provide a discharge path for the capacitor.

2.6.2 Parallel Connection

VI characteristic of a thyristor during forward conduction is such that a small change in voltage results in large variation in current. Thyristors connected in parallel will have equal voltage drops, therefore, their characteristics must be essentially identical for even sharing of load current. A slight mismatch in characteristics will reduce the current handling capability of the unit by a large amount. Fig. 2.46 shows the sharing of current between two

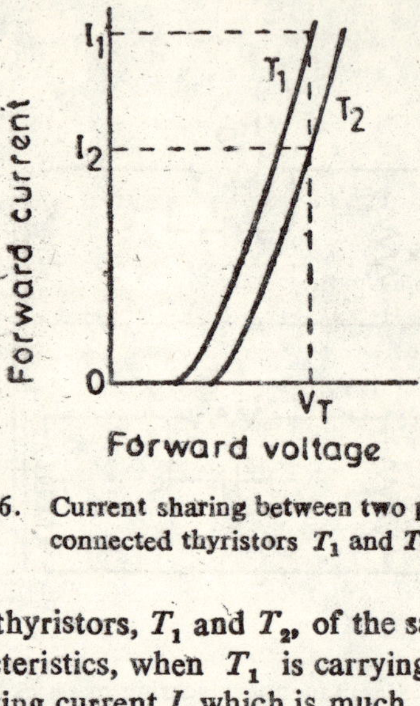

Fig. 2.46. Current sharing between two parallel connected thyristors T_1 and T_2.

parallel connected thyristors, T_1 and T_2, of the same type. Due to mismatch in characteristics, when T_1 is carrying rated current I_1 thyristor T_2 is carrying current I_2 which is much less than its rated value. Total rated current of the parallel unit is only $(I_1 + I_2)$ instead of $2I_1$.

If one of the thyristors has a large turn on delay, the forward voltage across it drops to a very low value when other thyristors in parallel have already turned on. For a given gate drive there is a minimum forward voltage called finger voltage for a thyristor to turn on. If the forward voltage is less than the finger voltage the thyristor will fail to turn on. The thyristors already on will now be required to carry the complete load current. They may be damaged due to overloading. Matched thyristors are generally available for direct parallel operation. The current rating of the combined unit is then around 10 per cent less than the sum of their individual ratings to allow for tolerances and thermal effects. The parallel thyristors are usually mounted on the same heat sink to maintain equal junction temperatures. This minimises current unbalance resulting from thermal differences between thyristors. Sharing of surge currents does not present any problem.

Unequal inductive effect on current carrying conductors may also cause uneven current distribution. When the thyristors are arranged as shown in Fig. 2.47(a), the middle one will have higher inductance due to higher flux linkage. More current will flow through the outer ones compared to the middle one. By having thyristors located symmetrically on the heat sink, as shown in Fig. 2.47(b), even distribution of current is obtained.

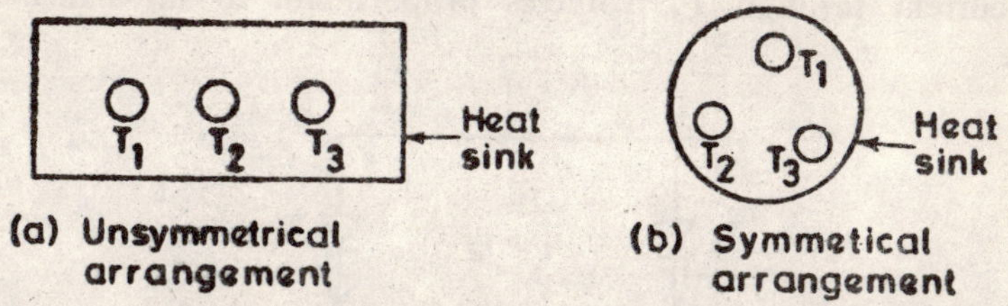

Fig. 2.47. Location of thyristors on heat sink for parallel operation.

Triggering requirements may differ quite widely between individual thyristors even when they are parallel matched. It is advisable to use gate signals with fast rise time and magnitude as large as the rating allows. This helps in reducing turn on dissimilarities between the individual units. The gate signals must be applied until the latching currents of all the thyristors have been exceeded. If gate pulse is removed before one or more thyristors have not latched into conduction, they will not conduct for the remainder of the cycle, thus overloading other parallel units. Another important problem related to triggering of parallel thyristors arises when the load current assumes low and high values during the conduction interval. Voltage drop across the unit at low current values may be insufficient to keep current through some of the thyristors above the holding current value. These thyristors will turn off. When the load current increases, rest of the thyristors will have to carry it. This problem can be overcome by using a gate signal lasting for full conduction interval. As long as gate current is maintained, the thyristor with a tendency to turn off at low load currents will switch into conduction again when the load current is increased.

When unmatched thyristors are to be connected in parallel, forced current sharing is necessary. For this external balancing series resistors or reactors are used. It is possible to get good sharing, to the extent that only 10% derating may be necessary. The disadvantage with the balancing resistors is of significant power loss typically equal to that of the device, as the voltage drop across the resistors would be comparable with the anode to cathode voltage drop across the thyristors. The reactors are more efficient, however they are expensive. Fig. 2.48 illustrates how two magnetically coupled reactors can be used to achieve uniform current sharing. If the current through thyristor T_1 tends to increase above the current through T_2, voltages proportional to unbalance in

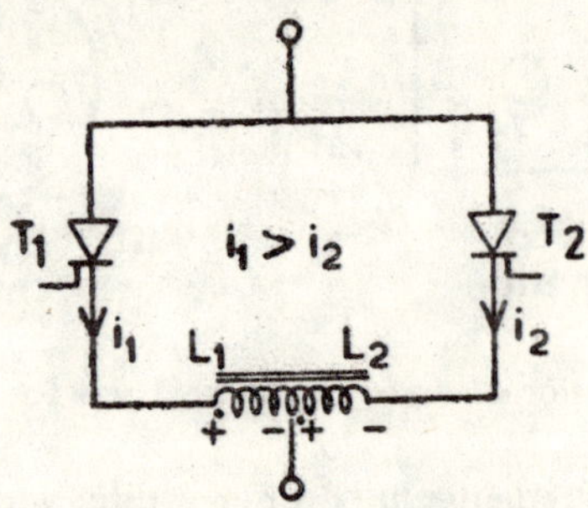

Fig. 2.48. Current equalisation using magnetically coupled reactors.

current and with the polarity as shown are induced. Voltage in reactor L_1 opposes the flow of current and voltage in L_2 boosts the current. As a result, the current unbalance is reduced. The most

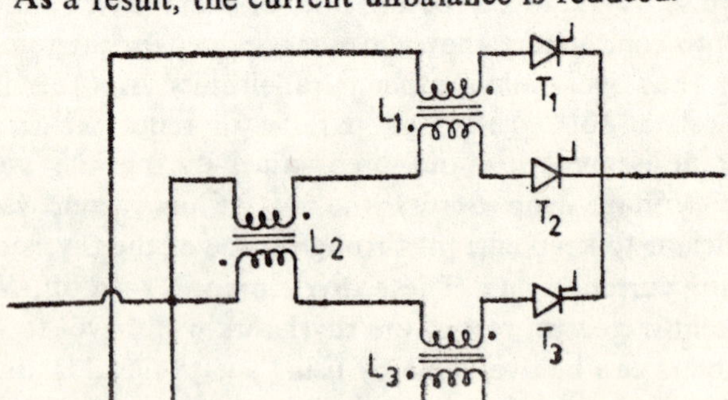

Fig. 2.49. Current equalising reactor connections for three thyristors in parallel.

important magnetic requirements of such reactors are high saturation and low residual flux densities in order to provide as much change in total flux in each cycle as possible. Arrangements of magnetically coupled reactors for three and four parallel connected thyristors are shown in Figs. 2.49 and 2.50 respectively.

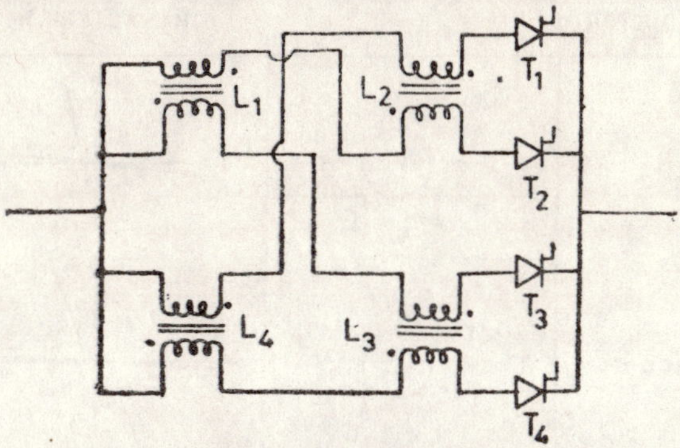

Fig. 2.50. Current equalising reactor connections for four thyristors in parallel.

2.7 TRIAC AND OTHER MEMBERS OF THYRISTOR FAMILY

2.7.1 Introduction

The silicon controlled rectifier (SCR) is by far the best known member of thyristor family. It has three terminals (anode, cathode, control gate) and is classified as a reverse blocking triode thyristor. The other members of thyristor family are silicon unilateral switch (SUS), light activated silicon controlled rectifier (LASCR), programmable unijunction transistor (PUT), silicon controlled switch (SCS), silicon bilateral switch (SBS), bidirectional triode thyristor (TRIAC). All these devices with the exception of the triac are generally low power devices which are widely used in low power control circuits. The silicon bilateral switch is a bidirectional thyristor which conducts current in either direction and so is a triac. However, the triac is commercially available in high power ratings. The name 'TRIAC' is an acronym that has been coined to identify the *TRI*ODE *AC* semiconductor switch which is triggered into conduction by a gate signal in a manner similar to the action of an SCR. But it differs from the SCR in that it can conduct in both directions of current flow in response to a positive or negative gate signal. Figure 2.51 shows different members of thyristor family along with their V-I characteristics and circuit symbols.

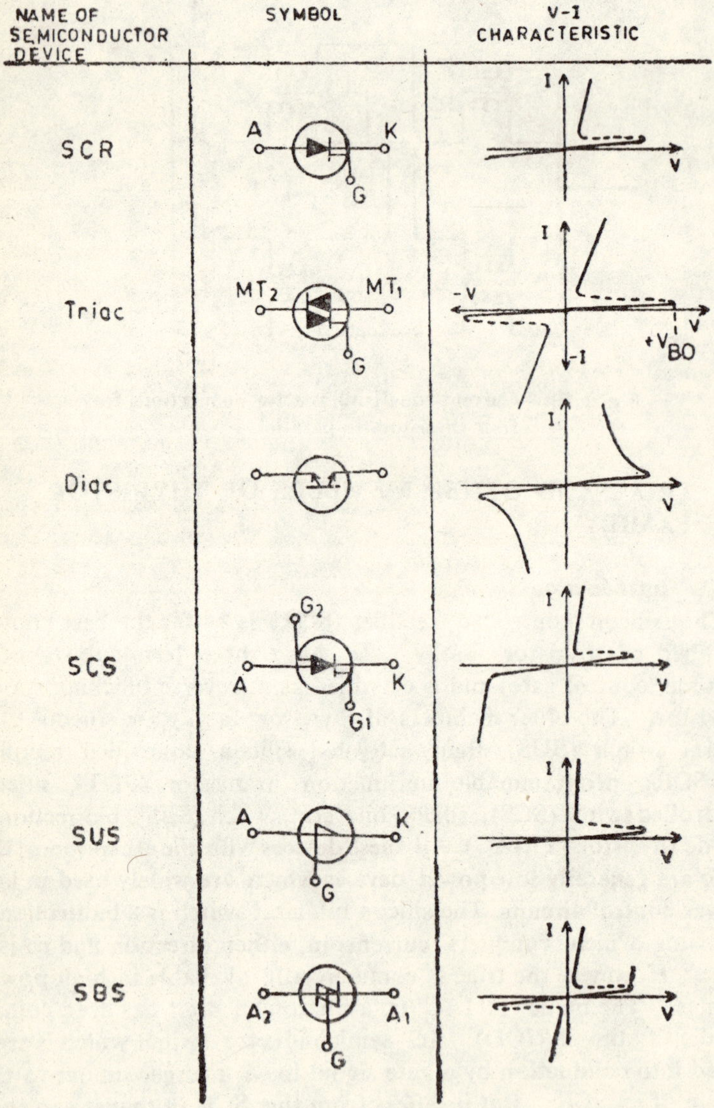

Fig. 2.51. (continued).

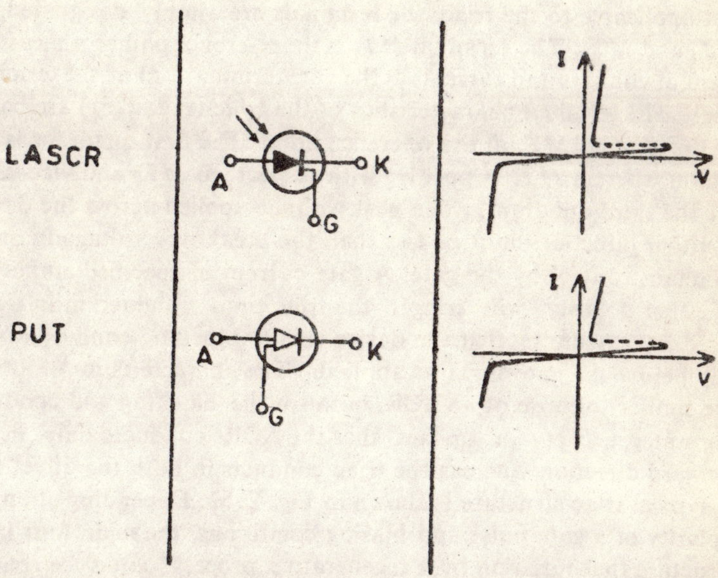

Fig. 2.51. Members of thyristor family.

Other types of thyristors which are finding widespread use in recent times are amplifying gate thyristor (AGT), reverse conducting thyristor (RCT), and gate turn off thyristor (GTO). These devices are commercially available in high power ratings. Asymmetrical thyristor and field controlled thyristor are new devices yet in the process of development. The amplifying gate thyristor is designed in such a way that it is capable of providing higher value of dI/dt with lower turn on time in comparison to that of a conventional thyristor. The reverse conducting thyristor has a diode built into its structure. The circuit symbol is equivalent to a thyristor and a diode connected in antiparallel. This finds application in inverters and choppers. The turn off time is usually quite small. The gate turn off thyristor can be easily turned off by the application of a negative gate current without reversing the anode to cathode voltage. It does not require any commutation circuitry in the main circuit. The power circuits particularly choppers and inverters are considerably simplified by using gate turn off thyristors.

2.7.2 Triac

Figure 2.51 shows the circuit symbol of a triac along with its terminal characteristics. Since the terms 'anode' and 'cathode' are not applicable to the triac, the terminals are simply designated by MT_1 and MT_2. The terminal MT_1 is the reference point for measurement of voltages and currents at the gate terminal and at the terminal MT_2. The terminal characteristics of the triac (Fig. 2.51) are based on the terminal MT_1 as the reference point. The first quadrant is the region wherein MT_2 is positive with respect to MT_1 and vice-versa for the third quadrant. The peak voltage applied across the device in either direction must be less than the breakover voltage in order to retain control by the gate. A gate current of specified amplitude of either polarity will trigger the triac into conduction in either quadrant, assuming that the device is in a blocking condition initially before the gate signal is applied. The characteristics of a triac are similar to those of an SCR, both in the blocking and conducting states, except for the fact that the SCR conducts only in the forward direction whereas the triac conducts in both the directions. A typical triac structure is shown in Fig. 2.52. Depending upon the polarity of a gate pulse and biasing conditions, the main four layer structure that turns on by a regenerative process could be one of

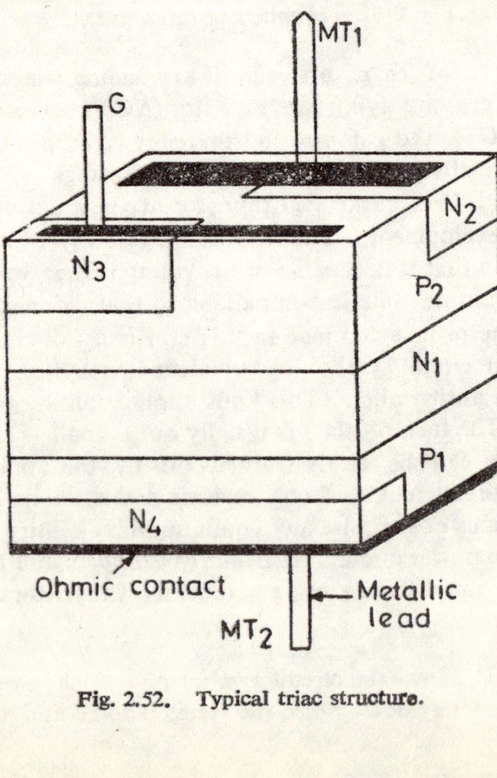

Fig. 2.52. Typical triac structure.

$P_1N_1P_2N_2$, $P_1N_1P_2N_3$, or $P_2N_1P_1N_4$ [1].

2.7.3 Triggering Mechanism of a Triac

As pointed out in section 2.7.1 the triac can be turned on with positive or negative gate current keeping the MT_2 terminal at positive or negative potential. Let us examine how the triac switches into conduction in each case.

2.7.3.1 *Main Terminal MT_2 Positive, Positive Gate Current*

The device operates in the first quadrant of *V–I* characteristics (Fig. 2.51). Fig. 2.53(a) (i) shows the cross-sectional view of the structure. The device turns on in the conventional manner as in the case of an SCR. However, in the case of a triac, the gate current requirement is higher for turn on at a particular voltage. This can readily be seen from the structure of a triac as shown in Fig. 2.52. The gate current flows from the gate lead to the terminal MT_1 through the P_2—N_2 junction. Because of ohmic contacts of gate and MT_1 terminals on the P_2-layer, some more gate current flows from the gate lead G to the main terminal MT_1 through the

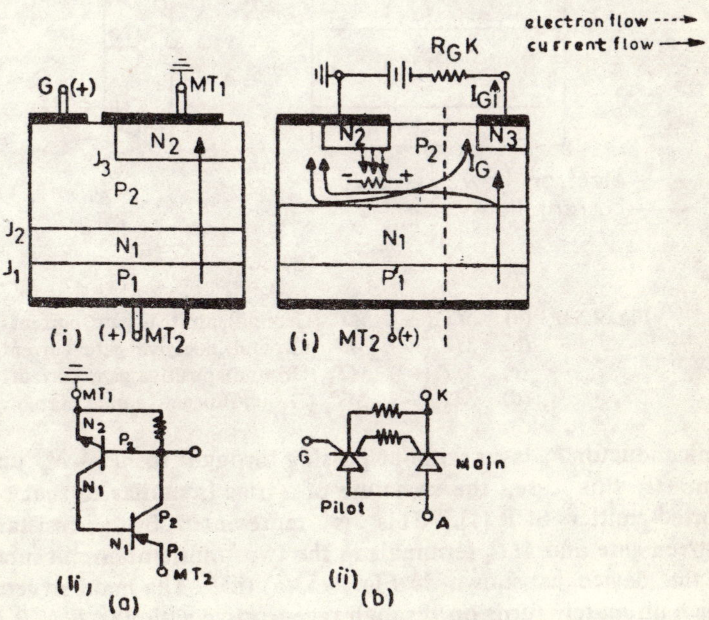

Fig. 2.53 (Continued)

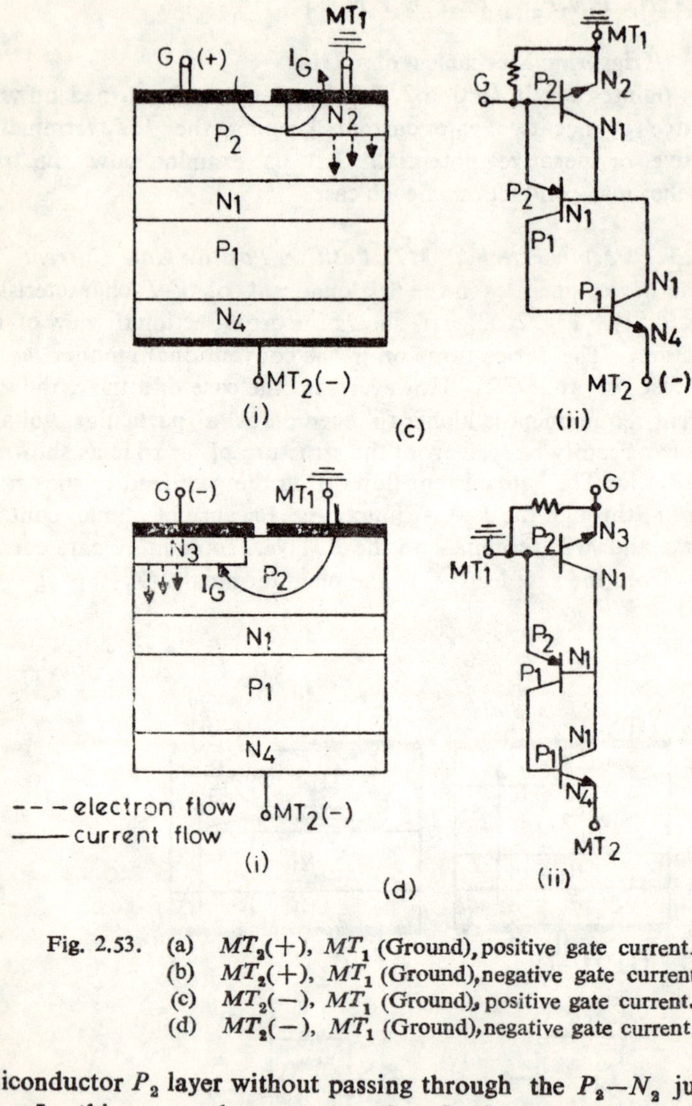

Fig. 2.53. (a) $MT_2(+)$, MT_1 (Ground), positive gate current.
(b) $MT_2(+)$, MT_1 (Ground), negative gate current.
(c) $MT_2(-)$, MT_1 (Ground), positive gate current.
(d) $MT_2(-)$, MT_1 (Ground), negative gate current.

semiconductor P_2 layer without passing through the P_2-N_2 junction. In this case, the structure of a triac is similar to that of a shorted-emitter SCR [1]. This is represented by a resistance between gate and MT_1 terminals in the two-transistor circuit model of the device as shown in Fig. 2.53(a) (ii). The main structure which ultimately turns on through regenerative action is $P_1N_1P_2N_2$. When the gate current flows across the P_2-N_2 junction, the P_2

layer is flooded with electrons. These electrons diffuse to the edge of the junction J_2, and are collected by the N_1 layer. As a result, the electrons build a space charge in the N_1 region and more holes from P_1 diffuse into N_1 to neutralize the negative space charge. These holes arrive at the junction J_2. They produce a positive space charge in the P_2 region which results in more electrons being injected from N_2 into P_2. This results in positive regeneration and ultimately the structure $P_1N_1P_2N_2$ conducts the external current. The device is more sensitive in this mode compared to the other modes of operation. This is the recommended method of triggering if the conduction is desired in the first quadrant.

2.7.3.2 *Main Terminal MT_2 Positive, Negative Gate Current*

A cross sectional view of the structure is shown in Fig. 2.53(b)(i). This structure is similar to a typical junction gate thyristor structure [1]. The main structure that conducts ultimately is $P_1N_1P_2N_2$. However, the initiation of conduction process is provided by the auxiliary structure $P_1N_1P_2N_3$. Initially, the gate current I_G forward biases the gate junction P_2-N_3 of the auxiliary $P_1N_1P_2N_3$ structure, and this structure turns on by normal regenerative action. As $P_1N_1P_2N_3$ turns on, the voltage drop across it falls, and the right hand section of the region P_2 moves towards anode potential. Since the left hand section of the region P_2 is clamped to the cathode potential, a transverse voltage gradient now exists across P_2 and current flows laterally through P_2. As the right hand edge of P_2N_2 becomes forward biased, electrons are injected at this point and the main structure turns on. The device auxiliary structure, $P_1N_1P_2N_2$ may be considered as a pilot SCR while the structure, $P_1N_1P_2N_2$ may be regarded as the main SCR both being built in one common structure. The anode current of the pilot SCR serves as a gate drive to the main SCR. Comparing the turn on action with the positive gate current as explained in section 2.7.3.1, the device is somewhat less sensitive and therefore requires more gate current in this mode.

2.7.3.3 *Main Terminal MT_2 Negative, Positive Gate Current*

In this mode, the device operates in the third quadrant when it is triggered into conduction. The turn on is initiated by remote gate control. The main structure that leads to turn on is $P_2N_1P_1N_4$ with N_2 acting as a remote gate as shown in Fig. 2.53(c) (i). The junction P_2N_2 is forward biased by the external gate current I_G. It injects electrons as shown by the dotted line. These electrons diffuse through the region P_2 and are collected by the junction P_2N_1. The electrons from N_2 collected by P_2N_1 junction cause an increase of current through the junction P_2N_1. The holes injected from P_2, diffuse through N_1 and arrive in P_1. As a result, a positive space charge builds up in the P_1 region. More electrons from N_4 diffuse into P_1 to neutralize the positive space charge. These electrons arrive at the junction J_2. They produce a negative space charge in the N_1 region which results in more holes being injected from P_2 into N_1. This regenerative process continues until the structure $P_2N_1P_1N_4$ completely turns on and conducts the current which is limited by the external load.

The turn on process can also be explained by the transistor analogy. Fig. 2.53c (ii) shows the transistor equivalent circuit. The two transistors $P_2N_1P_1$ and $N_1P_1N_4$ form the main structure while the remote layer N_2 in conjunction with the P_2N_1 form an additional transistor. Since the gate (base of $N_2P_2N_1$) is held at a positive potential with reference to the main terminal MT_1, electrons are injected from the emitter N_2. These are collected by the collector thereby increasing the base drive of transistor $P_2N_1P_1$. This initiates the regenerative action of the main structure $P_2N_1P_1N_4$ and ultimately the device switches into conduction. In the third quadrant, the device is less sensitive with the positive gate current.

2.7.3.4 *Main Terminal MT_2 Negative, Negative Gate Current*

Fig. 2.53 d (i) shows the cross section of main structure ($P_2N_1P_1N_4$) and its transistor equivalent circuit (Fig. 2.53d(ii)). N_3 acts as a remote gate. The external gate current I_G causes P_2N_3 junction to become forward biased, and electrons are injected as shown by the dotted lines. These electrons from N_3 collected by P_2N_1 cause an increase of current across P_1N_1. The structure turns on by the regenerative action. Looking at the structure by transistor circuit analogy, the remote transistor $N_3P_2N_1$ is properly biased to inject electrons to the collector N_1. This increases the base drive of $P_2N_1P_1$ initiating the regenerative action of the main structure $P_2N_1P_1N_4$. This leads to switching of the device into conduction. The device is more sensitive in this mode compared with the turn on by the positive gate current as discussed in section 2.7.3.3.

In all the modes of operation, the sensitivity of the triac is the greatest in the first quadrant with the positive gate current and also in the third quadrant with the negative gate current. The sensitivity is slightly lower in the first quadrant with the negative gate current. Further the device is much less sensitive in the third quadrant with the positive gate current. This particular mode of operation is not commonly used. In this context, the magnitude of the gate current required to turn on indicates the degree of sensitivity.

2.7.4 Commutation of Triacs

A triac is equivalent to a pair of thyristors connected in antiparallel. The output voltage waveform in an ac circuit with a triac is the same as that obtainable with a pair of thyristors connected in antiparallel. However, one important difference between them is that with SCRs each SCR has an entire half-cycle for its turn off, but the triac must turn off during the brief interval while the load current is passing through zero. For resistive loads this is fairly simple to accomplish since the time available for the triac to turn-off extends from the time the device current drops below the holding current until the reapplied voltage exceeds the value of the line voltage required to allow latching current. With inductive loads the task of commutating the triac becomes more difficult.

Fig. 2.54(a) shows the circuit of a triac with inductive load. The waveforms of triac voltage and current are also shown. For the sake of clarity, the waveform at the commutation point *A* is amplified, and shown in Fig. 2.54(b). It can be seen from the current waveform that the recovery current may act as a virtual gate current and turn the device on again. In addition there is a component to the reverse current which is due to the junction capacitance and the reapplied dv/dt. This component directly adds to the recovery current but does not appear until the triac begins to block a voltage of opposite polarity. As the rate of removal of the current $(-di/dt)$ decreases, the recovery current also decreases. This then implies that at lower values of di/dt, higher values of reapplied dv/dt are permissible for a given commutation capability. If by chance, the reapplied dv/dt is higher than the permissible value for a given (di/dt) rating, then additional protection circuits must be incorporated. The standard method is to use *RC* snubber such as R_s, C_s in Fig. 2.54(a). The values of R_s and C_s depend upon the load current, line voltage and reverse recovery charge of the triac.

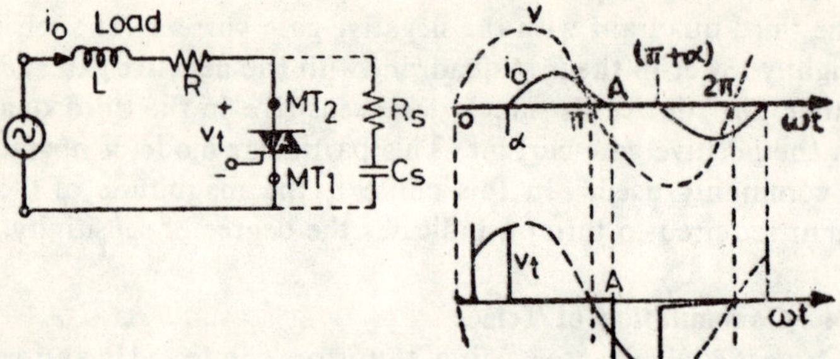

(a) Basic circuit and waveform.

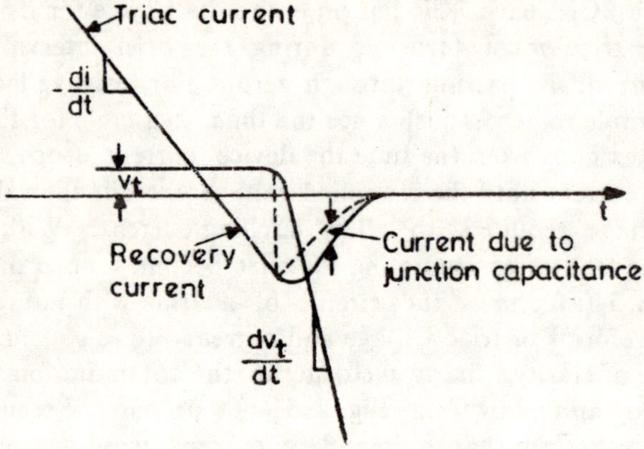

(b) Expanded commutation interval.

Fig. 2.54. Commutation of a triac.

REFERENCES

1. SCR Manual, 5th edition, Syracuse, N.Y. General Electric Company, 1972.
2. G.N. Revankar and P.K. Srivastava, 'Turn-off model of an SCR', IEEE Trans. IECI-22, No. 4, 1975, pp. 507–510.
3. F. Csaki et al., 'Power Electronics', Budapest: Akademiai Kiado, 1975.
4. F.E. Gentry et al., 'Semiconductor Controlled Rectifiers: Principles and applications of *p–n–p–n* devices', New Jersey; Prentice-Hall Inc., Englewood Cliffs, USA, 1964.
5. S.B. Dewan and A. Straughen, 'Power Semiconductor Circuits', Toronto: John Wiley & Sons, 1975.

6. G.K. Dubey, 'Classification of thyristor commutation methods', IEEE Industry Applications Society Annual Meeting, 1982, pp. 895–909.
7. N.W. Mapham, 'The classification of SCR inverter circuits', IEEE International Convention Record, part 4, 1964, pp. 99–105.
8. J.M.D. Murphy, 'Thyristor Control AC Motors', Oxford: Pergamon Press, 1973.
9. S.M. Sriraghavan, B.D. Pradhan and G.N. Revankar, 'An improved complementary impulse commutated inverter using saturable reactor', IEEE Trans. on IECI, vol. IECI-28, Feb. 1981, pp. 50–55.

3. LINE COMMUTATED AC TO DC CONVERTERS

3.1 INTRODUCTION

Many industrial applications require variable DC power. Some of these are:

1. Steel mills, paper and textile mills employing DC motor drives.
2. AC-fed traction system using DC traction motors.
3. Electro-chemical and electro-metallurgical processes.
4. Magnet power supplies.
5. Reactor controls.
6. Portable hand-tool drives.

Earlier AC to DC conversion was achieved using *MG* sets, Mercury-arc rectifiers, thyratrons. With the availability of high power thyristors, presently thyristor converters are used to convert AC to DC. The thyristors are phase-controlled to get variable direct voltage. The phase-controlled thyristor converters make use of natural reversal of AC source voltage for commutating the thyristors conducting earlier. This type of commutation is referred to as 'Line commutation' in section 2.5. The phase-controlled converters are simple and less expensive and are extensively used in industries. These converters are classified as two-quadrant if the output voltage can be made either positive or negative with a given polarity of current at the output. They are however classified as single-quadrant if the output voltage cannot be made negative with a given polarity of the output current. The two-quadrant converter is also known as a fully-controlled converter since it is obtained by replacing all diodes by thyristors in a conventional rectifier circuit arrangement. Section 3.2 deals with the two-quadrant converters.

The operation of the most commonly used 2-pulse mid-point converter circuit supplying an *R–L* load is described. Another classification of line commutated 2-quadrant converters is based on a circuit arrangement in the bridge configuration. The external performances of a converter circuit are defined. These performances are derived for a 2-pulse converter under the simplified assumption that the output current of the converter is continuous with an insignificant amount of ripple current [1]. The external performances of multi-pulse converters (up to 6 pulses) are given in a tabular form. Irrespective of number of pulses, the fundamental relations of all two-quadrant converters are derived neglecting output voltage and input current harmonic components. The output voltage and input supply current harmonics of all multi-pulse converters are also studied [1]. A detailed analysis of a 2-pulse converter for any finite load circuit parameters R, L and V_c (load source) is presented. The output current of a 2-pulse converter may be continuous or discontinuous depending upon the load circuit parameters and the firing angle. The regions of continuous and discontinuous current conduction are explained with the help of the operating diagram [2]. The different possible discontinuous conduction modes are also identified on the operating diagram of a 2-pulse converter [3], [4]. The output voltage and current waveforms corresponding to these discontinuous modes are also given. The operating diagram and the modes of operation of a three-phase fully controlled bridge converter are briefly explained. Lastly, the effect of source inductance is taken into account while deriving an expression for the output voltage of a converter [1]. The effect of the overlap angle with an increased load current on the output voltage of single-phase and three-phase bridge circuits is explained [5].

Many industrial applications require only one polarity of voltage and current at the output dc terminals. This therefore requires only one-quadrant operating characteristic in the $V_0 - I_0$ plane. Such a characteristic can be obtained from a two-quadrant converter by using one half of its control range. However an alternative approach is to modify the 2-quadrant converter circuit so that intrinsically it is capable only of 1-quadrant operation. Obviously this involves replacement of some thyristors in the 2-quadrant converter by uncontrolled rectifiers.

In some applications such as single-phase ac traction systems, it is desirable to improve the supply power factor without sacrificing the 2-quadrant operating characteristics of the converter circuit. The regeneration of load energy to the ac source may result in substantial improvement in the energy consumption. In such cases, the fully controlled converter circuits are invariably used. However, with a slight modification in the conventional triggering of the fully controlled circuit, it has been made possible to achieve rectification and inversion modes. Such converters are known as 'Fully-controlled converters with half-controlled characteristics' [6]. Section 3.3 deals with the study of half-controlled converters. The fundamental principles of all one-quadrant converters are explained. Some practical two-pulse and three-pulse half-controlled converters are described [1]. The possibility of 'Half-waving' effect with these converters is pointed out. The input and output performances are shown in a tabular form assuming constant load current. The output voltage harmonics and input line current harmonics are briefly studied. The fully-controlled converter with half-controlled characteristics is described. A detailed analysis of such a converter with actual load circuit parameters is presented. The operating diagrams showing continuous and discontinuous regions are explained. The discontinuous modes along with the voltage and current waveforms are presented. Minimum inductance estimation for continuous conduction is detailed [7].

The output voltage of positive or negative polarity with either positive or negative output current is often required in many applications. Two 2-quadrant converters connected in inverse parallel across the load can supply both voltage and current of either positive or negative polarity. Such a converter is familiarly known as dual converter [1], [8]. The basic operating features of a dual converter are explained in Section 3.4.

There are various ways of producing gating pulses for a three-phase 6-pulse thyristor converter [9]–[11]. These are reviewed in Section 3.5.

3.2. FULLY-CONTROLLED (TWO-QUADRANT) CONVERTERS

One possible way of classifying an AC to DC converter circuit is according to the number of pulses of current that pass through the load circuit in each cycle of the source voltage. Fig. 3.1 (a) illustrates a converter which gives two pulses of current in each input cycle. It is therefore called 2–pulse mid-point converter circuit configuration owing to the use of centre-tapped transformer. It is also known as single-phase full wave converter. Let us examine the operation of this converter circuit with an inductive (R–L) load. Let us make the following assumptions.

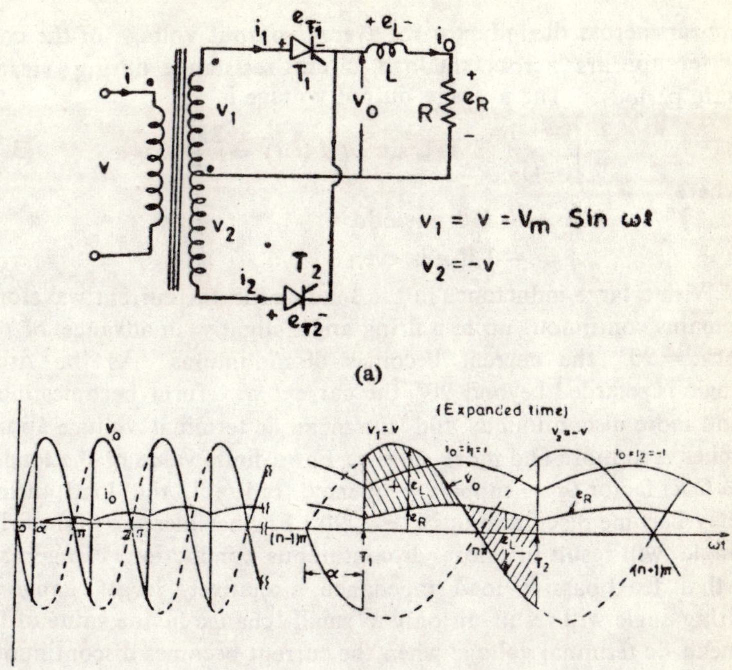

Fig. 3.1. Basic 2-pulse mid-point thyristor converter: (a) Circuit; (b) Typical waveforms.

1. The supply source is ideal. The effect of source impedance is studied in section 3.2.11.

2. The thyristors offer infinite resistance in the blocking state and zero resistance in the on state.

Let the thyristor T_1 be triggered at an angle α with reference to the voltage v_1 ($\alpha < \phi$, $\phi = \tan^{-1}\omega L/R$). Since the load is inductive, the current builds up with the initial value zero. At the end of one half-period ($\pi + \alpha$) the thyristor T_2 is triggered when the load current is greater than zero. The entire secondary voltage of the transformer appears across T_1, reverse biases it, and the thyristor T_1 is commutated. The load current now passes through the lower half of the secondary winding. Again after another half period the turn on of T_1 will commutate T_2.

During this transient period, the load current keeps on building. After the transient process, steady state waveform results. These waveforms are shown in Fig. 3.1 (b). Since no average voltage appears across the inductor the average output voltage of the converter appears across the load circuit resistance during steady state periods. The average output voltage is

$$\frac{1}{\pi}\int_{(n-1)\pi+\alpha}^{(n\pi+\alpha)} AV_m \sin \omega t \, d(\omega t) = \frac{2V_m}{\pi}\cos\alpha \qquad (3.1)$$

where

$A = +1$ if n is odd

$A = -1$ if n is even

With a large inductance in the load circuit the current waveform remains continuous up to a firing angle slightly in advance of 90°. At $\alpha = 90°$, the current becomes discontinuous. As the firing angle is retarded beyond 90°, the current waveform becomes more and more discontinuous and the mean dc terminal voltage approaches zero more and more closely. For a finite value of the load Q ($\omega L/R$) factor (ω = supply frequency, rad/sec), the load current may become discontinuous at $\alpha < 90°$. Further increase in the firing angle will result in more discontinuous conduction. In general, wth a fixed passive load inpedance, a relatively large change in firing angle will result in only a small change in the value of the mean dc terminal voltage when the current becomes discontinuous. With purely resistive loads, the load current is always discontinuous.

With an active oad such as a battery or a dc motor, the output current may be continuous or discontinuous. If the current is continuous, the output voltage is clearly defined and is dependent only upon the converter firing angle. The average output voltage becomes positive for $\alpha < 90°$ and negative for $\alpha > 90°$. The converter is thus inherently capable of providing two-quadrant operation in the V_0-I_0 plane. However, if the output current becomes discontinuous, the output voltage waveform and hence also its mean value is dependent both upon the firing angle and upon the load source. The detailed analysis of a fully controlled converter with actual load circuit parameters R, L and active load source V_c is presented in Section 3.2.8.

3.2.1. Single Phase Bridge Converter

Another common classification of line commutated 2-quadrant converters is based upon the circuit arrangement in a bridge configuration. Figure 3.2 (a) shows the single-phase converter circuit. This is also known as the fully-controlled converter or two-pulse bridge type converter. The waveforms of a single-phase bridge circuit are identical to those of a single-phase 2-pulse mid-point converter circuit.

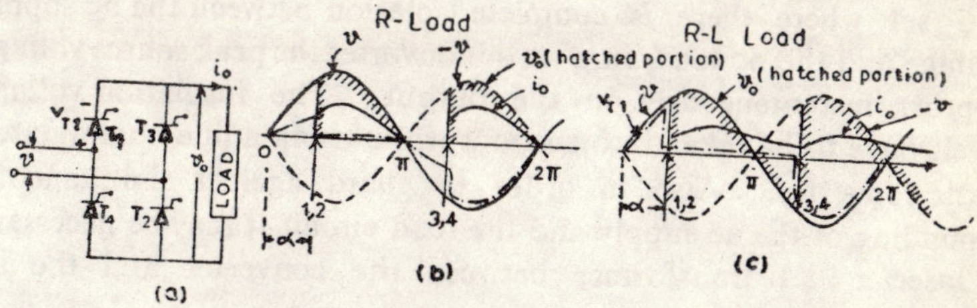

Fig. 3.2. Single phase bridge converter: (a) Circuit configuration. (b) Waveforms with R load. (c) Waveforms with R–L load.

The operation of the single-phase bridge converter circuit (Fig. 3.2. (a)) may be described as follows. Thyristors T_1, T_2 & T_3, T_4 are turned on at some delay angle α in each positive and negative half cycles of the supply voltage respectively. With a resistive load, the load current is given by V/R. When the supply voltage falls to zero, the current also goes to zero and the appropriate thyristors (T_1, T_2 in the positive half cycle, T_3, T_4 in the negative half-cycle) turn off by natural commutation. The supply voltage reverse biases the two thyristors which were conducting previously until the next pair of thyristors is turned on. With the turn-on of the other pair of thyristors, the supply voltage appears across each thyristor as a reverse bias. The chain waveform shown in Fig. 3.2. (b) shows the voltage across the thyristor T_1.

With an inductive load having sufficient inductance, the load current becomes continuous. Each diagonally connected thyristor pair conducts for one-half period of the supply frequency. When T_1 and T_2 are turned on together in the positive half cycle, the current flows from the source, through the thyristor T_1 to the load and back to the source through the thyristor T_2. By turning on thyristors T_3 and T_4 after one-half period from the instant of triggering thyristors T_1 and T_2, a negative voltage appears across the thyristors T_1 and T_2, reverse biases them and turns them off by the line commutation. The current flow through the ac source gets reversed. The load current is however unidirectional. The voltage across the thyristor T_1 is shown by the chain line waveform in Fig. 3.2. (c).

It may be necessary to use a transformer depending upon the supply voltage available and the load voltage required. Unlike in an *MG* set where there is complete isolation between the ac supply voltage and the dc load, in a static converter the peak source voltage appears instantaneously in the output. The insulation voltage levels have to be taken into account while using a line commutated static converter. Also in order to guard against simultaneous grounding of the ac supply and the load circuit it may be necessary to insert a 1 : 1 transformer between the converter and the ac mains.

3.2.2 Input and Output Performances of Line-Commutated Converters

The input voltage to a converter is generally sinusoidal while the input current is non-sinusoidal, i.e., harmonic currents are present in the ac lines. This has two main adverse effects. These harmonic currents conduct through the ac lines and interfere other equipment connected to the same lines. Furthermore, they cause Electro-magnetic interference (EMI) to nearby telephone and communication lines. Unlike an MG set, the static converter produces ripple in the output voltage which in turn generates ripple in the output current. The ripple current has several adverse effects. It is therefore necessary to define the external performances of a line-commutated thyristor converter.

3.2.2.1 *Input Displacement Factor* (*DSF or FPF*)

This is defined as the cosine of the input displacement angle ϕ_1 which is the angular displacement between the fundamental component of the ac supply current and the associated line to neutral voltage. In all line-commutated converters, the fundamental component is either in phase or lags behind the voltage by an angle which depends upon the firing angle. The input displacement factor is also known as the fundamental power factor (FPF).

3.2.2.2 *Input Power Factor* (*PF*)

This is defined as the ratio of the total mean input power to the total rms apparent power input to the converter. Since only the fundamental component contributes to the mean input power, the *PF* may be defined as

$$PF = \frac{V_1 I_1 \cos \phi_1}{V_{rms} I_{rms}} = \left(\frac{I_1}{I_{rms}}\right) \cos \phi_1 \tag{3.2}$$

where $V_1 = V_{rms}$ = Phase voltage

I_1 = Fundamental component of the supply current

ϕ_1 = Input displacement angle

I_{rms} = Supply rms current

3.2.2.3 *Input Current Distortion Factor*

It is defined as the ratio of the rms amplitude of the fundamental component to the total rms amplitude. It is a measure of the distortion present in the supply current.

3.2.2.4 *Input Harmonic Factor* (I_H)

It is the ratio of the total harmonic content to the fundamental component

$$I_H = (I^2_{rms} - I_1^2)^{1/2}/I_1 \tag{3.3}$$

3.2.2.5 *DC Voltage Ratio* (r)

It is defined as the ratio of the mean dc terminal voltage at a given firing angle α to the maximum possible mean dc terminal voltage.

3.2.2.6 *Voltage Ripple Factor* (K_V)

It is defined as the ratio of the net harmonic content of the output voltage to the average output voltage

$$K_V = \frac{(V^2_{0rms} - V^2_{0av})^{1/2}}{V_{0av}} \tag{3.4}$$

Sometimes, the peak-to-peak ripple is also considered as a measure of voltage ripple in the output voltage.

3.2.2.7 *Current Ripple Factor* (K_I).

This is defined as the ratio of the net harmonic content of the output current to the average output current.

$$K_I = \frac{(I^2_{0rms} - I^2_{0av})^{1/2}}{I_{0av}} . \tag{3.5}$$

Sometimes, the difference between the maximum and minimum

output current levels (peak-to-peak ripple) is taken as a measure of the output current ripple.

3.2.3 External Performance of a 2-Pulse AC-DC Converter

The different waveforms of a 2-pulse converter of Fig. 3.1 (a) are shown in Fig. 3.3 with the output comprising of a load resistance R, filter inductance L (infinitely large), and a DC voltage source, Vc. For the sake of clarity, the input source current and its fundamental component are shown separately in Fig. 3.4. Let us derive the input and output performances of a 2-pulse converter.

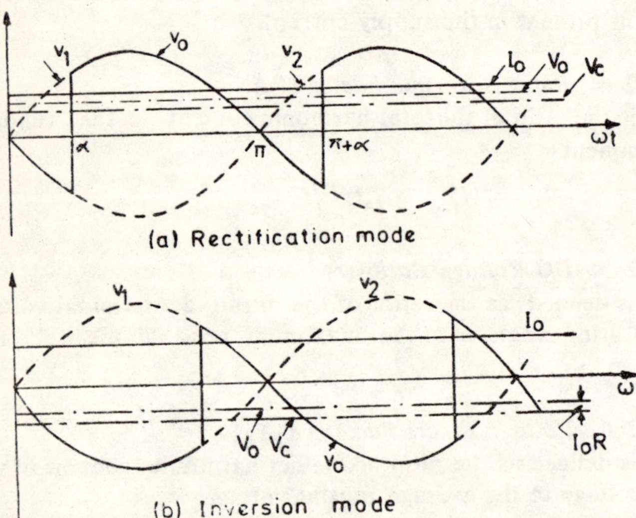

Fig. 3.3. Output waveforms of a 2 pulse converter with constant output current:
v_0=Instantaneous output voltage of converter;
V_0=Average output voltage;
I_0=Constant output current;
v_1, v_2=Transformer secondary voltages;
V_c=Output source voltage.

In general the supply current, i can be expressed by Fourier series. That is

$$= \sum_{n=1}^{\infty} (a_n \cos n\omega t) + b_n \sin n\omega t$$

$$= \sum_{n=1}^{\infty} C_n \sin (n\omega t + \phi_n) \tag{3.6}$$

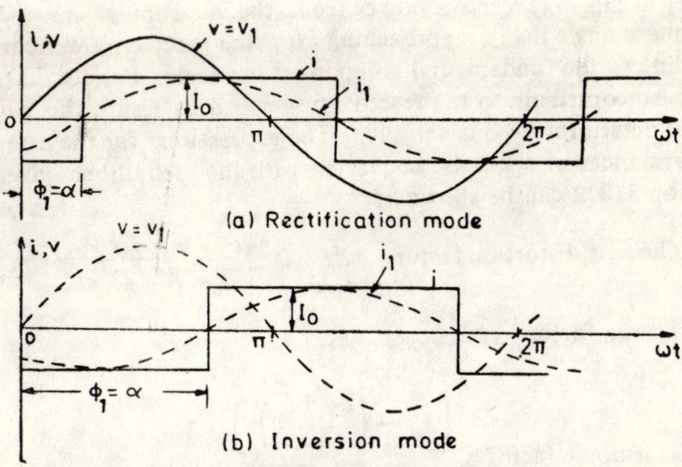

Fig. 3.4. Input waveforms of a 2-pulse converter.

where $\quad C_n = \sqrt{a_n^2 + b_n^2},\ \phi_n = \tan^{-1}\left(\dfrac{a_n}{b_n}\right)$

The fundamental component can be obtained by first determining the Fourier coefficients.

The Fourier coefficients of the current waveform, i assuming constant load current I_0 are

$$a_1 = \frac{2}{\pi}\int_{\alpha}^{(\alpha+\pi)} I_0 \cos \omega t \, d(\omega t) = -\frac{4}{\pi} I_0 \sin \alpha \tag{3.7}$$

$$b_1 = \frac{2}{\pi}\int_{\alpha}^{(\alpha+\pi)} I_0 \sin \omega t \, d(\omega t) = \frac{4}{\pi} I_0 \cos \alpha \tag{3.8}$$

$$C_1 = (a_1^2 + b_1^2) = \frac{4I_0}{\pi} \tag{3.9}$$

The fundamental component of the supply current and its phase angle with reference to the corresponding phase voltage are

$$I_1 = C_1/\sqrt{2} = 2\sqrt{2}\, I_0/\pi \tag{3.10}$$

$$\phi_1 = \tan^{-1}(-\tan \alpha) = -\alpha \tag{3.11}$$

The negative sign for ϕ_1 indicates that the current lags behind the voltage. For the inversion mode, the fundamental phase angle is greater than 90° ($\alpha > 90°$). Even though the output voltage source returns power to the supply source, the load is still a consumer of lagging reactive power from the ac supply line. With the phase angle delay approaching 90°, the reactive power corresponding to the fundamental component increases. Besides, harmonics also contribute to the reactive power. As a result, the power factor deteriorates considerably. The expressions for the external performances of a 2-pulse converter with the definitions given in Section 3.2.2 can be shown as:

$$\text{Current distortion factor} = \frac{I_1}{I_{rms}} = \frac{2\sqrt{2}\,I_0}{\pi I_0} = \frac{2\sqrt{2}}{\pi} \tag{3.12}$$

$$\text{Power factor} = \frac{2\sqrt{2}}{\pi}\cos\alpha \tag{3.13}$$

$$\text{Harmonic factor} = \frac{\left[I_0^2 - \left(\frac{2\sqrt{2}\,I_0}{\pi}\right)^2\right]^{1/2}}{(2\sqrt{2}\,I_0)/\pi} = \left(\frac{\pi^2}{8} - 1\right)^{1/2} \tag{3.14}$$

$$\text{DC Voltage ratio} = \frac{\frac{1}{\pi}\int_{\alpha}^{(\alpha+\pi)} V_m \sin\omega t \, d\,\omega t)}{\frac{1}{\pi}\int_0^{\pi} V_m \sin\omega t \, d\,(\omega t)} = \cos\alpha \tag{3.15}$$

$$\text{Voltage ripple factor} = \frac{\left[\frac{V_m^2}{2} - \left(\frac{2V_m}{\pi}\cos\alpha\right)^2\right]^{1/2}}{\frac{2V_m}{\pi}\cos\alpha} = \left(\frac{\pi^2}{8\cos^2\alpha} - 1\right)^{1/2} \tag{3.16}$$

$$\text{Maximum peak-to-peak ripple} = 2V_m \tag{3.17}$$

Because of a large ripple in the output voltage of a 2-pulse converter, the converter rating is limited to about 7.5 KW for operation on a single-phase ac supply system. For intermediate powers, a three-phase source is normally employed. And for large powers, a six-phase source has to be used.

3.2.4 Other Multi-Pulse Converters

Multi-pulse converters can be realized using either centre-tapped transformer type or bridge type converters. The centre-tapped transformer type converters are also known as mid-point converters owing to the use of neutral point of the transformer as one terminal of the DC output. In such converters, the conduction interval of each thyristor decreases with an increase in the pulse number. As a result, the form factor increases and the average current rating of the device decreases for a given rms current rating. To increase the conduction interval of the thyristors whereby higher average current through the thyristors can be allowed, two multipulse converters are connected using interphase reactors. This is equivalent to connecting two converters in parallel. In the bridge type converter configuration, the neutral wire of the transformer secondary is not required. A three-phase bridge converter provides six pulses of current in each cycle of the input voltage. A 12-pulse converter can be obtained by the series connection of the output terminals of two three-phase bridge converters. The input voltages to these converters are phase-shifted using star and delta connections of the transformer secondaries. Some commonly used multipulse converters are shown in Fig. 3.5.

3.2.4.1 *Three-pulse Converter*

Figure 3.5 (a) shows the circuit of a three-phase half-wave circuit or a three-pulse converter. The supply neutral and transformer neutral are solidly connected. This may be considered as the basic building block for higher multi-pulse converters. Each thyristor conducts for one-third of the supply cycle period. The earliest instant at which a thyristor in a phase is triggered into conduction is 30° with reference to the corresponding phase voltage. If the thyristors T_1 to T_3 were replaced by diodes D_1 to D_3, the diodes would conduct from 30° with respect to the phase voltages, v_A, v_B, v_C respectively. The firing angle α for a particular thyristor connected in a particular phase is therefore measured from 30° with respect to the corresponding phase voltage. The major drawback of the three-pulse converter is that there is a dc component in the supply currents. This component may result in core-saturation. The triplen harmonics in the supply current may also produce saturation in the transformer core thereby increasing the losses. The three-phase half-wave circuit is therefore used for low power applications only.

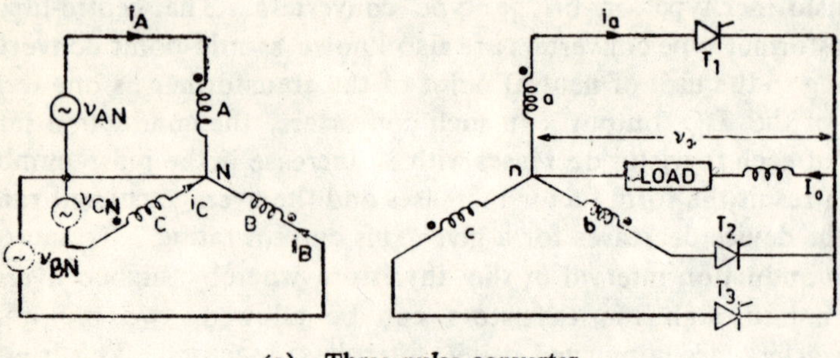

(a) Three-pulse converter.

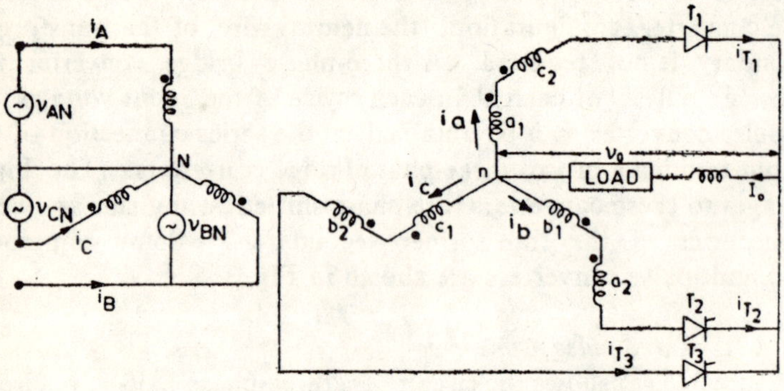

(b) Three-pulse converter with zig-zag secondary.

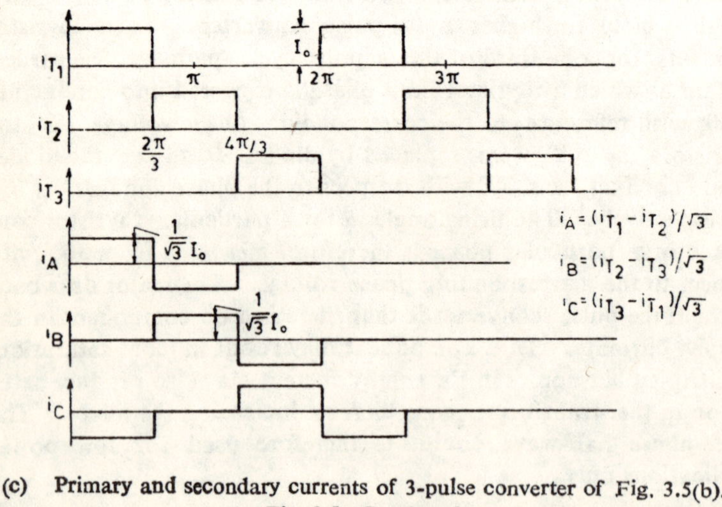

(c) Primary and secondary currents of 3-pulse converter of Fig. 3.5(b).

Fig. 3.5 (Continued)

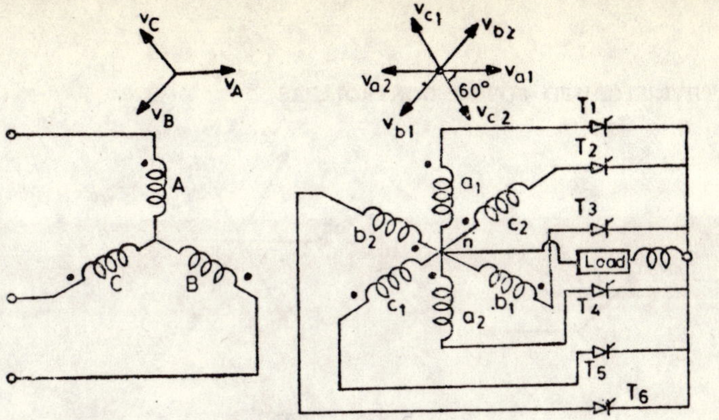

(d) 6-pulse mid-point converter.

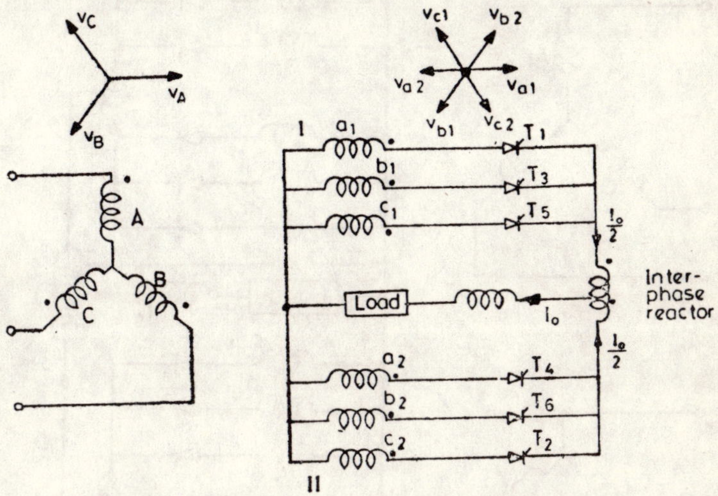

(e) 6-pulse converter with interphase reactor.

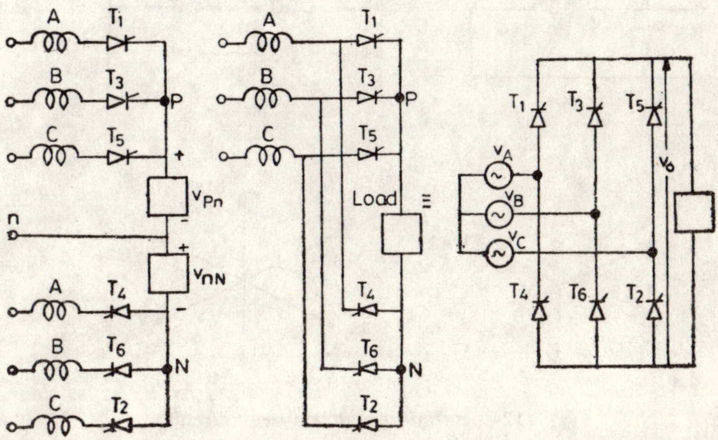

(i) (ii) (iii)

(f) Six-pulse bridge converter.

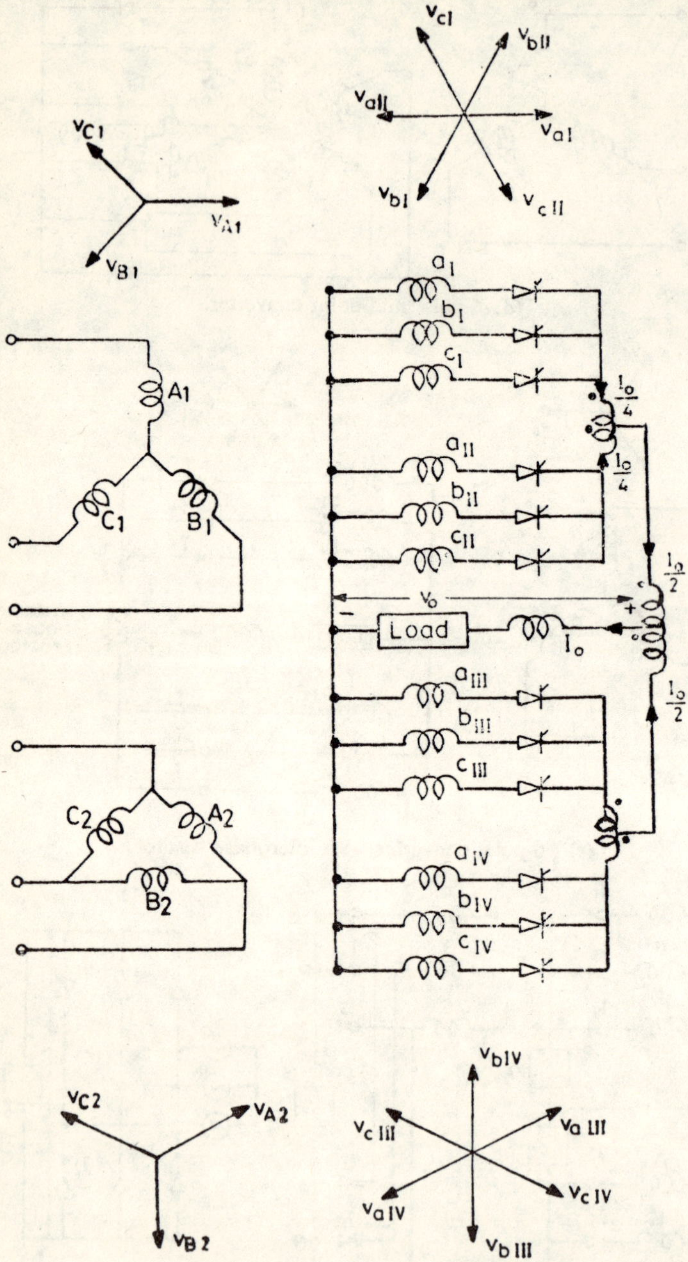

(g) 12-pulse mid-point converter circuit.

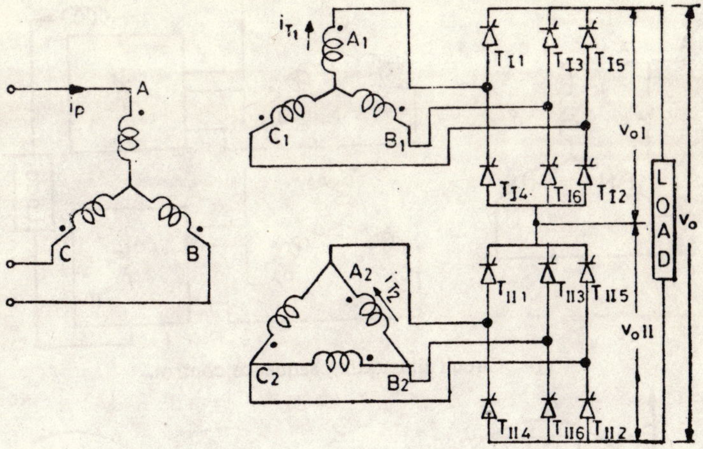

(h) 12-pulse converter using bridge converters.

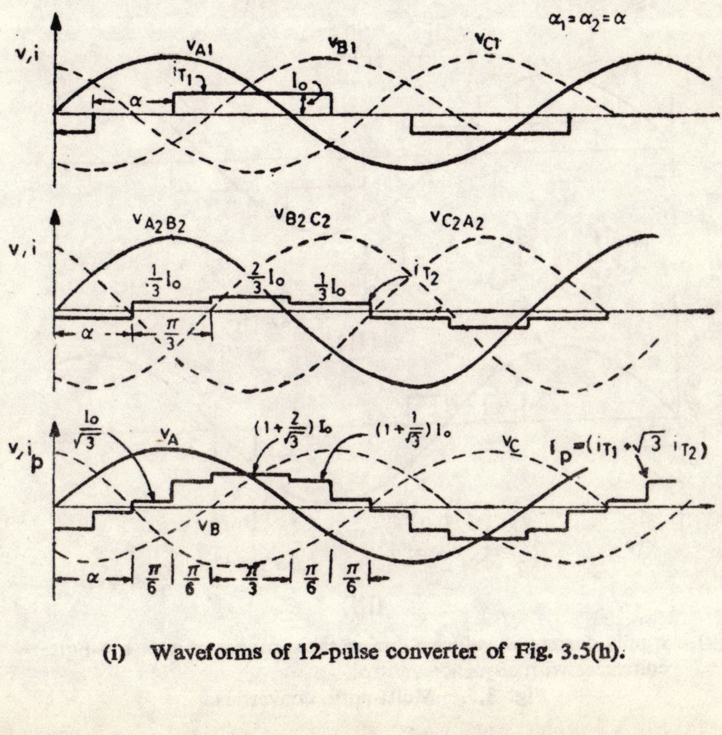

(i) Waveforms of 12-pulse converter of Fig. 3.5(h).

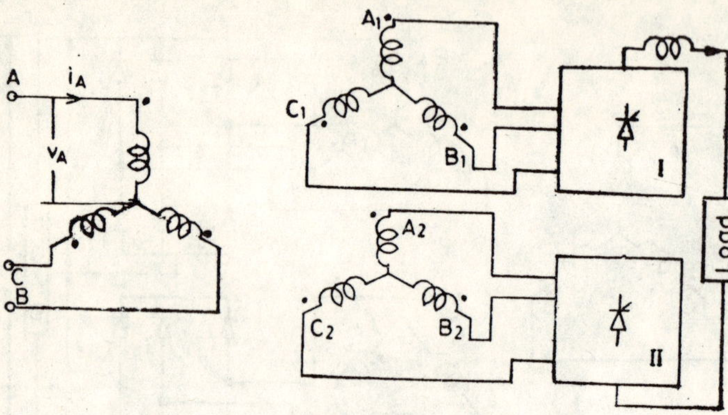

(j) Circuit illustrating sequence control.

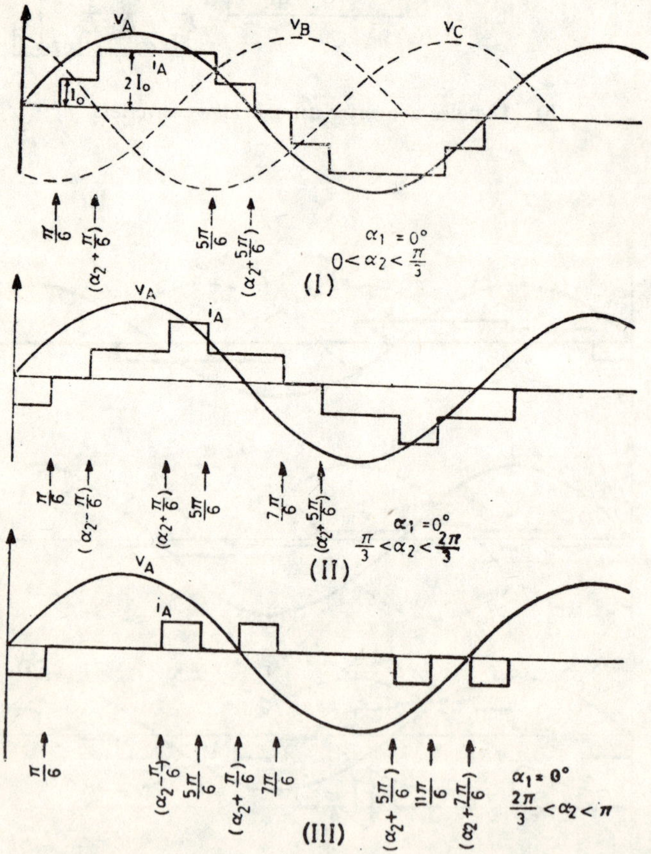

(k) Supply current waveforms during three ranges of α in a 6-pulse converter with sequence control.

Fig. 3.5. Multi-pulse converters.

With an isolated neutral point N and assuming 1 : 1 turns ratio of the transformer, the instantaneous current relations are

$$\left.\begin{aligned} i_A &= i_a - i_m \\ i_B &= i_b - i_m \\ i_C &= i_c - i_m \end{aligned}\right\} \begin{aligned} &i_m \text{ being the mean} && (3.18) \\ &\text{current in the} && (3.19) \\ &\text{secondary circuit} && (3.20) \end{aligned}$$

$$i_A + i_B + i_C = 0 \tag{3.21}$$

When T_1 conducts

$$i_b = i_c = 0 \tag{3.22}$$

$$i_a = i_{T1} \tag{3.23}$$

Equations (3.18) to (3.23) yields

$$i_A = \frac{2}{3}\, i_a \tag{3.24}$$

$$i_B = \frac{-1}{3}\, i_a$$

$$i_C = \frac{-1}{3}\, i_a \tag{3.25}$$

$$i_m = \frac{1}{3}\, i_a \tag{3.26}$$

The current i_m causes dc magnetization of the core. Further, during the conduction interval of T_1, the supply current i_A returns through the windings B and C. Since the secondary windings b and c are not conducting, the windings B and C offer a large impedance. As a result, the phase voltage v_A decreases and the corresponding secondary voltage also decreases. This causes a reduction in the output voltage [12]. This configuration with an isolated neutral is rarely used in practice.

3.2.4.2 *Three-pulse Converter with Zig-zag Connection of Transformer Secondary Windings*

Figure 3.5(b) shows the circuit of a three-pulse converter which eliminates the DC component in the supply currents. The primary and secondary currents are also shown in Fig. 3.5(c). Such a three-pulse converter can be used for high power applications compared with the basic converter of Fig. 3.5(a). However, it requires three more secondary windings. During the conduction interval of T_1, the phase A conducts in the positive direction with the supply current entering the dot terminal. During the conduction interval of T_2, the supply phase current i_A is reversed since the secondary current is entering the dot terminal. The phase A winding is coupled to the secondary windings a_1 and a_2. The turns ratio between the windings A, a_1 and a_2 is assumed $\sqrt{3} : 1 : 1$. The same is true for other windings also. The voltage ratio between the secondary and primary is 1 and the primary current in phase A is given by $\frac{1}{\sqrt{3}}(i_a - i_b)$.

3.2.4.3 *Six-pulse Converter*

The six-pulse converter circuit is shown in Fig. 3.5(d). The primary winding A is coupled to the secondary windings a_1 and a_2. The voltage induced in the winding a_2 is phase shifted by 180° with respect to the voltage induced in the winding a_1. Similar relations are valid for the windings B, b_1, b_2 and C, c_1, c_2. The voltages induced in the windings a_1, c_2, b_1, a_2, c_1 and b_2 are 60° apart. There is a considerable improvement in the distortion factor of a six-pulse converter over that of a three-pulse converter. Such converters are invariably used in high power applications. The secondary windings of the input transformer of Fig. 3.5(d) can be considered as six-phase. If the thyristors were replaced by diodes, the phase whichever is more positive among the six phases conducts for 60°. With controlled devices connected in each phase, the earliest instant at which a thyristor connected in a particular phase can be triggered into conduction is at 60° with reference to the corresponding phase voltage. The firing angle for a thyristor is correspondingly measured with reference to this point. In this six-pulse converter, each thyristor conducts (60°) for one sixth of the supply cycle period. The turn-on of an incoming thyristor should produce a reverse bias across the outgoing thyristor. Then only will the outgoing thyristor be line commutated reliably. Ideally, the control range of the firing angle α for any thyristor is 0 to 180° with continuous load current.

3.2.4.4 *Six-pulse Converter with Inter-phase Reactor*

Figure 3.5(e) shows an improved six-pulse mid-point converter using an interphase reactor. The output voltages of two 3-pulse converters are combined in an interphase reactor. Each 3-pulse group operates independently. As a result, each thyristor conducts for 120°. The average output voltage is the mean of the DC terminal voltages of either group. The output current is shared equally by the two 3-pulse converter groups. The two groups can be essentially considered to be operating in parallel. Of course, the input voltages to the two groups are phase shifted. A major advantage of this configuration is that the thyristors and the phases conduct for larger intervals giving better utilization of components apart from reducing the ripple in the output voltage and current.

3.2.4.5 *Six-pulse Bridge Converter*

In the above section, it has been shown that a six-pulse converter is derived by connecting the dc terminals of two 3-pulse converters in parallel. If we connect the dc terminals of two 3-pulse converters in series, the converter so formed is called a 6-pulse bridge converter which is shown in Fig. 3.5(f). In Fig. 3.5(f), the load voltage is divided into two equal values for the sake of developing the bridge converter by the series connection of two 3-pulse converters. The two converters are assumed to be symmetrical and the direct current flows from group I to group II. There is no current flowing through the neutral wire and as such no neutral wire is required. The secondary windings of the supply transformer can be connected either in wye or delta. The thyristors T_1, T_3, T_5 are called positive group since they are turned on when the supply voltages are positive. Similarly the thyristors T_4, T_6, T_2 which conduct when the supply voltages are negative form the negative group. Just as we defined the firing angle in a three-pulse converter, the firing angle for the thyristor T_1 is measured from 30° with reference to the positive half-cycle of the phase voltage v_A. In a similar manner the thyristors T_3 and T_5 are turned on taking the respective phase voltages v_B and v_C into account. The negative group thyristors T_4, T_6, T_2 are turned on 180° after the positive group thyristors T_1, T_3, T_5 are turned on respectively. In each input cycle, there are six pulses in the output voltage. That is, the output voltage is obtained by the conduction of two phases at any time such as *AB*, *AC*, *BC*, *BA*, *CA*, *CB* if we assume ideal source voltages. Each thyristor conducts for 120° and each phase conducts the positive current for 120° and the negative current for 120°. There is no dc magnetization of the transformer core. Because of the higher pulse number, the ripple in the output voltage is reduced and the distortion factor of the supply current improves significantly. It can be shown by the harmonic analysis of the supply current in each line that the lowest harmonic is 5 times the supply frequency and its amplitude is one-fifth of the fundamental.

If we assume negligible ripple in the output current, the external performances can be easily obtained for the multi-pulse converters discussed in the above sections. The performances can be derived as explained in section 3.2.3. Table 3.1 shows the external performances of some typical multi-pulse converters including single-phase converters. The rms phase voltage is taken as V while deriving these relations.

3.2.4.6 *Twelve-pulse Converter with Inter-phase Reactors*

Two 6-pulse converters are connected in parallel to form a 12-pulse converter. The 6-pulse converter is obtained by connecting two 3-pulse converters in parallel. Such an arrangement (Fig 3.5(g)) is known as parallel connection since four groups of three-pulse converters are connected in parallel. Group I and II with their double-star secondary windings coupled to a single star-connected primary winding constitute a 6-pulse output as explained in section 3.2.4.4. A similar 6-pulse converter group with a delta connected primary winding provides the same voltage waveform but shifted 30° with respect to the former. The parallel connection is made through interphase transformers which are basically centre-tapped iron cored reactors. Since the instantaneous output voltages of the two parallel-connected converters are not equal, there is a tendency for the circulating current to flow. These reactors are designed such that the maximum voltage difference between two converters is absorbed without saturating them. The current under this condition is limited to the magnetizing current of these reactors [1], [12].

3.2.4.7 *Twelve-pulse Converter with Bridge-type Converters*

Figure 3.5(h) shows the circuit configuration of a 12-pulse converter by the series connection of two 6-pulse bridge converters. Since the input voltages to these two converters are derived from wye and delta connected secondary windings of the input supply transformer, there is a phase difference of 30° between the two voltages. As a result, the resultant output voltage is the sum of the individual converter output voltages. Since each converter gives a 6-pulse output, the resultant voltage is a 12-pulse output. Because of the series connection, the same load current flows through the two converters. Figure 3.5(i) shows secondary and primary currents in the supply transformer. The primary of the supply transformer is wye-connected. It can as well be connected in delta. But the supply current waveform will be different in this case. However, the supply current waveform for a particular connection remains the same as the firing angle is varied as illustrated in Fig. 3.5(i). In the wye-connected secondary winding, the line currents and phase currents are the same. Assuming unity turns ratio, the reflected current in the primary is the same as that in the secondary. In the delta-connected secondary, the line currents are 120° width pulses with constant amplitude. The phase currents are $\pm \frac{2}{3}I_0$ or $\pm \frac{1}{3}I_0$. Whenever the phase A_2B_2 conducts in the positive direction, the current i_{T_2} is $\frac{2}{3}I_0$. It is however equal to $-\frac{1}{3}I_0$ when any one of the other two phases conducts in the positive direction. Similarly it is $-\frac{2}{3}I_0$ whenever the phase A_2B_2 conducts in the negative direction and $+\frac{1}{3}I_0$ when any one of the other two phases conducts in the reverse direction. From these considerations, the resultant phase current i_{T_2} can be easily obtained as shown in Fig. 3.5(i). Since the ratio of the voltages of delta connected windings to the wye-connected windings is $\sqrt{3}$, the reflected component of the secondary current in the primary winding is $\sqrt{3}\, i_{T_2}$. The resultant primary current is $(i_{T_1} + \sqrt{3}\, i_{T_2})$ as illustrated in Fig. 3.5(i). If we assume constant load current with negligible ripple and source impedance, the distortion factor does not change as the firing angle α is varied since the waveshape remains unchanged but shifts in phase. The displacement angle or fundamental power factor angle does vary as the firing angle is varied. Hence, the power factor deteriorates as it approaches 90° since the firing angle is the same as the fundamental phase angle for a fully controlled converter with a constant load current.

3.2.4.8 *Sequence Control*

Two 6-pulse bridge converters are supported from unshifted voltages as shown in Fig. 3.5(j). In sequence control, one converter is operated at a fully advanced angle and the other converter is controlled to provide a variable dc output voltage. In the rectifying mode of operation, for example the converter 1 is kept at full advance while the firing angle of the converter 2 is controlled in the range $0 < \alpha_2 < 180°$. The resultant output voltage varies from zero ($\alpha_2 = 180°$) to maximum ($\alpha_2 = 0°$). In the inverting mode of operation, the firing angle of the converter 2 may be kept at full retard and that of the converter 1 is controlled from 0 to 180° to vary the resultant output voltage from 0 to negative maximum. There is a significant reduction in the reactive power from the supply source if the two series connected converters are operated in this mode of sequence control. The resultant output voltage however has 6-pulse ripple. The supply current waveform has three different shapes in the ranges, 0 to 60°, 60° to 120° and 120° to 180° [8]. These are shown in Fig. 3.5 (k).

3.2.5 Fundamental Relations of Two-Quadrant Converters

The converters discussed in section 3.2.4 are called fully controlled since the devices used are all controlled type. They are also known as two-quadrant converters provided the output current is continuous. For firing angle less than 90°, the average output voltage and current are positive and the source supplies power to the load. The average output voltage becomes negative for firing angle α greater than 90° and the output current being still positive, the output power becomes negative. In this mode of operation, the load source supplies power to the ac supply source. This is known as inverter operation.

A simple equivalent circuit can be derived for all two-quadrant converters, which is independent of converter pulse number. The necessary assumptions for the derivation of the equivalent circuit are: 1. The input current harmonics are neglected. 2. The output voltage harmonics are neglected and the output voltage is assumed to be constant.

By Fourier analysis of the input current to the converter, the fundamental component can be determined. Let it be I_1. The output voltage of the converter at any firing angle α is given by

$$V_0 = V_{0max} \cos \alpha$$

where V_{0max} is the maximum output voltage of the converter for $\alpha = 0°$.

The DC terminal voltage ratio is, $\frac{V_0}{V_{0max}} = r = \cos \alpha$ (3.27)

Neglecting harmonic currents, the input current to the converter consists of two components, an in phase component, $I_1 \cos \alpha$ and a quadrature component, $I_1 \sin \alpha$.

$$I_P = I_1 \cos \alpha = I_1 r \tag{3.28}$$

$$I_Q = I_1 \sin \alpha = I_1 \sqrt{1 - r^2} \tag{3.29}$$

The normalized components are

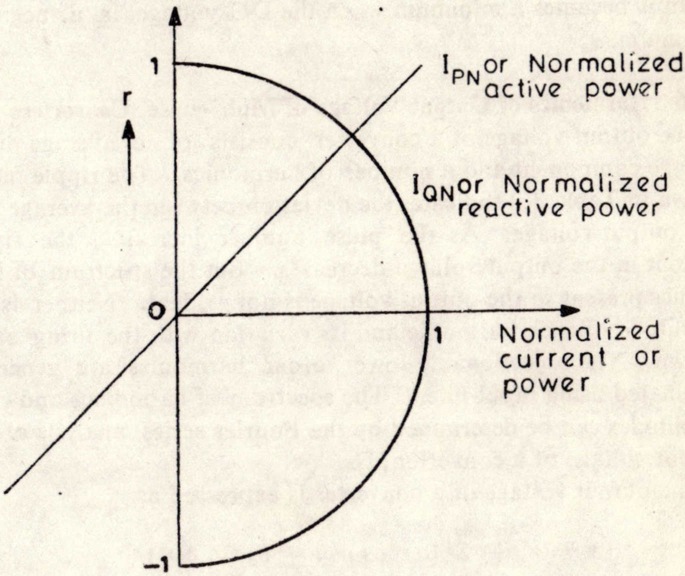

Fig. 3.6. Fundamental relations of an ideal two-quadrant converter.

$$I_{PN} = \frac{I_P}{I_1} = r \tag{3.30}$$

$$I_{QN} = \frac{I_Q}{I_1} = \sqrt{1 - r^2} \tag{3.31}$$

The expressions (3.30) and (3.31) also represent the normalized powers i.e.,

$$\text{Normalized active power} = \frac{V_0 I_0}{V_{0max} I_0} = \frac{V I_P}{V I_1} = r \tag{3.32}$$

$$\text{Normalized reactive power} = \frac{V I_Q}{V I_1} = \sqrt{1 - r^2} \tag{3.33}$$

where V is the rms supply voltage per phase.

I_0 is the output current.

The relations (3.32) and (3.33) showing the variations of active and reactive powers are plotted in Fig. 3.6. The active power increases with an increase in the DC voltage ratio in both quadrants. However, the reactive power requirement of the converter is minimum when the DC voltage ratio is maximum and then reaches to a maximum value when the DC voltage ratio is zero. Again it becomes a minimum when the DC voltage is a negative maximum.

3.2.6 Harmonics of Output Voltage of Multi-pulse Converters

The output voltage of a converter consists of an average direct voltage component and a number of harmonics. The ripple factor shown in Table 3.1 indicates the deviation between the average and rms output voltage. As the pulse number increases, the ripple content in the output voltage decreases. But the spectrum of harmonics present in the output voltage is not evident. Neither is the amplitude of each harmonic and its variation with the firing angle known. The objectionable lower order harmonics are generally eliminated using input filters. The spectrum of harmonics and their amplitudes can be determined by the Fourier series analysis of the output voltage of a converter [1].

The output voltage of a converter is expressed as

$$v_0 = V_0 + \sum_{n=1}^{\infty} (a_n \cos n\omega t + b_n \sin n\omega t)$$

$$= V_0 + \sum_{n=1}^{\infty} C_n \sin (n\omega t + \phi_n) \tag{3.34}$$

where V_0, a_n, b_n can be determined considering one fundamental period of the output voltage.

and $C_n = \sqrt{a_n^2 + b_n^2}$, $\phi_n = \tan^{-1}\left(\frac{a_n}{b_n}\right)$.

3.2.6.1 *Two-pulse Converter*

The solid curve of Fig. 3.7(a) shows the instantaneous output voltage of a single-phase converter assuming continuous output current. Since the waveform repeats twice in each supply cycle, this is also called the output voltage of a 2-pulse converter. The Fourier coefficients are determined considering one repetitive period of the output voltage waveform. The average output voltage is

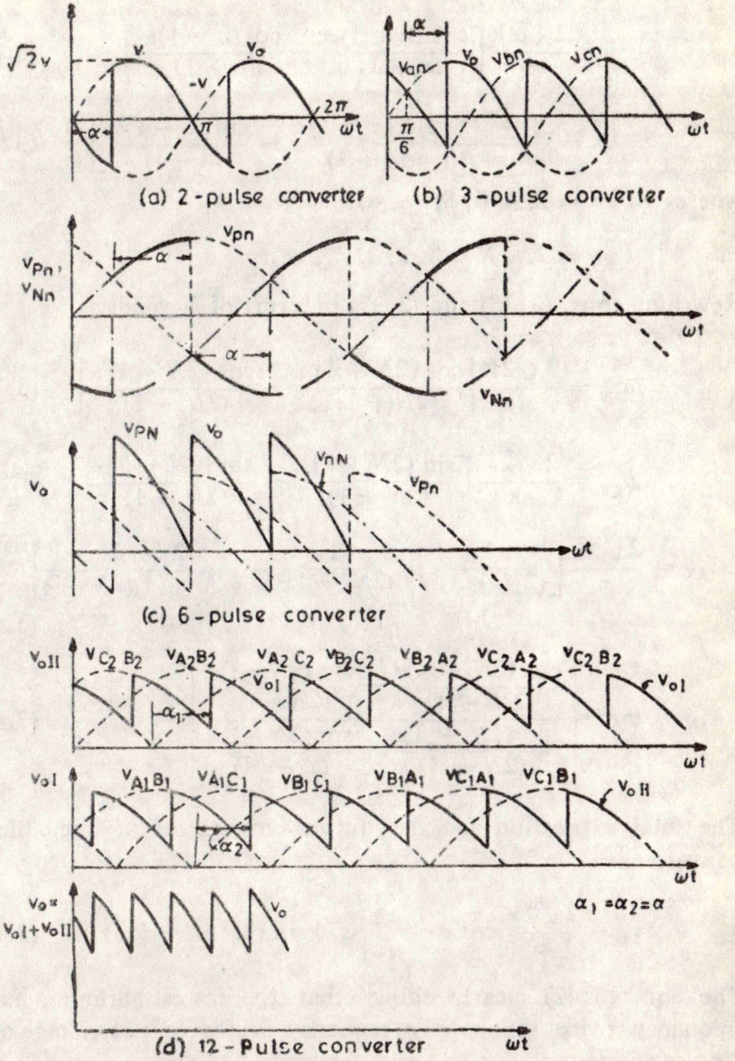

Fig. 3.7. Instantaneous output voltage of multi-pulse converters.

$$V_0 = \frac{1}{\pi}\int_{\alpha}^{(\pi+\alpha)} v_0\, d(\omega t) = \frac{1}{\pi}\int_{\alpha}^{(\pi+\alpha)} \sqrt{2}\, V \sin \omega t\, d(\omega t) = \frac{2\sqrt{2}V}{\pi}\cos\alpha \tag{3.35}$$

The Fourier coefficients are

$$a_n = \frac{2}{\pi}\int_{\alpha}^{(\pi+\alpha)} \sqrt{2}V \sin \omega t \cos n\omega t\, d(\omega t)$$

$$= \frac{2\sqrt{2}V}{\pi}\left[\frac{\cos(n+1)\alpha}{(n+1)} - \frac{\cos(n-1)\alpha}{(n-1)}\right] \tag{3.36}$$

$$b_n = \frac{2\sqrt{2}V}{\pi}\left[\frac{\sin(n+1)\alpha}{(n+1)} - \frac{\sin(n-1)\alpha}{(n-1)}\right] \tag{3.37}$$

where $n = 2, 4, 6, 8, \ldots, \infty.$

$= 2N, N = 1, 2, 3, \ldots, \infty.$

Rewriting eqns. (3.36) and (3.37) in terms of N yields

$$a_{2N} = \frac{2\sqrt{2}V}{\pi}\left[\frac{\cos(2N+1)\alpha}{(2N+1)} - \frac{\cos(2N-1)\alpha}{(2N-1)}\right] \tag{3.38}$$

$$b_{2N} = \frac{2\sqrt{2}V}{\pi}\left[\frac{\sin(2N+1)\alpha}{(2N+1)} - \frac{\sin(2N-1)\alpha}{(2N-1)}\right] \tag{3.39}$$

$$C_{2N} = \frac{2\sqrt{2}V}{\pi}\left[\frac{1}{(2N+1)^2} + \frac{1}{(2N-1)^2} - \frac{2\cos 2\alpha}{(2N+1)(2N-1)}\right]^{1/2} \tag{3.40}$$

$$\phi_{2N} = \tan^{-1}\left[\frac{\dfrac{\cos(2N+1)\alpha}{(2N+1)} - \dfrac{\cos(2N-1)\alpha}{(2N-1)}}{\dfrac{\sin(2N+1)\alpha}{(2N+1)} - \dfrac{\sin(2N-1)\alpha}{(2N-1)}}\right] \tag{3.41}$$

The final expression for the output voltage of a single-phase converter is

$$v_0 = \frac{2\sqrt{2}V}{\pi}\cos\alpha + \sum_{N=1}^{\infty} C_{2N}\sin(2N\omega t + \phi_{2N}) \tag{3.42}$$

The eqn. (3.42) clearly shows that the lowest harmonic has a frequency twice the supply frequency in the output voltage of a 2-pulse converter.

3.2.6.2 *3-Pulse Converter*

In Figure 3.7(b), the solid curve indicates the output voltage of a three-phase half-wave converter of Fig. 3.5(a). The average output voltage assuming continuous output current is

$$V_0 = \frac{3}{2\pi}\int_{(\alpha+\pi/6)}^{(\alpha+5\pi/6)} \sqrt{2}V \sin\omega t\, d(\omega t) = \frac{3\sqrt{6}V}{2\pi}\cos\alpha$$

The Fourier coefficients are

$$a_n = \frac{3}{\pi}\int_{\alpha+\pi/6}^{\alpha+5\pi/6} \sqrt{2}V \sin \omega t \cos n\omega t d(\omega t) \tag{3.43}$$

$$b_n = \frac{3}{\pi}\int_{\alpha+\pi/6}^{\alpha+5\pi/6} \sqrt{2}V \sin \omega t \cdot \sin n\omega t d(\omega t) \tag{3.44}$$

where $n = 3, 6, 9, \ldots = 3N$, $N = 1, 2, 3, \ldots, \infty$

Integrating eqns. (3.43) and (3.44) and expressing in terms of N, the Fourier coefficients can be shown to be

$$a_{3N} = \frac{3\sqrt{6}V}{2\pi}\left\{\frac{\cos\left[\frac{N\pi}{2} + (3N+1)\alpha\right]}{3N+1} - \frac{\cos\left[\frac{N\pi}{2} + (3N-1)\alpha\right]}{3N-1}\right\} \tag{3.45}$$

$$b_{3N} = \frac{3\sqrt{6}V}{2\pi}\left\{\frac{\sin\left[\frac{N\pi}{2} + (3N+1)\alpha\right]}{3N+1} - \frac{\sin\left[\frac{N\pi}{2} + (3N-1)\alpha\right]}{3N-1}\right\} \tag{3.46}$$

$$C_{3N} = \frac{3\sqrt{6}V}{2\pi}\left[\frac{1}{(3N-1)^2} + \frac{1}{(3N+1)^2} - \frac{2\cos 2\alpha}{(3N-1)(3N+1)}\right]^{1/2} \tag{3.47}$$

The output voltage of a 3-pulse converter is expressed as

$$v_0 = \frac{3\sqrt{6}V}{2\pi}\cos\alpha + \sum_{N=1}^{\infty} C_{3N}\sin(3N\omega t + \phi_{3N}) \tag{3.48}$$

where $$\phi_{3N} = -\frac{n\pi}{2} + \tan^{-1}\left[\frac{\frac{\cos(3N+1)\alpha}{3N+1} - \frac{\cos(3N-1)\alpha}{3N-1}}{\frac{\sin(3N+1)\alpha}{3N+1} - \frac{\sin(3N-1)\alpha}{3N-1}}\right]$$

Equation (3.48) shows that the output voltage of a three-pulse converter consists of, apart from the average DC voltage, harmonic components of frequency $3f$, $6f$, $9f$, etc. The lowest harmonic obviously has a frequency three times the sulppy frequency f.

3.2.6.3 *6-Pulse Converter*

Tne output voltage of a 6-pulse bridge converter can be easily obtained treating the bridge converter equivalent to two three-phase half-wave converters connected as shown in Fig. 3.5(f). The instantaneous output voltages v_{Pn} and v_{Nn} are shown in Fig. 3.7(c).

The output voltage due to the positive converter group may be represented by the eqn. (3.48). That is,

$$v_{Pn} = \frac{3\sqrt{6}V}{2\pi} \cos \alpha + \sum_{N=1}^{\infty} C_{3N} \sin (3N\omega t + \phi_{3N}) \quad (3.49)$$

Since the input voltages to the negative converter group are displaced by 180°, the output voltage due to negative converter group is obtained by replacing ωt by $(\omega t + 180°)$ in the eqn. (3.49). That is,

$$v_{nN} = \frac{3\sqrt{6}V}{2\pi} \cos \alpha + \sum_{N=1}^{\infty} C_{3N} \sin (3N(\omega t + \pi) + \phi_{3N}) \quad (3.50)$$

The resultant output voltage of a bridge converter is

$$v_0 = v_{PN} = v_{Pn} - v_{Nn} = v_{Pn} + v_{nN}$$

$$= \frac{3\sqrt{6}\, V \cos \alpha}{\pi} + \sum_{N=1}^{\infty} C_{6N} \sin (6N\omega t + \phi_{6N}) \quad (3.51)$$

where

$$C_{6N} = \frac{3\sqrt{6}V}{\pi}\left[\frac{1}{(6N-1)^2} + \frac{1}{(6N+1)^2} - \frac{2\cos 2\alpha}{(6N-1)(6N+1)}\right]^{1/2} \quad (3.52)$$

Equation (3.51) shows that the output voltage of a six-pulse converter consists of an average value plus a number of harmonics of the order $6f$, $12f$, $18f$ etc.

3.2.6.4 12-*Pulse Converter*

The circuit of a 12-pulse converter using bridge-type converters is shown in Fig. 3.5(h). The resultant output voltage is the sum of individual converter output voltages. If the phase voltage is taken as the reference and assuming unity turns ratio, the output voltage of converter II may be represented as given in eqn. (3.51). The input voltages to the converter I are displaced by 30° leading the input voltages to the converter II. Hence, the output voltage of converter I is obtained substituting $(\omega t + 30°)$ for ωt in the eqn. (3.51). The instantaneous output voltages of converter I and II are shown in Fig. 3.7(d). The expressions for the output voltages are

$$v_{0I} = \frac{3\sqrt{6}V}{\pi} \cos \alpha + \sum_{N=1}^{\infty} C_{6N} \sin (6N(\omega t + \pi/6) + \phi_{6N}) \quad (3.53)$$

$$v_{0II} = \frac{3\sqrt{6}V}{\pi} \cos\alpha + \sum_{N=1}^{\infty} C_{6N} \sin(6N\omega t + \phi_{6N}) \tag{3.54}$$

$$v_0 = v_{0I} + v_{0II} = \frac{6\sqrt{6}V}{\pi} \cos\alpha + \sum_{N=1}^{\infty} C_{12N} \sin(12N\omega t + \phi_{12N}) \tag{3.55}$$

where

$$C_{12N} = \frac{6\sqrt{6}V}{\pi}\left[\frac{1}{(12N-1)^2} + \frac{1}{(12N+1)^2} - \frac{2\cos 2\alpha}{(12N-1)(12N+1)}\right]^{1/2} \tag{3.56}$$

The different harmonics in the output voltage of a 12-pulse converter are $12f$, $24f$, $36f$ etc. In general, the harmonic components in the output voltage of P-pulse converter are PNf, $N = 1$, 2, 3,... and f is the supply frequency.

Equations (3.40), (3.47), (3.52) and (3.56) are used to determine the amplitudes of different harmonics present in the output voltage of 2, 3, 6 and 12-pulse converters. The relative amplitude is determined by dividing these expressions by the maximum possible output voltage in each case. The general expression for the relative amplitude of a particular harmonic component is

$$C_{PN} = \left[\frac{1}{(PN-1)^2} + \frac{1}{(PN+1)^2} - \frac{2\cos 2\alpha}{(PN-1)(PN+1)}\right]^{1/2} \tag{3.57}$$

where P is the pulse number. The variation of relative amplitudes of harmonics $2f$, $3f$, $4f$, $6f$, $12f$ and $24f$ as a function of the firing angle α is shown in Fig. 3.8. All integer multiples of $2f$, $3f$, $6f$, $12f$ are present in 2, 3, 6 and 12-pulse converters respectively. The harmonic amplitudes are maximum at $\alpha = 90°$ but minimum at both $\alpha = 0°$ and $180°$.

It must, however, be noted that the absolute value of a particular harmonic in a lower and a higher pulse converter will not be equal, although the relative amplitude in terms of the peak output voltage of the respective converter is the same. For example, the harmonic $12f$ in 2-pulse and 12-pulse converters have different values but the relative amplitudes in terms of the maximum output voltages of 2-pulse and 12-pulse converters are the same.

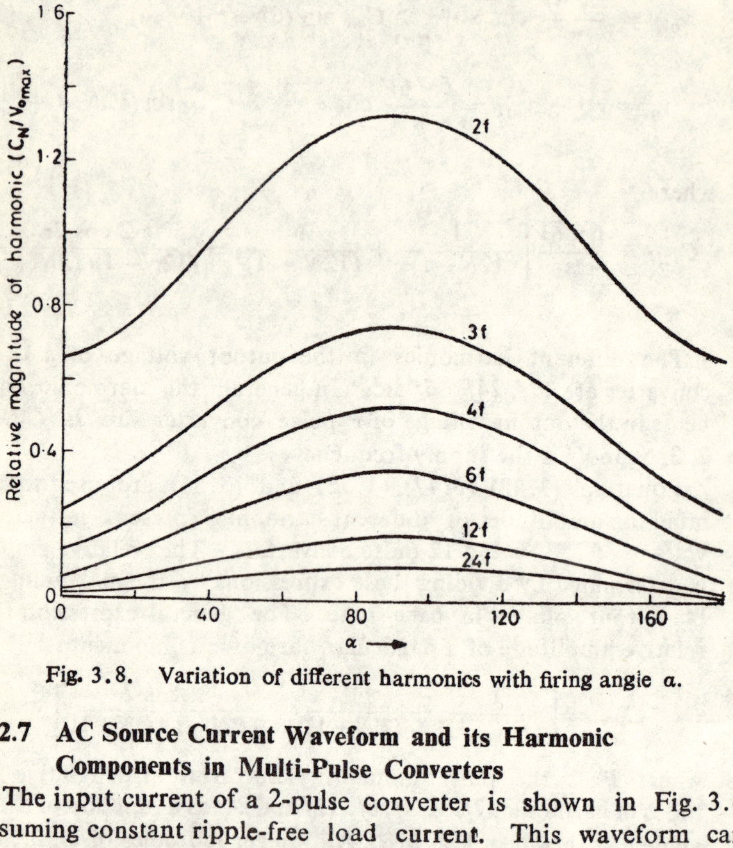

Fig. 3.8. Variation of different harmonics with firing angle α.

3.2.7 AC Source Current Waveform and its Harmonic Components in Multi-Pulse Converters

The input current of a 2-pulse converter is shown in Fig. 3.4 assuming constant ripple-free load current. This waveform can be analysed by Fourier series to determine different harmonics. Owing to half-wave symmetry, no even harmonics are present and neither is there any direct current component in the supply current. Apart from the fundamental, other harmonics are $3f$, $5f$, etc. The expression for the supply current can be shown to be

$$i = \frac{4I_0}{\pi} \sin(\omega t - \phi_1) + \Sigma \frac{4I_0}{n\pi} \sin(n\omega t - \phi_n) \qquad (3.58)$$

$$n = 2k + 1$$

$$k = 1, 2, \ldots, \infty$$

where I_0 is the load current, ϕ_n is the phase angle of the nth harmonic current.

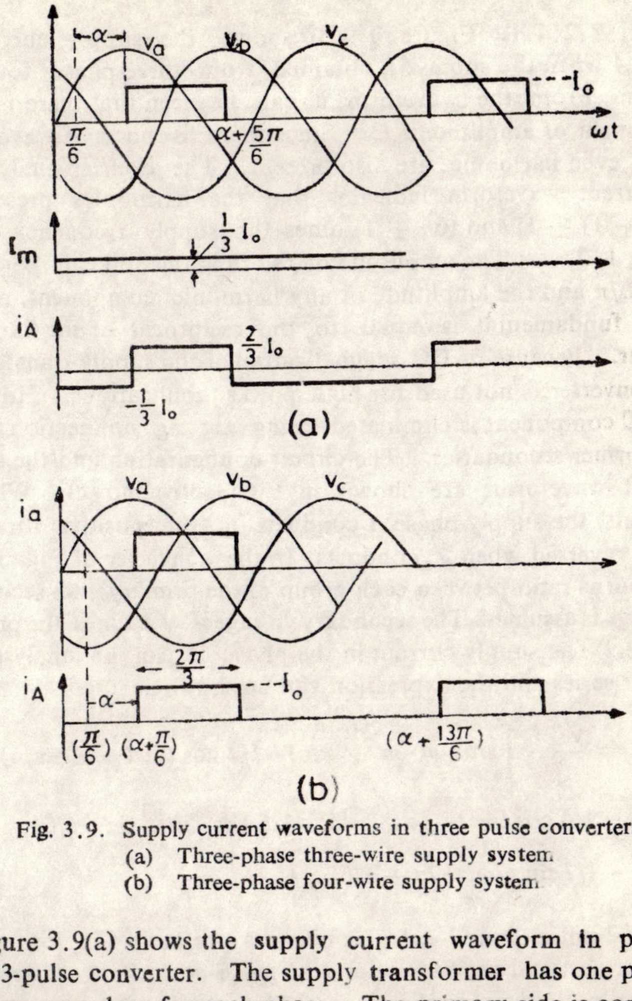

Fig. 3.9. Supply current waveforms in three pulse converters:
(a) Three-phase three-wire supply system
(b) Three-phase four-wire supply system

Figure 3.9(a) shows the supply current waveform in phase *A* for a 3-pulse converter. The supply transformer has one primary and one secondary for each phase. The primary side is connected in wye with isolated neutral. The turns ratio is assumed to be unity. During the interval when T_1 conducts, the reflected primary current enters the dot terminal of phase *A* and leaves through the dot terminals of phases *B* and *C*. Since T_2 and T_3 are not conducting during this interval, the phases *B* and *C* offer a large impedance to the flow of current. As a result, the phase *A* voltage decreases and consequently the output voltage also decreases. The 3-pulse converter supplied from a three-phase three-wire supply system is hardly used in practice as pointed out in

Section 3.2.4.1. Figure 3.9(b) shows the supply current in phase A when the supply is obtained from three-phase four-wire system. From the waveform it can be seen that there is a DC component of amplitude $I_0/3$. Since there is no half-wave symmetry, even harmonics are also present. The Fourier analysis of the current waveform indicates that the harmonics present are $[3(2n+1) \pm 1]$ and $(6n \pm 1)$ times the supply frequency where $n = 0, 1, 2, \ldots$. The amplitude of the fundamental can be shown to be I_0/π and the amplitude of any harmonic component, relative to the fundamental is equal to the reciprocal of the harmonic number. Because of DC magnetization of the supply transformer, this converter is not used for high power applications. However, the DC component is eliminated using zig-zag connection of the transformer secondaries. The circuit configuration and the supply current waveforms are shown in Fig. 3.5(b) and (c). When T_1 conducts, the supply phase A conducts in the positive direction. It gets reversed when T_2 conducts. In the converter of Fig. 3.5(b), unity turns ratio between each group of the primary and secondary windings is assumed. The secondary voltage is $\sqrt{3}$ times the primary voltage. The supply current in the phase A can be analysed by Fourier series and the expression can be shown as

$$i_A = \frac{\sqrt{3}\,I_0}{\pi}\Big[\sin(\omega t-\alpha+\pi/6)-1/2\cos(2\omega t-2\alpha-\pi/6)$$
$$-1/4\cos(4\omega t-4\alpha+\pi/6)-1/5\sin(5\omega t-5\alpha-\pi/6)$$
$$-1/7\sin(7\omega t-7\alpha+\pi/6)\ldots\Big] \tag{3.59}$$

The 2-pulse and 3-pulse converters are used for low power applications, and the 6-pulse and 12-pulse converters are generally used for high power applications. The supply current waveform of a 12-pulse converter is shown in Fig. 3.5(i). The secondary current of the converter group I or II also represents the supply current of a 6-pulse converter. Although the waveshapes in these two cases are different, the Fourier analysis of these current waveforms reveal certain common features. The harmonic components present in a 6-pulse converter are $(6n \pm 1)f$ where $n = 1, 2, \ldots, \infty$, and f is the supply frequency. The amplitude of a particular harmonic relative to the fundamental is equal to the reciprocal of the harmonic number. These observations also hold good for a

12-pulse converter with the exception that the harmonic components present are $(12n \pm 1)f$ where $n = 1, 2, \ldots, \infty$.

3.2.8 Detailed Analysis of a Two-Pulse Converter

In Section 3.2.4, the external performances of two-pulse and other multi-pulse converters are obtained assuming not only continuous but also constant load current. The simplifying assumption of constant load current (infinite load inductance) enables us to derive simple expressions for power factor, distortion factor, harmonic factor etc. This is justified when the ripple content in the load current is negligibly small. The actual shape of the load current depends upon the values of load circuit parameters and the control variable α. The load current may be continuous or discontinuous. In the case of continuous current operation the load voltage waveform comprises only portions of the ac supply voltage waveform and hence the mean value of this waveform is clearly defined, and is dependent only upon the converter firing angle. The determination of the steady state performance characteristics is relatively easy. In the case of discontinuous current operation, the load voltage waveform and its mean value is dependent upon both the triggering angle and load circuit parameters, that is, load phase angle and load source voltage. Different types of load voltage waveforms are encountered depending upon the triggering angle and load circuit parameters. The range of firing angle α over which the output current of a two-pulse converter is discontinuous for a given set of load circuit parameters is studied in this section.

Figure 3.10 (a) shows the equivalent circuit of a 2-pulse converter. The thyristor T_1 of Fig. 3.10 (a) is identical to thyristor T_1 of Fig. 3.1 (a) but it represents two thyristors connected in series in the single-phase bridge circuit of Fig. 3.2 (a). The thyristors in the mid-point converter and the bridge converter are assumed to be gated by long pulses. In the rectification mode, the firing pulses are applied from α to π and $(\pi + \alpha)$ to 2π. In the inverting mode, the firing pulses are maintained for about one-half cycle period, i.e. from α to $(\pi + \alpha)$ and $(\pi + \alpha)$ to $(2\pi + \alpha)$. The maintained pulse triggering improves the converter performance as can be seen later.

The source voltages are

$$v_{AN} = \sqrt{2}\, V \sin \omega t \tag{3.60}$$

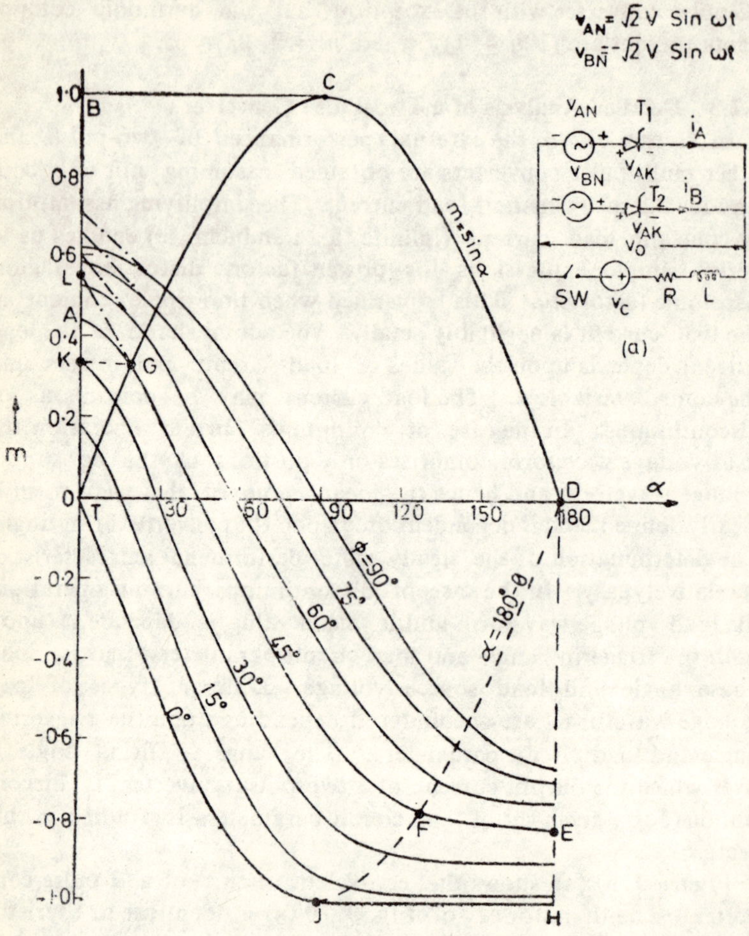

Fig. 3.10. Operating curves of a 2-pulse converter:
(a) Equivalent circuit of a 2-pulse converter.
(b) Continuous and discontinuous conduction of 2-pulse converter.

$$v_{BN} = \sqrt{2}\, V \sin(\omega t + \pi) = -\sqrt{2} V \sin \omega t \qquad (3.61)$$

Let the switch SW be closed and the thyristor T_1 be triggered at an angle α. The load current may be considered to consist of two components: one due to the ac source $\frac{\sqrt{2}\, V}{Z} \sin(\omega t - \phi)$ and

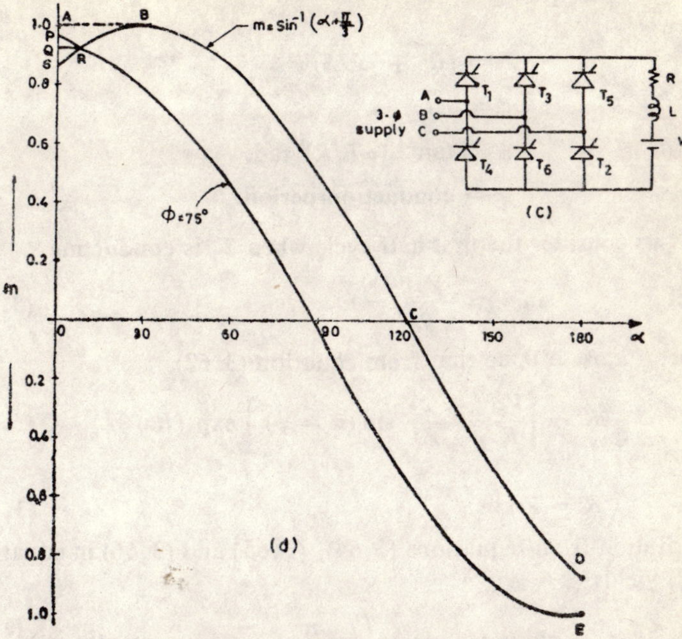

Fig. 3.10. Operating diagram of a three-phase fully controlled converter:
(c) Three-phase converter circuit.
(d) Operating diagram.

the other due to the direct emf $\left(-\frac{V_c}{R}\right)$. Each of these components has in turn a transient component. Let these be represented by a single exponential term K exp. $(-Rt/L)$.

The total current in the circuit is the sum of these three components. When T_1 is conducting

$$i_0(\omega t) = i_A(\omega t) = \frac{\sqrt{2}V}{Z} \sin(\omega t - \phi)$$

$$-\frac{V_c}{R} + K \exp(-Rt/L), \quad \alpha < \omega t < (\alpha + \gamma) \qquad (3.62)$$

After one half period, $T2$ conducts. The current is

$$i_0(\omega t') = i_B(\omega t') = \frac{\sqrt{2}V}{Z} \sin(\omega t' - \phi)$$

$$-\frac{V_c}{R} + K \exp(-Rt'/L), \quad \alpha < \omega t' < (\alpha + \gamma) \qquad (3.63)$$

where

$$Z = (R^2 + \omega^2 L^2)^{1/2}$$

$$\omega t' = \omega t - \pi$$

$$\phi = \tan^{-1}(\omega L/R) \text{ rad.}$$

$$\gamma = \text{conduction period.}$$

Let us consider the first half cycle when T_1 is conducting

Let
$$\sin\alpha = \frac{V_c}{\sqrt{2}V} = m \tag{3.64}$$

At $\omega t = \alpha$, $i_0 = 0$, so that from equation (3.62)

$$K = \left[\frac{V_c}{R} - \frac{\sqrt{2}V}{Z}\sin(\alpha - \phi)\right]\exp(R\alpha/\omega L) \tag{3.65}$$

Also

$$R = Z\cos\phi \tag{3.66}$$

Substitution from equations (3.64), (3.65) and (3.66) in equation (3.62) yields

$$\frac{Z}{\sqrt{2}V} i_0 = \sin(\omega t - \phi) - \left[\left(\frac{m}{\cos\phi} - B.\exp-(\omega t/\tan\phi)\right]\right. \tag{3.67}$$

where
$$B = \left\{\frac{m}{\cos\phi} - \sin(\alpha - \phi)\right\}\exp(\alpha/\tan\phi)$$

$\frac{Z}{\sqrt{2}V} i_0$ is the normalized value of the load current.

The normalized average value of the rectified current is given by

$$I_N = \frac{1}{\pi}\int_{\alpha}^{(\alpha+\gamma)} \frac{Z}{\sqrt{2}V} i_0 \, d(\omega t) \tag{3.68}$$

The normalized value of the rms output current is given by

$$I_{RN} = \left[\frac{1}{\pi}\int_{\alpha}^{(\alpha+\gamma)} \left(\frac{Z}{\sqrt{2}V} i_0\right)^2 d(\omega t)\right]^{1/2} \tag{3.69}$$

3.2.9 Operating Diagram of a Fully Controlled Converter

3.2.9.1 *Single-phase Fully Controlled Converter*

Equation (3.67) shows that the normalized instantaneous current is a function of variables, α, ϕ, and m. Figure 3.10 (b) shows continuous and discontinuous operating regions of a single-phase fully controlled converter. The curves are obtained [2] by solving eqn. (3.67) on a digital computer for various values of α, ϕ and m. Since the useful range of the firing angle is from 0 to 180°, the curves are therefore drawn for this range of the firing angle. A point (α, m) in this diagram for which $m > 1$ represents a condition in which no thyristors will turn on because they are always reverse biased. Similarly, a point (α, m) for which $m \leqslant -1$ represents a condition in which no thyristors will turn off since at no instant is a reverse voltage applied to them. Under such conditions short-circuit of the ac source through the thyristors results. Therefore the theoretical limits for m are $\pm$ 1. In Fig. 3.10 (b) the operating curves are drawn for a series of values of ϕ. For any chosen value of ϕ and normalized EMF m, the solid curves in the diagram define the values of firing angle α, at which the transition from continuous current operation to discontinuous current operation takes place as α is increased. The curves in the first quadrant correspond to rectification mode when the AC source supplies power to the load circuit. Those in the fourth quadrant correspond to inversion mode when the output source feeds power to the AC source. For a given load angle ϕ, any point lying below the curve represents continuous current operation and any point above the curve represents discontinuous current operation.

Let us consider the load angle $\phi = 45°$ as an example. *AGFE* is the boundary curve which separates the continuous and discontinuous current conduction regions. As pointed out earlier, the region above this curve corresponds to the discontinuous current operation and the region below this curve corresponds to the continuous current region. The continuous current region *AGFEHJTA* can be subdivided into two regions, namely *AGTA* and *TGFEHJT*. The region *AGTA* corresponds to $\alpha \leqslant \theta$, being the angle at which the instantaneous ac supply voltage equals the load source V_c, and the region *TGFEHJT* corresponds to $\alpha > \theta$. In the continuous current operation, the thyristors turn on at the instant of triggering irrespective of the firing angle setting. The load voltage waveform does not depend on the load source voltage V_c.

A point in the region *ABCDEFGA* represents a condition of discontinuous current operation, the subregion *ABCGA* corresponding to $\alpha \leqslant \theta$, and the subregion *GCDEFG* corresponds to $\alpha > \theta$. This discontinuous region is divided into four subregions resulting

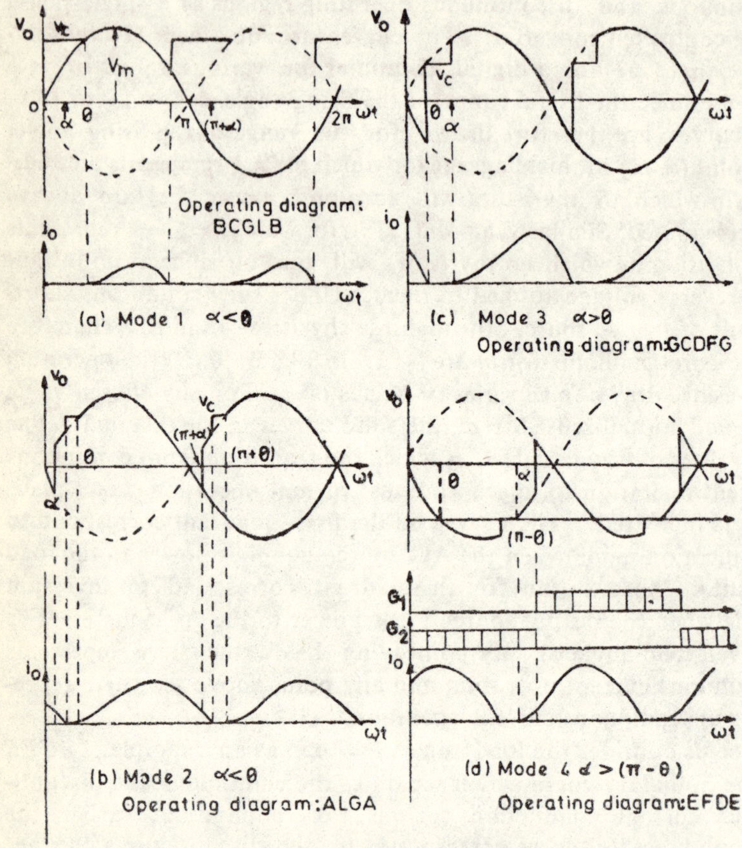

Fig. 3.11. Output voltage and current waveforms for discontinuous conduction modes:

(a)–(d): Single-phase fully controlled converter.

in four types of load voltage waveforms. Unlike in continuous current operation, the load voltage waveform and its mean value is dependent both upon the converter firing angle and load source V_c. The different possible discontinuous current conduction modes are illustrated in Fig. 3.11. To identify different modes on the operating diagram, the load angle $\phi = 45°$ is considered for illustration of the discontinuous modes in section 3.2.10.1.

3.2.9.2 *Three-phase Fully Controlled Converter*

A three-phase fully controlled converter (Fig. 3.10 (c)) can also

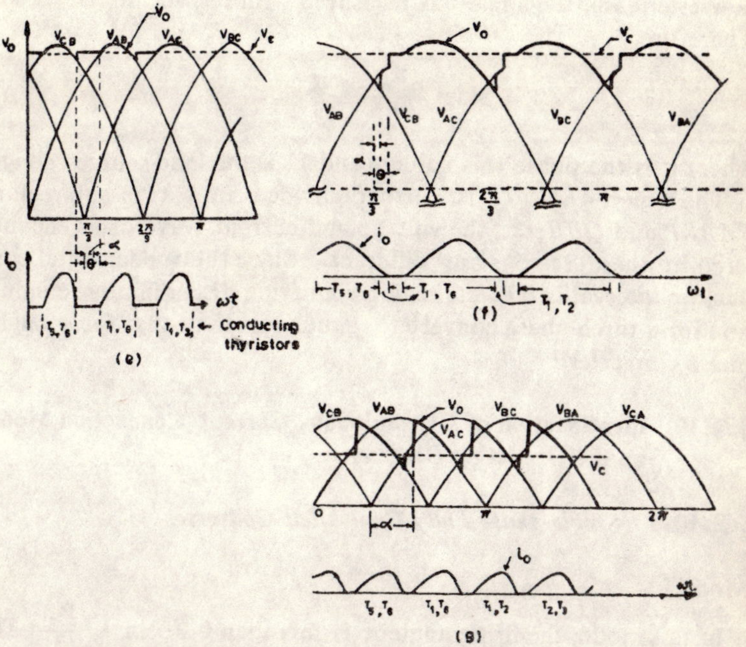

Fig. 3.11. **Three-phase fully controlled converter:** (e) Mode 1, (f) Mode 2, (g) Mode 3.

be analysed with different load circuit parameters along the lines discussed for a single-phase fully controlled converter in section 3.2.9.1. The datum for the firing angle α is $\omega t = 60°$ with respect to the line-to-line voltage. Taking this into account, the operating diagram is illustrated in Fig. 3.10 (d). In order to explain the continuous and discontinuous regions, $\phi = 75°$ is chosen as an example. As in the single-phase case, the range of α considered is 0 to 180°. *QRE* is the dividing curve separating the continuous and discontinuous regions. Any point below this curve represents an operation with continuous conduction while any point above this curve represents an operation signifying discontinuous conduction. The first quadrant indicates rectifier operation while the fourth quadrant relates to inverter operation of the converter. The discontinuous region *QABCDERQ* is further divided into sub-regions depending upon the load voltage and current waveforms. The discontinuous regions *QABRQ* and *RBCDER* correspond to $\alpha \leqslant \theta$ and $\alpha \geqslant \theta$ respectively. In the three-phase converter case, the angle θ is measured with respect to $\omega t = 60°$. Therefore,

$$\theta = \sin^{-1}\frac{V_c}{\sqrt{6}V} - \frac{\pi}{3}$$

where V is the phase rms voltage and V_c is the load source voltage.

The region $QABRQ$ is further divided into two sub-regions $PABRP$ and $QPRQ$ as the voltage and current waveforms encountered in these regions are different. Since the fundamental conduction interval is 60°, the region signifying discontinuous conduction for a three-phase converter is much less than that for a single-phase converter.

3.2.10 Identification of Discontinuous Current Conduction Modes on the Operating Diagram

3.2.10.1 *Single-phase Fully Controlled Converter*

Mode 1

In this mode, the firing angle α is less than $\theta = \sin^{-1}\frac{V_c}{\sqrt{2}V}$. The load voltage and load current waveforms along with supply voltage are shown in Fig. 3.11. The conduction angle is less than 180°. The thyristors turn on only at θ even though the firing pulses are applied at α since they are reverse biased from α to θ. This mode is identified by the region $BCGLB$ in the operating diagram of the converter (Fig. 3.10 (b)) for $\phi = 45°$. For a given load angle ϕ, the boundary curve GL can be determined from eqns. (3.62), (3.64) to (3.67) with the conditions that the current i_0 is zero at $\omega t = \theta$ and again at $\omega t = \theta + \gamma$. The resulting equation can be shown as

$$\exp\left(\frac{-(\pi + \alpha - \theta)}{\tan\phi}\right) = \frac{\dfrac{m}{\cos\phi} + \sin(\alpha - \phi)}{\dfrac{m}{\cos\phi} - \sin(\theta - \phi)} \tag{3.70}$$

This equation can be solved on a digital computer for different values of ϕ, m and α in order to obtain different boundary curves as illustrated in the operating diagram of the converter.

Mode 2

The firing angle α is less than θ in this mode also. The load voltage and current waveforms are shown in Fig. 3.11(b). In the operating diagram this mode occurs in the region *LGAL*. The load angle $\phi = 45°$ is considered for illustrating different modes on the operating diagram as before. When the incoming thyristors are fired, the load current is present but falls to zero since the load source is greater than the supply voltage ($\alpha < \theta$). The thyristors turn on again at $\omega t = \theta$ since they are triggered by long pulses. The necessary condition for this mode to occur is that the current must fall to zero before the AC supply voltage increases to the load source value, V_c. The boundary curve *GA* represents the condition that the current has become just continuous. That is, when the incoming thyristors are triggered at α, the current is present, but becomes zero again at θ, and then the current builds up again from θ onwards. From the above discussion, the necessary equation for determining the boundary curves for mode 2 can be shown as

$$\frac{m}{\cos\phi} = \frac{2\sin(\alpha - \phi)\exp\left(\dfrac{\alpha - \theta}{\tan\phi}\right)}{\exp\left(\dfrac{-\pi}{\tan\phi}\right) - 1} + \sin(\theta - \phi) \qquad (3.71)$$

Mode 3

In the operating diagram of Fig. 3.10(b), this mode is represented by the region *GCDFG* with the firing angle $\alpha \geqslant \theta$. The thyristors turn on at the instants that they are triggered, and the current builds up. It falls to zero before the incoming thyristors are triggered after one half-cycle. The waveforms of the load voltage and the load current are shown in Fig. 3.11(c). The necessary equation for computing the curve *GF* can be derived as

$$\alpha = \phi + \sin^{-1}\left\{\frac{m}{\cos\phi}\cdot\frac{1 - \exp\left(\dfrac{\pi}{\tan\phi}\right)}{1 + \exp\left(\dfrac{\pi}{\tan\phi}\right)}\right\} \qquad (3.72)$$

Mode 4

This mode occurs in the region *EFDE* in the operating diagram of the converter of Fig. 3.10(b) and arises during inversion only with the firing angle $\alpha > (\pi - \theta)$. The current through the outgoing thyristors first falls off to zero but because of maintained trigger pulses, the outgoing thyristors turn on again. They are line commutated when the incoming thyristors are turned on at the firing angle α. Since the current is present when the incoming thyristors are fired, the latching of the thyristors is fast, and this improves the response of the converter system [4]. The equation for determining the boundary *FE* representing mode 4 in the operating diagram of Fig. 3.10(b) can be shown as

$$\frac{m}{\cos\phi} = \sin(\theta+\phi) + \left\{\frac{2\sin(\alpha-\phi)\cdot\exp\left\{\dfrac{-(2\pi-\theta-\alpha)}{\tan\phi}\right\}}{1-\exp\left(\dfrac{-\pi}{\tan\phi}\right)}\right\} \tag{3.73}$$

$$\alpha > (\pi - \theta)$$

$$\gamma < \pi$$

Equations (3.70) to (3.73) can be solved on a digital computer to determine different modes for various values of ϕ, m and α. The horizontal line *GK* (Fig. 3.10 (b)) shows the boundary dividing the continuous and discontinuous regions if the thyristors are triggered by short pulses instead of by maintained long pulses. It is evident that by maintained pulse triggering the operating region of the converter is enhanced.

3.2.10.2 *Three-phase Fully Controlled Converter*

The analysis of a three-phase fully controlled converter reveals three discontinuous modes of operation corresponding to the three regions of the operating diagram.

Mode 1

In this mode, the load current falls to zero before the incoming thyristor is triggered. The region corresponding to this mode of operation is *ABRPA* of the operating diagram of Fig. 3.10(d) where $\alpha \leqslant \theta$. Typical waveforms of the output voltage and current in this mode are illustrated in Fig. 3.11(e). As in the case of a single-phase converter, it has been assumed that the thyristors are triggered with long pulses each of width 120°. Let us consider the operation of this circuit during the turn on interval of the thyristor T_1. Earlier to the turn on of the thyristor T_1, T_5 and T_6 conduct giving a pulse of the load current. Since the load current has become zero, both thyristors are turned off. Although a gate pulse is applied to the thyristor T_1 at $\omega t = \alpha + \pi/3$, T_1 cannot turn on since V_c being greater than v_{AB} reverse biases it. The earliest instant at which it turns on is at $\omega t = \pi/3 + \theta$. Therefore the thyristors T_1 and T_6 turn on at $(\pi/3 + \theta)$ and the load current rises, reaches a maximum and then decays to zero before the incoming thyristor T_2 is turned on. The load current repeats giving six pulses of current in each input cycle of the supply voltage.

Mode 2

As has been explained in section 3.2.9.2, the region in the operating diagram which represents this mode of operation is given by *PQRP* where $\alpha \leqslant \theta$. The load current is present when the incoming thyristor is turned on. Therefore, the turn on of the incoming thyristor commutates the outgoing thyristor by the line voltage. Since the load emf is greater than the supply voltage, the load current decays to zero at some instant say t_1 such that $\omega t_1 < (\pi/3 + \theta)$, and the thyristor is commutated naturally. It remains in the off condition till $\omega t = \pi/3 + \theta$ since it is reverse biased. As pointed out earlier, the thyristors being triggered by long pulses, the incoming thyristor turns on again at $\omega t = \pi/3 + \theta$. The current builds up again and continues to conduct the load current till the next commutation. Figure 3.11(f) shows the output voltage and current waveforms in mode 2 operation. In each cycle of the input voltage, six pulses of load current are obtained since there are six commutations.

Mode 3

This mode of operation is encountered in the region *RBDER* of the operating diagram of Fig. 3.10(d). In this region, $\alpha > \theta$ Typical waveforms of the output voltage and current are illustrated in Fig. 3.11(g). The incoming thyristor turns on at $\omega t = \alpha + \pi/3$ since the incoming thyristor is always forward biased. The load current becomes zero before the incoming thyristor is turned on.

3.2.11 Effect of Source Impedance on the Performance of Converters

The source impedance has been neglected in the derivation of average output voltages of different multipulse converters shown in Table 3.1 and hence during commutation of the outgoing thyristors, the transfer of current from the outgoing thyristors to the incoming thyristors is instantaneous. If the source impedance is resistive, then there will be a drop across this resistance and the average output voltage of a converter gets reduced by an amount equivalent to the average drop. Since the source resistance is usually small, the commutation angle during which the load current is transferred from the outgoing to the incoming thyristors is generally neglected. In most practical cases, the source resistance is small compared with the source reactance. In what follows, only source inductance is considered while analysing the converter performance. Further, the load inductance is assumed to be infinitely large so that the output current is virtually constant. The analysis of a converter taking the source impedance into account along with a finite load circuit inductance is quite involved [13].

3.2.11.1 *Two-pulse Converter*

Let us consider a single-phase 2-pulse converter as shown in Fig. 3.12(a) with a source inductance L_S. The load current is assumed constant. The waveforms of voltages and currents of the circuit of Fig. 3.12(a) are shown in Fig. 3.12(b). The thyristor T_2 is turned on at $\omega t = \pi + \alpha$. The commutation of T_1 begins at $(\pi + \alpha)$. Because of the presence of the source inductance L_S in series with each source, the current builds up gradually from zero to the full value of the load current I_0 through the thyristor T_2, while the current through the thyristor T_1 gradually reduces to zero from the initial value of I_0. During the commutation of T_1, writing down the voltage equation for the loop comprising v_{AN}, T_1, T_2 and v_{BN}, we get

$$L_S\left(\frac{di_{T1}}{dt}-\frac{di_{T2}}{dt}\right)=2\sqrt{2}\,V\sin\omega t \tag{3.74}$$

where $\qquad v_{AN} = -\,v_{BN} = \sqrt{2}\,V\sin\omega t$

With a constant load current,

$$i_{T1} + i_{T2} = I_0 \tag{3.75}$$

Differentiating eqn. (3.75)

$$\frac{di_{T1}}{dt} = -\frac{di_{T2}}{dt} \tag{3.76}$$

If the commutation interval is μ, then the current through T_2 builds up to I_0 during this interval.

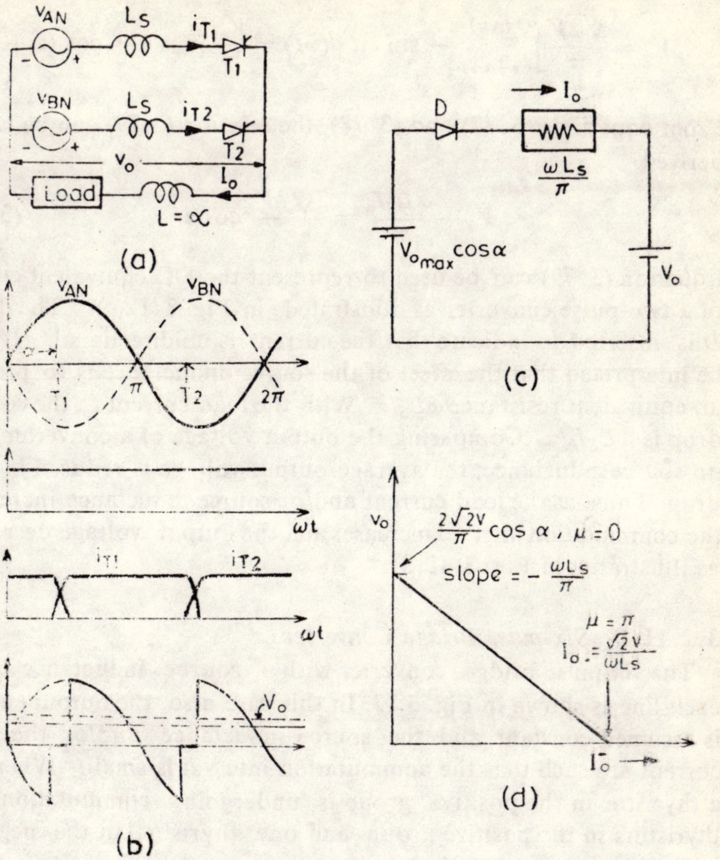

Fig. 3.12. Effect of source inductance in a single-phase FCC
(a) Equivalent circuit;
(b) Typical waveforms;
(c) DC equivalent circuit;
(d) Variation of output voltage with output current.

Therefore

$$i_{T_2} = \frac{1}{\omega L_S} \int_{(\pi+\alpha)}^{(\pi+\alpha+\mu)} -\sqrt{2}\, V \sin \omega t \, d(\omega t) = I_0$$

$$\frac{\sqrt{2}V}{\omega L_S} [\cos \alpha - \cos (\alpha + \mu)] = I_0 \tag{3.77}$$

During the interval μ the output voltage is zero. There are two commutations in each cycle. The average output voltage is

$$V_0 = \frac{\sqrt{2}V}{\pi}\int_{(\pi+\alpha+\mu)}^{(2\pi+\alpha)} -\sin\omega t\, d(\omega t) = \frac{V_m}{\pi}[\cos\alpha + \cos(\alpha+\mu)] \tag{3.78}$$

From equations (3.77) and (3.78), the relation (3.79) can be easily derived.

$$V_0 + \frac{\omega L_S I_0}{\pi} = \frac{2\sqrt{2}V}{\pi}\cos\alpha \tag{3.79}$$

Equation (3.79) can be used to represent the DC equivalent circuit of a two-pulse converter as illustrated in Fig. 3.12(c). The diode D is inserted to indicate that the current is unidirectional. It can be interpreted that the effect of the source inductance is to present an equivalent resistance $\omega L_S/\pi$. With the load current I_0, the voltage drop is $\omega L_S I_0/\pi$. Comparing the output voltage of a converter with no source inductance, the average output voltage is reduced by this drop. Thus, as the load current and/or source inductance increases, the commutation interval increases and the output voltage decreases as illustrated in Fig. 3.12(d).

3.2.11.2 *Six-pulse Bridge Converter*

The six pulse bridge converter with a source inductance L_S in each line is shown in Fig. 3.13. In this case also, the output current is assumed constant and the source inductance and/or the load current are such that the commutation interval is small. Whenever a thyristor in the positive group is undergoing commutation, two thyristors in the positive group and one thyristor in the negative

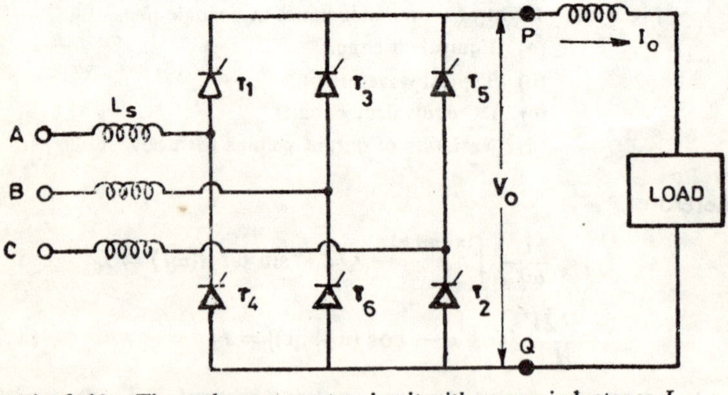

Fig. 3.13. Three-phase converter circuit with source inductance L_S in each line.

group conduct. Figure 3.14 shows the circuit configuration under this condition. After the commutation is completed, two thyristors, one in the positive group and the other in the negative group, but not in the same leg conduct. Similarly, whenever a thyristor in the negative group is undergoing commutation, three thyristors conduct followed by two thyristors. Thus three thyristors and two thyristors alternately conduct. For a 6-pulse converter, there are six commutations in each cycle of the supply voltage. The instantaneous waveform of the output voltage and the output current

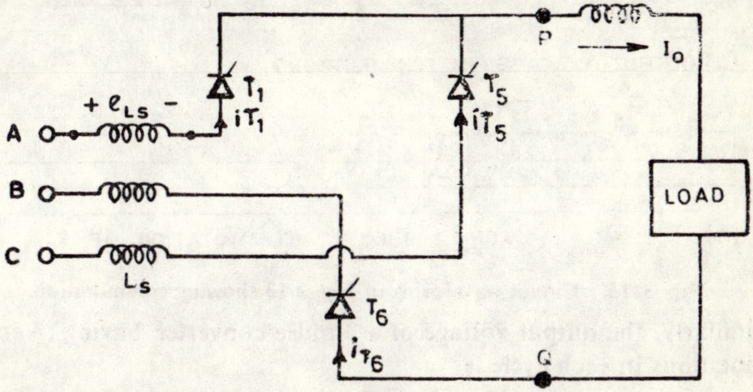

Fig. 3.14. Circuit configuration of Fig. 3.13 with conducting SCR's T_1, T_5 and T_6.

are shown in Fig. 3.15. Without undertaking a detailed analysis of the converter, the effect of the source inductance on the effective output voltage of a converter can be determined as follows.

Loss of Volt-sec due to each commutation $= Vt = L_S I_0$ (3.80)

Total loss of volt-sec in one cycle $= 6L_S I_0$. (3.81)

Average voltage lost due to 6 commutations $= 6L_S I_0 f = \dfrac{3\omega L_S I_0}{\pi}$ (3.82)

From Table 3.1, the average output voltage of a 6-pulse bridge converter neglecting source inductance $= \dfrac{3\sqrt{6}V}{\pi}\cos\alpha = V_{0_{max}}\cos\alpha$

where $\dfrac{3\sqrt{6}V}{\pi} = V_{0_{max}}$ for $\alpha = 0^\circ$.

The effective output voltage of the 6-pulse converter is

$$V_0 = V_{0_{max}}\cos\alpha - \frac{3\omega L_S I_0}{\pi} \tag{3.83}$$

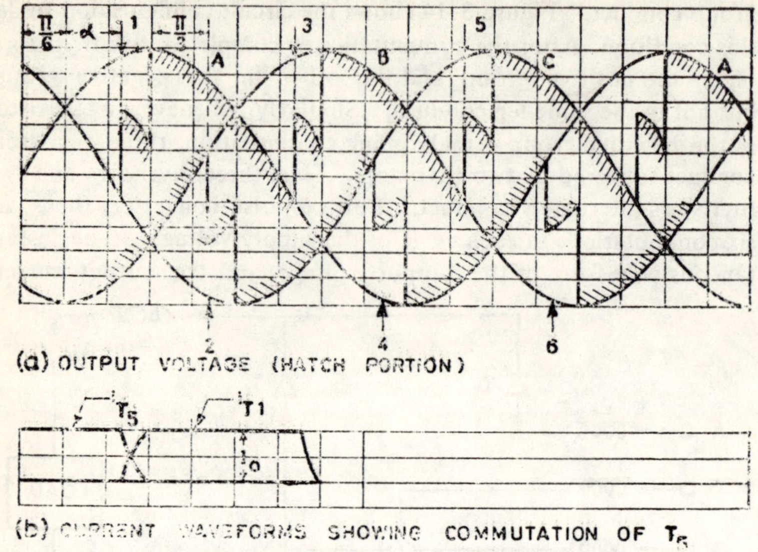

Fig. 3.15. Circuit waveforms of Fig. 3.13 showing commutation.

Similarly, the output voltage of a 3-pulse converter having 3 commutations in each cycle is

$$V_0 = \frac{3\sqrt{3}.\sqrt{2}V}{2\pi}\cos\alpha - \frac{3\omega L_S I_0}{2\pi} \tag{3.84}$$

3.2.12 Effect of Overlap on the Converter Performance

For a given source inductance, the output voltage of a 2-pulse

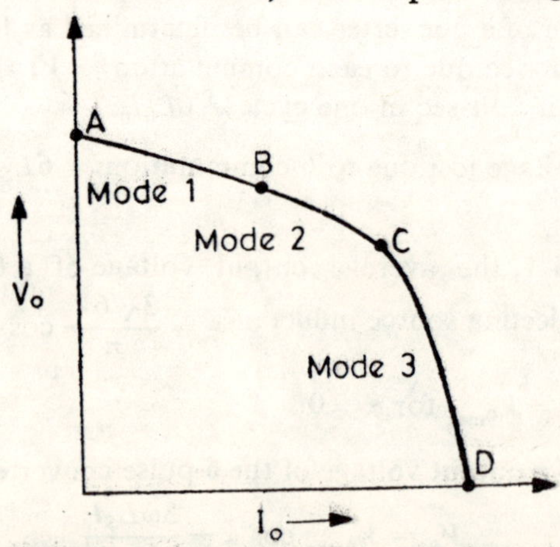

Fig. 3.16. Variation of the output voltage with the load current in a three-phase bridge converter.

converter, as given by eqn. (3.79), decreases linearly with an increase in the load current as depicted in Fig. 3.12(d) for the rectification mode. As the load current increases, the overlap angle also increases with reduced intervals of constant current and the output voltage decreases linearly. When the overlap angle becomes equal to π, the output voltage becomes zero and the converter is short-circuited. The alternating current becomes almost sinusoidal.

Unlike in a single-phase converter, several modes arise in a three-phase 6-pulse bridge type converter if the source inductance is taken into account. Since there are six commutations in each supply cycle, no overlap takes place for a given firing angle between two consecutive commutations as long as the overlap angle is less than 60°. The output voltage of the converter is given by eqn. (3.83) and it decreases linearly with an increase in the output current. This is called mode I operation of the converter, and is shown by the curve *AB* in Fig. 3.16. The conduction pattern of different thyristors in mode I over a period of one cycle is

1, 2 1, 2, 3 2, 3 2, 3, 4 3, 4 3, 4, 5
4, 5 4, 5, 6 5, 6 5, 6, 1 6, 1 6, 1, 2

where 1 to 6 are the thyristors T_1 to T_6 in the bridge circuit of Fig. 3.13. When the overlap angle reaches 60°, an overlap between two commutations takes place. Three thyristors conduction pattern prevails as the load current is increased. That is

1, 2, 3 2, 3, 4 3, 4, 5 4, 5, 6 5, 6, 1 6, 1, 2

The overlap angle remains constant at 60°. The incoming thyristor turns on after the outgoing thyristor is commutated. Due to this, an additional phase delay is introduced. This is known as Mode II operation of the converter. It is essential that the thyristors must be triggered by maintained long pulses of at least 30° duration for this mode to appear. During this mode, the average output voltage of the converter decreases more rapidly than in mode I as shown by the curve *BC* in Fig. 3.16. The instantaneous output voltage waveform during mode II operation of the converter is shown in Fig. 3.17(a).

When the total phase angle delay $(\alpha + \alpha') = 30°$, and if the load current is increased beyond that corresponding to the end of Mode II, an overlap in successive commutations takes place. Three thyristors and four thyristors alternately conduct. That is

1, 2, 3 1, 2, 3, 4 2, 3, 4 2, 3, 4, 5

3, 4, 5 3, 4, 5, 6 4, 5, 6 4, 5, 6, 1

5, 6, 1 5, 6, 1, 2 6, 1, 2 6, 1, 2, 3

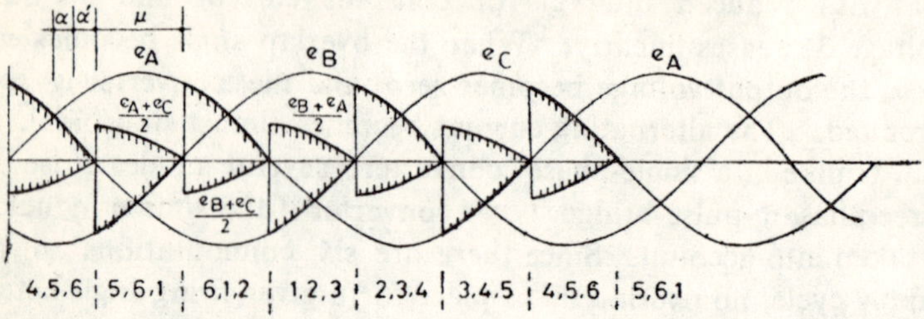

(a) Instantaneous output voltage during mode II (curve *BC* of Fig. 3.16).

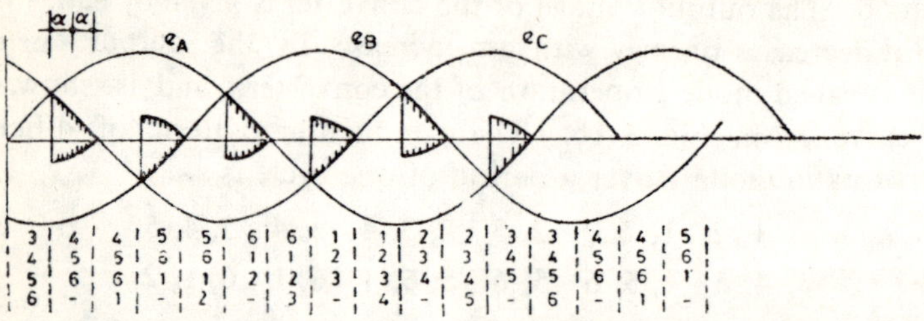

(b) Instantaneous output voltage during mode III (curve *CD* of Fig. 3.16).

Fig. 3.17. Instantaneous output voltages of three-phase converter taking the overlap angle into account.

Whenever four thyristors conduct, the output voltage is short-circuited. There will be six intervals during which the output voltage is short circuited in each cycle. This is called mode III operation of the converter. The variation of the output voltage with the load current is shown by the curve *CD* in Fig. 3.16. Because of zero output voltage intervals, the average output voltage falls off more rapidly than in the mode II. The output voltage waveform is shown in Fig. 3.17(b). The total overlapping interval can be divided into two periods, 4-thyristor conductor interval (u) and 3-thyristor conduction interval (μ). As the load current increases, the overlapping period increases with the total period ($u+\mu$) remaining at 60°. Finally, when u increases to 60°, the output voltage is short-circuited all the time with four thyristors conducting. That is, the conduction pattern of thyristors is

1, 2, 3, 4 2, 3, 4, 5 3, 4, 5, 6

4, 5, 6, 1 5, 6, 1, 2 6, 1, 2, 3

It has been assumed that the load current is ripple free in the foregoing discussion.

The reactive power requirement of a 2-quadrant converter is high. The power factor deteriorates considerably as α approaches 90° with a continuous load current. In some applications, the inverting operation may not be required. In such applications, the cost is reduced, and the power factor is improved by using half-controlled converters.

3.3 HALF-CONTROLLED (ONE-QUADRANT) CONVERTERS

In a 2-quadrant bridge converter, if half the total number of thyristors are replaced by diodes, the converter circuit so formed is called the 'Half-Controlled' converter. This name is derived owing to half the number of thyristors used in the circuit arrangement compared to the fully controlled converter. Half-controlled converters are also known as 'Semi-Converters'. The advantages of half-controlled converters over fully controlled converters are:

1. Since half the thyristors are replaced by diodes, the half-controlled converter costs less than the fully controlled converter.

2. The mean DC terminal voltage can be continuously controlled from maximum to virtually zero with an increased control range of the firing angle, α.

3. The reactive power requirement for the same mean DC output voltage is less than for a fully controlled converter specially for lower output voltages. As a result, the power factor is improved with half-controlled converters.

Half-controlled converters are widely used in main line A.C. traction where large DC motors are supplied from a single-phase AC supply. The supply power factor is an important quantity since the supply system is weak with low short-circuit capacity.

3.3.1 Fundamental Principles of All One-Quadrant Converters

The half-controlled converter can also be built using the basic building blocks of 2-quadrant converter and uncontrolled rectifier. It essentially consists of a controlled 2-quadrant converter connected in series with an uncontrolled converter as shown in Fig. 3.18. This is valid regardless of the pulse number. A general relationship can be derived between the firing angle, dc voltage ratio and displacement factor provided the following assumptions are made.

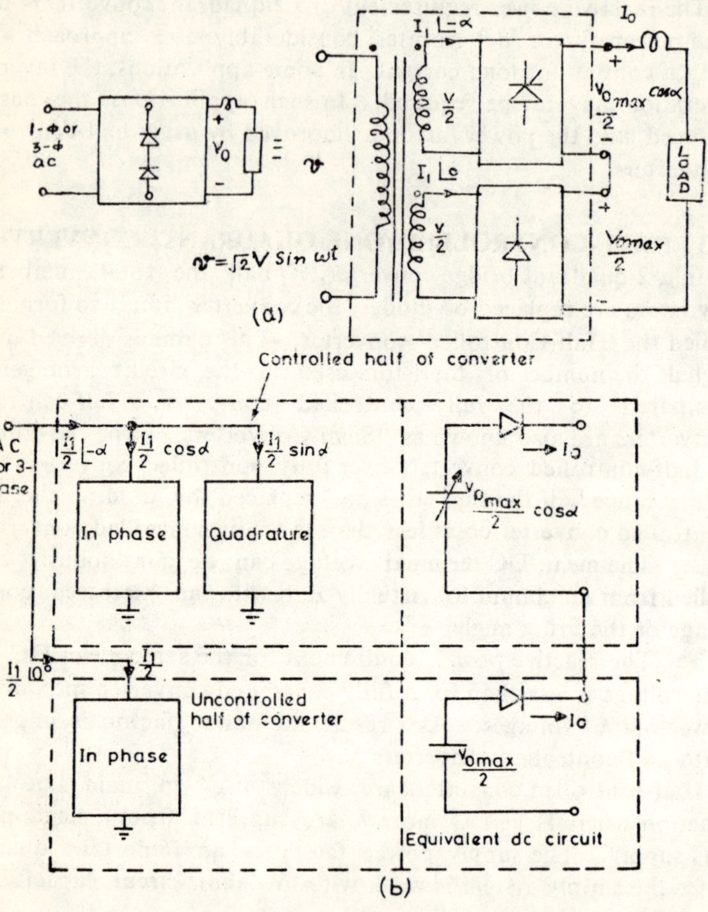

Fig. 3.18 (a) Block diagram representation of a half controlled converter consisting of a controlled converter in series with an uncontrolled converter.

(b) Equivalent circuit of neglecting harmonics in output voltage and input current.

1. The input AC harmonic currents are neglected.
2 The output voltage harmonics are neglected. That is, the

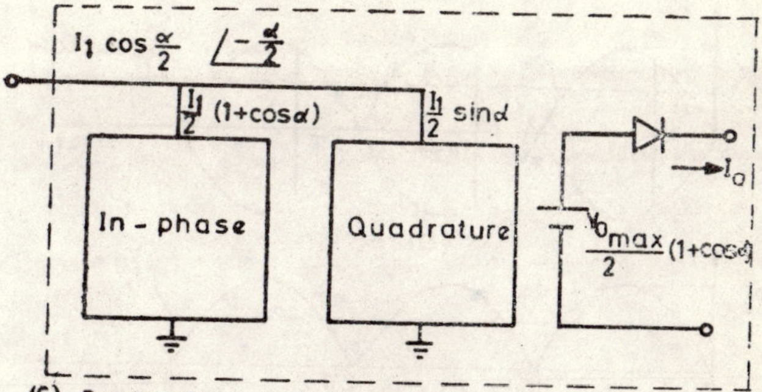

(C) Composite equivalent circuit of a half-controlled converter.

Fig. 3.18. Equivalent circuit representation of the half-controlled converter.

output voltage is assumed to be perfect dc. The output current is assumed to be continuous and constant.

These assumptions make it possible to represent the equivalent circuit of the converter independent of the pulse number as shown in Fig. 3 18(b). Fig. 3.19(a) and (d) represent the output voltage and the input current respectively for the fully controlled converter while Fig. 3.19(b) and (e) represent the same for the uncontrolled converter. In a similar way the waveforms for a multi-phase converter can be easily drawn. In the diagram of Fig. 3.19 the following relations can readily be obtained for the peak fundamental and peak output voltage of controlled and uncontrolled converters.

By Fourier analysis of the input current waveform of each converter, the rms fundamental component is given by

$$I_1 = \frac{I_{1\,max}}{\sqrt{2}} = \frac{2\sqrt{2}\,I_0}{\pi} \tag{3.85}$$

The maximum output voltage is

$$V_{0_{max}} = \frac{2\sqrt{2}\,V}{\pi} \tag{3.86}$$

The input current waveform of the half-controlled converter is obtained by summing the currents due to individual converter at the input side of the ac supply while the output voltage is obtained by the addition of the output voltages of controlled and uncontrolled

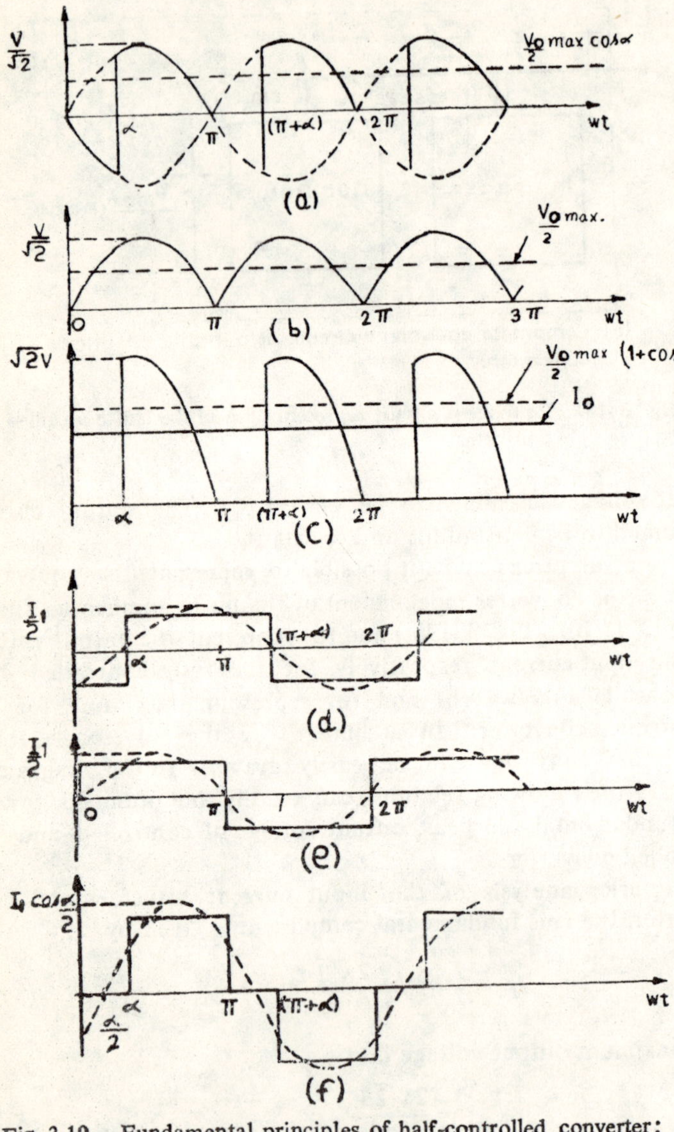

Fig. 3.19. Fundamental principles of half-controlled converter:

(a) Output voltage waveform of controlled half of converter.
(b) Output voltage waveform of uncontrolled half of converter.
(c) Resultant voltage of composite converter.
(d) Input current waveform of controlled converter.
(e) Input current waveform of uncontrolled converter.
(f) Input current waveform of composite converter.

converters. The resulting waveforms as depicted in Fig. 3.19(c) and (f) indicate the output voltage and the input current respectively of the 'Half-controlled' converter.

The equivalent circuit at the dc side consists of a controllable direct voltage source, $\frac{V_{0max}}{2}\cos\alpha$, representing the voltage of the controlled half of the converter, connected in series with a fixed direct voltage source, $\frac{V_{0max}}{2}$, representing the voltage of the uncontrolled half of the converter.

Thus, the DC terminal voltage ratio of the half-controlled converter is given by:

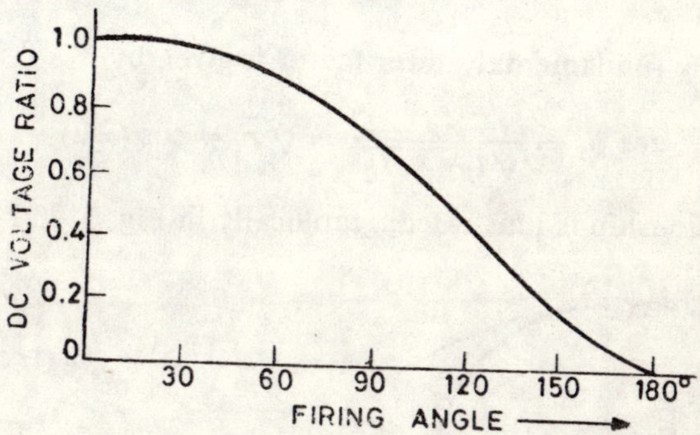

Fig. 3.20 (a) DC terminal voltage ratio vresus firing angle for half-controlled converter.

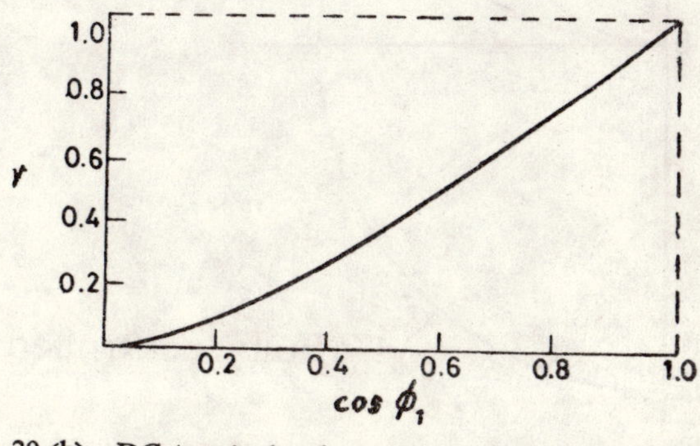

Fig. 3.20 (b) DC terminal voltage ratio versus the input displacement factor for the half-controlled converter.

$$r = \frac{V_0}{V_{0max}} = \tfrac{1}{2}(1 + \cos\alpha) \qquad (3.87)$$

This relationship is illustrated in Fig. 3.20(a).

The fundamental current at the AC side consists of an inphase-component, $I_1/2 \cos\alpha$, and a quadrature component, $I_1/2 \sin\alpha$, due to the controlled converter and a fixed amplitude inphase component, $I_1/2$ due to the uncontrolled converter.

Thus, the total inphase component of the current is given by

$$I_P = \frac{I_1}{2}(1 + \cos\alpha) = I_1 . r \qquad (3.88)$$

And the total quadrature component of the current is given by

$$I_Q = \frac{I_1}{2}\sin\alpha = I_1\sqrt{r - r^2} \qquad (3.89)$$

The supply fundamental power factor is given by

$$\cos\phi_1 = \frac{I_p}{\sqrt{I_P^2 + I_Q^2}} = \sqrt{r} = \cos(\alpha/2) \qquad (3.90)$$

This relationship is illustrated graphically in Fig. 3.20(b).

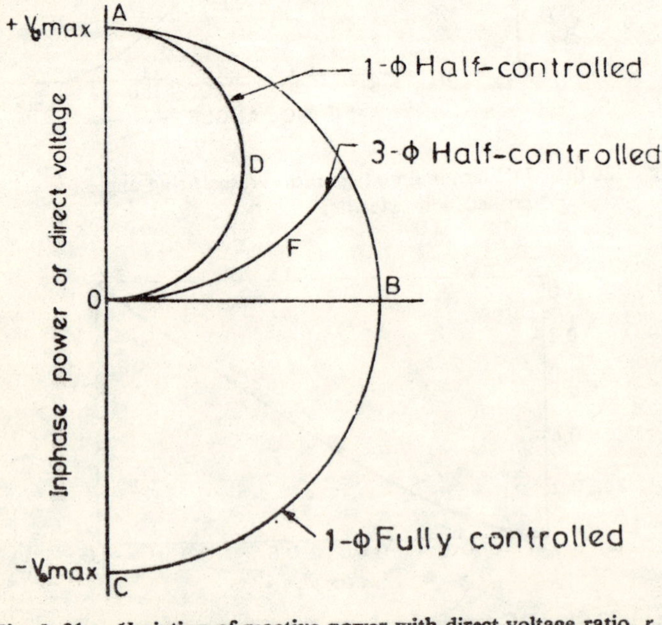

Fig. 3.21. Variation of reactive power with direct voltage ratio, r.

$$\text{Normalized active power} = \frac{VI_P}{VI_1} = \frac{V_0 I_0}{V_{0max} I_0} = r \tag{3.91}$$

$$\text{Normalized reactive power} = \frac{VI_Q}{VI_1} = \sqrt{r - r^2} \tag{3.92}$$

Similarly the normalized active and reactive powers for a fully controlled converter are shown to be r and $\sqrt{1-r^2}$ respectively in Section 3.2.5. For the sake of comparison, the reactive power variation for half-controlled and fully-controlled converters is shown in Fig. 3.21.

3.3.2 Half-controlled Two-pulse and Three-pulse Bridge Converters

Figure 3.22 shows two alternative arrangements of 2-pulse half-controlled bridge converters together with the pertinent voltage and current waveforms, for a particular firing angle α. The filter inductance, L_d is assumed to be sufficiently large as to produce continuous load current. The converter circuit with the thyristors T_1 and T_2 connected to form a common cathode as shown in Fig. 3.22(a) is known as the symmetrical configuration while the circuit with the thyristors T_1 and T_2 connected to the same AC terminal as shown in Fig. 3.22(b) is known as the asymmetrical configuration. The different waveforms of symmetrical and asymmetrical converter circuits are shown in Figs. 3.22(c) and (d) respectively.

Consider the symmetrical circuit configuration. T_1 is turned on at a firing angle α in each positive half cycle. T_1 and D_1 conduct from α to π. When the supply voltage reverses, the diode D_2 is forward biased since diode D_1 is already conducting. The diode D_2 then turns on, and the load current passes through D_2 and T_1. The supply voltage reverse biases D_1 and then turns it off. Thus the load current freewheels during the interval from π to $(\pi + \alpha)$ in each supply cycle. At the instant $(\pi + \alpha)$, a trigger pulse is applied to the thyristor T_2. Since it is forward biased, it turns on. Consequently the supply voltage reverse biases T_1 and then turns it off by the line commutation. The load current now flows through T_2 and D_2 and the supply current i is reversed. The supply current waveform, voltage across the thyristor T_1, output voltage and output current are shown in Fig. 3.22(c). Since the conduction intervals of thyristors and diodes are equal, this circuit is known as the 'symmetrical configuration'.

In the converter circuit of Fig. 3.22(b), T_1 and D_1 conduct from

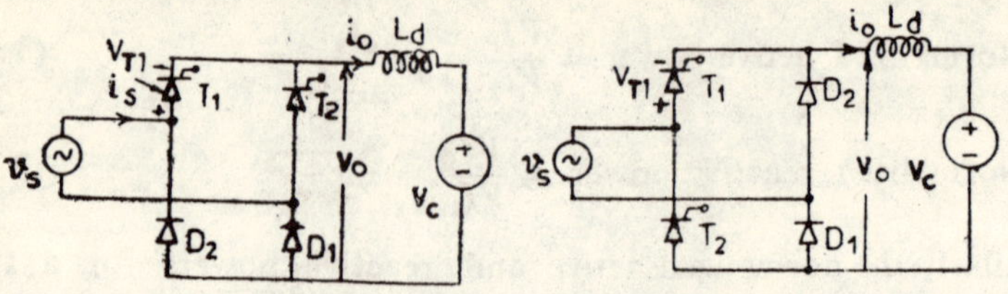

(a) Symmetrical configuration (b) Asymmetrical configuration

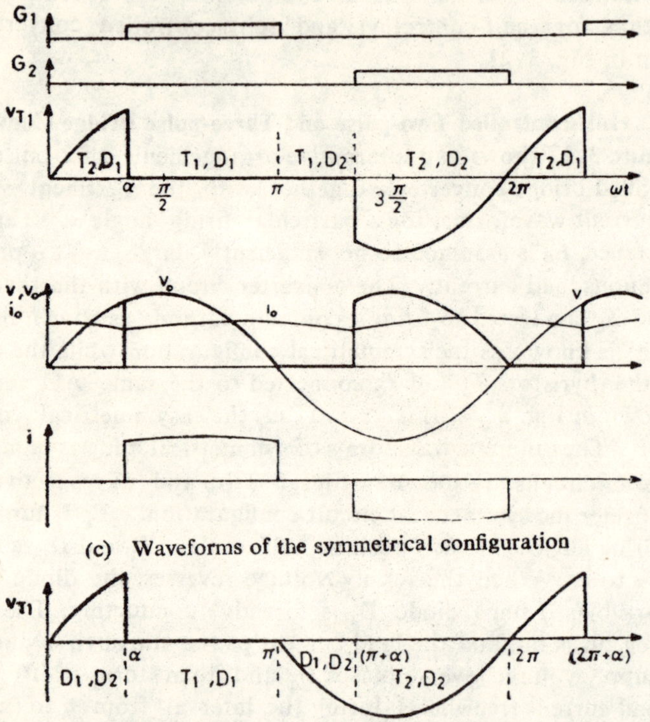

(c) Waveforms of the symmetrical configuration

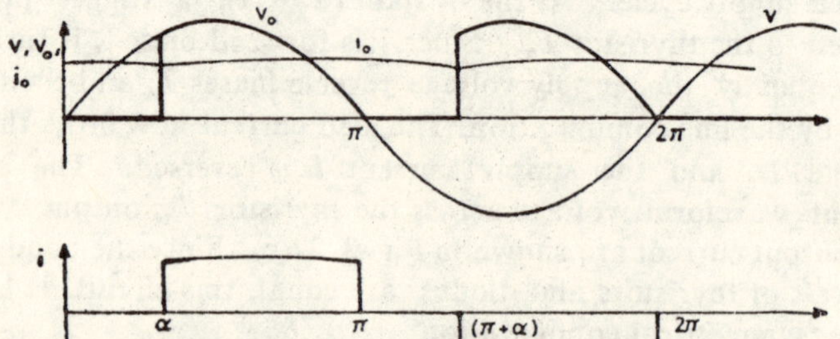

(d) Waveforms of the asymmetrical configuration

Fig. 3.22. Two-pulse half-controlled bridge circuits.

α to π, and T_2 and D_2 conduct from $(\pi + \alpha)$ to 2π in each supply cycle. The freewheeling action is provided by diodes D_1 and D_2 from 0 to α and from π to $(\pi + \alpha)$ in each supply cycle. The supply current waveform, voltage across the thyristor T_1, output voltage and current waveforms for a particular firing angle α are shown in Fig. 3.22(d). Due to unequal conduction intervals of thyristors and diodes, this circuit is known as the 'asymmetrical configuration'. The two converter circuit configurations are indistinguishable with regard to the output voltage and current waveforms as can be seen clearly from Fig. 3.22(c). However the internal conduction mechanism is different in the two converter circuits. Some essential differences between the symmetrical and asymmetrical converter circuit configurations may be summarised as follows.

1. For large firing angle delays, commutation failure may take place due to the limited time available in the symmetrical half-controlled converter circuit configuration if the current is assumed to be continuous. This may result in 'half-waving effect'. That is, the converter operates as an uncontrolled half-wave converter. This may lead to instability particularly when the converter supplies a motor load. The motor drive system may become oscillatory. The 'half-waving effect' may also take place if the firing pulses are inhibited due to faulty operation of the control circuit when the converter is in operation. However in the asymmetrical configuration, commutation failure does not take place even for a large firing angle delay. Also, the 'half-waving effect' does not arise

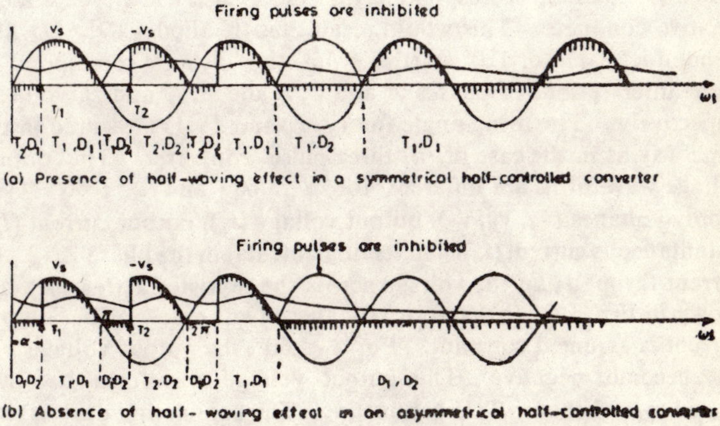

(a) Presence of half-waving effect in a symmetrical half-controlled converter

(b) Absence of half-waving effect in an asymmetrical half-controlled converter

Fig. 3.23. Waveforms illustrating half-waving effect.

with this circuit since there is an inherent freewheeling path for the load current through the diodes D_1 and D_2. When the trigger pulses are inhibited, the output current freewheels to zero. The time interval depends upon the electrical time constant of the circuit. The decay of the current is shown in Fig. 3.23 when the trigger pulse to T_2 is inhibited and the thyristor T_1 is conducting the load current. The nature of the waveforms shown in Fig. 3.23 would remain the same if the trigger pulses to either of the thyristors are inhibited provided the current is continuous.

2. Since the thyristors conduct for a longer interval in the symmetrical circuit, the thyristors used in this circuit must have a higher average current rating compared to those in the asymmetrical circuit.

3. Since the cathodes of the two thyristors are at the same potential in the symmetrical configuration, their gates can be connected and a single gate pulse can be used for triggering either thyristor. The particular thyristor which has the forward bias at the instant of firing will turn on. In the case of asymmetrical converter, separate pulses are to be used.

Just as we have two configurations of half-controlled converters in the case of single-phase, there are also two such configurations in the case of a three-phase converter circuit. Figure 3.24(a) shows a three-phase symmetrical half-controlled bridge circuit since the thyristors are connected to form a common cathode. Neglecting the source impedance, the commutation of each device is instantaneous. The diode corresponding to the phase whichever is more negative conducts. This would mean that the diodes D_1, D_2 and D_3 conduct each for 120° starting from the instants corresponding to the intersections of phases A and C, B and A, C and B voltages respectively. The firing angle for T_1, T_2 and T_3 is measured in the same way as in the case of a three-pulse converter. The output voltage waveforms are different for $\alpha > 60°$ and $\alpha < 60°$. The supply voltages (v_A, v_B, v_C), output voltage (v_0), output current (I_0), instantaneous current (i_A) and its fundamental in the phase $A(i_1)$, the current through and the voltage across the thyristor $T_1(i_{T1}, v_{T1})$ are shown in Fig. 3.24 (b) and (c) for these two cases. The output current is assumed constant. For $\alpha < 60°$, the output voltage will not become negative. The output voltage waveform becomes v_{AB} or v_{AC} whenever T_1 is conducting. The conducting devices are marked in Fig. 3.24(b) for one complete cycle. Each thyristor

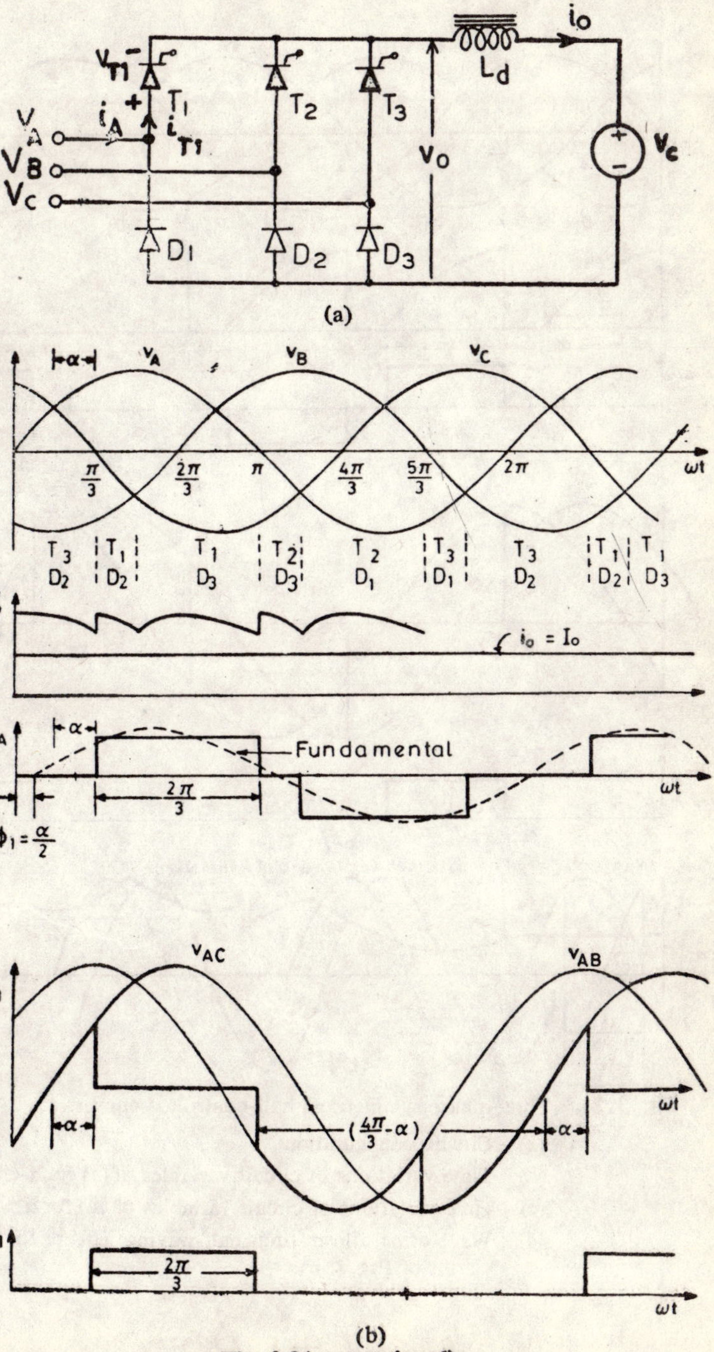

(b)

Fig. 3.24. (continued)

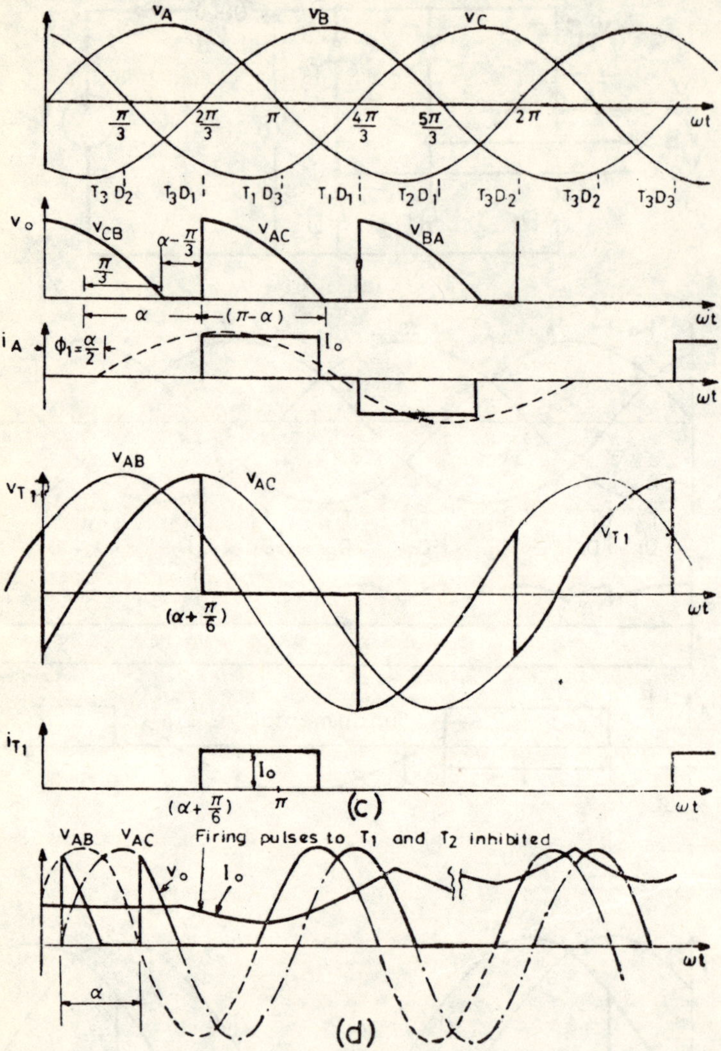

Fig. 3.24. Three-phase symmetrical half-controlled converter:

(a) Circuit configuration;

(b) Time variations of circuit variables of (a) for $\alpha < 60°$;

(c) Time variations of circuit variables of (a) for $\alpha > 60°$;

(d) Waveforms illustrating half-waving effect for the circuit of Fig. (a).

conducts for 120°. The output voltage waveform repeats for every 120°. It is therefore called a 3-pulse half-controlled converter.

For $\alpha > 60°$, the output voltage becomes zero during a part of the output voltage period because of freewheeling action. It can be noted from the waveforms that the freewheeling period is $\left(\alpha - \frac{\pi}{3}\right)$. The supply current flows for the period $(\pi - \alpha)$ in each half cycle. As α increases, the duration of the supply current pulse decreases. Therefore, the harmonic content in the source current increases as the firing angle α increases.

The half-waving effect is also possible in a three-phase converter. This is illustrated in Fig. 3.24(d) with the thyristor T_1 in conduction and the trigger pulses to T_2 and T_3 are inhibited. The current must be continuous as in the case of a single-phase converter for the half-waving effect to take place. For the condition shown in

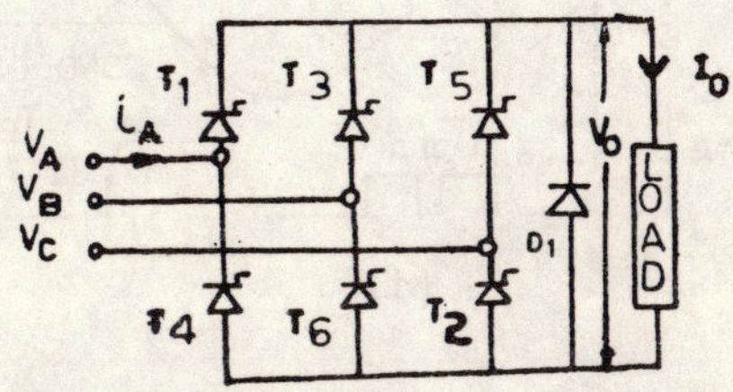

(a)

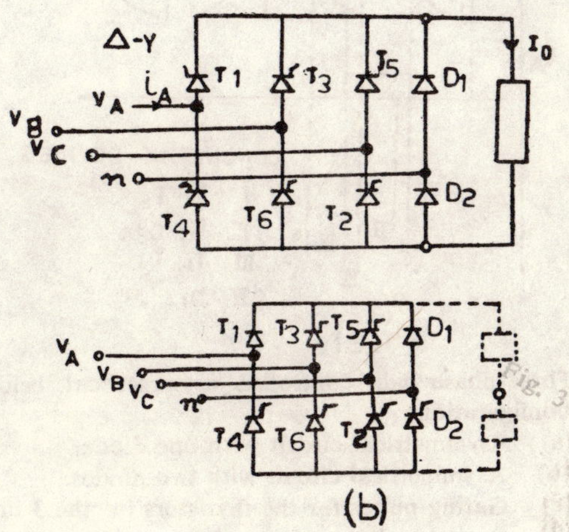

(b)

Fig. 3.25 (continued)

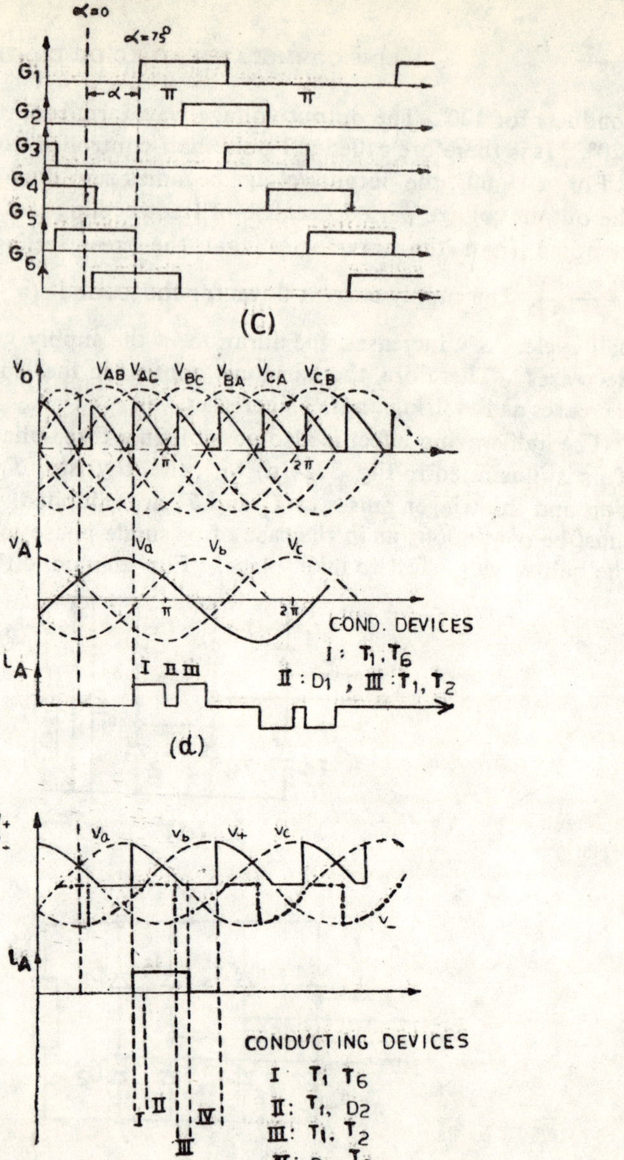

Fig. 3.25. Three-phase half controlled asymmetrical bridge circuit configuration:

(a) Asymmetrical circuit with one diode.

(b) Asymmetrical circuit with two diodes.

(c) Gating pulses for the thyristors in the 3 bridge circuit.

(d) Output voltage (V_0), phase voltage (V_A) and phase current (L_A) waveforms for the circuit (a).

(e) Same as (d) for the circuit (b).

Fig. 3.24(d) the output voltage becomes uncontrolled being v_{AB} or v_{AC} for 120° each. Prior to the half-waving effect, the waveforms are shown for $\alpha = 120°$ in Fig. 3.24(d). The ripple in the output current becomes quite appreciable. The thyristor current is the same as the load current since the thyristor conducts all the time. This half-waving effect can be eliminated by connecting a diode across the load circuit. Therefore a freewheel diode is usually connected across the output terminals when the symmetrical half-controlled converter is used to control a DC motor.

The asymmetrical configuration for a three-phase half-controlled bridge circuit can be arranged in two ways as shown in Fig. 3.25(a) and (b). The firing pulses for the thyristors T_1 to T_6 are shown in Fig. 3.25(c). The output voltage, supply voltage and current in phase A are shown in Fig. 3.25(d) for the asymmetrical bridge circuit configuration with one diode of Fig. 3.25(a). The freewheeling action takes place when α is more than 60° and the output voltage waveform is identical to that obtainable in the symmetrical half-controlled bridge of Fig. 3.24(a). However, the supply current waveform gets distorted in the case of an asymmetrical half-controlled bridge circuit. This is illustrated in Fig. 3.25(d). The control range for the firing angle α is from 0 to 120°.

The supply neutral must be available for the asymmetrical configuration with two diodes. This bridge circuit can be viewed as consisting of two three-phase half-wave circuits each with a freewheel diode connected across the output. They may be considered acting independently. The output voltage waveforms at the positive and negative terminals with reference to the neutral are shown by solid and chain lines in Fig. 3.25(e). Also shown in this figure is the supply current waveform in phase A for one half cycle along with the devices conducting. It may be noted that the freewheeling action commences earlier and hence the benefits of the freewheeling operation are obtained at an earlier point in the control range, as compared to the asymmetrical circuit of Fig. 3.25(a) with one freewheel diode. The major drawback with this circuit is that the neutral carries current when the firing angle is more than 30°. The circuit is also more expensive since it requires more number of devices. As in the case of single-phase asymmetrical bridge circuit, the 'half-waving' effect will not take place in a three-phase asymmetrical bridge circuit if the firing pulses are inhibited when the current is continuous. The three-phase asymmetrical bridge circuit is seldom used in actual practice since it involves more number of devices thereby making it uneconomical.

3.3.3 Input and Output Performances of Half-Controlled (One-Quadrant) Converters

If the output current is assumed constant, simple expressions can be derived for the external performances of all half-controlled converters. An expression for the output voltage is obtained by considering the output voltage waveform for one fundamental period. The displacement factor, distortion factor and power factor are obtained by the Fourier analysis of the input current knowing the conduction interval of each phase current. The input voltages are assumed sinusoidal with negligible source impedance. The expressions for different performances are derived following a similar procedure explained in Section 3.2.3 for fully-controlled converters. Table 3.2 shows these expressions for 2-pulse and 3-pulse half-controlled converters. The rms voltage per phase is taken as V while deriving these expressions.

3.3.4 Single-Phase Fully Controlled Converter with Optional Freewheeling

The half-controlled converter described earlier provides better power factor compared with a fully controlled converter. It is, however, not possible to operate the half-controlled converter in the inversion mode since no mean negative voltage can be developed across its output terminals. However, in traction locomotives, there may be significant energy saving if the load energy is regenerated to the ac source. With a slight modification in the conventional triggering of the thyristors in a fully controlled converter circuit (small diagram of Fig. 3.26) it is possible to incorporate freewheeling without sacrificing the inversion operation [6]. In this triggering scheme, one phase shifter controls the triggering angle or angle of connection of the main thyristor pairs T_1, T_2 and T_3, T_4 as described in Sec. 3.2.1. Also it involves another phase shifter which controls the freewheeling interval. This additional phase shifter controls the triggering angle of disconnection α_2 of the series-aiding thyristor pairs T_1, T_4 and T_2, T_3. Thyristors T_1 and T_2 are fired at some delay angle α_1 in the positive half cycle of the supply voltage and the current flows from the source through T_1 to the load and back to the source via T_2. T_1 and T_2

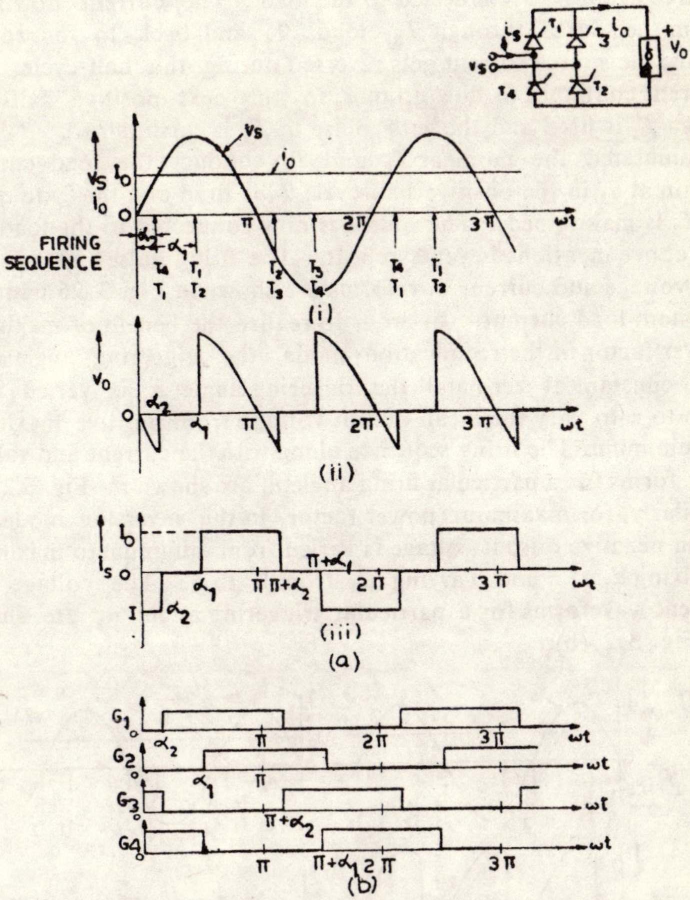

Fig. 3.26. Optional free-wheel control with variable α_1 and α_2:
(a) Voltage and current waveforms.
(b) Firing pulses.

continue to conduct up to the instant α_2 in the next half cycle when T_3 is fired and the gate pulse to T_2 is maintained. T_1 gets commutated by line commutation and the current is transferred to T_3. Thus, the source is disconnected from the load and the load current free-wheels through T_2 and T_3. The freewheeling of the load current continues up to α_1 in the negative half cycle when T_4 is fired and the gate pulse to T_3 is maintained. As T_4 turns on, T_2 gets commutated by the negative line voltage appearing across it and the source again gets connected to the load. The current now flows from the source through T_3, load, T_4 and back to the source. Thus, the source current gets reversed during this half-cycle. The current continues to flow up to α_2 in the next positive half-cycle when T_1 is fired and the gate pulse to T_4 is maintained. T_3 gets commutated the moment T_1 and T_4 conduct the load current. Again at α_1 in the positive half-cycle T_2 is fired and the gate pulse to T_1 is maintained. The source is now connected to the load and the above mentioned events repeat. The firing pulses along with the voltage and current waveforms are shown in Fig. 3.26 assuming constant load current. In order to realize the benefit of maximum power factor in the rectification mode, the triggering angle α_2 is held constant at zero and the triggering angle α_1 is varied from zero to π to vary the mean output voltage from positive maximum to minimum. The firing sequence along with the current and voltage waveforms for a particular firing angle α_1 are shown in Fig. 3.27(a). Similarly, for maximum power factor in the inversion mode, the mean negative output voltage is varied from minimum to maximum by fixing α_1 at π and varying α_2 from 0 to π. The voltage and current waveforms for a particular triggering angle α_2 are shown in Fig. 3.27(b).

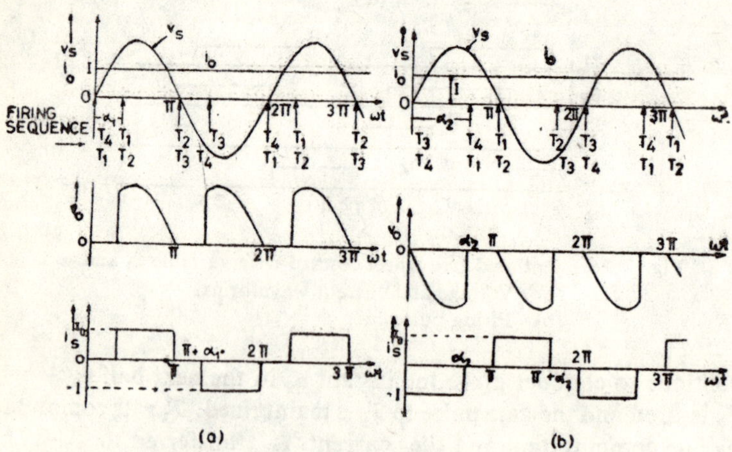

Fig. 3.27. Waveforms for optional free-wheel control:
(a) For rectification mode with $\alpha_2 = 0$.
(b) For inversion mode with $\alpha_1 = \pi$.

The variation of the reactive power with the direct voltage or in phase power can be derived as described in section 3.3.1. This is shown

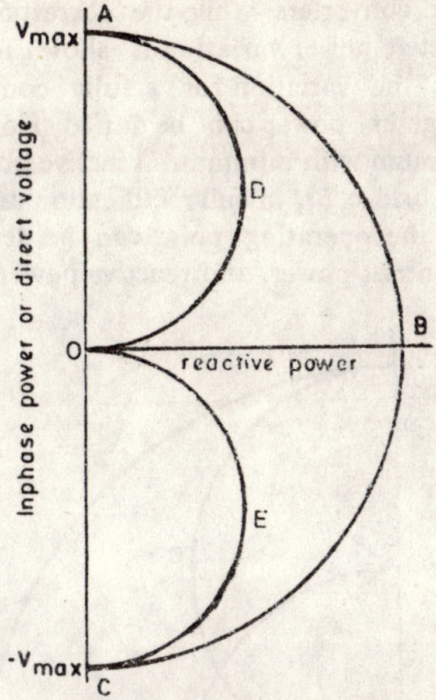

Fig. 3.28. Line power tool of fully controlled single-phase bridge using new optional flywheel mode.

in Fig. 3.28 for the preferred triggering strategy which offers maximum power factor. Locus *ODA* corresponds to the rectification mode while *OEC* corresponds to the inversion mode. Also shown in the Figure is the locus *ABC* for the fully controlled converter. A significant reduction in the reactive power especially at lower output voltages can be clearly noted. Further reduction in reactive power is obtained by connecting two converters in series, each converter operating with the above mentioned control strategy for maximum power factor. Figure 3.29 shows the series

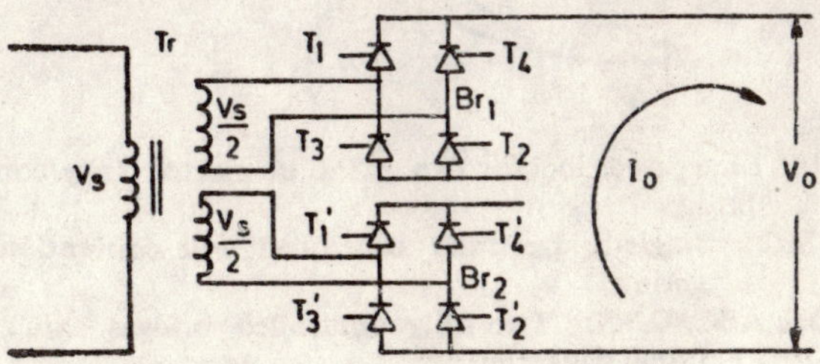

Fig. 3.29. Two series-connected fully controlled bridges for regenerative traction controller.

connection of two converters while the corresponding locus diagram for the reactive power variation is shown in Fig. 3.30. The locus *ABC* shows the variation for a fully controlled converter. The output voltage or power can be varied from zero to positive or negative maximum with minimum reactive power consumption. By operating the bridge Br_1 in full rectification and the bridge Br_2 in full inversion, the operating point can be set at *O* where the output voltage, output power, and reactive power are all zero. By

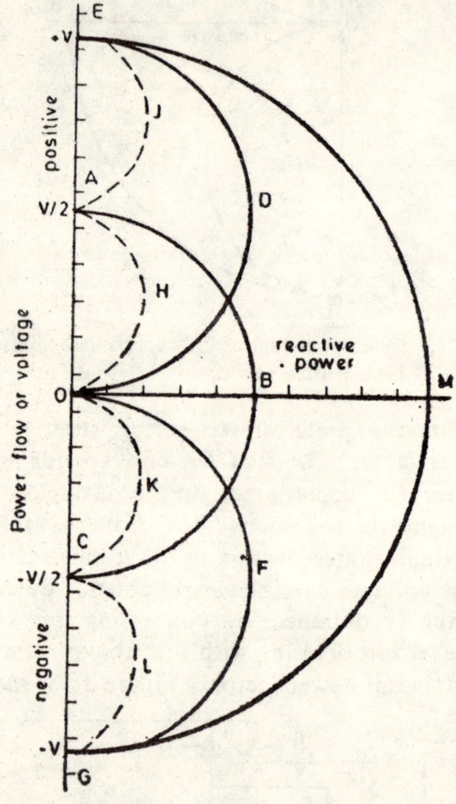

Fig. 3.30. Line power loci for two series connected fully controlled bridges.

Locus EMG: Single-bridge fully controlled with conventional triggering.

Loci ODE, ABC, OFG: Two fully controlled bridges with conventional triggering.

Loci EJA, AHO, OKC and CLG: Two fully controlled bridges with half controlled characteristics (modified triggering).

advancing the triggering angle α_2 at which the bridge Br_2 is operated in the freewheeling mode gradually, the output voltage also decreases gradually following the locus OKC. The result of this will be that the combined output of the two converters follows the locus OHA. Thus with Br_2 at O, the combined output will be at A and the fundamental power factor is unity at half the output voltage. Now operating the Br_2 in the rectification mode with $\alpha_2 = 0$, its locus follows OHA and the combined output of the two converters follows the locus AJE. The fundamental power factor is again unity at the full output voltage corresponding to the point E.

The inversion operation is the reverse of the above mode. Starting from the point O, the Br_1 phases back in the half-controlled mode along AHO for the combined output to reach $(-V/2)$ at the point C. Again operating Br_1 in the inversion mode (OKC), the combined output can traverse the locus CLG. If the operation is achieved by a single fully controlled converter with conventional triggering, the supply power locus would follow EMG. The improvement in power factor can thus clearly be seen with this new triggering strategy. Further improvement in power factor is possible if more number of bridges are connected in series. But this increases the initial investment on the transformers and semiconductor devices. A detailed study is required to determine the optimum number of converters to be connected in series so that the benefits derived owing to the improvement in the power factor will overweigh the investments on transformers and semiconductor devices and other protective devices.

3.3.5 Harmonics of Output Voltage and Input Current of Multi-Pulse Half-Controlled (One-quadrant) Converters.

The instantaneous output voltage of a single-phase half-controlled converter is shown in Fig. 3.22(c) and (d) assuming continuous load current. The waveform can be analysed by Fourier series. The Fourier coefficients are:

$$a_n = \frac{2\sqrt{2}V}{\pi}\int_{\alpha}^{\pi} \sin \omega t \cos n\,\omega t\, d\,(\omega t)$$

$$= \frac{-\sqrt{2}V}{2\pi}\left[\frac{\cos (n-1)\alpha}{(n-1)} - \frac{\cos (n+1)\alpha}{(n+1)} + (-1)^n\left(\frac{1}{n-1} - \frac{1}{n+1}\right)\right] \tag{3.93}$$

$$b_n = \frac{2\sqrt{2}V}{\pi}\int_\alpha^\pi \sin\omega t \sin n\,\omega t\, d(\omega t)$$

$$= -\frac{\sqrt{2}V}{2\pi}\left[\frac{\sin(n-1)\alpha}{n-1} - \frac{\sin(n+1)\alpha}{n+1}\right] \tag{3.94}$$

The expression for the output voltage is

$$v_0 = \frac{\sqrt{2}\,V(1+\cos\alpha)}{\pi} + \Sigma\,(a_n \cos n\omega t + b_n \sin n\omega t) \tag{3.95}$$

$$n = 2, 4, \ldots, 2N, \ldots, \infty$$

The circuit configuration of a three-pulse half-controlled bridge converter is shown in Fig. 3.24(a). The output voltage for a particular firing angle is shown in Fig. 3.24(b) and (c). The harmonic components in the output voltage can be determined in the same way as described for a 6-pulse fully controlled converter in Section 3.2.6.3. The voltage due to the positive group of three thyristors is given by the equation (3.49) whiie the voltage due to the uncontrolled negative group of three diodes is obtained by replacing ωt by $(\omega t + 180°)$ and substituting $\alpha = 0$ in eqns. (3.47) and (3.49). The resulting equations are

$$v_{Pn} = \frac{3\sqrt{6}V\cos\alpha}{2\pi} + \sum_{N=1}^{\infty} \frac{3\sqrt{6}V}{2\pi}\left[\frac{1}{(3N-1)^2} + \frac{1}{(3N+1)^2} - \frac{2\cos 2\alpha}{(3N-1)(3N+1)}\right]^{1/2} . \sin(3N\omega t + \phi_{3N}) \tag{3.95}$$

$$v_{nN} = \frac{3\sqrt{6}V}{2\pi} + \sum_{N=1}^{\infty} (-1)^{3N} \frac{3\sqrt{6}V}{2\pi}\left[\frac{1}{(3N-1)^2} + \frac{1}{(3N+1)^2} - \frac{2}{(3N-1)(3N+1)}\right]^{1/2} . \sin(3N\omega t + \phi_{3N}) \tag{3.95a}$$

The resultant output voltage of a 3-pulse half-controlled bridge converter is

$$v_0 = v_{Pn} + v_{nN} = \frac{3\sqrt{6}V}{\pi}\left\{\frac{1+\cos\alpha}{2} + \frac{1}{2}\left[\sum_{N=1}^{\infty}\left(\frac{1}{(3N-1)^2} + \frac{1}{(3N+1)^2} - \frac{2\cos 2\alpha}{(3N-1)(3N+1)}\right)^{1/2} + (-1)^{3N}\left(\frac{1}{(3N-1)^2} + \frac{1}{(3N+1)^2} - \frac{2}{(3N-1)(3N+1)}\right)^{1/2}\right] \sin(3N\omega t + \phi_{3N})\right\} \tag{3.96}$$

The input current waveform of single-phase and three-phase half controlled bridge converters are shown in Figs. (3.22) and (3.24). These waveforms can be analysed in the same way as described for the fully controlled converters. The amplitudes of these harmonics relative to the fundamental in these converters and other converters with freewheel diodes as a function of the firing angle can be referred [1]. These converters provide low content of lower order harmonics compared to fully controlled converters. But the net harmonic content increases in these converters due to the increase in the higher order harmonics. But these can be filtered out easily since the size of these filters decrease with an increase in the harmonic order.

3.3.6 Detailed Analysis of Single-Phase Fully-Controlled Converter with Half-Controlled Characteristics

The assumption of constant load current leads to simple expressions for the external performances. In general, the load current can be either continuous or discontinuous. In the case of continuous current operation the load voltage waveform comprises only portions of the AC supply voltage, and hence the mean value of this waveform is clearly defined, and is dependent only upon the firing angle. The computation of performance characteristics is trivial in this case. In the case of discontinuous current operation, the load voltage waveform and hence also its mean value is dependent both upon the triggering angles and load circuit parameters, i.e., load phase angle and load *EMF*. Different types of load voltage waveforms are encountered depending upon triggering angle and load circuit parameters. These are recognized here as 'Modes'. The expressions for the voltage and current necessary to determine the operating characteristics are different for different modes. Hence, a prior knowledge of the exact mode is required to determine the performance characteristics from the appropriate set of equations. In this section, the different modes that are possible for the fully controlled converter with half-controlled characteristics are recognized and then identified on the operating diagram of the converter in a similar way as explained in section 3.2.8. A graphical procedure making use of the operating diagrams is presented for the determination of minimum inductance for continuous conduction for both rectification and inversion operations [7].

3.3.6.1 *Determination of Operating Diagrams*

The operating diagrams are drawn separately for rectifier and inverter on the basis of computational results. For different load values of impedance angle ϕ, the load current is computed for various combinations of firing angles and normalized load *EMF*s to determine whether it is continuous or discontinuous. To obtain these results, it is necessary to compute the load current for only the first cycle of the supply voltage input to the converter circuit. The gate pulses are assumed to be of 180° width. The source impedance is neglected.

(a) RECTIFIER OPERATION

In the first half-cycle, the main thyristor pair T_1 and T_2 (Fig. 3.26) turn on at $\omega t = \alpha_{act}$ (α_{act} denotes the actual turning on instant being equal to α or θ as the case may be). Here, θ represents the instant when the instantaneous value of the supply voltage equals the

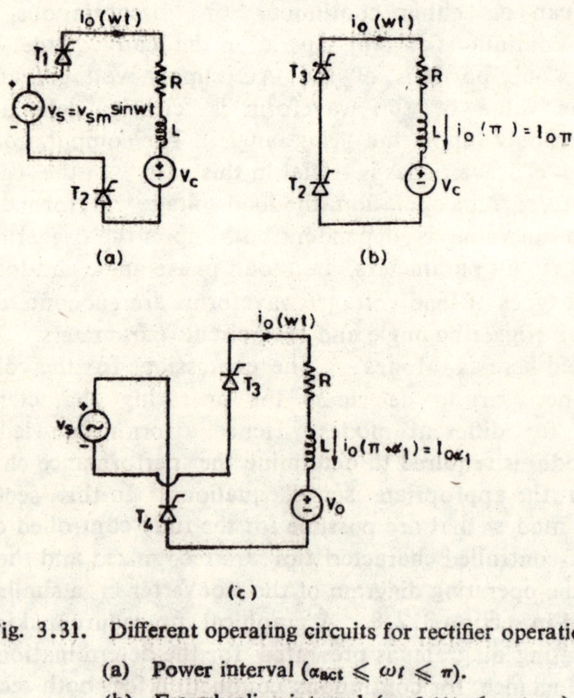

Fig. 3.31. Different operating circuits for rectifier operation:

(a) Power interval ($\alpha_{act} \leqslant \omega t \leqslant \pi$).

(b) Freewheeling interval ($\pi \leqslant \omega t \leqslant \pi + \alpha_1$).

(c) Power interval ($\pi + \alpha_1 \leqslant \omega t \leqslant 2\pi$).

load circuit *EMF*. The operating circuit is as shown in Fig. 3.31 (a). The load current during the period $\alpha_{act} \leqslant \omega t \leqslant \pi$ is given by

$$i_0(\omega t) = (V_{Sm}/Z) \sin(\omega t - \phi) - (V_c/R) + A.\exp(-Rt/L) \tag{3.97}$$

where
$$Z = [R^2 + (\omega L)^2]^{1/2},$$
$$V_{Sm} = \sqrt{2}V, \; V_c \text{ is load source,}$$
$$\phi = \tan^{-1}(\omega L/R).$$

A is a constant and can be evaluated from the initial condition,

$$i_0(\omega t) = 0 \text{ at } \omega t = \alpha_{act}.$$

Substituting this initial condition in the above equation and simplifying

$$\begin{aligned} i_0(\omega t) &= (V_{Sm}/Z) \sin(\omega t - \phi) - (V_c/R) \\ &+ [(V_c/R) - (V_{Sm}/Z) \sin(\alpha_{act} - \phi)] \\ &\exp[-R(\omega t - \alpha_{act})/\omega L] \end{aligned} \tag{3.98}$$

The normalized load current $i_{0N}(\omega t)$ is obtained by dividing the actual load current by a factor V_{Sm}/Z, i.e., 1 pu current. Also denoting the quantity V_c/V_{Sm} by the normalized load *EMF* m, eqn. (3.98) reduces to

$$\begin{aligned} i_{0N}(\omega t) &= \sin(\omega t - \phi) - (m/\cos\phi) + [(m/\cos\phi) - \sin(\alpha_{act} - \phi)] \\ &\times \exp[-(\omega t - \alpha_{act})/\tan\phi] \end{aligned} \tag{3.99}$$

The load current starts freewheeling when the series-aiding thyristors T_3 and T_2 are turned on at $\omega t = \pi$. Referring to the operating circuit of Fig. 3.31(b) for this period $\pi < \omega t \leqslant \pi + \alpha_1$ the normalized load current in a similar manner can be easily derived as

$$i_{0N}(\omega t) = -(m/\cos\phi) + [I_{0\pi N} + (m/\cos\phi)] \exp[-(\omega t - \pi)/\tan\phi] \tag{3.100}$$

where $I_{0\pi N}$ is the load current at $\omega t = \pi$.

The thyristors T_3 and T_4 are triggered at $\omega t = \pi + \alpha_1$. The operating circuit for this period, $\pi + \alpha_1 < \omega t \leqslant 2\pi$, is shown in Fig. 3.31 (c). The load current in its normalized form during this interval is given by

$$\begin{aligned} i_{0N}(\omega t) &= -\sin(\omega t - \phi) - (m/\cos\phi) + [(m/\cos\phi) + I_{0\alpha_1 N} \\ &- \sin(\alpha - \phi)] \cdot \exp[-(\omega t - \pi - \alpha_1)/\tan\phi] \end{aligned} \tag{3.101}$$

where $I_{0\alpha_1 N}$ is the initial normalized current for this mode.

The equations derived above for the load current do not hold good for the load impedance angle $\phi = 90°$ as the term $m/\cos\phi$ becomes infinite for this case. The expressions for the load current in this case are derived by putting $R = 0$ in the basic differential equations. Thus, for $\phi = 90°$, the expressions for the load current corresponding to eqns. (3.99–3.101) respectively are easily obtained as

$$i_{0N}(\omega t) = \cos\alpha_{act} - \cos\omega t - m(\omega t - \alpha_{act}) \quad \alpha_{act} \leqslant \omega t \leqslant \pi \quad (3.102)$$

$$i_{0N}(\omega t) = I_{0\pi N} + m(\pi - \omega t) \quad \pi \leqslant \omega t \leqslant \pi + \alpha_1 \quad (3.103)$$

and

$$i_{0N}(\omega t) = \cos\alpha_1 + \cos\omega t - m(\omega t - \pi - \alpha_1) + I_{0\alpha_1 N}$$
$$\pi + \alpha_1 \leqslant \omega t \leqslant 2\pi \quad (3.104)$$

(b) Inverter Operation

At the instant $\omega t = 0$ in the first half-cycle when the main thyristors T_3 and T_4 (Fig. 3.26) are triggered, the source is connected to the load and the operating circuit is as shown in Fig. 3.32(a). The load current during this period $0 \leqslant \omega t \leqslant \alpha_2$ is obtained as

$$i_0(\omega t) = (V_0/R) - (V_{Sm}/Z)\sin(\omega t - \phi) + A.\exp(-Rt/L) \quad (3.105)$$

where A is a constant and can be evaluated from the initial condition $i_0(\omega t) = 0$ at $\omega t = 0$. Thus

$$i_0(\omega t) = (V_c/R) - (V_{Sm}/Z)\sin(\omega t - \phi) - [(V_c/R)$$
$$+ (V_{Sm}/Z)\sin\phi].\exp(-R\omega t/\omega L) \quad (3.106)$$

By normalizing the load current, eqn. (3.106) reduces to

$$i_{0N}(\omega t) = (m/\cos\phi) - \sin(\omega t - \phi) - [(m/\cos\phi)$$
$$+ \sin\phi].\exp(-\omega t/\tan\phi) \quad (3.107)$$

The load current starts freewheeling when the series-aiding thyristors T_1 and T_4 are triggered at $\omega t = \alpha_2$. The operating circuit for the freewheeling interval $\alpha_2 \leqslant \omega t \leqslant \pi$, is shown in Fig. 3.32(b). Also the freewheeling action takes place through T_2 and T_3 during the interval, $(\pi + \alpha_2) \leqslant \omega t \leqslant 2\pi$. The equation for the load current in its normalized form during this period is easily obtained as

$$i_{0N}(\omega t) = (m/\cos\phi) + [I_{0\alpha_2 N} - (m/\cos\phi)].\exp[-(\omega t - \phi)/\tan\phi] \quad (3.108)$$

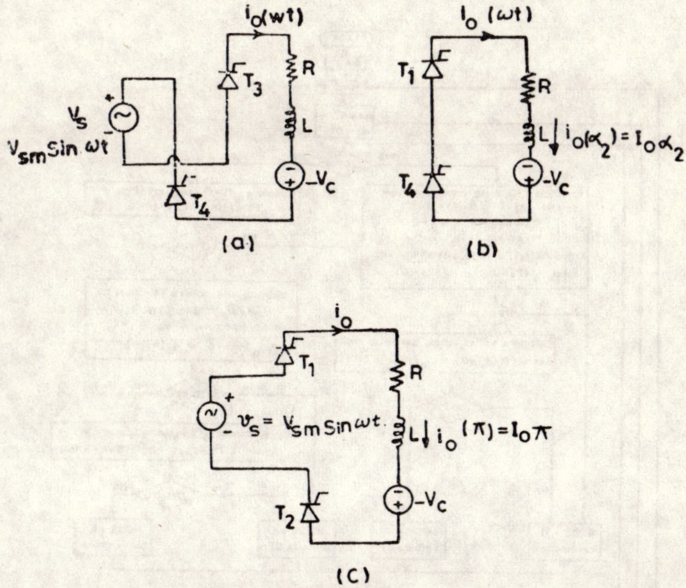

Fig. 3.32. Different operating circuits for inverter operation.
(a) Power interval ($3 \leqslant \omega t < \alpha_2$).
(b) Freewheeling interval ($\alpha_2 \leqslant \omega t \leqslant \pi$).
(c) Power interval ($\pi \leqslant \omega t \leqslant \pi + \alpha_2$).

where $I_{0\alpha_2N}$ is the initial current (normalized) in this mode.

Freewheeling of the load current continues till $\omega t = \pi$ when the thyristors T_1 and T_2 are triggered to connect the source to the load. The operating circuit for this period, $\pi \leqslant \omega t \leqslant \pi + \alpha_2$, is shown in Fig. 3.32 (c) and the expression for the normalized load current during this interval is obtained as

$$i_{0N}(\omega t) = (m/\cos\phi) + \sin(\omega t - \phi)$$
$$+ [I_{0\pi N} - (m/\cos\phi) - \sin\phi].\exp[-(\omega t - \pi)/\tan\phi] \qquad (3.109)$$

where $I_{0\pi N}$ is the initial current for this mode.

As pointed out in the case of rectification, eqns. (3.107)–(3.109) will not hold good when $\phi = 90°$. They are, however, to be derived from the first principles. These equations are:

$$i_{0N}(\omega t) = m\cdot\omega t - 1 + \cos\omega t \quad 0 \leqslant \omega t \leqslant \alpha_2 \qquad (3.110)$$

$$i_{0N}(\omega t) = I_{0\alpha_2N} + m(\omega t - \alpha_2) \quad \alpha_2 \leqslant \omega t \leqslant \pi \qquad (3.111)$$

$$i_{0N}(\omega t) = I_{0\pi N} + m(\omega t - \pi) - 1 - \cos\omega t \quad \pi \leqslant \omega t \leqslant \pi + \alpha_2 \qquad (3.112)$$

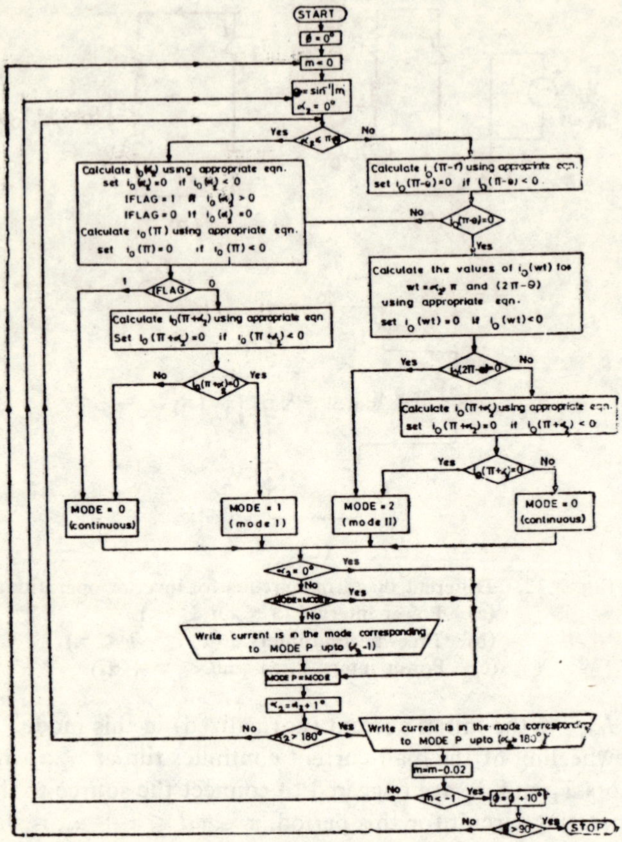

Fig. 3.33. Flow-chart of the computer program to obtain operating diagram and for identification of modes for rectifier operation.

The normalized equations are used to find the load current at different instants in the first cycle of the supply frequency. The flow-charts of the computer programs used to obtain data for constructing the operating diagrams are shown in Figs. 3.33 and 3.34 for rectifier and inverter operations respectively. In addition to the operating diagrams of the converter circuit, the different possible modes of current conduction are obtained using these flow-charts.

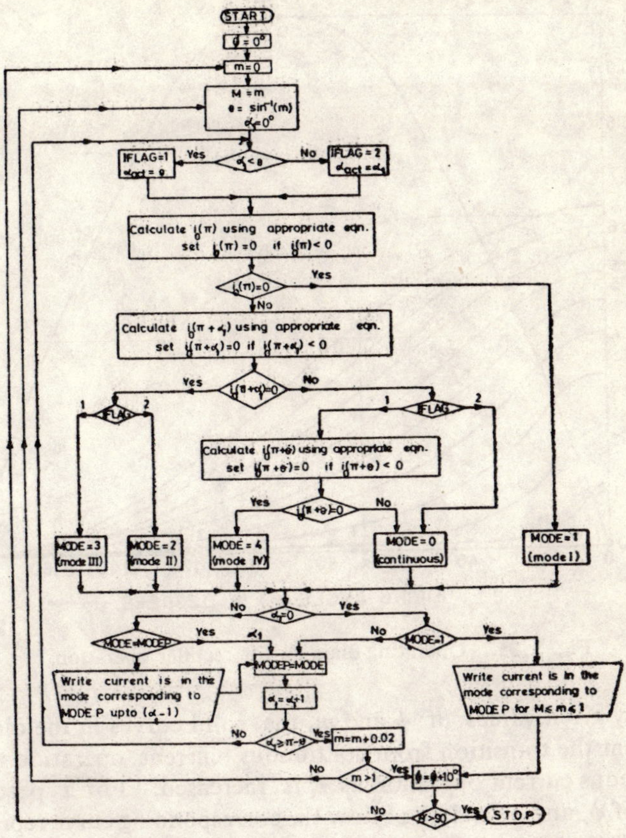

Fig. 3.34. Flow chart of the computer program to obtain operating diagram and for identification of modes for inverter operation.

3.3.6.2 *Classification of Modes and Their Identification on the Operating Diagrams.*

(a) RECTIFIER OPERATION

Apart from the continuous current mode, there are different modes of operation when the load current becomes discontinuous. The normalized equations for the load current derived earlier are used to construct the operating diagram. Figure 3.35 shows the operating diagram of the fully controlled converter with half-controlled characteristics in the rectifier operation. The operation of the converter is possible for any point below the boundary *PQE*.

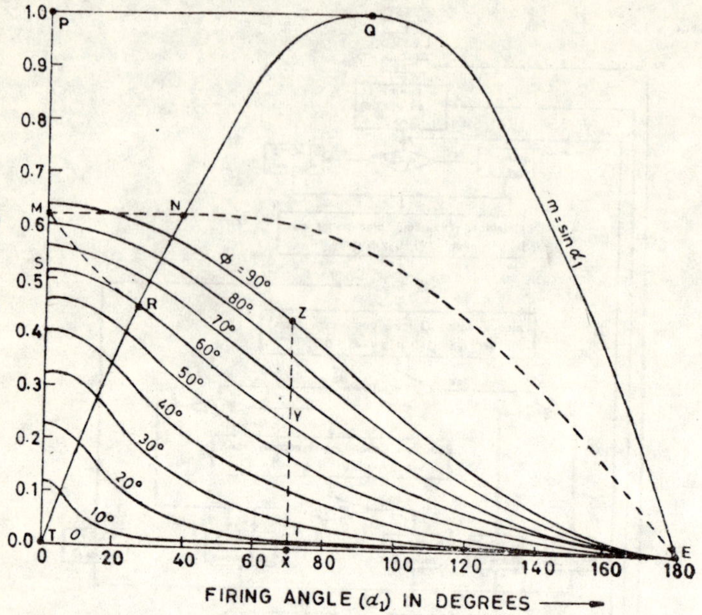

Fig. 3.35. Operating diagram for rectifier operation.

For any given values of ϕ and m, the solid curves in the diagram represent the transition from continuous current operation to discontinuous current operation as α_1 is increased. For a particular value of ϕ, any point lying below the corresponding curve represents continuous current operation, and any point above represents discontinuous current operation.

As an example, let us consider the load impedance angle $\phi = 60°$ in our discussion. *SRE* is the transition curve. Any point in the region *SRETS* represents operation with continuous current. The region *SRTS* corresponds to $\alpha_1 \leqslant \theta$ while the region *TERT* corresponds to $\alpha_1 > \theta$. In the continuous region the thyristors turn on at the instant that they are triggered whether the triggering angle $\alpha_1 > \theta$ or $< \theta$. A point in the region *PQERSP* respresents a state of discontinuous current operation. There are four distinct possible modes of operation in the discontinuous region.

Mode I: The load voltage and load current waveforms along with the supply voltage for this mode are shown in Fig. 3.36(a). The main thyristor pair turns on at α_1 or θ depending on whether $\alpha_1 \geqslant \theta$ or $\alpha_1 < \theta$. This mode is characterized by the load current falling to zero before the next half cycle of the supply voltage starts. This means that there is no freewheeling of the load current at all and $\beta \leqslant \pi$. Here β represents the extinction angle of the load current. This mode is identified by the region *PQENMP* in the operating diagram (Fig. 3.35).

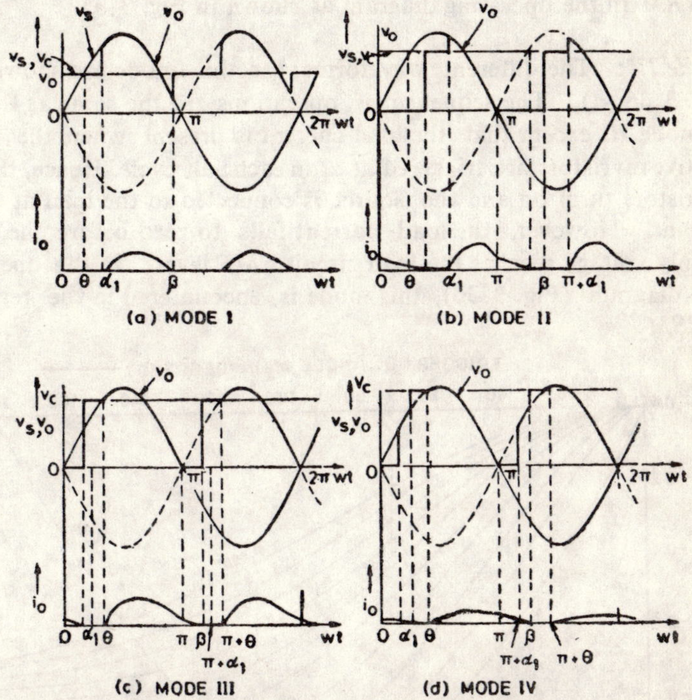

Fig. 3.36. Steady-state voltage and current waveforms for different modes of discontinuous current for rectifier operation.

Mode II: This mode is present for $\alpha_1 > \theta$. The thyristors turn on at α_1 and the load current builds up. At $\omega t = \pi$, the series-aiding thyristors are turned on and the load current starts freewheeling. The load current falls to zero before the other main thyristor pair is turned on at $\omega t = \pi + \alpha_1$ in the next half cycle, i.e., $\beta \leqslant \pi + \alpha_1$ for this mode. The associated waveforms are shown in Fig. 3.36

(b) and the region *NERN* represents this mode in the operating diagram of Fig. 3.35.

Mode III: This mode may occur for $\alpha_1 < \theta$. Although the firing pulses are applied at α_1 the main thyristor pair turns on only at $\omega t = \theta$. However, the load current falls to zero before the other main thyristor pair is triggered at $\omega t = \pi + \alpha_1$ in the next half cycle as illustrated in Fig 3.36 (c). This mode is identified by the region *MNRM* in the operating diagram as shown in Fig. 3.35.

Mode IV: The different waveforms for this mode are shown in Fig. 3.36 (d). The sequence of operations are the same as those in mode III except that the load current is present when the respective thyristors are triggered at α_1 in each half cycle. Hence, these thyristors turn on and the source is connected to the load at that instant. However, the load current falls to zero before the AC supply voltage reaches the load circuit *EMF* level. In the operating diagram (Fig. 3.35), this mode is encountered in the region *MRSM*.

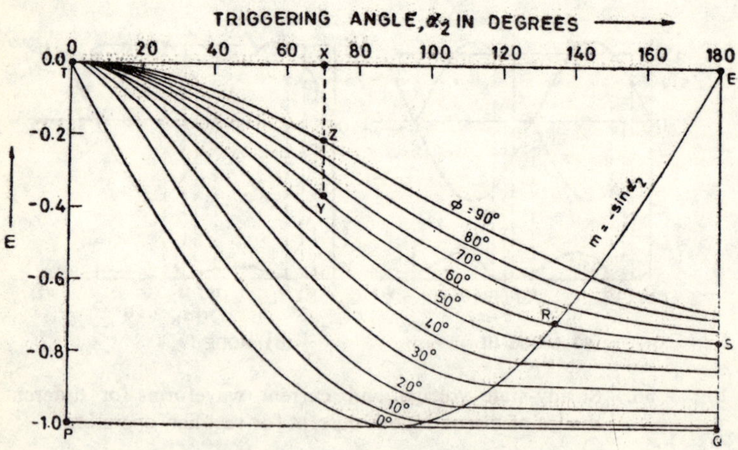

Fig. 3.37. Operating diagram for inverter operation.

(b) Inverter Operation

Figure 3.37 shows the operating diagram for inverter operation of the fully controlled converter with half-controlled characteristics. The inverter operation is possible at any point inside the region *TEQPT*. As in the case of rectifier operation for a given

ϕ, any point lying below the corresponding curve represents continuous current operation, and any point above represents discontinuous current operation. As an example, let us again consider $\phi = 60°$ in our discussion. *TRS* is the separating curve.

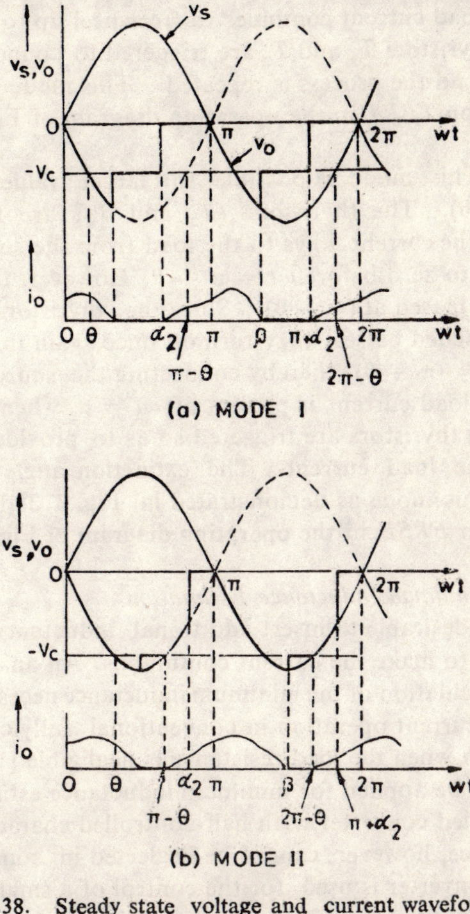

Fig. 3.38. Steady state voltage and current waveforms for different modes of discontinuous current for inverter operation.

Any point in the region *TRSQPT* represents a state of continuous current operation while that in the region *TESRT* represents a state of discontinuous current operation. In the discontinuous current region, there are two distinct possible modes.

Mode I: The associated waveforms for this mode are shown in Fig. 3.38. The supply source is connected to the load at $\omega t = 0$ when the main thyristors T_3 and T_4 are triggered. However, the load current falls to zero before $\omega t = \alpha_2$ when the freewheeling is initiated by triggering the corresponding series-aiding thyristor pair. The load current continues to freewheel up to $\omega t = \pi$ when the other thyristors T_1 and T_2 are triggered to connect the load to the source and the process is repeated. This mode is identified by the sub-region *TERT* in the operating diagram of Fig. 3.37.

Mode II: This mode is possible for larger values of α_2 when $\alpha_2 > (\pi - \theta)$. The thyristors ($T_3$ and T_4) are turned on at $\omega t = 0$ and the current flows to the load from the source. The load current falls to zero before $\omega t = \pi - \theta$. However, these thyristors are forward biased at $(\pi - \theta)$. Since the thyristors are triggered by 180° stretched pulses, they turn on once again in the same half cycle at $\omega t = (\pi - \theta)$ thereby connecting the source to the load. Thus, some load current is present at $\omega t = \alpha_2$ when the respective series aiding thyristors are triggered so as to provide freewheeling path for the load current. The extinction angle β is less than $(2\pi - \theta)$. This mode as demonstrated in Fig. 3.38(b) is identified by the region *ERSE* in the operating diagram of Fig. 3.37.

3.3.6.3 *Minimum Inductance Estimation*

It may be desirable to insert additional inductance in the load circuit so as to make the current continuous. An analytical method describes calculation of the minimum inductance necessary to achieve continuous current operation in conventional fully controlled converter system when the load resistance is negligible [13]. The same method may be applied for minimum inductance estimation for the fully controlled converter with half-controlled characteristics. The load resistance, however, cannot be neglected in some applications where the converter is used for the control of a small dc machine. A graphical method for estimating the minimum inductance necessary to achieve continuous current operation is presented for the fully controlled converter with half-controlled characteristics taking the load resistance also into account [3].

For example, in Fig. 3.35 for certain triggering angle α_1 indicated by the point X in the diagram, the ordinate XY gives the normalized value of the load circuit *EMF*, V_c/V_{Sm} at the limit of continuous current operation with $\phi = 60°$. The ordinate XZ gives the normalized value of the average output voltage V_0/V_{Sm}, at the limit of continuous current operation since the average voltage across a pure inductor ($\phi = 90°$) is zero. Therefore, for the triggering angle X, the ordinate YZ equals the normalized value of the average voltage drop, I_0R/V_{Sm}, across the load circuit resistance. This is represented by the point A in the plane whose coordinates are α_1 and I_0R/V_{Sm} as shown in Fig. 3.39 (a).

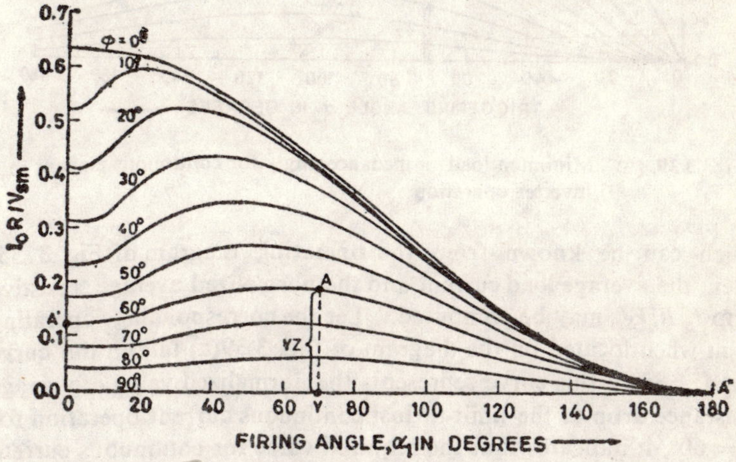

Fig. 3.39. (a) Minimum load impedance angle for continuous current in rectifier operation.

Similarly, the normalized average voltage drop across the resistance at the limit of continuous current operation for other values of triggering angle result in the curve $A'AA''$ in Fig. 3.39(a). Using a similar procedure the curves in Fig. 3.39(a) for other values of load impedance angle ϕ are obtained. The same method when used with the operating diagram shown in Fig. 3.37 for the inverter operation result in the curves shown in Fig. 3.39(b). The curves in Figs. 3.39(a) and 3.39(b) provide information regarding minimum inductance estimation for continuous current operation. As an example, let us consider the rectifier operation of a converter fed DC motor whose armature resistance R_m, and armature inductance L_m and hence armature circuit load angle ϕ_m, are known. For a particular firing angle α_1, and back $EMF\ V_c$, in the operating region of the motor, the load current may become discontinuous

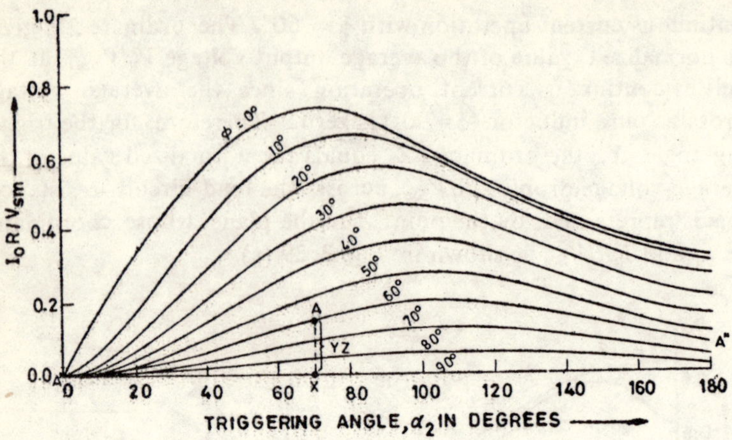

Fig. 3.39. (b) Minimum load impedance angle for continuous current in inverter operation.

which can be known from the operating diagram of Fig. 3.35. Then, the average load current and the normalized average resistive drop $I_0\ R/V_{Sm}$ may be computed. Let the corresponding operating point when located on the diagram of Fig. 3.39(a) fall on the curve $A'AA''$. Since this curve represents the normalized value of average resistance drop at the limit of just continuous current operation for $\phi = 60°$, it indicates that the required value for continuous current operation is 60°. If $\phi_m < 60°$, then the additional inductance L_{add}, to be connected in series with the motor armature to make the load current just continuous amounts to

$$L_{add} = (\tan 60° - \tan \phi_m).\ R_m/\omega \tag{3.113}$$

This gives the value of the additional inductance for continuous current for a particular operating point. If the region of operation of the converter-fed motor is known *a priori* for both rectification and inversion, the additional inductance can be calculated for different operating points in the entire region following the above procedure. Choosing the maximum value of these additional inductances would then assure converter operation in the continuous current mode.

3.4 FOUR-QUADRANT CONVERTER

Half-controlled and fully-controlled converters are discussed in Sections 3.3 and 3.2 respectively. All half-controlled converters provide single-quadrant operation only. They convert AC power to DC power which is controllable. The operation of all such

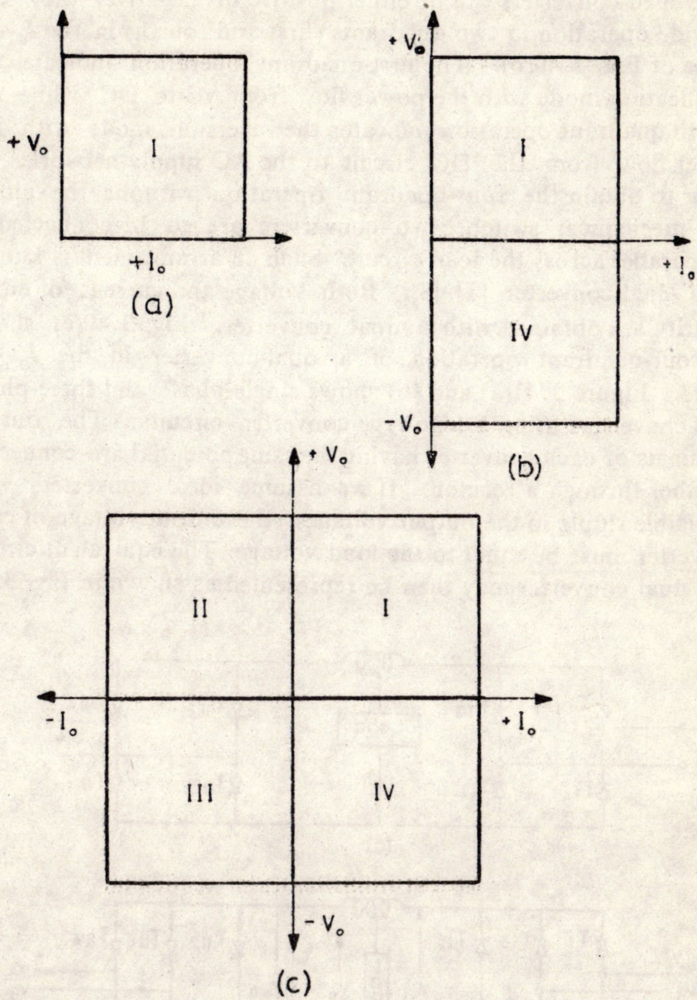

Fig. 3.40. Operating regions of converters in the $I_0 - V_0$ plane.

(a) Single-quadrant operation.
(b) Two-quadrant operation.
(c) Four-quadrant operation.

converters is confined to the first quadrant in the I_0-V_0 plane of Fig. 3.40(a) where V_0 and I_0 represent average output voltage and current respectively. Since the average output voltage of all fully-controlled converters can be either positive or negative, they can provide operation in two quadrants (first and fourth) in the I_0-V_0 plane of Fig. 3.40(b). The first quadrant operation indicates the rectification mode with the power flow from AC to DC while the fourth quadrant operation indicates the inversion mode with the power flow from the DC circuit to the AC supply network. In order to obtain the four-quadrant operation without the aid of any mechanical switch, two converters are to be connected in antiparallel across the load circuit. Such an arrangement is known as a dual converter [1], [8]. Both voltage and current of either polarity are obtained with a dual converter. Fig. 3.40(c) shows the four-quadrant operation of a dual-converter in the I_0-V_0 plane. Figure 3.41(a) and (b) shows single-phase and three-phase dual converters using bridge-type converter circuits. The output terminals of each converter having the same potential are connected together through a reactor. If we assume ideal converters with negligible ripple in the output voltages, the output voltage of each converter must be equal to the load voltage. The equivalent circuit of a dual converter may then be represented as shown in Fig. 3.42.

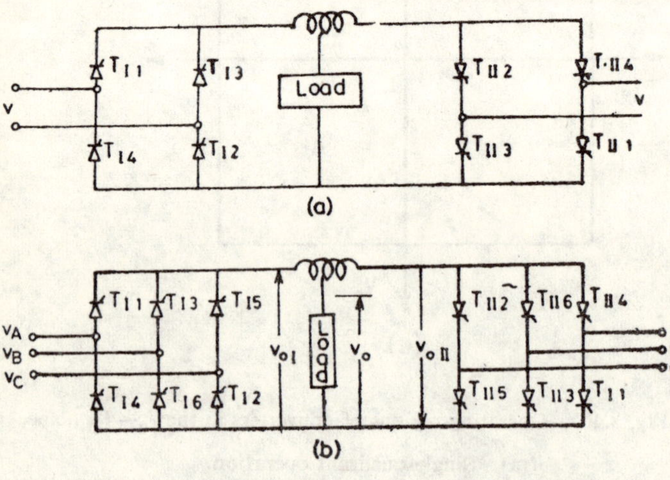

Fig. 3.41. Dual converter using bridge converters: (a) Single-phase, (b) Three-phase.

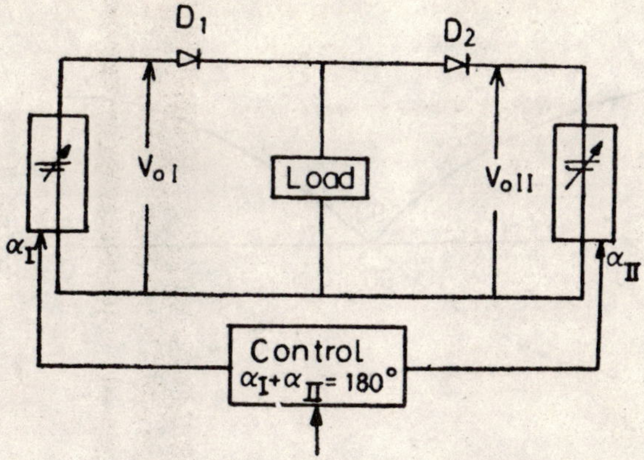

Fig. 3.42. Equivalent circuit of an ideal dual converter.

The diodes D_1 and D_2 are connected in series with the output voltage of converter *I* and converter *II* respectively to indicate the ability of each converter to conduct current in one direction only. However, the load current has the freedom to flow in either direction. From Fig. 3.41, the output voltages are

$$V_{0I} = V_{0max} \cos \alpha_I \tag{3.114}$$

$$V_{0II} = V_{0max} \cos \alpha_{II} \tag{3.115}$$

where the maximum output voltage V_{0max} is given in Table 3.1 for single-phase and three-phase converters. Equating the output voltages of converter *I* and converter *II*

$$V_{0I} = V_0 = -V_{0II} \tag{3.116}$$

Substituting for V_{0I} and V_{0II} from eqns. (3.114) and (3.115) in eqn. (3.116), we obtain

$$\cos \alpha_I + \cos \alpha_{II} = 0$$

or

$$\alpha_I + \alpha_{II} = 180° \tag{3.117}$$

From eqns. (3.114), (3.115) and (3.117), the variation of the output voltage of each converter with the firing angle is shown in Fig. 3.43.

Although the average output voltages of the converters are equal, the instantaneous voltages v_{0I} and v_{0II} are not equal. Due to this, a circulating current flows between the two converters. In a practical converter, a reactor is inserted between the two converters to limit the circulating current to a tolerable value. The reactor can,

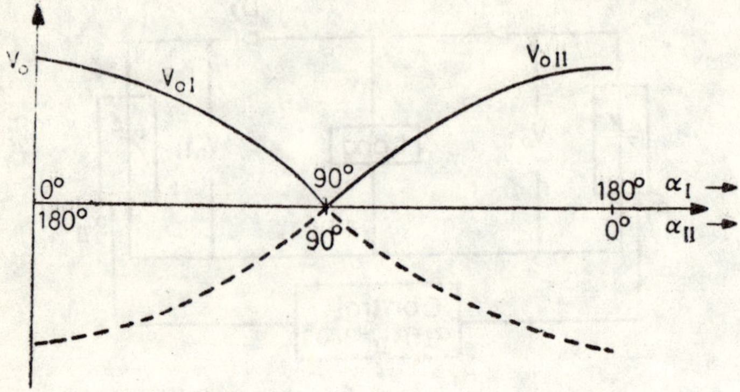

Fig. 3.43. Variation of terminal voltages of an ideal dual converter with firing angle.

however, be eliminated provided the converters are controlled such that no circulating current arises. The two most commonly used operations of a dual converter are (1) Non-circulating current operation (2) circulating current operation [1].

3.4.1 Non-Circulating Current Operation

In a dual converter with non-circulating current, only one converter operates at a time and carries the load current. The other converter is temporarily blocked from conducting by withdrawing firing pulses to the thyristors. Since only one converter operates at a time, no reactor is required between the converters. Let us suppose converter *I* of Fig. 3.41 is supplying the load current. If the load current is to be reversed, the firing pulses to converter *I* are either immediately blocked or the firing angle is first increased to a maximum value and subsequently the firing pulses are withdrawn from the converter *I*. The load current decays to zero after which the firing pulses to converter *II* are applied. The current will now build up through the load in the reverse direction. As long as the current is negative, it is supplied by the converter *II* and the converter *I* is non-operative since the firing pulses are withdrawn from it. The change over from one converter to another converter must take place after the load current has gone to zero. That is, the incoming converter is operated after the outgoing converter stopped conducting completely. Normally, after the current in one converter decays to zero, a delay period of 10 ms to 20 ms is introduced before the firing pulses are switched on to the other converter. This ensures reliable commutation of thyristors conducting earlier in the outgoing converter. Otherwise, a high circulating current will flow between the two converters. In the absence of any circulating current limiting reactor, a large circulating current may destroy thyristors in the converters. The load current, under steady state conditions, may be continuous or discontinuous. The control circuit for the dual converter is so designed as to give satisfactory operation both under continuous and discontinuous load current. The switching over from one converter to another converter must take place only from an external command.

The transfer characteristic between the output voltage and the firing angle α is linear so long as the current is continuous. However, with discontinuous load current, the transfer characteristic becomes non-linear. The control circuit becomes complex and the response is sluggish. Whenever a fast and frequent reversal of the load current is desired, the dual converter is operated in the circulating current mode.

3.4.2 Dual Converter with Circulating Current Operation

In this mode of operation, a controlled amount of circulating current is allowed to flow between converter *I* and converter *II* by connecting a reactor between the output terminals of the two converters. This circulating current keeps the two converters in a state of continuous conduction whether or not the load current is present. The transfer characteristic between the output voltage and the firing angle is linear and hence the response is very fast. The control circuit becomes relatively simple. The load current is free to flow in either direction at any time. The reversal of the load current is inherently a natural and smooth process. One converter acts as a rectifier while the other converter acts as a line commutated inverter. The sum of the firing angles of the two converters is maintained at 180° as in the case of the idealized dual-converter discussed in section 3.4. If the load current is to be reversed, the roles of the two converters are interchanged. That is, the converter which was operating as a rectifier is now operated as an inverter while the converter which was operating as an inverter is made to operate as a rectifier. The current sensing is not required and the normal delay period of 10 ms to 20 ms as in case of a circulating current-free operation is eliminated. Thus, the dual converter with the circulating current provides a fast response. The main drawbacks of this scheme are: (1) A reactor is required to limit the circulating current flowing between the two converters. The size of this reactor becomes significant at high power levels. (2) Because of the losses in the circulating current reactor, the efficiency and power factor are low. (3) The converters are to be rated for a current higher than the load current depending upon the circulating current (normally 15% to 20% full load current) flowing between the two converters. If the load current is to be reversed quite frequently and a fast response is to be a primary consideration, the dual converter with the circulating current scheme is the obvious choice in the four quadrant operation of a dual converter.

3.5 GATING CIRCUITS FOR A THREE-PHASE FULLY CONTROLLED THYRISTOR CONVERTER

3.5.1 Gating Pulse Requirements

The operation of a three-phase bridge converter (Fig. 3.5(f)) has been explained in Section 3.2.4.5. There are six thyristors T_1 to T_6 which have to be triggered into conduction in a particular sequence so as to make the supply currents balanced. If the output current is assumed constant, the output voltage at any instant is equal to one of the six line to line voltages: v_{AB}, v_{AC}, v_{BC}, v_{BA}, v_{CA}, v_{CB}. This requires that two thyristors one in the positive group (T_1, T_3, T_5) and one in the negative group (T_4, T_6, T_2) be gated together simultaneously. In order to vary the average value of the output voltage, the gate pulses to thyristor pairs are to be controlled. The earliest instant at which each thyristor can be turned on is taken as the datum for measuring the firing angle for each thyristor. These instants are the same as those at which the diodes connected in the thyristor positions would normally conduct. Obviously the instants at which the phase voltages cross over are the reference points for the measurements of firing angles. For example, the instant at which the positive values of v_{an} and v_{cn} intersect is taken as the reference point for the measurement of the firing angle for the thyristor T_1. The reference points for other thyristors are likewise fixed. Fig. 3.44 shows phasor relationships between line voltages and gating signals. The line voltages are displaced from each other by 60°. The thyristors are gated in pairs forming six groups,

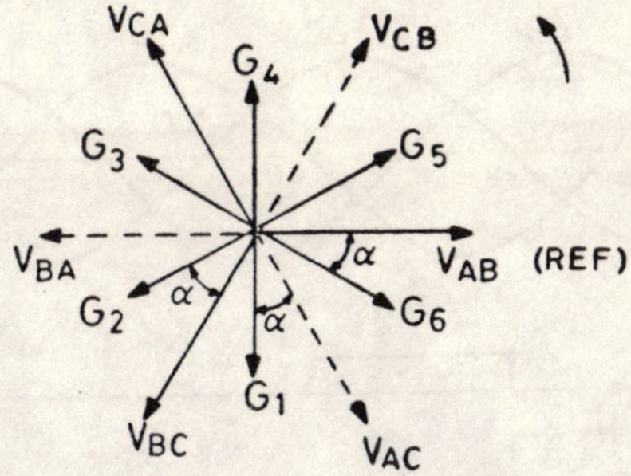

Fig. 3.44. Phasor relationship of line (control) voltages & gating signals.

i.e., (T_1, T_2), (T_2, T_3), (T_3, T_4), (T_4, T_5), (T_5, T_6), (T_6, T_1). The interval for each group is 60° and each thyristor conducts for two periods. The firing angle for the thyristor T_1 is measured from $\omega t = 30°$ with reference to the phase voltage v_{an}. Or it is measured from $\omega t = 60°$ with reference to the zero crossing of the line voltage v_{AB} or with reference to the zero crossing of v_{AC}. In a similar manner, the firing delay angles for T_2, T_3, T_4, T_5, T_6 are measured from 60° with reference to the zero crossings of v_{AC}, v_{BC}, v_{BA}, v_{CA} and v_{CB} respectively. The vector marked G_1 shows the starting point of gating signal G_1. Similarly the vectors G_2, G_3, G_4, G_5, G_6 are delayed by α degrees with respect to V_{BC}, V_{BA}, V_{CA}, V_{CB} and V_{AB} respectively. Figure 3.45 shows the time variation of the line voltages and the gating signals. For the sake of clarity, only three line voltages v_{AB}, v_{BC}, and v_{CA} are shown. If the gating signal G_1 to T_1 is terminated at $\omega t = \pi$ and other gating signal G_n terminated 60° behind G_{n-1} for $n = 2, 3, 4, 5, 6$, then this scheme works satisfactorily for $0° < \alpha \leqslant 60°$. However for $\alpha > 60°$, only one thyristor is gated at a time. Though this may work satisfactorily in the steady state, it may lead to difficulties in starting the converter at $\alpha \geqslant 60°$. An alternative scheme is to extend the pulse by 60° beyond the $(180° - \alpha)$ point. This may not be quite satisfactory in the inverting region particularly when α approaches 180°. Therefore it may be desirable to have maintained gate pulses as this may enhance the operating region of the converter if the current has a tendençy to become discontinuous. From theee considerations, gate pulses of 120° width are considered in this control circuit over the entire range of the firing angle. However,

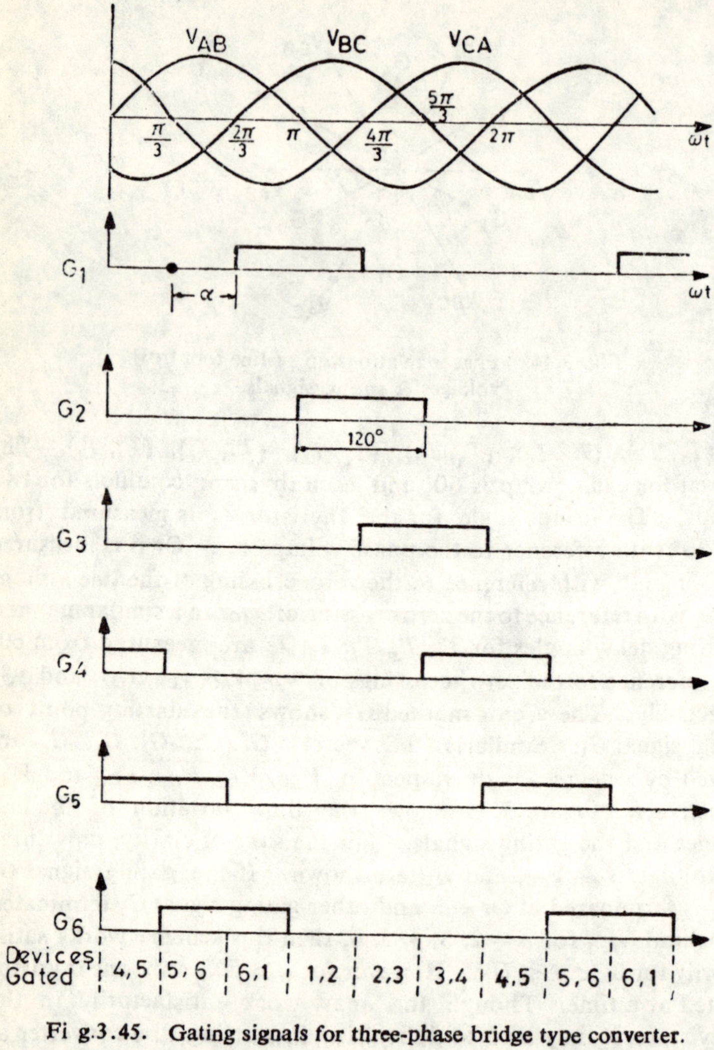

Fig.3.45. Gating signals for three-phase bridge type converter.

in order to reduce the size of pulse transformers, they are usually modulated by a high frequency of about 10 KHz.

Thus, the problem of generating the gating pulses for a three-phase thyristor bridge converter reduces to designing a 'pulse generator' to generate signals G_1 to G_6 of Fig. 3.45. A set of transformers are used to obtain isolated, low level 'synchronizing' signals which are in phase with the six voltages V_{AB} to V_{CB}. The pulse generator uses these signals as reference for generating the firing delay angles. It should be noted that the gating signals are sensitive to the phase sequence of the synchronizing signal. Therefore, care must be taken in connecting the ac lines to the synchronizing transformer in the correct phase sequence.

3.5.2. Schemes of Generating Gating Pulses

There are numerous approaches of designing gate pulses at the required delay angle. The purpose of this section is not to give an exhaustive treatment of generating gate pulses but rather mention broad schemes most commonly used in the firing pulse generator. These are:

1. Cosine wave crossing control [1].
2. Ramp comparator approach.
3. Equidistant pulse firing scheme [9], [10].
4. Digital firing scheme [11].

3.5.2.1 *Cosine Wave Crossing Control*

The cosine wave crossing control technique is explained with the help of Fig. 3.46. The synchronizing line voltage v_{ab} is integrated and compared with a control voltage V_c to obtain the gating pulse commencing at the required delay angle, α. The dc control voltage is varied from V_m to $-V_m$ to vary α from 0 to π. A clock pulse is produced at the output of the comparator. This clock pulse sets a flip-flop. A similar block is used taking the synchronizing line voltage v_{bc}. The clock-pulse obtained from this channel is used to clear the flip-flop in the former channel and also to set another flip-flop. Since the synchronizing signals are 120° apart in phase, the former flip-flop is cleared after 120°. Similar circuits are used to generate pulses for other thyristors. The output of each flip-flop goes through another block consisting of pulse isolation, amplification and high frequency modulation before it is actually applied between gate and cathode terminals of each thyristor.

The delay angle α is related to the control voltage V_c by the equation

$$V_m \cos \alpha = V_C \tag{3.118}$$

The average output voltage of a 6-pulse bridge converter is given by

$$V_{dc} = 1.35\ V_{l-l} \cos \alpha = 1.35\ V_{l-l} \left(\frac{V_C}{V_m}\right) \tag{3.119}$$

where V_{l-l} is the line voltage input to the converter.

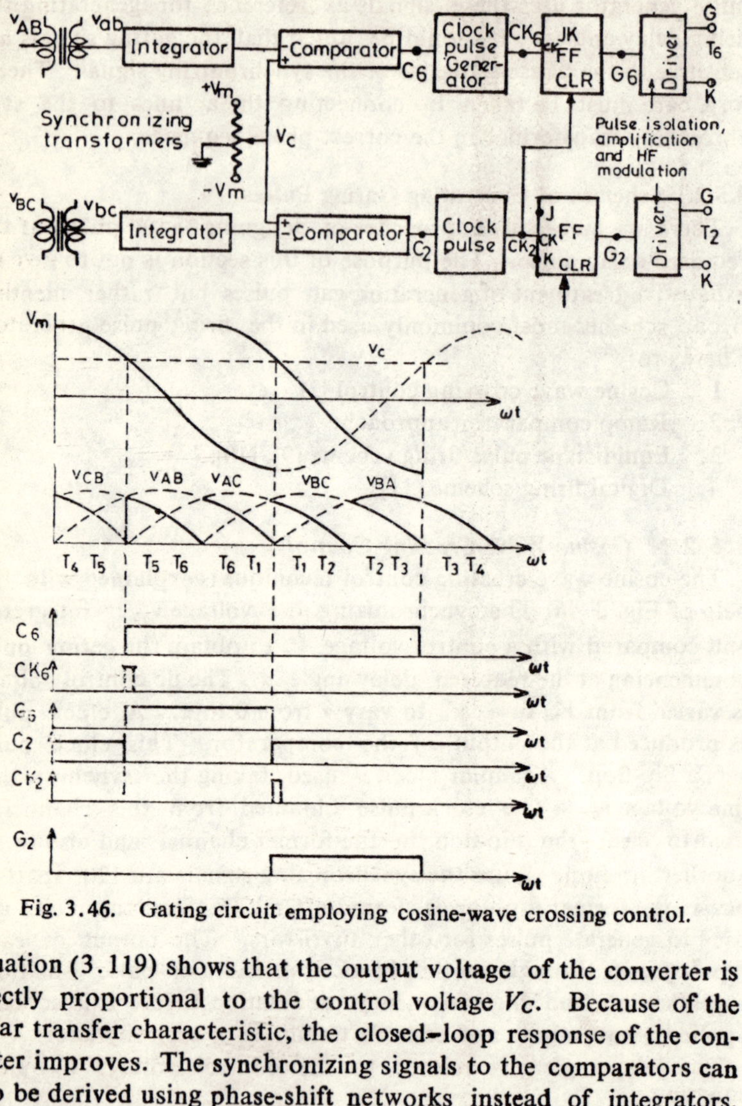

Fig. 3.46. Gating circuit employing cosine-wave crossing control.

Equation (3.119) shows that the output voltage of the converter is directly proportional to the control voltage V_C. Because of the linear transfer characteristic, the closed-loop response of the converter improves. The synchronizing signals to the comparators can also be derived using phase-shift networks instead of integrators. It is advantageous to use integrator since it attenuates any high frequency noise that may be present in the synchronizing voltage. The disadvantages of this scheme are:

1. The circuit is sensitive to line voltage variations. Adjustments are therefore required to equalise all line voltages. Otherwise the firing angles will be unequal for the thyristors.

2. Since the firing pulse is produced individually for each thyristor, it requires a large number of components.

The second disadvantage can be overcome employing cosine timing wave multiplexing instead of individual timing control. This scheme is illustrated in Fig. 3.47. The six cosine wave signals are compared using a single comparator. Six clock pulses are therefore produced. These are passed through a ring counter which gets reset after a count of 6. The output of the ring counter goes through decoding and pulse isolation, amplification and high

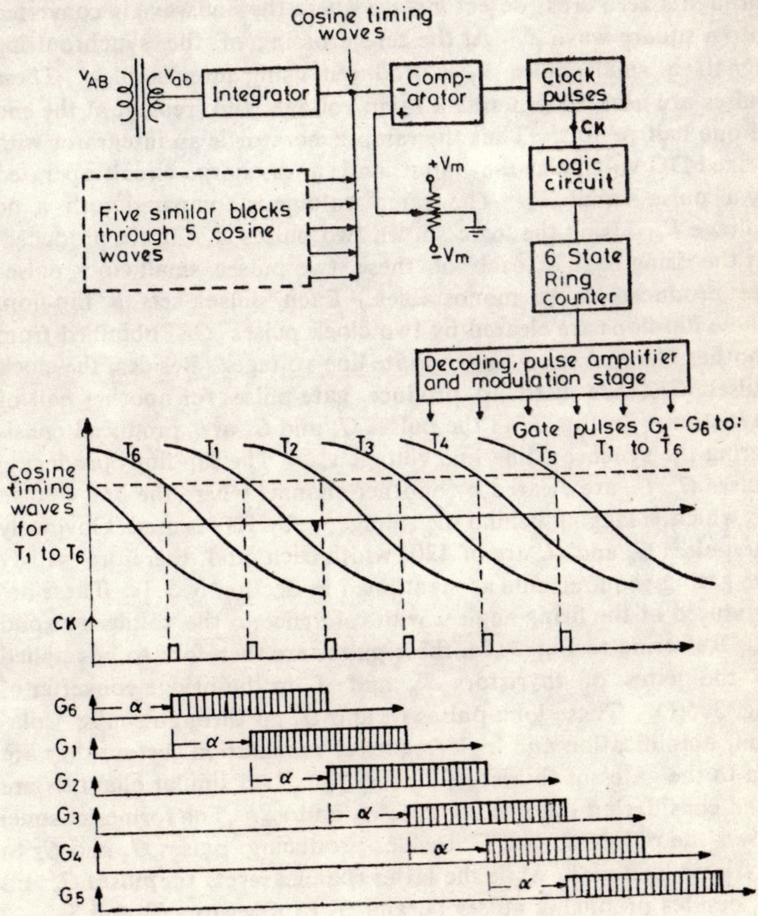

Fig. 3.47. Gating circuit showing time multiplexing cosine wave crossing control.

frequency modulation stages before they are fed to the gates of six thyristors. It can clearly be seen that the number of components required in this scheme is considerably reduced. However, it still suffers from the drawback of unequal firing angles due to supply voltage and frequency variations.

3.5.2.2 *Ramp-Comparator Approach*

Figure 3.48 shows the block diagram of this method along with waveforms at different points. The appropriate line voltage is obtained through a synchronizing transformer. This signal passes through a zero cross detector stage where the sine wave is converted into a square wave Z. At the zero crossing of the synchronizing signal, a small pulse V_t is produced using monostables. These pulses are used to generate a ramp voltage and reset it at the end of one half period. Thus the ramp generator is an integrator with a fixed DC voltage at the input and a discharge circuit operated by a pulse signal V_t. The ramp voltage is compared with a dc voltage V_c. Using the logic shown two pulses Z_6, Z_3 are produced. At the rising edge of each of these two pulses, small clock pulses are produced using monostables. Each pulse sets a flip-flop. These flip-flops are cleared by two clock pulses CK' obtained from another channel using appropriate line voltage. Besides, the clock pulses CK' are used to produce gate pulses for another pair of thyristors. In Fig. 3.48 the pulses G_6 and G_3 are produced considering the synchronizing line voltage v_{ab}. The flip-flops producing pulses G_6, G_3 are cleared by another channel where the line voltage v_{bc} which is lagging behind the voltage v_{ab} by 120° is used. Obviously the pulses G_6 and G_3 are of 120° width each and therefore satisfy the gating requirements as mentioned in Section 3.5.1. These are produced at the firing angle α with reference to the voltages v_{ab} and v_{ba}. Referring to Fig. 3.44, these pulses are therefore to be applied to the gates of thyristors T_6 and T_3 in the bridge converter of Fig. 3.5(f). These long pulses G_6 and G_3 go through pulse isolation, amplification and high frequency modulation before they are fed to the gates of thyristors T_6 and T_3. Two similar channels are used considering the line voltages v_{bc} and v_{ca}. The former channel resets the pulses G_6 and G_3 besides producing pulses G_2 and G_5 to thyristor T_2 and T_5 while the latter channel resets the pulses G_2 and G_5 besides producing pulses G_4 and G_1 to thyristors T_4 and T_1.

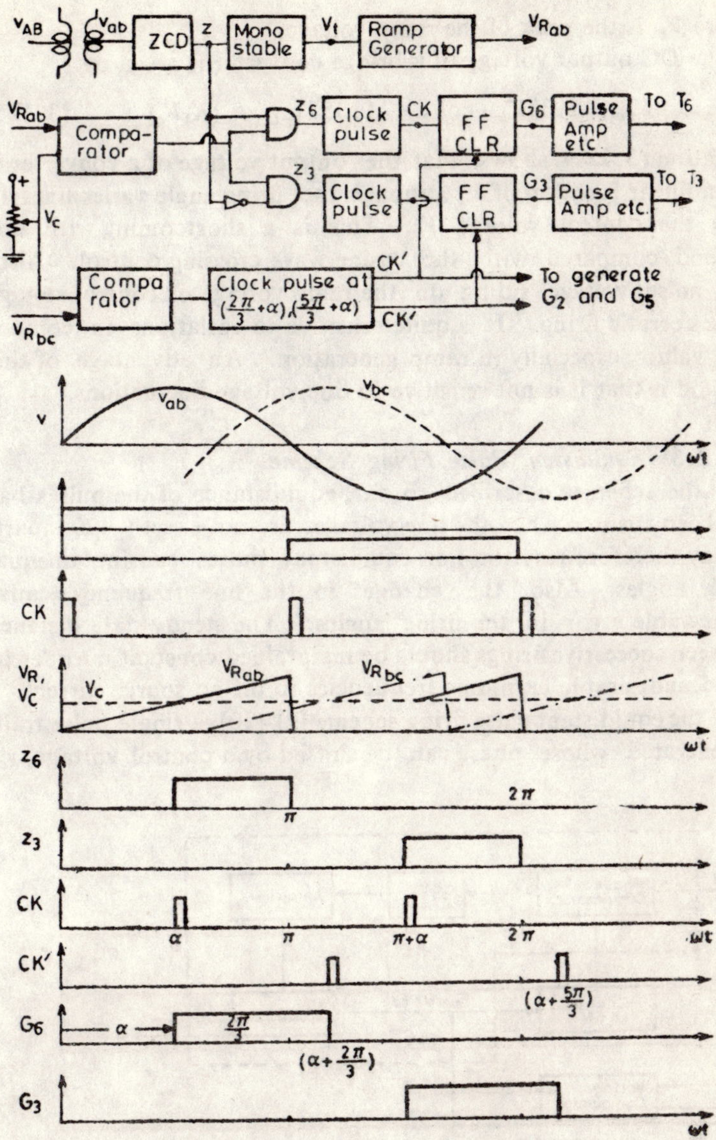

Fig. 3.48. Gating circuit employing ramp-comparator scheme.

From Fig. 3.48, the delay angle α is given by

$$\alpha = \left(\frac{V_c}{V_m}\right)\cdot\pi = K_1 V_c \tag{3.120}$$

where V_m is the peak of the ramp voltage.

The DC output voltage of a bridge converter is given by

$$V_{dc} = 1.35 V_{l-l} \cos \alpha = 1.35 V_{l-l} \cos (K_1 V_c) \tag{3.121}$$

Equation (3.121) shows that the output voltage of a converter is not a linear function of V_c although the firing angle varies linearly with the control voltage V_c. This is a shortcoming with this method compared with the cosine wave crossing control. Also, any noise voltage riding on the ramp voltage or control voltage causes erratic firing. It is quite sensitive to variations in component values, especially in ramp generation. An advantage of this method is that it is not sensitive to line voltage fluctuations.

3.5.2.3 *Equidistant Pulse Firing Scheme*

In the schemes described so far, equidistance of the pulses has not been ensured when the line voltages are not exactly 120° apart. It may therefore result in non-equidistant pulses causing unequal firing angles. Also, the changes in the line frequency cause appreciable error in the firing angles. The steady state distance between successive firings should be maintained constant in order to avoid undesirable harmonic frequencies in the ac source current.

In the equidistant pulse firing scheme [9]–[10], a single pulse train is generated whose phase can be shifted by a control voltage, v_c.

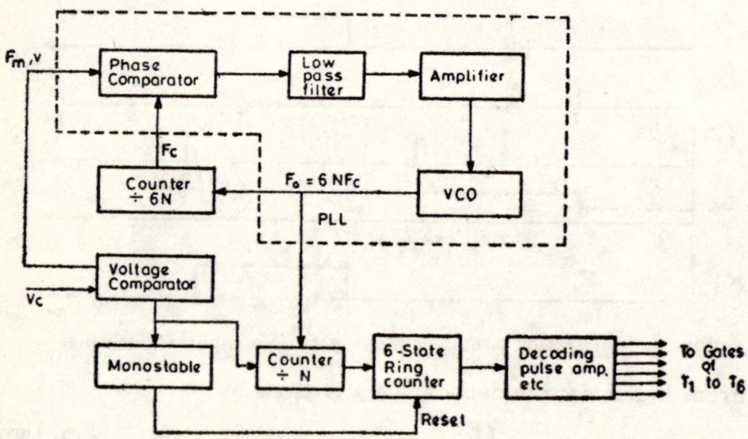

Fig. 3.49. Equidistant pulse firing scheme using phase locked loop (PLL).

A ring counter with a coding circuit is used to steer this train of pulses in six different paths through pulse amplifier and isolation stages.

Figure 3.49 shows the block diagram of an equidistant pulse firing scheme. The block enclosed by the dotted line represents the circuit of a phase locked loop (PLL). It consists of a highly stable and linear voltage controlled oscillator (VCO), phase comparator, a low pass filter and an amplifier. The PLL maintains the output fraquency of the VCO at exactly $6N$ times the mains frequency. In the event that the mains frequency F_m changes, then the output frequency of the VCO also changes. The frequency F_c tracks the mains frequency F_m. The output frequency of the VCO, F_0 is divided by $6N$ to obtain F_c which is compared with and maintained at the same frequency as F_m. There may exist a phase difference between F_m and F_c. But this should be as small as possible to minimise errors in firing angles. The maximum error that may be introduced in the equidistant spacings of the six consecutive pulses can be shown to be $360°/N$ where N is a frequency multiplying factor. If N is set at 600, then the maximum error in the equidistant spacing is 0.6°. The phase control of the pulse train is obtained by comparing the synchronizing supply voltage with a control voltage in a comparator circuit. The comparator resets the $\div N$ counter. At the same time, a small clock pulse is produced at the output of the comparator using a monostable circuit. This pulse resets the ring counter at the end of count 6. The output of the ring counter is steered through appropriate blocks to the gates of thyristors.

The main disadvantage of this scheme is that it requires a minimum delay of at least one period to effect any correction in the firing angle. In the schemes described earlier, the minimum delay is only one-sixth of a period.

3.5.2.4 *Digital Firing Scheme*

Figure 3.50(a) gives the block diagram of a digital firing scheme and Fig. 3.50(b) depicts the relevant waveforms at different points.

The digital PLL is used to multiply the mains frequency f by 6.2^{N+1} (consisting of $a \div 2^{N+1}$ counter and $a \div 6$ ring counter). The ring counter generates the necessary three-phase synchronizing waveforms which are spaced at 60°. Also, the VCO output is divided by $a \div 6$ ring counter (R_2) to generate a frequency of

$2^{N+1}f$. This is used to clock three sets of N-bit down counters, 1, 2, and 3. The preset enable (PE) points of these down counters is made high at the zero crossings of V_{AB}, V_{BC}, and V_{CA}. These zero crossings are obtained by differentiating and adding the appropriate synchronizing waveforms obtained from the output of the ring counter R_1. The digital control input is used to adjust the firing angle. The carry outputs of the down counters give out pulses after the counting is completed. This information along with the synchronizing waveforms is fed to generate phase-delayed 180° wide pulses at the outputs of the D-flip-flops. These outputs are properly gated to obtain the necessary 120° wide pulses.

The response time of this scheme in the worst case is one-sixth of the time period as each down counter is loaded independently by the appropriate synchronizing waveform.

The resolution of the firing angle is $180°/2^N$. With the above scheme an accurate and precise firing delay control is possible. The circuit is insensitive to waveform distortion and transients. The correction of the firing angle takes place within one-sixth of the time period.

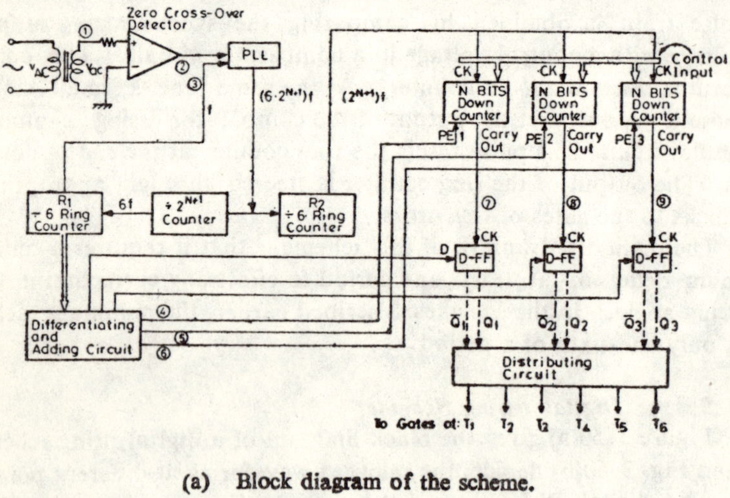

(a) Block diagram of the scheme.

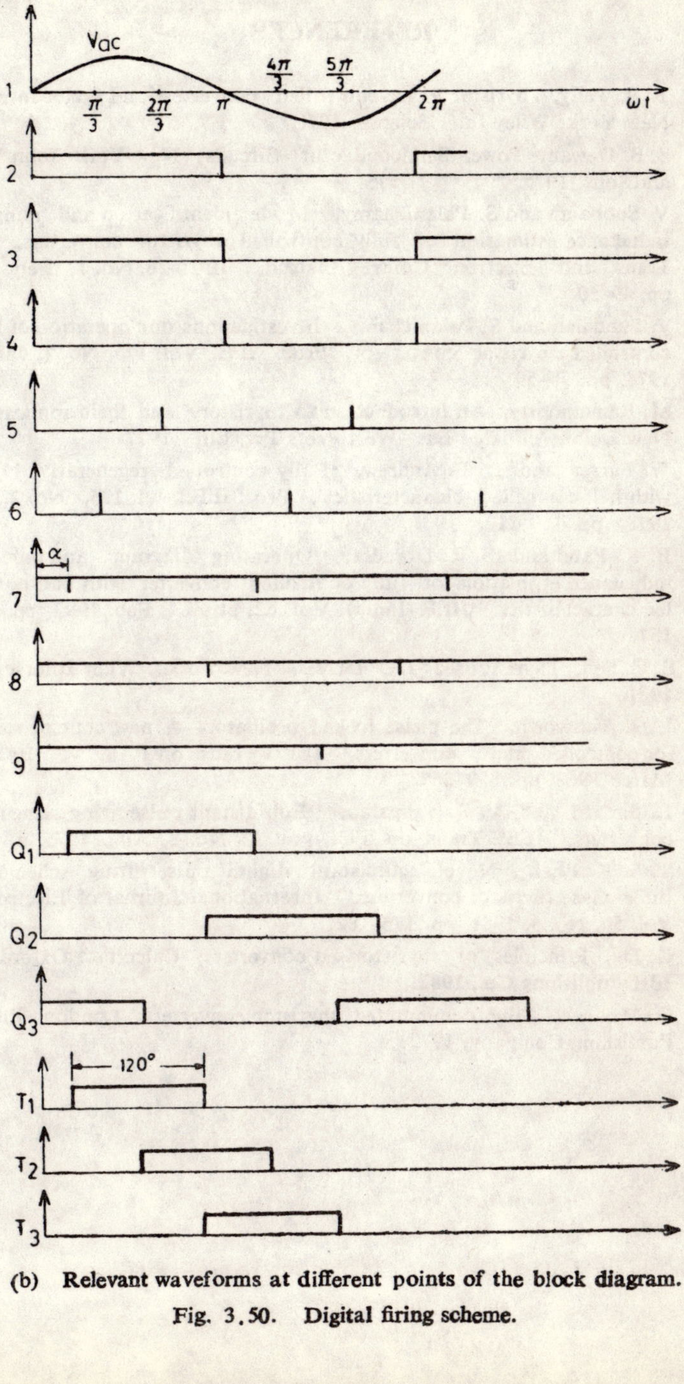

(b) Relevant waveforms at different points of the block diagram.

Fig. 3.50. Digital firing scheme.

REFERENCES

1. B.R. Pelly, Thyristor Phase-controlled converters and cycloconverters, New York: Wiley-Inter Science, 1971.
2. S.B. Dewan, Power Semiconductor Circuits, New York: John Wiley and Sons, 1975.
3. V. Subbaiah and S. Palanichamy, 'Mode identification and minimum inductance estimation for fully controlled thyristor converters, 'IEEE Trans. Ind. Electron. Contrl. Instrum., IECI-26, No. 1, Feb. 1979, pp. 49–50.
4. V. Subbaiah and S. Palanichamy, 'Investigations on operation of fully-controlled thyristor converters', Proc. IEE, Vol. 125, No. 1, January 1978, pp. 58–59.
5. M. Ramamoorty, An introdnction to thyristors and their applications, New Delhi, Affiliated East-West Press Pvt. Ltd., 1977.
6. W. Farrer and D. F. Andrew, 'Fully controlled regenerative bridges with half-controlled characteristics', Proc. IEE, vol. 125, No. 2, Feb. 1978, pp. 109–112.
7. B. K. Patel and S. R. Doradla, "Operating diagrams and minimum inductance estimation for fully controlled converter with half-controlled characteristics," JIEE (India), Vol. 62, pt EL4, Feb. 1982, pp. 150–157.
8. P. C. Sen, Thyristorised DC Drives, New York: Wiley Interscience 1981.
9. I. N. Ainsworth, "The phase locked oscillator—A new control scheme for controlled static converters," IEEE Trans. on PAS, vol. PAS-87, March 1968, pp. 859–865.
10. R. Simard and V. Rajagopalan, "Equidistant pulse firing scheme for converters," IEEE Trans. on IECI, vol. 22, No. 3, Aug. 1975, p. 475.
11. S. A. K. Bhat, "Novel equidistant digital pulse firing scheme for three-phase thyristor converters," International Journal of Electronics, Vol. 50, No. 3, 1981, pp. 175–182.
12. G. De, Principles of thyristorised converters, Calcutta: Oxford and IBH Publishing Co., 1982.
13. G. Möltgen, Line commutated thyristor converters, London: Pitman Publishing Company, 1972.

4. CYCLOCONVERTERS

4.1 INTRODUCTION

Cycloconverters are basically AC to AC power converters where the alternating voltage at supply frequency is converted directly to a lower frequency without going through any intermediate DC power conversion stage. These were developed in the 1930s using grid controlled mercury arc rectifiers primarily for application in the AC traction. The three-phase AC power supply frequency was converted to single-phase supply at a lower frequency mainly to reduce losses in the AC series motors used in the AC traction system. The earlier schemes using cycloconverters were not very popular owing to technical and economical reasons. The advent of high power thyristors has revived the interest in the cycloconverter circuits. Presently cycloconverters are used for the speed control of AC motors, in HVDC for interconnecting two power grids operating at two different frequencies, VSCF (variable frequency input and constant frequency output) generating systems such as used in Aircraft power supplies, mobile power supplies and space vehicles. Recently cycloconverters are being actively considered for high speed rapid transit systems with power generating stations operating at high frequency.

The basic principle of a cycloconverter can be explained with the help of an equivalent circuit which is similar to that of a dual converter discussed in Chapter 3. If each two-quadrant converter is represented by a voltage source of the desired frequency f_0 and consider the output power is provided by the alternating voltage and current at the desired frequency f_0, the circuit so configured represents the equivalent circuit of the ideal cycloconverter. This is shown in Fig. 4.1. The diodes connected in series with each voltage source shows the unidirectional conduction of each two quadrant converter. If the ripple in the output voltage of each converter is neglected, then it becomes ideal and represents the desired output voltage. The firing angle for the positive and negative converters is continuously modulated so as to obtain a voltage of the desired waveshape at frequency f_0. In the equivalent circuit of Fig. 4.1, the desired voltage (V_r) is shown as sinusoidal. The load current can flow freely to and fro from the load through the positive and

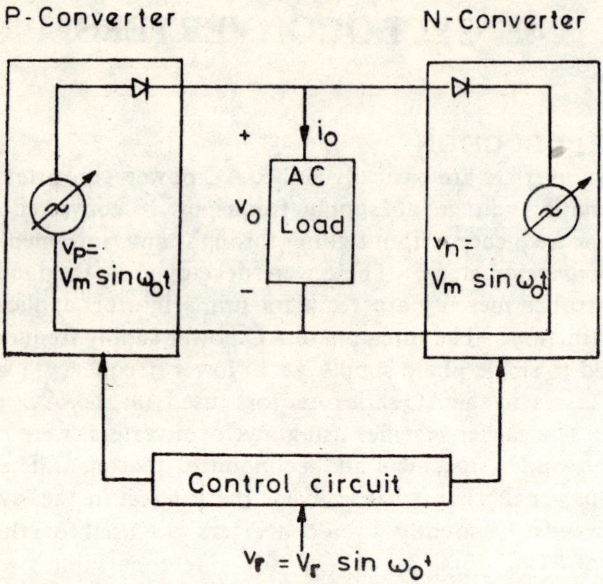

Fig. 4.1. Equivalent circuit of the cycloconverter.

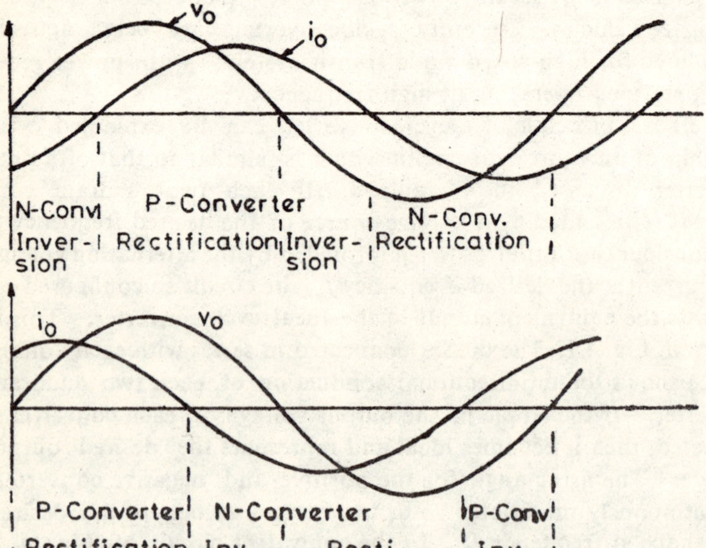

Fig. 4.2. Ideal output voltage and current waveforms of a cycloconverter.

negative converters. It is possible to supply loads of any power factor through the cycloconverter circuit. It will be discussed in later sections how the load current is reflected on to the AC supply system. Figure 4.2 shows the output voltage and current waveforms using converters of infinite pulse number for lagging and leading power factor loads where the conduction intervals of positive and negative converters are also marked. Also shown in the figure are the intervals during which each converter operates in the rectification and inversion modes. The positive converter operates whenever the load current is positive with the negative converter remaining idle during this period. In a similar manner, the negative load current is supplied by the negative converter with the positive converter remaining idle during this period. The method by which the firing angle for each converter is regulated so as to obtain the output voltage of the desired waveform has been discussed in section 4.4. Some basic cycloconverter circuits are reviewed in the next section.

4.2 PRACTICAL CYCLOCONVERTER CIRCUITS

4.2.1 Single-Phase to Single-Phase Cycloconverter Circuit

Figure 4.3 shows the power circuit of a single-phase to single-phase cycloconverter employing a centre tapped transformer and bridge type converter configurations. Although the single-phase to single-phase cycloconverter is rarely used in practice, nevertheless the operating principle of a cycloconverter can be easily understood with the help of the single-phase circuit [1]. If each transformer secondary is replaced by a voltage source, then the equivalent circuit as shown in Fig. 4.3(c) is obtained. The bridge configuration also is represented by the same equivalent circuit if it is recognized that $v_{An} = v$ and $v_{Bn} = -v$. Instead of one thyristor in the centre-tap transformer type, two thyristors come in series with each voltage source in the bridge configuration. The equivalent circuit of Fig. 4.3(c) is considered for describing the operation of the single-phase to single-phase cycloconverter. The thyristors T_{P_1} and T_{P_2} represent the positive group while T_{N_1} and T_{N_2} represent the negative group as indicated in the equivalent circuit of the cycloconverter shown in Fig. 4.1. Since the positive load current is obtained by the conduction of thyristor T_{P1} or T_{P2}, these form the positive group. So is the case with the negative

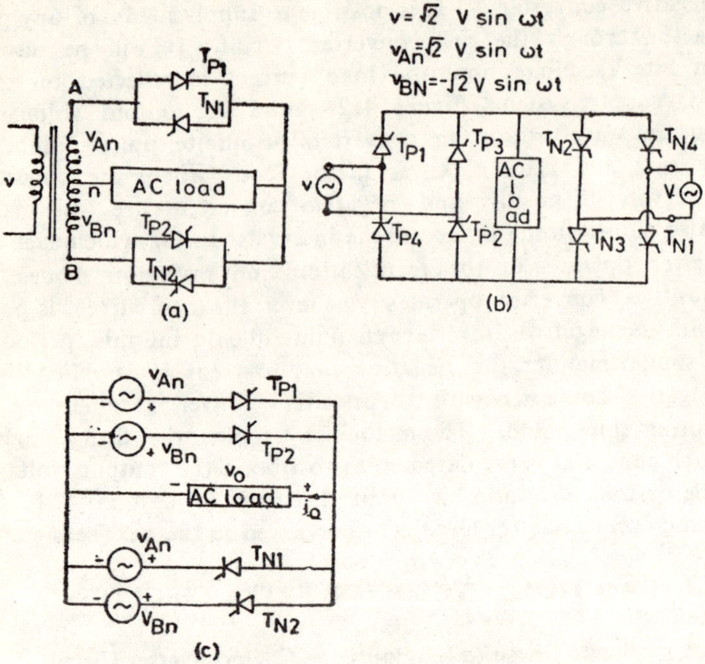

Fig. 4.3. Single-phase to single-phase cycloconverter circuit:
(a) Centre tapped transformer type.
(b) Bridge type.
(c) Equivalent circuit of both types.

group which conduct the negative load current. As in a converter circuit, the line commutation is employed to turn off a conducting thyristor. By turning on a thyristor in a particular group, a reverse voltage must appear across the conducting thyristor. Then only will the thyristor conducting earlier be commutated. By turning T_{P1} on, T_{P2} gets commutated and vice-versa. In a similar way, the commutation of thyristors in the negative group is achieved. If the current is discontinuous, natural commutation takes place since the current through a thyristor goes to zero before the incoming thyristor is turned on. The thyristors T_{P1}, T_{N2} or T_{P2}, T_{N1} can not be turned on simultaneously since this leads to the short circuiting of secondary voltage sources v_{An} and v_{Bn}. Whenever the load current is to be reversed from a positive value to a negative value, the changeover should take place from T_{P1} to T_{N1} or T_{P2} to T_{N2} in order to avoid a short-circuit of the supply voltage. Similarly, whenever the load current is changing from a negative value to a positive value, the change over must be from T_{N1} to T_{P1} or T_{N2} to T_{P2}. Figure 4.4 shows the waveforms of the cycloconverter of Fig. 4.3(c) when the load current is discontinuous assuming *R-L* load, and

$$\phi < \alpha < \pi, \ \phi \text{ being the load phase angle.}$$

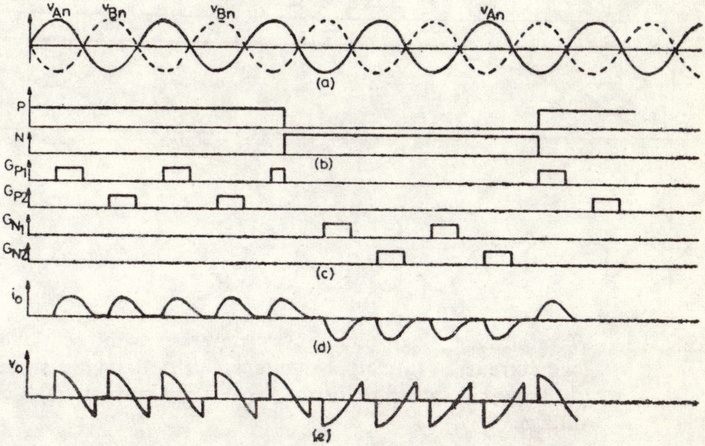

Fig. 4.4. Waveforms of single-phase cycloconverter with discontinuous load current: (a) Supply voltages, (b) Blanking signals, (c) Trigger pulses to thyristors, (d) Load current, (e) Load voltage.

The blanking signals (Fig. 4.4(b)) determine the desired output frequency. The gating signals (Fig. 4.4(c)) are applied to the *P*-group thyristors whenever the blanking signal *P* is present. In a similar way, the gating signals are applied to the *N*-group thyristors whenever the blanking signal *N* is present. Fig. 4.5 shows the waveforms of the same circuit variables as in Fig. 4.4 when the current is continuous i.e.,

$$0 < \alpha < \phi$$

From the above discussion and from the waveforms shown in Figs. 4.4 and 4.5 the following points may be noted:

1. The average output voltage waveform is far from sinusoidal since the firing angle is held constant. In this case, the average voltage is of square wave shape.

2. The current waveform may not repeat after every cycle if

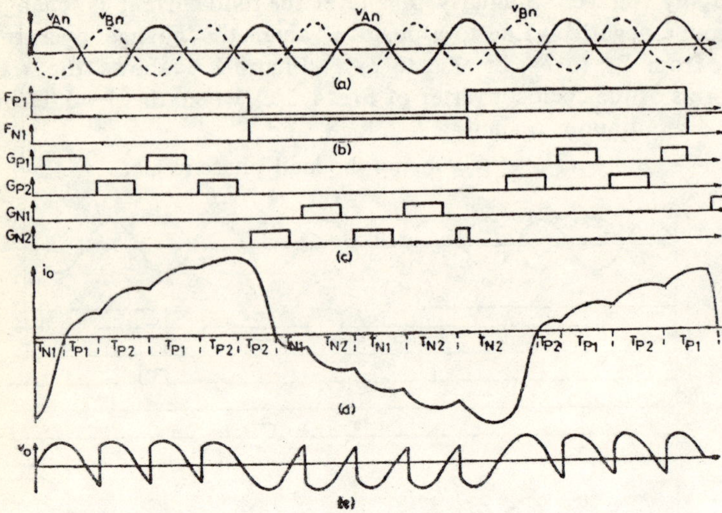

Fig. 4.5. Waveforms of single-phase cycloconverter with continuous load current: (a) Supply voltage, (b) Blanking signals, (c) Trigger pulses to thyristors, (d) Load current, (e) Load voltage.

the output frequency is not an integer submultiple of the supply frequency. It may not have half-wave symmetry. This introduces even harmonics apart from producing subharmonic components.

3. In order to eliminate short-circuits, some design features have to be incorporated in the control circuit: the trigger pulses are applied to the thyristors in a particular group only when the blanking signal is present in that particular group; the blanking signal is generated after the current in the outgoing group has fallen to zero; the interval between the instant at which the current in the outgoing group becomes zero and the instant of application of blanking signal to the incoming group must exceed the turn-off time of the thyristors in the outgoing group.

4. Some circulating current is allowed to flow from the positive converter to the negative converter. In this case, a circulating current limiting reactor is connected between the positive and negative converters. The design features mentioned in step 3 are not built into the control circuit under this condition. The operation is analogous to the circulating current operation of a dual converter.

5. In order to obtain reasonably good sinusoidal or square

output voltage waveforms using line commutated two-quadrant converters and eliminating the possibility of short-circuit of the supply voltages, the output frequency is limited to a value that is much lower than the supply frequency. By using *P*-group and *N*-group converters of higher pulse number, the output voltage of improved waveform can be realized.

4.2.2 Three-Phase to Single-Phase and Three-Phase to Three-Phase Cycloconverter Circuits

The cycloconverter circuits using converter groups of 3 and 6 pulses are shown in Figs. 4.6 and 4.7 respectively. The inter-group reactor shown dotted in these figures is required only when the cycloconverter is operated with the circulating current scheme where some circulating current is allowed to flow between the two converters bypassing the load circuit. The cycloconverter circuit of Fig. 4.6(a) shows the single-phase output from the three-phase supply sources. In order to derive three-phase output, three such circuits are connected as shown in Fig. 4.6(b). The loads are connected to form a wye connection. It is desirable to connect the load neutral to the supply neutral when the loads are unbalanced, otherwise the load neutral can be left isolated. Since each converter is known as a mid-point converter, this circuit is also called as the three-pulse mid-point cycloconverter. By using mid-point converters of higher pulse number, cycloconverters of higher pulse number can be constructed. Fig. 4.7 shows 6-pulse cycloconverter using bridge converters. If the input voltages are of appropriate magnitude each converter group can be supplied directly from the supply source without the necessity of supply transformers. However, the output phases are to be isolated as shown in Fig. 4.7(a) since there is no common terminal between the source and the load. This is generally possible whenever the cycloconverter supplies a three-phase AC motor the phases of which can be isolated. If the load is to be connected to form a common terminal as shown in Fig. 4.7(b), then the input supply for each converter group must be obtained from an independent source. This arrangement requires an input transformer with three secondaries, each supplying independently one positive and negative converter group. Though this is not indicated in Fig. 4.6(a), it is rather understood when the three phases are connected in the wye configuration.

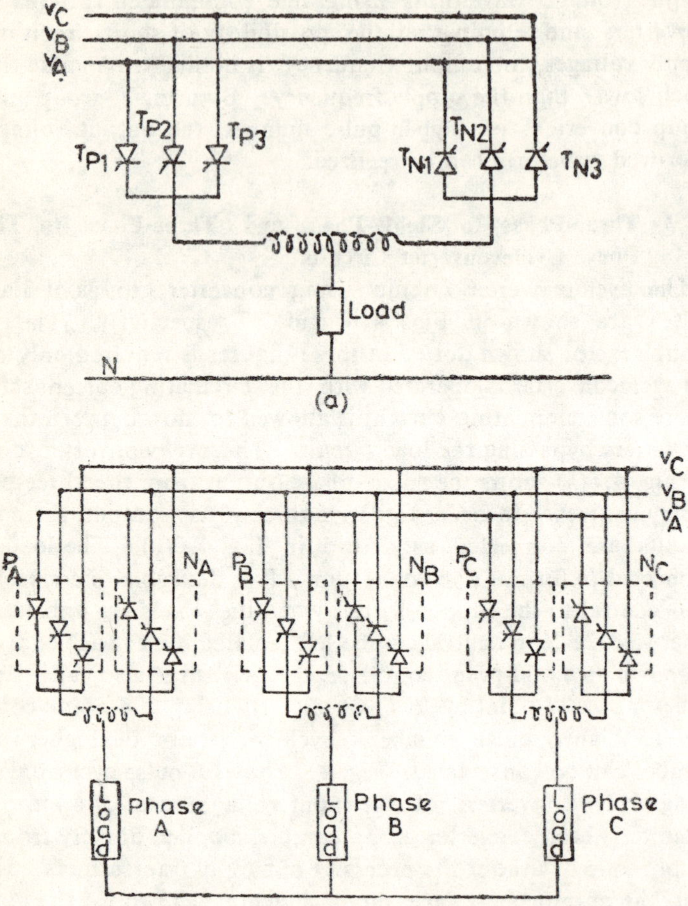

Fig. 4.6. Cycloconverter circuit using 3-pulse converter groups:
(a) Three phase to single phase cycloconverter.
(b) Three phase to three phase cycloconverter.

4.3 OUTPUT VOLTAGE

The study of a single-phase to single-phase cycloconverter in section 4.2.1 shows that the output voltage is fabricated by switching appropriate segments of the input voltages across the output. Therefore, the waveform of the output voltage is very much dependent upon the control logic that selects appropriate segments of the input voltages. The cycloconverter circuits of

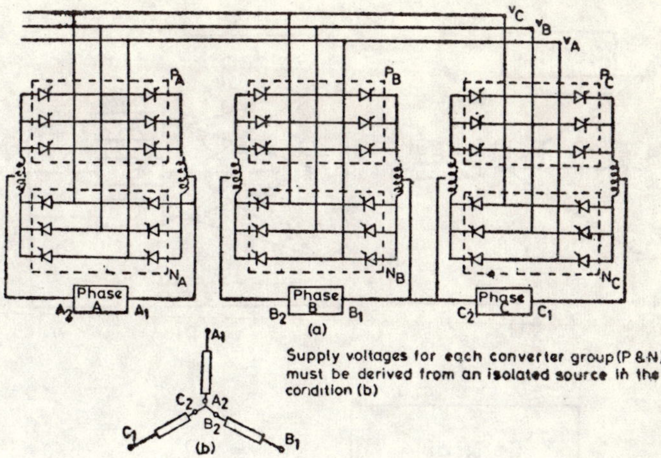

Fig. 4.7. Three-phase to three-phase cycloconverter circuit using 6 pulse bridge converters: (a) Independent loads, (b) Star connected loads.

Figs. 4.6 and 4.7 are represented as switching circuits in Fig. 4.8 replacing the thyristors by simple switches [2]. The oppositely directed arrows indicate whether the supply voltage is connected to the load circuit through positive group thyristors or negative group thyristors. The block labelled 'control' operates the closing and opening of these parallel connected switches. The desired output voltage waveform is set by the reference signals. The control circuit monitors the reference signals switching different segments of input voltages so as to produce an output voltage waveform that is identical to the reference signal. The average of the finite segments represents the desired output waveform. Using converters of higher pulse number in the positive and negative groups, the waveform can approach close to the reference signal waveform.

The exact wave shape of the output voltage mainly depends upon the following factors.

1. Converter pulse number.
2. Ratio between output and input frequencies.
3. Relative level of the output voltage.
4. Load displacement angle.

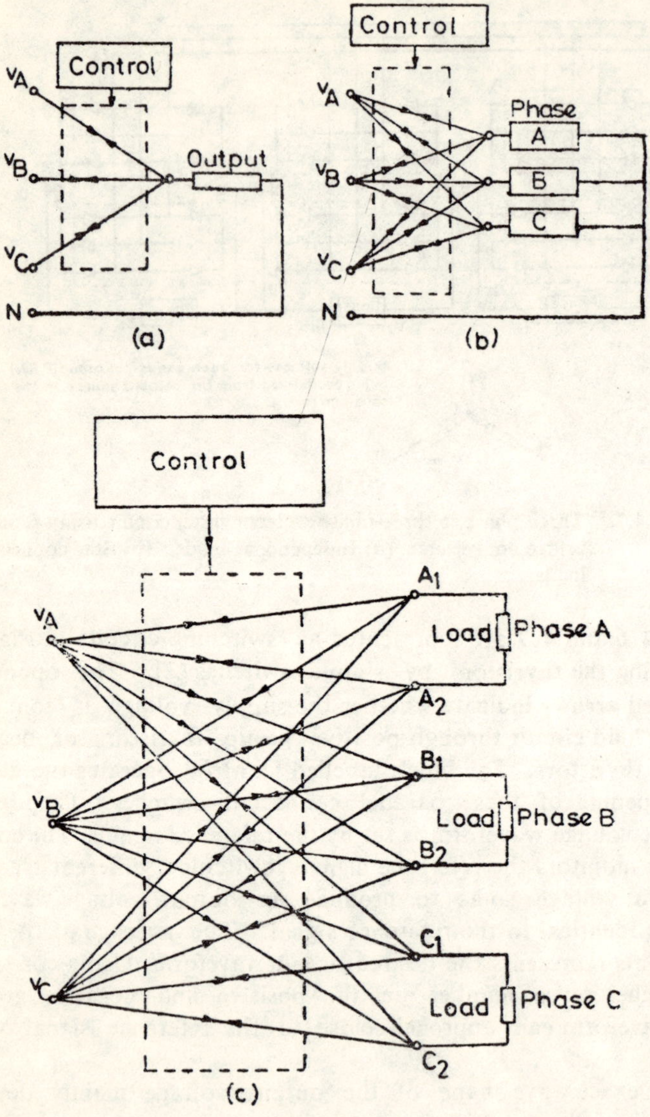

Fig. 4.8. Switching diagrams of cycloconverter circuits of Figs. 4.6 and 4.7.

5. **Modulation technique which controls the firing pulses to the thyristors.**

4.3.1 Circulating Current Scheme

In this scheme, some circulating current is permitted to flow between the two converters. The output voltage of the positive converter is constructed by changing the firing angle in successive pulses so as to obtain the desired waveform. The heavy lines in Fig. 4.9(a) shows the output voltage of the positive converter of a 3-pulse cycloconverter. The ratio of the output frequency to the input frequency is taken as 1/6. The average DC output voltage of a three pulse converter for a given firing angle as discussed in chapter 3 is given by the expression

$$V_{dc} = \frac{3\sqrt{6}\,V_{ph}}{2\pi}\cos\alpha = V_{dc_{max}}\cos\alpha$$

where V_{ph} is the phase voltage.

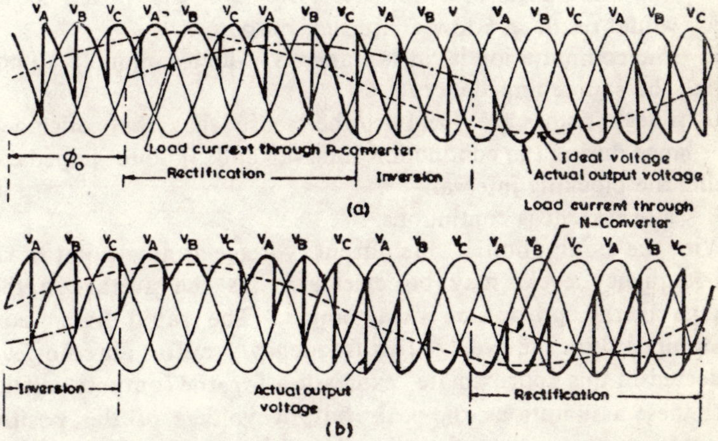

Fig. 4.9. Waveforms of 3-pulse cycloconverter with circulating current operation:
(a) Output voltage of positive converter
(b) Output voltage of negative converter.

The thyristors T_{P_1}, T_{P_2} and T_{P_3} of Fig. 4.6(a) are turned on in sequence. The phase voltages are connected to the output terminals in such a way that the average of the voltage segments conforms to the desired waveshape of the output voltage. If the firing angle is varied from $\pi/2$ to 0, 0 to $\pi/2$, $\pi/2$ to π, and π to $\pi/2$ with suitable intermediate steps, one cycle of low frequency signal is superimposed on the average output voltage. The low frequency is solely determined by the rate at which the firing angle is varied in successive steps and is independent of the supply frequency. Since the line commutation is employed, the incoming voltage must be always greater than the outgoing voltage. Then only will the outgoing thyristor be reverse biased when the incoming thyristor is turned on. For the positive converter, the incoming voltage is known as the positive type voltage. The instantaneous output voltage of the positive converter which comprises positive type voltage at each commutation can clearly be noted in Fig. 4.9(a). Also shown a dotted curve superimposed on the voltage segments is the low frequency alternating voltage of the desired frequency which in this particular case is one-sixth the supply frequency.

The expression for the rms output voltage of the positive converter can be derived with the following assumptions.

1. The firing angle delay in successive cycles is varied slowly which would result in a low frequency output.
2. The commutation is instantaneous which amounts to neglecting the source impedance.
3. The thyristors are ideal in the sense that they offer zero impedance during the conduction interval and infinite impedance during the blocking interval.
4. The current is continuous.

With the assumption 1, the output voltage at any point of the low-frequency cycle may be calculated as the average output voltage for the appropriate delay angle. The rapid fluctuations superimposed on the average low-frequency waveform are however neglected in this approximate expression for the output voltage. With these assumptions, the peak output voltage of the positive converter must correspond to the zero firing angle delay. If V_{op} is the rms output voltage of the positive converter, then for a three-pulse converter

$$\sqrt{2}\, V_{op} = \frac{3\sqrt{6}\, V_{ph}}{2\pi}$$

or

$$V_{op} = \frac{3\sqrt{3}\, V_{ph}}{2\pi} \tag{4.1}$$

where V_{ph} is the rms phase voltage.

This approach can be extended to multi-pulse three-phase to three-phase cycloconverters. The output voltage V_{do} of a multi-pulse converter for any firing angle α is

$$V_{do} = \sqrt{2}\, V_{ph} \left(\frac{m}{\pi}\right) \sin\left(\frac{\pi}{m}\right) \cos\alpha \tag{4.2}$$

where m is the number of pulses.

The maximum DC voltage is

$$V_{do_{max}} = \sqrt{2}\, V_{ph} \left(\frac{m}{\pi}\right) \sin\left(\frac{\pi}{m}\right) \tag{4.2a}$$

If V_{op} is the rms voltage per phase of the positive converter, then

$$\sqrt{2}\, V_{op} = \sqrt{2}\, V_{ph} \left(\frac{m}{\pi}\right) \sin\left(\frac{\pi}{m}\right) = V_{do_{max}}$$

or

$$V_{op} = V_{ph} \left(\frac{m}{\pi}\right) \sin\left(\frac{\pi}{m}\right) \tag{4.3}$$

If the negative converter is controlled such that the relation $\alpha_N = 180° - \alpha_p$ (α_P = firing angle of the positive converter, α_N = firing angle of the negative converter) is satisfied, the output voltage of the negative converter can be obtained along the lines explained for the positive converter. Since the thyristors in the negative converter are connected in series with the supply voltage in a direction opposite to that in the positive converter, the incoming voltage must be always lower than the outgoing voltage. Then only will the outgoing thyristor be reverse biased when the incoming thyristor is turned on. Therefore for the negative converter, the incoming voltage being always lower than the preceding voltage is called negative-type voltage. This can clearly be seen from the instantaneous output voltage waveform of the negative converter shown in Fig. 4.9 (b). The equations (4.1) to (4.3) are valid for the negative converter also. The rms output voltage V_0 of the cycloconverter is

$$V_0 = V_{op} = V_{on} \tag{4.4}$$

The firing angle of the negative group cannot be increased to 180°. This would mean zero turn off time to the outgoing thyristor. Therefore commutation failure is sure to occur under this condition. In practice, the maximum firing angle will be somewhat less than 180° by an amount depending upon commutation overlap and turn-off time specification of the device. Consequently, the delay angle of the positive group can not be reduced below a certain finite value, $\alpha_{P\,min}$. The maximum output voltage is reduced by a factor $\cos\alpha_{P\,min} = r_1$. Therefore, the rms output voltage is,

$$V_0 = V_{0N} = V_{0P} = r_1 V_{ph} \left(\frac{m}{\pi}\right) \sin\frac{\pi}{m} \qquad (4.5)$$

Although the rms output voltages of the desired low-frequency signal of the positive and negative converters are equal, the actual waveforms as illustrated in Fig. 4.9 (a) and (b) are different. The actual output voltage of the cycloconverter at the mid-point of the circulating current limiting reactor (Fig. 4.6 (a)) is obtained

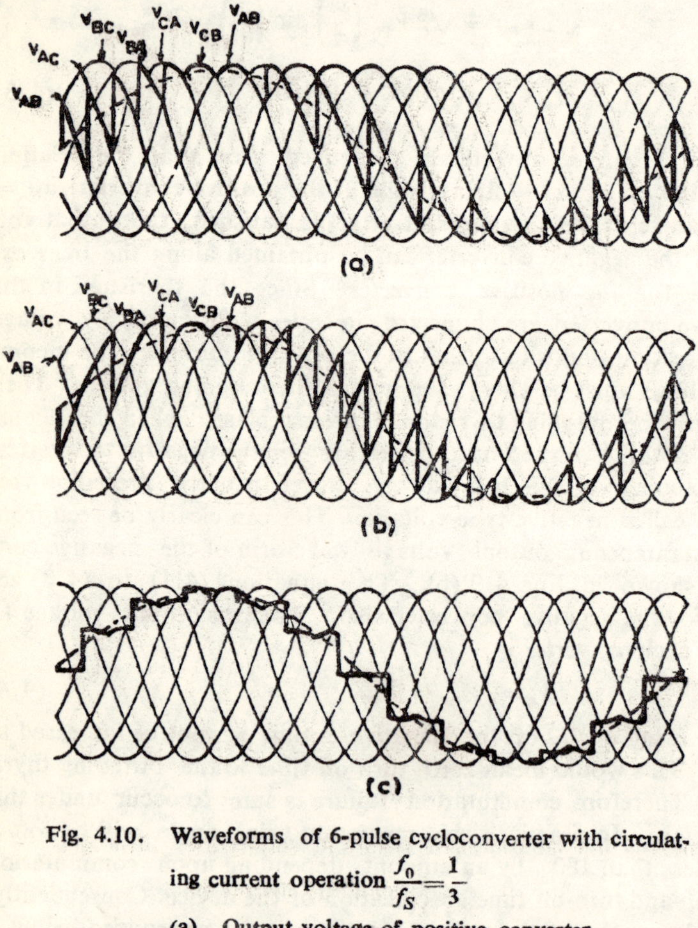

Fig. 4.10. Waveforms of 6-pulse cycloconverter with circulating current operation $\frac{f_0}{f_S} = \frac{1}{3}$

(a) Output voltage of positive converter.
(b) Output voltage of negative converter.
(c) Output voltage at the mid-point of circulating current reactor.

by taking the average of the instantaneous output voltages of the positive and negative converters.

That is,

$$v_0 = \frac{v_{0P} + v_{0N}}{2}$$

The load current depends upon load voltage, load impedance and also back *EMF*, if any. Since the voltage impressed across the load circuit has some rapid fluctuations superimposed on the average low-frequency waveform, the current waveform also will have some fluctuations. But these fluctuations will be of much lower amplitude since the load impedance has the filtering effect. As a result, the load current waveform can be assumed to be almost sinusoidal. The chain curve shown in Fig. 4.9 indicates the load current. Apart from this current, some additional circulating current flows through each converter. This is, however, not shown in the figure. The rectification and inversion intervals of both converter groups are also marked in Fig. 4.9.

The output voltage waveform can be improved significantly if converters of higher pulse number are employed. Fig. 4.10 illustrates the output voltages of positive and negative 6-pulse converters. The average output voltage which is about 1/3 times the supply frequency is also shown. The voltages as can be seen from Fig. 4.10 approach more closely to the low frequency signal than for a three-pulse cycloconverter. Further improvement is possible with 12-pulse converters. But this increases the cost and complexity. In general, the 6-pulse cycloconverter of Fig. 4.7 provides reasonably good waveforms for output to input frequency ratio not exceeding about 0.5.

4.3.2 Non-Circulating Current Operation

In the non-circulating current operation, only one converter operates at a time. The positive converter is gated when the load current is positive and the negative converter is blocked temporarily during this period by inhibiting the firing pulses to the thyristors as explained in Section 4.1. The reverse is true during the negative half-cycle of the load current. With this operation no circulating current flows between the two converters and the need for the circulating current limiting reactor is eliminated. The load voltage is the same as the output voltage of the conducting group.

The output voltage is fabricated using positive type and negative type voltages. The positive type voltages appear when the load current is instantaneously positive and the negative type voltages appear when the load current is instantaneously negative. Fig. 4.11 shows the output voltage (Heavy lines) of a three-pulse cycloconverter with the non-circulating current operation. The load current lagging the mean output voltage (dotted curve) at an angle 60° is shown in Fig. 4.11 (b). The rectification and inversion intervals of positive and negative converters are also indicated. As explained for the single-phase to single-phase cycloconverter in section 4.2.1, there is a possibility of short-circuit of supply voltages at the crossing over points of the load current if proper care is not taken in the control circuit. Let us examine the fabrication process of the output voltage.

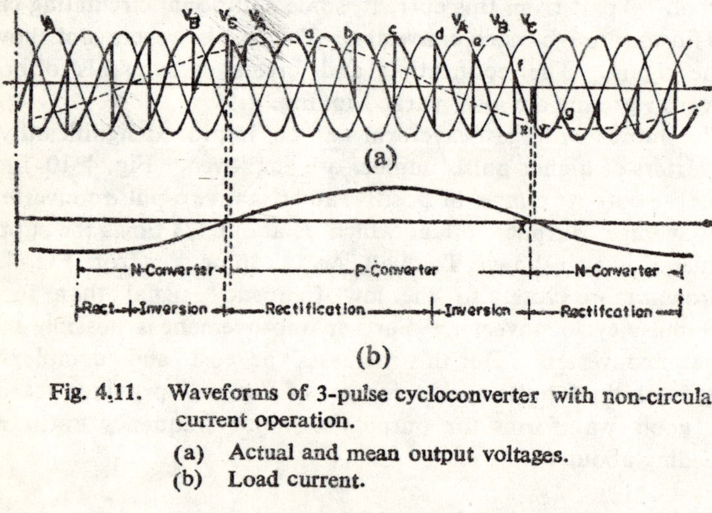

Fig. 4.11. Waveforms of 3-pulse cycloconverter with non-circulating current operation.
(a) Actual and mean output voltages.
(b) Load current.

The average voltage is reduced by retarding the firing angle gradually as indicated by the points *a, b, c, d, e* and *f* in Fig. 4.11 (a). In the presence of the positive load current extending beyond the point *g*, the thyristor T_{P_2} would continue to conduct up to *g*' where the thyristor T_{P_3} would have taken over from T_{P_2}. However, T_{P_2} turns off naturally since the load current has become zero at the transition point *X*. From this point onwards, the thyristors in the negative group converter conduct the load current. If the first incoming thyristor in the negative group after the crossing over point *X* happens to be one connected to a phase other than the previously conducting phase preceding the crossing over point X, the short-circuit of these two phase voltages will take place. In the absence of any current limiting reactor, a high current will flow destroying the thyristors. In order to eliminate such a possibility, a small pause is introduced after the current in the positive group has fallen to zero. This is illustrated in Fig. 4.11(a) at each crossing-over point of the load current. In the waveforms illustrated in Fig. 4.11 (a), this possibility does not exist since the first thyristor in the negative group after the crossing over point belongs to the same previously conducting phase B. Therefore, the incoming thyristor T_{N_2} can safely be turned on at the point X without causing any short-circuit of supply voltages. However, the small delay is essential since the short-circuit may take place at some other output frequency. Figure 4.12 illustrates the output voltage and current waveforms of the 6-pulse bridge type non-circulating current cycloconverter for the output to input frequency ratio of 1/5 showing pauses at the crossingover points of the load current which is shown lagging behind the mean output voltage. There is a significant improvement in the waveshape of the raw output voltage of the six-pulse cycloconverter

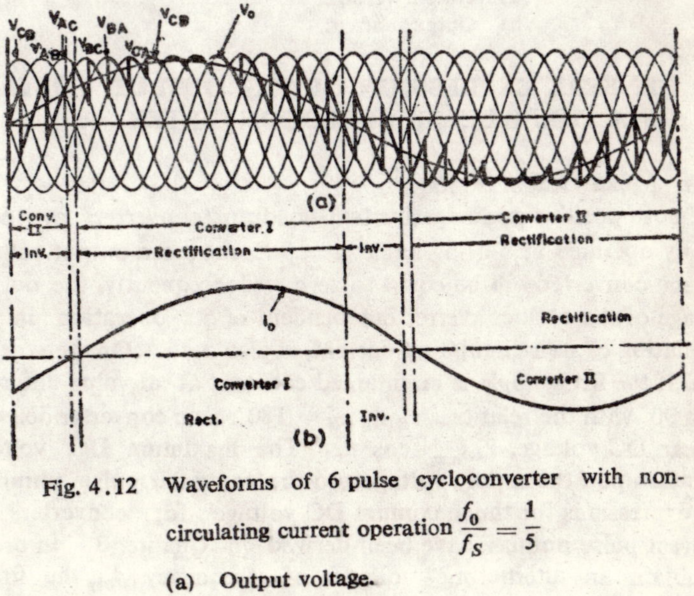

Fig. 4.12 Waveforms of 6 pulse cycloconverter with non-circulating current operation $\frac{f_0}{f_S} = \frac{1}{5}$

(a) Output voltage.
(b) Output current.

compared to that of the three-pulse cycloconverter. In order to see how the waveforms compare between 3-pulse and 6-pulse converters for the same frequency ratio 1/6 and the load current of the same lagging power factor, Fig. 4.13 is illustrated. The pauses at the crossing-over points are however not shown since the same phases conduct in the vicinity of the crossing over points. The improvement in the raw output voltage waveform can clearly be seen.

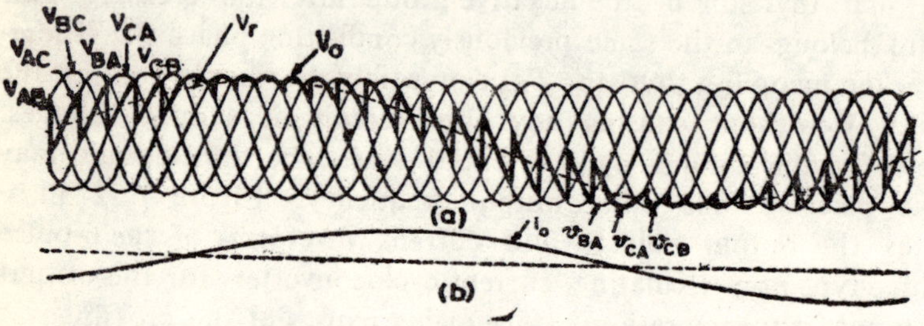

Fig. 4.13. Waveforms of 6 pulse cycloconverter with non-circulating current operation $\frac{f_0}{f_S} = \frac{1}{6}$

(a) Output voltage.
(b) Output current.

4.4 GENERAL EXPRESSION OF THE OUTPUT VOLTAGE OF A THREE-PULSE CYCLOCONVERTER

4.4.1 Mean Output Voltage

If both positive and negative two-quadrant converters are constantly operated at a firing angle, $\alpha = 90°$, the mean output voltage of each converter will be equal to zero. Consequently, the output voltage of the cycloconverter independent of its operation in the circulating or non-circulating current, is also zero. On the other hand if the firing angle is maintained constant at a value different from 90° with the relation, $\alpha_P + \alpha_N = 180°$, each converter delivers a mean DC voltage, $V_{do_{max}} \cos \alpha$. The maximum DC voltage depends upon the supply voltage and the converter pulse number. The expressions for the maximum DC voltage for converters of different pulse number have been derived in Chapter 3. In order to obtain an alternating voltage of frequency f_0, the firing angle also has to be continuously oscillated. It may be noted from

Figs. 4.9 and 4.10 that the firing angle of the positive converter is oscillated to and fro about the quiescent point corresponding to the 90° firing angle. This control process is also carried out for the negative converter in a direction opposite to that of the positive converter. Let us suppose that the firing angle is oscillated to and fro about the quiescent point according to some function say $f(\theta_0)$ which would depend upon the output frequency. Thus the value of $f(\theta_0)$ oscillates to and fro about the 90° quiescent point. The rate at which it is oscillated is solely determined by the frequency of the output voltage. Theoretically $f(\theta_0)$ can vary between $+\pi/2$ and $-\pi/2$. Since the firing angles are modulated equally in the negative converter but in a sense opposite to that of a positive converter, the corresponding modulating function for the negative converter is $-f(\theta_0)$. Taking the modulating function into account, the appropriate segments of input voltages are considered while deriving an expression for the output voltages of positive and negative converters. Taking the average of these output voltages, an exact expression for the output voltage of the cycloconverter can be obtained. This expression contains the desired alternating component at the defined output frequency f_0 and other harmonic components which depend upon input and output frequencies. In the case of a cycloconverter with the noncirculating current operation, in order to obtain expressions for the output voltage, suitable switching functions are to be introduced. These switching functions take into account the load displacement angle and have amplitude unity or zero. During the half-cycle at the output frequency, f_0 when the load current is positive, the switching function for the positive converter will have unity and the switching function corresponding to the negative converter will be zero. The output voltage of the cycloconverter will be the same as that of the positive converter. In a similar manner, the output voltage of the cycloconverter will be the same as that of the negative converter whenever the output current is negative since the switching function for the negative converter will have unity amplitude while that of the positive converter will have zero amplitude. The output voltage at the desired frequency f_0 of the three-pulse cycloconverter with the circulating current or non-circulating current operation can be shown to be [3]

$$v_0 = \frac{3\sqrt{3}V_{ph}}{2\pi} \sin f(\theta_0)$$

where V_{ph} is the rms phase voltage and $f(\theta_0)$ is an arbitrary modulating function. As mentioned in section 4.3, there are several factors which contribute to the waveshape of the output voltage. Dominant among these factors are the modulation function and reference voltage. Some commonly employed modulation functions are:

1. Linear modulation
2. Triangular wave modulation
3. Square wave modulation
4. Sinusoidal modulation.

Each modulation function has some advantages over the other [2]. The linear modulation technique produces unwanted harmonic components which are of much higher order than the desired output frequency. The triangular modulation produces a DC component at discrete ratios of output to input frequencies over the whole output frequency range. Also it generates harmonic components of order lower than the desired output frequency. As a result, the ratio of the output frequency to the input frequency is limited to about $\frac{1}{3}$ for three-pulse converters. The square wave modulation fabricates the output voltage waveform which is predominantly square wave in shape. The sine wave modulation contains odd harmonic components of which the third harmonic has significant amplitude. It also generates harmonic components of order lower than the desired output frequency and as such the comments made for the triangular modulation are also relevant for the sine wave modulation.

4.4.2 Output Voltage of a Cycloconverter with Cosinusoidal Modulating Signals and Sinusoidal Reference Voltages

In order to produce the output voltage waveform with minimum total harmonic distortion, the sinusoidal reference voltage and cosinusoidal modulating voltage signals are preferred over other reference and modulating signals [3], [8]. The principle of the cosinusoidal modulation has been explained in section 3.5.2. An expression will be derived for the output voltage using three-pulse converters. Fig. 4.14 shows cosine modulating signals along with the desired reference voltage, and switching functions for thyristors in the positive group while Fig. 4.15 illustrates the same waveforms for thyristors in the negative group. These switching functions

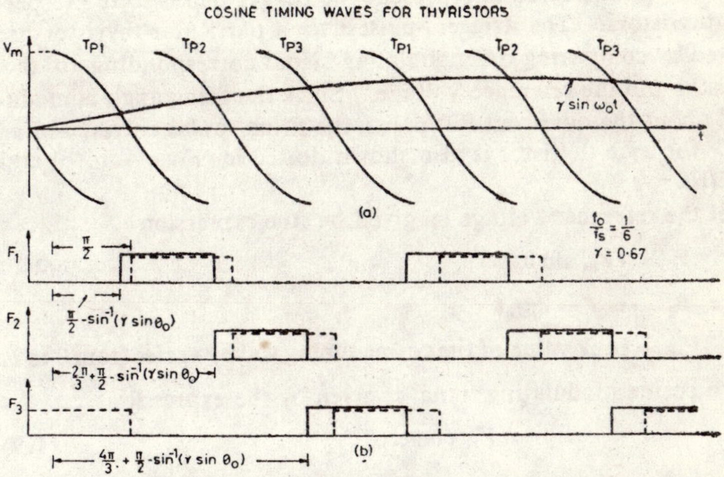

Fig. 4.14. Modulation scheme for a three-pulse cycloconverter:

(a) Sinusoidal reference wave and cosine modulating signals for T_{P_1} to T_{P_3}.

(b) Switching functions for positive group thyristors.

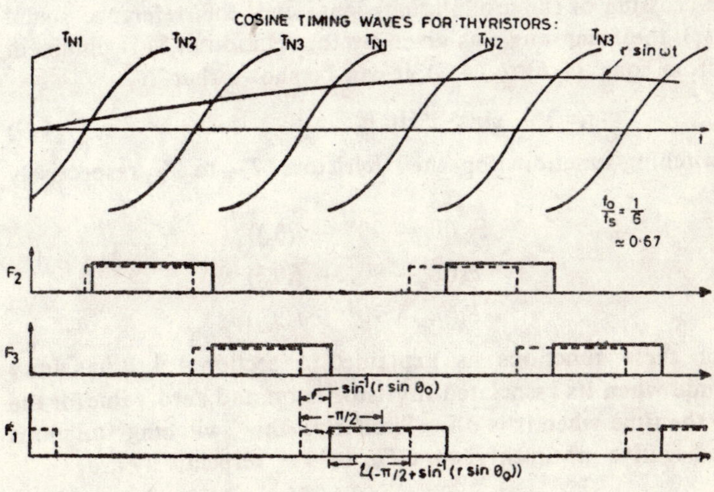

Fig. 4.15. Modulation Scheme:

(a) Sinusoidal reference wave and cosine timing waves for T_{N1} to T_{N2}

(b) Switching functions for negative group thyristors.

can also be considered to represent the trigger pulses to the respective thyristors. The trigger pulses to a particular thyristor are derived by comparing the modulating signal corresponding to this thyristor and the reference voltage. Since the firing angle is modulated about the quiescent 90° point, the firing pulse corresponding to 90° for each thyristor is also shown dotted in Figs. 4.14(b) and 4.15(b).

Let the reference voltage be given by the expression

$$v_r = rV_m \sin \theta_0 \tag{4.6}$$

where $\theta_0 = \omega_0 t = 2\pi f_0 t$

V_m = peak value of the cosine timing wave r varies from 0 to 1.

The cosine modulating signal is given by the expression

$$v_c = V_m \cos \alpha \tag{4.7}$$

where α depends upon the reference signal and cosine time wave. In order to obtain the positive voltage, the firing angle is advanced about the quiescent point by $f(\theta_0)$. The firing angle α is related to $f(\theta_0)$ as follows

$$\alpha = 90° - f(\theta_0) \tag{4.8}$$

At the crossing of the modulating signal and the reference signal ($v_r = v_c$), the firing angle as given by the relation (4.8) is obtained. Thus from eqns. (4.6) to (4.8), it can be shown that

$$f(\theta_0) = \sin^{-1} r \sin \theta_0 \tag{4.9}$$

The switching functions for the thyristors T_{P1} to T_{P3} respectively are:

$$F_1(\theta_s - 90° + f(\theta_0))$$

$$F_2(\theta_s - 90° + f(\theta_0))$$

$$F_3(\theta_s - 90° + f(\theta_0))$$

Each of these functions as explained in section 4.4.1 has unity amplitude when its associated thyristor is on and zero value for the rest of the time when it is off. Similarly, the switching functions for the negative group thyristors T_{N1} to T_{N3} respectively are:

$$F_1(\theta_s + 90° - f(\theta_0))$$

$$F_2(\theta_s + 90° - f(\theta_0))$$

$$F_3(\theta_s + 90° - f(\theta_0))$$

In this case also, each switching function has unity amplitude whenever its corresponding thyristor is on and zero for the rest of the period. The output voltages of positive and negative converters after substituting for $f(\theta_0)$ in the switching functions are:

$$\begin{aligned} v_{op} = {} & V_{ph} \sin(\omega_s t)\cdot F_1(\theta_s - 90° + \sin^{-1}(r \sin \theta_0)) \\ & + V_{ph} \sin(\omega_s t - 120°)\cdot F_2(\theta_s - 90° + \sin^{-1}(r \sin \theta_0)) \\ & + V_{ph} \sin(\omega_s t + 120°)\cdot F_3(\theta_s - 90° + \sin^{-1}(r \sin \theta_0)) \end{aligned} \tag{4.10}$$

$$\begin{aligned} v_{0N} = {} & V_{ph} \sin(\omega_s t)\cdot F_1(\theta_s + 90° - \sin^{-1}(r \sin \theta_0)) \\ & + V_{ph} \sin(\omega_s t - 120°)\cdot F_2(\theta_s + 90° - \sin^{-1}(r \sin \theta_0)) \\ & + V_{ph} \sin(\omega_s t + 120°)\cdot F_3(\theta_s + 90° - \sin^{-1}(r \sin \theta_0)) \end{aligned} \tag{4.11}$$

4.4.2.1 *Circulating Current Scheme*

Replacing each switching function by its Fourier series and simplifying eqns. (4.10) and (4.11) and taking their average, the maximum output voltage ($r = 1$) of a 3-pulse cycloconverter can be shown as

$$\begin{aligned} v_0 = \frac{3\sqrt{6}V_{ph}}{2\pi} \Big\{ & \sin \theta_0 + \tfrac{1}{4} [\sin(3\theta_s + 2\theta_0) + \sin(3\theta_s - 2\theta_0)] \\ & + \tfrac{1}{8} [\sin(3\theta_s + 4\theta_0) + \sin(3\theta_s - 4\theta_0)] \\ & + \tfrac{1}{10} [\sin(6\theta_s + 5\theta_0) - \sin(6\theta_s - 5\theta_0)] \\ & + \tfrac{1}{14} [\sin(6\theta_s + 7\theta_0) - \sin(6\theta_s - 7\theta_0)] + \ldots \Big\} \end{aligned} \tag{4.12}$$

4.4.2.2 *Non-circulating Current Scheme*

As explained in section 4.3.2, the positive and negative converters conduct during the positive and negative half cycles of the load current respectively. The load current which is assumed to be sinusoidal is displaced with respect to the mean sinusoidal reference voltage by the load displacement angle, ϕ_0. In order to obtain a general expression for the output voltage of a cycloconverter with the non-circulating current operation, it is necessary to introduce two switching functions for the positive and negative converters. They are:

$$\left.\begin{aligned} F_P(\theta_0) &= 1 \quad \phi_0 < \omega_0 t < (\phi_0 + \pi) \\ &= 0 \quad (\phi_0 + \pi) < \omega_0 t < (\phi_0 + 2\pi) \\ F_N(\theta_0) &= 1 \quad (\phi_0 + \pi) < \omega_0 t < (\phi_0 + 2\pi) \\ &= 0 \quad \phi_0 < \omega_0 t < (\phi_0 + \pi) \end{aligned}\right\} \tag{4.13}$$

The output voltage is,

$$v_0 = v_{op} \cdot F_p(\theta_0) + v_{on} \cdot F_N(\theta_0) \tag{4.14}$$

The Fourier series expansion of $F_p(\theta_0)$ and $F_N(\theta_0)$ are:

$$F_p(\theta_0) = -\frac{1}{2} + \frac{2}{\pi}\Big[\sin(\theta_0 + \phi_0) + \frac{1}{3}\sin 3(\theta_0 + \phi_0) + \frac{1}{5}\sin 5(\theta_0 + \phi_0) + \frac{1}{7}\sin 7(\theta_0 + \phi_0) + \ldots\Big] \tag{4.15}$$

$$F_N(\theta_0) = \frac{1}{2} - \frac{2}{\pi}\Big[\sin(\theta_0 + \phi_0) + \frac{1}{3}\sin 3(\theta_0 + \phi_0) + \frac{1}{5}\sin 5(\theta_0 + \phi_0) + \frac{1}{7}\sin 7(\theta_0 + \phi_0) + \ldots\Big] \tag{4.16}$$

From eqns. (4.10) to (4.16) and after replacing F_1, F_2 and F_3 by their Fourier series and after necessary algebraic simplification, the output voltage of a 3-pulse cycloconverter with the non-circulating current operation with $r = 1$ and $\phi_0 = 0$, can be shown as follows:

$$\begin{aligned} v_0 = \frac{3\sqrt{3}V_{ph}}{2\pi}\Big\{\Big(&\sin\theta_0 + \frac{1}{4}[\sin(3\theta_s + 2\theta_0) + \sin(3\theta_s - 2\theta_0)] \\ &+ \tfrac{1}{8}[\sin(3\theta_s + 4\theta_0) + \sin(3\theta_s - 4\theta_0)] \\ &+ \tfrac{1}{10}[\sin(6\theta_s + 5\theta_0) - \sin(6\theta_s - 5\theta_0)] \\ &+ \tfrac{1}{14}[\sin(6\theta_s + 7\theta_0) - \sin(6\theta_s - 7\theta_0)] + \ldots\Big) \\ &+ \frac{1}{\pi}\Big(\cos 3\theta_s + \frac{1}{3}[\cos(3\theta_s + 2\theta_0) + \cos(3\theta_s - 2\theta_0)] \\ &- \tfrac{1}{5}[\cos(3\theta_s + 4\theta_0) + \cos(3\theta_s - 4\theta_0)] \\ &- \tfrac{1}{4}[\cos(3\theta_s + 6\theta_0) + \cos(3\theta_s - 6\theta_0)] \\ &- \tfrac{1}{15}[\cos(3\theta_s + 8\theta_0) + \cos(3\theta_s - 8\theta_0)] \end{aligned}$$

$$
\begin{aligned}
&-\tfrac{2}{105}\,[\cos(6\theta_s+\theta_0)-\cos(6\theta_s-\theta_0)]\\
&+\tfrac{6}{85}\,[\cos(6\theta_s+3\theta_0)-\cos(6\theta_s-3\theta_0)]\\
&+\tfrac{18}{175}\,[\cos(6\theta_s+5\theta_0)-\cos(6\theta_s-5\theta_0)]\\
&-\tfrac{15}{245}\,[\cos(6\theta_s+7\theta_0)-\cos(6\theta_s-7\theta_0)]\ldots\Big)\Big\}
\end{aligned}
\tag{4.17}
$$

Equations (4.12) and (4.17) reveal that the wanted component and certain distortion terms are the same in circulating and non-circulating current cycloconverters. There are additional distortion terms in the non-circulating current cycloconverter. The harmonic distortion terms are thus more complex in a cycloconverter with the noncirculating current operation than with the circulating current operation.

4.5 CONTROL CIRCUIT

The main function of the control circuit is to produce trigger pulses in a particular sequence and feed them to the gates of positive and negative group of thyristors so as to generate a voltage of the desired wave shape at the output terminals of a cycloconverter. Figure 4.16 shows a block diagram of the control circuit of a cycloconverter with the non-circulating current operation [4]. The same control circuit is applicable to a cycloconverter with the circulating current operation but the block designated as 'converter group selection' will not be present in this case. The power circuit basically consists of a pair of two-quadrant converters connected in antiparallel across the load as discussed in section 4.2. The control circuit can be arranged broadly in four functional blocks.

1. Synchronizing circuit. 2. Reference voltage sources. 3. Logic and triggering circuit. 4. Converter group selection circuit. In what follows the function of each block is explained.

4.5.1 Synchronizing Circuit

The function of the synchronizing circuit is to derive low voltage signals to the control circuit which operates at low voltages. These low voltage signals must be synchronized to the voltages supplied to the main power circuit. Step-down transformers may be used for this purpose with filter circuit to avoid waveform distortion if any. Phase shifting network may also be required while deriving modulating voltages at the supply frequency. The

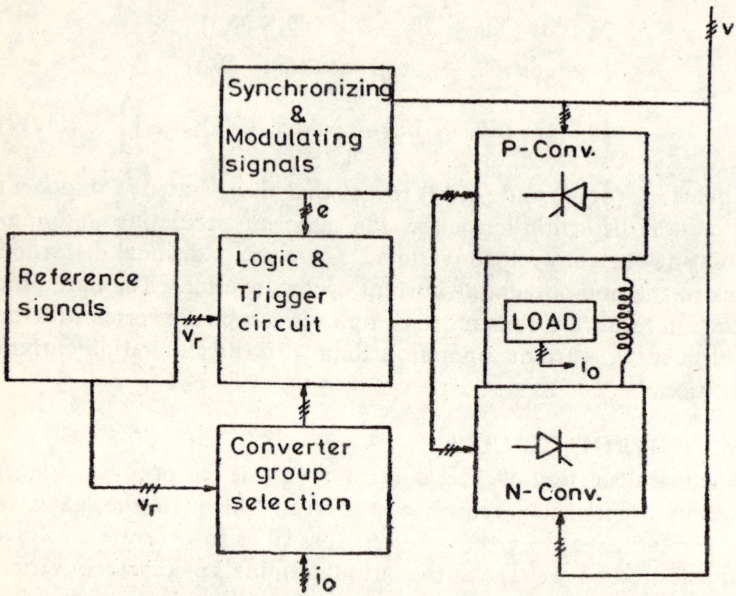

Fig. 4.16. Block diagram illustrating control scheme for a cycloconverter with non-circulating current operation.

modulating voltages are compared with the reference voltages to determine the instants at which the firing pulses are to be produced to the thyristors in the two converter groups.

4.5.2 Reference Voltage Signals

In most cases it is required to control the output voltage and frequency of a cycloconverter. In the case of a three-phase to three-phase cycloconverter, ability to control the phase sequence of the output voltages may also have to be built into the system. The purpose of the reference voltages is to faithfully reproduce the output voltages, the waveforms of which approach more closely to the set reference waveforms. Besides, output voltage, frequency and phase sequence (in case of polyphase output) are controlled from the reference voltage signal block. There are various ways of implementing a three-phase variable-frequency, variable-voltage sine wave reference voltage sources. The designs are based on purely analogue [5], hybrid digital-analogue [6] and purely digital [7] approaches. In the design using analogue approach, a fixed high frequency oscillator and resistor, capacitor phase shifting networks are used to generate a three-phase sine wave. A variable frequency oscillator is also provided. The output frequency of the latter is controlled by a DC voltage. Three mixers, one for each phase, are used to combine the fixed and variable frequencies. The output of each mixer contains both sum and difference of input frequencies. The phase information is preserved in the mixers. Figure 4.17 shows the block diagram of the reference voltage signal block. In this case, the *UJT* is used as a relaxation oscillator at a variable frequency to trigger a three stage ring counter. The three-phase square wave output at a frequency f_d of the ring counter is used to modulate a single-phase fixed frequency (f_c) variable amplitude sinusoidal voltage in a three-phase full wave transistor chopper. The output of the transistor chopper

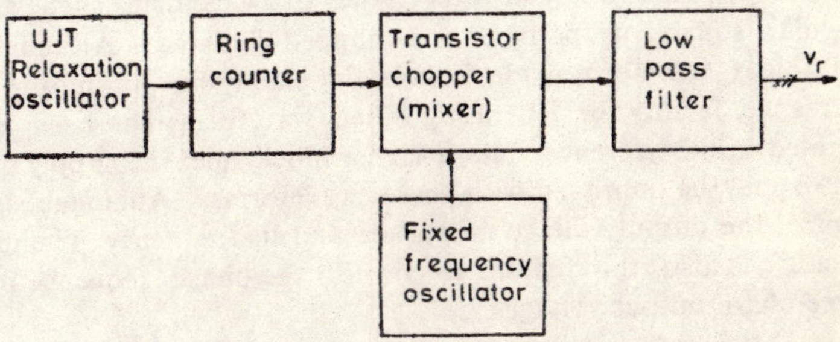

Fig. 4.17. **Block diagram showing variable voltage and variable frequency three-phase reference source.**

is a square wave with half-wave symmetry. It generates a fundamental and a series of odd harmonics. If all higher order harmonics, are neglected it can be shown that the sum (f_c+f_d) and the difference ($f_c - f_d$) of fixed frequency and chopper fundamental frequency (f_d) would result at the output of the chopper. By using low pass-filters, the frequency component ($f_c - f_d$) is selected and the component ($f_c + f_d$) is eliminated. Finally the reference signals can be shown to be

$$v_{ra} = rV_m \sin 2\pi (f_c - f_d) t \tag{4.18}$$

$$v_{rb} = rV_m \sin (2\pi (f_c - f_d) t - 120°) \tag{4.19}$$

$$v_{rc} = rV_m \sin (2\pi (f_c - f_d) t + 120°) \tag{4.20}$$

where V_m is the peak of the cosine modulating wave.

Equations (4.18) to (4.20) show that the amplitude and frequency of the reference waves are controlled by varying r and f_d respectively while the phase sequence of three-phase output is controlled by setting f_d greater or less than f_c. For example, if $f_c = 1000$ Hz, $f_d = (1000 - 30)$ Hz the output frequency can be varied up to a maximum of 30 Hz. The same range is maintained if $f_d = (1000 + 30)$ Hz but now the phase sequence is reversed. The amplitude is controlled by varying r from 0 to 1.

In the hybrid digital-analogue design [6], the digital logic was used to determine the output frequency and phase sequence, and hybrid circuitry was used to determine the output amplitude and waveform.

A versatile three-phase oscillator [7] was developed based on the digital approach. The oscillator provides not only three-phase sinusoidal output but also other types of waveforms such as triangular, square, trapezoidal and clipped sinusoid. Although the sinusoidal waveform is ideal from the standpoint of motor performance, facility for generating other waveforms has been incorporated in the reference voltage signal block since the supply power factor may be improved with these waveforms. Analogue signals control the output voltage amplitude and its frequency. The digital circuit has also the feature to reverse the phase sequence of the three-phase output voltages.

4.5.3 Logic and Triggering Circuit

Figure 4.18 shows the block diagram of the main logic and triggering circuit for the phase A output of a cycloconverter using the reference voltage, v_{ra}. Two similar blocks, one with the reference voltage v_{rb} and the second with the reference voltage v_{rc} are used for the other two phases of the output in the case of a three-phase to three-phase cycloconverter circuit. The reference voltage, v_{ra} is compared with the modulating voltages e_a, e_b and e_c corresponding respectively to supply voltages, v_A, v_B and v_C. Short pulses are produced at the outputs of comparators. These short pulses act as clock pulses to the flip-flops. The outputs of the flip-flops are fed to appropriate thyristors. In each channel the 'And' gate before the block consisting of pulse amplification, isolation and high frequency modulation stage has two inputs, one coming from the flip-flop and the other coming from the blanking circuit. The blanking circuit is essential in a cycloconverter circuit with the

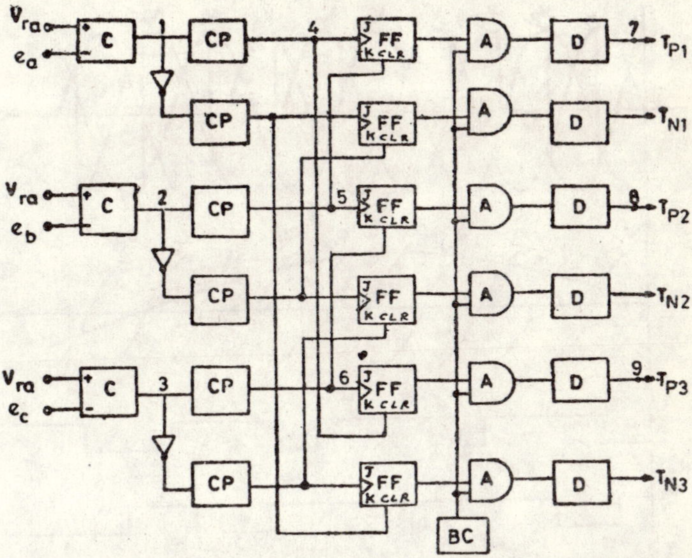

Fig. 4.18. Block diagram of logic and trigger circuit:

V_{ra}	:	ref. voltage for phase and output.
$e_u, e_b e_c$	:	modulating signals.
C	:	Comparator.
CP	:	Clock pulse generator.
FF	:	JK flip-flop, $J = 1$, $K = 0$.
A	:	And gate.
D	:	Driver circuit consisting of pulse isolation amplification and high frequency modulation.
BC	:	Group selection and blanking circuit logic.
T_{P1}–T_{N3}	:	Gating pulses to thyristors in 3-pulse cycloconverter.

non-circulating current operation as this steers the firing pulses to the appropriate converter group depending upon whether the current is positive or negative. With the circulating current operation, no blanking circuit is needed. Hence this input may be set at the logic '1' permanently. The block diagram of the logic circuit is general in the sense that the reference signals and modulating signals could be of any waveshape as mentioned earlier. In most applications, sinusoidal reference signals and cosinusoidal modulating signals are commonly employed. The basic principle of the cosinusoidal modulation is the same as that of cosine wave

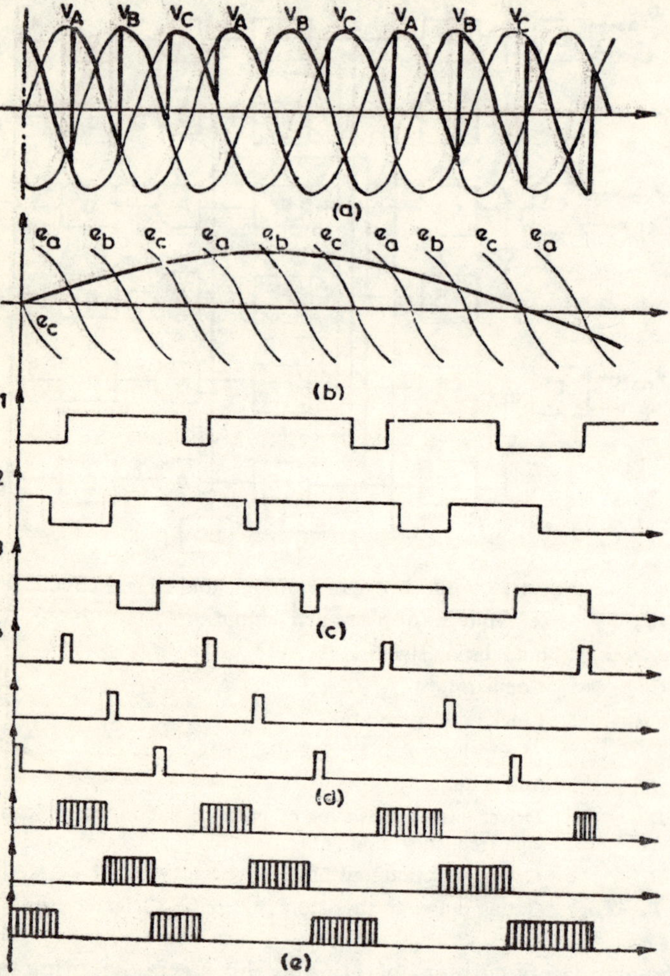

Fig. 4.19. Typical waveforms of logic circuit of Fig. 4.18 for one phase only:

(a) Supply voltage of 3-pulse cycloconverter showing output voltage of one half cycle.

(b) Modulating signals for positive converter and reference voltage (100%).

(c) Outputs of comparators used in the positive group.

(d) Clock-pulses for the positive group.

(e) Gate pulses to thyristors in the positive group of 3-pulse cycloconverter shown in Fig. 4.6(a).

crossing-pulse timing control which is explained in Chapter 3. Fig. 4.19 illustrates typical waveforms at different points of the logic and triggering circuit of Fig. 4.18 assuming sinusoidal reference voltages and consinusoidal modulating signals. The waveforms are drawn for one half-cycle of the output voltage for the positive converter with unity power factor load. The same procedure can be followed for developing the output voltage for the other half cycle of the positive converter. For the negative converter, different modulating signals have to be taken into account. This is taken care of by using an inverter at the output of each comparator (Fig. 4.18).

The generation of the output voltages of the positive and negative converters using cosinusoidal modulating signals for a 3-pulse cycloconverter is illustrated in Fig. 4.20 for one complete cycle. The output voltage is controlled by varying the amplitude of the sinusoidal reference signal while the output frequency is controlled by controlling the frequency of the reference signal. In this diagram, the outputs of clock pulse generators and comparators are not given. The trigger pulses are, however indicated. Figure 4.21 shows the waveforms of the output voltage of a positive converter for a six-pulse cycloconverter for the two cases of 50% and 100% of reference voltage. The actual raw output voltage waveform improves with a reduction in the output frequency and an increase in the depth of modulation i.e., with increasing value of the reference voltage. This can be clearly seen from the waveforms depicted in Fig. 4.21.

4.5.4 Converter Group Selection

Figure 4.22 shows the block diagram of the converter bank selection [4]. The load current is allowed to flow through the *P*-converter or the *N*-converter through suitable logic. There are various ways of sensing the load current. These have been discussed in Chapter 9. *D* is the delay during which period the firing pulses to both converters are inhibited. The delays are not introduced in some control schemes where a small circulating current is permitted during the cross-over instants of the fundamental load current only. The scheme is still recognized as the non-circulating current operation since during a major portion of the output cycle, it operates in the non-circulating current mode.

This circuit is an essential part of the control scheme of a

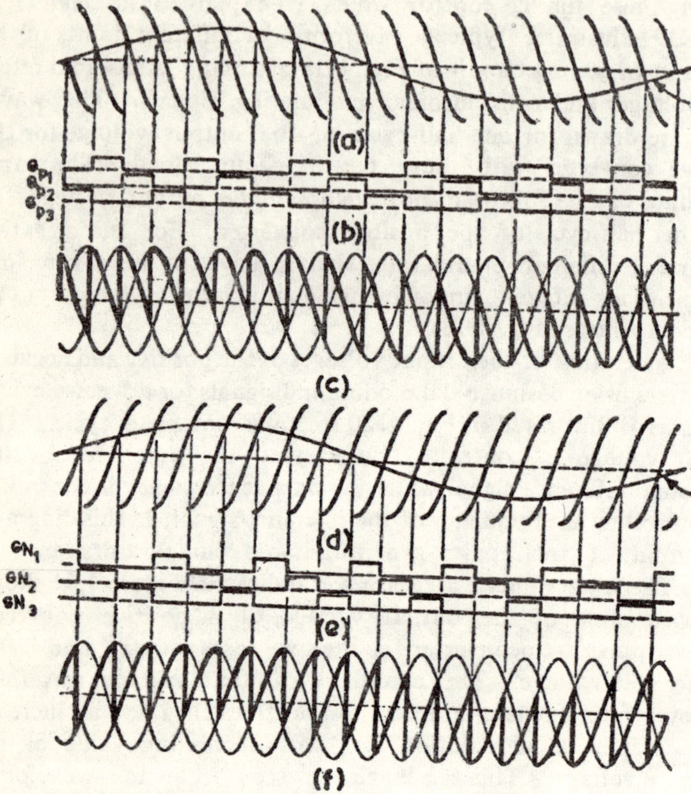

Fig. 4.20. Fabrication of the output voltages of the positive and negative converters for one complete cycle of a 3-pulse circulating current cycloconverter:

(a) Cosinusoidal modulating and sinusoidal (v_r) reference signals for the positive converter.

(b) Gate pulses to thyristors in the positive converter.

(c) Output voltage of the positive converter.

(d) Cosinusoidal modulating and sinusoidal (v_r) reference signals for the negative converter.

(e) Gate pulses to thyristors in the negative converter.

(f) Output voltage of the negative converter.

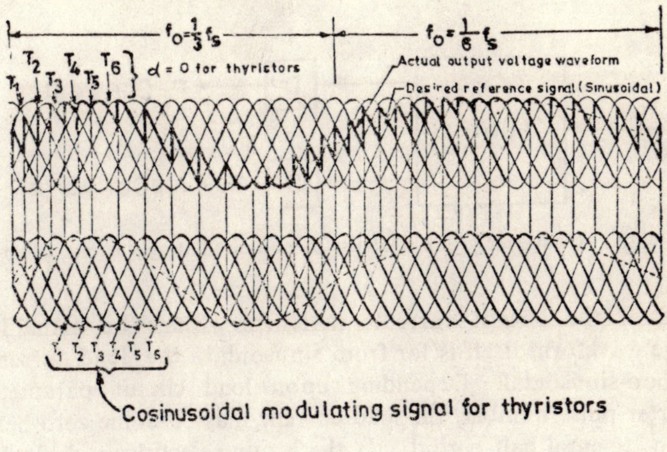

(a)

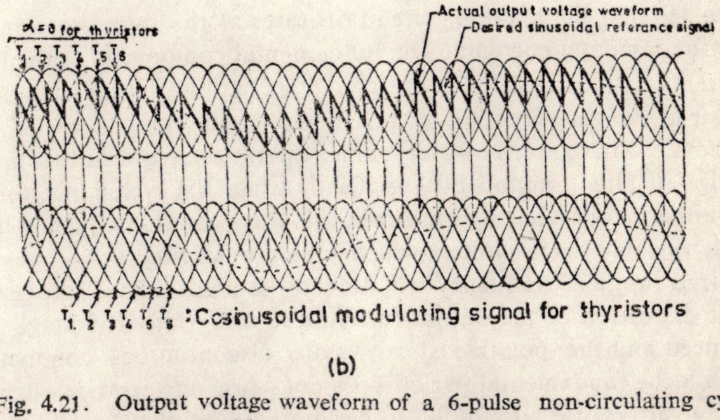

(b)

Fig. 4.21. Output voltage waveform of a 6-pulse non-circulating cycloconverter with varying output frequency:
(a) 100% output voltage.
(b) 50% output voltage.

cycloconverter with the non-circulating current operation. The purpose of this circuit is to see that only one converter operates at a time depending upon the polarity of the current. As explained earlier, the positive converter is operated when the load current is positive and the opposite is true for the negative converter. The converter group selection is not straight forward primarily due to non-ideal

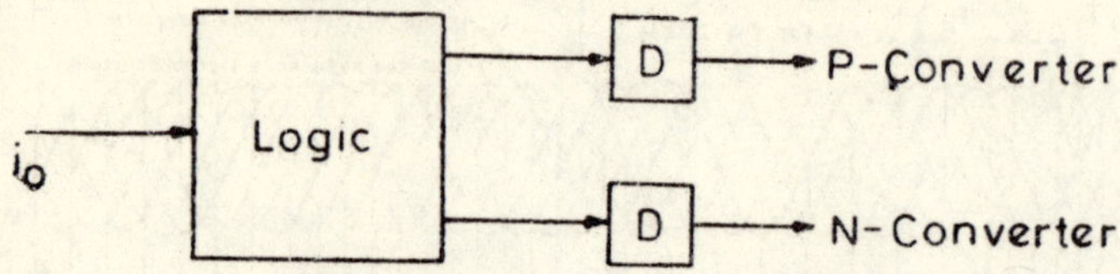

Fig. 4.22. Converter group selection in non-circulating current scheme.

nature of the output current waveform. Since the actual load voltage waveform itself is far from sinusoidal, the load current is also non-sinusoidal. Depending upon load circuit parameters, converter pulse number, the load current may become zero before the fundamental half-period. If the group selection and blanking circuit were to operate at each current zero instant, it may cause erratic switching of converters. The result of this is to further distort the output voltage. One possible solution to this problem is to see that the blanking circuit operates at the zero crossings of the fundamental current. The fundamental component of the load current is extracted and the converter bank-selection is made to occur at the zero-crossing of this fundamental component of the current. There are, however, some operational difficulties in the design of filter components specifically when the cycloconverter is required to operate over a range of the output frequency and with variable load. Filters which operate satisfactorily over the desired range of frequency will have to be used. Thus the envelope distortion of the output current and the output voltage are reduced and the possible steady state discontinuous conduction within the fundamental period does not cause any erratic switching of converters. Because of filters, certain amount of phase shift may be introduced between the zero-crossings of the fundamental output current and the actual load current. In order to eliminate waveform distortion, certain amount of circulating current may be allowed to flow during these short overlap periods in some control schemes. The presence of the circulating current reactor is a must in such a design. However, if no circulating current is permitted to flow, some distortion in the output voltage is to be tolerated. This distortion arises due to the delays introduced at the zero-crossings of the load current to ensure turn off of thyristors in the outgoing group before the thyristors in the incoming group are turned on.

Another approach to the converter group selection is based on the closed-loop control of the output voltage [3]. In this scheme, the firing pulses are permanently applied to both converters. Also a bias voltage is introduced between their voltage transfer characteristics as a result of which, the circulating current between the two converters is reduced. The closed-loop control scheme is such that it compensates the bias voltage of the converter in operation thereby produing its voltage at the output terminals, while the other converter is blocked because of the non-compensation of its bias voltage. At the current cross-over points an error exists between the reference voltage and the mean output voltage. The error voltage in the closed-loop control scheme is such that it automatically adopts the incoming converter and blocks the outgoing converter. In order to prevent excessive circulating current, the bias voltage is increased. This introduces some cross over distortion and the output voltage waveform is somewhat distorted. More information with regard to detailed operation of this particular scheme can be obtained from the references [2], [3].

4.6 HARMONICS, SUPPLY CURRENT WAVEFORM AND POWER FACTOR

4.6.1 Harmonice

4.6.1.1 *Necessary Distortion Terms*

The main factors which significantly influence the waveshape of the output voltage have been listed in section 4.3. Obviously the harmonics also depend upon the same factors. In a phase controlled converter, the firing angle is held constant for a given output voltage. Apart from the DC voltage, the spectrum of harmonics are integral multiples of supply frequency. In the case of a cycloconverter, the firing angle is modulated 'to and fro' about the 90° quiescent point. The harmonic spectrum, as seen from section 4.4.., is different from that of a converter. The theoretical expressions in section 4.4.2 for the output voltage reveal that the harmonic distortion components have frequencies which are sums or differences between multiples of the output and input frequencies. Strictly speaking, these are not harmonic frequencies since they are not multiple of the input frequency alone. However, for the sake of convenience, they are generally referred as harmonic distortion components. The general theoretical expression for the output voltage of a cycloconverter with the non-circulating current operation shows [3] the following spectrum of harmonic frequencies for 3-, 6- and 12-pulse cycloconverters.

3-pulse : $f_{0H} = | 3(2P-1) f_s \pm 2nf_0 |$

and $f_{0H} = | 6Pf_s \pm (2n+1) f_0 |$

6-pulse : $f_{0H} = | 6Pf_s \pm (2n+1) f_0 |$

12-pulse : $f_{0H} = | 12Pf_s \pm (2n+1) f_0 |$

where

f_{0H} is the harmonic component in the output voltage,

f_s is the supply frequency,

f_0 is the output frequency,

P is any integer from 1 to infinity,

n is any integer from 0 to infinity.

It is very interesting to note that the spectrum of harmonics present in the output voltage is dependent upon the ratio of output to input frequencies. For certain ratios, the order of the harmonic may be less than or equal to the desired output frequency. All such harmonics are known as subharmonics since they are submultiples of supply frequencies. The successive cycles of the output voltage may not be the same. The harmonic components whose order is more than the desired output frequency are known as superharmonic components. The subharmonics are potentially objectionable since the filtering of these harmonics poses problems. As the ratio f_0/f_s increases, the subharmonic spectrum increases, and the output voltage waveform deteriorates. Therefore the ratio f_0/f_s is limited to a value at which the output voltage of tolerable waveform is obtained. The superharmonics may not be as objectionable as the subharmonics since their amplitudes are relatively less. In our discussion of cycloconverters, the source impedance has been neglected. If the source impedance is also taken into account, there is a further distortion in the output voltage which may generate some more harmonic frequencies. Since all these harmonics are inherently present depending upon the voltage ratio, they are generally referred to as 'Necessary distortion terms' [3].

In the case of circulating current operation, the same spectrum is present in the output voltage. Unlike in the non-circulating current operation where the harmonic components extend up to infinity, the harmonic frequencies terminate at a definite term in the circulating current operation. The termination points for 3-, 6- and 12-pulse converters are:

3-pulse : $n \leqslant 3(2P - 1) + 1, (2n + 1) \leqslant (6P + 1)$

6-pulse : $(2n + 1) \leqslant (6P + 1)$

12-pulse : $(2n + 1) \leqslant (12P + 1)$

As a guideline, the ratios of output to input frequencies are limited to 0.3, 0.5 and 0.75 for 3, 6 and 12 pulse cycloconverters.

4.6.1.2 *Unnecessary Distortion Terms*

These distortion terms arise if the relationship between the reference voltage and the mean output voltage is not linear. The spectrum of frequencies generally consists of integral multiples of desired output frequency. For output frequencies much less than the input frequency, the transfer characteristic between output and reference voltages is almost linear with the sinusoidal reference voltages and cosinusoidal modulating waves. These unnecessary distortion terms are absent under such conditions. These are, however, present with other type of control signals.

4.6.1.3 *Practical Distortion Terms*

These arise when there are practical imperfections in the control circuits. Non-linear characteristics of thyristors also contribute to the practical distortion terms. The terms arising by the former are reduced by employing integral control or end-stop control in the closed-loop control schemes [1]–[3]. The practical distortion terms are quite significant specifically at relatively low levels of output voltage.

4.6.2 Supply Current Waveform

It has been pointed out in section 4.3 that the cycloconverter circuit is basically a switching circuit. An output voltage of the desired frequency is fabricated by switching segments of supply voltages selectively across the load circuit. As pointed out earlier the exact waveshape of the output voltage depends upon several factors. The amplitude of the load current depends upon relative level of the output voltage and load impedance. Using sinusoidal reference signals and cosinusoidal modulating signals, the mean output voltage becomes almost sinusoidal if the converters of large pulse number are employed. The output current also becomes sinusoidal and its phase with respect to the mean output voltage is solely determined by the load phase angle. In this section, it will be examined how the load current is reflected on the input supply system. Cycloconverters supplying single-phase output as well as three-phase output are considered. Although the load current may be assumed to be sinusoidal, the supply current is more complex being made up of pulses. If the thyristors are assumed as ideal switches with no losses, the net instantaneous power at the input terminals must always be equal to the net instantaneous power at the output terminals. For a single-phase cycloconverter supplying a single-phase load, the fluctuations of the instantaneous output power is transmitted directly to the AC supply system. The instantaneous power can be easily shown to be

$$p_0 = V_0 I_0 \cos \phi_0 - V_0 I_0 \cos (2\omega_0 t + \phi_0)$$

where

V_0 is the rms output voltage,

I_0 is the rms output current,

ω_0 is the output frequency in rad/sec,

ϕ_0 is the load phase angle.

Because of fluctuations of the instantaneous output power, the harmonic components of the input current have frequencies, $(f_s + 2f_0)$ and $(f_s - 2f_0)$. These harmonic components are present independent of the pulse number of the converter since they are produced due to the transfer of instantaneous power from the input to the output. The actual output voltage waveforms as depicted in Figs. 4.9 to 4.13 are not sinusoidal. Neither the actual output current is sinusoidal. The actual voltage waveforms depend upon the pulse number of the constituent converters. Evidently, there are other input harmonic currents which will depend upon the pulse number. In the case of a cycloconverter supplying a balanced three-phase load, the net instantaneous power is the sum of three-individual fluctuating instantaneous powers. The resultant power can be easily shown to be a constant, $3V_0 I_0 \cos \phi_0$, Consequently, the beat frequency harmonic components $(f_s + 2f_0, f_s - 2f_0)$ of the input current are absent in this case. The other harmonic currents which are dependent upon the pulse number are however present. Thus for a cycloconverter supplying a balanced three-phase load, the net harmonic content in the input current is much reduced as compared to that of a single-phase load. In general, the supply current consists of three components:

1. Inphase component, 2. Quadrature component and 3. Harmonic component.

The inphase component mainly depends upon the mean or active power output. In a phase controlled converter, the relative level of the output voltage decides the quadrature component. Since the output DC voltage is controlled by the firing angle the quadrature component depends upon the firing angle. This component vanishes when the firing angle is zero. However, in the case of a cycloconverter, each output voltage cycle is formed by the oscillation of the firing angle. The mean value of this hypothetical firing angle will always be more than zero even for the maximum output voltage. As the level of the reference voltage decreases, the net average firing angle over one cycle increases. The supply current of a cycloconverter always consists of a quadrature component which is a function of the voltage fabrication technique. So long as the thyristors are line commutated, this quadrature component is always of lagging nature quite independent of whether the load displacement factor is lagging, leading or unity. The factors which contribute to the presence of harmonics, more correctly distortion components, has been explained in section 4.6.1.

4.6.3 Input Power Factor

The input performances of a cycloconverter such as input displacement angle, displacement factor or fundamental power factor, supply current distortion factor, supply power factor are defined similar to those for a converter. The expressions given in section 3.2.2 for a converter are therefore valid for a cycloconverter also. However, the harmonic factor is relatively complex for a cycloconverter owing to the fact that the harmonic frequencies are not simple multiples of supply frequency alone. Each distortion component will have to be considered [2].

The rms inphase component of the current at the input of the cycloconverter is obtained by equating the real components of the input and output power. The output power P_0 is

$$P_0 = q.r.S.V_{0p} I_0 \cos \phi_0$$

where q is the number of output phases

r is the voltage ratio factor

S represents the number of 3-pulse converters connected in series.

I_0 is the load current

ϕ_0 is the load displacement angle

Substituting for V_{0p} from eqn. (4.1),

$$P_0 = q.r.S.\frac{3\sqrt{3}\,V_{ph}}{2\pi} I_0 \cos \phi_0 \quad (4.21)$$

The real input power P_s is

$$P_s = 3V_{ph}I_1 \cos \phi_s = 3V_{ph}I_p \quad (4.22)$$

where I_1 is the fundamental component and ϕ_s is the input displacement angle.

Equating eqns. (4.21) and (4.22), the relation for I_p is

$$I_p = q.r.s.\frac{\sqrt{3}}{2\pi} I_0 \cos \phi_s \quad (4.23)$$

In a similar manner, the quadrature component I_ϕ in the AC line due to the load current I_0 is

$$I_{Q1} = q.r.s\,\frac{\sqrt{3}}{2\pi} I_0 \sin \phi_s$$

In addition to this component I_{Q_1} which is a function of the load displacement factor and output voltage ratio, there is another component I_{Q_2} which arises due to the modulation of the firing angle in the fabrication process of the output voltage. The sum of these two components represents the net quadrature component in the AC line at the supply frequency. This resultant quadrature current is always of lagging nature at the input supply of a line commutated cycloconverter for the load current of leading, lagging, or unity displacement factor. In order to obtain an expression for the resultant quadrature component, a detailed analysis of the input current is required. This is beyond the scope of the book. The interested reader can refer the reference [3] for an expression for the resultant quadrature current $I_Q = I_{Q1} + I_{Q2}$. The input displacement factor is thus given by

$$\cos \phi_s = \frac{I_p}{\sqrt{I_p^2 + I_Q^2}} \tag{4.24}$$

From eqn. (4.23) and the relevant expression for I_{Q2} [3], the input displacement factor can be determined from the expression (4.24). Figure 4.23 shows the variation of the input displacement factor with

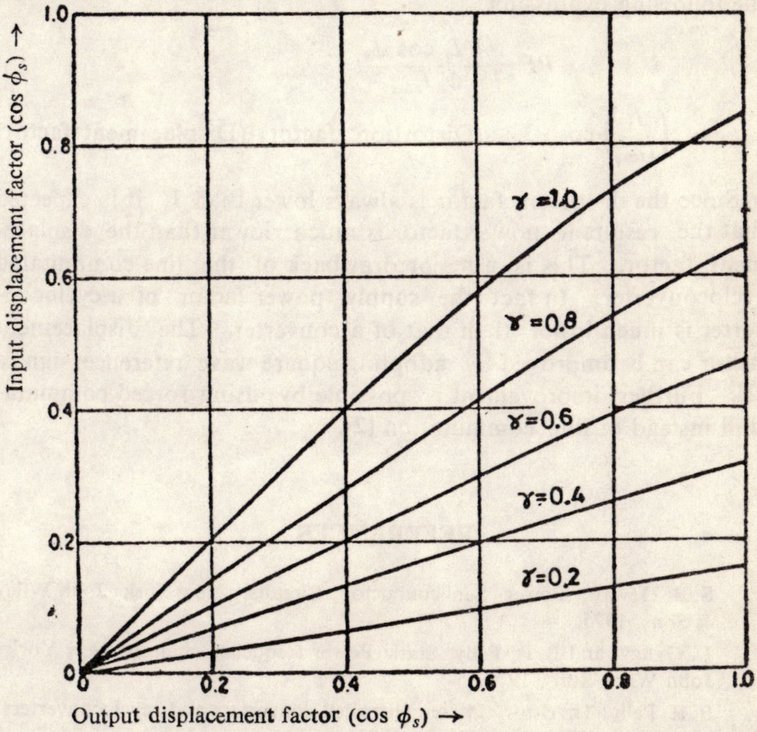

Fig. 4.23. Supply displacement factor versus output displacement factor for various output voltage ratios.

the output displacement factor for different output voltage ratios. These curves are general and as such are applicable for a phase-controlled cycloconverter of any pulse number, supplying both single-phase and balanced three-phase loads. The sinusoidal reference voltages and the cosine modulating signals have been considered while determining these curves. In a phase controlled converter supplying DC load, the maximum displacement factor is unity for maximum DC output voltage. However, in the case of a cycloconverter, the maximum supply displacement factor at the maximum output voltage with an output displacement factor of unity is only 0.843. The displacement factor decreases with a reduction in the output voltage ratio. It increases with an increase in the output displacement factor. The resultant power factor (*PF*) is given by the following expression

$$PF = \frac{V I_1 \cos \phi_s}{V I_{\text{rms}}}$$

$$= \left(\frac{I_1}{I_{\text{rms}}}\right) \cos \phi_s = \text{(Distortion factor)} \cdot \text{(Displacement factor)}$$

Since the distortion factor is always lower than 1, it is expected that the resultant power factor is much lower than the displacement factor. This is a major drawback of the line-commutated cycloconverter. In fact the supply power factor of a cycloconverter is much lower than that of a converter. The displacement factor can be improved by adopting square wave reference signals [2]. Further improvement is possible by using forced commutation instead of line commutation [2].

REFERENCES

1. S.B. Dewan, Power Semiconductor Circuits, New York: John Wiley & Sons, 1975.
2. L. Gyugyi and B.R. Pelly, Static Power frequency changes, New York: John Wiley Sons, 1976.
3. B.R. Pelly, Thyristor Phase-controlled converters and cycloconverters, New York: Wiley-interscience, 1971.
4. A.K. Chattopadhyay and T.J. Rao, "Simplified control electronics for a practical cycloconverter," Int. Journal of Electronics, Vol. 43, No. 2, 1977, pp. 137–150.
5. S.K. Datta, "A novel three-phase oscillator for the speed control of AC motors," IEEE Trans. on Industry and General applications, Vol. IGA-7, No. 2, Jan.-Feb. 1971, pp. 61–68.

6. J.P. Walden and F.G. Turnbull, "Adjustable voltage and frequency poly-phase sine wave generator," IEEE Industrial Application society conference Record, 1974, pp. 1015–1020.

7. D.J. Clarke and P.C. Sen, "A versatile three-phase oscillator," IEEE Transactions on Industrial Electronics and control Instrumentation, Vol. IECI–24, No. 1, Feb. 1977, pp. 57–60.

8. R. Dirr, J. Neuffer, W. Schluter and H. Waldmann, "New Electronic control equipment for Doubly-fed induction motors of high rating," Siemens Review, 3, 1972, pp. 121–126.

5. A.C. REGULATORS

AC regulators are used to get variable ac voltage from the fixed mains voltage. Some of the important applications of ac regulators are: domestic and industrial heating, induction heating in metallurgical industries, induction motor speed control for fan and pump drives, transformer tap changers in utility systems, static reactive power compensators, lighting control etc. Earlier, autotransformers, transformer with taps and magnetic amplifiers were employed in these applications. Now thyristor and triac ac regulators have replaced them in most of the applications because of high efficiency, compact size, flexibility in control etc. They are also suitable for closed loop control because of low control power and fast response. The main disadvantage of these controllers is the presence of harmonics in supply current and load voltage, particularly at lower output voltages.

The present chapter will describe single phase and three phase ac regulators and their performance.

5.1 SINGLE PHASE AC REGULATORS

Figure 5.1 shows three configurations of single phase ac regulators. Figure 5.1 (a) employs two thyristors connected in anti-parallel. Sometimes circuit of 5.1 (b) is employed to reduce the cost. In this circuit isolation between control and power circuit is not required. In the case of circuit of Fig. 5.1 (a) isolation is a must because the cathodes of two thyristors are not connected to a common point. Circuit of Fig. 5.1 (c), which employs triac, is used in low power applications where the load is either resistive

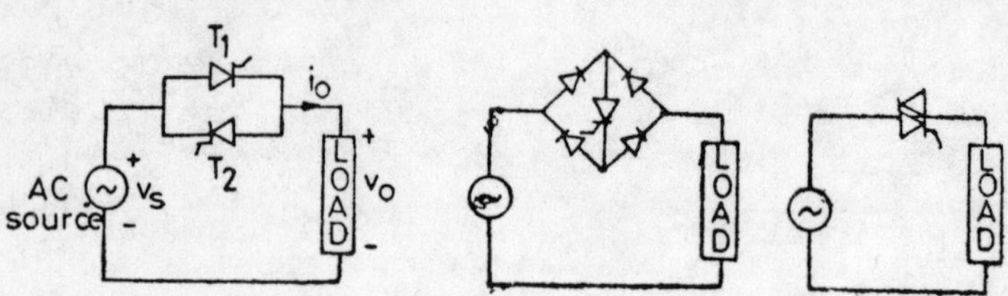

Fig. 5.1. Single phase AC regullaors.

or only has a small inductance. In this case also control circuit need not be isolated from power circuit.

5.1.1 Single Phase AC Regulator with Resistance (R) Load

Figure 5.2 shows an ac regulator feeding a R load. Two thyristors connected in antiparallel have been employed. Gate pulses, load voltage and load current waveforms are also shown in the figure. Thyristor T_1, which is forward biased during the positive half cycle, is triggered at an angle α. It conducts from α to π applying source voltage to the load.

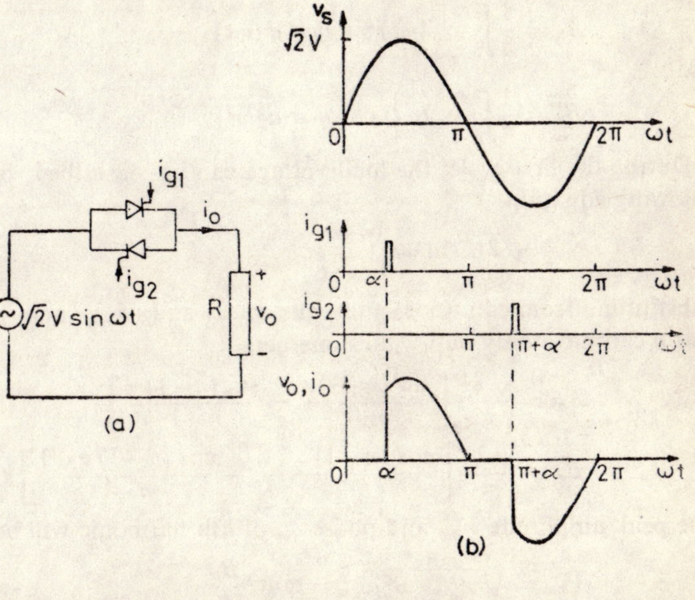

Fig. 5.2. AC regulator with R load.

At π current falls to zero and then T_1 is subjected to a reverse bias and turns off. Thyristor T_2, which is forward biased during negative half cycle, is triggered at $(\pi + \alpha)$ and conducts up to 2π. After commutation each thyristor is subjected to a reverse bias of duration π, for example T_1 is subjected to a reverse bias equal to supply voltage from π to $(\pi + \alpha)$ and equal to the drop across T_2 (conducting) from $(\pi + \alpha)$ to 2π. Since, for all the values of α, each thyristor is reverse biased for a duration of π, there is no restriction on the maximum value of α. It can be controlled up to π. As α is varied from 0 to π the rms value and the fundamental component of the load voltage change from V to 0.

Figure 5.2 (b) shows that the load voltage v_o is non-sinusoidal. It can be described by a Fourier series. Because of the half wave symmetry, the dc component and the even harmonics will be absent, and therefore, for this case v_o can be represented by the following series.

$$v_0 = \sum_n^{\infty} A_n \sin n\omega t + \sum_n^{\infty} B_n \text{ cost } n\omega t \tag{5.1}$$

where $n = 1, 3, 5, \ldots$

$$A_n = \frac{1}{\pi}\int_0^{2\pi} v_0(\omega t) \sin n\omega t \, d(\omega t) \tag{5.2}$$

and

$$B_n = \frac{1}{\pi}\int_0^{2\pi} v_0(\omega t) \cos n\omega t \, d(\omega t) \tag{5.3}$$

During the first cycle, the load voltage can be described by the following equation

$$v_0 = \sqrt{2}V \sin \omega t \Big|_{\alpha,\ \pi+\alpha}^{\pi,\ 2\pi} \tag{5.4}$$

Substituting from eqn. (5.4) into eqns. (5.2) and (5.3) and noting that n can have only odd values one gets

$$A_n = \frac{\sqrt{2}V}{\pi}\left[\frac{\sin(n+1)\alpha}{n+1} - \frac{\sin(n-1)\alpha}{n-1}\right] \tag{5.5 a}$$

$$B_n = \frac{\sqrt{2}V}{\pi}\left[\frac{\cos(n+1)\alpha - 1}{n+1} - \frac{\cos(n-1)\alpha - 1}{n-1}\right] \tag{5.5b}$$

The peak amplitude V_{nm} and phase ϕ_n of nth harmonic will be

$$V_{nm} = \sqrt{A_n^2 + B_n^2},\ \phi_n = \tan^{-1}\frac{B_n}{A_n} \tag{5.6}$$

Equations (5.5) and (5.6) can be used to evaluate the magnitude of harmonics for which $n = 3, 5 \ldots$. It cannot be used to calculate the fundamental component because substitution of $n = 1$ in eqns. (5.5 a) and (5.5 b) leads to undefined expressions. The fundamental component, which has the same frequency as the supply voltage, can however be calculated from eqns. (5.2) and (5.3) by substituting $n=1$ and for v_o from eqn. (5.4). This yields,

$$A_1 = \frac{\sqrt{2}V}{\pi}\left[\frac{\sin 2\alpha}{2} + (\pi - \alpha)\right] \tag{5.7a}$$

$$B_1 = \frac{\sqrt{2}V}{\pi}\left[\frac{\cos 2\alpha - 1}{2}\right] \tag{5.7b}$$

The amplitude (V_{1m}) and the phase (ϕ_1) of the fundamental can now be calculated as follows:

$$V_{1m} = \sqrt{A_1^2 + B_1^2}$$

$$= \frac{\sqrt{2}V}{\pi}\left[\{\tfrac{1}{2}\sin 2\alpha + (\pi - \alpha)\}^2 + \left\{\frac{\cos 2\alpha - 1}{2}\right\}^2\right]^{1/2} \tag{5.8a}$$

$$\phi_1 = \tan^{-1}\left(\frac{B_1}{A_1}\right) = \tan^{-1}\left[\frac{\cos 2\alpha - 1}{\sin 2\alpha + 2(\pi - \alpha)}\right] \tag{5.8b}$$

In some applications, the fundamental component is the only useful component and the harmonics are the undesirable components which adversely affect load performance. It is one of the major limitations of the ac regulator that the variation in the output voltage of supply frequency is obtained with the simultaneous introduction of the harmonics. As an example, in the case of an induction motor useful torque is mainly produced by the fundamental and the harmonics mainly contribute to the losses.

In some applications both the fundamental and the harmonics produce useful power. For example in the case of a heater, heating effect is produced both by the fundamental and the harmonics. In such applications rms value of the output voltage V_0 is of interest. V_0 is described by the following equation,

$$V_0 = \left[\frac{1}{2\pi}\int_0^{2\pi} v_0^2(\omega t)\, d(\omega t)\right]^{1/2} \tag{5.9}$$

Substituting from eqn. (5.4) gives

$$V_0 = V\left[\frac{1}{\pi}\left\{\frac{1}{2}\sin 2\alpha + (\pi - \alpha)\right\}\right]^{1/2} \tag{5.10}$$

Fundamental component, nth harmonic and the rms value of load current are obtained by dividing V_{1m}, V_{nm} and V_0 (eqns. (5.8a), (5.6) and (5.10)) by load resistance R.

In addition to the limitation described above, other important limitations of *ac* regulator are the introduction of harmonics in the source currents and poor power factor. Since the load and source currents have identical waveforms, eqns. (5.5) to (5.10) when divided by R provide expressions for the appropriate quantities of the load and source currents. Power factor is evaluated as given below.

Basic definition of power factor is

$$P.F. = \frac{\text{Real power}}{\text{Apparent power}} \tag{5.11a}$$

If it is now assumed that the source voltage remains sinusoidal in spite of drawing non-sinusoidal current from it

$$P.F. = \frac{VI_1 \cos \phi_1}{VI_{rms}} = \frac{I_1}{I_{rms}} \cos \phi_1 \tag{5.11b}$$

where I_1 = rms fundamental source current
I_{rms} = rms value of source current
ϕ_1 = phase angle between I_1 and V and is given by eqn. (5.8b)

Equation (5.11b) gives the definition of power factor when the source voltage is sinusoidal but the current is non-sinusoidal. For the present case one can derive an expression for power factor directly from eqn. (5.11a) as follows.

$$\text{Real power} = \frac{V_0^2}{R}$$

$$\text{Apparent power} = V \cdot \left(\frac{V_0}{R}\right)$$

Substituting in eqn. (5.11a) gives

$$P.F. = \frac{V_0}{V} \tag{5.12a}$$

Substituting for V_0 from eqn. (5.10) yields

$$P.F. = \left[\frac{1}{\pi}\left\{\frac{1}{2}\sin 2\alpha + (\pi - \alpha)\right\}\right]^{1/2} \tag{5.12b}$$

The maximum values of rms output voltage and current are produced at $\alpha = 0$ and are given by V and V/R respectively. Since at $\alpha = 0$, harmonics are absent, these are also the maximum values of fundamental rms voltage and current. If these quantities are used to normalise various voltages and currents, then from eqns. (5.12a) and (5.10),

$P.F.$ = Normalised output voltage
= Normalised source (or load) current (5.13)

Normalised fundamental component of load current, normalised values of harmonics in load current and power factor are plotted against normalised rms load current in Fig. 5.3 [3]. For low values of the load current the power factor is poor and appreciable amount of harmonics is present. For example at a per unit value of 0.4 of the load current, third harmonic becomes comparable to the fundamental and at a p.u. value of 0.2, fifth and seventh harmonics also have comparable values. Since the source current and load current waveforms are identical, the curves of harmonic components and fundamental of Fig. 5.3 are also applicable to the source current.

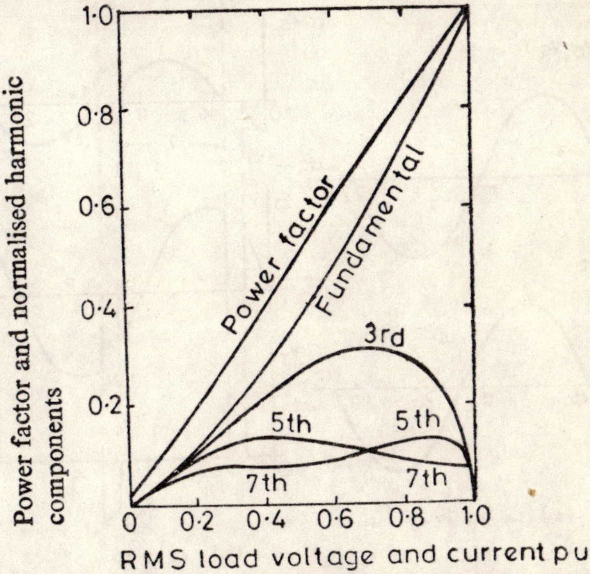

Fig. 5.3. Power factor and normalised harmonic components versus RMS load voltage and current, p.u., for 1 phase ac regulator with *R* load.

5.1.2 Single Phase AC Regulator with Resistance (R)–Inductance (L) Load

Single phase ac regulator with *R-L* load is shown in Fig. 5.4(a). Source voltage, gate current, load current and voltage, and thyristor voltage waveforms are shown in Fig. 5.4(b). The gate signal now consists of a sustained pulse of duration $(\pi-\alpha)$ or a pulse train of the same duration. The reason for using a sustained pulse will be explained later. Current flows through the load

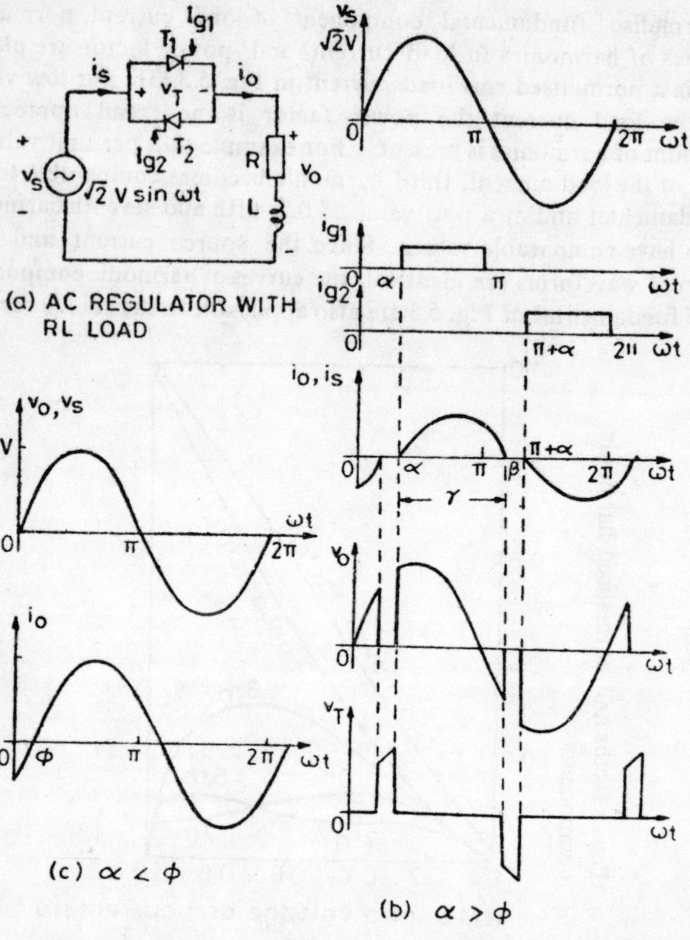

Fig. 5.4. Single phase ac regulator with *R-L* load.

when thyristor T_1 is fired at an angle α. Due to the presence of a inductance, the load current does not become zero at π, but continues to flow until β which is called extinction angle. The load is subjected to the source voltage from α to β. At β thyristor T_1 turns off due to a reverse bias and the voltage across T_2 jumps from a small voltage equal to the thyristor on state voltage to the source voltage, subjecting T_2 to a large rate of change of positive voltage. This may turn on thyristor T_2. This is prevented by employing a proper snubber circuit. Thyristor T_2 is triggered at $\pi + \alpha$ as before. In drawing the waveforms of Fig. 5.4(b) it has been assumed that $(\pi + \alpha) > \beta$. The expression of load current and β are calculated as follows.

For $\alpha \leqslant \omega t \leqslant \beta$,

$$L\frac{di_0}{dt} + Ri_0 = v_0 \tag{5.14}$$

with the initial condition $i_0(\alpha) = 0$ and

$$v_0 = \sqrt{2}V \sin \omega t \tag{5.15}$$

Solution of eqn. (5.14) is of the form

$$i_0 = \frac{\sqrt{2}V}{Z} \sin (\omega t - \phi) + A \exp [-(R/L)t] \tag{5.16}$$

where

$$Z = (R^2 + \omega^2 L^2)^{1/2} \quad \text{and} \quad \phi = \tan^{-1} \frac{\omega L}{R} \tag{5.17}$$

Substituting the initial condition $i_0 = 0$ at $\omega t = \alpha$ gives

$$i_0 = \frac{\sqrt{2}V}{Z}[\sin (\omega t - \phi) - \sin (\alpha - \phi) \exp \{(R/L)(\alpha/\omega - t)\}] \tag{5.18}$$

The load current i_0 again falls to 0 at angle β. Substituting this condition in eqn. (5.18) yields

$$\sin (\beta - \phi) = \sin (\alpha - \phi) \exp \{(R/L)(\alpha - \beta)/\omega\} \tag{5.19}$$

Extinction angle β can be determined by the solution of this transcendental equation.

The duration for which the current flows is called conduction angle. If the conduction angle is denoted by γ then,

$$\gamma = \beta - \alpha \tag{5.20}$$

From eqns. (5.19) and (5.20) one can obtain a relationship between γ and α for a given value of ϕ. γ versus α curves for various values of ϕ are shown in Fig. 5.5.

Figure 5.5 shows that as α is decreased the conduction angle increases. For $\gamma < \pi$, current i_{T1} flows and T_1 remains on from α to $(\alpha + \gamma)$, and current remains zero and T_1 stays off from $(\alpha + \gamma)$ to $(\pi + \alpha)$. Similarly from $(\pi + \alpha)$ to $(\pi + \alpha + \gamma)$ current i_{T2} flows and T_2 remains on and T_1 stays off. If α is progressively decreased to make γ just equal to π then T_1 will be on from α to $(\pi + \alpha)$ and T_2 will be on from $(\pi + \alpha)$ to $(2\pi + \alpha)$. Thus at no

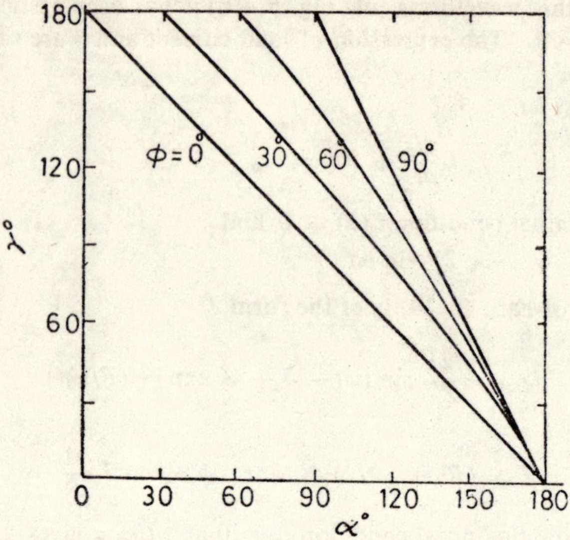

Fig. 5.5. γ vs α curves for AC regulator with *RL* load.

time current will be zero and all the time source will be connected to the load. The load will be subjected to a sinusoidal voltage and the circuit condition will be similar to the one which is obtained when the load is connected directly across the source, provided the thyristor voltage drop is ignored. Now let us imagine a load, with a load phase angle ϕ, connected to a sinusoidal voltage source. When in steady state, the load current will be sinusoidal and lag behind source voltage by an angle ϕ, the current will be positive from ϕ to $(\pi + \phi)$ and negative for the rest of the cycle. If this condition is compared with the ac regulator case under consideration, one can conclude that in the case of ac regulator when α becomes equal to ϕ, γ will become equal to π. This can also be verified from eqns. (5.19) and (5.20). Substituting $\alpha = \phi$ in eqn. (5.19) gives,

$$\sin(\beta - \phi) = 0$$

which has a solution

$$\beta - \phi = \pi.$$

Substituting from eqn. (5.20) gives

$$\alpha + \gamma - \phi = \pi$$

or

$$\gamma = \pi$$

Let us now consider the operation of ac regulator for $\alpha \leqslant \phi$. Suppose that the ac regulator is initially working in steady state for $\alpha = \phi$. Let α be decreased now. When a gate pulse is given to T_1 at α, it is reverse biased due to the drop of T_2. T_2 will continue to conduct until $\omega t = \phi$, and therefore, T_1 will remain reverse biased until this time. Thus T_1 will be able to turn on only at $\omega t = \phi$. As before T_1 will conduct from ϕ to $(\pi + \phi)$ and T_2 from $(\pi + \phi)$ to $(2\pi + \phi)$ and the load voltage and current waveforms will remain the same. Thus reducing α below ϕ does not bring any change in the operation of the ac regulator. Further the control on the output voltage can only be obtained up to $\alpha \geqslant \phi$. Load voltage and current waveforms for $\alpha < \phi$ are shown in Fig. 5.4(c). It should also be noted that reducing α below ϕ does not change the value of γ. It remains equal to π. Because of this reason, the angle γ is held constant at π for all values of α less than ϕ as depicted in Fig. 5.5.

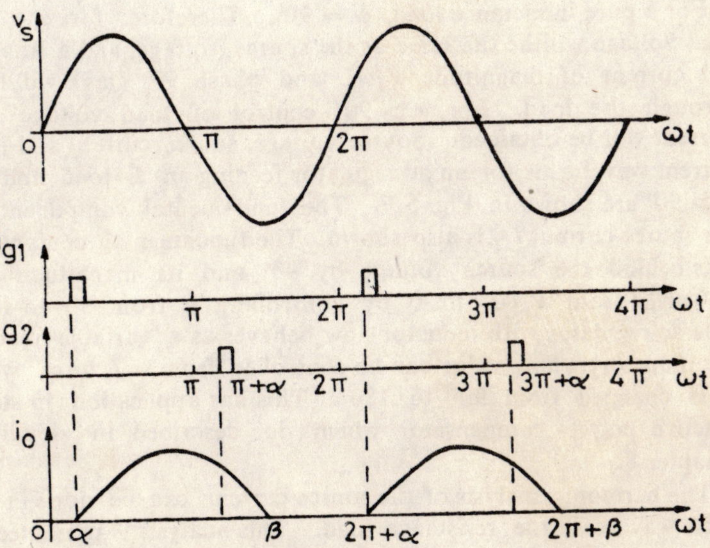

Fig. 5.6. Operation of AC regulator feeding *RL* load for $\alpha < \phi$ and with narrow gate pulses.

The reason for using a sustained gate pulse with a *RL* load can also be explained now. For $\alpha < \phi$, thyristor T_1 will be reverse biased when the gate pulse is applied at α, and therefore, will not turn on. If a narrow gate pulse is used then the thyristor will fail to turn on. However, if a sustained pulse is used, it will turn on at ϕ when it is forward biased. This requirement is further explained with the help of Fig. 5.6. The operation of the circuit is initiated with $\alpha < \phi$. Let T_1 be turned on first. It will turn on at α and the load current will start flowing. The load current will still be flowing at $(\pi + \alpha)$ when T_2 is given the gate pulse. If a narrow gate pulse is used, then T_2 will fail to fire. In the next half cycle T_1 will again conduct. The circuit will essentially work as a halfwave rectifier. To eliminate this problem sustained gate pulse is used with a RL load. In Figure 5.4 (b) it is shown to be present from α to π. It is not necessary to keep it from α to π. The minimum duration should be equal to ϕ plus the angle required for the current to reach the latching current value.

5.1.3 Single Phase AC Regulator with Inductance (L) Load

For a pure inductance load, $\phi = 90°$. Therefore, for $\alpha < 90°$ load voltage will be the same as the source voltage, and a sinusoidal current of magnitude $V/\omega L$ and phase 90° (lag) will flow through the load. For $\alpha > 90°$ control of load voltage and current will be obtained. Source voltage, source current and load current waveforms for an ac regulator feeding an L load and for $\alpha > 90°$ are shown in Fig. 5.7. The fundamental component of the source current i_{S1} is also shown. The fundamental component lags behind the source voltage by 90° and its magnitude can be varied from $V/\omega L$ to 0 by controlling α from 90° to 180°. The ac regulator with inductor now behaves as a variable reactor (or inductor) whose value can be controlled from ωL to ∞ when α is changed from 90° to 180°. This has application in static reactive power compensator which is described in detail in Chapter 8.

The harmonic analysis of the source current can be done in the same way as for the resistance load. This analysis is presented in Chapter 8. In this case the fundamental component of the load current is the only useful component, therefore, variation of the amplitude of the harmonics in p.u. (amplitude of a harmonic/ maximum amplitude of the fundamental, $\sqrt{2}V/\omega L$) is plotted against the per unit amplitude of the fundamental in Fig. 5.8. This figure when compared with Fig. 5.3 shows that the magnitudes of

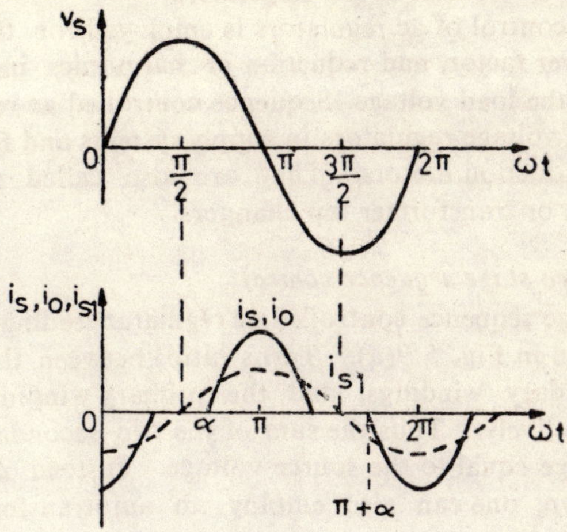

Fig. 5.7. Waveforms of AC regulator of Fig. 5.4(a) with L load (R=0).

all the harmonics are lower with an inductance load compared to a resistance load. With RL load the magnitudes of harmonics lie in between their values with R and L loads.

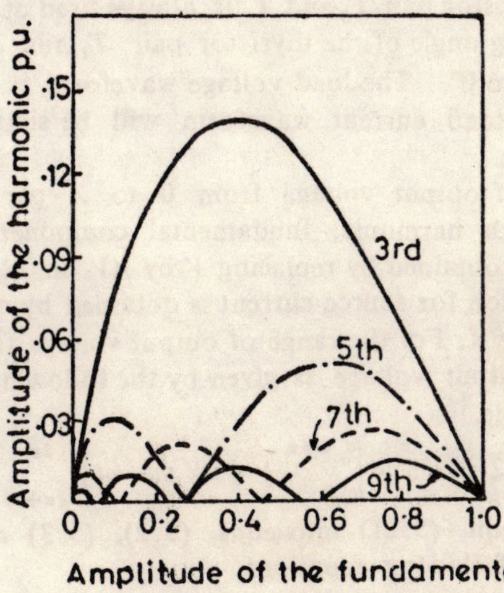

Fig. 5.8. Variation of harmonic currents with the fundamental reactive current.

5.1.4 Sequence Control of AC Regulators

Sequence control of ac regulators is employed for the improvement of power factor, and reduction of harmonics in the source current and the load voltage. Sequence controlled ac regulators are employed as voltage regulators in supply systems and for the speed control of induction motors. They are also called synchronous tap changers or transformer tap changers.

5.1.4.1 *Two stage sequence control*

A two stage sequence controlled ac regulator feeding a resistance load is shown in Fig. 5.9(a). Turns ratios between the lower and upper secondary windings, and the primary winding are A and $(1-A)$ respectively. Thus the sum of the two secondary voltages gives a voltage equal to the source voltage. Instead of the transformer shown, one can also employ an autotransformer with a tapping at A.

(a) RESISTANCE LOAD

For the control of output voltage from 0 to A per unit, the firing angle of thyristor pair T_3 and T_4 is controlled from 180° to 0°. The load voltage and current waveform is identical to that shown in Fig. 5.2(b). For the control of output voltage from A per unit to 1 per unit, thyristor pair T_3 and T_4 is always fired at zero firing angle and the firing angle of the thyristor pair T_1 and T_2 is controlled from 180° to 0°. The load voltage waveform is shown in Fig. 5.9(b). The load current waveform will be similar to that of the load voltage.

For the range of output voltage from 0 to A per unit, the expressions for nth harmonic, fundamental component and rms output voltage are obtained by replacing V by AV in eqns. (5.5) to (5.10). Expression for source current is obtained by multiplying the load current by A. For the range of output voltage from A per unit to 1, the output voltage is given by the following equation during the first cycle.

$$v_0 = \sqrt{2}VA \sin \omega t \Big|_{0,\,\pi}^{\alpha \quad \pi+\alpha} + \sqrt{2}V \sin \omega t \Big|_{\alpha,(\pi+\alpha)}^{\pi,\, 2\pi} \tag{5.21}$$

Substituting from eqn. (5.21) into eqns. (5.2), (5.3) and (5.9), one can derive the following equations.

$$A_n = \frac{\sqrt{2}V}{\pi}\left[(1-A)\left\{\frac{\sin (n+1)\,\alpha}{n+1} - \frac{\sin (n-1)\,\alpha}{n-1}\right\}\right] \tag{5.22}$$

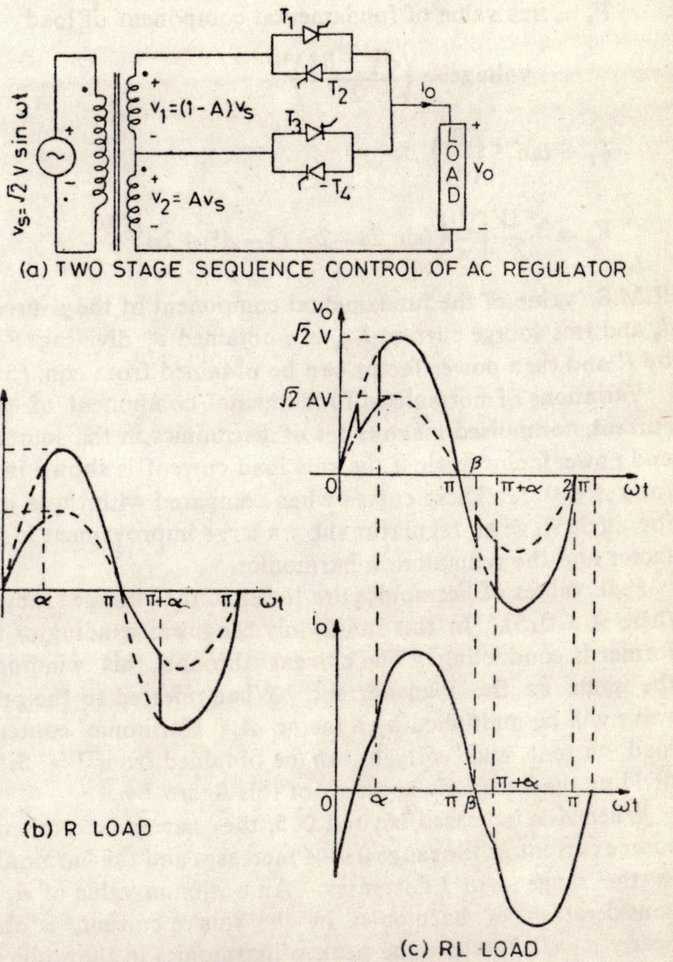

Fig. 5.9. Sequence control of AC regulator.

$$B_n = \frac{\sqrt{2}V}{\pi}\left[(1-A)\left\{\frac{\cos(n+1)\alpha}{n+1} - \frac{\cos(n-1)\alpha}{n-1}\right\} - \frac{1}{n+1} + \frac{1}{n-1}\right] \tag{5.23}$$

$$A_1 = \frac{\sqrt{2}V}{2\pi}\left[(\sin 2\alpha - 2\alpha)(1-A) + 2\pi\right] \tag{5.24}$$

$$B_1 = \frac{\sqrt{2}V}{2\pi}\left[(\cos 2\alpha - 1)(1-A)\right] \tag{5.25}$$

V_1 = rms value of fundamental component of load

$$\text{voltage} = \left(\frac{A_1^2 + B_1^2}{2}\right)^{1/2} \tag{5.26 a}$$

$$\phi_1 = \tan^{-1}\left(\frac{B_1}{A_1}\right) \tag{5.26 b}$$

$$V_0 = \frac{\sqrt{2}V}{2}\left[\frac{1}{\pi}\left\{(\sin 2\alpha - 2\alpha)(1 - A^2) + 2\pi\right\}\right]^{1/2} \tag{5.27}$$

R.M.S. value of the fundamental component of the source current I_1 and rms source current I_{rms} are obtained by dividing V_1 and V_0 by R and then power factor can be obtained from eqn. (5.11 b).

Variations of normalised fundamental component of the source current, normalised magnitudes of harmonics in the source current, and power factor against the rms load current is shown in Fig. 5.10 for a $A = 0.5$. These curves when compared with those of Fig. 5.3 for single stage ac regulator show a large improvement in the power factor and the reduction in harmonics.

Peak values of harmonics are lower in the voltage range 0 to A (here $A = 0.5$). In this range only the lower winding of the transformer is conducting. The current through this winding will be the same as the load current. When referred to the primary, its value will be multiplied by a factor A. Harmonic content in the load current and voltage can be obtained from Fig. 5.10 (b), by dividing the harmonic contents of this figure by A.

When A is increased beyond 0.5, the harmonic content in the source current in the range 0 to A increases and the harmonic content in the range A to 1 decreases. An optimum value of A, from the consideration of harmonics in the source current, is obtained at nearly $A = 0.7$, where the peak of harmonics in the whole range of operation will be minimum. When the control of output voltage is required only in a limited range (greater than 0.5 to 1) the value of A is chosen such that with the thyristor pair T_3 and T_4 fully on, the output voltage is equal to the minimum voltage to be delivered.

For the improvement of power factor and reduction of harmonics, some of the methods described in Section 8.2 for the converter case, can be used for an ac regulator also.

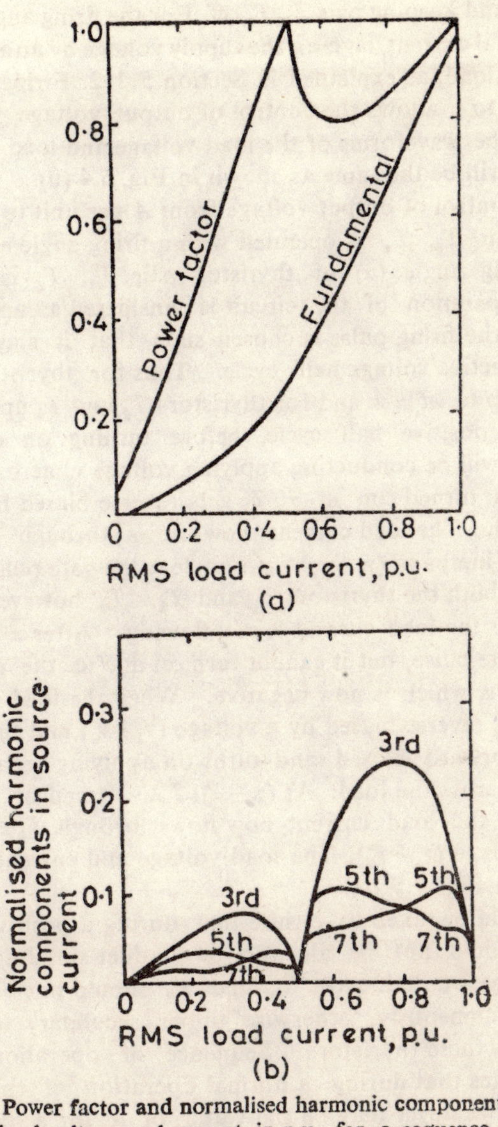

Fig. 5.10. Power factor and normalised harmonic components versus RMS load voltage and current, in p.u., for a sequence controlled AC regulator with *R* load.

(b) *R–L* Load

As in the case of *R* load, the variation of output voltage from 0 to *A* per unit is obtained by controlling the firing angle of thyristor pair T_3, T_4, and keeping pair T_1, T_2 off. For the firing angles less than ϕ, a sinusoidal current, lagging the supply voltage by an angle ϕ, flows through the load, as explained in Section 5.1.2. Firing angle variation from ϕ to π allows the control of output voltage from A per unit to 0. The waveforms of the load voltage and load current, and gate pulses will be the same as shown in Fig. 5.4 (b).

For the control of output voltage from A per unit to 1 per unit, thyristor pair T_3, T_4 is operated with a firing angle equal to zero and the firing angle (α) of thyristor pair T_1, T_2 is controlled. Now the operation of the circuit is considered assuming that the duration of the firing pulse is chosen such that it stays until the end of respective voltage half cycle. Thus for thyristors T_1 and T_3 it stays up to $\omega t = \pi$ and for thyristors T_2 and T_4 up to $\omega t = 2\pi$. During the positive half cycle, before turning on of T_1 at α, thyristor T_3 will be conducting applying voltage v_2 across the load. When T_1 is turned on at α, T_3 gets reverse biased by voltage v_1 and turns off. The load current now flows through T_1 and the load voltage jumps to (v_1+v_2). At angle π the gate pulses are withdrawn from both the thyristors T_1 and T_3. T_1 however continues to conduct as the load current is still flowing. After π thyristor T_4 is given a gate pulse, but it cannot turn on due to the reverse bias applied by v_1 which is now negative. When the load current falls to zero, T_1 is reverse biased by a voltage (v_1+v_2) and turns off. T_4 now gets forward biased and turns on applying negative voltage equal to v_2 across the load. At $(\pi+\alpha)$, T_2 is turned on. This turns off T_4 and the load current now flows through T_2 and the load voltage jumps to (v_1+v_2). The load voltage and current waveforms are shown in Fig. 5.9 (c).

Care should be taken to ensure that during a positive half-cycle T_1 and T_4 should not be allowed to conduct simultaneously and during a negative half-cycle T_3 and T_2 should not be allowed to conduct simultaneously, otherwise upper secondary winding will be shorted by these thyristors. Sequence of operation described above indicates that during a normal operation of the circuit in steady state, whatever may be the value of α, for a positive half cycle, turning on of T_1 will be preceded by the conduction of either T_2 or T_3, and the turning off of T_1 will be only in the negative half cycle. Thus there is no possibility of T_1 and T_4 conducting together in the positive half cycle. In the same way one can also explain that there is no possibility of T_3 and T_2 conducting simultaneously during the negative half cycle during a normal steady state operation. However, the possibility of T_1 and T_4 conducting simultaneously during a positive half cycle, and T_3 and T_2 conducting during a negative half cycle does exist under certain transient conditions. For example let us consider that the ac regulator is in operation giving A per unit output voltage. Here T_1 and T_2 are not fired and T_3 and T_4 are working with $0 < \alpha < \phi$. Now with a view to increase the output voltage to 1 per unit, gate pulses are released to thyristors T_1 and T_2 with a firing angle $\alpha \rightarrow 0$. Let T_4 be conducting when this is done. The load current is now negative. At the beginning of the positive half cycle T_4 is still conducting and T_1 is to be turned on. This will lead to a short circuit of the transformer secondary by T_1 and T_4. This type of situation can be avoided by gradually increasing the output voltage. As a further safety measure the firing angle of thyristor pair T_1 and T_2 can be restricted to values above ϕ. Alternatively one can have the firing circuit such that T_1 is turned on only after the load current has become positive and similarly T_2 is turned on only when the load current has become negative.

5.1.4.2 *Multistage Sequence Control and Sinusoidal Voltage Regulator*

Further improvement in power factor and reduction in harmonics is obtained by using more than two stages of sequence control. Figure 5.11 shows an n stage sequence control. Each transformer winding has a voltage equal to v_s/n. Variation of voltage from say $(n-2)v_s/n$ to $(n-1)v_s/n$ is obtained by firing thyristor pair 3 at $\alpha = 0$ and controlling the firing angle of thyristor pair 2. Thus by choosing any two appropriate adjacent pair of thyristors, the voltage can be controlled from v_s/n to v_s.

If continuous voltage control over a wide range with low harmonic content and improved power factor, is required, then a large number of stages will have to be used. An alternative circuit requiring less number of stages is shown in Fig. 5.12. It employs $(m+1)$ stages $0, 1, \ldots, m$ with the transformer voltage of stages 1 to m being in geometric progression with a common ratio of 2. Thus if v is the voltage of secondary number 1, then the voltage of mth secondary is $2^{m-1}v$. The circuit uses two sets of thyristors, named as control thyristors (T_C) and by-pass thyristors (T_B). Thyristor pairs of first m stages are controlled to get gate

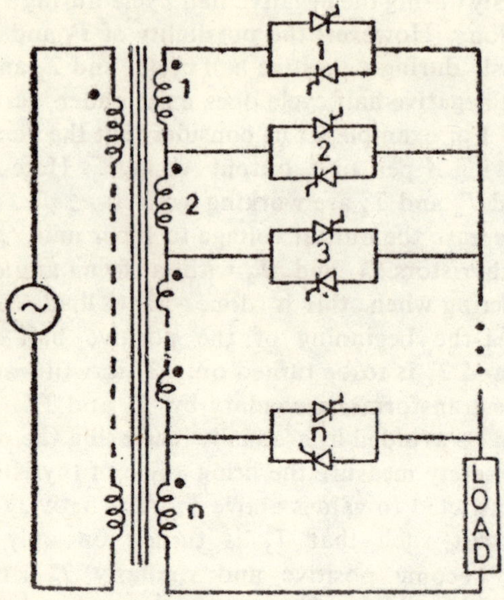

Fig. 5.11. Multistage sequence control of AC regulator.

pulses either throughout the cycle or not at all. Thus essentially they are made to work as switches which are either on throughout the cycle or remain off throughout the cycle. When the control thyristor pair of a particular stage is on and the bypass thyristor pair is off, then the load current is made to flow through the transformer of that stage, thus the voltage of this transformer will appear across the load. On the other hand, when the bypass thyristor pair of a particular stage is on and the control thyristor pair is off, that stage will be simply bypassed and will not contribute any voltage across the load. Thus with this arrangement of thyristors, any series combination of transformer secondaries 1 to m may be achieved and the load voltage can be varied from v to $(2^m-1)v$ in discrete steps equal to v. It should be noted that the sequence control of Fig. 5.11 does not have this flexibility of choosing any series combination of transformer secondaries. This feature is made available in the present circuit by the use of bypass thyristors.

Stage 0 is phase controlled and its voltage can be continuously varied from 0 to v. It is provided to obtain continuous variation between the discrete values of output voltage. This winding will certainly introduce harmonics but their magnitude will be small,

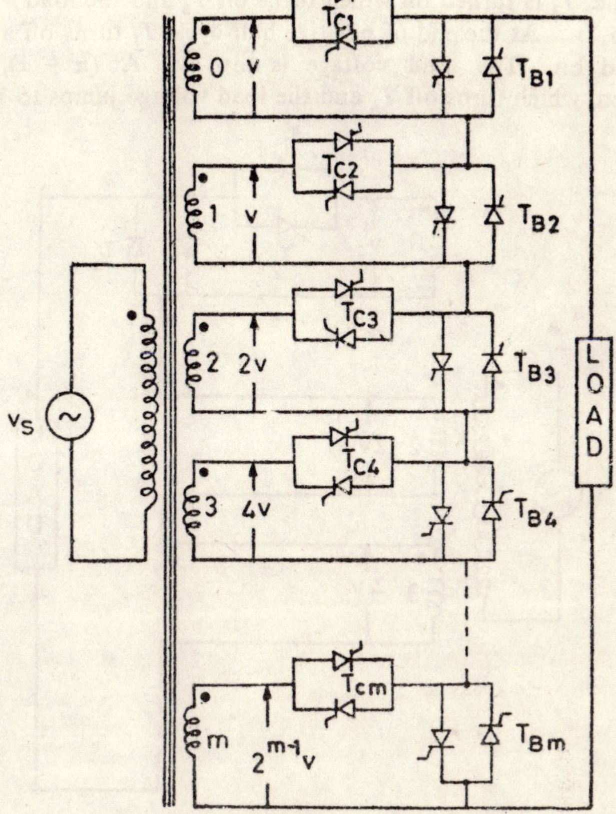

Fig. 5.12. Sinusoidal voltage AC regulator.

because the contribution of this winding to the load voltage is only a small fraction of the total voltage.

The operation of the circuit of Fig. 5.12 can now be elaborated by a specific example of a resistance load. Let the load needs a stepless variation of voltage from $6v$ to $7v$. Stages 2 and 3 together can be used to get a voltage $6v$ and the variation from $6v$ to $7v$ is obtained by using stage 0. For this, the by-pass thyristors of all the stages except those of stages 0, 2 and 3 are given the continuous gate pulses to keep them on all the time. The control thyristor pairs of stages 2 and 3 are also given the continuous gate pulses to keep them on all the time. The circuit now reduces to the one shown in Fig. 5.13. During a positive half cycle thyristor T_3 is turned on at $\omega t = 0$. A voltage equal to $6v$ appears across the load. At α, T_1 is turned on which turns off T_3 and the load voltage jumps to $7v$. At the end of positive half-cycle T_1 turns off and T_4 is turned on. The load voltage is now $6v$. At $(\pi + \alpha)$, T_2 is turned on, which turns off T_4 and the load voltage jumps to $7v$.

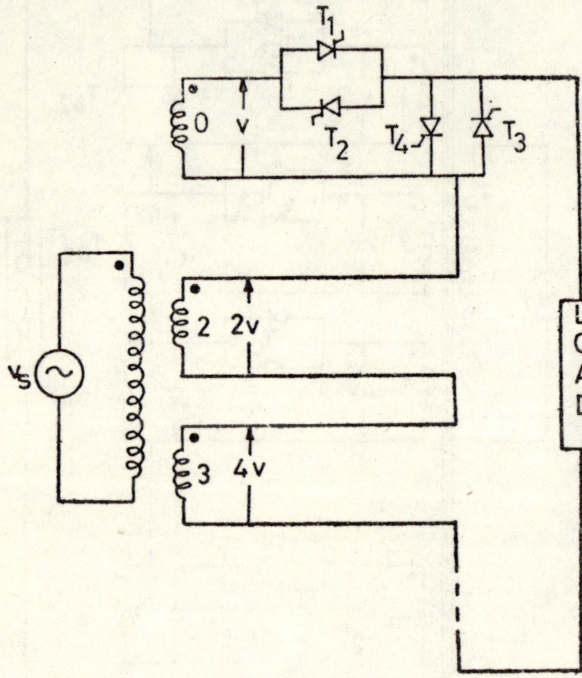

Fig. 5.13. Operation of circuit of Fig. 5.12.

5.2 THREE-PHASE AC REGULATORS

Some three-phase ac regulator circuits are shown in Fig. 5.14. In order to explain the method of analysis, important waveforms and gate circuit requirements, one particular circuit is considered. However, the important performance features and characteristics of all the circuits will be presented.

5.2.1 Important Performance Features [1, 5-8]

Circuits 1 and 2 can be regarded to be a combination of three single phase ac regulators, which operate independent of each other. Therefore, the normalised fundamental and harmonic components of the load current versus the rms load current curves of Fig. 5.3

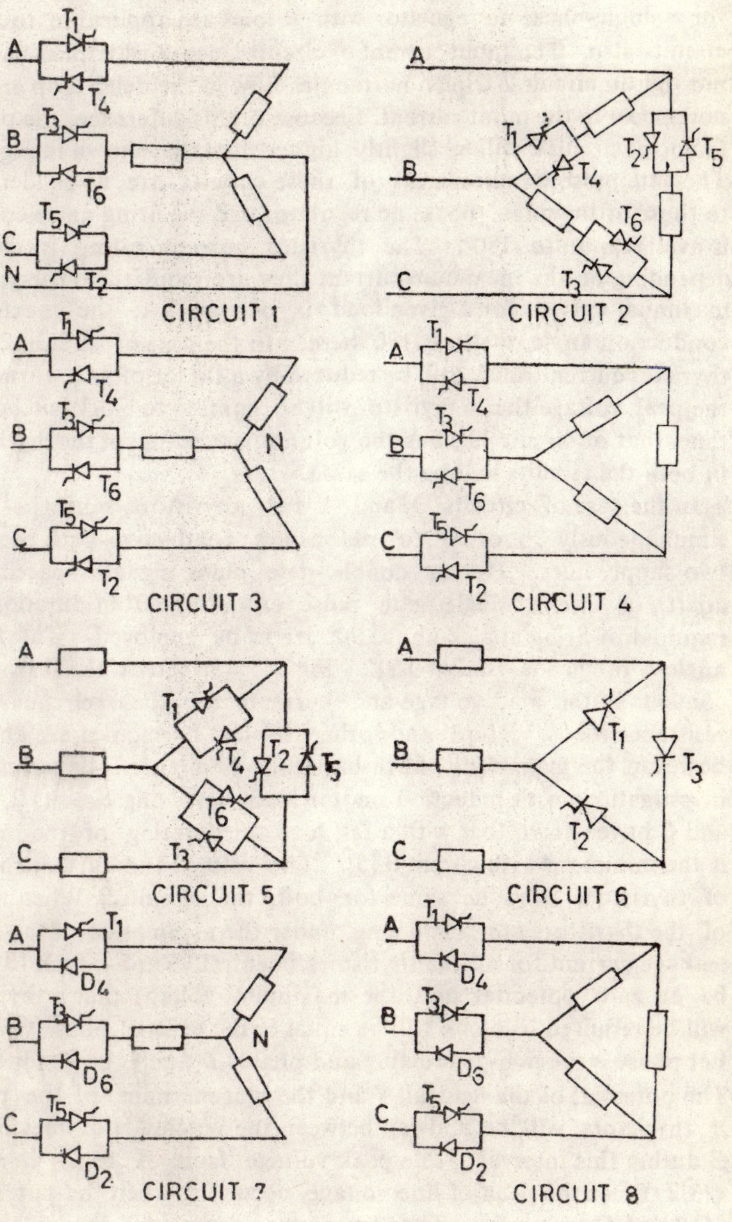

Fig. 5.14. Three-phase AC regulators.

for a single-phase ac regulator with R load are applicable to these circuits also. The input current of circuit 1 contains triplen harmonics but in circuit 2 triplen harmonics flow in the delta loop and do not appear in the input current. Because of this difference, the power factor of circuit 2 will be slightly higher than that shown in Fig. 5.3. The gate pulse requirements of these circuits are also identical to those of the single phase ac regulator and the firing angle control is available up to 180°. The thyristor current rating is chosen depending on the maximum current they are required to carry. The maximum current for a given load is obtained at the maximum conduction angle, which is 180° here. In the case of circuit 2 the thyristor current rating will be reduced by a factor of $\sqrt{3}$. However, the peak voltage that a thyristor will be required to block will be $\sqrt{3}$ times that of circuit 1. Thus the volt ampere ratings of the thyristors in both the circuits will be the same.

In the case of circuits 3 and 4 two thyristors must be fired simultaneously in order to establish a conductive path between two supply lines. Hence, double gate pulse signals spaced 60° apart, or else a single gate pulse exceeding 60° in duration are required, if firing angles above 90° are to be employed. The firing angle range is only 150°. Fig. 5.15 shows the harmonic contents in the load voltage and currents for these circuits with resistance loads. Third and other triplen harmonics are absent, however, the magnitude of 5th harmonic is higher. Experimental investigations with induction motor loads involving circuits 2, 3, 4, and 6 have shown that with a fan load the derating of the motor is the minimum with circuit 3 [5]. The voltage and current ratings of thyristors are the same for both the circuits. When none of the thyristors are couducting, under the assumption of the same leakage current for all the thyristors, neutral N of circuit 3 will be at zero potential and the maximum voltage that a thyristor will be required to block will be equal to the peak of phase voltage. Let phase A be non-conducting and phases B and C be conducting. The potential of the neutral N and the inner terminal of the phase A thyristors will be midway between the potentials of lines B and C during this interval. The peak voltage from A to N is, then $\sqrt{3}/2$ times the peak of line voltage, occurring when the potentials of B and C are equal. Thus for normal operation, the maximum voltage that a thyristor is required to block is $\sqrt{3}/2$ times the peak of line voltage. Under fault condition the voltage across thyristors of circuit 3 may be still higher. Let us imagine that due to device failure thyristor in one of the phases is rendered conductive, then full line to line voltage is applied across the blocking thyristors of the other two phases. If the thyristors should be designed to withstand this condition then the voltage rating of the thyristors of this circuit will be the same as that of circuit 2 while the current rating will be equal to that of thyristors of circuit 1. Thus voltampere ratings of thyristors of this circuit will be $\sqrt{3}$ times that of circuits 1 and 2. What has been said about the voltage ratings of circuit 3 is also true for circuit 4.

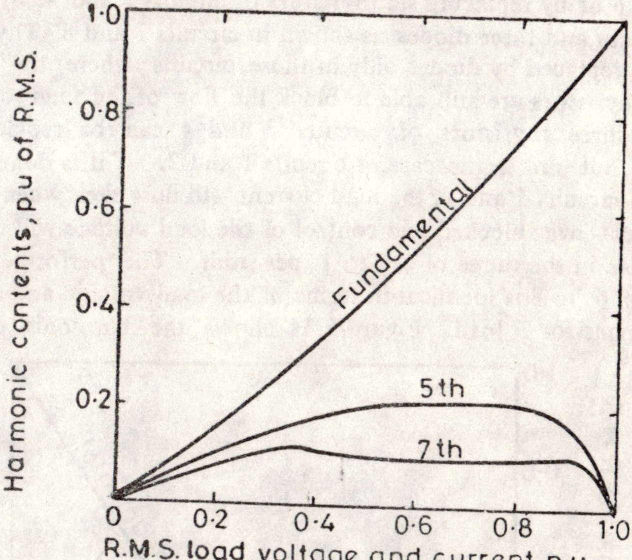

Fig. 5.15. Harmonic content of load voltage and current for 3-phase AC regulators 3, 4 and 5 of Fig. 5.14 for *R* load.

Comparison of circuit 5 with circuit 3 will show that the load currents are the same in both the circuits, only the thyristor currents are different. Harmonic content in the load voltage and current for circuit 5 with a *R* load will be the same as shown in Fig. 5.15 for circuit 3. The important difference between these two circuits is that in the case of circuit 5 current can flow between two lines even when one thyristor is conducting. Thus each thyristor requires only one firing pulse per cycle. The voltage and current ratings of the thyristors of this circuit are nearly the same as that of circuit 2. A substantial amount of triplen harmonic current flows around the inner delta such that with a motor load the thyristors conduct for a maximum of 120° each cycle, whereas maximum conduction angle for circuit 2 is 180°. The average thyristor current is reduced (due to low value of conduction angle) to a minimum but the rms value is slightly increased over that in circuit 2. Both these circuits and also circuit 6 have the drawback that all the six load terminals must be accessible.

Cost of 3-phase ac regulators can be reduced by replacing six thyristors in circuit 5 by only three thyristors as shown in circuit 6 or by replacing six thyristors of circuits 3 and 4 by three thyristors and three diodes as shown in circuits 7 and 8. Thyristors can be replaced by diodes only in those circuits where the remaining thyristors are still able to block the flow of the load current. Thus, three thyristors of circuits 3 and 4 can be replaced by diodes but not in the case of circuits 1 and 2. If it is done in the case of circuits 1 and 2, the load current will flow even when all the thyristors are blocked and control of the load voltage will only be available in the range of 0.5 to 1 per unit. The performance of circuits 6 to 8 is identical in terms of the load voltage and current waveforms for *R* load. Figure 5.16 shows the harmonic content

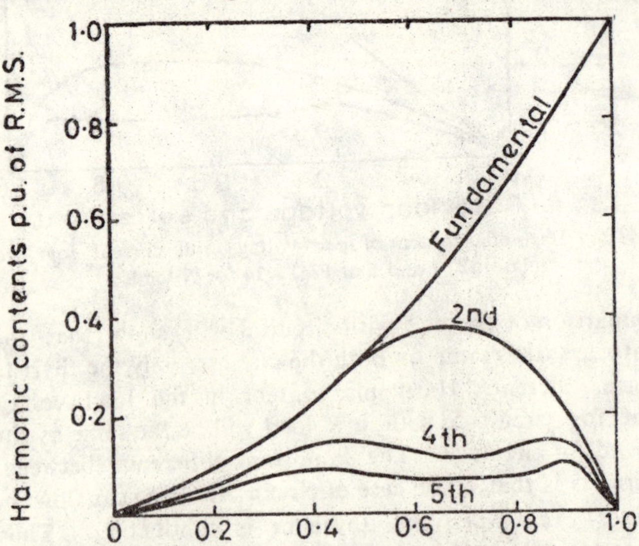

Fig 5.16. Harmonic content of load voltage and current for 3-phase AC regulators 7 and 8 of Fig. 5.14 for *R* loads.

in the load voltage and current for R loads for these three circuits. The main drawback of these circuits is the large harmonic content in the output voltage. In addition to odd harmonics, even harmonics are also present. Predominant harmonic is the 2nd compared to 3rd in the case of circuits 1 and 2, and 5th in the case of circuits 3, 4 and 5. When used with an induction motor load, due to the presence of low frequency harmonics, efficiency is low and considerable derating of the motor takes place, especially at low speeds. Use of diodes in circuits 7 and 8 may give the impression that the dc component might be present but it is not so. In all these circuits variation of firing angle up to 210° is required to get zero voltage and current at the load terminals. Circuits 7 and 8 require three pulses per cycle, one for each thyristor and displaced by 120°. In the case of circuit 6, each thyristor must be provided with two consecutive gate pulses in each cycle, with a phase displacement of 120°.

Though circuit 6 uses only three thyristors as compared to six in the case of circuit 5, each thyristor used in this circuit must be rated approximately two to three times the current rating of a thyristor used in circuit 5, depending on the type of load. Since the voltage ratings are the same, total volt-ampere rating of the thyristors of circuit 6 is higher than that of thyristors of circuit 5, and it is nearly the same as that of circuit 2. When derating of the motor is also taken into account, particularly when speed is to be controlled in wide limits, circuit 5 may not be economical.

In those low power applications where harmonics may not adversely affect the load, for example in industrial heating, and induction motor control applications requiring only limited variation of speed, the circuits 6 to 8 may be used. For large power applications circuits 1 to 5 may be employed.

In addition to the circuits shown in Fig. 5.14, asymmetrical circuits involving back to back connected thyristors in one or two lines and with Y or Δ connected loads have also been used. In such circuits excessively large currents flow in two lines. This renders them unsuitable for most applications.

5.2.2 Analysis of Three-phase AC Regulators

In the case of circuits 1 and 2 of Fig. 5.14, each antiparallel connected thyristor pair appears either directly across a line and supply neutral or directly across two lines. The load current in each phase can also be controlled independent of the other phases. These circuits can therefore, be analysed in the same way as a single-phase ac regulator and all the data presented earlier for harmonics etc. is applicable to these circuits also.

Circuits 3 to 8 of Fig. 5.14 are such that the load current in a phase cannot be controlled independent of the other phases, and therefore, they cannot be analysed as single phase regulators.

In the case of circuits which allow the load current to flow even with one thyristor on, for example circuits 1 and 2, and 5 to 8, of Fig. 5.14, each thyristor requires only a single pulse per cycle. In the case of circuits 3 and 4, ability of a thyristor of any pair to supply the load current depends on the status of thyristors of other pairs. For example, T_1 when fired will be able to supply load current only when at least one of the thyristors from among T_2 and T_6 is on. If T_6 has already ceased to conduct and T_2 is reverse biased then T_1 will be able to conduct with T_2 if it is given a sustained gate pulse or a second gate pulse, when it gets forward biased. Therefore in these circuits each thyristor is either given a sustained gate pulse or two pulses per cycle. This will be further explained later in this section.

In order to illustrate the method of analysis of three phase ac regulators, example of circuit 3 Fig. 5.14 will be taken. The analysis will aim at determining thyristor gate pulse requirement and the load voltage and current waveforms. As the analysis with a *RL* load is relatively difficult, the analysis will be presented only for a *R* load. This circuit with a *R* load is redrawn in Fig. 5.17(a).

The phase voltages and the line voltages of the source are drawn in Fig. 5.17(b). The origin has been chosen to coincide with zero of v_{AB}. Thus

$$v_{AB} = \sqrt{2}V \sin \omega t \tag{5.28}$$

The line voltages v_{BC} and v_{AB} lag behind v_{AB} by $2\pi/3$ and $4\pi/3$ radians respectively. And

$$v_{AN} = \frac{\sqrt{2}V}{\sqrt{3}} \sin\left(\omega t - \frac{\pi}{6}\right) \tag{5.29}$$

The phase voltages v_{BN} and v_{CN} lag behind v_{AN} by $2\pi/3$ and $4\pi/3$ radians respectively.

For symmetrical operation of the circuit, the firing pulses of the thyristors in the three branches must have the same sequence and phase displacement as do the source voltages. Thus the firing

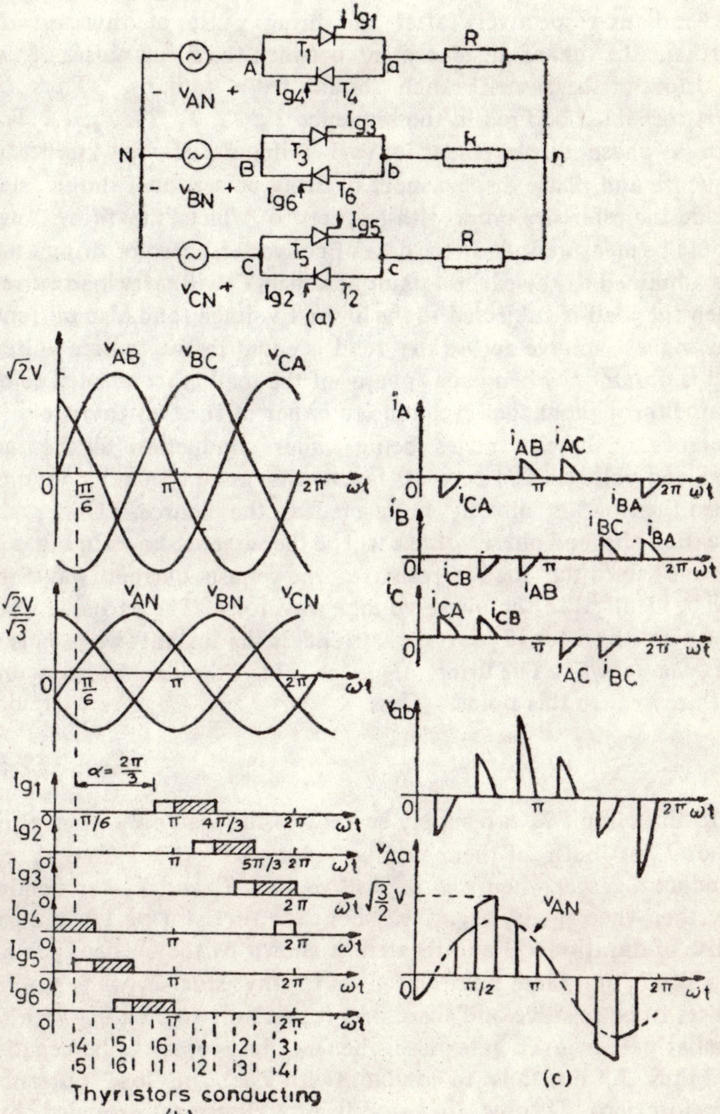

Fig. 5.17. Three-phase AC regulator with R load and its waveforms for $\alpha = 120°$.

pulses to thyristors T_3 and T_5 should be delivered $2\pi/3$ and $4\pi/3$ radians respectively after the firing pulse of thyristor T_1. Further, the phase displacement between the firing pulses of two thyristors of the same branch should be π radians. Thus the thyristors must be fired in the sequence T_1, T_2, T_3, T_4, T_5, T_6, T_1... with a phase displacement of $\pi/3$ radians. Having known the sequence and phase displacement of firing pulses, one should next decide the reference point with respect to which the firing angle should be measured. As in the case of converters, zero of firing angle α is obtained as the earliest point at which T_1 will carry load current when the load is subjected to the highest voltage (and also current). The highest voltage across the load is equal to the source voltage and is obtained when each phase of the load is connected to the source throughout the cycle due to either of the two thyristors of each of the three branches being under conduction all the time. The load voltages and currents in this case are the same as obtained when the load is directly connected to the source. Under this situation the load phase voltage will be the same as shown in Fig. 5.17 (b) and since the load is resistive, the phase current waveform will be identical to the phase voltage waveform. The earliest instant at which thyristor T_1 carries current is the instant when i_A is 0+ i.e., $\omega t = \pi/6$. The firing angle should therefore be measured with respect to this point. Thus

$$\alpha = \omega t - \frac{\pi}{6} \tag{5.30}$$

In this circuit T_1 can supply current to the load only when either T_6 or T_2 or both of them conduct along with it. T_1 and T_6 can conduct together when v_{AB} is positive and T_1 and T_2 can conduct together when v_{CA} is negative. Let each thyristor be given a gate pulse of duration $\pi/3$ and located as shown by the shaded parts in Fig. 5.17 (b). Gate pulse is given to thyristor T_1 when v_{AB} just ceases to be positive and therefore, it just fails to conduct with T_6. Similarly T_2 is given gate pulse when v_{CA} just ceases to be negative, and thus, T_2 just fails to conduct with T_1. The load current is therefore zero. If now thyristor firing pulses are extended by a small amount to the left, without changing the end points, then T_6 and T_1, and T_1 and T_2 will conduct together for the duration by which the gate pulses have been extended to the left. Thus the shaded part of the gate pulse represents the highest value of firing angle, which is $5\pi/6$, here. The load voltage and current can be controlled by extending the firing pulses to the left i.e. by reducing the firing angle.

If narrow gate pulses are chosen, instead of those shown in Fig. 5.17 (b), control of the load voltage and current cannot be realised. For the load current to flow, two thyristors, which are forward biased, must receive gate pulses simultaneously. If only one thyristor at a time, receives a gate pulse it will not turn on due to the absence of the load current.

As an alternative to gate pulses shown in Fig. 5.17 (b), one can also use two narrow gate pulses per thyristor per cycle situated $\pi/3$ rad. apart. Another alternative is to use a single gate pulse of fixed duration but greater than $\pi/3$ rad.

The waveforms of the load current, load voltage and thyristor voltage for $\alpha = 2\pi/3$ rad. are shown in Fig. 5.17 (c). Only two thyristors conduct at a time. Line current i_A consists of two pulses in each half cycle. Pulse i_{AB} is produced due to the simultaneous conduction of thyristors T_6 and T_1 and pulse i_{AC} is produced due to the simultaneous conduction of thyristors T_1 and T_2. And

$$i_{AB} = \frac{v_{AB}}{2R} \text{ and } i_{AC} = \frac{v_{AC}}{2R} \tag{5.31}$$

The line currents i_B and i_C have waveforms similar to that of i_A but displaced by $2\pi/3$ and $4\pi/3$ rad. respectively. The load phase voltage waveforms are identical to the line current waveforms in shape and are given by

$$v_{an} = Ri_A,\ v_{bn} = Ri_B \text{ and } v_{cn} = Ri_C \tag{5.32}$$

The load line voltage waveforms can be obtained from eqn. (5.32). For example

$$v_{ab} = v_{an} - v_{bn} = Ri_A - Ri_B \tag{5.33}$$

Fig. 5.17 (c) also shows the voltage waveform across the thyristor pair T_1 and T_4. When none of the thyristor is in conduction, the pair will be subjected to a voltage v_{AN}. When thyristors of the other two branches are in conduction the voltage across the pair v_{Aa} is obtained as follows:

Since current in phase *an* of the load is zero and thyristor pairs T_3, T_6 and T_2, T_5 are in conduction

$$v_{an} = v_{Bb} = v_{Cc} = 0 \tag{5.34}$$

As the load phases *bn* and *nc* carry the same current

$$v_{bn} = -v_{cn} \tag{5.35}$$

For the mesh consisting of the load branches *bn* and *cn*

$$v_{BC} = v_{bn} + v_{nc} \tag{5.36}$$

From eqns. (5.35) and (5.36)

$$v_{bn} = v_{nc} = \frac{v_{BC}}{2} \tag{5.37}$$

For the mesh consisting of the load phases *an* and *cn*

$$v_{Aa} = -v_{an} - v_{nc} - v_{cC} - v_{CA} \tag{5.38}$$

From eqns. (5.34) and (5.38)

$$v_{Aa} = -v_{nc} - v_{CA}.$$

Substituting the expressions for the line voltages v_{BC} and v_{CA} gives,

$$v_{Aa} = -\frac{\sqrt{2}V}{2}\sin(\omega t - 2\pi/3) - \sqrt{2}V\sin(\omega t - 4\pi/3)$$

$$= -\frac{\sqrt{3}}{2}[\sqrt{2}V\cos(\omega t + \pi/3)] \tag{5.39}$$

In this particular case ($\alpha = 120°$), the regulator operates in two modes, the mode when none of the thyristors conduct and the mode when two thyristors conduct at a time. When none of the thyristors conduct, $v_{Aa} = v_{AN}$ and when two thyristors conduct at a time, excluding T_1 and T_4, v_{Aa} is giving by eqn. (5.39). When either T_1 or T_4 conducts $v_{Aa} = 0$. Thus

$$v_{Aa} = v_{AN} \text{ for } 0 \leqslant \omega t \leqslant \frac{\pi}{6};$$

$$\frac{\pi}{3} \leqslant \omega t \leqslant \frac{\pi}{2};$$

$$\frac{2\pi}{3} \leqslant \omega t \leqslant \frac{5\pi}{6};$$

$$\pi \leqslant \omega t \leqslant \frac{7\pi}{6};$$

$$\frac{4\pi}{3} \leqslant \omega t \leqslant \frac{9\pi}{6};$$

$$\frac{5\pi}{3} \leqslant \omega t \leqslant \frac{11\pi}{6}.$$

$$v_{Aa} = -\frac{\sqrt{3}}{2}\left[\sqrt{2}V\cos\left(\omega t + \frac{\pi}{3}\right)\right]$$

$$\text{for } \frac{\pi}{2} \leqslant \omega t \leqslant \frac{2\pi}{3};$$

$$\frac{3\pi}{2} < \omega t \leqslant \frac{10\pi}{6}.$$

and

$$v_{Aa} = 0 \text{ for } \frac{\pi}{6} \leqslant \omega t \leqslant \frac{\pi}{3};$$

$$\frac{5\pi}{6} \leqslant \omega t \leqslant \pi;$$

$$\frac{7\pi}{6} \leqslant \omega t \leqslant \frac{4\pi}{3};$$

$$\frac{11\pi}{6} \leqslant \omega t \leqslant 2\pi \tag{5.40}$$

The thyristors in branch 1 are subjected to a maximum voltage at $\omega t = 2\pi/3$ and $5\pi/3$. The magnitude of this voltage is $(\sqrt{3}/2)$ times the peak of the line voltage.

For $\alpha = 0$, throughout the cycle, three thyristors conduct simultaneously. For the range of α, $0 < \alpha \leqslant \pi/3$ a cycle consists of alternate intervals involving simultaneous conduction of three thyristors and two thyristors. Relative duration of these intervals depends on the value of α. At $\alpha = \pi/3$, the interval involving simultaneous conduction of three thyristors ceases. For α, $\pi/3 \leqslant \alpha \leqslant \pi/2$ a cycle consists of only the interval in which two thyristors conduct simultaneously. For α, $\pi/2 < \alpha < 5\pi/6$, a cycle consists of alternative intervals when two thyristors conduct simultaneously and none of the thyristors conduct. At $\alpha = 5\pi/6$, which is the maximum value of α, none of the thyristor conducts. The interval in which three thyristors conduct simultaneously and the interval in which two thyristors conduct at a time are designated as Mode I and Mode II respectively. The mode involving conduction of none of the thyristors will be designated as Mode III of operation.

Figures 5.18 (b) and (c) show the waveforms for $\alpha = \pi/6$. It consists of Mode I and Mode II. When in Mode II, current flows through two phases of the load, with each phase sharing half of the line voltage. The line current for this interval is obtained by

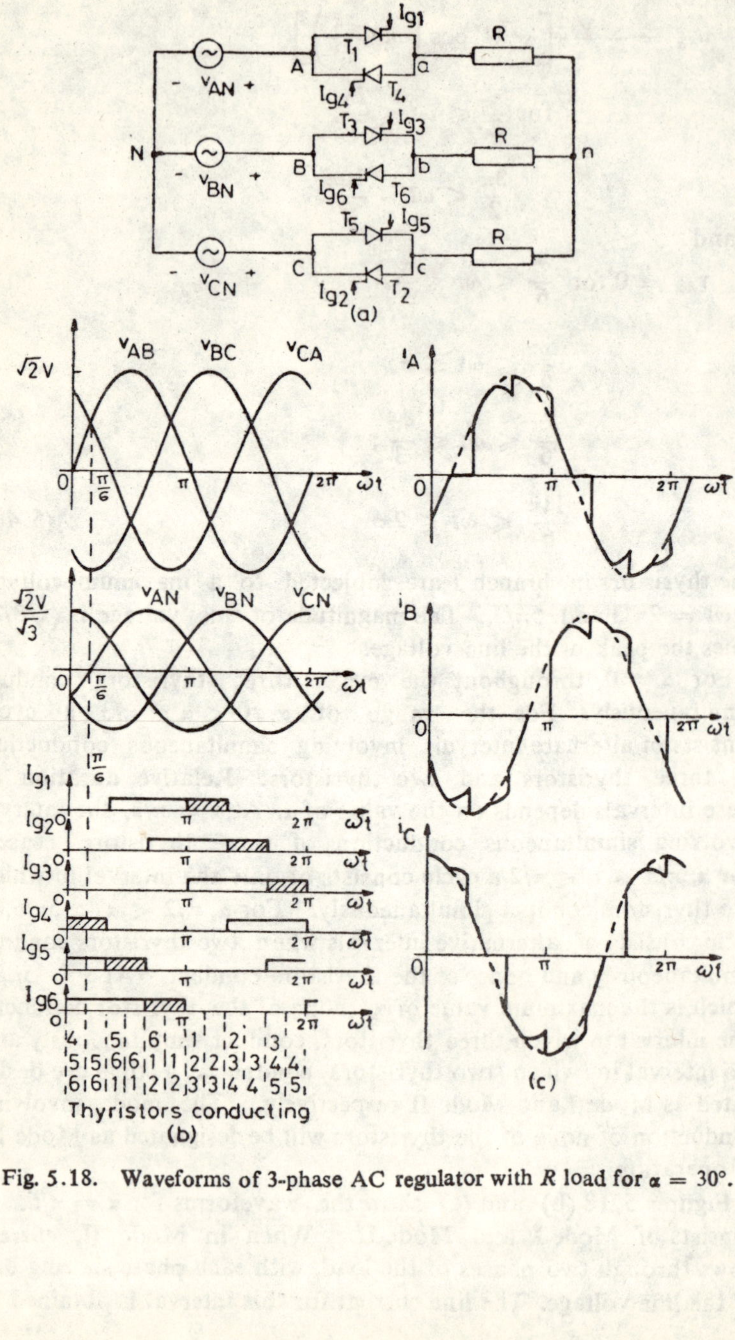

Fig. 5.18. Waveforms of 3-phase AC regulator with R load for $\alpha = 30°$.

dividing the load phase voltage by R. When in Mode I, since three thyristors conduct simultaneously, all the three supply terminals are connected to the load. The line and phase voltage waveforms of the load are identical to those of the source. The load current waveforms (i.e. i_A, i_B and i_C) are identical to the phase voltage waveforms. When in Mode II, the thyristor voltage waveforms are obtained in the same way as for $\alpha = 2\pi/3$ above. When in Mode I, each nonconducting thyristor is shorted by a conducting thyristor and therefore, reverse bias on thyristors is equal to thyristor on state voltage.

At $\alpha = 0$, only mode I is present and the line current waveforms are given by the dotted lines of Fig. 5.18 (c), which is the same as the waveforms of Fig. 5.17 (b).

Once the waveforms are known, the rms values and harmonic content of the load voltage and current can be evaluated. The harmonic content in the load voltage and current is shown in Fig. 5.15.

The above analysis is presented for a R load. With R-L loads, the analysis becomes difficult because the conduction continues even after the instant at which the driving voltage (phase or line voltage of the source) becomes zero. Due to the absence of analytical expressions for load currents, the analysis is done using numerical methods. As in the case of 1-phase circuits, the harmonic content reduces with the increase of ϕ.

5.2.3 Firing Circuits for Three-phase AC Regulators

In 3-phase ac regulators one needs generation of equidistant firing pulses with a phase difference of either 60° or 120° and of appropriate duration. In some cases one may also employ a second set of pulses displaced with respect to the first either by 60° or 120°. These requirements are explained in Sec. 5.2.1. Thus the firing pulse requirements of ac regulators are not very much different from those of converters. Using the same approaches as described for converters in Chapter III one can develop the firing circuits for ac regulators.

REFERENCES

1. W. Shepherd, 'Thyristor control of ac circuits', Lockwood Staples Ltd., England, 1975.
2. S. B. Dewan and A. Straughen, 'Power semiconductor circuits', John Wiley & Sons, New York, 1975.

3. K. A. Krishnamurthy, G.K. Dubey and G.N. Revankar, 'A solid state synchronous tap charger permitting a selective reduction of line harmonics', Int. J. Electronics, vcl. 45, No. 2, 193-202, 1978.
4. N. Mohan, 'Improvement in energy efficiency of induction motors by means of voltage control', IEEE Trans. on PAS, 1982.
5. D. A. Paice, 'Induction motor speed control by stator voltage control', IEEE Trans. Power Apparatus and Systems, vol. PAS-87, pp. 585–590, 1968.
6. W. McMurray, 'A comparative study of symmetrical three-phase circuits for phase-controlled ac motors', IEEE Trans., Industry Applications, vol. IA–10, pp. 403–411, 1974.
7. E. D. Spooner, 'Three phase three-thyristor voltage control scheme', IEEE Trans. Industry Applications, vol. IA–11, pp. 478–482, 1975.
8. S. Rahman and W. Shepherd, 'Thyristor and diode controlled variable voltage drives for 3-phase induction motors', Proc. IEE, vol. 124, pp. 784-790, 1977.

6. CHOPPERS

The choppers are circuits which convert power between a constant voltage d.c. input and a variable voltage d.c. output. The directions of the output voltage and current are restricted according to the configurations of chopper circuits and form a basis for their classification.

In this chapter different chopper configurations using ideal switches are discussed. The relationships between the input and the output voltages under steady state are derived for a *PWM* chopper. The calculations of the ripple voltage and the current based on the Fourier Analysis of the output voltage are given, for continuous current operation. The principle of Time Domain Analysis is discussed through a simple example. Specific practical thyristor choppers using the voltage and the current commutation are discussed. The gating requirements of *PWM* chopper are described and a practical logic and control circuit is given.

6.1 CHOPPER CONFIGURATIONS USING IDEAL SWITCHES

The d.c. choppers consist of power switches and diodes connected in suitable circuit configurations. The circuit cofigurations can be considerably simplified by assuming that the power switches and the diodes are ideal. The resulting ideal chopper circuits are described in this section.

The ideal switches and diodes are assumed to have the following characteristics.

1. The switches have zero resistance in the closed state and infinite resistance in the open state.

2. The switches can be closed and opened instantaneously, independent of the current through them.

3. In the closed state the switches conduct current in one direction only. The resistance to reverse conduction is infinite.

4. The diodes have zero forward and infinite reverse resistance. The threshold voltage is zero.

Due to the unidirectional conduction in the ideal switches and the diodes, the directions of the instantaneous output currents and the polarities of the output voltages are restricted. Consequently, the operation of the ideal choppers is restricted to certain quadrants of the output current–output voltage plane. This property forms a basis for the classification of the choppers in the Single Quadrant, Two Quadrant and the Four Quadrant configurations. The structure of these choppers is discussed in the following.

6.1.1 Single Quadrant Choppers, Type A and Type B

Figure 6.1 shows the Type *A* and Type *B* single quadrant choppers. The *Type A* chopper of Figure 6.1(a) operates by periodic opening and closing of the switch S_1. The switch S_1 conducts only in the direction shown by the arrow i.e. downwards. Similarly, the current through diode D_1 flows upwards only, as shown by the arrow. Thus it is clear that output current i_0 can only be positive. Similarly, diode D_1 limits the output voltage v_0 to be positive. The region of operation in the v_0-i_0 plane is therefore the first quadrant shown hatched in Fig. 6.1(b). The

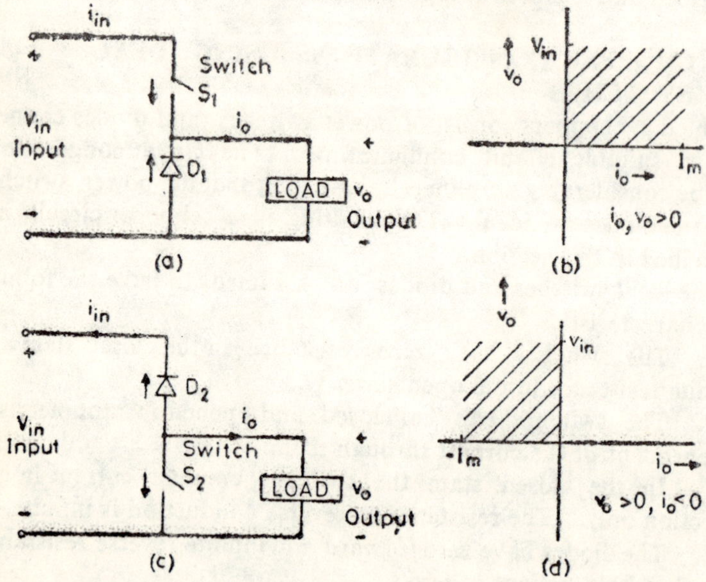

Fig. 6.1. Single quadrant choppers using ideal switches: (a) Type *A* or motoring chopper; (b) Operating diagram for (a); (c) Type *B* or regenerative chopper; (d) Operating diagram for (c). Hatched areas show permissible operating regions.

maximum instantaneous voltage v_0 is limited to V_{in} when S_1 is closed. The maximum output current I_m for all configurations is determined by the ratings of the switches and diodes. The power flow in this chopper is from the input voltages source V_{in} to the load connected at the output. It is also called the 'Step Down Chopper' as the average value of v_0 is always lower than the input voltage V_{in}.

Figure 6.1(c) shows the *Type B* single quadrant chopper in which the current i_0 is restricted to negative values by the switch S_2 and the diode D_2. However, the voltage v_0 is restricted to be positive and lower than V_{in}. A negative value of v_0 will render both S_2 and D_2 open. The region of operation of the chopper in the v_0—i_0 plane is shown hatched in Fig. 6.1(d). It is also called a 'Step Up Chopper' as the power can be transferred only from a lower voltage v_0 to a higher voltage V_{in}. The reversed power flow i.e. from the load to the source can take place only if the load is active i.e. capable of providing continuous power output. A common example of reversed power flow is the regenerative braking in d.c. drives. The d.c. motor in this case acts, for a short duration, like a generator and the kinetic energy stored in the system is returned to the input d.c. source. Other applications are transfer of energy stored in large inductors and stepping up battery voltages to higher values.

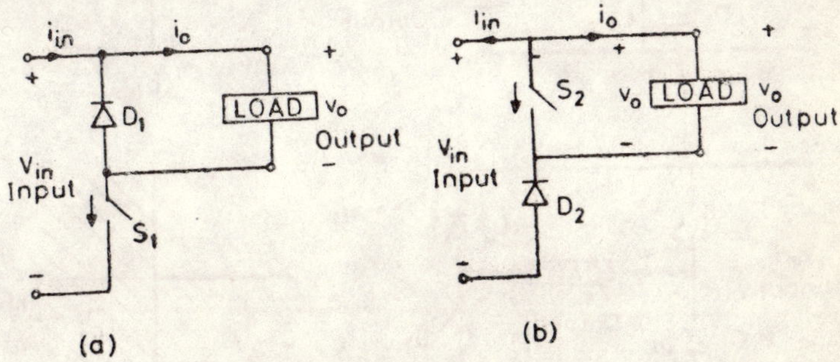

Fig. 6.2. Alternative configurations for the single quadrant choppers of Fig. 6.1: (a) Type *A* chopper; (b) Type *B* chopper.

The single quadrant choppers shown in Fig. 6.1 have a common negative terminal between the input and the output. Fig. 6.2 shows alternative configurations which have a common positive terminal between the input and the output. Otherwise, the configurations of Type *A* and Type *B* choppers shown in Fig. 6.2 are functionally equivalent to the corresponding types shown in Fig. 6.1. A comparison of Fig. 6.1 and 6.2 shows that the Type *A* chopper of Fig. 6.1(a) is identical in structure to the Type *B* chopper of Fig. 6.2(b). Therefore, the changeover from Type *A* to Type *B* chopper may be obtained by simply reconnecting the load to the appropriate terminals with correct polarity. This change can be accomplished by a double pole, double throw switch or a contactor. Similarly it can be seen that the Type *A* chopper of Fig. 6.2(a) is identical in structure to the type *B* chopper of Fig. 6.1(c). The above mentioned changeover between Type *A* and Type *B* choppers is possible only if the load is not required to have a fixed common terminal between the input and the output.

6.1.2 Two Quadrant Choppers, Type C and Type D

The two quadrant chopper configurations can be obtained by

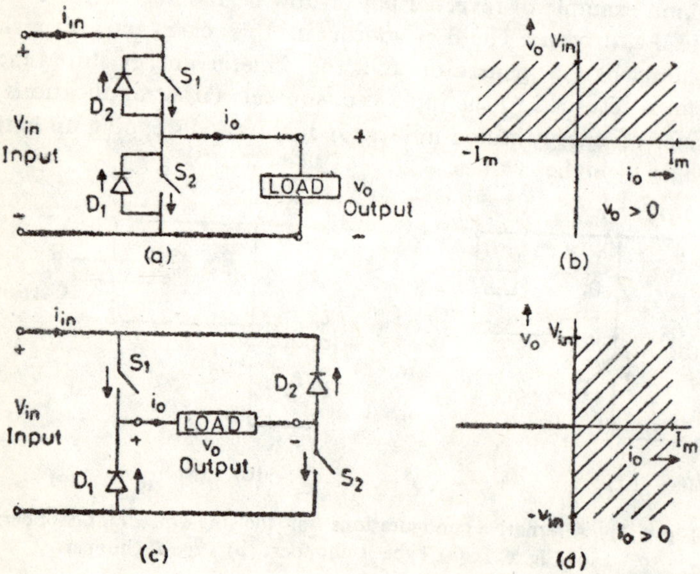

Fig. 6.3. Two quadrant choppers using ideal switches: (a) Type *C* chopper; (b) Operating diagram for (a); (c) Type D chopper; (d) Operating diagram for (a). Hatched area shows permissible operating regions.

connecting the single quadrant choppers of Type A and Type B (Fig. 6.1) in parallel. Figure 6.3 (a) shows the *Type C* chopper. The output voltage v_0 of the Type C chopper is restricted to be positive due to diode D_1. The output current i_0 however can be positive or negative. If i_0 is positive either S_1 or D_1, but not both, is conducting. Similarly if i_0 is negative either S_2 or D_2 is conducting. Figure 6.3 (b) shows the operating region for the Type C chopper. The maximum value of v_0 is limited to V_{in} while the rating of the switches and the diodes limit the magnitude of the current i_0 to I_m. The Type C chopper shown in Fig. 6.3 (a) has a common negative terminal between the input and the output. The load could also be connected between the junction of S_1, S_2

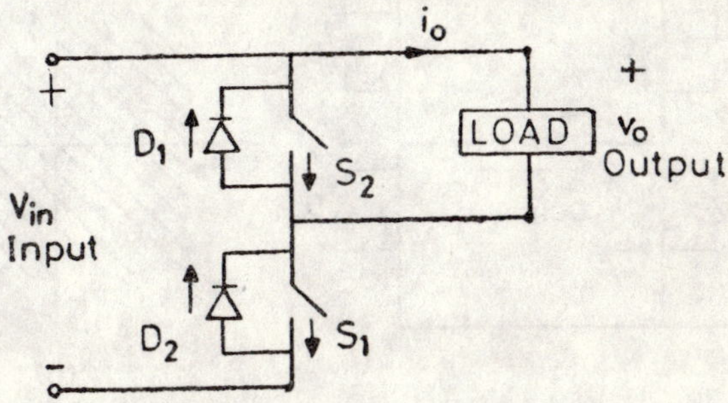

Fig. 6.4. Alternative configuration for the type C chopper.

and the positive terminal of input. This alternative configuration is shown in Fig. 6.4. The configuration of Fig. 6.4 is equivalent to connecting the Type A and Type B choppers of Fig. 6.2 in parallel. The configuration of Figure 6.4 has a common positive terminal between the input and the output. The labels of switches and diodes are interchanged between Figs. 6.3 (a) and 6.4 to preserve common control signals. In the Type C choppers of Figs. 6.3 (a) and 6.4, a simultaneous closure of S_1 and S_2 shorts the input d.c. source and must be prevented by suitable protection circuits.

Another basic two quadrant chopper configuration is shown in Fig. 6.3 (c). In this *Type D* chopper the load is connected differentially between the Type A and the Type B chopper. The Type D chopper of Fig. 6.3 (c) can be reduced to a Type $A(B)$ chopper by fixing one of the switches, i.e. either S_1 or S_2, in the closed (open) position. The direction of the output curren t i_0 in the Type D chopper is restricted to be positive because S_1, D_1, S_2, D_2 conduct only in the directions shown by arrows in Fig. 6.3 (c). A reversal of i_0 renders all devices open. The output voltage V_0 however can be positive or negative. Figure 6.3 (d) shows by hatched area the operating regions of the Type *D* chopper. It should be noted from Figs. 6.3 (b) and (d) that the Type *C* chopper operates in the first and the second quadrant while the Type *D* chopper operates in the first and the fourth quadrant.

6.1.3 Four Quadrant Chopper, Type E

Figure 6.5 shows the circuit configuration of a Type *E* four quadrant chopper. The load is connected differentially between

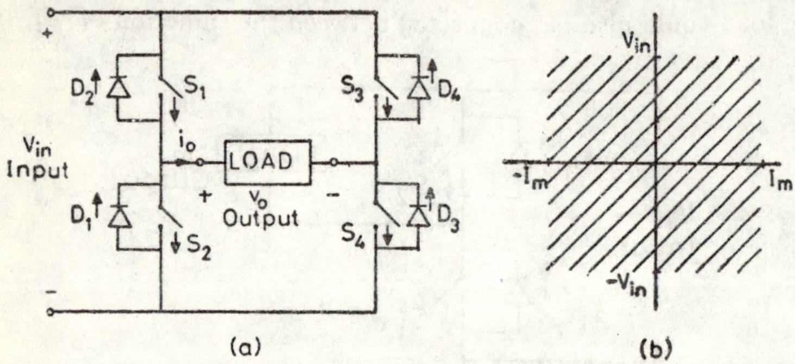

Fig. 6.5. Four quadrant chopper using ideal switches: (a) Type *E* chopper, (b) Operating diagram for (a). Hatched area shows permissible operating regions.

two Type *C* choppers. The Type *E* chopper can also be thought of a parallel combination of two Type *D* choppers (Fig. 6.3) with opposite polarities of current i_0. The operating region of Type *E* chopper is shown by hatched area in Fig. 6.5 (b). The Type *E* chopper can be reduced to a Type *D* chopper by fixing one of the cross-pair of switches S_1, S_4 or S_2, S_3 in the open position. Similarly it can be reduced to a Type *C* chopper by fixing any one of switches in the closed position and keeping its complement in the open position to avoid shorting the d.c. bus.

6.1.4 Isolated Choppers

In the chopper configurations mentioned above the input source and the load have a common terminal. In some applications complete conductive isolation between the source and the load is required. An example of such application is the multiple output switching power supply. In such cases the input d.c. is inverted to a high frequency ac in the range of 400 Hz to 20 KHz. A high frequency transformer provides the isolation and the required voltage ratio. The secondary output of the transformer is rectified to obtain the dc output. The discussion given in the rest of the chapter excludes the isolated choppers.

6.2 CONTROL STRATEGIES

The chopper configurations shown in Figs. 6.1 to 6.5 operate from a fixed dc input voltage V_{in}. The average value of the output voltage, v_0, is controlled by periodic opening and closing of the switches. The various control strategies for operating the switches are described in this section. The operation is illustrated by the waveforms of the Single Quadrant choppers.

6.2.1 Pulse Width Modulation (PWM)

In this strategy the switches are turned on at a constant chopping frequency. Figure 6.6 shows the waveforms of the Type *A* and Type *B* chopper operating under *PWM* control. The output voltage v_0 of the Type *A* chopper (Figure 6.1a) is given by,

$$\begin{aligned} v_0 &= V_{in} \text{ when } S_1 \text{ is closed} \\ &= 0 \text{ when } S_1 \text{ is open} \end{aligned} \tag{6.1}$$

In equation (6.1) it has been assumed that the output current i_0 is continuous. Figure 6.6 (a) shows the resulting waveforms of v_0, i_0 and i_{in} for the Type *A* chopper, of Figure 6.1 (a).

A rectangular train of voltage pulses can be generated at the output by periodically closing switch S_1. The dc component of the output voltage v_0, denoted by V_0, is given by

$$V_0 = V_{in}\,(T_{on}/T) \tag{6.2}$$

In eqn. (6.2) T_{on} is the on-time of the switch S_1 and T is the chopping period. The reciprocal of T is the chopping frequency f. The average or the dc voltage V_0 is varied by keeping T constant and varying the pulse width T_{on}. Ideally the ratio (T_{on}/T) can be varied between zero and unity. Therefore, the average output voltage V_0 can be varied between zero and V_{in}. The ratio (T_{on}/T) is often called the Duty Cycle of the chopper.

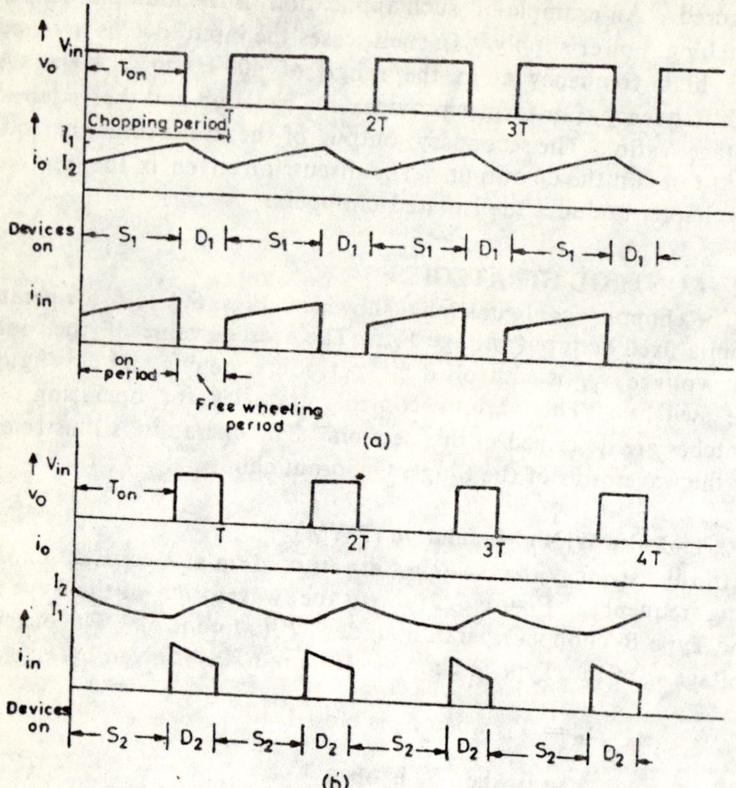

Fig. 6.6. Steady state waveforms for choppers of Fig. 6.1: (a) Type *A*; b) Type *B*.

For the Type *B* chopper of Fig. 6.1 (c), the output voltage V_0 is given by,

$$v_0 = 0 \quad \text{when } S_2 \text{ is closed}$$
$$= V_{in} \text{ when } S_2 \text{ is open} \tag{6.3}$$

In eqn. (6.3) it has been assumed that the output current i_0 is continuous. Figure 6.6 (b) shows the waveforms of v_0, i_0 and i_{in} for the Type *B* chopper of Fig. 6.1 (c) operating under *PWM* control. The average or the dc output voltage is given by

$$V_0 = V_{in} \{1 - (T_{on}/T)\} \tag{6.4}$$

The value of V_0 is varied by keeping T constant and varying T_{on}.

6.2.2 Constant Pulse Width, Variable Frequency

In the *PWM* strategy low values of V_0 are obtained by reducing the on-time T_{on} of the switch. In practical switches T_{on} cannot be reduced below a minimum value, T_{min}. Under these conditions the ratio (T_{on}/T) is reduced by keeping T_{on} constant at T_{min} and increasing T. Hence, in general the output voltage V_0 given by eqns. (6.2) and (6.4) can also be varied by keeping the pulse width T_{on} constant and varying the chopping period T.

6.2.3 Current Limit Control (CLC)

In this strategy the switch S_1 of Type *A* chopper (Fig. 6.1a) is turned on when the output current i_0 equals a preset value I_2. The switch S_1 is kept closed till i_0 increases to another preset value I_1. The switch S_1 is turned off when i_0 equals I_1 and is kept off till the current i_0 decays to I_2 by freewheeling through D_1. The switch S_1 is turned on when i_0 equals I_2 and the cycle is repeated. A similar strategy can be used with switch S_2 in the Type *B* chopper. In steady-state, the current i_0 is thus limited to a maximum of I_1 and a minimum of I_2 as shown in Fig. 6.6. In this strategy the chopping frequency and the pulse width are dependent on the load parameters and are not controlled explicitly. An advantage of the current limit control is that the ripple $(I_1 - I_2)$, in current i_0 can be regulated to a small, predetermined value for a wide range of loads.

6.2.4 Variable Pulse Width and Frequency

The output voltage V_0 of the chopper is determined by the ratio T_{on}/T, (eqn. 6.2, 6.4). Thus both the pulse width T_{on} and the frequency, $1/T$, may be varied to obtain a particular value of (T_{on}/T). In some choppers T_{on} is not explicitly controlled but is determined by the load. The chopping frequency is then varied to obtain the required value of V_0.

6.3 STEADY STATE OPERATION UNDER PWM CONTROL

The Pulse Width Modulation is the most commonly used strategy for the control of dc choppers. In this section, the steady-state operation of the Type *A*, *B*, *C* and *D* chopper under *PWM* control and continuous output current is described. The expressions for the output voltage harmonics are given. Two different *PWM* control strategies for the Type *D* chopper are discussed and the corresponding steady-state waveforms are given.

6.3.1 Type A Chopper

The operation of the Type *A* chopper has been described in section 6.2.1. The steady-state waveforms for the chopper are given in Fig. 6.6(a). The average output voltage V_0 is given by eqn. (6.2). The waveform of the instantaneous output voltage, v_0 is shown in Fig. 6.6(a). As the instantaneous load current i_0 is assumed to be continuous, the waveform of v_0 is completely defined independent of the load parameters and is periodic. Therefore v_0 can be resolved into the following Fourier Series.

$$v_0 = V_0 + \sum_{h=1}^{\infty} v_h, \tag{6.5}$$

where,
$$v_h = \left(\frac{2V_{in}}{\pi}\right)\cdot\frac{1}{h} \sin(h\pi\cdot D)\cdot\sin\left(\frac{2\pi h}{T}t + \phi_h\right) \tag{6.6}$$

$$V_0 = V_{in}\cdot D,$$

$$D = (T_{on}/T) \tag{6.7}$$

The average output voltage V_0 is controlled by varying the duty cycle, *D*. The amplitude of the harmonics is given by eqn. (6.6) and varies with the order *h* and the ratio (T_{on}/T). The maximum rms value of the *h*th harmonic v_{hr}, is given by,

$$V_{hr} = 0.45\, V_{in}/h \tag{6.8}$$

Figure 6.7(a) shows the variation of the r.m.s. harmonic voltages with the ratio (T_{on}/T). It is seen that the harmonic voltages decrease rapidly as the order *h* increases. The harmonic current in the load is given by,

$$i_h = V_h/Z_h \tag{6.9}$$

where Z_h is the impedance of the load at the harmonic frequency h.f. If the load resistance is negligible compared to its inductance, then i_h is proportional to $(1/h^2)$. Thus the harmonic current decreases as the order of the harmonic increases. An alternative measure of the harmonic content of a waveform which does not require the calculation of the individual harmonics, is the a.c. ripple voltage V_r.

It is defined as,
$$V_r = \sqrt{V_{rms}^2 - V_0^2} \tag{6.10}$$

In the eqn. (6.10) V_{rms} is the rms value and V_0 is the average value of the waveform. Using this definition it can be shown that

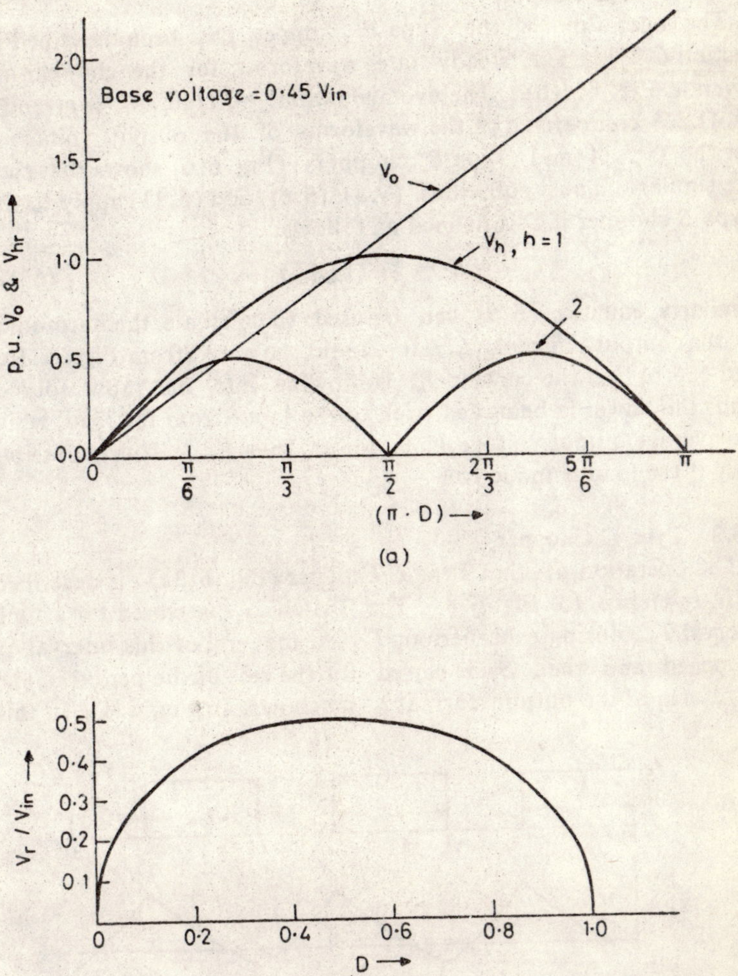

Fig. 6.7 The variation of (a) Harmonic voltages, (b) AC ripple voltage for the type *A*, *B*, *C*, *D* choppers.

the ac ripple in the output voltage for Type *A* chopper is given by,

$$V_r = V_{in} \cdot \sqrt{D - D^2} \tag{6.11}$$

Figure 6.7(b) shows the variation of the per unit ripple voltage (V_r/V_{in}) with the duty cycle D.

6.3.2 Type B Chopper

The operation of the Type *B* chopper has been described in section 6.2.1. The steady-state waveforms for the chopper are given in Fig. 6.6(b). The average output voltage, V_0, is given by (6.4). A comparison of the waveforms of the output voltage v_0 for the Type *A* and Type *B* choppers (Fig. 6.6) shows that they are similar. Thus equations (6.5), (6.6) and (6.11) apply to the Type *B* chopper if *D* is defined as follows.

$$D = 1-(T_{on}/T) \tag{6.12}$$

Similarly, equation (6.9) can be used to calculate the harmonics in the output current, i_0. It should be noted from Figs. 6.1(c) and 6.6(b) that the current i_{in} is flowing into the input source. Thus the power is being fed back to the input from the load, while the average output voltage V_0 is lower than V_{in}. This is possible only if the load is inductive.

6.3.3 Type C Chopper

The operation of the Type *C* chopper (Fig. 6.3a) is described with reference to Fig. 6.8. The switch S_1 is closed for a time interval T_{on}, during each period, T. At the end of this interval S_1 is opened and then S_2 is closed for the rest of the period. The waveform of the output current i_0 is shown in Fig. 6.8. In this

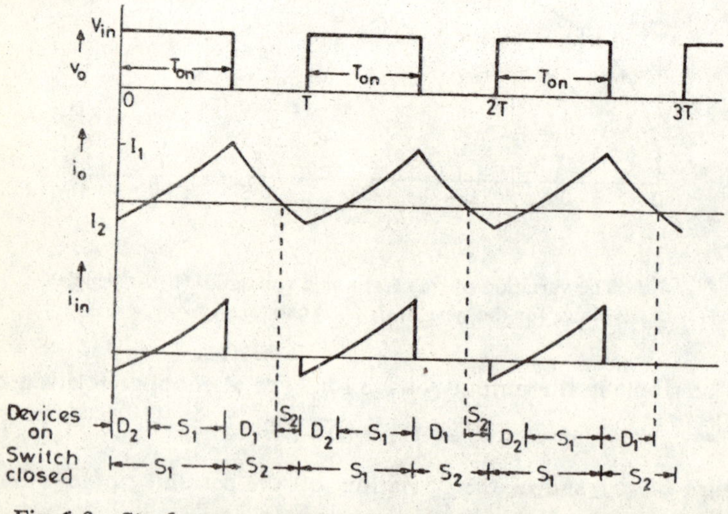

Fig. 6.8. Steady-state waveforms for the type *C* chopper of Fig 6.3(a).

chopper the output current is continuous for active loads. Therefore, the output voltage v_0 can be defined independent of the load. In this respect, it is different from the Type A and the Type B choppers where the current i_0 may be discontinuous for low values of (T_{cn}/T). The waveform of the voltage v_0 in Fig. 6.8 is identical to that of Type A chopper with continuous current shown in Fig. 6.6(a). Therefore eqn. (6.5) to (6.11) apply to Type C chopper. Figure 6.7 gives the average, ripple and the harmonic voltages for Type C chopper. As the Type C chopper can provide bidirectional current (Fig. 6.8b) each of the currents I_1 and I_2 may independently be either positive or negative. The average value of i_0 and i_{in} can also be either positive or negative.

6.3.4 Type D Chopper

The circuit configuration of the Type D chopper is shown in Fig. 6.3(c). If the output current i_0 is assumed to be continuous, the instantaneous output voltage v_0 is given by the following equation,

$$\begin{aligned} v_0 &= +V_{in} \text{ when } S_1 \text{ and } S_2 \text{ are closed,} \\ &= -V_{in} \text{ when } S_1 \text{ and } S_2 \text{ are open,} \\ &= 0 \text{ when } S_1 \text{ is open and } S_2 \text{ is closed,} \\ &= 0 \text{ when } S_1 \text{ is closed and } S_2 \text{ is open.} \end{aligned} \tag{6.13}$$

Equation (6.13) shows that v_0, the instantaneous output voltage, can be modulated between $+V_{in}$ and $-V_{in}$ or between $\pm V_{in}$ and zero. Thus there are two strategies for the Pulse Width Modulation (PWM) control of the Type D chopper.

(a) Single Polarity Modulation

In this strategy the switches S_1 and S_2 are closed for a time T_{on} during each period, T. However, there is a phase difference of $T/2$ seconds between the closing of S_1 and S_2 as shown in Fig. 6.9. In this figure the states of the switches are represented by the waveforms of S_1 and S_2. S_1 (S_2) equals one when the switch S_1 (S_2) is closed and zero when it is open. The resulting waveform of v_0 (eqn. 6.13) modulates between $+V_{in}$ and zero when T_{on} is greater than $T/2$ as shown in Fig. 6.9 (a). When T_{on} is exactly equal to $T/2$, v_0 is zero. When T_{on} is reduced below $T/2$, the waveform of v_0 modulates between zero and $-V_{in}$ as shown in Fig. 6.9 (b).

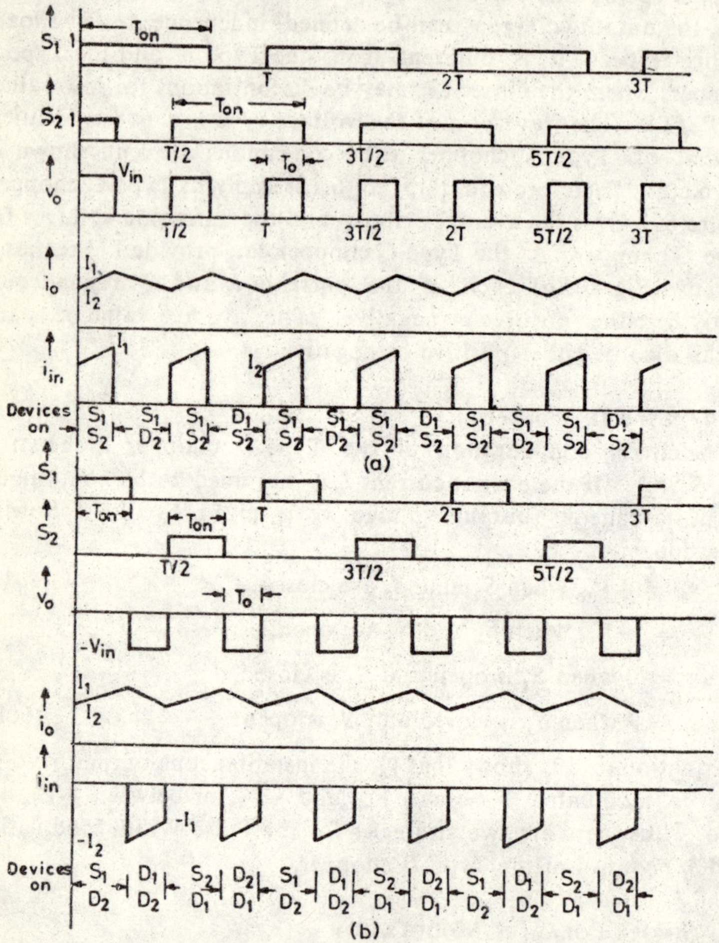

Fig. 6.9. Steady state waveforms for type *D* chopper with single polarity modulation:
(a) $T_{on} > T/2$; (b) $T_{on} < T/2$.

Thus the average value of the output voltage V_0 is given by,

$$V_0 = V_{in} \cdot D \tag{6.14}$$

where $$D = 2\left(\frac{T_{on}}{T} - 0.5\right) \tag{6.15}$$

The average output voltage V_0 can be varied between $+V_{in}$ and $-V_{in}$ by controlling the duty cycle of the switch (T_{on}/T).

The waveform of the output voltage v_0 for Type D chopper can be resolved into the following Fourier Series.

$$v_0 = V_0 + \sum_{h=1}^{\infty} v_h, \quad \text{where,} \tag{6.16}$$

$$v_h = \frac{2V_{in}}{\pi}\frac{1}{h}\sin(h\pi D)\cdot\sin\left(\frac{4\pi h}{T}t + \phi_h\right) \tag{6.17}$$

$$D = 2\left(\frac{T_{on}}{T} - 0.5\right) \tag{6.18}$$

A comparison of the eqns. (6.17) and (6.6) shows that the amplitudes of the harmonics in the Type D and the Type A chopper are equal. For the same duty cycle of the switch i.e. T_{on}/T, the frequency of the harmonics in the Type D chopper is twice that of the Type A chopper. The variation of the rms values of harmonics for the Type D chopper is given by Fig. 6.7 (a) if the abscissa is taken to be the modulus $|\pi D|$. The a.c. ripple in the output voltage is given by the following equation.

$$V_r = V_{in}\sqrt{|D| - D^2} \tag{6.19}$$

where D is given by (6.15) and $|D|$ represents the modulus of D. Figure 6.7 (b) shows the variation of the ripple voltage if the abscissa is taken to be $|D|$.

(b) Dual Polarity Modulation

In this strategy the switches S_1 and S_2 are turned on simultaneously for a duration T_{on} during each period. The switches are turned off simultaneously at the end of the on period. Figure 6.10 shows the states of the switches and the resulting waveform of v_0 obtained by applying eqn. (6.13). The current i_0 (Fig. 6.3c) has been assumed to be continuous. The waveform of v_0 modulates between $+V_{in}$ and $-V_{in}$. The average value of the output voltage is given by eqns. (6.14) and (6.15). The amplitudes of the harmonics are double the value given by equation (6.6). Similarly the ripple voltage is double the value given by eqn. (6.19). Therefore Fig. 6.7 (a) and (b) apply to this strategy with an increase in the ordinates by a factor of two and with abscissa equal to $|\pi D|$ and $|D|$ respectively. The dual polarity modulation therefore increases the harmonics in the output voltage by a factor of two compared to the single polarity modulation. However it is simpler to implement and may have a faster transient response than the single polarity modulation.

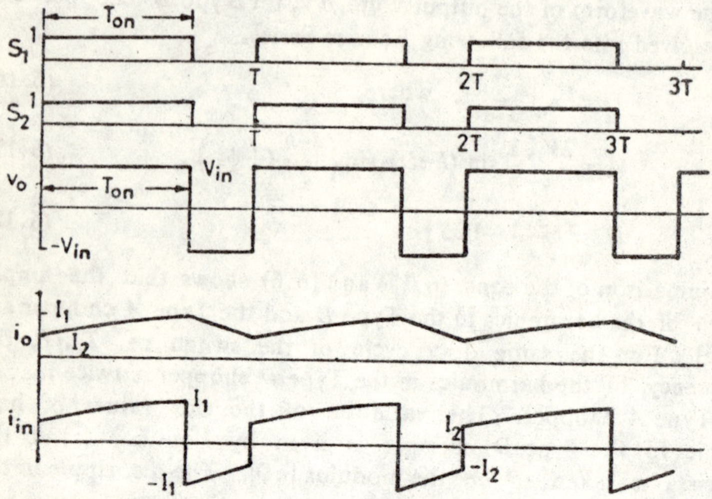

Fig. 6.10. Steady state waveforms for type *D* chopper with dual polarity modulation.

6.3.5 Discontinuous Current Operation

The results presented in the sections 6.3.1 to 6.3.4 are derived on the assumption that the load current is continuous. If the load inductance is small or the chopping period *T* is large (Fig. 6.6 a) the load current may fall to zero during the freewheeling period. The output voltage during the interval in which the output current is zero, can be defined only if the nature of the load is known. For example, if the load is a dc motor the instantaneous output voltage, when the output current is zero, is equal to the back *EMF* of the motor. The operation in the discontinuous mode increases the ripple in the output current. The calculation of the harmonics and the ripple in the output voltage and current can be done by detailed simulation of the particular system under consideration. The analysis of dc drives under discontinuous conduction is discussed in detail in Section 6.4.

6.3.6 Input Current Harmonics

Figures 6.6, 6.8 to 6.10 show the waveform of the current i_{in} drawn from the dc source. It is seen that the current i_{in} is discontinuous in all types of choppers. The current i_{in} is unidirectional in Type *A* and Type *B* choppers and is bidirectional in Type *C* and

Type D choppers. The voltage source V_{in} must therefore be capable of supplying pulsed current and also be capable of absorbing energy when the chopper is regenerating ($i_{in} < 0$). Some practical sources, i.e. batteries, operate more efficiently if the current drawn from them is ripple free. In dc traction systems the input current of the chopper is supplied through the third rail and the track. The harmonics in the input current can therefore interact with the communication circuits connected through the track. For these reasons a filter is often connected between the dc voltage source V_{in} and the input terminals of the chopper. Figure 6.11 shows a type A chopper with a LC filter at the input. If the output current i_0 of any the choppers is assumed to be *ripple free* then the waveforms of the chopper input current i_{in} are similar to the waveforms of the output voltage v_0 (Figs. 6.6, 6.9, 6.10). Therefore the harmonics expressed by eqns. (6.6) and (6.17) are also applicable to i_{in} if V_{in} is replaced by the average output current, I_0. If the output current i_0 has significant ripple then the harmonics of i_{in} can only be determined by Fourier Analysis of the waveforms given in Figs. 6.6, 6.8 to 6. 10.

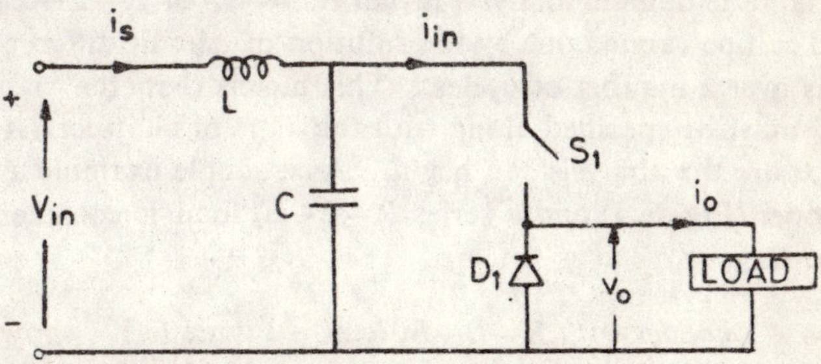

Fig. 6.11. Type A chopper with LC input filter.

The chopper input current i_{in} contains components at frequencies $h.f_c$ where f_c is the chopping frequency and h is an integer. The frequency f_c is the reciprocal of the period T in Fig. 6.6. If the load contains an ac source of frequency f_0, then the current i_{in} would also contain components at frequencies f_0 and $(hf_c \pm f_0)$. The filter components L and C (Fig. 6.11) must be selected such that its cut off frequency, $1/\sqrt{LC}$, is much lower than f_0 and f_c. In systems where the voltage V_{in} is obtained from an ac source through a rectifier there is a further possibility of obtaining beat frequencies in the source current i_s (Fig. 6.11). For example, if the ac input at 50 Hz is rectified to obtain V_{in} then i_s contains a 300 Hz component. If the chopper frequency is close to 300 Hz e.g. 310 Hz, then beats of frequency 10 Hz are obtained in the source current i_s [1]. The reflection of i_s in the ac lines via the rectifier can generate further harmonics in the ac input line.

6.4 TIME DOMAIN ANALYSIS

The calculation of harmonic voltages and currents as given in section 6.3 provides valuable information about the frequencies and magnitudes of different components. Often it is also necessary to calculate the instantaneous values of the load voltages and currents. This enables the determination of the peak voltages and currents, on the basis of which the components like thyristors can be selected. Further, the frequency domain analysis of section 6.3 assumes that the chopper is operating in the steady state. The transition from the initial state to the steady-state is of importance in some applications of choppers e.g. dc traction drives. For these reasons, a time domain analysis of the chopper and its load is required.

The time domain analysis for an R, $R-L$ or $R-L-E_b$ type of load can be carried out by the solution of circuit differential equations over a number of cycles. This means that the nature of the load must be specified along with the type of chopper (A, B, C, or D) before the analysis can begin. As a simple example a Type 'A' chopper (Fig. 6.1) and a series $R-L-E_b$ load is considered in the following.

Type A Chopper with $R-L-E_b$ load: Figure 6.12 shows the two modes of circuit operation. In the on period (Mode (i)) the load is connected to the supply V_{in} through the switch S_1. In the off period, (Mode (ii)), the load is short circuited through D_1. The differential equations for the load current i_0 with time t equal to zero at the start of each mode are as follows:

$$\begin{aligned} &\text{Mode (i):} \quad di_0/dt + (R/L)i_0 = (V_{in} - E_b)/L \\ &\text{Mode (ii):} \quad di_0/dt + (R/L)i_0 = -E_b/L \end{aligned} \tag{6.20}$$

Equation (6.20) can be solved repeatedly for each period. For a constant chopping period T and a constant on-period, T_{on}, equal to $T/2$, the waveform for i_0 is as shown in Figure 6.12(d). It is seen that with the initial condition $i_0 = 0$ @ $t = 0$, the current

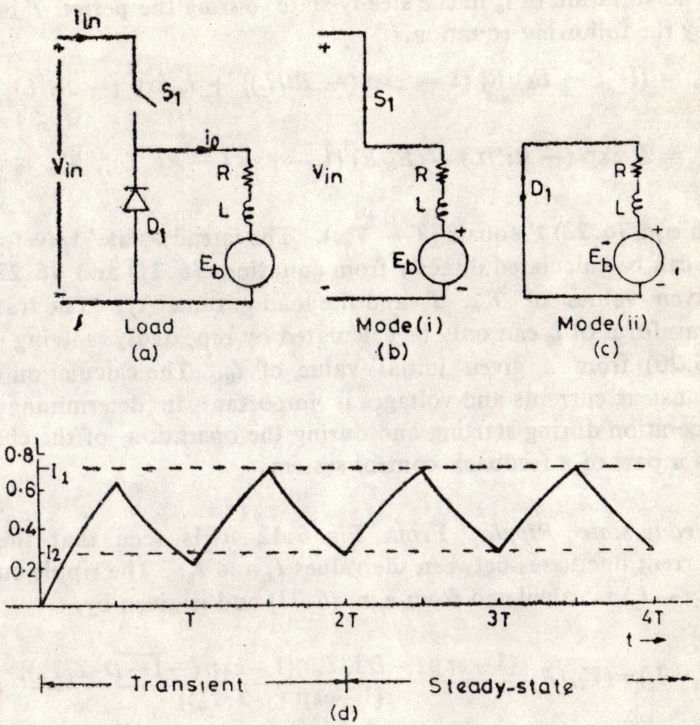

Fig. 6.12. Type A chopper with $R-L-E_b$ load: (a) circuit diagram; (b) and (c) Equivalent circuits; (d) Waveform of i_0.

i_0 builds up to a waveform which repeats itself in every period T. This 'steady-state' waveform can also be directly determined by solving eqns. (6.20) under the boundary condition,

$$(i_0@t = 0) = (i_0@t = T) = I_2$$

where T is the time period of the chopper. For a given on period T_{on}, the solution of (6.20), gives,

$$I_2 = \frac{V_{in}}{R}\left\{1 - \exp\left(\frac{R}{L}\,T_{on}\right)\right\}\left\{1 - \exp\left(\frac{R}{L}\,T\right)\right\}^{-1} - \frac{E_b}{R}$$

The peak current I_1 is given by

$$I_1 = \frac{V_{in}}{R}\left\{1 - \exp\left(-\frac{R}{L}\,T_{on}\right)\right\}\left\{1 - \exp\left(-\frac{R}{L}\,T\right)\right\}^{-1} - \frac{E_b}{R} \quad (6.21)$$

The variation of i_0 in the steady-state during the period T is given by the following equation.

$$i_0 = \{(V_{in} - E_b)/R\}\{1 - \exp(-Rt/L)\} + I_2 \exp(-Rt/L), \quad 0 \leqslant t \leqslant T_{on}$$

$$= I_1 \exp(-Rt'/L) - (E_b/R)\{1 - \exp(-Rt'/L)\}, \quad T_{on} \leqslant t \leqslant T \tag{6.22}$$

In eqn. (6.22) t' equals $(T - T_{on})$. The 'steady-state' waveform of i_0 can be calculated directly from equations (6.21) and (6.22) for given values of T_{on}, T and the load parameters. The transient waveform of i_0 can only be calculated by repeatedly solving eqns. (6.20) from a given initial value of i_0. The calculation of the transient currents and voltages is important in determining their operation during starting and during the operation of the chopper as a part of a feedback control system.

Steady-state Ripple: From Fig. 6.12 it is seen that the load current fluctuates between the values I_1 and I_2. The ripple current $(I_1 - I_2)$ is calculated from eqn. (6.21) and is given by,

$$(I_1 - I_2) = (V_{in}/R) \cdot \frac{\{1 - \exp(-DT/T_a)\}\{1 - \exp(-\overline{1-D} \cdot T/T_a)\}}{\{1 - \exp(-T/T_a)\}} \tag{6.23}$$

where $D = T_{on}/T$ and $T_a = L/R$.

The ripple current given by above equation is independent of back *EMF* E_b. Figure 6.13 shows the per unit ripple current $(I_1 - I_2)/(V_{in}/R)$ as a function of the duty cycle D and the ratio (T/T_a). The p.u. ripple current is maximum when D equals 0.5. As the inductance L is increased, (T/T_a) reduces. The currents I_2 and I_1 tend to the average value $(V_{in} - E_b)/R$ and the ripple current $(I_1 - I_2)$ tends to zero.

Limit of Continuous Conduction: The output current i_0 in Type A chopper is constrained to be positive by the switch S and the free-wheeling diode D (Fig. 6 1a). Thus the equation (6.20) for mode (ii) is valid only till the current i_0 fails to zero. Since the current cannot reverse it stays at zero. The limit of continuous conduction is reached when the current I_2 given by (6.21) goes to zero. Equating I_2 to zero, the duty cycle (T_{on}/T) at the limit of continuous conduction is given by the following equation.

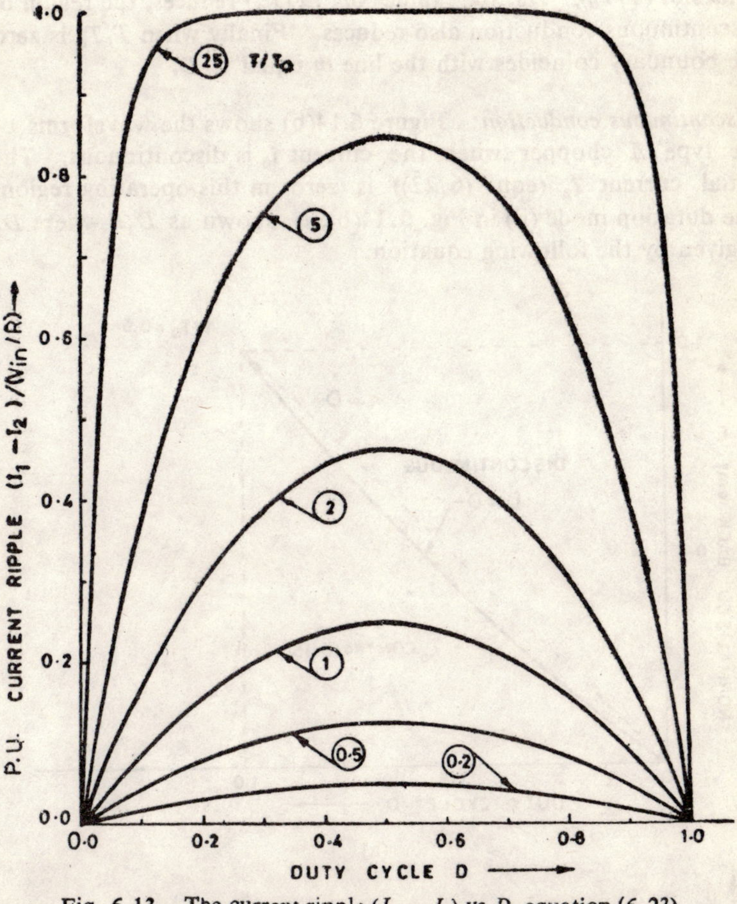

Fig. 6.13. The current ripple $(I_1 - I_2)$ vs D, equation (6.23).

$$D' = \left(\frac{T_{on}}{T}\right) = \left(\frac{T_a}{T}\right) \ln\left[1 + m\left\{\exp\left(T/T_a\right) - 1\right\}\right] \tag{6.24}$$

where $m = E_b/V_{in}$, $T_a = L/R$.

Figure 6.14(a) shows the variation of D' when (T/T_a) equals 0.5. The operation of the type *A* chopper is limited to the region defined by the *D* axis, and the lines, *D* equal to one and *m* equal to one. The curve of D' divides the operating region into two parts. Above the curve the output current is discontinuous and below the curve it is continuous. Similar curves can be drawn for other values of (T/T_a). As the value of (T/T_a) reduces, the region of discontinuous conduction also reduces. Finally when T/T_a is zero the boundary coincides with the line m equal to D.

Discontinuous conduction: Figure 6.14(b) shows the waveforms of the type A chopper when the current i_0 is discontinuous. The initial current I_2 (eqn. (6.22)) is zero in this operating region. The duration mode (ii) in Fig. 6.14(b), is shown as D_2T where D_2 is given by the following equation.

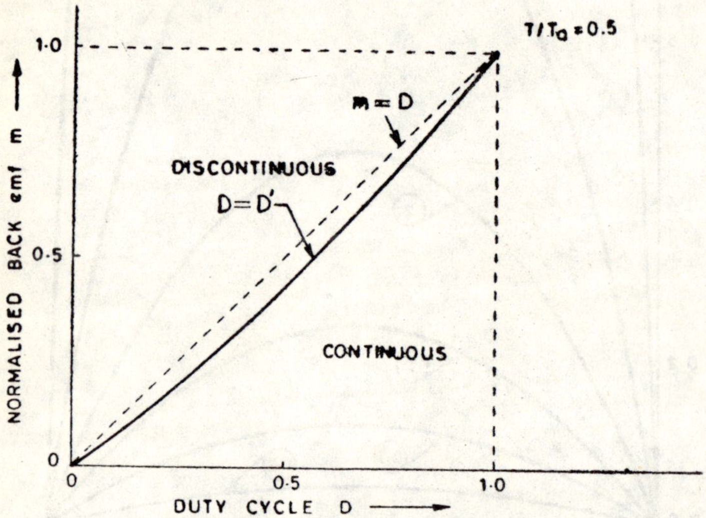

(a)

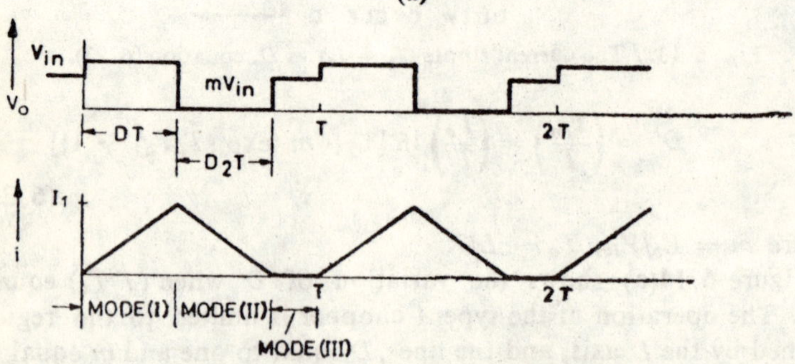

(b)

Fig. 6.14. (a) Limit of continuous conduction in one quadrant chopper.
(b) Discontinuous current operation in one quadrant chopper.

$$D_2 = \left(\frac{T_a}{T}\right) \ln\left[\frac{1}{m} + \left(1 - \frac{1}{m}\right) \cdot \left\{\exp\left(-DT/T_a\right)\right\}\right] \quad (6.25)$$

where $D = T_{on}/T$, $on = E_b/V_{in}$, $T_a = L/R$.

The transient analysis given above is valid for a type A chopper. Similar analysis can be carried out for the type B chopper (Fig. 6.1c) operating in the second quadrant. An interesting symmetry relationship exists between the p.u. currents of type A and type B choppers supplying the $R-L-E_b$ load. The currents I_2 and I_1 given by eqn. (6.21) can be normalised by dividing both sides by (V_{in}/R). The p.u. currents are numerically the same if in the first quadrant the duty cycle is D_1 and p.u. back EMF is m_1 and the corresponding values in the second quadrant are $(1-D_1)$ and $(1-m_1)$. The duty cycle of the switch is equal in both quadrants. A boundary curve separating the regions of discontinuous and continuous current operation similar to Fig. 6.14(a) can be obtained. It however lies in the triangle bounded by the lines D equal to zero, m equal to one and m equal to D.

6.5 MULTIPHASE CONFIGURATIONS USING IDEAL SWITCHES

From the considerations given in Section 6.3.6 it is seen that when the choppers are used for high power circuits, e.g. in d.c. traction, the problem of filtering the input current is severe. An elegant method for reducing the filter size is to operate several choppers in parallel. Such an arrangement is called a multiphase chopper. Figure 6.15(a) shows the circuit diagram of a multiphase chopper showing three Type A choppers connected in parallel. The output of each Type A chopper is connected to a series reactor L and the load. The reactor L is chosen to be large so that each chopper can be assumed to operate independently.

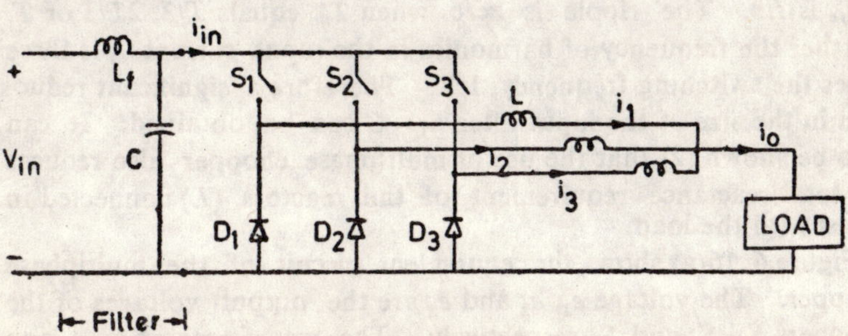

Fig. 6.15. (a) Circuit diagram of multiphase chopper.

The load current i_0 can be assumed to be ripple free. If the value of i_0 is I, the chopper currents i_1, i_2, i_3 are also ripple-free and each is equal to $I/3$. The switches S_1, S_2 and S_3 are closed for a period T_{on} when t equals zero, $T/3$ and $2T/3$ respectively as shown in Fig. 6.15(b). The waveform of the input current i_{in} can be obtained by adding the currents i_1, i_2 and i_3 for the on period of the respective chopper.

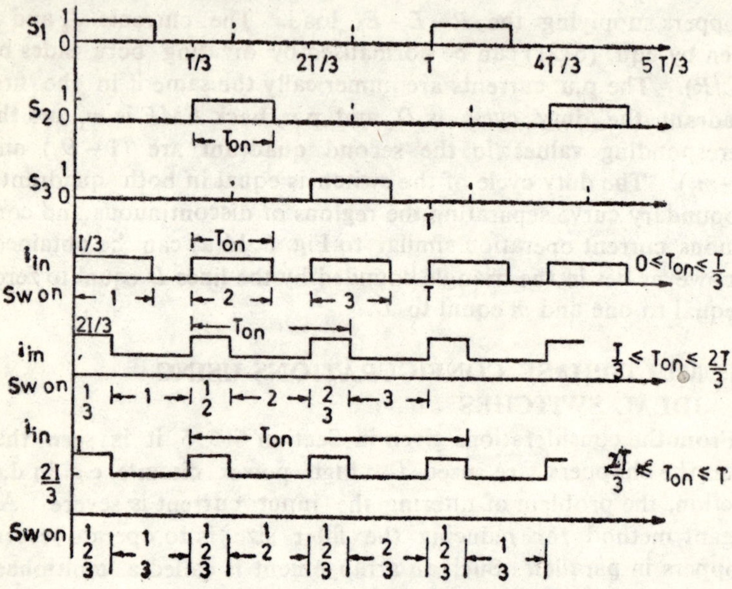

Fig. 6.15. (b) Waveforms of i_{in} for multiphase chopper of Fig. 6.13. Currents i_1, i_2, i_3 are assumed ripple-free.

Figure 6.15(b) shows the waveforms of the input current i_{in} for three different values of T_{on}. It is seen that the maximum ripple in i_{in} is $I/3$. The ripple is zero when T_{on} equals $T/3$, $2T/3$ or T. Further the frequency of harmonics in the input current i_{in} is three times the switching frequency, $1/T$. Therefore a significant reduction in the size of the input filter $L_f - C$ can be obtained. It can also be shown [2] that the use of multiphase chopper also reduces the total reactance requirement of the reactors (L) connected in series with the load.

Figure 6.16(a) shows the equivalent circuit of the multiphase chopper. The voltage e_1, e_2 and e_3 are the output voltages of the choppers S_1, S_2 and S_3 respectively. The waveforms of e_1, e_2 and

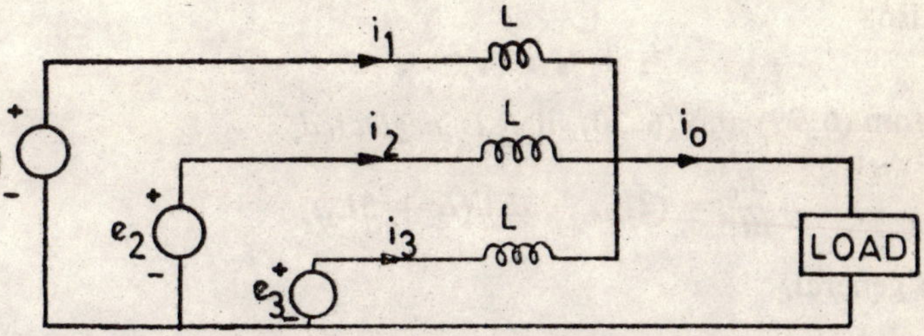

Fig. 6.16. (a) Equivalent circuit for multiphase chopper of Fig. 6.15a.

e_3 are similar to v_0 in Fig. 6.6(a). The harmonics of e_1, e_2 and e_3 are therefore given by eqns. (6.6) and (6.7). The voltages e_1, e_2 and e_3 have a phase difference in time equal to $T/3$ or 120 degrees. The harmonics in i_0 can be calculated by superposition of currents through the load due to e_1, e_2 and e_3.

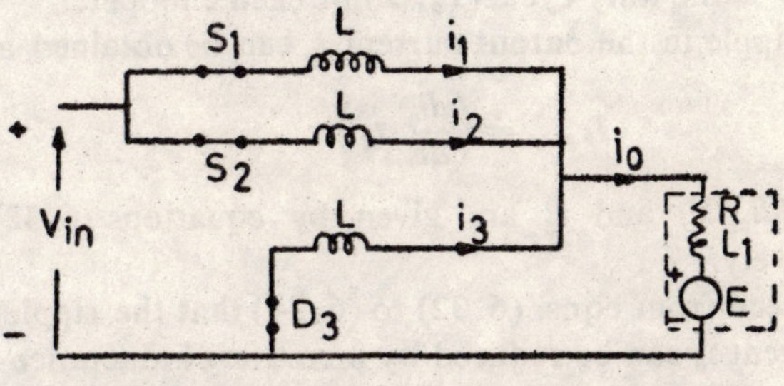

Fig. 6.16. (b) Equivalent circuit of chopper Fig. 6.15a: S_1 and S_2 on, S_3 off.

A time domain analysis (section 6.4) of the multiphase chopper can be carried out using the equivalent circuits for each mode. Figure 6.16(b) shows the equivalent circuit when S_1 and S_2 are on and S_3 is off. For this mode, assuming a series R, L_1 and E load,

$$V_{in} = L\frac{di_1}{dt} + L_1\frac{di_0}{dt} + i_0R + E : S_1 \text{ on} \tag{6.26}$$

$$V_{in} = L\frac{di_2}{dt} + L_1\frac{di_0}{dt} + i_0R + E : S_2 \text{ on} \tag{6.27}$$

$$O = L\frac{di_3}{dt} + L_1\frac{di_0}{dt} + i_0R + E : S_3 \text{ off} \tag{6.28}$$

Adding equation (6.26) to (6.28)

$$2V_{in} = L\frac{d}{dt}(i_1 + i_2 + i_3) + 3L_1\frac{di_0}{dt} + 3i_0R + 3E \tag{6.29}$$

Also,

$$i_1 + i_2 + i_3 = i_0 \tag{6.30}$$

From (6.29) and (6.30), if R *is neglected,*

$$\frac{di_0}{dt} = (2V_{in} - 3E)/(L + 3L_1) \tag{6.31}$$

In general,

$$\frac{di_0}{dt} = (n\,V_{in} - mE)/(L + mL_1) \tag{6.32}$$

where n is the number of conducting switches and m is the total number of switches. The time t_0 for which the current i_0 is increasing is given by,

$$t_0 = \{D - (n - 1)/m\}\ T \tag{6.33}$$

where D is the duty cycle (T_{on}/T) of each chopper.

The ripple in the output current i_0 can be obtained as,

$$I_{r\,p-p} = \left(\frac{di_0}{dt} \cdot t_0\right) \tag{6.34}$$

where (di_0/dt) and t_0 are given by equations (6.32) and (6.33) respectively.

It is seen from eqns. (6.32) to (6.34) that the ripple in the output current i_0 can be reduced by increasing inductance L and the total number of phases m.

6.6 CHOPPER CIRCUITS USING THYRISTOR SWITCHES

In the preceding sections, the existence of an ideal switch which can be closed and opened at required instants of time is assumed. The process of opening the switch is also termed as *'commutation'*. In practice the switches are devices in which the flow of current can be started and interrupted by an external electric signal. The switches can be mechanical, arc-based or solid state devices. The solid state switches are presently the most prevalent, due to the advantages of low loss, fast response and longer operating life. The two common types of devices used for power switching are the 'Transistor' and the 'Thyristor'. The power transistor has the advantage that it can be turned on and off by the base drive circuit. The thyristor, on the other hand, when used in the d.c. choppers, requires a separate *commutation circuit* for turn-off. However, the thyristor is available in much higher voltage and current ratings.

The thyristor also has a regenerative turn on, which reduces the gate drive requirements. Thus except for low voltage applications, the thyristor is the preferred power switch.

The process of turn-off or commutation of a conducting thyristor requires the reduction of the anode current to zero, followed by the application of a reverse voltage across the device. The reverse voltage for few microseconds is necessary to regain the forward blocking capability. The various commutation circuits used in the d.c. choppers differ from one another in the manner in which the above mentioned requirements are met. Two broad categories of the commutation circuits are the current commutation and the voltage commutation circuits. The salient features of these commutation circuits are described below.

Current commutation: In this type of commutation the conducting thyristor is turned-off by injecting a pulse of reverse current through it. The current pulse is usually generated by an *L–C* circuit and is initiated by gating an auxiliary thyristor. The auxiliary thyristor may be self-commutated. The reverse current pulse is of sufficient magnitude to reduce the anode current to zero. A distinctive feature of the current commutation is that the reverse voltage across the device is applied through a diode connected in anti-parallel to the thyristor. Since this is limited to about one volt, the turn-off time of the device increases in comparison to the voltage commutation. The device manufacturers specify the turn-off time with reverse diode for such applications.

Voltage commutation: In this type of commutation the conducting thyristor is turned off by impressing a large reverse voltage across it. The reverse voltage is usually applied by switching a previously charged capacitor. The commutation is initiated by gating an auxiliary thyristor. The application of reverse voltage across the conducting thyristor reduces the anode current to zero rather abruptly. The reverse voltage then assists in completing the turn-off process. The turn off time in this commutation is smaller than the current commutation due to the increased reverse voltage.

In the commutation methods mentioned above the load current is diverted to an auxiliary path before its transfer to the freewheeling diode. The inductance in this auxiliary path stores some energy which is recovered or dissipated by additional circuitry.

The trapped energy problem is more severe in voltage commutation rather than current commutation. A large number of chopper circuits have been designed using the voltage and the current commutation principles. A few circuits which are in common use are discussed in the following sections. The operation of the circuits has been explained in detail and the design equations are given. A comparison of several types of chopper circuits is given in references [3] and [4].

6.6.1 Current Commutated Switch

In this section a practical thyristor switch is described. The commutation (or turn-off) of the thyristor is effected by injecting a reverse current through it. The energy for commutation is

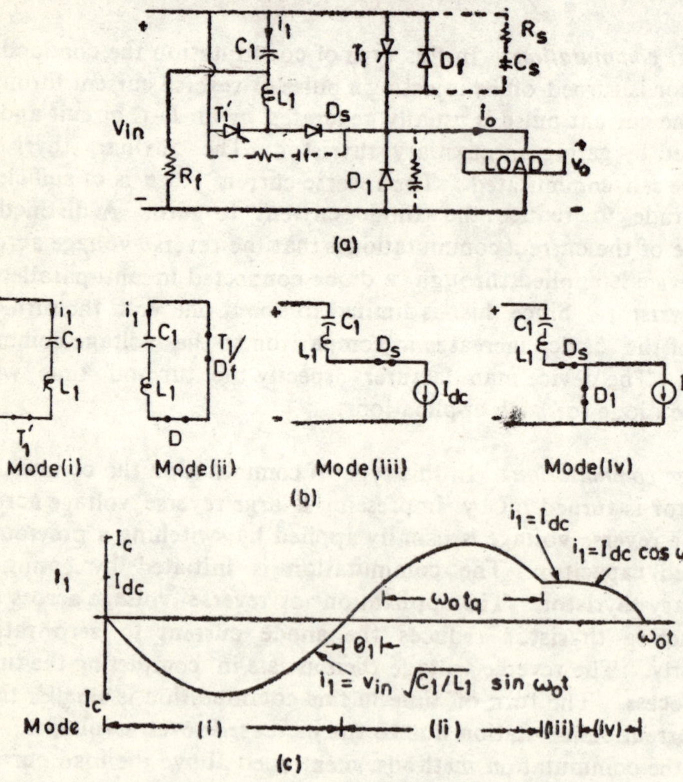

Fig. 6.17. (a) Circuit diagram; (b) Equivalent circuits, (c) Capacitor current for a practical type A chopper $\omega_0 = 1/\sqrt{L_1 C_1}$

stored in a capacitor and is released at required instant by gating an auxiliary thyristor. The circuit elements for the protection of the thyristors and diodes are also described.

Figure 6.17 shows the circuit of the thyristor switch and a freewheeling diode, connected as type *A* chopper configuration (Fig. 6.1a). Thyristor T_1 is the power switch which is turned on by application of the gate-cathode signal. The turn-off of T_1 is initiated by gating thyristor T_1'. The commutation circuit consisting of capacitor *C*, inductor *L*, diodes D_s and D_f operates in the following modes.

Mode (*i*): It is assumed that at the start of this mode the commutation capacitor *C* (Fig. 6.17a) is charged to the supply voltage V_{in} through resistor R_f. The resistor R_f is assumed to be large so that it can be considered to be an open circuit during the commutation.

When thyristor T_1' is gated, it short-circuits the $L_1 - C_1$ branch and a sinusoidal pulse of current, i_1, flows through L_1, C_1 and T_1'. Thyristor T_1' is automatically turned off when the current i_1 goes to zero.

Mode (*ii*): The current i_1 reverses and flows through C_1, L_1, D_s and thyristor T_1, as shown in Fig. 6.17(c). When current i_1 equals the load current the net current flowing through thyristor T_1 is reduced to zero. The peak of the sinusoidal pulse current, i_1 is higher than the load current and the excess of i_1 over the load current, flows through the diode D_f. This mode ends when i_1 starts falling and is equal to the load current.

Mode (*iii*): A constant current equal to load current now flows through C_1, L_1, D_s and the load, till the capacitor voltage v_{cl} is equal to the supply voltage V_{in}.

Mode (*iv*): At this point, the diode D_1 is forward biased and load voltage is reduced to zero. Due to the trapped current in the inductor L_1 (equal to the load current), the capacitor C_1 is overcharged i.e. v_{cl} rises above V_{in}. The overcharge on C_1 is dissipated through R_f in the freewheeling (off) interval of the chopper. Figure 6.18 shows detailed waveforms of the circuit currents and voltages.

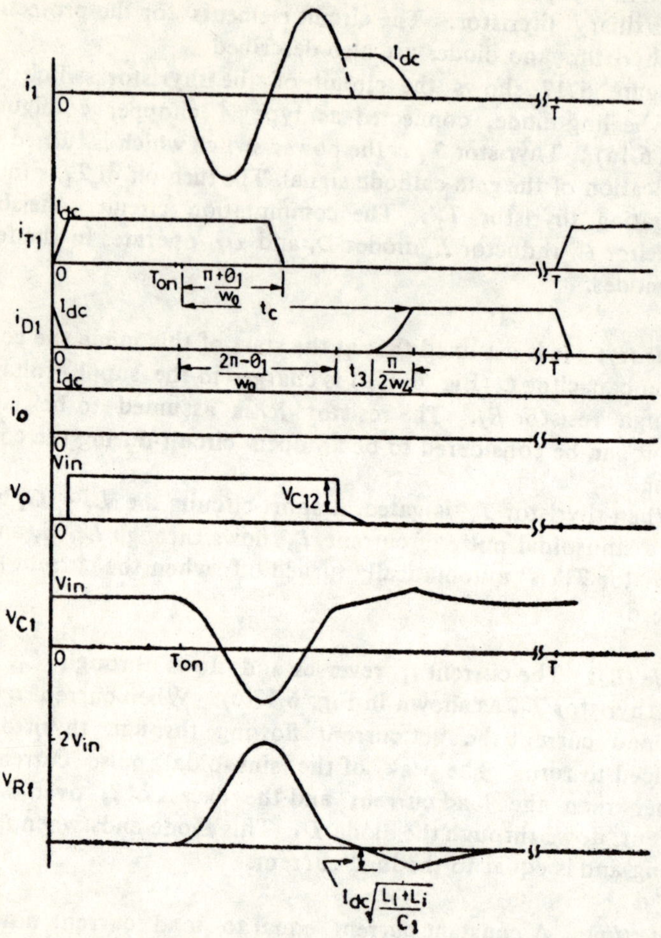

Fig. 6.18. Circuit waveforms for practical thyristor switch.

Protective elements: The diodes and thyristors are shunted by a $R_s - C_s$ network to limit the rate of rise of forward voltage across the thyristors. These elements are shown in dotted lines in Fig. 6.17 (a). The *di/dt* across the thyristor T_1 is limited by the inductance of the circuit layout. If this is not sufficient a small inductor can be connected in series with T_1 to limit *di/dt*.

Resistor R_f is not essential for the commutation modes but is connected to limit the voltage across the capacitor C_1. It also provides the initial charge on the capacitor when the power is turned on. Therefore a special starting sequence is not required for this chopper.

Turn off time: The current through thyristor switch T_1 falls to zero when the peak capacitor current I_c equals the load current I_{dc} (Fig. 6.17c). The excess of capacitor current in mode (ii) over the load current flows through the diode D_f. During this time, t_q in Fig. 6.17 (c), the voltage across the thyristor T_1 is equal to, the voltage drop across diode D_f. The diode drop reverse biases T_1 and provides the necessary turn off time for T_1 to regain its forward blocking capability. The peak capacitor current I_c is given by,

$$I_c = V_{in}\sqrt{\frac{C_1}{L_1}} \tag{6.35}$$

The turn off time t_q for a given load current I_{dc} is given by,

$$t_q = (\pi - 2\theta_1)\sqrt{L_1 C_1} \tag{6.36}$$

where

$$\theta_1 = \sin^{-1}(I_{dc}/I_c) \tag{6.37}$$

It is seen from equations (6.36) and (6.37) that the turn off time t_q reduces as I_{dc} is increased. Thus the (I_{dc}/I_c) ratio must be maintained below a critical value to obtain the necessary turn off time Similarly for given values of L_1, C_1, V_{in} and t_q there is a limit on the magnitude of load current I_{dc} that can be successfully commutated.

Peak capacitor voltage: The capacitor voltage V_{cl} is charged to the supply voltage V_{in} at the end of mode (iii). During mode (iv) the energy stored in the inductor L_1 is transferred to the capacitor C_1. The capacitor voltage at the end of mode (iv) is given by,

$$V_{c4} = V_{in} + I_{dc}\sqrt{L_1/C_1} \tag{6.38}$$

The voltage V_{c4} is also the peak forward voltage across thyristors T_1 and T_1'.

Power loss in R_f: The peak voltage across the capacitor is higher than the supply voltage V_{in}. Therefore, the capacitor discharges through resistor R_f during the remaining interval in the chopping period T. A part of the energy stored in the capacitor due to overcharge is returned to the supply and a part is dissipated in the resistor R_f. If the chopping period T is greater than $3R_fC_1$, it is shown in [5], that the power dissipated in R_f is,

$$P_{R_f} = \frac{1}{2} C_1 V_{in}^2 / x^2 T \tag{6.39}$$

where $x = I_c / I_{dc}$.

Effect of circuit resistance: The equivalent circuits shown in Figure 6.17(b) show a purely oscillatory L_1C_1 circuit. In practical circuits there is a small resistance associated with the inductor L_1, the conducting devices and the connecting wires. The quality factor, Q, of the oscillatory circuits in modes (i) and (ii) is usually of the order of 6 to 10. The main effect of a low Q is reduction in the peak capacitor current from the ideal value I_c and a small reduction in the over charge on the capacitor.

Operation with resistance R_f: In an ideal circuit the discharge resistance R_f is essential to reduce the peak capacitor voltage from V_{c4} (eqn. 6.38) to V_{in} at the beginning of each commutation. An ideal circuit with infinite Q when operated without R_f will continuously build up the voltage across the capacitor. In a practical circuit, however, the build up of capacitor voltage is limited to about $3V_{in}$ due to finite value of Q. This effect can be utilised, especially for low values of V_{in} to obtain larger values of peak capacitor current I_c. Since the overcharge on the capacitor increases with larger load current, this process automatically increases the commutation capability at higher load current.

Design equations: The design steps of the current commutated chopper of Fig. 6.17(a) are explained. The design equations are based on the analysis of the chopper operation during the modes (i) to (iv) shown in Fig. 6.17(b).

Step 1: Choose a value of the ratio $x = I_c/I_{dc}$ (Fig. 6.17c). I_{dc} is the maximum current that the chopper is required to commutate and must be specified from the load considerations. The ratio x is usually chosen to be between 1.4 and 3.0.

Step 2: Calculate,

$$G = \pi - 2 \sin^{-1} (1/x) \tag{6.40}$$

Step 3: Calculate,

$$L_1 = (V_{in}t_q)/(x \cdot G \cdot I_{dc}) \quad H \tag{6.41}$$

$$C_1 = (x \cdot I_{dc}t_q)/G \cdot V_{in} \quad F \tag{6.42}$$

where t_q is the required turn off time for the thyristor T_1.

Step 4: Calculate peak capacitor current I_c (Eqn. 6.35), peak capacitor voltage V_{c4} (Eqn. 6.38) and resistor loss P_{R_f} (Eqn. 6.39). The peak current I_c and voltage V_{c4} are also peak ratings of thyristors T_1 and T_1'.

Step 5: Choose R_f such that chopping period T is much greater than $3R_fC_1$.

6.6.2 Voltage Commutated Switch

In this section a practical thyristor switch is described which uses the voltage commutation principle. The commutation (or turn off) of the thyristor switch is effected by impressing a reverse voltage across it. The reverse voltage is obtained from charge stored on a capacitor and is impressed by using an auxiliary thyristor.

Figure 6.19 shows the circuit diagram of the Type A chopper. Thyristor T_1 is the main power switch corresponding to S_1 in Fig. 6.1a. Thyristor T_1' is the auxiliary thyristor. The circuit operates in the following modes.

Mode (*i*): It is assumed that prior to this mode T_1 has been conducting and that the capacitor C_1 is charged to V_{in} volts in the direction shown in Fig. 6.19(b). The gating of T_1' by signal G_1' starts this mode. Thyristor T_1 is almost instantaneously turned off. The load current, assumed to be constant during this mode, now flows through C_1 and T_1'. The capacitor C_1 discharges from $-V_{in}$ to $+V_{in}$, almost linearly. The output voltage V_0 jumps to $2V_{in}$ at the start of this mode and falls linearly to zero. The

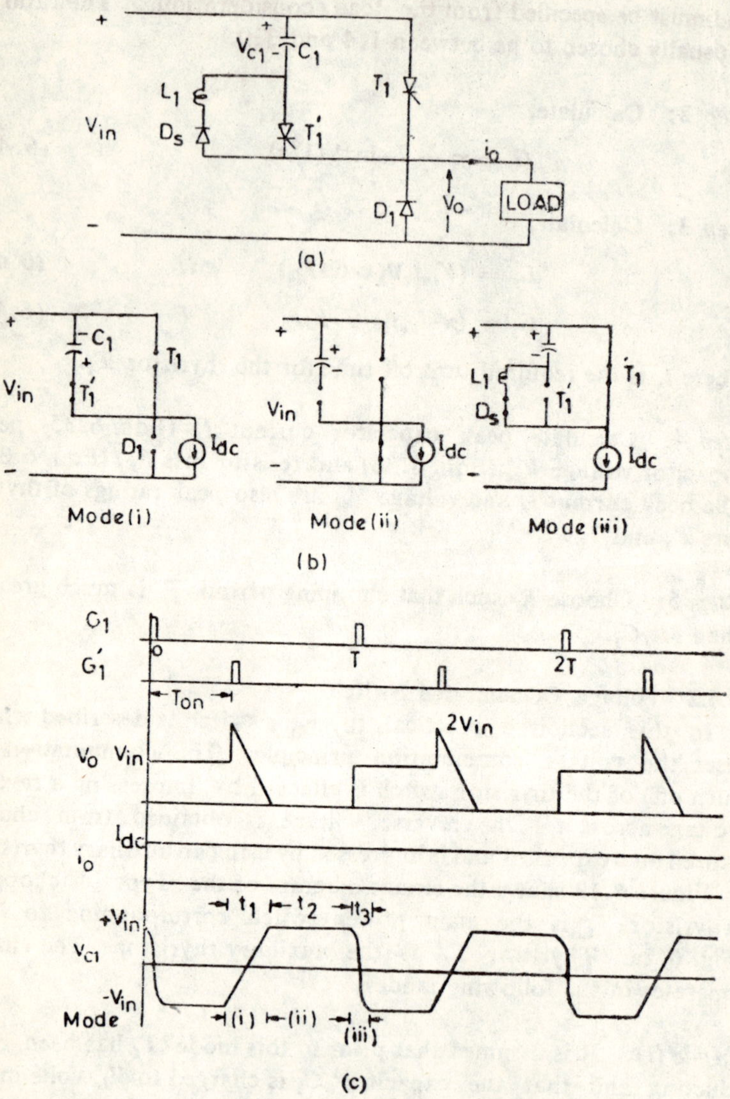

Fig. 6.19. (a) Circuit diagram.
(b) Equivalent circuits.
(c) Waveforms for type *A* voltage commutated chopper.

duration of this mode is given by,

$$t_1 = 2V_{in}C/I_{dc} \tag{6.43}$$

The mode ends when diode D_1 is forward biased and is conducting. The capacitor is charged to $+V_{in}$ at the end of mode. The turn off time for T_1 is $t_1/2$.

Mode (ii): The load current freewheels through diode D_1 in this mode. The duration of the mode is $(T - T_{on} - t_1)$.

Mode (iii): This mode occurs when the main thyristor T_1 is gated. An oscillatory circuit is formed with $L_1C_1T_1$ and D_s. The capacitor voltage changes sinusoidally from $+V_{in}$ to $-V_{in}$. The duration of this mode is given by,

$$t_3 = \pi\sqrt{L_1C_1} \tag{6.44}$$

The turn off time for T_1' is $t_3/2$.

Design equations: The components L_1 and C_1 can be calculated for given values of load current I_{dc} and thyristor turn off times t_{q1} and t_{q1}'.

From Eqn. (6.43) and Eqn. (6.44)

$$t_{q1} = V_{in}C_1/I_{dc} \tag{6.45}$$

$$t_{q1}' = \frac{\pi}{2}\sqrt{L_1C_1} \tag{6.46}$$

From equations (6 45) and (6.46),

$$C_1 = (I_{dc}\, t_{q1}/V_{in}) \tag{6.47}$$

and $$L_1 = (2/\pi \cdot t'_{q1})^2/C_1 \tag{6.48}$$

The peak current for thyristor T_1 is,

$$i_{Tp} = I_1 + V_{in}\sqrt{\frac{C_1}{L_1}} \tag{6.49}$$

The peak voltage across thyristors T_1 and T_1' is,

$$(V_{T1})_p = \pm V_{in} \tag{6.50}$$

The peak current through thyristor T_1' is,

$$i_{T2p} = V_{in}\sqrt{C_1/L_1} \tag{6.51}$$

The peak voltage across the freewheeling diode D_1 is,

$$V_{om} = 2V_{in} \tag{6.52}$$

Important features of voltage commutated chopper: (1) The average load voltage is given by

$$V_0 = (V_{in}\, T_{on}/T) + (V_{in} \cdot t_1/T)$$

$$= \frac{V_{in}}{T} T_{on} + \frac{2V_{in}C}{I_{dc}} \tag{6.53}$$

The effective on period of the chopper is therefore larger than the period T_{on} and varies with the load current.

(2) Since the charge on the capacitor must be reversed when T_1 is turned on, the minimum on period of the chopper is $(t_3 + t_1)$. Thus T must be increased if low values of V_0 are to be obtained.

(3) At start, T'_1 must be gated before T_1 so that a charge is established on C_1. Further if T_1 fails to commutate, C_1 is completely discharged. The commutation therefore cannot be resumed in the next cycle and T_1 must then be turned off by some other means, e.g. switching off V_{in}.

(4) The circuit cannot be operated at no load. The voltage commutation circuit described above and others of the same kind, usually result in larger commutation components (L_1C_1) than the current commutation circuit of section 6.6.1. However, it must be noted that the device manufacturers specify different turn-off times for the two types of commutations. Since the reverse voltage across the thyristor is only a diode drop ($\simeq 1V$) in current commutation, the corresponding turn off time is longer.

6.6.3 Common Chopper Circuits

A large number of chopper circuit configurations are presently being used in d.c. drives. In this section a few common circuits are discussed. The operation of these circuits is similar to the current and voltage commutation circuits analysed in sections 6.6.1 and 6.6.2 Therefore only a qualitative discussion of the operation of Type *A* chopper configurations is given and similarities with the basic commutation circuits are established.

Modified current commutation: A basic modification to the circuit of Fig. 6.17(a) can be made by using a diode in parallel with T'_1 in place of the series diode D_s. Figure 6.20 shows one such circuit. This circuit operates in the four modes shown in Fig. 6.17(b), when T'_1 is gated to commutate T_1. A fifth mode is added when the overcharge on the capacitor causes D_p to conduct. The capacitor is thus left with a voltage lower than V_{in} by the amount $I_{dc}\sqrt{L_1/C_1}$. Further when T_1 is turned on to apply power to the load the capacitor C_1 reverses charge by oscillation through T_1, D_p and L_1. Thus T_1 carries the extra current pulse over and above the load current. Reference [6] gives the design of the circuit of Fig. 6.20 in detail.

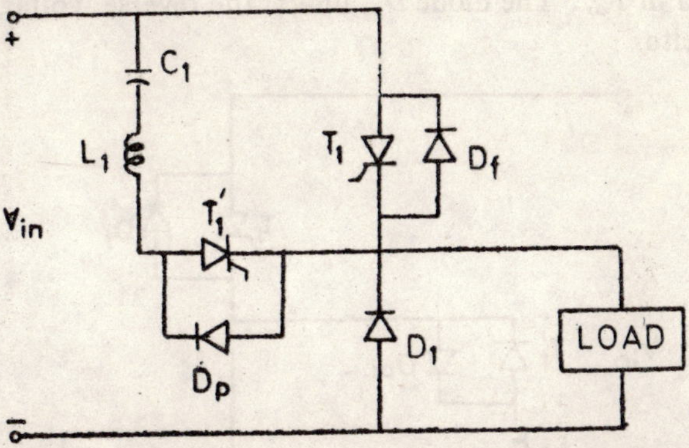

Fig. 6.20. Type *A* chopper using modified current commutation.

Figure 6.21 shows another circuit in which the commutating inductor L_1 is placed in series with freewheeling diode. This configuration therefore eliminates the overcharging of capacitor due to L_1. The trapped energy in the source inductance may still result in a slight under charging of the capacitor. The inductor L_1 also

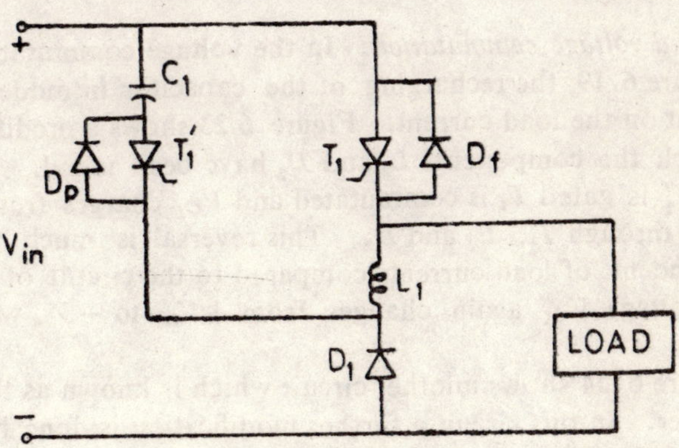

Fig. 6.21. Type *A* chopper using modified current commutation.

reduces the rate of rise of current when T_1 is turned on to very low values. However, as L_1 is carrying the load current during the freewheeling period, its rating is considerably increased.

Figure 6.22 shows another current commutated chopper in which the commutation components are placed across the load. The capacitor C_1 now charges to a value $2V_{in}$ when T_1 is turned on. When T_1' is gated T_1 is turned off by the current flowing through C_1, L_1, T'_1, T_1/D_f and the d.c. supply. The energy trapped in L_1 is dissipated in R_d. The diode D_{p2} limits the reverse voltage across the capacitor.

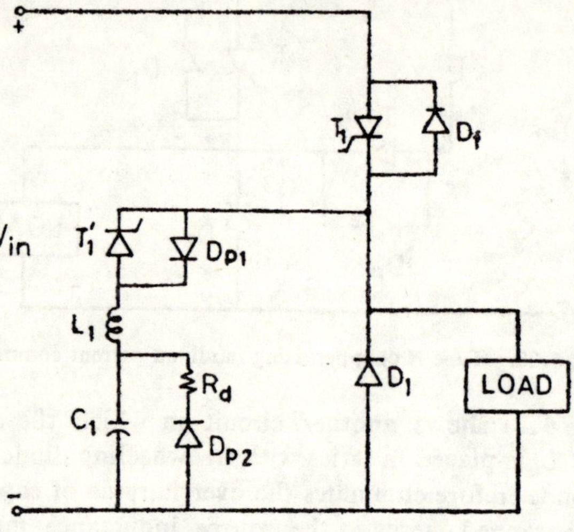

Fig. 6.22. Type *A* chopper using modified current commutation.

Modified voltage commutation: In the voltage commutation circuit of Figure 6.19, the recharging of the capacitor in mode (i) is dependent on the load current. Figure 6.23 shows a modified circuit in which the components L_2 and D_2 have been added. Therefore, when T_1' is gated T_1 is commutated and V_{C1} charges from $-V_{in}$ to $+V_{in}$ through T_1', L_2 and D_2. This reversal is much faster and independent of load current compared to the circuit of Fig. 6.19. The voltage V_{C1} again changes from $+V_{in}$ to $-V_{in}$ when T_1 is gated.

Figure 6.24 shows another circuit which is known as the Mazda Chopper. In this circuit a further modification is done by providing a separate thyristor T_2 for independent reversal of the capacitor voltage from $+V_{in}$ to $-V_{in}$. This eliminates the extra current pulse in T_1 at turn on.

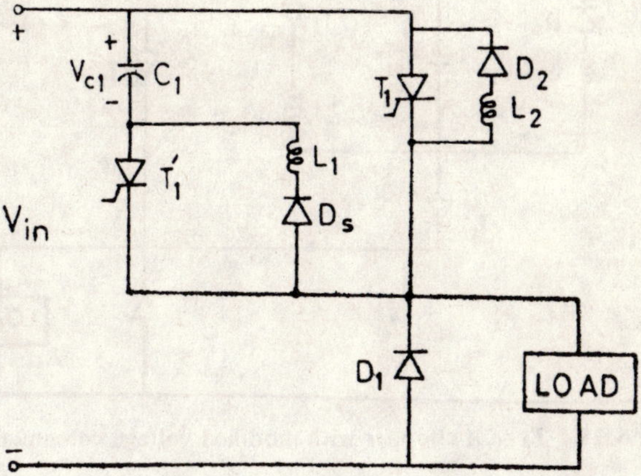

Fig. 6.23. Type *A* chopper with modified voltage commutation.

A significant drawback of the chopper circuits of Figs. 6.19, 6.23 and 6.24 is that the on period cannot be reduced below a minimum value. Figure 6.25 shows a circuit known as Farrer Chopper which overcomes this drawback. Here the thyristor T_1' can be turned on at a time when V_{C1} is negative but lower than V_{in}. The recharging time to $+V_{in}$ is thus reduced. The peak output voltage is also reduced from the maximum value of $2V_{in}$.

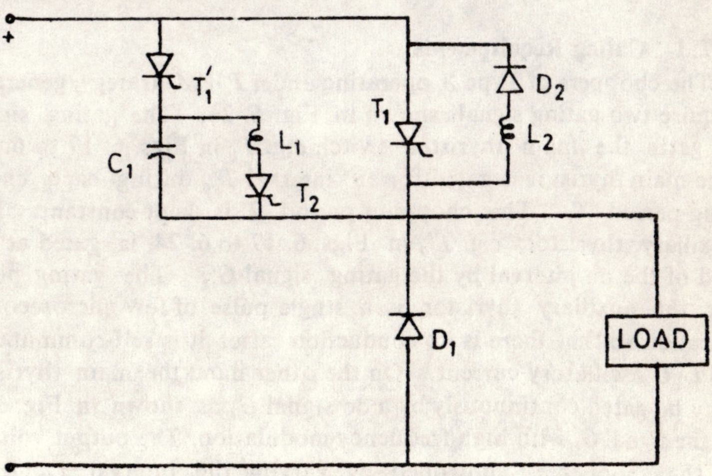

Fig. 6.24. Type *A* chopper with modified voltage commutation.

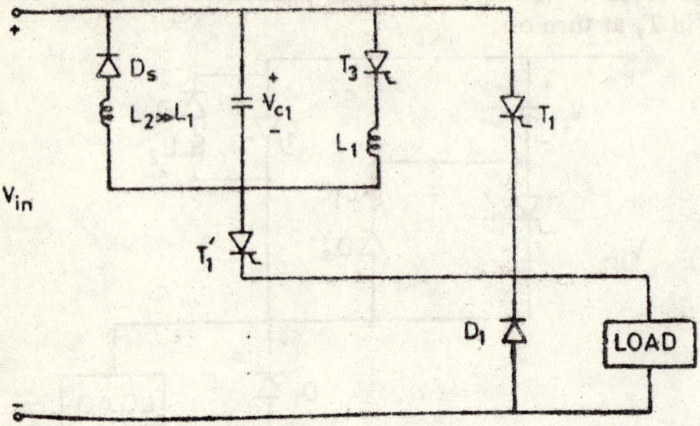

Fig. 6.25. Type A chopper with modified voltage commutation.

6.7 GATING AND CONTROL CIRCUITS

The operation of the chopper circuits described in section 6.6 assumes that the main and the auxiliary thyristors are gated at appropriate instants of time. The general structure of the gating or firing circuit has been given in Chapter 2. The present section discusses the implementation of the 'Pulse Generator' (Fig 2.19) for providing the gating pulses to a Type A chopper under the *PWM* control strategy (section 6.2.1). The implementation is based on the integrated circuit components described in Chapter 10.

6.7.1 Gating Requirements

The choppers of Type A operating under *PWM* strategy generally require two gating signals shown in Fig. 6. 26. The gating signal G_1 gates the main thyristor switch e.g. T_1 in Figs. 6. 17 to 6. 24. The main thyristor is gated for an interval T_{on} during each chopping period T. The chopping period T is kept constant. The auxiliary thyristor, e.g. T'_1 in Figs. 6. 17 to 6. 24, is gated at the end of the on interval by the gating signal G'_1. The gating pulse for the auxiliary thyristor is a single pulse of few microseconds duration so that there is no conduction after it is self-commutated by L–C oscillatory current. On the other hand the main thyristor may be gated continuously by a dc signal G_1 as shown in Fig. 6.26 or the signal G_1 with high frequency modulation. The output voltage of the chopper is controlled by varying the interval T_{on} from nearly zero to T seconds.

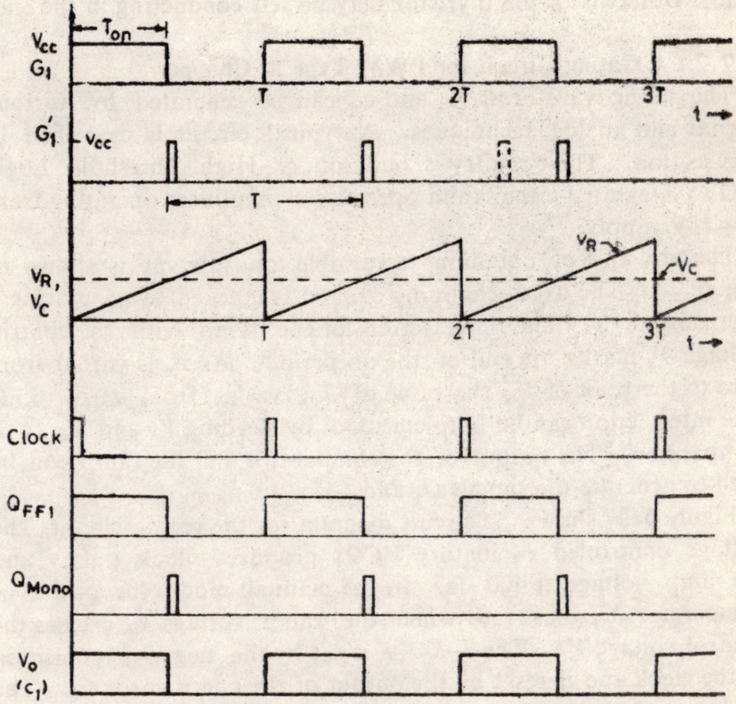

Fig. 6.26. Gating and control signals for type *A* PWM chopper Symbols refer to Fig. 6.27.

Another common requirement in choppers is the instantaneous current limit protection. This protection acts to commutate the main thyristor if the instantaneous load current exceeds a preset limit. Therefore the signal G'_1 may contain a pulse which is asynchronous with the regular commutation pulses. Such a pulse is shown dotted in Fig. 6. 26. The turn-on following the asynchronous commutation should be prevented till the load current has fallen below a safe limit.

Some commutation circuits may require special starting and stopping sequences. For example the circuit of Fig. 6.19 requires that at the start of the operation T'_1 be gated before T_1, so that a charge can be built on the commutation capacitor C_1. Otherwise gating of T'_1 will fail to commutate T_1 in the very first cycle. The stopping sequence for all choppers must include the provision to block to the gating signals after the last commutation has taken place. Otherwise some thyristor may be left conducting in the end.

6.7.2 A Gating Circuit for PWM Type A Chopper

The gating waveforms G_1 and G_1' can be generated by various digital and analog techniques. A typical circuit is described in this section. The circuitry is built up of High Threshold Logic (*HTL*) integrated circuits and operational amplifiers operating from a ± 15V supply.

The principle of obtaining a variable on interval is shown in Fig. 6. 26. The fixed chopping period is marked by generating a ramp signal V_R. The intersection of the ramp with the control voltage V_c marks the end of the on period. As V_c is varied from zero to the peak of V_R, the value of T_{on} varies from zero to T. The intersection can be implemented by feeding V_c and V_R to a comparator. The output of the comparator and the ramp can be used to generate the signals G_1 and G_1'.

Figure 6.27 shows the circuit diagram for the gating circuit. The voltage controlled oscillator (*VCO*) produces clock pulses and the ramp voltage at 400 Hz. In the manual mode, the output of comparator C_1 goes low when the ramp voltage V_R crosses the control voltage V_c. The *J–K F/F* is set by the negative transition of the clock and is reset by the output of the comparator C_1. The resetting of *F/F*1 triggers the monostable, Mono 1 which produces the narrow pulse G_1'. The output of the *JK F/F*1 is modulated by the high frequency oscillator (*HFC*) to generate G_1. T_{on} is varied by potentiometer P_1.

Comparator C_2 provides the instantaneous current limit. The instantaneous load current is compared with a limit set by potentiometer P_2. The output of C_2 is normally high. When the current limit is exceeded it goes low thereby resetting *JK F/F*1 and generating the commutation pulse G_1'.

The start/stop signal is set by a toggle switch and after debouncing fed to the *JK F/F*2. The status of the start/stop switch is checked after the commutation pulse has been produced. Thus the gating signals are blocked if required only after the main thyristor has been commutated.

For closed loop control of the output current the input to the comparator C_1 is derived from a current regulator rather than the potentiometer P_1.

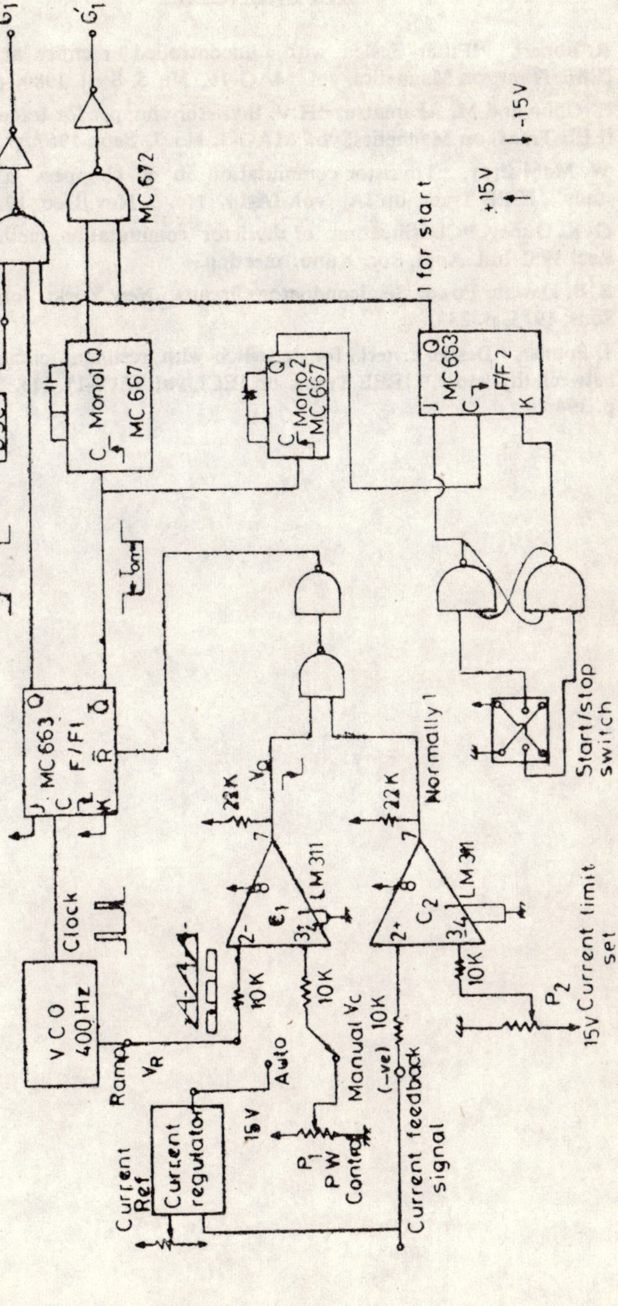

Fig. 6.27. Logic and control circuit for type *A* PWM chopper.

REFERENCES

1. R. Bonert, "Filter design with uncontrolled rectifier and chopper", IEEE Trans. on Magnetics, vol. MAG-16, No. 5, Sept. 1980, p. 928.
2. E. Ohno and M. Akamatsu, "H.V. thyristor chopper for traction control", IEEE Trans. on Magnetics, vol. MAG-3, No. 3, Sept. 1967, p. 232.
3. W. McMurray, "Thyristor commutation in dc choppers—a comparative study", IEEE Trans. on IA, vol. IA-14, No. 6, Nov./Dec. 1978, p. 547.
4. G.K. Dubey, "Classification of thyristor commutation methods", Conf. Rec. 1981 Ind. Appl. Soc. Annu. meeting.
5. S.B. Dewan, Power Semiconductor Circuits, New York: John Wiley and Sons. 1975, p. 283.
6. I. Ponner, "Design criteria for dc switch with resonant circuit connected between thyristors," IEEE Trans. on IECI, vol. IECI-17, No. 5, Aug. 1970, p. 394.

7. INVERTERS

Inverters are used to convert dc power into ac power at desired output voltage and frequency. Some of the important applications of inverters are in variable frequency ac drives, induction heating, standby aircraft and uninterruptible power supplies, high voltage dc transmission etc. The line commutated inverters are described in chapter III. The present chapter deals with the inverters with forced commutation and load commutation. The forced commutated inverters can be classified into voltage source inverters and current source inverters. Voltage source inverters are described in sections 7.1 and 7.2. Current source inverters and load commutated inverters (series inverters) are described in sections 7.3 and 7.4 respectively. Gating circuit requirements of inverters are presented in section 7.5.

7.1 VOLTAGE SOURCE BRIDGE INVERTERS

Inverters fed by a dc source with small or negligible internal impedance are called voltage source inverters. Inverters employing thyristors connected in bridge configuration are called bridge inverters.

7.1.1 Basic Bridge Inverter Circuits—Control Strategy and Output Voltage Waveforms

Two configurations of 1-phase bridge inverters are shown in

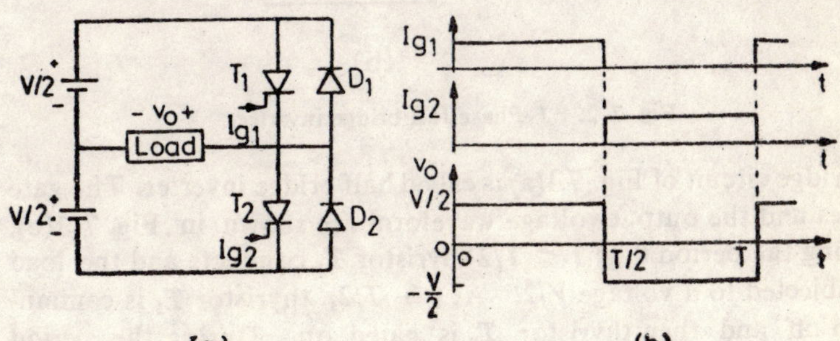

Fig. 7.1. Half bridge inverter.

Figs. 7.1(a) and 7.2(a). Their gate pulse requirements and output voltage waveforms are shown in Figs. 7.1(b) and 7.2(b) respectively. Commutation of thyristors is not being considered at this stage. For simplicity it is assumed that each thyristor conducts for the duration indicated by its gate pulse and gets commutated as soon as this pulse is removed.

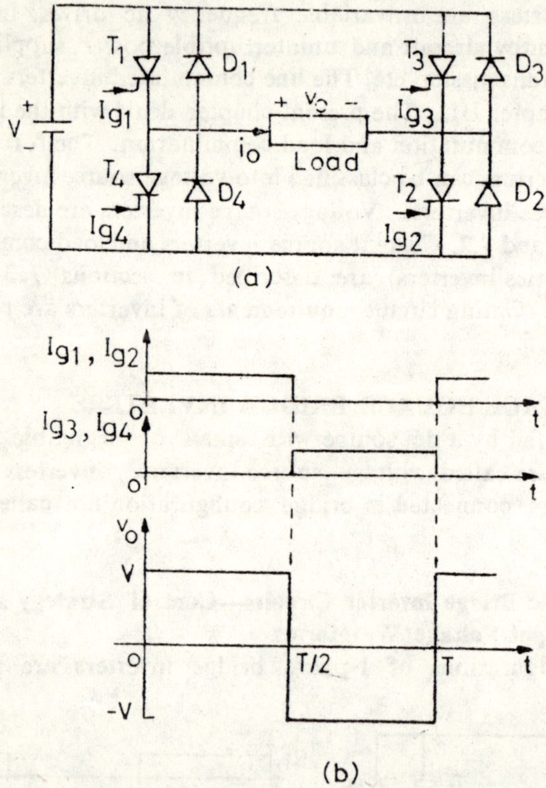

Fig. 7.2. 1-Phase full bridge inverter.

Bridge circuit of Fig. 7.1(a) is called half bridge inverter. The gate pulses and the output voltage waveform are shown in Fig. 7.1(b). During the period $0 \leqslant t \leqslant T/2$ thyristor T_1 conducts and the load is subjected to a voltage $V/2$. At $t = T/2$, thyristor T_1 is commutated off and then thyristor T_2 is gated on. During the period $T/2 \leqslant t \leqslant T$ thyristor T_2 conducts and the load is subjected to a voltage $-V/2$. Thus the load voltage waveform is an alternating rectangular voltage waveform of frequency $(1/T)$. By controlling T one can control the frequency of the inverter output voltage.

For a pure resistive load, thyristors will be enough. But load may be inductive or capacitive. Furthermore, even a resistive load will have some inductance or capacitance. Thus for all types of loads, current i_0 will not necessarily reverse at the same instant as does the load voltage v_0. The diodes D_1 and D_2 connected in anti-parallel with the thyristors will then permit load current to flow. These diodes are called feedback diodes.

The main disadvantage of the half bridge inverter is that it requires a three wire dc source. This limitation is eliminated in the full bridge inverter simply called bridge inverter shown in Fig. 7.2(a). The number of thyristors and diodes required in this case is twice that of the half bridge inverter. However, this is not a disadvantage because the amplitude of output voltage and therefore output power is also twice that of the half bridge. The gate pulse requirement and the output voltage waveform are shown in Fig. 7.2(b). Here also diodes have the same function as explained above. In both these inverters care is taken to ensure that two thyristors in the same branch (in series across the source) do not conduct simultaneously otherwise short circuit will take place across the source.

Three phase bridge inverter configuration is shown in Fig. 7.3. The load may be in star or delta and is connected across terminals

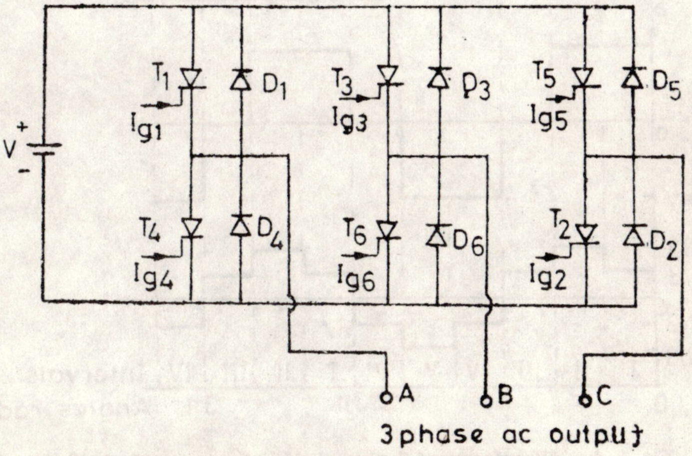

Fig. 7.3. Three-phase bridge inverter.

A, *B* and *C*. Thyristors are fired in the sequence T_1, T_2, T_3, T_4, T_5, T_6, T_1, ... If the inverter ac output voltage period *T* is divided into 2π radians then each thyristor is fired with an interval of $\pi/3$.

Two patterns are employed for the gating and commutation of thyristors. According to one pattern, each thyristor conducts for a duration π and according to another each thyristor conducts for a period of $2\pi/3$.

Figure 7.4 shows the various waveforms for the pattern in which each thyristor conducts for a duration of π. Gate signals I_{g1} to I_{g6} are given with an interval of $\pi/3$ and each one lasts for 180°. Like in the case of 1-phase inverter, the gate pulses indicate that the respective thyristor conducts for the duration of its gate pulse i.e. it is

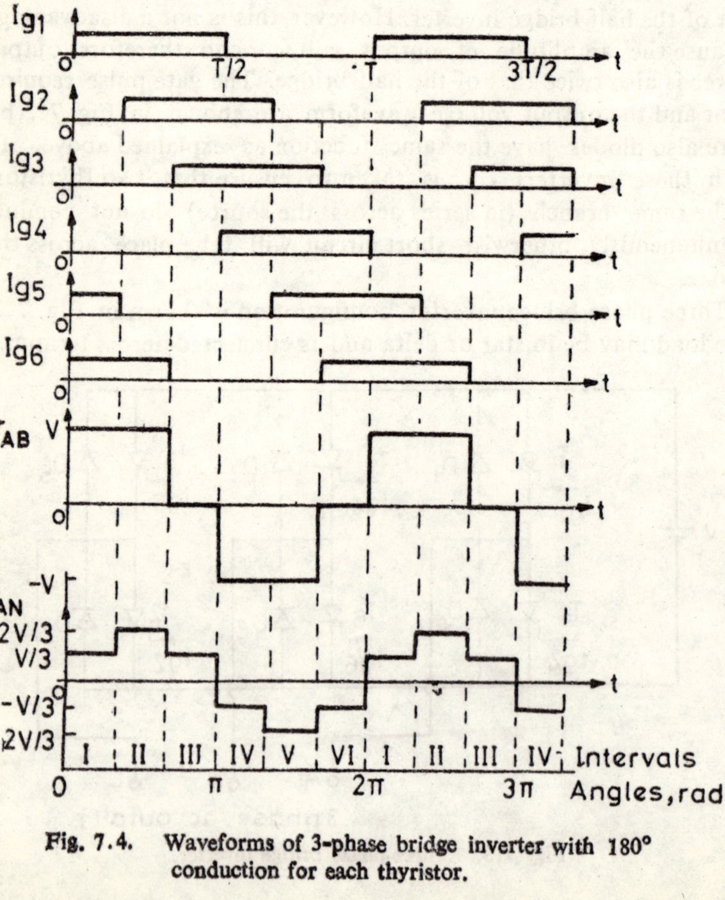

Fig. 7.4. Waveforms of 3-phase bridge inverter with 180° conduction for each thyristor.

turned-on on the application of the gate pulse and commutated as soon as the gate pulse is withdrawn Looking at the gate pulses indicates that 6 commutations per cycle are required. One period of inverter operation has been divided into six intervals. In each interval thyristors conduct according to the following pattern.

Interval	Thyristors conducting	Interval	Thyristors conducting
I	1,5, 6	IV	2,4,3
II	1,2,6	V	5,4,3
III	1,2,3	VI	5,4,6

Thus in this gating and commutation pattern, three thyristors conduct at a time. Once the conducting thyristors are identified, line voltage waveforms can be easily drawn. In Fig. 7 4, line voltage v_{AB} has been plotted. Other line voltage waveforms v_{BC} and v_{CA} will be displaced by 120° and will have similar waveforms. These output voltages can be described by the following Fourier series

$$v_{AB} = \sum_{n=1,3,5,\ldots}^{\infty} \frac{4V}{n\pi} \cos\frac{n\pi}{6} \sin n\left(\omega t + \frac{\pi}{6}\right) \tag{7.1}$$

$$v_{BC} = \sum_{n=1,3,5,\ldots}^{\infty} \frac{4V}{n\pi} \cos\frac{n\pi}{6} \sin n\left(\omega t - \frac{\pi}{2}\right) \tag{7.2}$$

$$v_{CA} = \sum_{n=1,3,5,\ldots}^{\infty} \frac{4V}{n\pi} \cos\frac{n\pi}{6} \sin n\left(\omega t - \frac{\pi}{6}\right) \tag{7.3}$$

Equations (7.1) to (7.3) show that the tripplen harmonics are zero in the line voltages. These line voltages have well defined values during all the six intervals, and therefore, their waveforms are independent of load, which may be linear or non-linear, balanced or unbalanced and may have any combination of resistance, inductance and capacitance. If load circuit is delta connected the phase and line currents can be calculated using equations (7.1) to (7.3). If the load is star connected and linear, super-position theorem can be used to obtain the phase currents and voltages.

In Fig. 7.4, phase voltage v_{AN} has also been drawn for a star connected resistance load connected across terminals ABC. This voltage is drawn by considering equivalent circuits of the inverter and load valid for various intervals, as an example, the load equivalent circuits for intervals I, II and III are shown in Fig. 7.5. From these figures it can be seen that

Interval		Phase voltage	
I	$(0 < \omega t < \pi/3)$	$v_{AN} = V/3$	
II	$(\pi/3 < \omega t < 2\pi/3)$	$v_{AN} = 2V/3$	
III	$(2\pi/3 < \omega t < \pi)$	$v_{AN} = V/3$	(7.4)

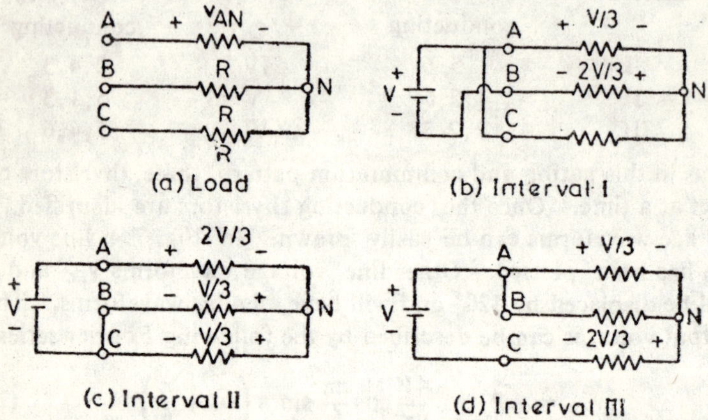

Fig. 7.5. Inverter load equivalent circuits.

Similarly, one can obtain other phase voltages v_{BN} and v_{CN}, which will be lagging behind v_{AN} by $2\pi/3$ and $4\pi/3$ rad respectively.

Figure 7.6 shows various waveforms for the gating and commutation pattern in which each thyristor conducts for $2\pi/3$ radians. Like in the previous case here also gate pulse indicates the conduction period of each thyristor. Six commutations per cycle are required in this case also. In Fig. 7.6 one period of inverter operation has been divided into six intervals. In each interval thyristors conduct according to the following pattern.

Interval	Thyristors conducting	Interval	Thyristors conducting
I	1, 6	IV	4, 3
II	2, 1	V	5, 4
III	3, 2	VI	6, 5

Thus in this gating and commutation pattern, two thyristors conduct at a time, one from among T_1, T_3 and T_5 and one from the group T_2, T_4 and T_6. Through the conduction of two thyristors,

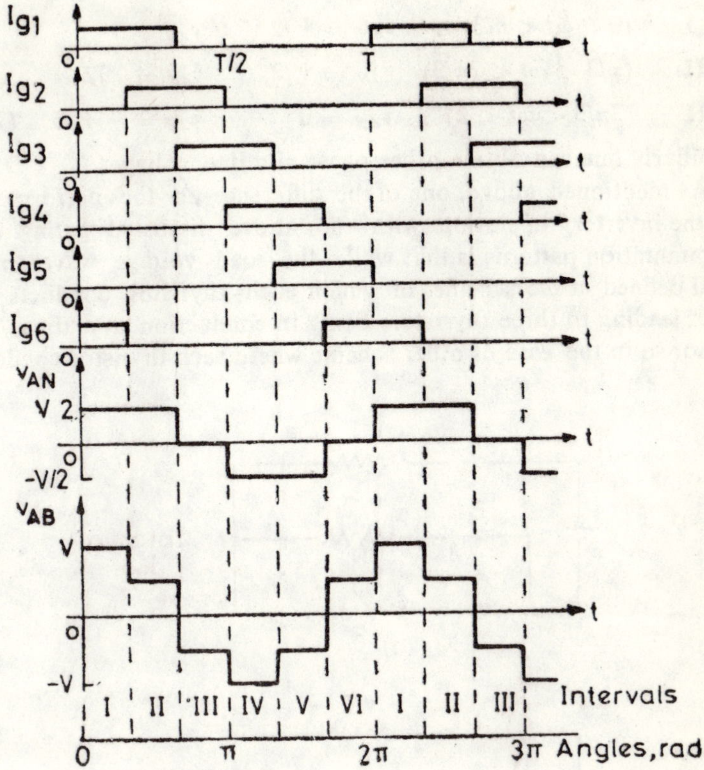

Fig. 7.6. Waveforms of 3-phase bridge inverters with 120° conduction for each thyristor.

two of the load terminals get connected to dc supply terminals and the third terminal remains floating during any of the above six intervals. The potentials of the terminals connected to the dc source are well defined but the potential of the floating terminal at any instant will depend on the nature of the load and will not remain constant during the duration ($\pi/3$) of the interval. The load voltage waveforms are, however, well defined for a balanced star connected resistance load (Fig. 7.5(a)) and can be obtained by considering equivalent circuits of the inverter load combination valid for various intervals. As an example equivalent circuits valid for intervals I, II and III are shown in Fig 7.7. From these equivalent circuits one can note the following

Interval		Phase voltage	Line voltage	
I	$(0 < \omega t < \pi/3)$	$v_{AN} = V/2$	$v_{AB} = V$	
II	$(\pi/3 < \omega t < 2\pi/3)$	$v_{AN} = V/2$	$v_{AB} = V/2$	
III	$(2\pi/3 < \omega t < \pi)$	$v_{AN} = 0$	$v_{AB} = -V/2$	(7.5)

Similarly one can obtain other phase and line voltages.

As mentioned above, one of the difference in the performance of the inverters operating with the above discussed gating and commutation patterns is that while the load voltage waveform is well defined in the scheme in which each thyristor conducts for 180° leading to three thyristors being in conduction at a time, this is not so in the case of other scheme where each thyristor conducts

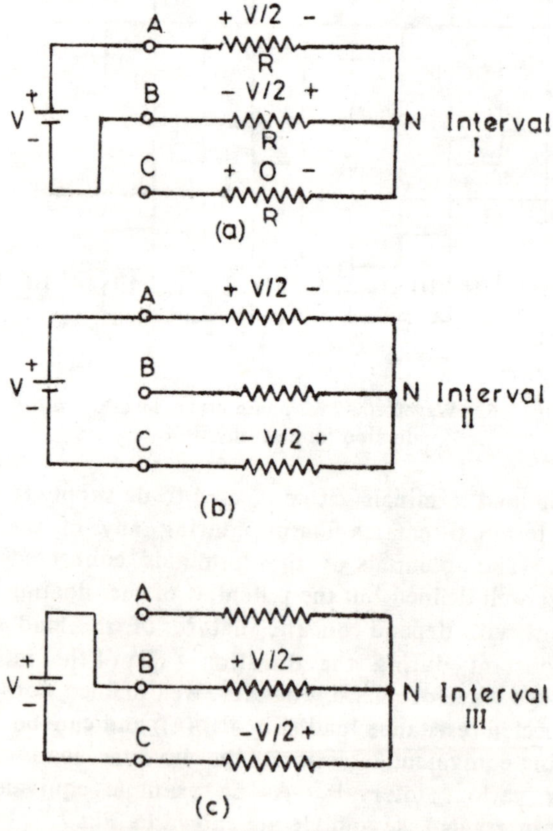

Fig. 7.7. Single-phase bridge inverter equivalent circuits.

for 120° and so at a time only two thyristors are on. There are two other important differences which are as follows:

Examination of the gating and commutation pattern involving conduction of each thyristor for 180° will show that the turning on of any thyristor in the series branches (T_1, T_4), (T_3, T_6) and (T_5, T_2) is followed immediately after the commutation of the thyristor in that branch. Thus enough time may not be provided for the commutation of thyristor and two thyristors in series may simultaneously conduct, resulting into short circuit of the source by these thyristors. On the other hand in the gating and commutation pattern, requiring 120° conduction for each thyristor, there is an interval of $\pi/3$ between the commutation and the turning on of the series thyristors. Thus enough time is made available for the outgoing thyristor to commutate before the thyristor in series is turned on. Therefore, commutation is more reliable and the possibility of two series thyristors conducting simultaneously is much less.

The second important difference is in relation to utility of thyristors. The comparison of the two schemes is done using a figure of merit termed as utility factor (*UF*) and is defined as

$$UF = P_0/P_T \tag{7.6}$$

where P_0 is the rated output power of the inverter and P_T is the measure of total power handling capability of the thyristors employed in the inverter, and is defined as

$$P_T = NV_{\text{DRM}} I_{\text{rms}} \tag{7.7}$$

where N = number of thyristors

V_{DRM} = repetitive peak forward voltage

and I_{rms} = rated RMS forward current

It is usual to calculate utility factor for ideal resistive loads for the purpose of comparison of the two control schemes.

Inverter with 180° conduction: If resistance per phase is R, then for a star connected load,

$$P_0 = 3\left(1/\pi \int_0^\pi \frac{v_{AN}^2}{R}\, d(\omega t)\right) \tag{7.8}$$

Substituting from eqn. (7.4) gives

$$P_0 = 2V^2/3R \tag{7.9}$$

Thyristor rating will be chosen such that the rms current flowing through it will be equal to its rms current rating I_{rms}. During a cycle, current of any phase is shared between two thyristors in series, for example for phase A, thyristor T_1 carries load current during the positive half cycle and thyristor T_4 during the negative half cycle. Therefore the RMS value of load phase current under the rated conditions, I_p is given by

$$I_p = \sqrt{2}\, I_{rms} \tag{7.10}$$

Load power P_0 can also be written as

$$P_0 = 3I_p^2 R = 6I_{rms}^2 R \tag{7.11}$$

From eqns. (7.9) and (7.11)

$$I_{rms} = V/3R \tag{7.12}$$

The repetitive peak forward voltage rating that a thyristor should have for this load, $V_{DRM} = V$. Thus,

$$P_T = 6V\frac{V}{3R} = \frac{2V^2}{R} \tag{7.13}$$

From eqns. (7.6), (7.9) and (7.13)

$$UF = \frac{P_0}{P_T} = \frac{1}{3} = 33.33 \text{ per cent} \tag{7.14}$$

This indicates that the rated power output of the inverter in this case is only 33.33 per cent of the combined maximum power rating of the six thyristors.

Inverter with 120° *conduction*: For a star connected load with a per phase resistance of R, P_0 will be given by eqn. (7.8). Substituting from eqn. (7.5) in eqn. (7.8) yields

$$P_0 = \frac{V^2}{2R} \tag{7.15}$$

Here also the relation between I_{rms} and I_p is given by eqn. (7.10) and P_0 in terms of I_{rms} is given by eqn. (7.11). From eqns. (7.11) and (7.15)

$$I_{rms} = \frac{1}{2\sqrt{3}}\frac{V}{R} \tag{7.16}$$

Now from eqn. (7.7)

$$P_T = 6V\frac{1}{2\sqrt{3}}\frac{V}{R} = \sqrt{3}\,\frac{V^2}{R} \tag{7.17}$$

and from eqn. (7.6)

$$UF \simeq \frac{1}{2\sqrt{3}} \simeq 28.9 \text{ per cent} \qquad (7.18)$$

In this case the rated inverter output is 28.9 per cent of the combined maximum power rating of 6 thyristors.

Comparison of eqns. (7.14) and (7.18) will show that the utility factor is lower in the case of 120° conduction scheme.

7.1.2 Single Phase Bridge Inverter Load Current Waveforms and Commutation Requirements

The load voltage waveform is shown in Fig. 7.2(b). It is independent of the nature of the load and is given by

$$\begin{aligned} v_0 &= V \qquad 0 < t < T/2 \\ &= -V \qquad T/2 < t < T \end{aligned} \qquad (7.19)$$

The load current waveform will depend on the nature of the load. Feedback diodes provide continuity in load current and a path for flow of reactive energy. As an example, let us consider an RL load. Let T_3 and T_4 be conducting initially. The load current i_0 will be flowing in the negative direction. Now at $t = 0$, T_3 and T_4 are turned off. The load current will now flow through diodes D_1 and D_2 against the supply voltage. If now gate pulses are applied to thyristors T_1 and T_2 they will not turn on as they are reverse biased by the drops of D_1 and D_2. As soon as the current through D_1 and D_2 falls to zero value, T_1 and T_2 will turn on provided the gate pulse is present and this is being ensured by using a gate pulse of duration π. Now the current will start flowing in the positive direction reaching a maximum value at the instant of commutation of T_1 and T_2. Similarly minimum will occur at the instant of commutation of T_3 and T_4. In steady state absolute values of maximum and minimum will be the same.

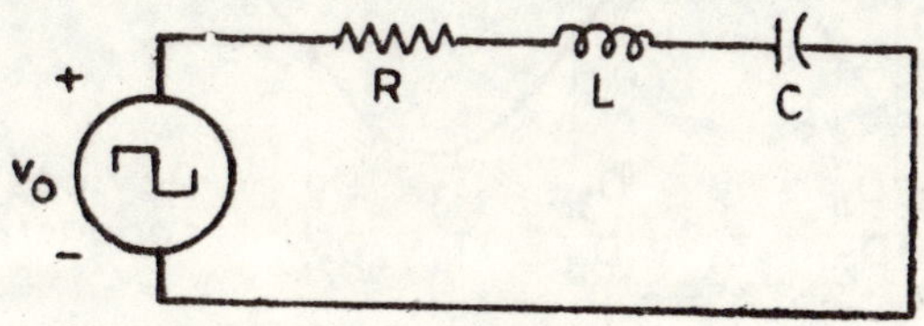

Fig. 7.8. Single-phase bridge inverter equivalent circuit.

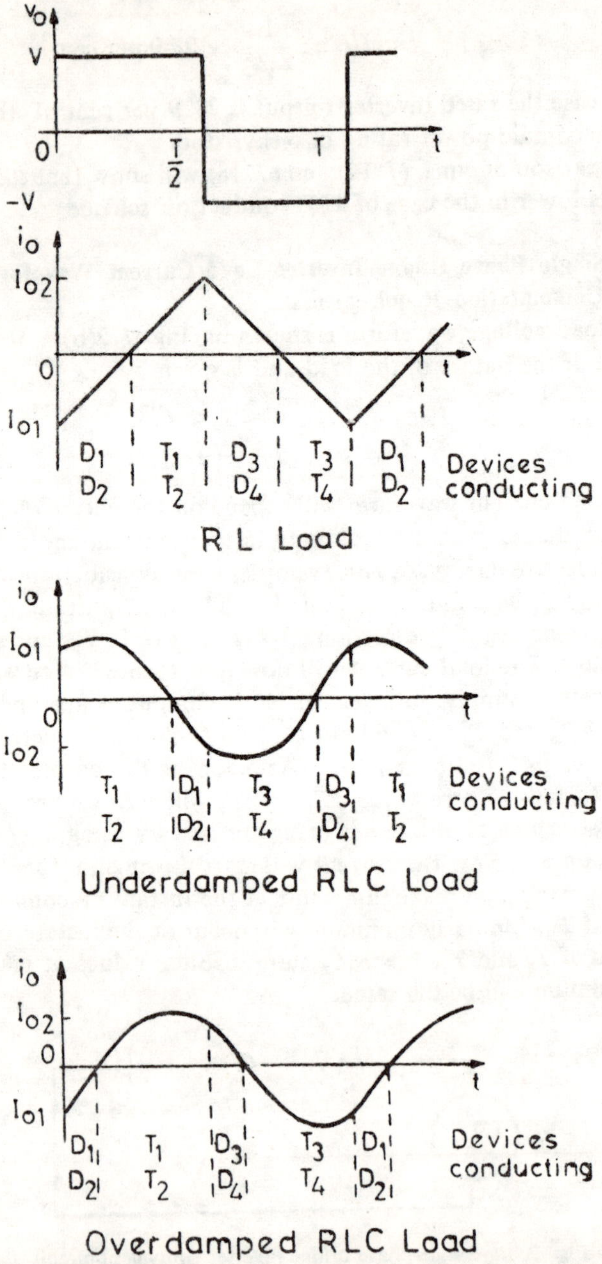

Fig. 7.9. Load voltage and current waveforms of 1-phase bridge inverter.

Since the current continuity is being ensured and since the load voltage waveform is independent of the nature of the load, one can obtain an equivalent circuit of the bridge inverter consisting of a square wave generator feeding the load. In Fig. 7.8 inverter equivalent circuit has been drawn for a general *RLC* load. The operation is described by the following differential equations.

For $O < t < T/2$

$$V = Ri_0 + \frac{Ldi_0}{dt} + \frac{1}{C}\int_0^t i_0 dt + V_{C1} \tag{7.20}$$

with the assumed initial conditions

$$V_c(O) = V_{C1} \text{ and } i_0(O) = I_{O1} \tag{7.21}$$

and for $T/2 < t < T$,

$$-V = Ri_0 + L\frac{di_0}{dt'} + \frac{1}{C}\int_0^{t'} i_0 dt' + V_{C2} \tag{7.22}$$

with the assumed initial conditions

$$v_0(O) = V_{C2} \text{ and } i_0(O) = I_{O2} \tag{(7.23)}$$

where $t' = t - T/2$.

Differentiating eqn. (7.20) gives

$$\frac{d^2i_0}{dt^2} + \frac{R}{L}\frac{di_0}{dt} + \frac{1}{LC}\, i_0 = 0 \tag{7.24}$$

Differentiating equation (7.22) will give a similar equation with t replaced by t'. These second order equations can be solved using initial conditions of eqns. (7.20) and (7.23). In steady state $|I_{O1}| = |I_{O2}|$ with opposite signs.

The load current waveforms for a *RL* load and underdamped and overdamped *RLC* loads are shown in Fig. 7.9. Various devices conducting during various intervals are also indicated.

In the case of *RL* load and overdamped *RLC* load, the thyristors carry forward current at the time of commutation (at $t = 0$ for T_3 and T_4 and $t = T/2$ for T_1 and T_2) and therefore, can be commutated only by using forced commutation. In the case of underdamped *RLC* load, the load current, which is also the current flowing through thyristors, becomes zero before the end of respective half cycle. The thyristors stop conducting. Next instant the load current reverses and is carried by diodes connected in antiparallel to them, applying reverse bias across the thyristors. If the duration of the reverse bias is larger than the thyristor turn off time, thyristors will get commutated and no additional arrangement will be required for their commutation. This method of commutation, known as self commutation or load commutation or natural commutation, is in fact employed in inverters used for high frequency induction heating.

Some of the commonly used commutation circuits for bridge inverters will be described in the following sections.

7.1.3 Modified McMurray Half Bridge Inverter

Modified McMurray Half-Bridge inverter is shown in Fig. 7.10(a). T_1 and T_2 are main thyristors and D_1 and D_2 are main diodes. The commutation circuit consists of auxiliary thyristors T_{A_1} and T_{A_2}, diodes D_{A_1} and D_{A_2} and inductance L and capacitance C. The operation of the circuit will be described for a lagging *pf* load with the following simplifying assumptions.

(1) The load circuit time constant is sufficiently large to permit the assumption of a constant current during the commutation interval.

(2) Thyristors and diodes are ideal switches.

(3) L and C are ideal elements having no resistance.

Initially T_1 is conducting, supplying current to a predominantly inductive load, other thyristors and diodes are not conducting. The commutation capacitor C has been charged previously to a voltage V with the polarity indicated in the figure. The commutation of T_1 is initiated at $t = 0$ by turning on auxiliary thyristor T_{A_1}. The inverter equivalent circuit for this mode (I) of operation is shown in Fig. 7.10(b). The value of the load current at this instant is I_0 and according to assumption (1) this will be assumed constant during the commutation of T_1. Capacitor current i_c flows through the loop consisting of L, T_1, T_{A_1} and C. As I_0 is constant a rise in the capacitor current results in a corresponding decrease of the current through thyristor T_1. Waveforms of i_c, I_0, i_0 and device currents are shown in Fig. 7.11. At time $t = t_1$, i_c equals I_0 and therefore, thyristor T_1 current becomes zero. Further increase in i_c causes a current equal to $(i_c - I_0)$ to flow through diode D_1. i_c, which has a sinusoidal waveform, reaches the maximum value I_m and then decreases. At $t = t_2$, i_c again equals I_0 and the current through D_1 falls to zero value. During the period $(t_2 - t_1) = t_c$, diode D_1 conducts applying a reverse bias equal to its voltage drop

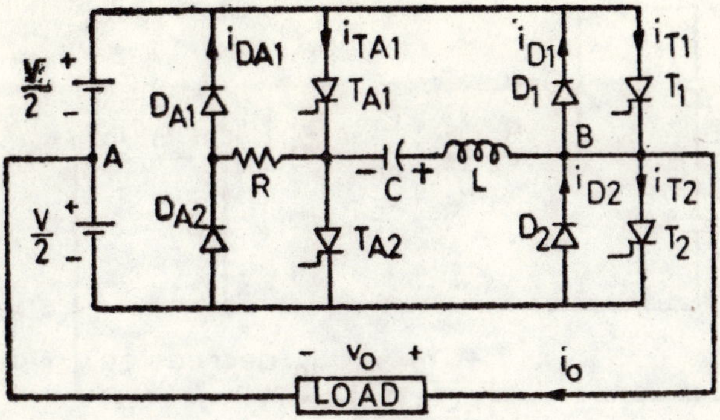

(a) Inverter circuit

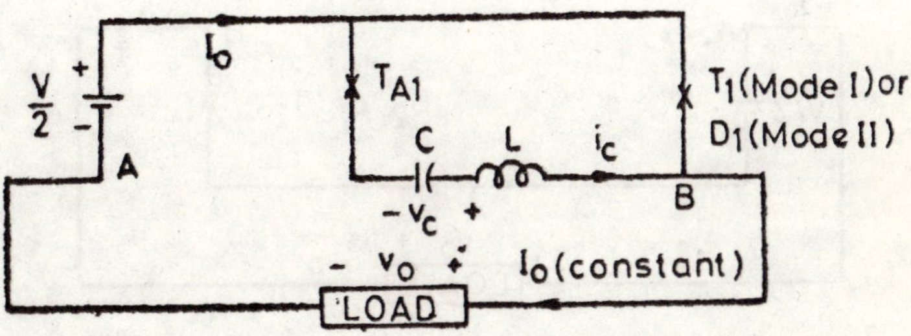

(b) Equivalent circuit, Mode I & II

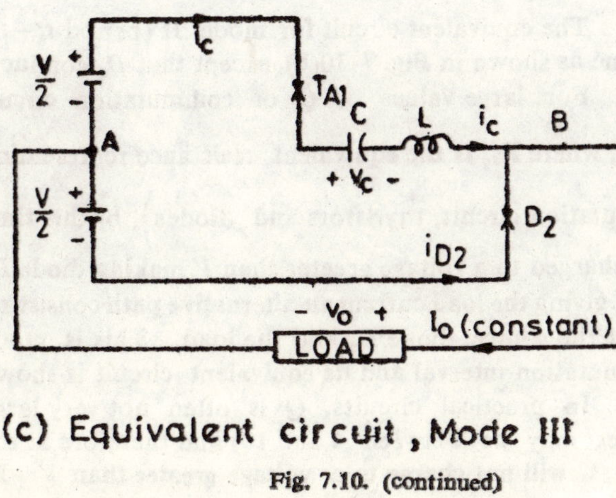

(c) Equivalent circuit, Mode III

Fig. 7.10. (continued)

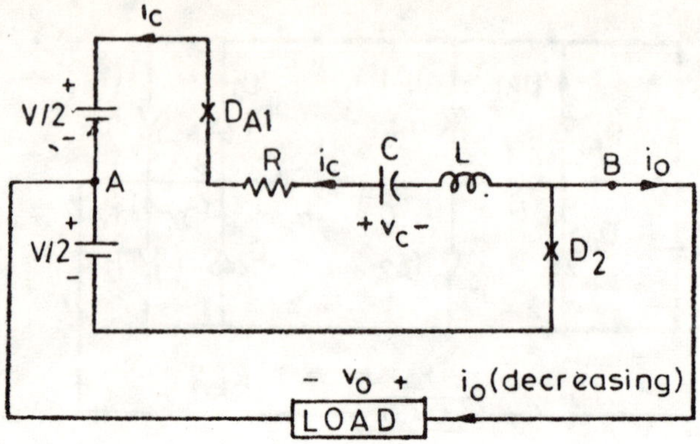

(d) Equivalent circuit, Mode IV

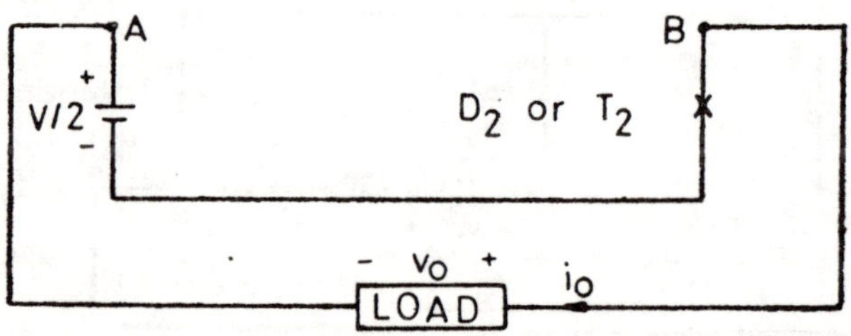

(e) Equivalent circuit, Mode V

Fig. 7.10. McMurray inverter.

to thyristor T_1. The equivalent circuit for mode II (period t_2-t_1) remains the same as shown in Fig. 7.10(b), except that D_1 conducts instead of T_1. For large values of Q of commutation circuit $\left(Q = \sqrt{\frac{L}{C}}\Big/R_{eq}\right.$, where R_{eq} is the equivalent resistance representing losses in commutation circuit, thyristors and diodes$\Big)$, by this time C would have charged to a voltage greater than V making diode D_2 forward biased, giving the load current an alternative path consisting of lower part of the source, diode D_2 and the load. This is mode III of the commutation interval and its equivalent circuit is shown in Fig. 7.10(c). In practical circuits, Q is often not very large (practical values may be between 4 and 10) and therefore at the end of Mode II, C will not charge to a voltage greater than V. In that case, an additional mode consisting of constant current charging of C by load current I_0 will occur before Mode III.

During Mode III, i_c will decrease and i_{D_2} will increase. Thyristor T_2 is given trigger pulse $\pi\sqrt{LC}$ seconds after thyristor T_{A_1}. It will not turn on at this instant due to reverse bias of conducting diode D_2. Ultimately at $t = t_3$, i_c will fall to zero and Mode IV will start. This is the end of the commutation interval and the load current is no more assumed constant. The equivalent circuit for this mode is shown in Fig. 7.10(d). At t_3 entire load current is transferred to D_2. Because the load current has to flow against the supply voltage, it will be decreasing. Meanwhile C, which would

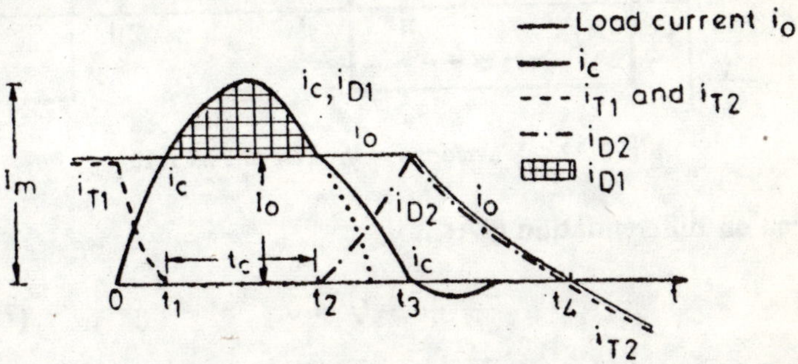

Fig. 7.11. Waveforms during commutation and load current reversal.

have charged to a voltage higher than the source voltage V, with left hand plate positive, will discharge through the damped RLC circuit to a voltage V through the path consisting of R, diode D_{A_1}, source, diode D_2, L and C. The reverse bias due to R and D_{A_1} across T_{A_1} will commutate it. When i_c falls to zero, Mode V will start. At some time t_4, decreasing load current will become zero. Thyristor T_2 will get forward biased and trigger pulse already applied earlier will turn it on and the load will be subjected to negative voltage and current. The load voltage waveform is shown in Fig. 7.12.

Design of Commutation Circuit: McMurray [1] has described a method for selection of optimum commutating capacitance and inductance. For Modes I and II of the commutation interval, equation for capacitor current i_c is

$$L\frac{di_c}{dt} = V - \frac{1}{C}\int i_c\, dt \tag{7.25}$$

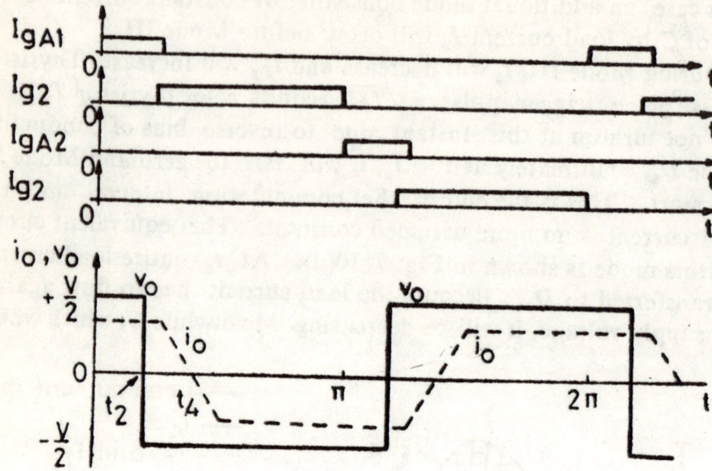

Fig. 7.12. Waveforms of inverter of Fig. 7.10.

which on differentiation gives,

$$\frac{d^2 i_c}{dt^2} + \frac{1}{LC} i_c = 0 \tag{7.26}$$

with initial conditions

$$i_c(0) = 0 \quad \text{and} \quad \frac{di_c}{dt}(0) = V/L \tag{7.27}$$

Solving eqn. (7.26) yields

$$i_c = I_m \sin \omega_0 t \tag{7.28}$$

where $I_m = V\sqrt{C/L}$ and $\omega_0 = \dfrac{1}{\sqrt{LC}}$ (7.29)

From Fig. 7.11

$$t_1 = \frac{1}{\omega_0} \sin^{-1}\left(\frac{I_0}{I_m}\right)$$

and

$$t_2 = \frac{1}{\omega_0}\left(\pi - \sin^{-1}\left(\frac{I_0}{I_m}\right)\right)$$

Therefore,

$$t_c = t_2 - t_1 = \frac{1}{\omega_0}\left(\pi - 2\sin^{-1}\left(\frac{I_0}{I_m}\right)\right) \tag{7.30}$$

The circuit turn off time t_c should be greater than the thyristor turn of time. This condition however can be realised by various combinations of values of C and L and commutation pulses as shown in Fig. 7.13. The optimum commutating pulse is the one which achieves the required turn-off time with the least amount of energy of the capacitor, $\frac{1}{2}CV^2$. From eqn. (7.30) on rearranging terms

$$\cos\frac{\omega_0 t_c}{2} = \frac{I_0}{I_m} = \frac{1}{x} \text{ (say)} \tag{7.31}$$

or

$$\frac{t_c}{\sqrt{LC}} = 2\cos^{-1}\frac{1}{x} = g(x) \tag{7.32}$$

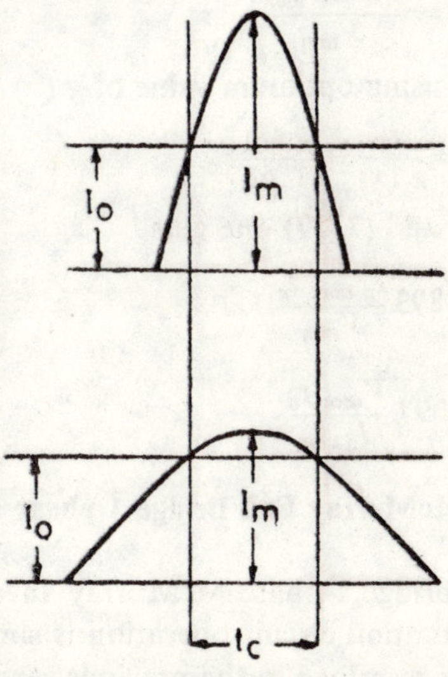

Fig. 7.13.

The energy w that the commutating circuit must provide in order to commutate the thyristor is

$$w = \frac{1}{2}CV^2 = \frac{1}{2}LI_m^2 \tag{7.33}$$

$$= \frac{1}{2}\sqrt{LC}\,VI_m$$

$$= \frac{1}{2}\frac{t_c}{2\cos^{-1}(1/x)}V\,x\,I_0 \tag{7.34}$$

Expressing eqn. (7.34) in normalised form as a function of the parameter x gives,

$$\frac{W}{VI_0t_c}=\frac{x}{4\cos^{-1}(1/x)}=h(x) \qquad (7.35)$$

On plotting $h(x)$ and $g(x)$ against x from eqns. (7.32) and (7.35) it is found that the normalised commutation energy $h(x)$ has a minimum value of 0.446 when $x = 1.5$ and $g(x) = 1.68$. The values of commutating circuit are now calculated as follows:

The design is done for the worst condition which consists of minimum supply voltage V_{min} and maximum load current $I_{0\,max}$.

From eqn. (7.29) and (7.31) using optimum value of x=1.5.

$$V_{min}\sqrt{C/L}=x\,I_{0\,max}=1.5\,I_{0\,max}$$

or

$$\sqrt{C/L}=\frac{1.5\,I_{0\,max}}{V_{min}} \qquad (7.36)$$

From eqn. (7.32), using optimum value of $g(x) = 1.68$

$$\sqrt{LC}=\frac{t_c}{g(x)}=\frac{t_c}{1.68} \qquad (7.37)$$

From eqns. (7.36) and (7.37) one gets

$$C=0.893\,\frac{I_{0\,max}\,t_c}{V_{min}} \qquad (7.38)$$

and

$$L=0.397\,\frac{V_{min}\,t_c}{I_{0\,max}} \qquad (7.39)$$

7.1.4 Modified McMurray Full Bridge 1-phase and 3-phase Bridge Inverters

A modified full bridge 1-phase McMurray inverter is shown in Fig. 7.14. Commutation circuit operation is similar to that of the half bridge inverter described in the previous section.

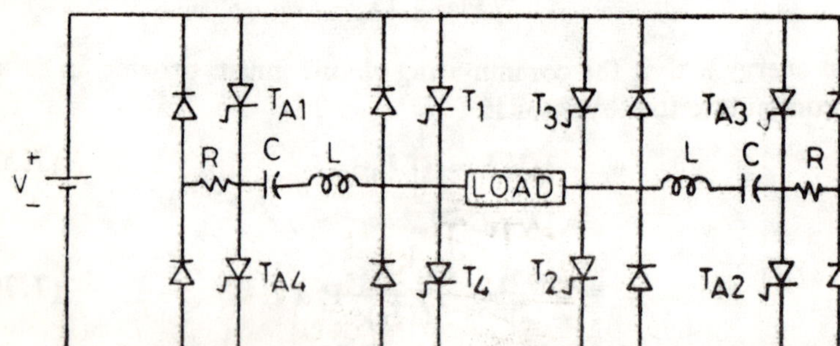

Fig. 7.14. Full bridge modified McMurray inverter.

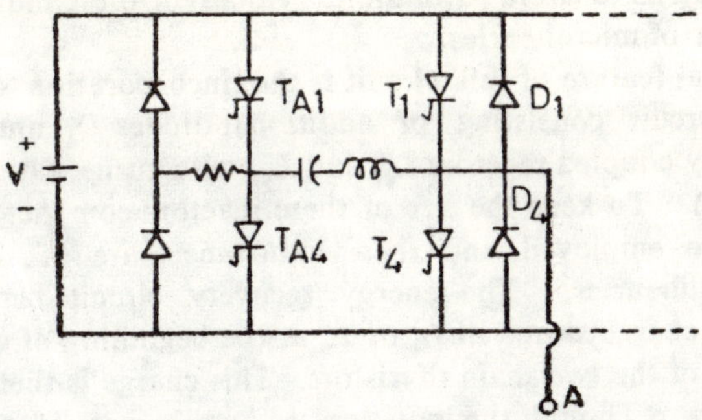

Fig. 7.15. Three-phase modified McMurray inverter.

Three-phase bridge inverter using McMurray commutation method is shown in Fig. 7.15. Only the leg supplying phase A, consisting of thyristors T_1 and T_4 along with their commutation circuits, is shown in the figure. Other two legs feeding phases B and C have the similar circuits and therefore have not been shown.

7.1.5 Modified McMurray Bedford Half-Bridge Inverter [2]

The circuit is shown in Fig. 7.16. It uses less number of thyristors compared to the Half-Bridge McMurray inverter. However the numbers of inductors and capacitors are higher. T_1 and T_2 are main thyristors, and D_1 and D_2 are feedback diodes. Commutation circuit employs capacitors C_1 and C_2 and magnetically coupled inductors L_1 and L_2 having the same inductance L and the same number of turns. Inductors L_1 and L_2 are wound on a core with

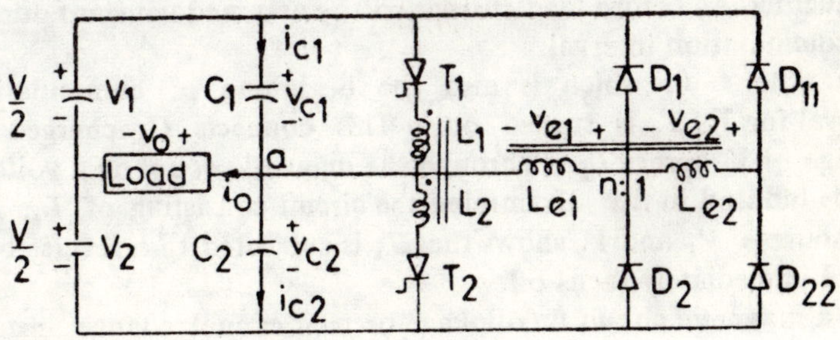

Fig. 7.16. Modified McMurray-Bedford half bridge inverter.

an air gap so as to avoid saturation. Values of these inductors are of the order of microhenries.

Additional feature of this circuit is the incorporation of energy recovery circuit consisting of additional diodes D_{11} and D_{22} and magnetically coupled reactors L_{e1} and L_{e2} with turns ratio of $n:1$ (with $n<1$). To keep the size of these reactors compact iron core reactors are employed and their inductances are of the order of few millihenries. The energy recovery circuit permits the recovery of charge stored in L_1 or L_2 at the beginning of commutation of any of the two main thyristors. This charge is then fedback to the source. Though the inductors L_1 and L_2 are of the order of microhenries, the recovery of charge from them helps in improving the efficiency and frequency of operation of the inverter.

The operation of this circuit will also be considered with the same simplifying assumptions as described in the previous section.

To study the operation of the circuit, let us consider the operation from the instant of turning on of thyristor T_2. Prior to turning on T_2, T_1 is conducting load current. As this is the end of positive half cycle, the rate of change of load current will be very small, and therefore, drop across L_1 can be neglected and C_1 can be assumed to have zero voltage, C_2 will then have voltage V. Capacitor currents will also be zero. Energy recovery transformer will be feeding the energy back to the source only when secondary voltage V_{e2} is either greater than V or less than $-V$. Since the ratio of turns of L_{e1} and L_{e2} is $n:1$ this statement implies that for energy recovery v_{e1} should either be greater than nV or less than $-nV$. In the situation under consideration D_1 is reverse biased and v_{e1} is zero. Thus none of the diodes is conducting and the currents through L_{e1} and L_{e2} are also zero. The current through L_2 is also zero as T_2 is not conducting. As before load current will be assumed constant during the commutation interval.

At time $t=0$, which is also the beginning of commutation interval for T_1, T_2 is turned on. This connects C_2 charged to voltage $+V$ across L_2. Through the magnetic coupling a voltage $+V$ is induced in L_1. Examining the circuit consisting of L_2, L_1, T_1, sources V_1 and V_2 shows that T_1 is subjected to a reverse bias V and immediately turns off.

In a magnetic circuit flux linkage or *mmf* cannot change instantaneously. In a single reactor current also cannot change instantaneously because flux linkage or *mmf* value depends totally on its value. In a magnetically coupled circuit current in a coil can change provided total flux linkage or *mmf* does not change. In the present case when T_1 turns off current through L_1 drops to zero and in order to maintain *mmf* at the same value an equal value of current is induced in L_2. The equivalent circuit of the inverter for this mode I of commutation interval is given in Fig. 7.17 (a). Load current is assumed to be constant at I_0 during the commutation interval, where I_0 is the value of the load current at the beginning of mode I.

For the loop consisting of V_1, C_1, C_2 and V_2

$$\frac{1}{C}\int_0^t i_{c1}\,dt + \int_0^t i_{c2}\,dt + V = V \tag{7.40}$$

which gives

$$i_{c1} = -i_{c2} \tag{7.41}$$

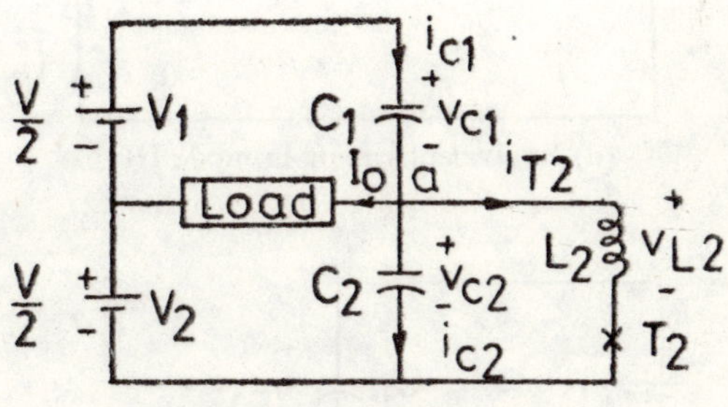

(a) Equivalent circuit, mode I

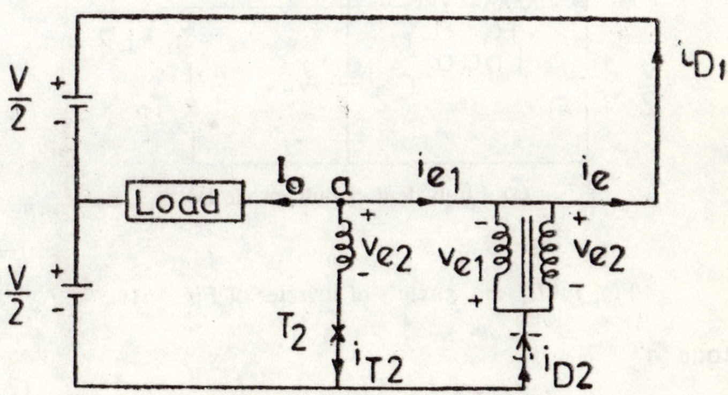

(b) Equivalent circuit, mode II

Fig. 7.17. (continued)

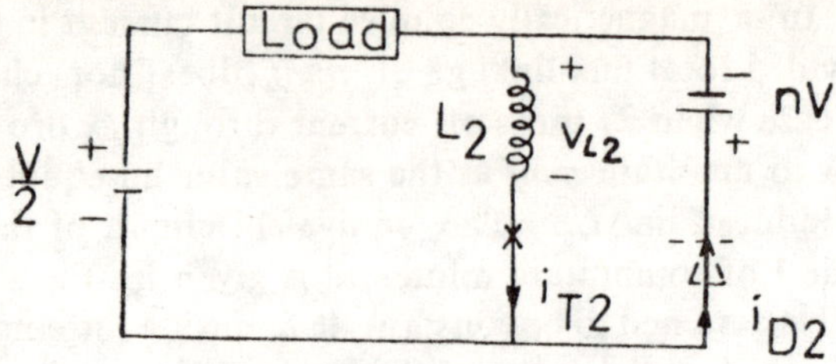

(c) Equivalent circuit of (b) above

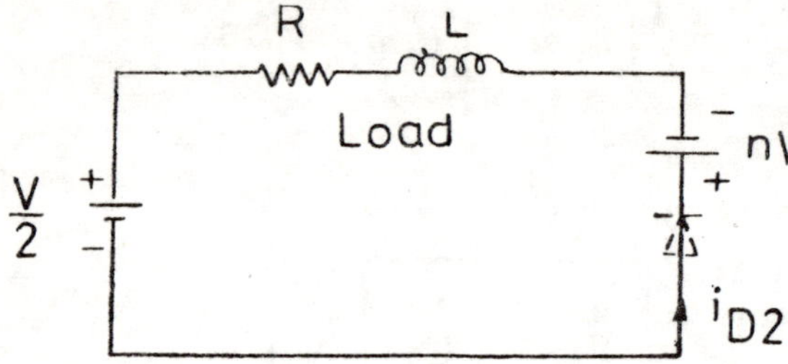

(d) Equivalent circuit in mode III

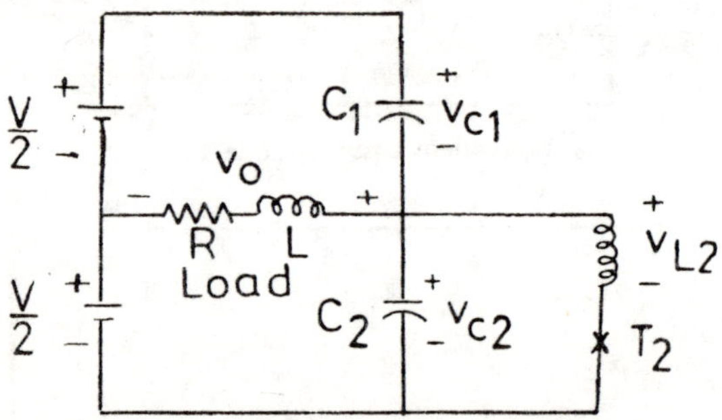

(e) Equivalent circuit, mode IV

Fig. 7.17. Eq. circuits of inverter of Fig. 7.16.

For node '*a*'

$$I_0 + i_{T2} + i_{c2} = i_{c1} \tag{7.42}$$

From eqns. (7.41) and (7.42)

$$i_{C_1} = \frac{I_0 + i_{T2}}{2} \text{ and } i_{C_2} = -\frac{I_0 + i_{T2}}{2} \tag{7.43}$$

From the loop comprising C_2 and L_2

$$L\frac{di_{T2}}{dt}=\frac{1}{C}\int_0^t i_{C2}\,dt+V \tag{7.44}$$

Substituting for i_{C2} from eqn. (7.43) and differentiating gives

$$\frac{d^2i_{T2}}{dt^2}+\frac{di_{T2}}{2LC}+\frac{I_0}{2LC}=0 \tag{7.45}$$

This is a second order differential equation with the following initial conditions

$$i_{T2}(0)=I_0 \text{ and } \frac{di_{T2}}{dt}(0)=\frac{V}{L} \tag{7.46}$$

Its solution is obtained to be

$$i_{T2}=2I_0\cos\omega_0 t+\frac{V}{\omega_0 L}\sin\omega_0 t-I_0 \tag{7.47}$$

where $\quad \omega_0=\dfrac{1}{\sqrt{2LC}}$

Now

$$v_{L2}=L\frac{di_{T2}}{dt}=-2\omega_0 L\,I_0\sin\omega_0 t+V\cos\omega_0 t \tag{7.48}$$

From the loop comprising of L_1, T_1, L_2 and the source, and noting that $v_{L1}=v_{L2}$

$$v_{T1}=V-2v_{L2}$$

Substituting from (7.48)

$$v_{T1}=V+4\omega_0 L\,I_0\sin\omega_0 t-2V\cos\omega_0 t \tag{7.49}$$

Thyristor T_1 will be subjected to a reverse bias as long as v_{T1} is less than zero. Therefore from eqn. (7.49) circuit turn off time t_{C1} of thyristor T_1 is given by

$$\cos\omega_0 t_{C1}-\frac{2\omega_0 L\,I_0}{V}\sin\omega_0\,t_{C1}=\tfrac{1}{2} \tag{7.50}$$

Mode I comes to an end at $t=t_1$ when $v_{c1}=nV$. At this instant diodes D_{11} and D_2 start conducting and energy recovery starts. From eqn. (7.48) the value of t_1 is obtained as follows:

$$-2\omega_0 L\,I_0\sin\omega_0 t_1+V\cos\omega_0 t_1=nV \tag{7.51}$$

The equivalent circuit of Mode II is shown in Fig. 7.17 (b). Since diodes D_2 and D_{11} are conducting, voltages v_{e1} and v_{e2} are clamped at nV and V respectively. Therefore v_{L2} and v_{C2} are also clamped at $-nV$ and the load voltage is clamped at $-(V/2)(2n+1)$. In the equivalent circuit, the transformer formed by L_{e1} and L_{e2} has a connection in which the secondary is connected across a voltage V. When the voltage is transferred to the primary side, the equivalent circuit part consisting of the transformer and the secondary voltage V can be replaced by a source of voltage $-nV$ in series with a diode and the equivalent circuit shown in Fig. 7.17 (c) is obtained. The diode is incorporated to emphasize that the circuit permits the flow of current only in one direction.

Now

$$L\frac{di_{T2}}{dt'} = -nV \tag{7.52}$$

where t' is the time measured from the beginning of this mode, i.e.

$$t' = t - t_1.$$

The initial value $i_{T2}(0) = I_{T2}$ can be obtained from eqn. (7.47) by substituting $t = t_1$. Using this initial condition, the solution of eqn. (7.52) yields

$$i_{T2} = I_{T2} - \frac{nV}{L}t' \tag{7.53}$$

Equation (7.53) shows that i_{T2} decreases linearly. From node a,

$$i_{e1} = -(i_{T2} + I_0) \tag{7.54}$$

$$= \frac{nVt'}{L} - I_{T2} - I_0 \tag{7.55}$$

Now $\quad i_{D11} = i_{e2} = -n\, i_{e1}$

$$i_{D2} = i_{e2} - i_{e1} = -(n+1)\, i_{e1} \tag{7.56}$$

This mode comes to an end at $t' = t_1'$ when i_{T2} becomes zero. From eqn. (7.53)

$$t_1' = \frac{L}{nV} I_{T2} \tag{7.57}$$

Let

$$t_1 + t_1' = t_2 \tag{7.58}$$

This is also the end of the commutation interval. Now onwards the assumption of constant load current is removed.

Soon after i_{T_2} falls to a zero value it will tend to reverse. This will be prevented by thyristor T_2 and therefore i_{T_2} and v_{L_2} will remain zero until T_2 is able to conduct again.

Since D_2 and D_{11} are still conducting in Mode II, voltages v_{e1}, v_{e2}, v_{c1} and v_{c2} are all still clamped to their values in Mode II, however i_{T_2} and v_{L_2} have fallen to zero values. Equivalent circuit for Mode III is shown in Fig. 7.17 (d).

From the equivalent circuit

$$\frac{di_{D2}}{dt''} + \frac{R}{L}\, i_{D2} = -\frac{V}{2L}(1 + 2n) \tag{7.59}$$

where t'' is the time measured from the beginning of mode III and is given by

$$t'' = t - t_2 \tag{7.60}$$

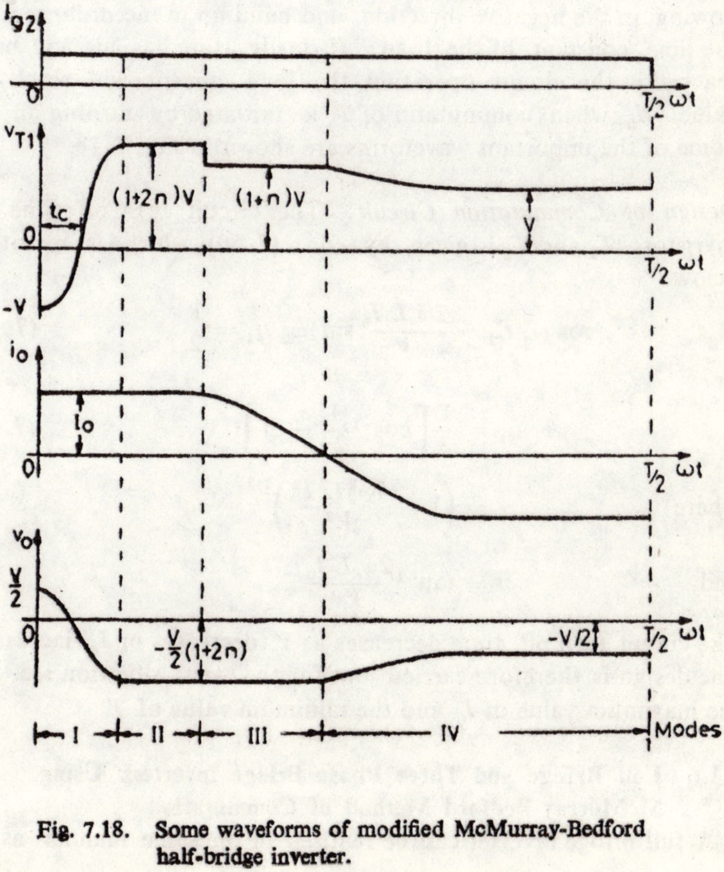

Fig. 7.18. Some waveforms of modified McMurray-Bedford half-bridge inverter.

Initial value of i_{D2} is obtained from eqns. (7.55) and (7.56) as

$$i_{D2} = (n + 1)\, I_0 \tag{7.61}$$

This mode is over at $t'' = t_1'$ when $i_{D2} = 0$.

Let
$$t_2 + t'' = t_3 \tag{7.62}$$

Equivalent circuit for mode IV is shown in Fig. 7.17 (e). All circuit currents are momentarily zero. Thyristor T_2 is forward biased and will conduct again if gate pulse is present. This explains the reason for using a gate pulse of duration of 180°. At the beginning of this mode C_2 and C_1 are charged to voltages $-nV$ and $(n + 1)\,V$ respectively. Further transients in the commutation circuit will occur until C_1 and C_2 are charged to voltages of V and 0 respectively. The load current will start flowing in the negative direction and build up in accordance with the time constant of the load. If steady state has already been reached in the circuit operation, the load current will reach the value $-I_0$ when commutatin of T_2 is initiated by turning on T_1. Some of the important waveforms are shown in Fig. 7.18.

Design of Commutation Circuit: The circuit turn off time of thyristors T_1 and T_2 is given by eqn. (7.50) which is rewritten below:

$$\cos \omega_0 t_{c1} - \frac{2\omega_0 L\, I_0}{V} \sin \omega_0 t_{c1} = \frac{1}{2} \tag{7.50}$$

or

$$t_{c1} = \frac{1}{\omega_0}\left[\cos^{-1}\frac{1}{2\gamma} - \theta\right] \tag{7.63}$$

where
$$\gamma = \left(1 + \frac{4\omega_0^2 L^2 I_0^2}{V^2}\right)^{1/2}$$

and
$$\theta = \tan^{-1}\frac{2\omega_0 L\, I_0}{V}$$

The circuit turn off time decreases as V decreases or I_0 increases. The design is therefore carried out for the worst situation i.e. for the maximum value of I_0 and the minimum value of V.

7.1.6 Full Bridge and Three Phase Bridge Inverters Using McMurray-Bedford Method of Commutation

A full bridge inverter can be realised in the same manner as is done for McMurray inverter. Three phase bridge inverter with the commutation circuit but without energy recovery arrangement is shown in Fig. 7.19. Connections for only phase *A* are shown. Similar connections are employed for phases *B* and *C*.

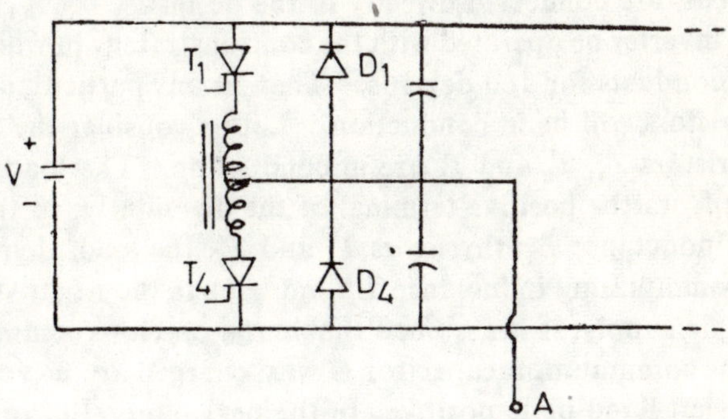

Fig. 7.19. 3-Phase bridge inverter using McMurray-Bedford commutation method.

7.1.7 Three-Phase Bridge Inverter with DC Side Commutation

A three phase bridge inverter with dc side commutation is shown in Fig. 7.20[4]. The commutation circuit, consisting of thyristors T_{C1} to T_{C4}, diodes D_{C1}, and D_{C2}, inductors L_1 and L_2 and capacitor C, is enclosed by dotted lines. It alternately commutates the

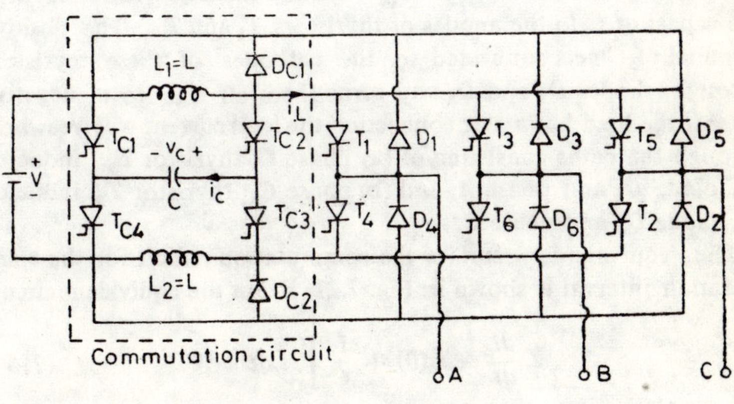

Fig. 7.20. 3-Phase bridge inverter with DC side commutation.

thyristors connected to the positive inverter bus and those connected to the negative inverter bus.

The six thyristors of the inverter, T_1 to T_6, are connected to the dc supply through the commutation reactors L_1 and L_2. The six inverter diodes D_1 to D_6, which provide a path for returning reactive load current, are connected directly to the dc bus.

Let the inverter be operated with the control strategy in which each thyristor conducts for 180 degrees. Then at any particular instant, three thyristors will be in conduction. Let us consider the interval when thyristors T_1, T_2 and T_3 are in conduction. The load current is flowing from the positive terminal of the dc supply, to the commutating inductance L_1, thyristors T_1 and T_3, the load, thyristor T_2 and the commutating inductance L_2, and back to the negative terminal of the dc supply. It is assumed that in the previous commutation interval the commutation capacitor C was charged to a voltage V with the left hand plate positive. In the next interval of operation of the inverter thyristors T_2, T_3 and T_4 must conduct, for which first T_1 should be turned off and then T_4 should be turned on. In the present circuit all the conducting thyristors connected to the positive bus i.e. T_1 and T_3 will be turned off first. Then T_3 is turned on again with T_4 to start the next interval of inverter operation. The commutation of thyristors T_1 and T_3 is carried out in the following manner:

The commutation of T_1 and T_3 is initiated by turning on thyristors T_{C_2} and T_{C_4}. At this instant C is charged to a voltage $-V$. The turn on of thyristors T_{C_2} and T_{C_4} connects the negative terminal of the capacitor C to the anodes of thyristors T_1 and T_3. The positive terminal of C gets connected to the cathodes of these thyristors through diodes D_4 and D_6 and turns them off due to a negative bias. If the load has a star connection the load current will freewheel through the paths consisting of (a) phase C, thyristor T_2, inductor L_2, diode D_4 and phase A, and (b) phase C, thyristor T_2, inductor L_2, diode D_6 and phase B.

The equivalent circuit of the commutation circuit for the commutation interval is shown in Fig. 7.21. From the equivalent circuit

$$L\frac{di_c}{dt} - v_c(0) + \frac{1}{C}\int_0^t i_c dt = V \tag{7.64}$$

with the initial conditions $i_c(0) = I_L$, where I_L is the current through the inductor at $t = 0$.

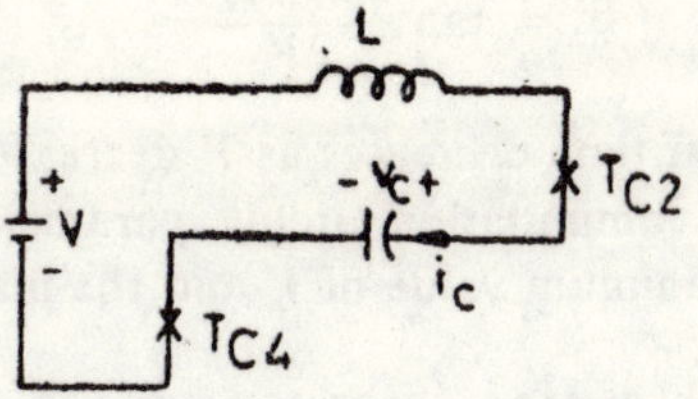

Fig. 7.21. Equivalent circuit of the commutation circuit of inverter of Fig. 7.20.

Solving eqn. (7.64) gives

$$i_c = \frac{2V}{\omega L} \sin \omega_0 t + I_L \cos \omega_0 t \tag{7.65}$$

where $\omega_0 = \sqrt{1/LC}$, the natural frequency of oscillation of the commutation circuit.

Now from eqn. (7.64)

$$v_c = V - L\frac{di_c}{dt}$$

$$= V - 2V \cos \omega_0 t + \omega_0 L I_L \sin \omega_0 t \tag{7.66}$$

The capacitor voltage, from its initial value $(-V)$ will come to zero and then the capacitor will charge in the positive direction. When the capacitor is charged to V volts, diode D_{C1} gets forward biased and the energy left in the inductor is dissipated in the loop consisting of L_1 and D_{C1}. The commutation capacitor is now charged with a positive voltage and is ready for the commutation of thyristor T_2 at the end of the next interval in which thyristors T_2, T_3 and T_4 conduct.

Design of commutation circuit: Equating v_c to zero in eqn. (7.66) gives an expression for circuit turn-off time t_c

$$\cos \omega_0 t_c - \frac{LI_L\omega_0}{2V} \sin \omega_0 t_c = -\frac{1}{2} \tag{7.67}$$

or

$$t_c = \frac{1}{\omega_0}\left[\cos^{-1}\left(-\frac{1}{2\gamma'}\right) - \theta'\right] \tag{7.68}$$

where $$\gamma' = \left(1 + \frac{\omega_0^2 L^2 I_L^2}{4V^2}\right)^{1/2}$$

and $$\theta' = \tan^{-1}\frac{\omega_0 L I_L}{2V} \tag{7.69}$$

The circuit turn off time decreases as V decreases or I_L increases. The values of the commutation circuit parameters are, therefore, selected for the minimum value of V and the maximum value of I_L.

7.1.8 Regenerative Braking of Voltage Source Inverter Fed Drive

When the main supply is ac usually a converter is used to supply dc voltage to the inverter, as shown in Fig. 7.22. The converter is

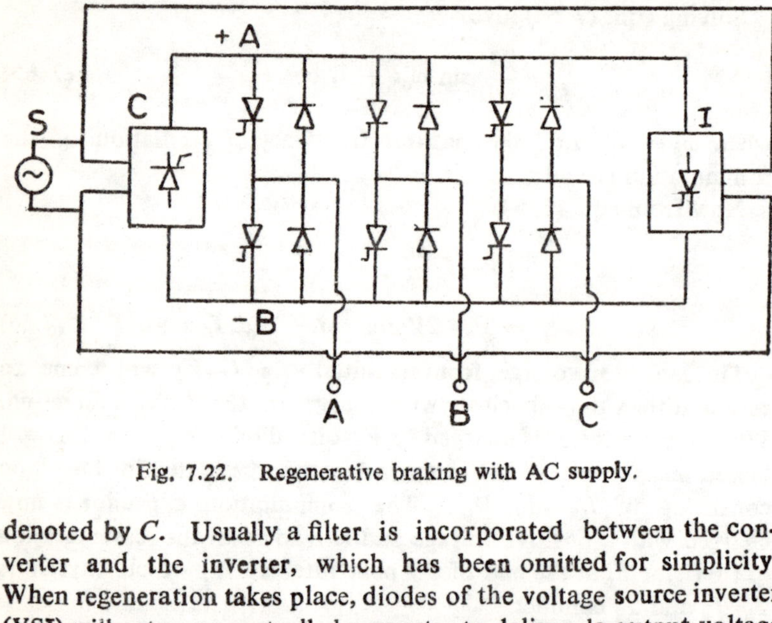

Fig. 7.22. Regenerative braking with AC supply.

denoted by C. Usually a filter is incorporated between the converter and the inverter, which has been omitted for simplicity. When regeneration takes place, diodes of the voltage source inverter (VSI) will act as uncontrolled converter to deliver dc output voltage with terminal A positive with respect to B. A line commutated inverter I connected to these terminals and operating with $\alpha > 90°$ can be used to feed this power to the ac supply. Thus if regeneration is required two converters connected in antiparallel have to be used.

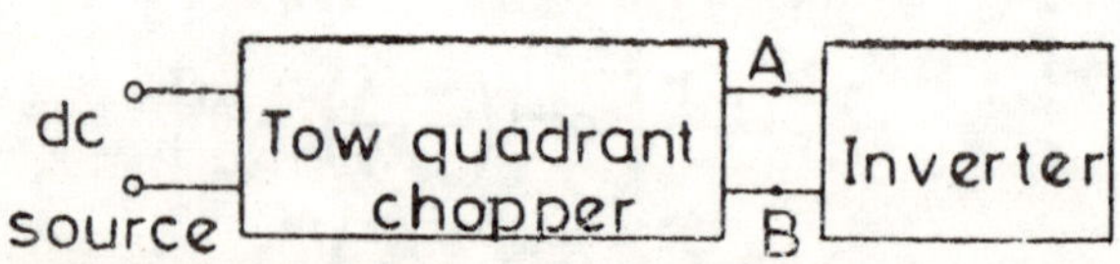

Fig. 7.23. Regenerative braking with DC source.

When main source is dc a two quadrant chopper capable of giving positive voltage and current in either direction can be employed for regeneration, as shown in Fig. 7.23.

7.2 VOLTAGE CONTROL AND HARMONIC REDUCTION IN INVERTERS

7.2.1 Voltage Control

In most inverter applications, the control of output voltage is required because of any one of the following reasons:

(i) In some applications a constant ac voltage is required. To compensate for voltage variation and regulation of dc source, and regulation of inverter itself some means of voltage control is required, as in case of standby and uninterruptible power supplies.

(ii) In motor control applications, control of circuit voltage is required along with frequency in order to avoid saturation of the motor magnetic circuit.

The methods of voltage control can be classified under the following three broad categories:

(a) control of dc voltage supplied to the inverter,

(b) control of voltage delivered by the inverter,

(c) control of voltage within the inverter using time ratio control techniques.

7.2.1.1 *Control of DC Voltage Supplied to the Inverter*

The dc voltage supplied to the inverter can be controlled by a chopper when the source is dc and by a fully controlled converter or by an uncontrolled converter followed by a chopper or by an ac regulator followed by an uncontrolled converter (Fig. 7.24) when the source is ac. The main advantages of voltage control schemes where the dc supply to the inverter is controlled, are as follows:

(i) Inverter output voltage waveform and its harmonic content are not appreciably changed as the output voltage is controlled.

(ii) If the dc input to the inverter is varied to compensate for source voltage fluctuations, the inverter can be designed for a very limited voltage range. Such an inverter is most efficient, both in terms of power loss and component utilisation.

However, this method of voltage control has the following disadvantages:

(i) The power delivered by the inverter is handled by two to

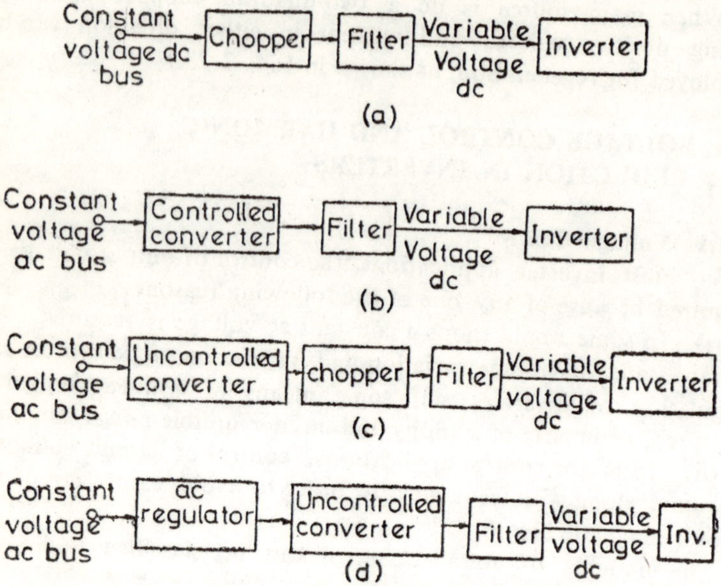

Fig. 7.24. Voltage control by controlling DC input voltage.

three power converters. For example in Fig. 7.24(b) it is handled by controlled converter and inverter and in Fig. 7.24(c) and (d) it is handled by uncontrolled converter, chopper and inverter, and by ac regulator, uncontrolled converter and inverter respectively. More stages of conversion not only make control costly, they also increase losses and reduce efficiency.

(ii) As shown in Figs. 7.24(a) to (d) filter circuits are required for obtaining dc voltage with permissible ripple content. Use of filter circuits increase cost, weight and size, reduces efficiency, and makes the transient response slow.

(iii) The commutation capacitor voltage in many inverters is proportional to the dc input voltage. The duration of reverse bias on the thyristor undergoing commutation decreases as the commutation capacitor voltage reduces. Thus the commutation capability decreases as the dc voltage is reduced. Therefore, the control of the dc input voltage is not desirable when a large variation in the output voltage is required and when high load current is necessary at the reduced output voltage. This limitation is overcome in some inverter circuits by using an auxiliary dc supply for charging commutation capacitors, this however increases the cost of the inverter and causes greater voltage stress on some of the thyristors.

7.2.1.2 *Control of Voltage Delivered by the Inverter*

Inverter output voltage can be controlled by introducing ac regulator between the inverter and the load. This method has the drawback that it gives a high harmonic content in the output voltage at lower values of the load voltage. This method is generally not used except for very low power applications.

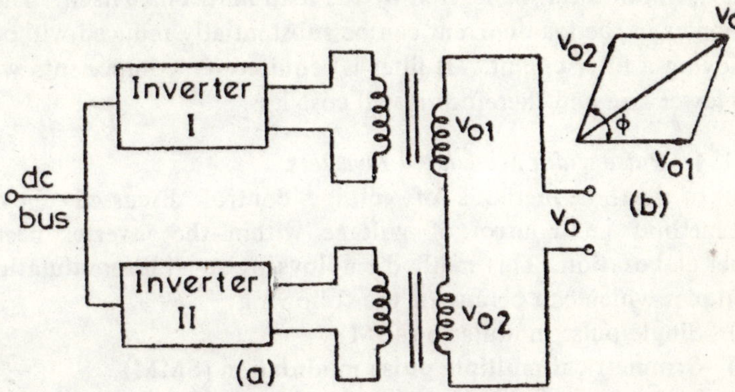

Fig. 7.25. Multiple inverter connection for output voltage control.

Multiple inverters can be efficiently used to control inverter output voltage, if the power requirement is high. The inverter outputs are summed through transformer secondaries. The resulting total output voltage is controlled by varying the phase angle of the gate trigger signals between the individual inverters. Fig. 7.25 shows a basic single phase configuration of this scheme. Two inverters generate output voltages of the same frequency and displaced by an angle ϕ in phase (Fig. 7.25(b)). The output voltage will be maximum when $\phi = 0$ and zero when $\phi = 180°$.

An important advantage of this scheme is that even at low output voltages, the relative harmonic content does not increase.

7.2.1.3 *Control of Voltage within the Inverter Using Time Ratio Control Techniques*

One of the most promising means of controlling the inverter

output voltage is to incorporate time ratio control within the inverter using a suitable modulation scheme. These inverters are called pulse width modulated inverters. The advantages of this method of control are:

(i) It is possible to control the output voltage without significantly adding to the total number of power circuit components of the inverter.

(ii) It is also possible to substantially reduce or eliminate (along with voltage control) lower order harmonic frequencies. The high order harmonics can be filtered by the load inductance itself. Thus harmonics in the load current can be substantially reduced without employing a filter circuit. If filter is required its components will have lower size and therefore it will cost less.

7.2.1.4 *Pulse width Modulated Inverters*

Out of the three methods of voltage control discussed above, third method i.e. control of voltage within the inverter needs further elaboration. This method employs pulse-width modulation techniques which can be any of the following:

(a) Single pulse modulation (SM)

(b) Symmetrical multiple pulse modulation (SMM)

(c) Multiple pulse modulation with selective reduction of harmonics (MMSR)

(d) Sinusoidal pulse modulation (Sin M).

(a) *Single pulse modulation (SM)* [1]: The output voltage waveform of a single-phase bridge inverter is shown in Fig. 7.26(a). The waveform after modulation takes the shape of Fig. 7.26(b). Instead of a pulse of 180° duration, a pulse of width $2d$ located symmetrically around 90° is generated. Control of voltage is realised by varying the pulse width $2d$. The waveform of v_0 can be described by the following Fourier series

$$v_0 = \sum_n^{\infty} A_n \sin n\omega t + \sum_n^{\infty} B_n \cos n\omega t \tag{7.70}$$

where $n = 1, 3, 5, \ldots$ and

$$A_n = \frac{2}{\pi}\int_0^{\pi} V \sin n\omega t \, d(\omega t) = \frac{2V}{\pi}\int_{(\pi/2-d)}^{(\pi/2+d)} \sin n\omega t \, d(\omega t)$$

$$= \frac{4V}{n\pi} \sin nd \tag{7.71}$$

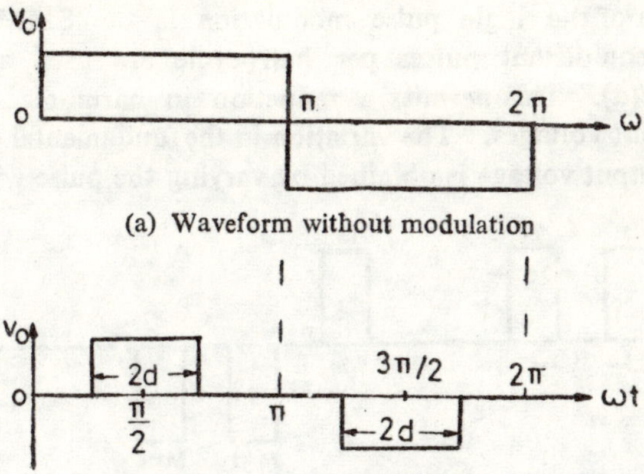

(a) Waveform without modulation

(b) Modulated waveform

Fig. 7.26. Single pulse modulation.

and

$$B_n = \frac{2V}{\pi} \int_{(\pi/2-d)}^{(\pi/2+d)} \cos n\omega t \, d(\omega t) = 0 \tag{7.72}$$

Thus

$$v_0 = \sum_{n=1,3,5}^{\infty} \left(\frac{4V}{n\pi} \sin nd \right) \sin n\omega t \tag{7.73}$$

The maximum value of fundamental component $(v_0)_{1m}$ occurs when $d = \pi/2$ or v_0 is a square wave and is given by

$$(v_0)_{1m} = \frac{4V}{\pi} \tag{7.74}$$

Thus the maximum rms value of the fundamental is 90.4 per cent of the dc supply voltage.

The ratios of $(v_0)_{nm}$ and $(v_0)_{1m}$, where $(v_0)_{nm}$ is the amplitude of nth harmonic component, have been plotted in Fig. 7.30 for $n = 1$, 3, 5, 7 and 9. From these curves it may be observed that when the fundamental component is reduced to nearly 0.3, the amplitude of the 3rd harmonic becomes equal to that of the fundamental; when the fundamental component is reduced to nearly 0.15 all the four harmonics ($n = 3$, 5, 7 and 9) become comparable to the fundamental. Thus in this method of voltage control, substantial amount of harmonic content is introduced in the output voltage, particularly at low values of the output voltage.

(b) *Symmetrical multiple pulse modulation* (*SMM*) [2]: Natural extension of the single pulse modulation is the SMM in which several equidistant pulses per half cycle are used as shown in Fig. 7.27(a). This permits a reduction in harmonic content at low output voltages. The variation in the fundamental component of the output voltage is obtained by varying the pulse width $2d$.

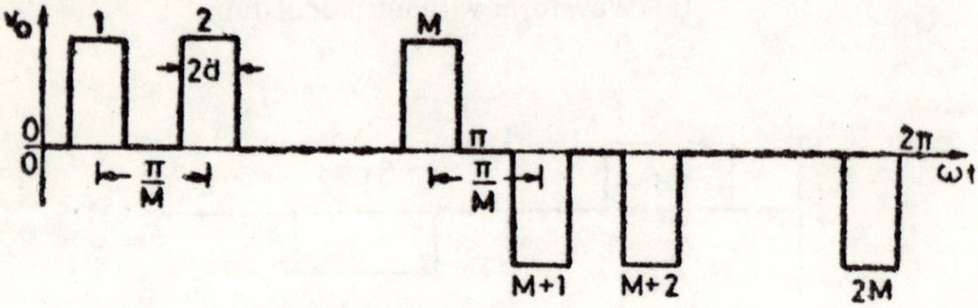

(a) Symmetrical multipulse modulation with *M* pulses per half cycle

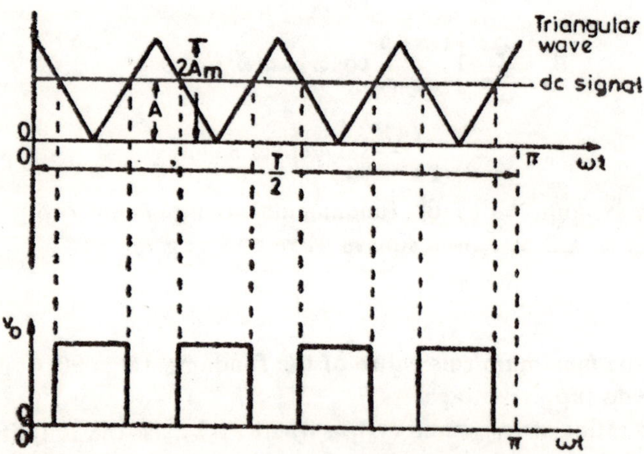

(b) Four pulses per half cycle

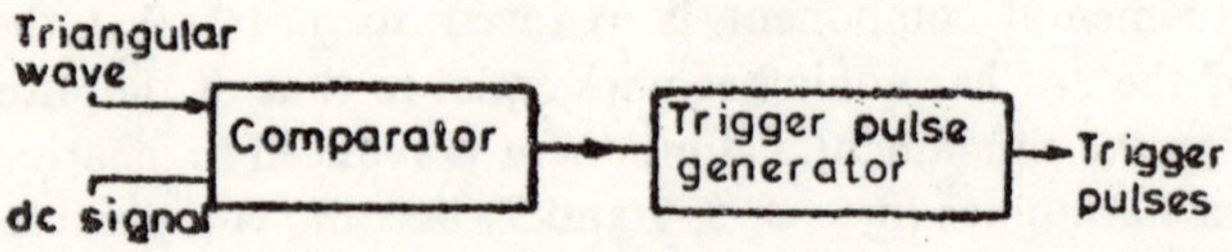

(c) Production of pulses

Fig. 7.27. Symmetrical multipulse modulation.

This type of modulation is achieved by comparing a dc control signal of variable magnitude A with a triangular wave of magnitude

A_m (with a direct component A_m) in a comparator as shown in Fig. 7.27 (b) and (c). The pulse width is changed by varying the magnitude of the dc control signal. The triggering pulses, which are generated at the points of intersections of the two waves, are

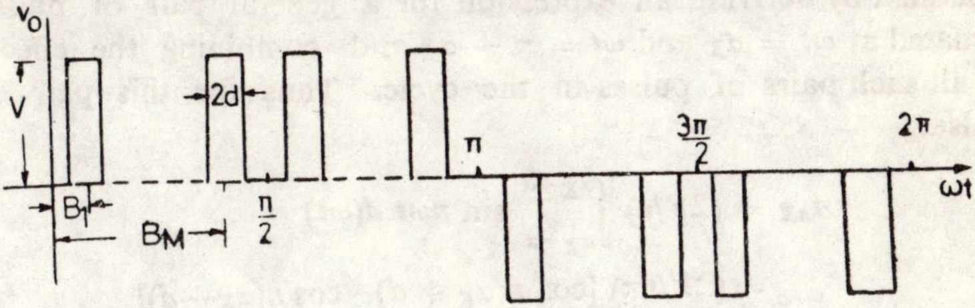

Fig. 7.28. Selected harmonic elimination method.

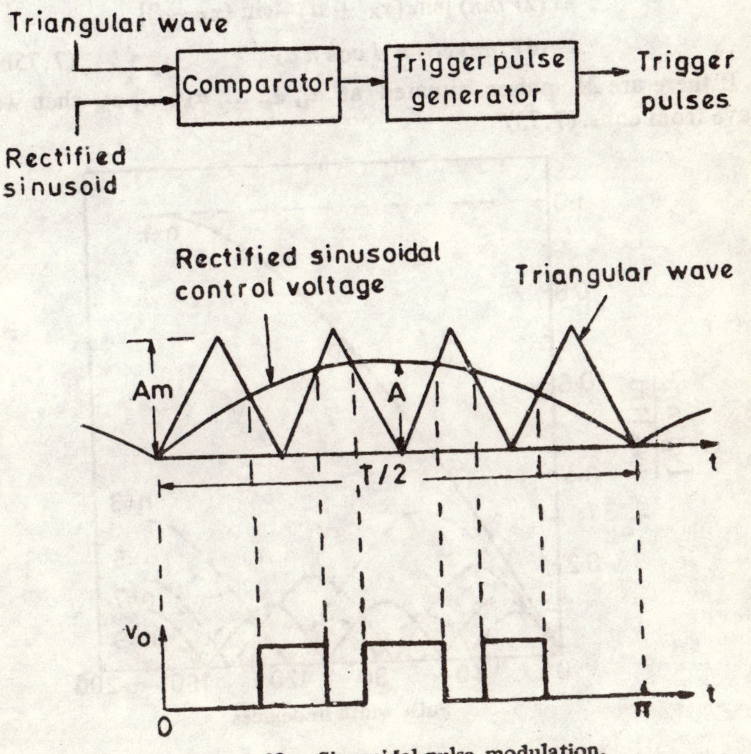

Fig. 7.29. Sinusoidal pulse modulation.

used to trigger the thyristors of the inverter in such a way that the inverter output voltage has a pulse of duration for which the dc control signal exceeds the triangular wave.

Expressions for various harmonics in the output voltage are obtained by deriving an expression for a general pair of pulses situated at $\omega t = \alpha_K$ and $\omega t = \pi + \alpha_K$ and combining the effects of all such pairs of pulses in the cycle. Thus for this pair of pulses,

$$A_{nK} = (2V/\pi) \int_{\alpha_K - d}^{\alpha_K + d} \sin n\omega t \, d(\omega t)$$

$$= (2V/n\pi) [\cos n(\alpha_K + d) - \cos n(\alpha_K - d)]$$

$$= (4V/n\pi) \sin nd \sin n\alpha_K \tag{7.75a}$$

$$B_{nK} = (2V/\pi) \int_{\alpha_K - d}^{\alpha_K + d} \cos n\omega t \, d(\omega t)$$

$$= (2V/n\pi) [\sin (\alpha_K + d) - \sin (\alpha_K - d)]$$

$$= (4V/n\pi) \cos n\, d \cos n\, \alpha_K \tag{7.75b}$$

If there are M pulses situated at $\alpha_1, \alpha_2, \ldots, \alpha_K, \ldots, \alpha_M$ then we have from eqns. (7.75)

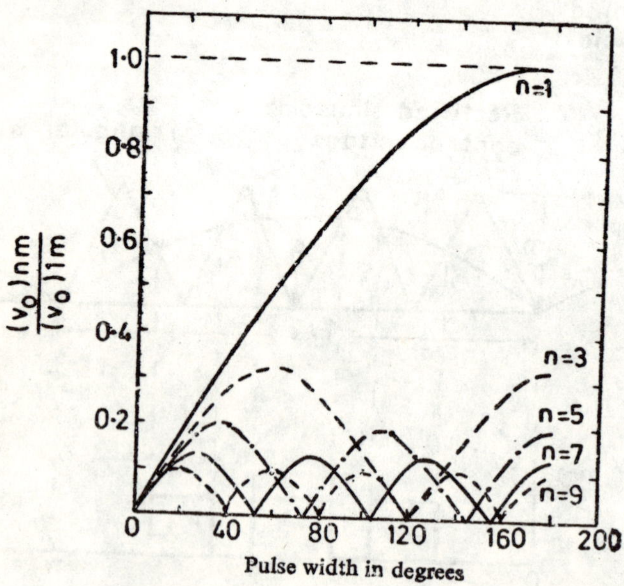

Fig. 7.30. Harmonic content of single pulse modulation.

$$A_n = \frac{4V}{n\pi} \sin nd \sum_{K=1}^{M} \sin n\alpha_K$$

$$B_n = \frac{4V}{n\pi} \cos nd \sum_{K=1}^{M} \sin n\alpha_K \tag{7.76}$$

Amplitudes of first, third, fifth and seventh harmonics as a ratio of the maximum value of fundamental (i.e. $(v_0)_{nm}/(v_0)_{1m}$) are plotted against the pulse width expressed as a ratio of distance between two adjacent pulses, i.e. $(2d/(\pi/m))$ in Fig. 7.31 for $M = 3$ and $M = 10$. The figure shows that considerable reduction in lower order harmonics is achieved as the number of pulses per half cycle (i.e. M) is increased. For example with $M = 10$ substantial reduction in 3rd, 5th and 7th harmonics is achieved in the complete range of the output voltage. If the magnitudes of higher harmonics are obtained it will be found that they increase as M is increased. However, this is not a disadvantage because such harmonics produce negligible current in the load circuit with appreciable inductance. In the absence of appreciable inductance in the load circuit, a cheap filter circuit consisting of small inductance and capacitance will be able to filter them out.

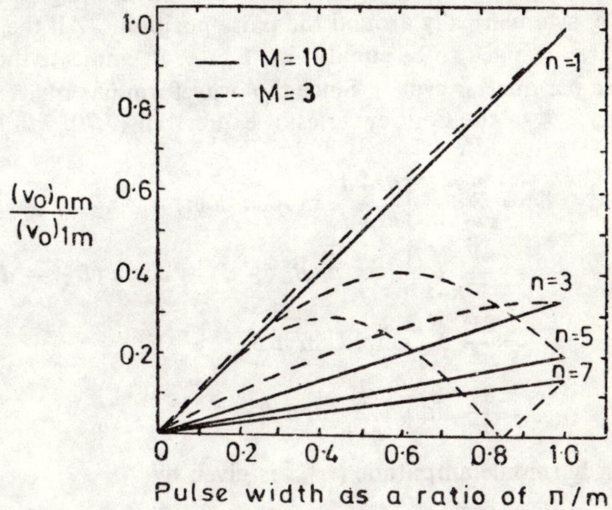

Fig. 7.31. Harmonic content of symmetrical multipulse modulation.

The advantage of this scheme over single pulse modulation is that the voltage control is achieved with a simultaneous reduction of lower order harmonics. The disadvantage is that the number of pulses per half cycle is large, which requires frequent turning on and turning off of thyristors; this increases switching losses and special inverter grade thyristors are required which are costly.

(c) *Multiple pulse modulation with selective elimination of harmonics (MMSR)* [6]: A method of selective harmonic elimination is illustrated in Fig. 7.28. The principle is based on selecting suitable M pulse positions per quarter cycle to eliminate any M harmonics and obtaining output voltage variation by controlling the pulse width $2d$ symmetrically around the pulse position. All the pulse-widths are arranged to be equal. $B_1, B_2, \ldots, B_M$ indicate the pulse positions per quarter cycle. Since the waveform has quarter-wave symmetry, B_n of the Fourier series of expression (7.70) will be zero and

$$A_n = \sum_{K=1}^{M} \frac{4V}{\pi} \int_{B_K-d}^{B_K+d} \sin n\,\omega t\; d(\omega t)$$

$$= \frac{4V}{\pi} \sum_{K=1}^{M} \frac{1}{n} [\cos n\,(B_K + d) - \cos n\,(B_K - d)]$$

$$= \frac{8V}{n\pi} \sum_{K=1}^{M} \sin nd \sin n\,B_K$$

$$= \frac{8V}{n\pi} \sin nd \sum_{K=1}^{M} \sin n\,B_K$$

Now nth harmonic amplitude $(v_0)_{nm}$ is given by

$$(v_0)_{nm} = \frac{8V}{n\pi} \sin nd \sum_{K=1}^{M} \sin n\,B_K \tag{7.77a}$$

and the fundamental component is given by

$$(v_0)_{1m} = \frac{8V}{\pi} \sin d \sum_{K=1}^{M} \sin B_K \tag{7.77b}$$

where n = harmonic numbers to be eliminated and M = total number of harmonics to be eliminated.

M harmonics can be eliminated by equating eqn. (7.77a) to zero. Thus

$$\sum_{K=1}^{M} \sin n\,B_K = 0 \tag{7.78}$$

Eqn. (7.78), yields M non-linear algebraic equations with M unknowns and they have to be solved for $B_1, B_2, \ldots, B_M$.

To illustrate the procedure, elimination of third, fifth and seventh harmonics is considered. As three harmonic components have to be eliminated, three pulses per quarter cycle are used. Let these pulses be situated at B_1, B_2 and B_3. From eqn. (7.77a) we get

$$(v_0)_{nm} = \frac{8V}{n\pi} \sin nd\,(\sin nB_1 + \sin nB_2 + \sin nB_3) \tag{7.79}$$

For elimination of 3rd, 5th and 7th harmonics $n = 3, 5$ and 7. Therefore from eqn. (7.78)

$$\sin 3B_1 + \sin 3B_2 + \sin 3B_3 = 0 \tag{7.80}$$

$$\sin 5B_1 + \sin 5B_2 + \sin 5B_3 = 0 \tag{7.81}$$

$$\sin 7B_1 + \sin 7B_2 + \sin 7B_3 = 0 \tag{7.82}$$

These non-linear equations can be solved by using Newton-Raphson numerical technique. The solution for above set of equations is as follows:

$$B_1 = 33°,\ B_2 = 62° \text{ and } B_3 = 78°$$

Table I gives the values of B computed for elimination of different harmonics throughout the voltage control range. The values of B are given as integral values whereas the exact values of B will be fractions. Hence, the magnitudes of harmonics to be eliminated will be of negligible quantity instead of being exactly zero.

Table I Values of B for elimination of different harmonics

Harmonics to be eliminated	Approx. values of B
5th	$B_1 = 36°$
5th and 7th	$B_1 = 33°$, $B_2 = 69°$
5th, 7th and 11th	$B_1 = 17°$, $B_2 = 36°$, $B_3 = 52°$
3rd	$B_1 = 60°$
3rd and 5th	$B_1 = 42°$, $B_2 = 78°$
3rd, 5th and 7th	$B_1 = 33°$, $B_2 = 62°$ and $B_3 = 78°$

Variation of fundamental and harmonic contents with half pulse width d are shown in Fig. 7.32. As the integral values are employed for B_1, B_2 and B_3, as given in Table I, harmonics 3rd and 7th are not exactly zero but their magnitudes are negligible. Variation of higher harmonics (9th to 15th) is also shown in the figure. Compared to fundamental, they are also low throughout the range of output voltage.

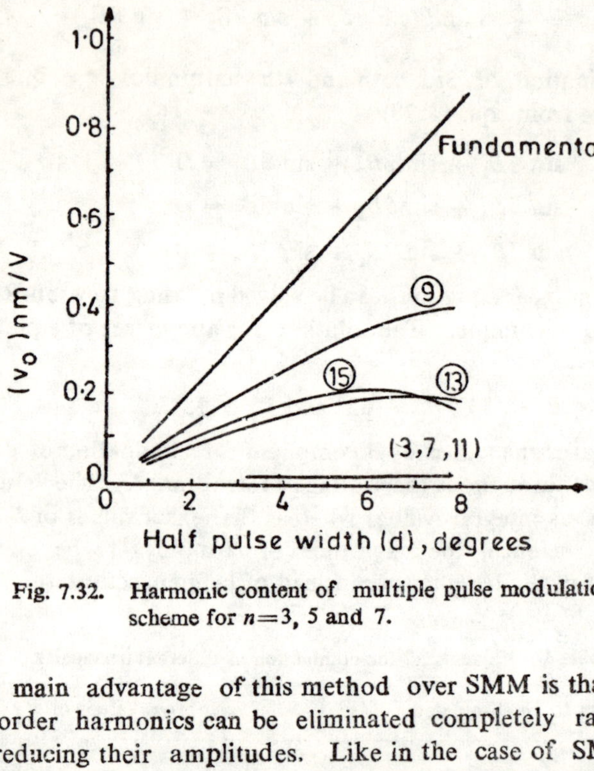

Fig. 7.32. Harmonic content of multiple pulse modulation scheme for $n=3$, 5 and 7.

The main advantage of this method over SMM is that certain lower order harmonics can be eliminated completely rather than only reducing their amplitudes. Like in the case of SMM, here also switching losses are large and inverter grade thyristors are required, which are costly. Since the pulses are not uniformly distributed, they will overlap at some pulse width for which the fundamental output voltage is much less than its value in the rectangular wave of Fig. 7.26 (a). Beyond this value of the pulse width the harmonics will not be eliminated. If the pulse width is restricted to this value, inverter output voltage gets derated considerably. Actual derating depends on the harmonics being eliminated. In the case of elimination of 3rd, 5th and 7th harmonics, the maximum available fundamental component is only 67 per cent of the fundamental component of the unmodulated square wave.

(d) *Sinusoidal pulse modulation* (*sin M*) [9]: In this modulation scheme also several pulses per half cycle are used as in the case of SMM and MMSR. However, the pulse width (instead of being same for all the pulses) is made sinusoidal function of the angular position of the pulse in the cycle. This type of modulation is realised by comparing a control signal consisting of a rectified sinusoidal wave of variable amplitude A and frequency $f = 1/T$ and a triangular wave, of fixed amplitude $A_m/2$ and frequency f_m with a direct component $A_m/2$, in a comparator as shown in Fig. 7.29. The frequency of the triangular wave decides the number of pulses in the output voltage in each half cycle. Therefore, the period of the triangular wave is chosen such that integral number of triangular pulses are produced in a cycle of the rectified sinusoidal control voltage. In the example considered here, there are four cycles of the triangular wave in a cycle of the rectified sinusoidal control voltage. The comparator output is high when the sinusoidal wave has magnitude higher than the triangular, otherwise it is low. The output of the comparator is processed through the firing pulse generator which controls the turn on and commutation of the thyristors of the inverter such that the output voltage has a pulse duration for which the rectified sinusoidal control voltage exceeds the triangular voltage, otherwise it is zero. Triangular wave can be timed to have either its zero or its peak coincident with the zero of the sinusoid. For a ratio f_m/f equal to m there will be $(m - 1)$ pulses in the former case and m pulses in the later case. Both the approaches provide nearly similar harmonic content for m higher than 5.

The ratio A/A_m is called the modulation index. The output voltage is varied by controlling the modulation index (i.e. A). Variation of various harmonics with the modulation index is shown in Fig. (7.33). The following are the important features of this modulation technique.

(i) Amplitude of mth harmonic and its integral multiples is zero.

(ii) The harmonic content as a per unit of fundamental is higher for lower values of modulation index except for harmonics $(2m-3)$ and $(2m + 3)$. The predominant harmonics are $(2m-1)$ and $(2m + 1)$. The next lower content harmonics are $(4m \pm 1)$ and $(6m \pm 1)$.

(iii) For a modulation index up to 1, harmonics lower than

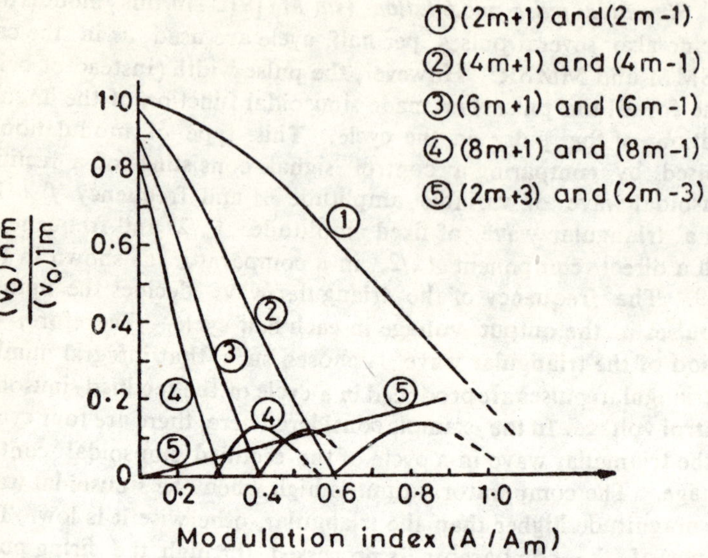

Fig. 7.33. Harmonic content in sinusoidal pulse width modulation.

those listed in (ii) are absent. However, any increase of modulation index beyond this value will introduce lower frequency harmonics.

Due to these reasons the modulation index may not be increased beyond 1, which restricts the maximum available output voltage from the inverter. Even if this restriction is not imposed, the output voltage (fundamental) will always be less than that available with unmodulated rectangular wave. This causes derating of the inverter output voltage. Exact derating depends on the number of pulses per half cycle.

Comparative study of sinusoidal multiple pulse modulation (sin M) with selective reduction of harmonics (MMSR) has been given in [6]. Sin M is essentially a harmonic reduction method and MMSR is a harmonic elimination method; thus in this method the harmonics to be eliminated become zero in the whole range of the output voltage control, which is not so with sinusoidal modulation. It has been shown in [6] that for a few pulses per half cycle MMSR is superior to Sin M.

7.2.1.5 *Implementation of Pulse Width Modulation in Single Phase Bridge Inverters*

(a) ***Single-phase full bridge inverter:*** The output voltage waveforms of Figs. 7.26 to 7.29 have three levels V, 0 and $-V$. In the bridge inverter of Fig. 7.2 (a) these waveforms are realised as follows. During a positive half cycle $+V$ is obtained by firing T_1 and T_2. Zero level of the output voltage for a general load is obtained by turning off T_2 and firing T_3. Since gate pulses are available for both T_1 and T_3, depending on the nature of the load inductive or capacitive, the load current flows through one of these thyristors and the diode connected in parallel with the other thyristor. In the negative half cycle, the output voltage level $-V$ is obtained by turning on thyristors T_3 and T_4. The zero level of the output voltage for a general load is obtained by turning off T_3 and firing T_2. Depending on the nature of the load, the load current flows through one of the two thyristors (T_2 or T_4) and the diode connected in parallel with the other thyristor.

(b) *Single phase half bridge inverter*: In the case of a half bridge inverter, zero level of output voltage cannot be obtained, only positive and negative levels are possible. Therefore, in this case the output voltage is allowed to reverse instead of making it zero. A modulation method involving equal pulse widths and single reversal per half cycle is shown in Fig. 7.34. The output voltage is controlled by controlling the width $2d$. One can also employ more than one reversal per half cycle. Usually this approach is not employed in practice due to large harmonic content.

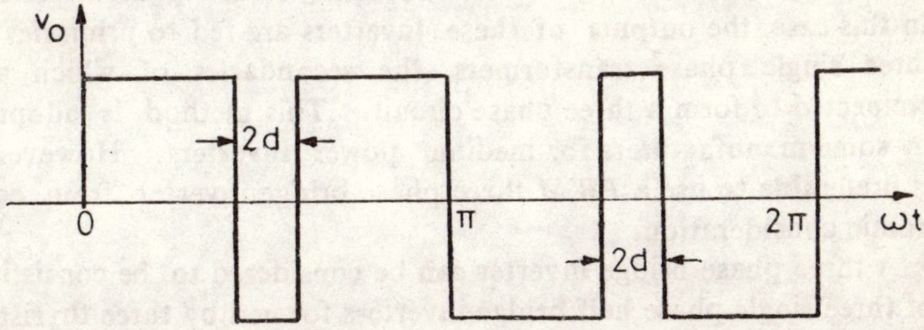

Fig. 7.34. Pulse width modulation in half bridge inverters.

Sinusoidal pulse width modulation shown in Fig. 7.35 is often used with half bridge inverters. A sinusoidal control signal is compared with a triangular wave. During positive half cycle, the inverter thyristors are operated to produce positive going pulses for the duration in which the sinusoidal control signal is higher than the triangular, otherwise negative going pulses are produced.

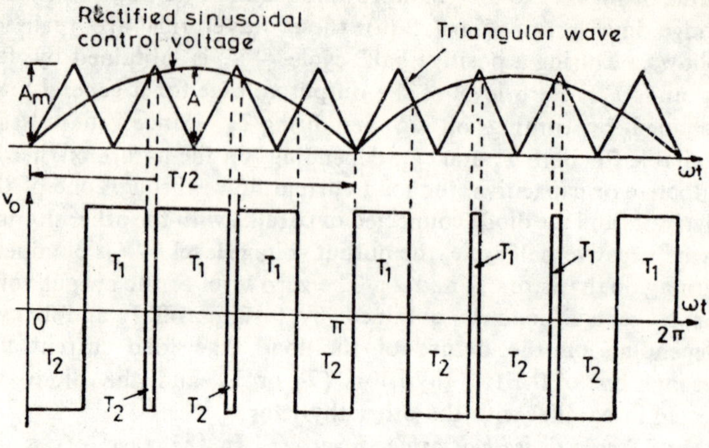

Fig. 7.35. Sinusoidal pulse modulation in half bridge inverters.

Thyristors which conduct during different intervals are marked in the figure.

7.2.1.6 *Implementation of Pulse Width Modulation in Three Phase Bridge Inverters*

It is possible to use three single phase pulse width modulated inverters phase shifted by 120° for obtaining three phase inverters. In this case, the outputs of these inverters are fed to primaries of three single phase transformers, the secondaries of which are connected to form a three phase circuit. This method is adopted by some manufacturers for medium power inverters. However it is preferable to use a *PWM* three phase bridge inverter from economic consideration.

A three phase bridge inverter can be considered to be consisting of three single phase half bridge inverters formed by three thyristor pairs T_1 and T_4, T_3 and T_6, and T_5 and T_2 with a common dc supply (Fig. 7.3). For sinusoidal modulation three sinusoidal control voltages displaced by 120° are compared with a common triangular wave. The intersections of the sinusoidal waves with the triangular wave determine the switching instants of these thyristor pairs. The load voltages produced by these three half bridge inverters with respect to the centre point of the dc supply, 0, are shown in Fig. 7.36. One line voltage V_{BC}, which is the difference of voltage V_{BO} and V_{CO}, is also shown in the figure.

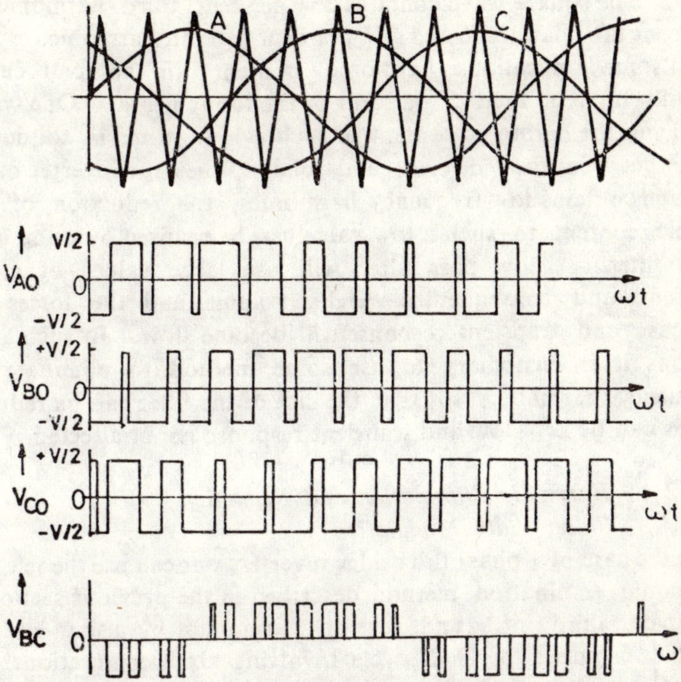

Fig. 7.36. Sinusoidal pulse with modulated waveforms of a three-phase bridge inverter.

Harmonic content can be evaluated in the same manner as in the case of single phase inverters. Some of the important points about this modulation method for *m* triangular pulses per cycle of sinusoidal voltage are [9].

(i) *m*th harmonic and its integral multiples are not present.

(ii) The predominant harmonics are $(2m-1)$ and $(2m+1)$. The other lower amplitude harmonics are $(m-2)$, $(m+2)$, $(3m-1)$, $(3m+1)$, $(4m-1)$ and $(4m+1)$.

(iii) Increasing *m* will shift the dominant harmonic to a higher frequency and

(iv) Harmonic content is higher at lower values of modulation index.

7.2.2 Harmonic Reduction

In the case of single phase bridge inverters at the maximum output voltage condition, squarewave voltage is delivered to the load. The square wave contains $33\frac{1}{3}$ per cent third harmonic, 20 per cent fifth harmonic and $14\frac{2}{7}$ per cent seventh harmonic.

In some applications harmonic content in the load current should be less than 5 per cent of the fundamental. One way of reducing the harmonic content to such a low value is to employ filter between the inverter and load. When the inverter output voltage contains low frequency harmonics, the reduction of harmonic content to such a low value can be realised by using a low pass filter. A low pass filter will use large values of circuit elements and consequently weight, volume and the losses will increase and transient response will become slow. In such applications it is customary to use some method to eliminate low frequency harmonics so that the size of the filter can be reduced, losses can be kept low and transient response is not affected.

7.2.2.1 *Harmonic Elimination and Reduction by Pulse Width Modulation*

In the case of 1 phase full bridge inverters one can use the selective harmonic elimination method described in the previous section to eliminate third, fifth and seventh harmonics by using suitably located 6 pulses per half cycle involving six commutations. In general, one can employ $2m$ pulses per half cycle to eliminate m lowest harmonics. This method however has a serious limitation that due to uneven distribution of pulses the harmonic elimination is achieved at the expense of considerable derating of the inverter. One can also employ the sinusoidal pulse width modulation to eliminate and reduce lower order harmonics. If the value of A in Fig. 7.29 is kept less than $2A_m$, with m pulses per half cycle, one can eliminate or substantially reduce all the harmonics up to $(2m-1)$. In this case also the derating of the inverter will take place but it will be less than for the selective harmonic elimination. Both these methods have the drawback of large commutations per half cycle. A method which can be used for harmonic elimination but not for voltage control is described below.

A modulation method which can be used both for half-bridge and full-bridge inverters, for the elimination of any two unwanted lower order harmonics, is shown in Fig. 7.37. It employs four extra commutation per cycle instead of one for the unmodulated square-wave. Because of the quarter wave symmetry, B_n in eqn. (7.70) will be zero and

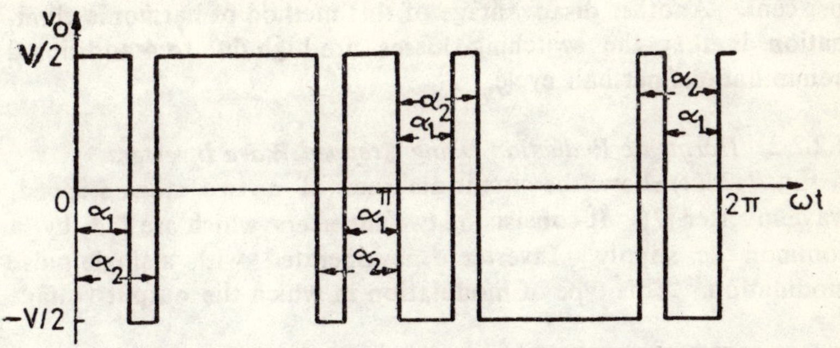

Fig. 7.37. Harmonic elimination using 4 extra commutations per cycle.

$$A_n = \frac{4}{\pi}\cdot\frac{V}{2}\left[\int_0^{\alpha_1} \sin n\,\omega t\, d(\omega t) - \int_{\alpha_1}^{\alpha_2} \sin n\,\omega t\, d(\omega t) + \int_{\alpha_2}^{\pi/2} \sin n\,\omega t\, d(\omega t)\right]$$

$$= \frac{2V}{\pi}\left[\frac{1 - 2\cos n\,\alpha_1 + 2\cos n\,\alpha_2}{n}\right] \tag{7.83}$$

Third and fifth harmonics have the lowest frequency and the largest magnitude among the various harmonics present in a square-wave. For the elimination of these harmonics

$$A_3 = 0 \text{ and } A_5 = 0$$

From eqn. (7.83),

$$1 - 2\cos 3\alpha_1 + 2\cos 3\alpha_2 = 0$$

$$1 - 2\cos 5\alpha_1 + 2\cos 5\alpha_2 = 0$$

These equations are solved numerically to obtain α_1 and α_2 subjected to the constraint

$$0° < \alpha_1 < 90°;\ 0 < \alpha_2 < 90°,\ \alpha_1 < \alpha_2$$

which gives

$$\alpha_1 = 23.62° \text{ and } \alpha_2 = 33.6°$$

The harmonic analysis of the waveform for these values of α_1 and α_2 indicates that 7th, 9th and 11th harmonics are respectively 29.6, 48.6 and 36.4 per cent of the fundamental and 3rd and 5th harmonics are absent. The fundamental component is, however, 83.9 per cent of the fundamental component of the unmodulated rectangular wave. Thus the inverter is derated by more than 16 per cent. Another disadvantage of this method of harmonic elimination is that the switching losses are high due to 4 additional commutations per half cycle.

7.2.2.2 *Harmonic Reduction Using Stepped Wave Inverters*

Fig. 7.38(a) shows the circuit diagram of a two stage stepped-wave inverter [7]. It consists of two inverters which are fed by a common dc supply. Inverter *I* is operated with a single pulse modulation. This type of modulation in which the output voltage

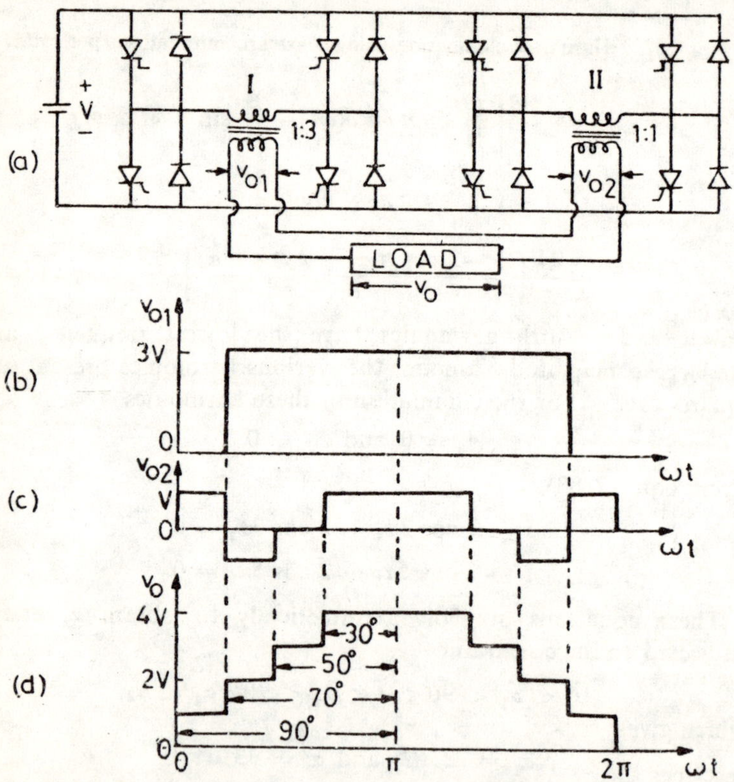

Fig. 7.38. Stepped wave inverter.

has only two levels during any half cycle, positive and zero during positive half cycle and negative and zero during negative half cycle, is called two level modulation. Inverter *II* is operated with 3 level multiple pulse modulation. It is a three level modulation method because during each half cycle the output voltage has positive, zero and negative values. These inverters are connected to a common load through transformers having turns ratios of 1 : 3 and 1 : 1 respectively. The output voltage waveforms of individual inverters and the load voltage waveform are shown in Figs. 7.38(b) to (d). Load voltage has 4 steps per quarter cycle. The output voltage has no tripplen harmonics. Fifth, seventh and eleventh harmonics are 2.4, 4.04 and 2.5 per cent of fundamental respectively. Total harmonic distortion is 11.75 per cent.

One can employ more than two stages. By using three stages one can get 13 steps per quarter cycle [8].

7.3 CURRENT SOURCE INVERTER

The current source inverter converts power between an adjustable current source and a single or three phase ac load. Since the input power is usually supplied at a constant voltage, the inverter system consists of the two stages described in the following.

(1) The 'current source' which receives power from a fixed voltage ac or dc bus and supplies an adjustable dc current at the output terminals. This power conversion stage uses either a controlled rectifier bridge or a dc chopper. A current feedback loop and a *L*-filter is employed to obtain a ripple-free dc current from the bridge or the chopper. The current source is sometimes referred to as the 'front end' of the inverter.

(2) The 'current source inverter' (CSI) is connected in cascade with the 'current source' (CS). It converts the dc current at the input to an ac current at the output terminals. The frequency of the ac current is adjusted by the control circuit of the CSI The amplitude of the ac current can be controlled only indirectly by controlling the magnitude of the dc input current produced by the current source.

In this section the basic circuit configurations of the single phase and the three phase current source inverter are given. Section 7.3.1 describes the operation of the current source inverter using ideal switches. The block diagrams of the practical system operating from the ac and dc bus are also given. Section 7.3.2 analyses the single phase capacitor commutated CS Inverter for different loads and operating frequencies. Section 7.3.3 analyses the three phase auto sequential CS inverter (ASCI) with a counter *EMF* load.

The limit frequency for no overlap operation is derived. Operation at higher frequencies is discussed qualitatively. Section 7.3.4 describes some inverter configurations other than the widely used ASCI.

7.3.1 Circuit Configurations Using Ideal Switches

In this section the basic circuit configurations of a single phase and a three phase current source inverter using ideal switches are described. The block diagrams of the practical inverters operating from an ac and dc bus are given. The characteristics of the ideal switches have been given in section 6.1.

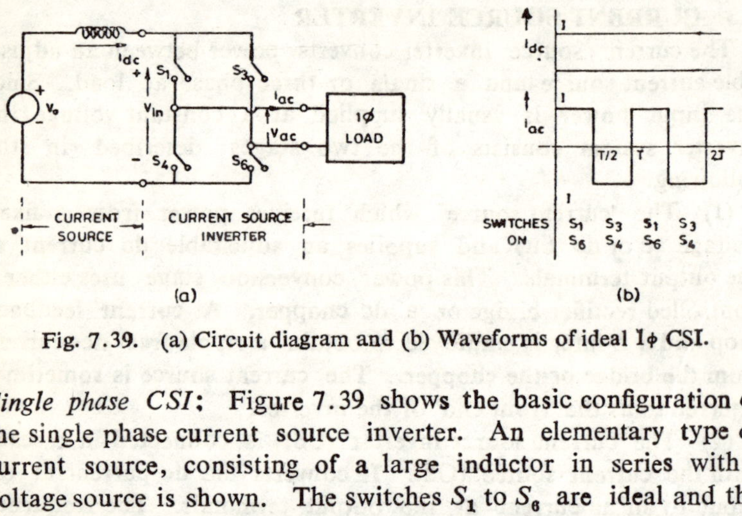

Fig. 7.39. (a) Circuit diagram and (b) Waveforms of ideal 1φ CSI.

Single phase CSI: Figure 7.39 shows the basic configuration of the single phase current source inverter. An elementary type of current source, consisting of a large inductor in series with a voltage source is shown. The switches S_1 to S_6 are ideal and the assumptions given in Chapter 6, Section 6.1 are applicable. The switch pairs S_1, S_6 and S_3, S_4 are closed alternately at a constant frequency. The resulting output current is an ac square wave whose amplitude is equal to the dc input current.

Figure 7.39 shows that the single phase current source inverter supplies the load with a defined current waveform. Therefore the waveform of the load voltage v_{ac} (Fig. 7.39a) depends upon the type of load. In general, voltage v_{ac} consists of a component at the inverter frequency f, equal to $1/T$, and other components at multiples of inverter frequency.

The input voltage v_{in} in Figure 7.39(a) is a waveform of frequency $2f$. It can be derived from v_{ac} by the following relationship.

$$v_{in} = v_{ac} \text{ when } S_1 \text{ and } S_6 \text{ are on,}$$
$$= -v_{ac} \text{ when } S_3 \text{ and } S_4 \text{ are on.}$$

The average value of v_{in} is equal to the source voltage V_0. The product $(V_0 \cdot I)$ is the input power of the inverter.

The switches S_1 to S_6 conduct current only in one direction, downwards in Fig. 7.39(a). The input current i_{dc} is therefore unidirectional. Thus the power is supplied from the source to the load when the average value of v_{in} is positive. The power is regenerated from the load to the source when the average value of v_{in} is negative. The source voltage V_0 must therefore be adjusted in response to the load to a positive or a negative value, while the source current i_{dc} is always positive.

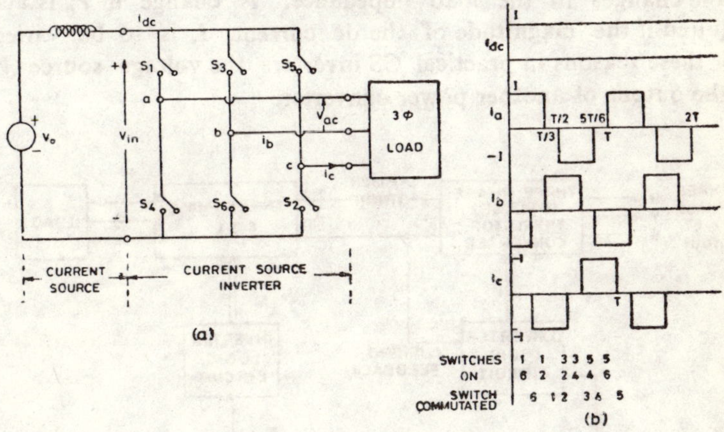

Fig. 7.40. (a) Circuit diagram and (b) waveforms of ideal 3ϕ CSI.

Three phase CSI: Figure 7.40 shows the basic configuration of the three phase CS inverter. The switches S_1 to S_6 are closed periodically in the pairs shown in Fig. 7.40. Each switch conducts for 120 degrees and is commutated when the next switch in the top (1, 3, 5) or bottom (4, 6, 2) group is turned on. The resultting three phase output current is a 120 degree wide square wave of amplitude I. The frequency of the output current is controlled by separate logic circuit, not shown in Fig. 7.40, which closes and opens the switches in the required sequence.

The three phase CSI supplies 120 deg. wide square wave to the load. The line voltages V_{ab}, V_{bc} and V_{ca} (Fig. 7.40) depend upon the load and contain a fundamental component at frequency f, equal to $1/T$, and its harmonics. The input voltage v_{in} can be derived from the segments of line voltages corresponding to the pair of conducting switches. For example, v_{in} equals v_{ab} when S_1 and S_6 are conducting. As there are six pairs of switches (Fig. 7.40b) v_{in} is a waveform of frequency $6f$. The average value of v_{in} is equal to the source voltage V_0. The value of voltage V_0 is adjusted in response to the load. A positive value of V_0 corresponds to power flow from the source to the load. A negative value of V_0 corresponds to power flow from the load to the source.

Practical current source inverters: In the inverter circuits of Fig. 7.39 and 7.40 the source voltage V_0 must be changed in response to the changes in the load impedance. A change in V_0 is also required if the magnitude of the dc current, I, is to be varied. For these reasons in practical CS inverters the voltage source V_0 is the output of another power converter.

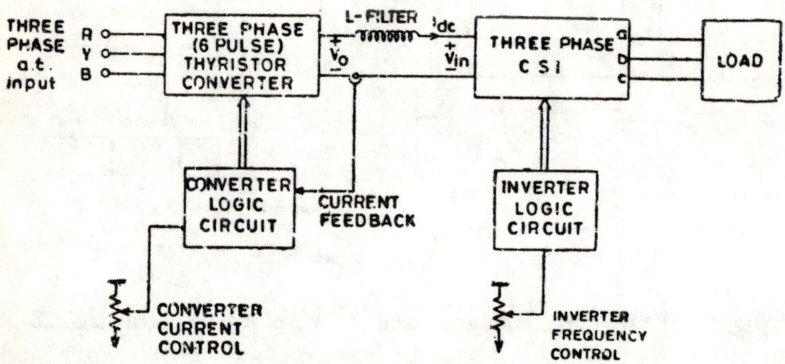

Fig. 7.41. Block diagram of a particle 3ϕ CS inverter system with AC input.

Figure 7.41 shows the block diagram of a practical system operating from a three phase ac supply. A three phase (6 pulse) controlled rectifier bridge provides the average output voltage which can be $+ve$ or $-ve$. A current feedback loop is used to adjust V_0 continuously to regulate the average value of i_{dc}. A large L-filter is used so that the currents due to harmonics in v_0 and v_{in} are negligible. Thus the current i_{dc} is continuous and ripple free.

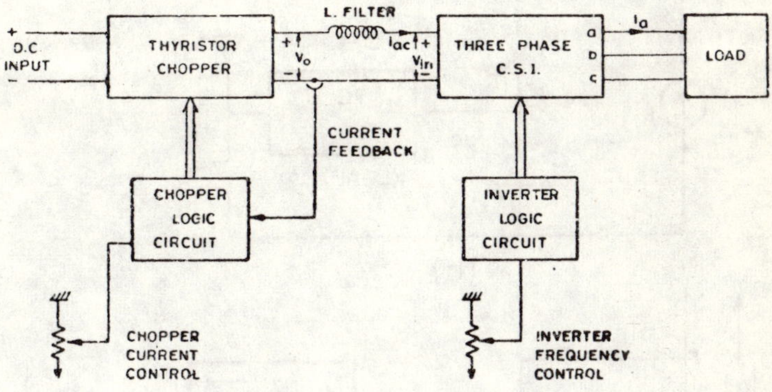

Fig. 7.42. Block diagram of a practical 3ϕ CS inverter system with DC input.

Figure 7.42 shows the block diagram of a practical system operating from a dc bus. In this system a thyristor chopper of configuration '*D*' (Fig. 6.3c) provides a positive or negative average output voltage V_0. A current feedback loop is used to regulate the average value of current i_{dc}.

In practical inverters, the switches S_1 to S_6 of Fig. 7.39 and 7.40 are replaced by thyristors. In addition, a commutation circuit is used to turn off the conducting switch at appropriate instant of time. Due to the finite commutation time of the switch, the current waveforms in a practical inverter have finite rise and fall times. The practical inverters therefore have a maximum operating frequency beyond which the load current is not the square waveform shown in Fig. 7.39 and 7.40. A detailed analysis of the operation of the single phase and the three phase CSI circuit is given in sections 7.3.2 and 7.3.3 respectively.

7.3.2 Single Phase Capacitor Commutated Current Source Inverter

In this section, a single phase CSI circuit is described and analysed in detail. The circuits shown in Figs. 7.43 and 7.45 (a) are similar to the ideal single phase inverter of Fig. 7.39(a). A majority of practical loads have lagging power factor. Therefore the thyristor switches must be force-commutated. In the circuits described in this section a capacitor is used to store charge for commutation.

Section 7.3.2.1 discusses the single phase capacitor commutated

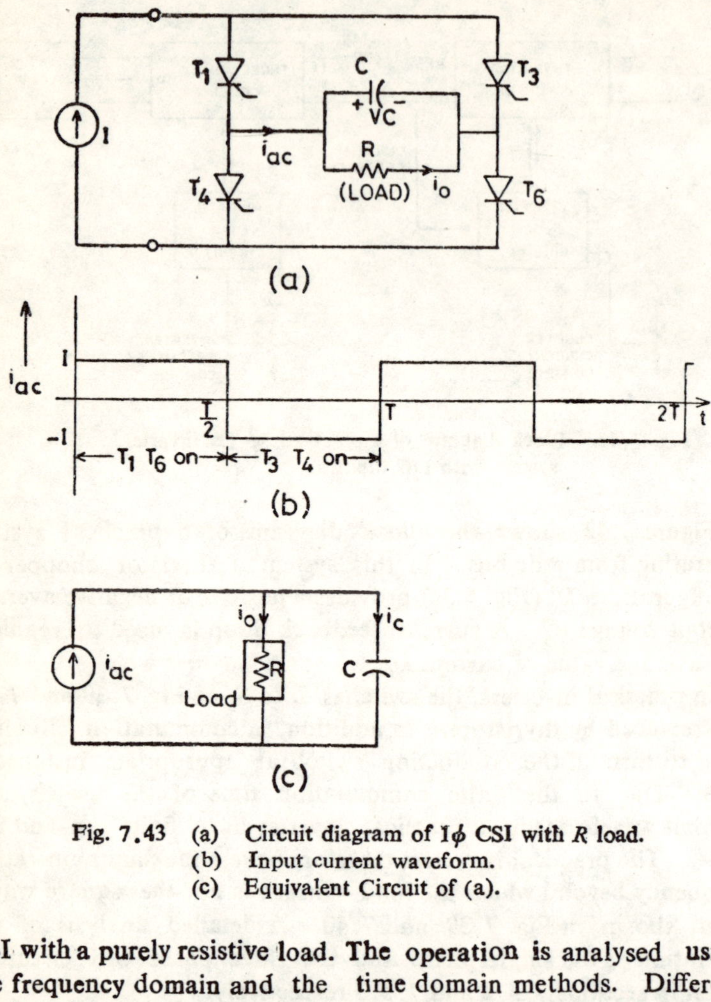

Fig. 7.43 (a) Circuit diagram of 1ϕ CSI with R load.
(b) Input current waveform.
(c) Equivalent Circuit of (a).

CSI with a purely resistive load. The operation is analysed using the frequency domain and the time domain methods. Different modes of operation are established and design criteria are given.

Section 7.3.2.2 discusses the single phase C.S.I. with a purely inductive load. The circuit modifications to prevent discharge of the commutation capacitors is described. The circuit has been analysed in detail and different operating regions over the frequency range are established.

Section 7.3.2.3 discusses the single phase CSI with general R-L load. The conditions for different modes of operation are described.

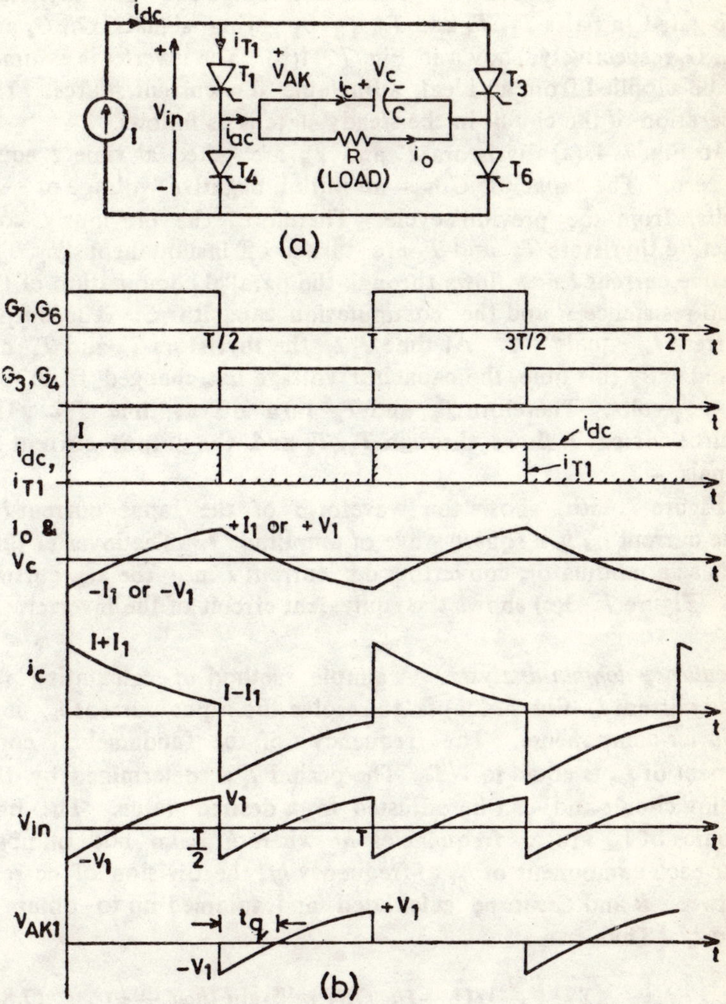

Fig. 7.44. (a) Circuit diagram.

(b) Waveforms for 1ϕ CSI with resistive load and $T/2RC = 2.5$

7.3.2.1 *Single Phase C.S.I. with Resistive Load*

Figure 7.43 shows the circuit diagram of a single phase CSI supplying a resistive load. The thyristors T_1, T_6 and T_3, T_4 form the power switches equivalent to S_1, S_6 and S_3, S_4 in Fig. 7.39(a).

The capacitor C stores charge for commutation. The thyristors are gated in pairs T_1, T_6 and T_3, T_4 by gating signals, G_1, G_6 and G_3, G_4 respectively, shown in Fig. 7.44(b). The inverter is assumed to be supplied from an ideal, adjustable d.c. current source. The operation of the circuit in the steady state is as follows.

In Fig. 7.43(a) thyristors T_1 and T_6 are gated at time t equal to zero. The capacitor C has an initial negative voltage of $-v_1$ volts, from the previous cycle. Therefore, the previously conducting thyristors T_3 and T_4 are turned off instantaneously. The source current I now flows through the parallel combination of the load resistance R and the commutation capacitor C. The output current i_{ac} equals $+I$. At time $T/2$, the thyristors T_3 and T_4 are gated. By this time, the capacitor voltage has changed from $-v_1$ to $+v_1$ volts. Therefore T_1 and T_6 turn off at time $T/2$. The source current I flows through T_3, T_4 and the output current i_{ac} equals $-I$.

Figure 7.43(b) shows the waveform of the input current i_{ac}. The current i_{ac} is a square wave of amplitude I. The inverter thus acts as a modulator, converting d.c. current I into the a.c. current i_{ac}. Figure 7.43(c) shows the equivalent circuit of the inverter.

Frequency domain analysis: A simple method of calculating the load current i_0 (Fig. 7.43c) is to resolve the input current i_{ac} into Fourier components. The frequency f of the fundamental component of i_{ac} is equal to $1/T$. The period T is determined by the gating circuit and can be adjusted to a desired value. The harmonics of i_{ac} are at frequencies hf where h is an odd number. For each component of i_{ac} at frequency hf, the division of current between R and C can be calculated and summed up to obtain i_0 and i_c. Thus,

$$i_0 = \sum_{h=1,3,5\ldots} \left(\frac{4I}{h\pi}\right) \cdot \{1 + (h\omega CR)^2\}^{-1/2} \cdot \sin(h\omega t - \phi) \qquad (7.84)$$

where $\phi = \tan^{-1}(\omega CR)$, $\omega = 2\pi/T$.

It is seen from Fig. 7.43(c) that for higher harmonics of i_{ac}, the load R is by-passed by the commutation capacitor. Therefore, the waveform of i_0 depends upon the relative values of R and reactance X_c at the inverter frequency f. Figure 7.44(b) shows the typical waveforms of the currents and voltages across different circuit components.

Time domain analysis: An alternative method of analysing the inverter operation is to write the Kirchhoff's voltage equation for $0 \leqslant t \leqslant T/2$ and solve the resulting equation for load current i_0. If at $t = 0$, $i_0 = -I_1$, $v_c = -I_1R = -v_1$, then,

$$i_0R = -v_1 + \frac{1}{C}\int_0^t (I - i_0)\,dt,\ 0 < t \leqslant \frac{T}{2} \tag{7.85}$$

The solution of (7.85) with above mentioned initial conditions gives,

$$i_0 = I(1 - e^{-t/RC}) - I_1e^{-t/RC},\ 0 < t \leqslant T/2 \tag{7.86}$$

Under steady-state, current i_0 reaches the value $+I_1$ at $T = T/2$ (Fig. 7.44). Therefore,

$$i_0(t = T/2) = +I_1 \tag{7.87}$$

From equations (7.86) and (7.87)

$$I_1 = I\cdot\frac{1 - e^{-T/2RC}}{1 + e^{-T/2RC}} \simeq I \text{ if } T \gg 2RC \tag{7.88}$$

Substituting for I_1 from (7.88) in (7.86)

$$i_0 = I(1 - A\cdot e^{-t/RC}) \text{ where } A = 2(1 + e^{-T/2RC})^{-1} \tag{7.89}$$

The turn off time t_q provided by the circuit to each thyristor is calculated from the equation,

$$v_c(t = t_q) = i_0(t = t_q) = 0 \tag{7.90}$$

From equations (7.90) and (7.86),

$$t_q = RC \ln\left(\frac{2}{1 + e^{-T/RC}}\right) \tag{7.91}$$

$$\simeq 0.69\ RC \text{ if } T \gg 2RC$$

The average value of the input v_{in} can be calculated from the equation,

$$V_{in} = \frac{2}{T}\int_0^{T/2} i_0R\,dt = IR\left[1 - \frac{4RC}{T}\cdot\left(\frac{1 - e^{-T/2RC}}{1 + e^{-T/2RC}}\right)\right] \tag{7.92}$$

The input power is $(V_{in}I)$ and this power is supplied to the load resistor R.

Design criteria: It is seen from equation (7.91) that the turn off time provided by the circuit reduces as the inverter frequency, $1/T$, is increased. Thus there is a maximum frequency beyond which the inverter will fail to commutate.

It is seen from equation (7.88) that at high values of inverter frequency, $1/T$, the peak load current I_1 reduces from the maximum value of I. Thus at low values of the inverter frequency, $1/T$, the waveform of i_0 is close to a square while at high values it is close to a sinusoid. If a square waveshape of load current is desired then the ratio $(T/2RC)$ must be higher than the 5.0. Using this criterion.

$$C = t_q'/0.69\ R \tag{7.93}$$

where t_q' is the turn off time of the thyristors used in the circuit. The maximum operating frequency of the inverter is then given by,

$$f_{max} = 0.069/t_q' \tag{7.94}$$

In order to obtain a square waveshape of load current the inverter must be operated at frequencies lower than the limit given by equation (7.94). On the other hand, if a sinusoidal waveshape of load current is desired then the frequency domain analysis provides a simple criterion. The value of capacitive reactance X_c at the lowest harmonic frequency $3f_{min}$ must be small ($<50\%$) compared to R. Therefore,

$$C \geqslant 1/(10\ Rf_{min}) \tag{7.95}$$

The inverter should be operated at frequencies *higher* than f_{min} in order to obtain the sinusoidal waveshape. The upper limit on frequency is determined by the turn off time which can be calculated from (7.91). It can be shown using eqn. (7.91) that the ratio $(t_q/0.5T)$ varies from 13.8% when $T/2RC$ equals 5.0 to 38% when $(T/2RC)$ equals 1.0 and approaches 50% at low values of $(T/2RC)$.

7.3.2.2 *Single Phase CSI with Purely Inductive Load*

The inverter circuit of Figure 7.43 (a) is not suitable for highly inductive loads due to the reduction in the voltage across the commutation capacitor. For example, if the constant current flows through purely inductive load, the voltage v_c (Fig. 7.43a) becomes zero. Then the commutation of the thyristors is not possible. A modified circuit for the CS inverter is shown in Figure 7.45 (a). In this circuit diodes have been added to prevent the commutation

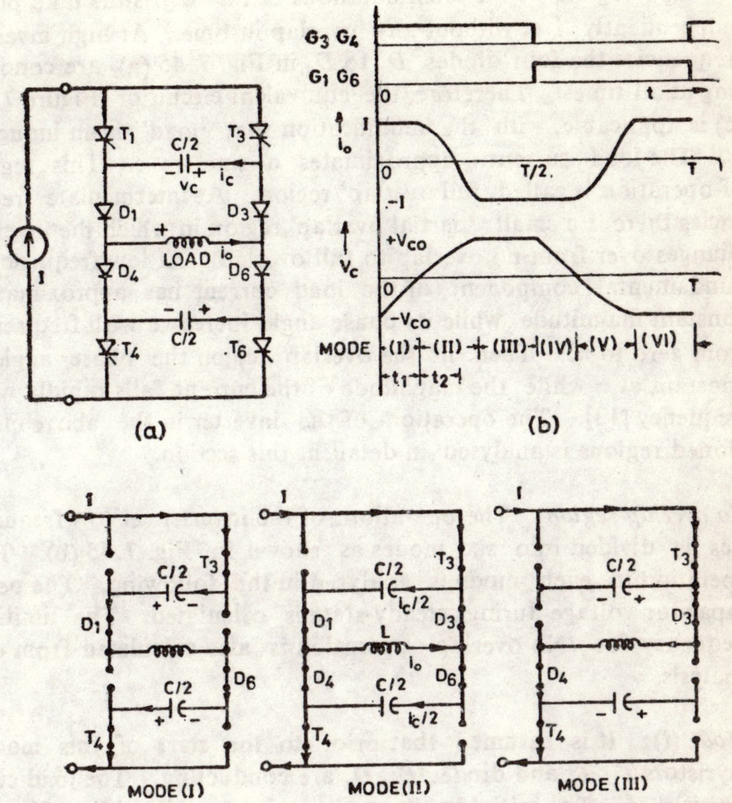

Fig. 7.45. (a) Circuit diagram, (b) Waveforms, (c) Equ. circuits for 1ϕ CSI with inductive load. 'No overlap' region.

capacitors from discharging into the load. The capacitor C of Fig. 7.43 (a) has been split into two groups for the commutation of the top thyristors T_1 and T_3 and the bottom thyristors T_4 and T_6. The thyristors are gated in pairs T_1, T_6 and T_3, T_4 by signals G_1, G_6 and G_3, G_4 shown in Fig. 7.45 (b). The inverter gating period T is controlled by the gating circuit and it can be adjusted to a desired value.

The operation of the CS Inverter with inductive load is dependent upon the inverter frequency f i.e. ratio $1/T$. At low frequencies the load current i_0 (Fig. 7.45) approximates the square wave of the ideal inverter. This region of operation is called the 'no overlap' region, as the commutations of the thyristors take place independently i.e. without any overlap in time. At high inverter frequencies the four diodes, D_1 to D_6 in Fig. 7.45 (a), are conducting at all times. Therefore, the equivalent circuit of Figure 7.43 (c) is applicable, with the modification that 'load' is an inductor L. The load current i_0 approximates a sine wave. This region of operation is called 'full overlap' region. At intermediate frequencies there is a small 'partial overlap' region in which the inverter changes over from no overlap to full overlap. At low frequencies, fundamental component of the load current has approximately constant magnitude while its phase angle increases with frequency from zero to π. Then in the overlap region the phase angle is constant at π while the magnitude of the current falls rapidly with frequency [13]. The operation of the inverter in the above mentioned regions is analysed in detail in this section.

No overlap region: The operation of the inverter at low frequencies is divided into six modes as shown in Fig. 7.45 (b). The operation in each mode is analysed in the following. The peak capacitor voltage during steady-state is calculated. The limiting frequency for 'No overlap' operation is also calculated from the analysis.

Mode (i): It is assumed that prior to the start of this mode, thyristors T_1, T_6 and diodes D_1, D_6 are conducting. The load current is $+I$. The initial voltage on each capacitor is $-V_{c0}$, as shown in Fig. 7.45 (c).

At the start of this mode, thyristors T_3 and T_4 are gated. The initial charge on the capacitors, reverse biases thyristors T_1 and T_6 and turns them off. The input current, I, now flows through devices T_3, $C/2$, D_1, load, D_6, $C/2$ and T_4. The load current is therefore unchanged during this mode. The diodes D_3 and D_4 are reverse biased during this mode due to the reverse voltage on the capacitors. The constant current I flowing through the capacitors linearly charges them during this mode, as shown by the waveform of v_c in Fig. 7.45(b). The mode ends when v_c reduces to zero, since the diodes D_3 and D_8 start conducting.

The variation of voltage v_c during this mode is given by,

$$v_c = -v_{c0} + \frac{2}{C} I.t \tag{7.96}$$

The duration of mode (i), t_1, is derived from above equation by equating v_c to zero. Thus,

$$t_1 = \frac{1}{2}\frac{CV_{c0}}{I} \tag{7.97}$$

Mode (ii): In this mode, all diodes D_1, D_3, D_4 and D_6 are conducting. As seen from Figure 7.45(c) the capacitors are now effectively connected in parallel with the inductive load. The load current i_0 and total capacitor current i_c are related by the following equations:

$$(I + i_0) = i_c \tag{7.98}$$

$$-L\frac{di_0}{dt} = \frac{1}{C}\int_0^t (I + i_0)\, dt \tag{7.99}$$

The above equations are solved with the initial conditions $i_0 = I$, $i_0' = 0$ @ $t = 0$, to get

$$i_0 = I(2\cos\omega_0 t - 1) \tag{7.100}$$

and

$$i_c = 2I\cos\omega_0 t,\ \omega_0 = 1/\sqrt{LC} \tag{7.101}$$

The capacitor voltage v_c is given by

$$v_c = (2I\sin\omega_0 t/\omega_0 C) \tag{7.102}$$

The mode ends when the capacitor current i_c goes to zero, since its reversal is prevented by diodes D_1 and D_6. The duration of mode (ii), t_2, is obtained by equating i_c (eqn. 7.101) to zero. Therefore,

$$t_2 = \pi/2\,\omega_0 \tag{7.103}$$

The capacitor voltage v_0 at the end of mode (ii) is obtained from equation (7.102) by putting $\omega_0 t$ equal to $\pi/2$. Thus,

$$v_c = 2I/\omega_0 C \text{ @ } t = t_2 \tag{7.104}$$

The load current i_0 at the end of mode (ii) is equal to $-I$ i.e. the load current has reversed the direction during this mode.

Mode (iii): The commutation process has been completed at the end of mode (ii). The input current I now flows through devices T_3, D_3, the load inductor, D_4 and T_4. The mode continues till the thyristor pair T_1, T_6 is turned on.

Mode (iv): Thyristors T_1 and T_6 are gated to turn off T_3 and T_4.

This mode is similar to mode (i). The thyristors T_1, T_6 and diode D_3, D_5 are conducting. The duration of this mode is given by equation (7.97).

Mode (*v*): This mode is similar to mode (ii). The thyristors T_1, T_6 and all the diodes are conducting. The duration of this mode is given by equation (7.103).

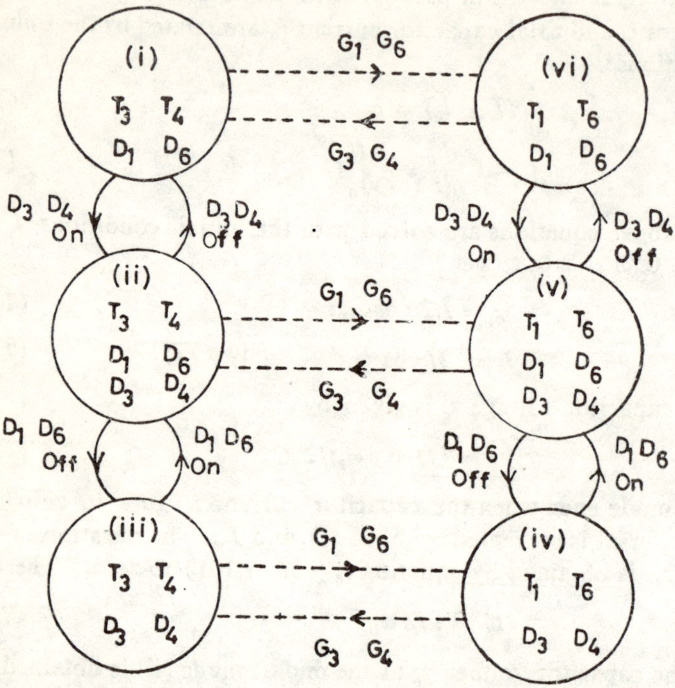

Fig. 7.46. Mode transition for the CSI of Fig. 7.45(a).

Mode (*vi*): This mode is similar to mode (iii). The commutation process started at the beginning of mode (iv) has been completed at the end of mode (v). The current I now flows through devices T_1, D_1, the load inductor, D_6 and T_6.

Steady-state operation: Figure 7.46 shows the transition diagram for the six modes of the single phase current source inverter. The devices conducting in a particular mode are placed inside a circle e.g. T_3, T_4, D_1 and D_6 in mode (i). The six circles represent the six modes. The inverter changes mode when a diode pair changes from on to off or vice-versa. For example, if diode pair D_1D_6 turns off the inverter transits from mode (ii) to (iii) or mode (v) to (iv). The transition between modes can also be brought about by turning on a thyristor pair. For example if during mode (i) gating signals G_1G_6 are applied to the thyristors T_1T_6 the inverter transits to mode (vi) provided the thyristors are forward biased and turn on.

During steady-state operation in the "No Overlap region" the inverter follows the mode sequence (i), (ii), (iii), (iv), (v), (vi) during each period. At the start of each period in the steady-state the capacitor voltage changes from $-V_{c0}$ to $+V_{c0}$ during modes (i) and (ii) as shown in Fig. 7.45(b). Thus, the capacitor voltage at the end of mode (ii), given by equation (7.104), is equal to V_{c0}. The peak capacitor voltage is therefore given by,

$$V_{c0} = 2I/\omega_0 C \tag{7.105}$$

The duration of mode (i), t_1, can be calculated from eqns. (7.97) and (7.105) as,

$$t_1 = 1/\omega_0 \tag{7.106}$$

From equations (7.103) and (7.106) it is seen that durations t_1 and t_2 are dependent only on the commutation frequency ω_0. The total commutation time is given by,

$$t_c = (t_1 + t_2) = \left(1 + \frac{\pi}{2}\right)\sqrt{LC} = 2.57/\omega_0 \tag{7.107}$$

The duration of mode (iii) is given by

$$t_3 = \left(\frac{T}{2} - t_c\right) = \left(\frac{T}{2} - \frac{2.57}{\omega_0}\right) \tag{7.108}$$

It is seen from equation (7.108) that the inverval t_3 reduces as the time period T is decreased. At the limiting value T_{NL}, t_3 reduces to zero. T_{NL} is given by,

$$T_{NL} = 5.14/\omega_0$$

or

$$f_{NL} = 1/T_{NL} = 0.19/\sqrt{LC} \tag{7.109}$$

Thus the analysis given above is valid for inverter frequencies $f \leqslant f_{NL}$. At frequencies higher than f_{NL} mode (ii) of the commutation is not completed before the next pair of thyristors is turned on. The inverter then operates in the 'Partial overlap' region.

Partial overlap region: In this region the inverter frequency f i.e. $1/T$, is higher than the value f_{NL} given by equation (7.109). The commutation of one pair of thyristors therefore starts when it is in mode (ii) or (v) (Fig. 7.46). The inverter operates in the mode sequence (ii), (i), (ii); (v), (iv), (v) in the steady-state. The modes (iii) and (vi) are absent in the Partial Overlap region. The commutation of a particular thyristor pair now starts with load current I_0 which is less than the source current I.

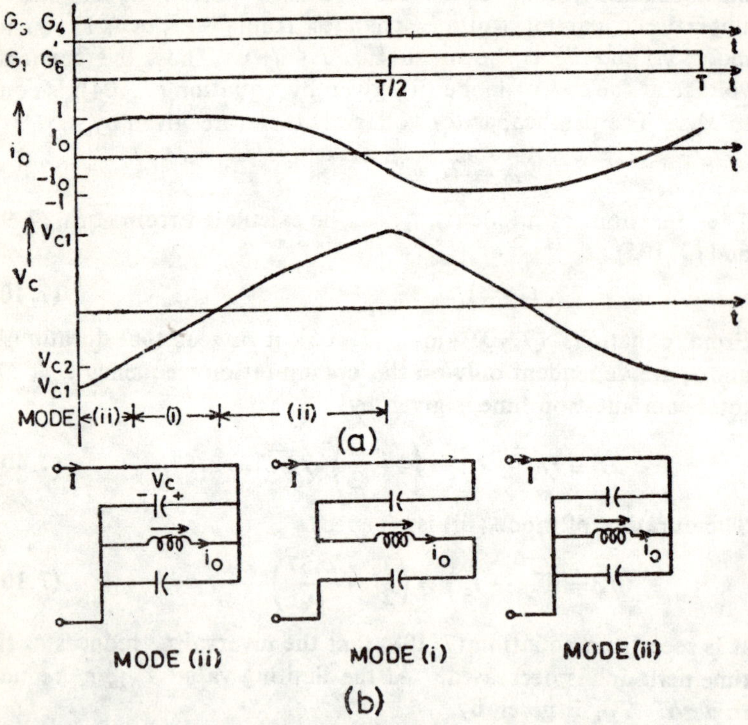

Fig. 7.47 1φ CSI with inductive load (Fig. 7.45a):
(a) Waveforms.
(b) Equivalent circuits in partial overlap region.

Figure 7.47(a) shows typical waveforms for operation in partial overlap. The circuit operates in three modes per half cycle and the corresponding equivalent circuits are given in Fig. 7.47(b). A detailed analysis of this region can be carried out using the circuit differential equations [10]. From such an analysis it has been shown that the initial current I_0 reduces as the inverter frequency is increased. The peak capacitor voltage V_{c1} also reduces from the value given by eqn. (7.105) as the frequency is increased. The limit of this region is reached when I_0 reduces to zero and mode (ii) (Fig. 7.47) disappears completely. It can be shown [10] that the end of the region, when the inverter frequency f equal f_{PL} is given by,

$$f_{PL} = 0.24/\sqrt{LC} \tag{7.110}$$

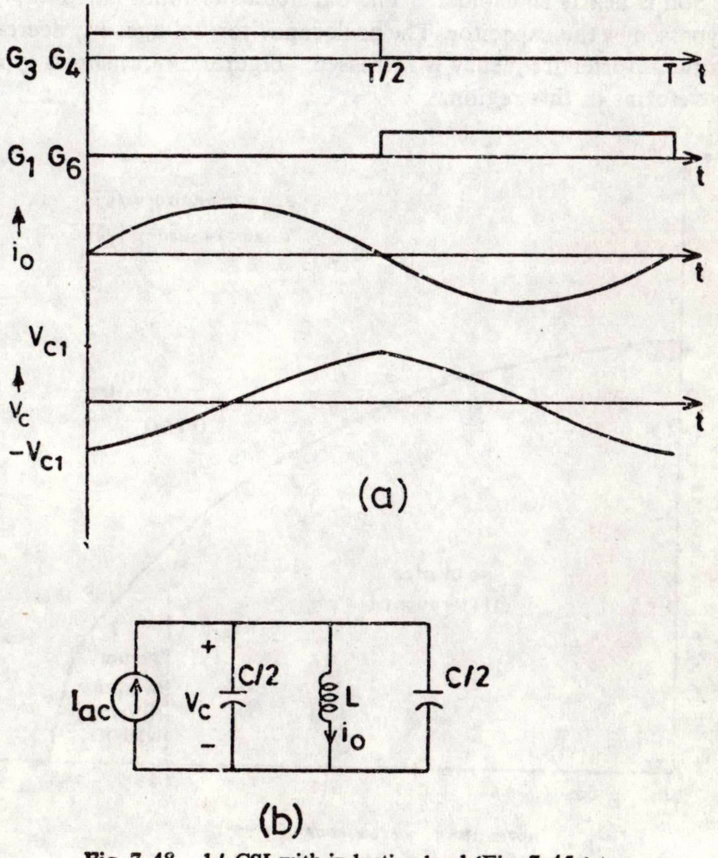

Fig. 7.48. 1ϕ CSI with inductive load (Fig. 7.45a):
(a) Waveforms.
(b) Equ. circuit in full overlap region.

The peak capacitor voltage V_{c1} reduces from V_{c0} (eqn. 7.105) at the beginning of the region to $0.866\ V_{c0}$ at the end of the region.

Full overlap region: If the inverter frequency is increased to values higher than f_{PL} (eqn. 7.110) the diodes conduct continuously. The inverter operates in the mode sequence (ii), (v) in the steady-state in this region. The equivalent circuit for this region is shown in Fig. 7.48(b). The capacitors are now in parallel with the load. The operation in this region can be analysed using the frequency domain analysis of section 7.3.2.1. The current I_{ac} in Fig. 7.48(b) is the square wave shown in Fig. 7.43(b). The load current i_0 in this region is nearly sinusoidal. The harmonics of input current i_{ac} are bypassed by the capacitor. The peak capacitor voltage, V_{c1} decreases as the inverter frequency is increased. Figure 7.48(a) shows typical waveforms in this region.

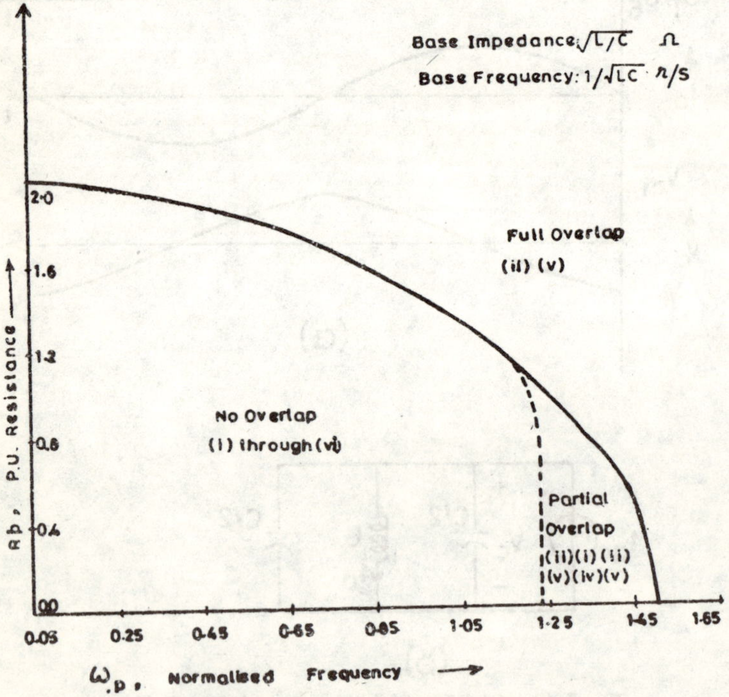

Fig. 7.49. Regions of various mode sequences for single phase current source inverter with R—L load.

7.3.2.3 *Single Phase CSI with R—L Load*

The current source inverter with $R-L$ load can be analysed in the same manner as the inverter with purely inductive load discussed in the preceding section. The circuit equations are more complex due to the appearance of a damping factor. The operation of the inverter can be simulated by using the circuit differential equations and the mode transition diagram of Fig. 7.46. The operation with $R-L$ load gives different mode sequences at different R/L ratios. Figure 7.49 shows mode sequences as a function of frequency for different values of resistance R and a constant inductance L. This diagram shows that for values of $R\sqrt{C/L}$ less than unity, the inverter operates in the 'No overlap', 'Partial Overlap' and the 'Full Overlap' regions. At higher values of $R\sqrt{C/L}$ the 'Partial Overlap' region is nearly absent while the 'No Overlap' region reduces rapidly. The inverter operates in the 'Full Overlap' region for all frequencies if $R\sqrt{C/L}$ exceeds two [10]. In this region all the four diodes conduct continuously and are thus redundant. The circuit of Fig. 7.43 can be used for such operation. The operation in the 'Full Overlap' region can be analysed by using the equivalent circuit of Fig. 7.48(b) with L replaced by the $R-L$ load. The frequency domain analysis given in section 7.3.2.1 can be applied to obtain the circuit waveforms. A detailed analysis of a current source inverter equivalent to the one discussed here, but using centre tapped transformer is given in [11].

7.3.2.4 *Single Phase CSI with L-Back EMF Load*

The current source inverter can be used to supply active loads, e.g. a single phase induction motor. In such applications the load is modelled as an ac counter *EMF* in series with an inductance, L. The resistance is usually small and can be neglected. The frequency of the counter *EMF* is equal to the frequency of the inverter gating circuit. However, the phase angle of the back *EMF* is dependent upon the load and is not under the control of the inverter, unless additional feedback is used. Figure 7.50 shows the gating signals for the thyristors, the ideal output current and the back *EMF*. The angle α determines the value of the back *EMF* at the time of commutation and thus influences the commutation time. The circuit of Fig. 7.45(a) with a back *EMF* e in series with load inductance L can be analysed by writing circuit differential equation for each of the six modes shown in Fig. 7.46. The phase angle α of the back

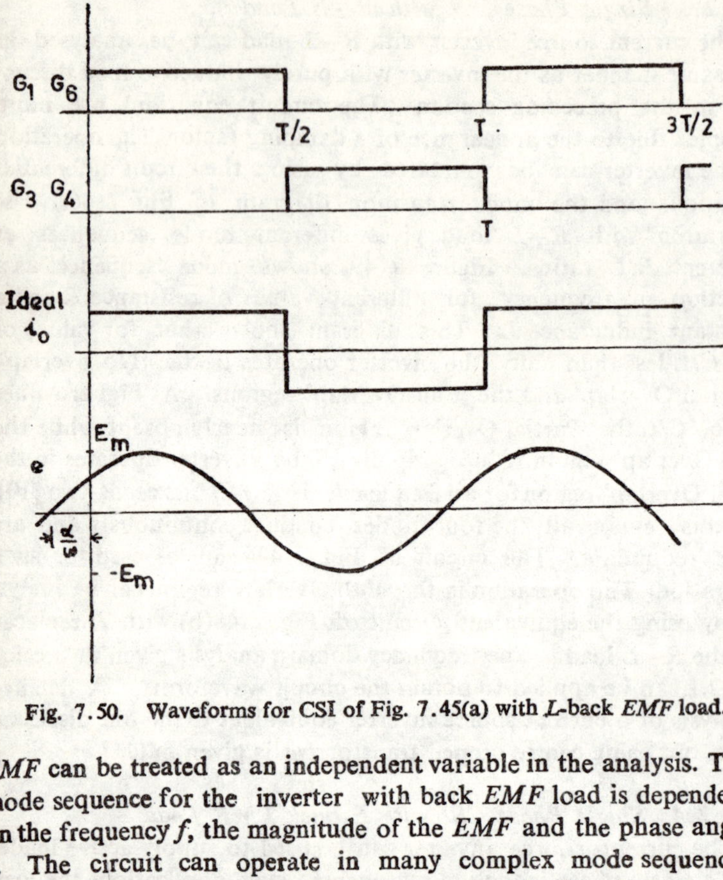

Fig. 7.50. Waveforms for CSI of Fig. 7.45(a) with *L*-back *EMF* load.

EMF can be treated as an independent variable in the analysis. The mode sequence for the inverter with back *EMF* load is dependent on the frequency *f*, the magnitude of the *EMF* and the phase angle α. The circuit can operate in many complex mode sequences besides the usual three corresponding to the No Overlap, Partial Overlap and Full Overlap regions [12], [29].

7.3.3 Three Phase Capacitor Commutated Current Source Inverter

In this section a three phase CSI circuit is described and analysed in detail. The circuit, shown in Fig. 7.51, is similar to ideal inverter (Fig. 7.40) with thyristors T_1 to T_6 working as switches S_1 to S_6 respectively. The commutation capacitors are divided into two groups and connected in a delta across the legs of the inverter. The commutation capacitors are separated from the output by blocking diodes D_1 to D_6. The thyristors are gated in a sequence so that only two thyristors are conducting at a time, as per the sequence shown in Fig. 7.40(b). The circuit shown

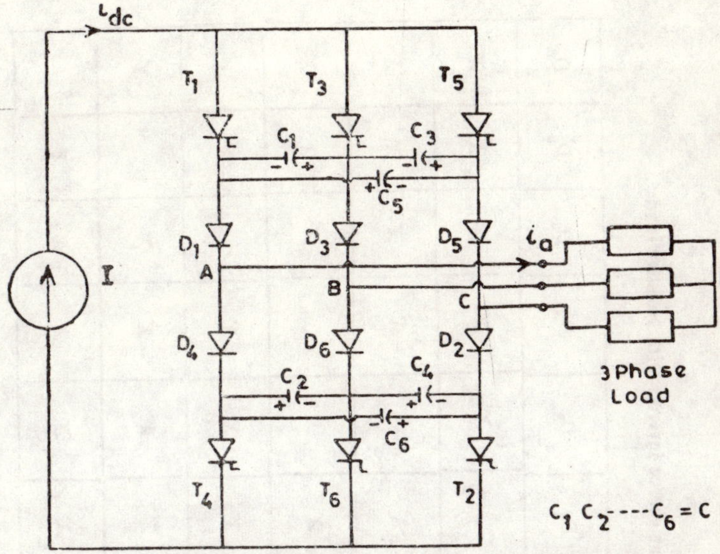

Fig. 7.51 Circuit diagram of three-phase capacitor commutated CS inverter.

in Fig. 7.51 is often called Auto Sequential Current Source Inverter (ASCI).

7.3.3.1 *Three-phase CSI with Resistive Load*

The three-phase current source inverter shown in Fig. 7.51 can be used to supply a balanced three phase resistive load. The operation of the three phase CSI in the steady-state can be divided into six modes in each cycle. The thyristors are gated in pairs of two for sixty degrees i.e. one-sixth of the period. Table 7.1 shows the conducting devices in each of the six modes. In each mode two thyristors and five diodes conduct for all inverter frequencies. The remaining devices are off. Equivalent circuits can be drawn for each mode based on the conducting devices given in Table 7.1. Figure 7.52 shows equivalent circuits for modes (i) and (ii). Similar equivalent circuits can be drawn for the other four modes. The calculation of the inverter performance can be done only by detailed simulation of the circuit using the equivalent circuits or other methods of circuit simulation. Figure 7.53 shows the waveforms obtained by circuit simulation. It is seen from Fig. 7.53 that the phase current is approximately the 120 degree square wave

Table 7.1. Conducting devices (X) in a three-phase current source inverter with resistive load

Mode	T_1	T_2	T_3	T_4	T_5	T_6	D_1	D_2	D_3	D_4	D_5	D_6
(i)					X	X		X	X	X	X	X
(ii)	X					X	X		X	X	X	X
(iii)	X	X					X	X		X	X	X
(iv)		X	X				X	X	X		X	X
(v)			X	X			X	X	X	X		X
(vi)				X	X		X	X	X	X	X	

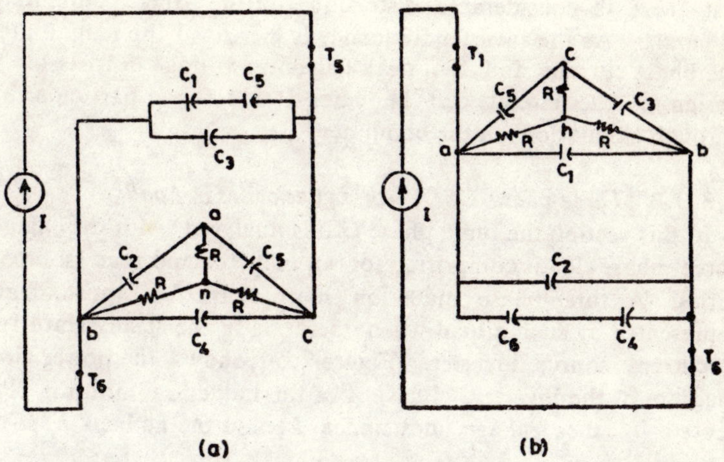

Fig. 7.52. Equivalent circuits for three-phase CSI of Fig. 7.51; (a) Mode (i) and (b) Mode (ii).

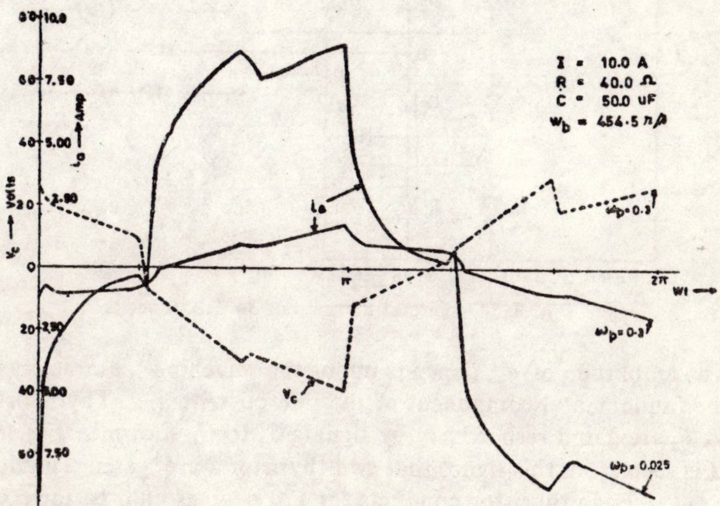

Fig. 7.53. Waveforms for phase current and capacitor voltage of the three-phase CSI with resistive load.

but there is considerable distortion during the commutation intervals. As the inverter frequency is increased the magnitude of the phase current and the peak capacitor voltage decreases. The design of the circuit can be carried out for a particular load through simulation on the computer.

7.3.3.2 *Three-phase CSI With L-Back EMF Load*

In this section the three phase CSI is analysed with a balanced three phase load comprising of an inductor and a ac source in series. A three-phase induction motor can be approximately represented by such a load when operating in the steady-state from a current source inverter. Figure 7.54 shows the power circuit diagram of the inverter. Phase *A* of the induction motor is represented by the leakage inductance L_1 and the ac back *EMF* e_{an}.

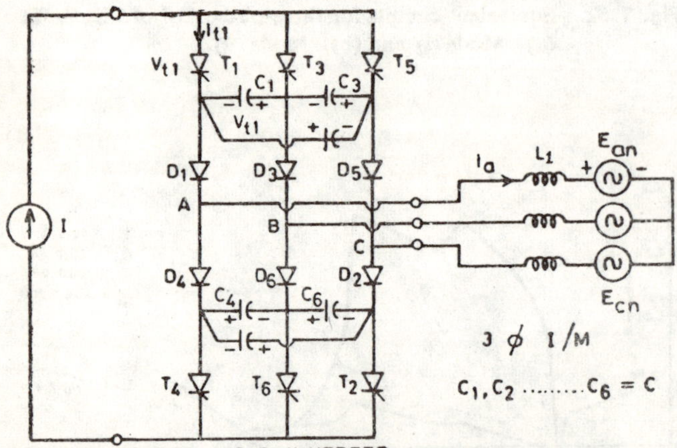

Fig. 7.54. Circuit diagram of 3ϕ C.S. inverter.

The amplitude of e_{an} depends upon the machine parameters and the fundamental component of the line current i_a. Thyristors T_1 to T_6 are gated respectively by signals G_1 to G_6 shown in Fig. 7.56. It is seen from this figure that two thyristors are gated simultaneously. Each thyristor conducts for 120 degrees and is then commutated. The thyristors can be divided into the top group comprising of T_1, T_3, T_5 and the bottom group comprising of T_2, T_4, T_6. Each thyristor is commutated by the stored charge on the capacitors when the next thyristor in the same group is gated. Thus

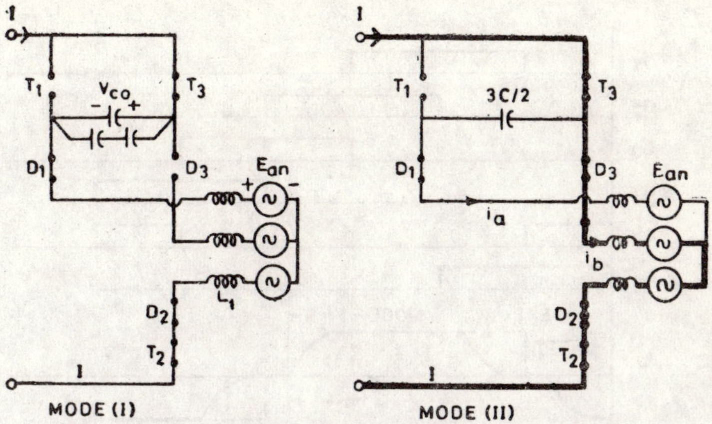

Fig. 7.55. Equivalent circuits for commutation modes of CSI.

there are six commutations during each period of the output current and the commutations alternate between the top and the bottom groups.

The operation of the CS inverter with inductive load can be divided into several modes depending on the state of the thyristors and diodes. The mode sequence is dependent on the frequency of the inverter and the load parameters. At low frequencies the commutation interval T_c is smaller than one-sixth of the inverter period T. Thus the commutations proceed independently. This is the 'No Overlap' region. As the inverter frequency is increased the commutations proceed simultaneously in the top and the bottom groups of thyristors while T_c is less than $T/3$. This is the 'Partial Overlap' region. As the inverter frequency is increased further the commutations take place continuously in the top and the bottom groups and T_c equals $T/3$. This is the 'Full Overlap' region. The operation of the inverter supplying induction motors is usually restricted to the 'No Overlap' region. The operation in this region is analysed in detail in the following. The operation in the other regions is qualitatively explained.

No overlap region: The operation of the inverter in each period of T seconds can be divided into six intervals of $T/6$ seconds each. The conducting thyristors in each mode are shown in Fig. 7.56. A commutation occurs in each of the six intervals. For example in Fig. 7.56 when t equals $T/2$ thyristor T_3 is gated and thyristor T_1

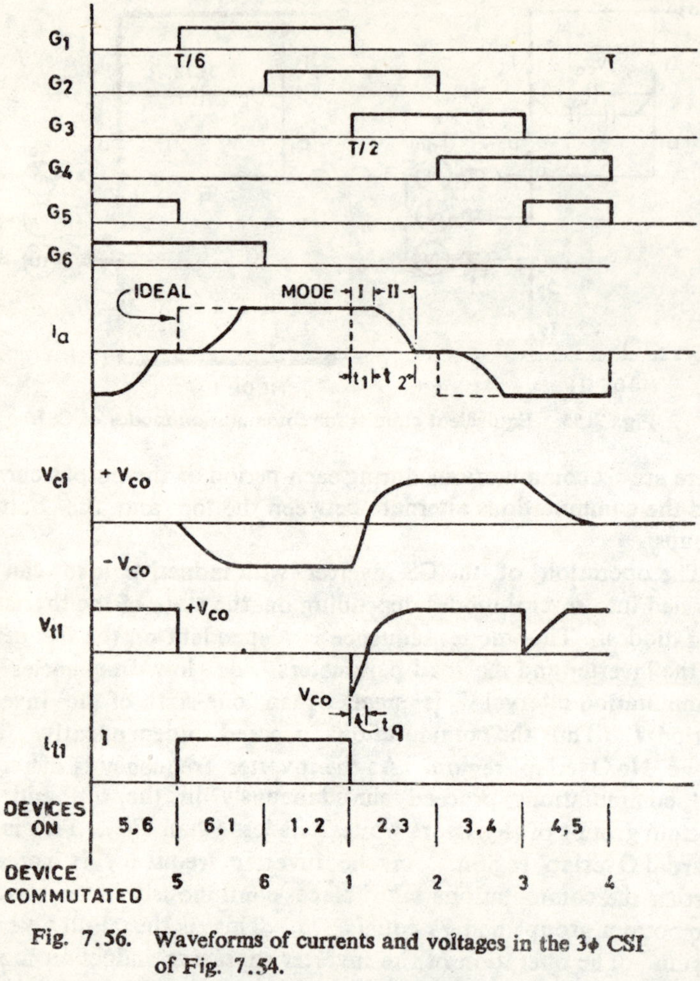

Fig. 7.56. Waveforms of currents and voltages in the 3ϕ CSI of Fig. 7.54.

is commutated. Each interval of $T/6$ seconds can be further subdivided into three modes in which two or three diodes conduct. Table 7.2 shows the conducting devices in various modes in the no overlap region by a cross sign. The other devices are off. At low frequencies of operation the time duration of modes (i) and (ii) is small compared to the duration of the interval, i.e. $T/6$. As the frequency is increased the time duration of the modes (i) and (ii) remains nearly constant while that of mode (iii) reduces. At the

Interval	Mode	T_1	T_2	T_3	T_4	T_5	T_6	D_1	D_2	D_3	D_4	D_5	D_6
I	(i)					X	X				X	X	
	(ii)					X	X				X	X	X
	(iii)					X	X					X	X
II	(i)	X					X					X	X
	(ii)	X					X	X				X	X
	(iii)	X					X	X					X
III	(i)	X	X					X					X
	(ii)	X	X					X	X				X
	(iii)	X	X					X	X				
IV	(i)		X	X				X	X	X			
	(ii)		X	X				X	X	X			
	(iii)		X	X					X				
V	(i)			X	X				X	X			
	(ii)			X	X				X	X	X		
	(iii)			X	X					X	X		
VI	(i)				X	X				X	X		
	(ii)				X	X				X	X	X	
	(iii)				X	X					X	X	

limit of the no overlap region the duration of mode (iii) reduces to zero. The analysis of the commutation during interval IV following turn on of thyristor T_3 is given below. The modes under analysis occur when t equals $T/2$ in Fig. 7.56. The operation is assumed to be in the steady-state. Prior to the gating of thyristor T_3 the input current I is flowing through devices T_1, D_1, phase A, phase C, D_2 and T_2. In the steady-state the initial voltages on the top group of capacitors are

$$v_{c1} = - V_{c0},\ v_{c3} = 0,\ v_{c5} = + V_{c0} \tag{7.111}$$

The reference directions of the capacitor voltages are given in Fig. 7.54. The values of the steady-state peak capacitor voltage, v_{c0}, is calculated later in this analysis.

The commutation period starts with gating of thyristors T_3 and is divided into the following three modes.

Mode (*i*): The equivalent circuit during this mode is shown in Fig 7.55. The thyristor T_1 is turned off almost instantaneously by the stored charge on the capacitors C_1 and C_5. The input current I, now flows through T_3, equivalent capacitor of $3c/2$, D_1, phases A, C, D_2 and T_2. Due to the stored charge on capacitors C_1, conduction of diode D_1 reverse biases diode D_3. This flow of input current I, discharges the capacitor C_1 and the mode ends when v_{d3}, the voltage across D_3, is reduced to zero.

In order to simplify the analysis, the back emf sources e_{an}, e_{bn} and e_{cn} are assumed to remain constant during the commutation period, i.e. modes (i) and (ii) [14]. The constant values, denoted by E_{an}, E_{bn} and E_{cn}, are equal to the inst. values at $t = T/2$. The diode voltage v_{d3} is given by

$$v_{d3} = - V_{c0} + \frac{1}{c} \cdot \frac{2}{3} \cdot I \cdot (t - T/2) + (E_{an} - E_{bn}) \tag{7.1.2}$$

The time interval t_1 of mode (i) is obtained by equating v_{d3} to zero. Therefore,

$$t_1 = 1.5\, C\, (E_{bn} - E_{an} + V_{c0})/I \tag{7.113}$$

The ac line current does not change during this mode.

Mode (*ii*): The diodes D_1 and D_3 conduct during this mode. The equivalent circuit of this mode is given in Fig. 7.55. The diode

D_1, phase A, phase B, diode D_3 and capacitor $3C/2$ form an oscillatory circuit. The current i_a now decreases from its initial value I while current i_b increases from the initial value of zero.

The differential equation governing the variation of i_a during this mode can be shown to be,

$$\frac{d^2 i_a}{dt'^2} + \frac{1}{3L_1 C} i_a = 0 \tag{7.114}$$

where $t' = t - \left(\frac{T}{2} + t_1\right)$.

Solving (7.114) with the initial conditions, $i_a = I$, $I_a' = 0$ @ $t' = 0$ gives,

$$i_a = I \cos \omega_0 t' \tag{7.115}$$

Then, $$i_b = I - i_a = I(1 - \cos \omega_0 t') \tag{7.116}$$

where $\omega_0 = \frac{1}{\sqrt{3L_1 C}}$, is the frequency of oscillation in mode (ii).

The reversal of current i_a is prevented by diode D_1. Therefore, this mode ends when i_a goes to zero. The duration of mode (ii), t_2, is given by

$$t_2 = \pi/2\,\omega_0 \tag{7.117}$$

Mode (iii): The commutation process is completed at the end of mode (i) and (ii). The input current now flows through devices T_3, D_3, phase B and C, D_2 and T_2 as shown by dark line in the equivalent circuit of mode (ii) in Fig. 7.55.

Steady-state operation: Figure 7.56 shows the waveforms of line current i_a, voltage across C_1, v_{c1}, thyristor voltage v_{t1} and current i_{t1} during steady-state operation. The capacitor voltage v_{c1} is given by,

$$\begin{aligned} v_{c1} &= -V_{c0} \text{ @ } t = T/2 \\ &= (E_{bn} - E_{an}) \text{ @ } t = T/2 + t_1 \\ &= + V_{c0} \text{ @ } t = T/2 + t_1 + t_2 \end{aligned} \tag{7.118}$$

Therefore, peak capacitor voltage is given by,

$$V_{c0} = (E_{bn} - E_{an}) + \frac{2}{3}\frac{I}{\omega_0 C} \tag{7.119}$$

The last term is the integral of current, $(2/3\ i_a)$, through C_1, during mode (ii). The peak capacitor voltage V_{c0} is also the peak reverse voltage of the thyristors. It is dependent on the back *EMF*, $(E_{bn}-E_{an})$ which is a function of the machine parameters, slip freq., inverter frequency and the fundamental component of the line current.

The waveform of v_{t1} in Fig. 7.56 shows the turn off t_q of the thyristor. It is given by,

$$t_q = 1.5\ C\ V_{c0}/I \tag{7.120}$$

The turn off time in the steady-state is minimum when the machine is at standstill, i.e. $E_{bn} = E_{an} = 0$ and is given by,

$$t_q \text{ min} = 1.73\ \sqrt{L_1 C} \tag{7.121}$$

The turn off time given eqn. (7.121) is sufficiently large so that non-inverter grade devices can be used. However, it has been found experimentally that the turn off time during the transient operation may be one half of the value given by (7.121). This factor must also be considered in the selection of the thyristors for C.S.I.

Operating frequency range: The total time t_c for each commutation in the inverter is the sum of intervals t_1 and t_2 given by eqns. (7.113) and (7.117) respectively. Note that the gating of a thyristor immediately turns off the previously conducting thyristor. But the commutation is not complete till the transfer of current between the phases is over. From eqns. (7.113), (7.117) and (7.119), the commutation time t_c under steady-state is given by

$$t_c = \frac{3C}{I}(E_{bn} - E_{an}) + \frac{1}{\omega_0} + \frac{\pi}{2}\cdot\frac{1}{\omega_0} \tag{7.122}$$

where $\omega_0 = (3L_1C)^{-1/2}$.

There are six identical commutations during each cycle (Fig. 7.56). The commutations can be independent only if $t_c < (T/6)$ where T is the inverter period. Thus, given the circuit and load parameters, the limit of '*No overlap*' operation is given by

$$f_{NL} = (1/T) = (1/6t_c) \tag{7.123}$$

The frequency f_{NL} is influenced by the magnitude of the ac back *EMF* $(F_{bn} - E_{an})$ and the oscillation frequency ω_0.

Partial and full overlap region: The operation at frequencies above f_{NL} is complex [15, 16] because more than two diodes may conduct at the same time. For frequencies between f_{NL} and $2f_{NL}$, the commutations in the top (T_1, T_3, T_5) and bottom (T_2, T_4, T_6) groups of Fig. 7.54 proceed simultaneously. If only two diodes conduct at a time such an operation is stable. Under some conditions this condition is not satisfied even for frequencies below f_{NL} [17]. For frequencies higher than $2f_{NL}$ the operation of the inverter motor system is complex. A part of the input dc current bypasses the motor due to conduction of two diodes in the same leg of the inverter. The motor torque is therefore reduced and the operation may even be unstable. Therefore the commutation capacitor in a motor load should be chosen such that 'No overlap' operation takes place throughout the operating frequency range.

Protective components: The circuit shown in Fig. 7.54 does not include the protective components normally required for proper operation of the inverter. It is seen from Fig. 7.56 that the thyristor current i_{t1} rises instantaneously at turn on. A small inductor is connected in series with each thyristors to limit the di/dt at turn on. The value of di/dt with a series inductance of L_t is given by

$$di/dt = V_{c0}/2L_t \tag{7.124}$$

where V_{c0} is given by eqn. (7.119)

In addition to the series inductance, each thyristor and diode in Fig. 7.54 is provided with a parallel R–C network (also called snubber) provides a path for the recovery current when the particular device is turned off. The snubber resistor is chosen to be such that the oscillation due to recovery current is critically damped.

The current source inverter is inherently protected against a short circuit at the output terminals due to the current regulator in the front end current source. But, the occurrence of an open circuit at the output causes very high voltages across the devices. Therefore, voltage detection circuit is provided at the output terminals. A sudden rise in the output voltage causes simultaneous gating of all thyristors and thus provides a low impedance path for the current.

7.3.3.3 *Three-phase CSI with Purely Inductive and R–L Load*

The operation of the current source inverter with purely inductive load can be analysed from the results of section 7.3.3.2 by reducing the ac back *EMF* to zero. Thus eqns. (7.113), (7.117), (7.119), (7.122) and (7.123) are also valid for the purely inductive load if E_{an} and E_{bn} are reduced to zero.

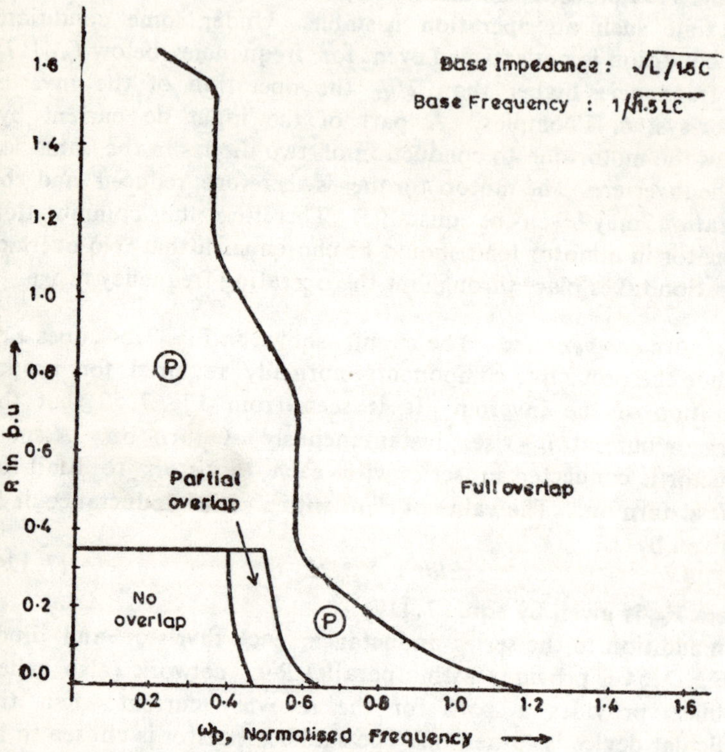

Fig. 7.57. Regions of various mode sequences in a three-phase CS inverter with R–L load.

The operation of the inverter with *R–L* load is highly complex and several mode sequences are possible. The mode sequences are dependent on the *R*/*L* ratio and the inverter frequency. Figure 7.57 shows the regions of various mode sequences when the load resistance is increased keeping the load inductance constant. These mode sequences are discussed briefly below.

No overlap: At low values of inverter frequency and load resistance the mode sequence is as shown in Table 7.2. In each interval of $T/6$ seconds either two or three diodes are conducting. As seen from Fig. 7.54 this sequence lasts for resistance values less than approximately 0.35 p.u. and inverter frequencies less than approximately 0.45 p.u. An analysis of delta connected R–L load in this region is given in [18].

Partial overlap: In this region the mode (ii) in the top group of thyristors overlaps with the mode (ii) in the bottom group of thyristors at each commutation. Thus in each interval there are modes during which either three or four diodes are conducting.

Pseudo random: The region for these sequences are marked by *P* in Fig. 7.57. In this region the following diode conduction patterns are possible in each interval.

(a) Two, three and five diodes conduct.
(b) Three, four and five diodes conduct.
(c) Four and five diodes conduct.

Thus, each interval may have up to three modes during which two, three, four or five diodes and two thyristors conduct. In some mode patterns the repetition is after half a cycle rather than one-sixth of a cycle. Hence the name 'Pseudo-Random'.

Full overlap: At high frequencies the inverter operates in a sequence in which five diodes are conducting simultaneously in each interval. This sequence is shown in Table 7.1 and is the same as the one for a purely resistive load.

7.3.4 Miscellaneous CSI Configurations

The Auto Sequential Current Source Inverter ASCI has been analysed in detail in section 7.3.3. While it is a very widely used inverter, other configurations have been developed to obtain better performance. A few such inverters are briefly described in this section.

Figure 7.58 shows a modified ASCI. In this circuit the inductors L_c, resistors R_c, diodes D_1' to D_6' and the commutation capacitors form six independent loops through the thyristors T_1 to T_6. At the start of each commutation a quick reversal of the capacitor voltage polarity takes place through the loops mentioned above. For example, when thyristor T_3 is gated thyristor T_1 turns off and the top group of capacitors oscillate through the loop T_3,

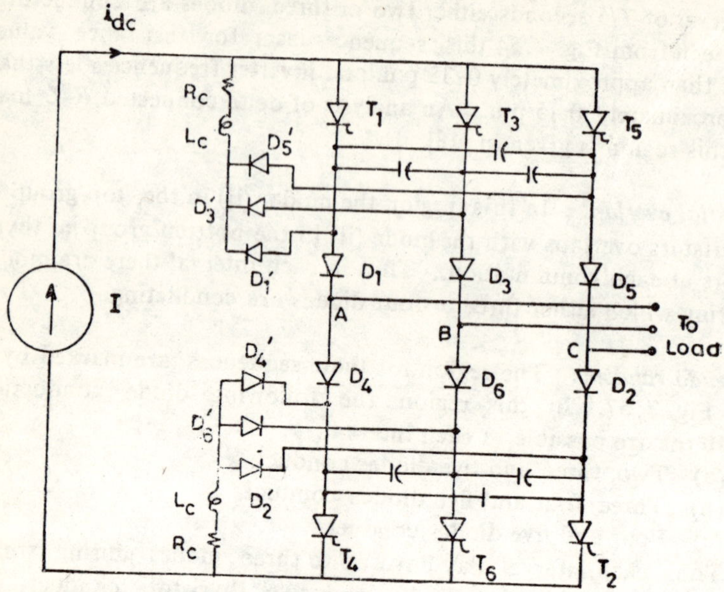

Fig. 7.58. **Modified auto sequential current source inverter with reset circuit.**

D_3', L_c, R_c. Thus, the oscillation modifies the mode (i) of commutation described in section 7.3.3.2. The net result is a reduction in the variation of commutation interval with load current. A detailed analysis of this type of commutation scheme is given in [19].

Figure 7.59(a) shows a current source inverter in which the commutation capacitor is connected between the neutral of the load and the input terminals. This inverter is also sometimes referred as 'Third Harmonic Auxiliary Commutated Inverter' (THACI). The commutation in this inverter proceeds in two modes as in ASCI. Auxiliary thyristor T_p is turned on to commutate a thyristor in the top group, for example T_3. The resulting equivalent circuit is shown in Fig. 7.59(b). The commutation capacitor charges linearly and reverses polarity. The next incoming thyristor T_1 is then gated at a suitable instant of time. The resulting equivalent circuit is shown in Fig. 7.59(c). In this mode the L–C oscillation turns off thyristor T_p thereby completing the commutation. The modes (i) and (ii) shown in Fig. 7.59(b) and (c) are similar to the modes of ASCI described in section 7.3.3.

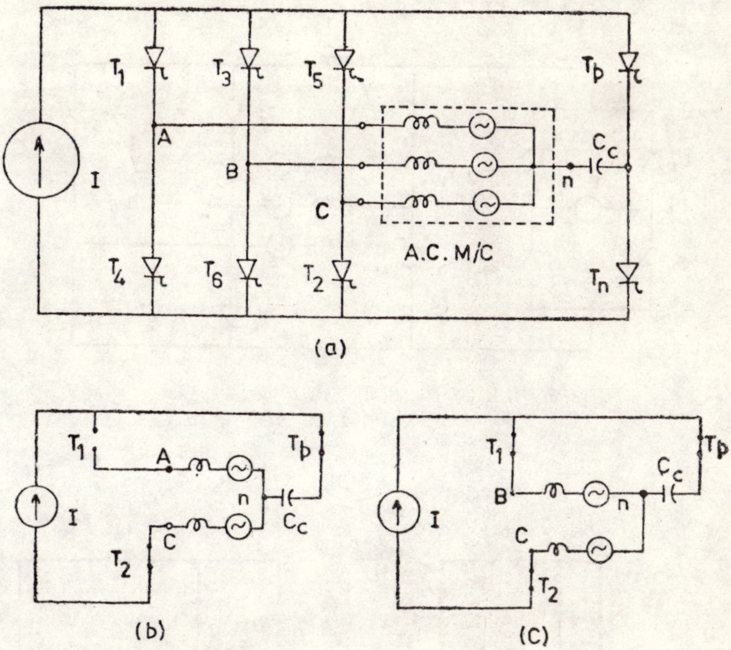

Fig. 7.59. (a) Current source inverter with commutation through load neutral.
(b) Equivalent circuit for mode (i).
(c) Equivalent circuit for mode (ii).

A detailed analysis of this inverter and its comparison with ASCI is given in [20].

Figure 7.60(a) shows a current source inverter with coupled reactor commutation. In this circuit there are no commutation capacitors. L_1 and L_2 are two closely coupled reactors. FT_1 and FT_2 are two freewheeling thyristors. S_1 and S_2 are two ideal choppers. The DC source E is used to effect the commutation.

At start, assume that thyristors T_1 and T_6 are the only devices that are on. Consequently the current flows through $D_2, -E, D_1, L_1, T_1$, Phase A, Phase B, T_6 and L_1'. The current is maintained by the reactor L_1 and L_1' while source E is opposing it. The transfer of current to thyristor pair T_1, T_2 occurs in the following modes.

Mode (*i*): Freewheeling thyristor FT_1 is turned on. The current now flows through FT_1, L_1, T_1 and Phase A. Thyristor T_6 turns

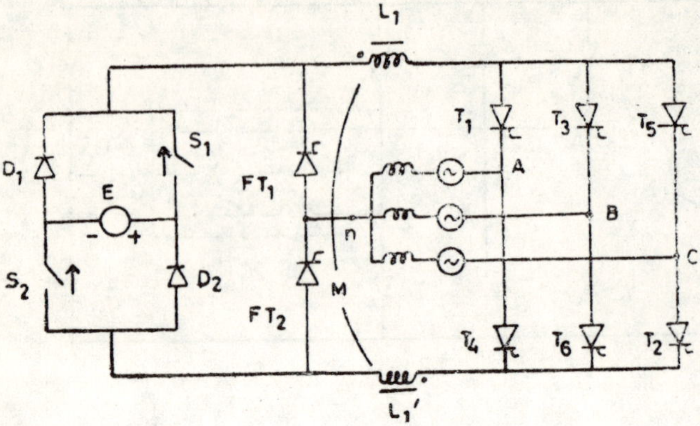

(a)

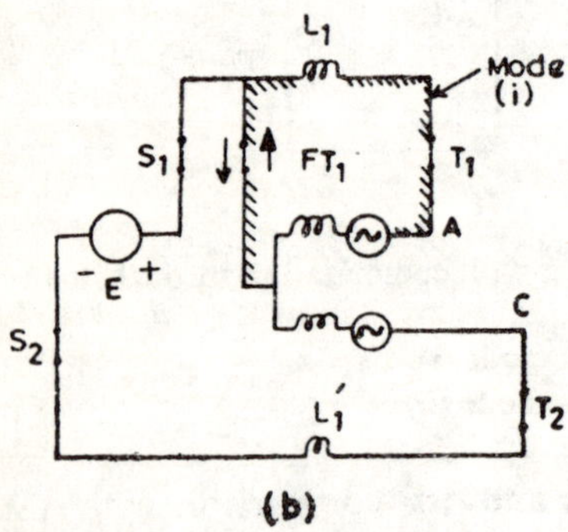

(b)

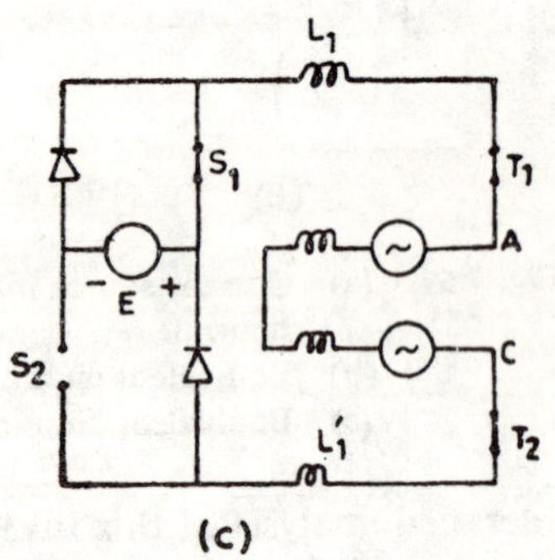

(c)

Fig. 7.60. (a) Current source inverter with coupled reactor commutation (CRC).
(b) Equivalent circuit for mode (i) and (ii).
(c) Equivalent circuit for mode (iv).

off as current through it is reduced to zero. Energy stored in L_1' increases the current in the above mentioned loop. Figure 7.60(b) shows the equivalent circuit of the mode by the hatched boundary.

Mode (ii): The incoming thyristor T_2 and switches S_1 and S_2 are turned on. A new loop consisting of S_2, E, S_1, FT_1, Phase C, T_2, L_1' is formed which reduces the net current flowing through thyristor FT_1 and after some time FT_1 turns off (Fig. 7.60(b)).

Mode (*iii*): The current now flows through S_2, E, S_1, L_1, T_1. Phase A, Phase C, T_2, L_1'. The current has been transferred to thyristors T_1 and T_2. In this mode source E is building up current in reactors L_1 and L_1'. This is similar to mode (iii) of ASCI discussed in Section 7.3.3.2. The equivalent circuit of this mode is same as Fig. 7.60(b) with FT_1 off.

Mode (*iv*): The current build up in reactors L_1 and L_1' can be controlled by introducing freewheeling. To start freewheeling S_2 is turned off. The current in S_2 now transfers to D_2. The equivalent circuit of this mode is shown in Fig. 7.60(c).

Mode (*v*): To retard the current flow even further S_1 is now turned off. The current in S_1 now transfers to D_1. The current now flows through D_2, $-E$, D_1, L_1, T_1, Phase A, Phase C, T_2 L_1. The situation now is similar to that prevailing before the start of mode (i) with thyristor T_6 replaced by T_2. The equivalent circuit of this mode is same as Fig. 7.60(c) except that S_1 is off and current flows through D_1 and D_2. To start the commutation of T_1, the freewheeling thyristor FT_2 is turned on. A mode sequence similar to the one described above takes place during commutations of T_1.

It is seen from above description that in this inverter the

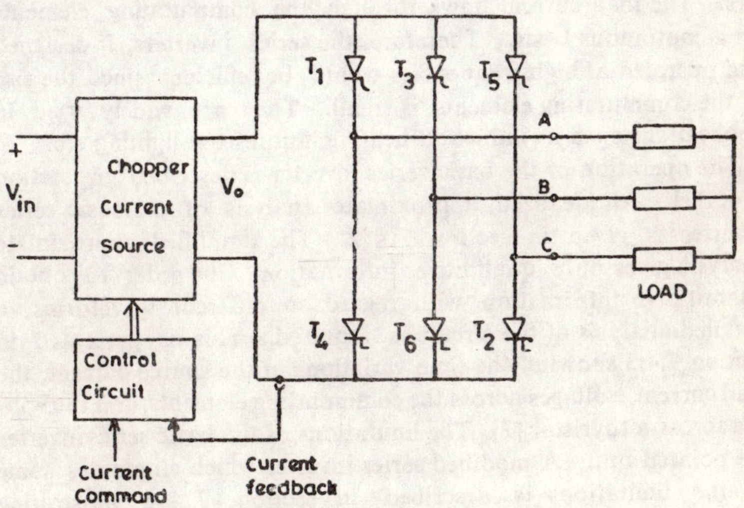

Fig. 7.61. Current source inverter with chopper commutation.

functions of input current source and commutation are mixed. The amplitude of current can be adjusted by controlling the duration of modes (ii) and (iii) in each period. Detailed analysis and waveforms of this inverter are discussed in [21].

Figure 7.61 shows another scheme in which the function of commutation is completely carried out by the front end current source. Thus, the commutation capacitors are not required; the filter reactor is included in the circuitry of the input current source. In order to effect commutation in such an inverter the input current is reduced to zero so as to commutate both the conducting thyristors, e.g. T_1 and T_6. The next pair, e.g. T_1 and T_2 is then gated and the input dc current is increased from zero to the full value. Clearly the input current source must have fast transient response to operate such an inverter at acceptable frequencies. An experimental system using a fast chopper operating from dc input is described in [22]. Such a system can be used in traction applications where the input power is obtained from a dc bus.

7.4 SERIES INVERTERS

Inverters in which the commutating elements are permanently connected in series with the load resistance and the load commutation as described in Section 2.5 is made use of to effect the turn off of the previously conducting thyristors are called series inverters. The load current flows through the commutating elements on a continuous basis. Therefore, the series inverters, if designed and operated at high frequencies would be efficient since the size of the commutating elements is small. They are widely used in applications such as induction heating, fluorescent lighting etc.

The operation of the basic series inverter is described in section 7.4.1[1]. An idealized, approximate analysis of the basic series inverter is given in section 7.4.2. The simplified approximate analysis gives only qualitative information. In order to obtain quantitative information with regard to different waveforms, a detailed analysis of the circuit is required. This is presented in section 7.4.3 showing the time variations of the source current, the load current, voltages across the commutating elements, and the voltage across a thyristor [2]. The limitations of the basic series inverter are pointed out. A modified series inverter which eliminates some of the limitations is described in section 7.4.4 illustrating waveforms of different circuit variables [1]. A three-phase series inverter can be built using three single-phase basic series inverters. If the output frequency for each phase is to be greater than the circuit resonant frequency, then the coupled inductors are required [1]. This is briefly mentioned in section 7.4.5. Finally, the principle of time-sharing for developing high frequency inverters is briefly explained [23]-[24] in section 7.4.6.

7.4.1 Basic Series Inverter

The basic series inverter is shown in Fig. 7.62. L and C are the commutating elements. The load consists of the resistance R. The commutating elements are selected snch that the current flowing through the series connected elements R, L, and C is underdamped. The thyristors T_1 and T_2 are turned on alternately at the intervals required to give the necessary output frequency f. The two thyristors should not be turned on simultaneously as this will lead to the short-circuit of the DC source and ultimate destruction of the devices. It must be ensured that prior to the turn on of one thyristor, the other thyristor is commutated. If the output frequency is selected such that it is always lower than the circuit ringing frequency f_r then the possibility of the short circuit of the DC source can be avoided. That is

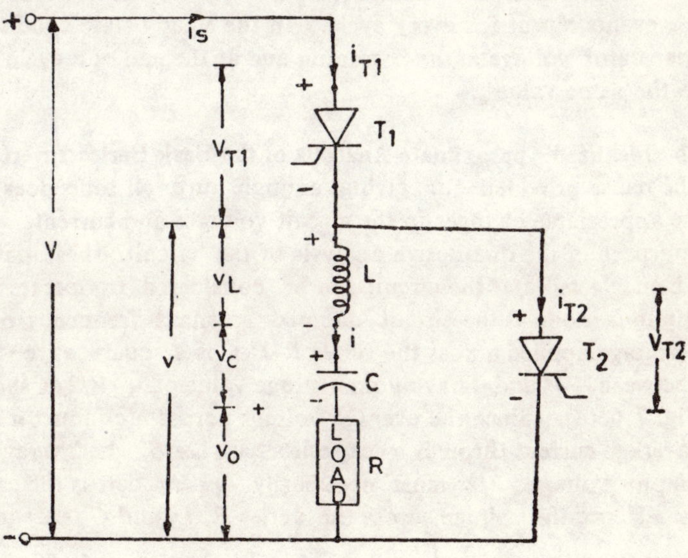

Fig. 7.62. Basic series inverter.

$$f < f_r \quad \text{or} \quad \frac{\pi}{\omega} - \frac{\pi}{\omega_r} = t_{q\,min}$$

where $t_{q\,min}$ is the minimum turn off time available for the turn off of either thyristor and, ω and ω_r being output and circuit resonant angular frequencies respectively. When T_1 is turned on, the supply voltage is applied across the series R, L, C circuit and a pulse of load current flows in the direction marked. The capacitor voltage builds up in the positive direction as shown in Fig. 7.62. The load current builds up to a peak value and then decreases to zero since the response is underdamped. When the current falls below the holding current level, then the thyristor T_1 turns off. After a short pause ($t_{q\,min}$) sufficient to cause the turn off of T_1, the second half cycle of the output current commences when T_2 is triggered into conduction. The capacitor which has been charged with the polarity as shown in the figure at the commencement of the second half cycle discharges into the load circuit giving a negative pulse of current. The current builds up to the negative maximum and then decreases to zero. The capacitor voltage reverses to some value depending upon the values of R, L, and C at the end of the second half cycle. Again, after sufficient time to cause the successful turn off of T_2, T_1 is triggered into conduction. These events repeat for every cycle. In the steady state condition, the capacitor voltage at the beginning and at the end of a cycle will have the same values.

7.4.2 Idealized Approximate Analysis of the Basic Series Inverter

The pause provided for giving enough turn off time does not cause appreciable changes in the circuit voltages and currents. For the purpose of the qualitative analysis of the circuit, these pauses can be neglected and the circuit can be considered to operate in a continuous mode at the circuit damped resonant frequency only. The voltage applied across the series R–L–C is a square wave varying between $+V$ and 0 having an average value of $+ V/2$ as shown in Fig. 7.63(a). Since the average voltage across an inductor and the average current through a capacitor are zero, the average of the input voltage $V/2$ must necessarily appear across the capacitor. Hence the voltage across the series R, L, and C is a square wave varying between $+\frac{V}{2}$ and $-\frac{V}{2}$. The Fourier analysis of this

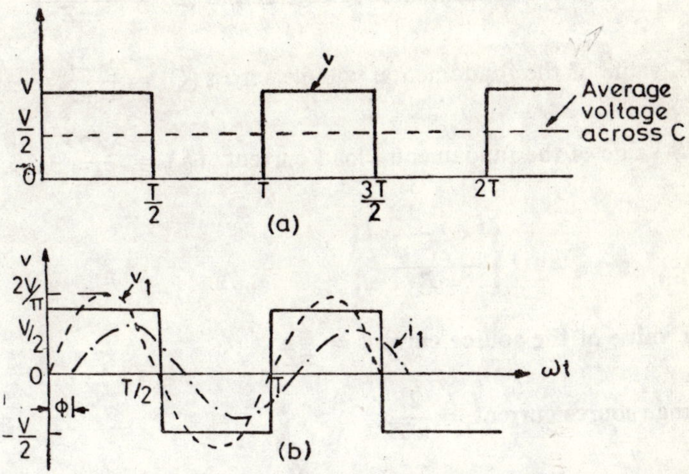

Fig. 7.63. Idealized waveforms of the basic series inverter

$$\text{Inverter frequency} = \frac{1}{2\pi}\left(\frac{1}{LC} - \frac{R^2}{4L^2}\right) = \frac{1}{T}$$

(a) Instantaneous voltage across series connected elements R, L, C

(b) Fundamental voltage and current through R, L and C.

square wave gives the fundamental and harmonics having 3, 5, 7, 9, 11,... times the fundamental frequency. The inductor offers a large impedance at the harmonic frequencies, and therefore the current due to higher harmonics can be neglected. Figure 7.63(b) shows the fundamental voltage across and the current through the series R–L–C circuit. Considering only the fundamental component of the voltage, the following expressions can be easily derived.

RMS value of the fundamental voltage

$$\text{across the series } R\text{–}L\text{–}C \text{ Circuit } (V_1) = \frac{\sqrt{2}V}{\pi}$$

RMS value of the fundamental voltage

$$\text{across the load resistance } (R) = \frac{\sqrt{2}VR}{\pi Z}$$

where
$$Z = \left[R^2 + \left(\omega L - \frac{1}{\omega C}\right)^2\right]^{1/2}$$

RMS value of the fundamental voltage across (L) $= \dfrac{\sqrt{2}V\omega L}{\pi Z}$

RMS value of the fundamental voltage across (C) $= \dfrac{\sqrt{2}V}{\omega c\pi Z}$

RMS value of the fundamental load current (I_1) $= \dfrac{\sqrt{2}V}{\pi Z}\angle\phi$

where $\phi = \tan^{-1}\left\{\dfrac{\left(\omega L - \dfrac{1}{\omega c}\right)}{R}\right\}$

Peak value of the source current $= \dfrac{2V}{\pi Z}$

Average source current $= \dfrac{2V}{\pi^2 Z}$

Fundamental RMS source current $= \dfrac{V}{\pi Z}$.

Output power $= \dfrac{\sqrt{2}VR}{\pi Z} \cdot \dfrac{\sqrt{2}V}{\pi Z} \cos\phi = \left(\dfrac{\sqrt{2}V}{\pi Z}\right)^2 R\cos\phi$

For low damping ratios, the capacitor voltage and the inductor voltage balance each other since $\omega L \simeq 1/\omega C$. In order to obtain a fairly sinusoidal voltage across the load resistance R, the ratio $\omega L/R$is limited to 2 to 5. In any case, the minimum value should not be below $\frac{1}{2}$ since this is the critical damping ratio.

7.4.3 Detailed Analysis of a Basic Series Inverter

Let the thyristor T_1 be triggered into conduction at $t = 0$ with the thyristor T_2 in the blocking condition. Writing down the KVL for the loop consisting of V, T_1, L, C, and R

$$V = Ri + L\frac{di}{dt} + \frac{1}{c}\int idt + v_C(0) \quad 0 < t < \frac{T}{2} \tag{7.125}$$

The initial conditions at $t = 0$ are

$$i = i(0) \tag{7.126}$$

$$\frac{di}{dt} = \frac{V - v_C(0)}{L} \tag{7.127}$$

$$v_C = V_C(0) \tag{7.128}$$

Solving eqn. (7.125) with the initial conditions given in eqns. (7.126) to (7.128), the expression for the load current is

$$i = \frac{V - V_C(0)}{\omega_r L} e^{-\xi t} \sin \omega_r t \tag{7.129}$$

where $\xi = \frac{R}{2L}$, $\omega_0 = \frac{1}{\sqrt{LC}}$

$$\omega_r = \left(\frac{1}{LC} - \frac{R^2}{4L^2}\right)^{1/2} = (\omega_0^2 - \xi^2)^{1/2}$$

The voltage across the inductor L is

$$v_L = L\frac{di}{dt} = \frac{\omega_0}{\omega_r}(V - V_C(0))\, e^{-\xi t} \cos(\omega_r t + \psi) \tag{7.130}$$

where $\psi = \tan^{-1}\frac{\xi}{\omega_r}$.

The voltage across the capacitor C is

$$v_C = V - v_L - iR \tag{7.131}$$

Substituting for i and v_L from eqns. (7.129) and (7.130) in eqn. (7.131), the expression for the capacitor voltage is

$$v_C = V - \frac{\omega_0}{\omega_r}(V - V_C(0))\, e^{-\xi t} \cos(\omega_r t - \psi) \tag{7.132}$$

$$i_{T_1} = i,\; v_{T_1} = 0 \qquad 0 < t < \frac{\pi}{\omega_r} \tag{7.133}$$

$$i_{T_2} = 0,\; v_{T_2} = V \qquad 0 < t < \frac{\pi}{\omega_r} \tag{7.134}$$

The current i falls to zero at the end of one-half period (π/ω_r) of the circuit resonant frequency. The capacitor voltage at the end of this period can be determined from eqn. (7.132) by substituting π for $\omega_r t$. Let this be $v_C(\pi/\omega_r) = V_{C\,max}$. Assuming ideal thyristors, the capacitor voltage remains constant from π/ω_r to $T/2$, T being the period at the inverter operating frequency.

$$i_{T_1} = 0,\; v_{T_1} = (V - V_{C\,max}) \qquad \frac{\pi}{\omega_r} < t < \frac{T}{2} \tag{7.135}$$

$$i_{T_2} = 0,\; v_{T_2} = V_{C\,max} \qquad \frac{\pi}{\omega_r} < t < \frac{T}{2} \tag{7.136}$$

This capacitor voltage $v_c\,(T/2)$ is taken as the initial voltage impressed around the loop comprising T_2, L, C, and R when T_2 is triggered into conduction at $t = T/2$. The thyristor T_1 is assumed to be in the off condition when T_2 is turned on. The expressions for i, v_L and v_C are obtained by putting $V=0$ and replacing t and $V_C(0)$ by t' and $V_C(T/2)$ respectively in eqn. (7.129) to (7.132).

$$i = \frac{-V_C(T/2)}{\omega_r L} \exp(-\xi t') \sin \omega_r t' \tag{7.137}$$

$$v_L = -\frac{\omega_0}{\omega_r} V'\left(\frac{T}{2}\right) \exp(-\xi t') \cos(\omega_r t' + \psi) \tag{7.138}$$

$$v_C = \frac{\omega_0}{\omega_r} V_C\left(\frac{T}{2}\right) \exp(-\xi t') \cos(\omega_r t' - \psi) \tag{7.139}$$

$$i_{T_2} = -i,\ v_{T_2} = 0 \qquad 0 < t' < \frac{\pi}{\omega_r} \tag{7.140}$$

$$i_{T_1} = 0,\ v_{T_1} = V \qquad 0 < t' < \frac{\pi}{\omega_r} \tag{7.141}$$

Assuming that the load current has half-wave symmetry, it can be written that

$$i(t) = -i(t')$$

The result of equating eqns. (7.129) and (7.137) is

$$V - V_C(0) = V_C\left(\frac{T}{2}\right) \tag{7.142}$$

Substituting for $V_C(T/2)$ from eqn. (7.142) in eqns. (7.137) to (7.139), we get,

$$i = -\frac{(V - V_C(0))}{\omega_r L} \exp(-\xi t') \sin \omega_r t' \tag{7.143}$$

$$v_L = -\frac{\omega_0}{\omega_r} \cdot (V - V_C(0)) \exp(-\xi t') \cos(\omega_r t' + \psi) \tag{7.144}$$

$$v_C = \frac{\omega_0}{\omega_r} (V - V_C(0)) \exp(-\xi t') \cos(\omega_r t' - \psi) \tag{7.145}$$

Also

$$i_{T_2} = 0,\ v_{T_2} = V_C(0) \qquad \frac{\pi}{\omega_r} < t' < \frac{T}{2} \tag{7.146}$$

$$i_{T_1} = 0,\ v_{T_1} = V + V_C(0) \qquad \frac{\pi}{\omega_r} < t' < \frac{T}{2} \tag{7.147}$$

Eqns. (7.129), (7.130), (7.132) and (7.137) to (7.139) can be used to determine the time variations of i, v_L and v_C for both positive and negative half cycles in the basic series inverter of Fig. 7.62. Under steady state, the initial capacitor voltage and the final capacitor voltage at the end of second half period $\left(t=\frac{T}{2}+\frac{\pi}{\omega_r}\right)$ will be equal. From eqn. (7.145), we get

$$v_C = V_C(0) = \frac{\omega_0}{\omega_r}(V - V_C(0)) \exp\left(\frac{-\xi\pi}{\omega_r}\right) \cos(\pi - \psi)$$

or

$$V_C(0) = \frac{-V \exp\left(-\xi\frac{\pi}{\omega_r}\right)}{\left(1-\exp\left\{-\frac{\xi\pi}{\omega_r}\right\}\right)} \tag{7.148}$$

From eqns. (7.142) and (7.148), the capacitor voltage at the beginning of the second half period is

$$V_C\left(\frac{T}{2}\right) = \frac{V}{\left(1-\exp\left\{-\frac{\xi\pi}{\omega_r}\right\}\right)} \tag{7.149}$$

The variation of the capacitor voltage with various values of the load resistance is shown in Fig. 7.64. The circuit resonant frequency is assumed to be constant. For low values of the load resistance, the circuit resonant frequency may be approximated to be equal to the inverter frequency. Higher values of the capacitor voltage result and the load voltage and current waveforms remain almost sinusoidal for changes in the load resistance. On the other hand, when the load resistance is large with a fixed value of L, the capacitance will have to be changed considerably so as to keep the circuit still underdamped for changes in the load resistance. Under this condition, the capacitor charges to lower voltages as illustrated in Fig. 7.64.

Typical waveforms of capacitor voltage v_C, voltage across the inductor v_L, load current, source current, voltage v_{T_2} across the thyristor T_2 are shown in Fig. 7.65 when the inverter frequency is less than the circuit ringing frequency. The distortion of the load current can clearly be seen from these waveforms. The basic series inverter has the following drawbacks:

1. The maximum inverter frequency is limited to a value that is slightly less than the circuit resonant frequency. If the inverter

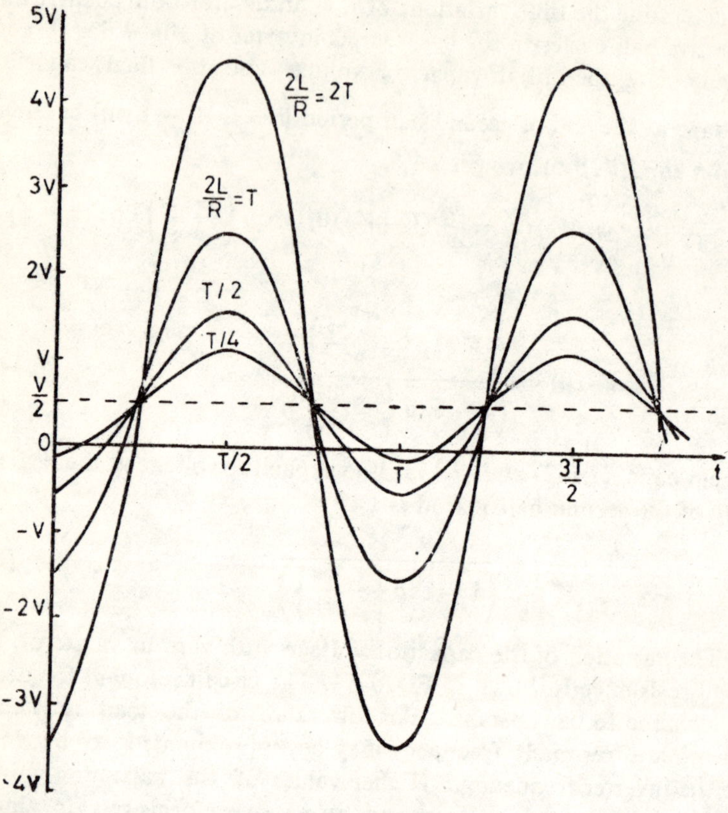

Fig. 7.64. Variation of the capacitor voltage with the load resistance R at constant resonant frequency.

frequency exceeds the circuit ringing frequency, the dc source will be short-circuited.

2. For output frequencies much less than the circuit ringing frequency, the load voltage is distorted.

3. The source current flows only during the period when the thyristor T_1 is conducting. This results in a large ripple in the source current and the peak current rating of the source increases.

The inverter can be operated at a frequency higher than the circuit ringing frequency and the source current ripple is reduced by modifying the circuit configuration of the basic series inverter as discussed in the following section.

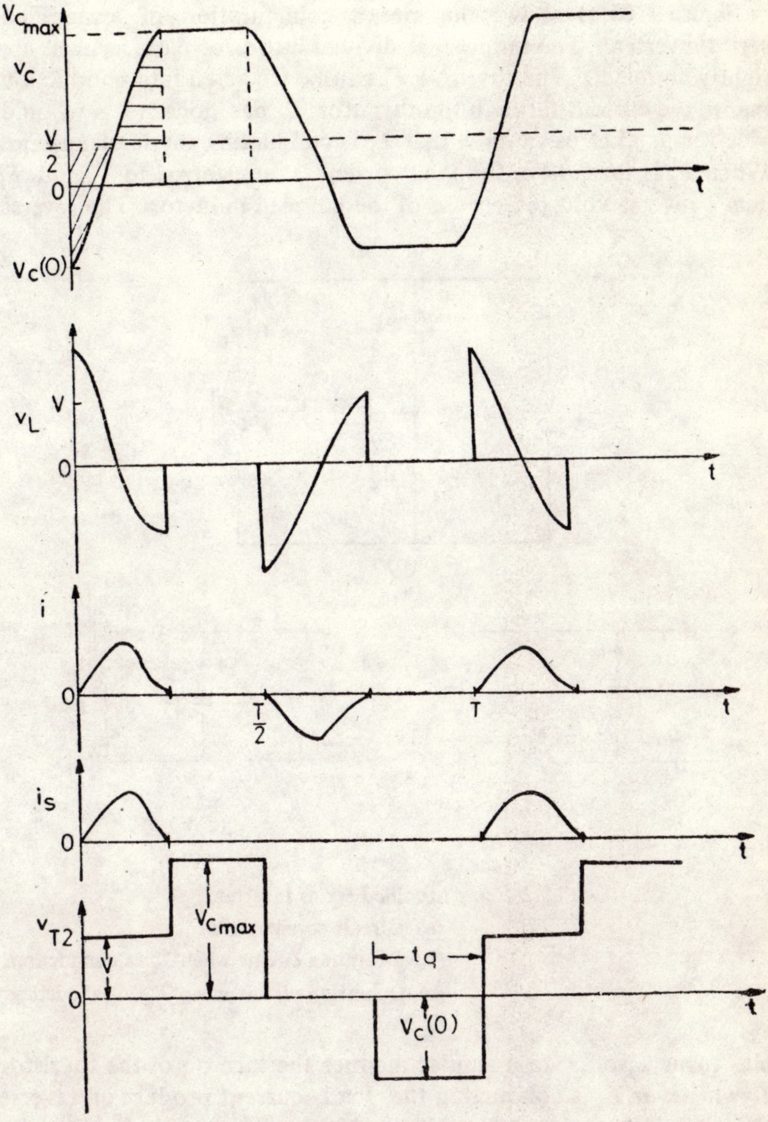

Fig. 7.65. Time variations of different voltages and currents shown in the basic series inverter at a frequency less than the circuit resonant frequency.

7.4.4 Modified Series Inverter

Figure 7.66(a) shows the circuit configuration of a modified series inverter. The inductor is divided into two parts which are tightly coupled. The thyristor T_2 can be triggered into conduction before the current through the thyristor T_1 has gone to zero and vice-versa. Let us suppose that T_1 is conducting the load current. When T_2 is turned on, the load current is transferred to T_2 and T_1 gets a reverse voltage because of the coupled inductor. The reverse

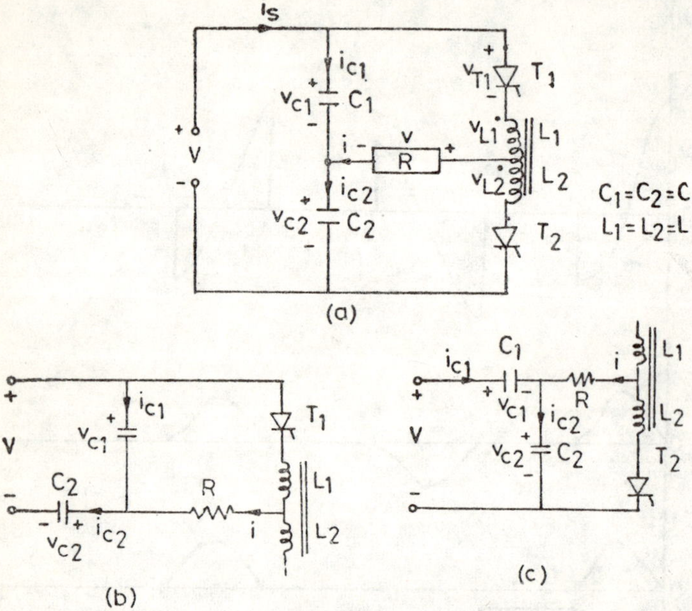

Fig. 7.66. Modified series inverter.
(a) Circuit configuration.
(b) Operating circuit when T_1 is conducting.
(c) Operating circuit when T_2 is conducting.

bias turns T_1 off. In a similar manner the turn on of the thyristor T_1 whenever T_2 is conducting the load current produces a reverse bias across T_2 and thus turns it off. Although the type of commutation employed in this circuit is not the load commutation strictly speaking it is still called as a series inverter owing to the connection of the commutation elements in series with the load. An important requirement is that the incoming thyristor should be turned on only when the current through the previously conducting thyristor has passed the peak value. Then only will the outgoing thyristor get a reverse bias and be commutated reliably. The maximum

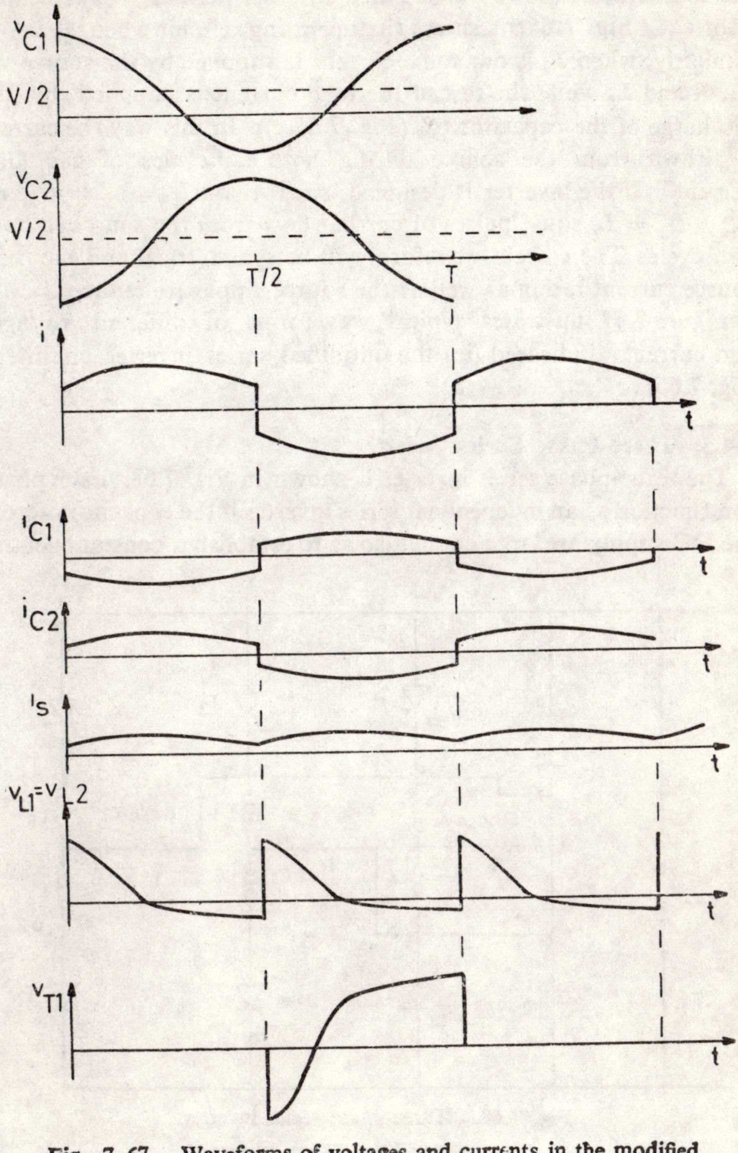

Fig. 7.67. Waveforms of voltages and currents in the modified series inverter circuit of Fig. 7.66.

frequency of the modified inverter is decided on this consideration only.

When T_1 is conducting, a part of the load current is supplied from the source while the rest is supplied by the discharge of the capacitor C_1. Fig. 7.66(b) shows the operating circuit when T_1 is on. Similarly, when T_2 is on, some current is supplied by the source via C_1, R and L_2 while the rest of the load current is supplied by the discharge of the capacitor. C_2 (Fig. 7.66(c)). In this way the current is drawn from the source during both half-cycles of the load current. If the inverter is designed such that $C_1 = C_2 = C$ and $L_1 = L_2 = L$, equal pulses of current flow from the source in both half-cycles. The current waveform will be symmetrical and the peak source current rating as well as the source ripple are reduced.

Figure 7.67 illustrates typical waveforms of different voltages and currents indicated in the modified series inverter circuit of Fig. 7.66.

7.4.5 Three-Phase Series Inverter

The three-phase series inverter is shown in Fig. 7.68. Each phase can function as an independent series inverter if the capacitors across the DC supply are large enough so as to establish a constant neutral

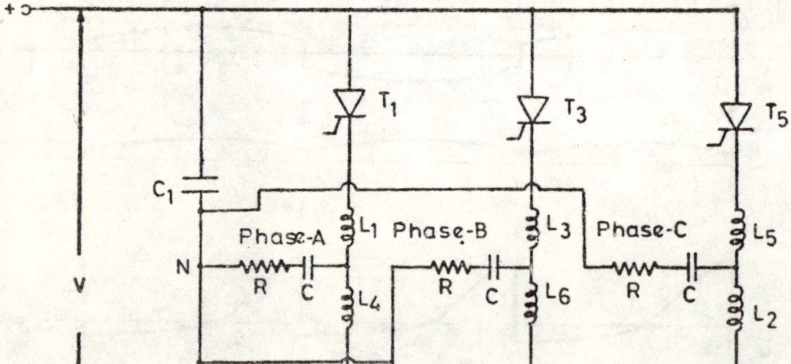

Fig. 7.68. Three-phase series inverter.
$L_1 = L_2 = \dots = L_6 = L$

voltage. The elements L, C and R connected in series with each thyristor are designed such that the current response is underdamped as in the basic series inverter. If coupled inductors similar to those in the modified inverter were used in each leg of the inverter then T_4 can be turned on before the current in T_1 has gone to zero as explained in section 7.4.4. Similarly the thyristors in the other legs can likewise be turned on. However, if the inductors in each leg are not coupled, one thyristor in a leg can be turned on only when the current through the other thyristor in the same leg has decreased to zero. The thyristors are triggered in the sequence T_1, T_2,..., T_6 as explained in section 3.5. In a three-phase bridge converter, the synchronization with the AC input voltages is essential while deriving the gating pulses to the thyristors. However, in this case, no such synchronization is necessary. A three-phase reference oscillator is essential in order to derive gating pulses to the six thyristors. The frequency of the oscillator is six times the output frequency. The triggering scheme explained in section 7.5 can be used to establish gating pulses to the six thyristors.

7.4.6 Time-Sharing High Frequency Inverter

The output frequency of the series inverter is limited to the circuit resonant frequency. It is however possible to extend the range of the inverter frequency slightly above the circuit resonant frequency by using a coupled inductor. But, ultimately the device turn-off time specification limits the maximum possible frequency at which a series inverter can be designed and operated. Many applications require quite substantial amounts of power at frequencies well in excess of 10 KHz One method of designing an inverter which can operate at a high frequency with thyristors having normal turn-off time specifications is to employ the principle of time-sharing where the thyristor turn off time is not a basic limiting factor. In this time-sharing inverter, several thyristor sub-circuits are operated sequentially into a common load circuit so that the fundamental frequency of operation of each of the thyristors is a sub-multiple of the output frequency. The principle of the time-sharing is not a new concept as such. Earlier, the voltage source inverters employing the time-sharing principle were used [25]. But the current source time-sharing inverter is relatively recent. Because of several advantages associated with the current source, this type of inverter is preferred over the voltage source type. Figure 7.69 shows the circuit configuration of the inverter employing the time-sharing principle.

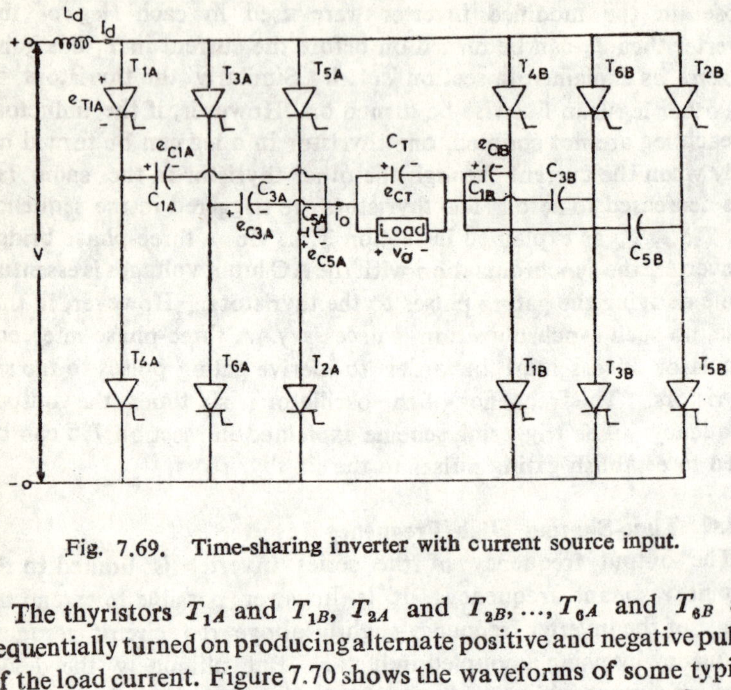

Fig. 7.69. Time-sharing inverter with current source input.

The thyristors T_{1A} and T_{1B}, T_{2A} and T_{2B}, ..., T_{6A} and T_{6B} are sequentially turned on producing alternate positive and negative pulses of the load current. Figure 7.70 shows the waveforms of some typical circuit variables indicated in Fig. 7.69. Although the waveform of the load current is shown having sharp transition edges, such sharp edges give rise to large values of di/dt which are detrimental to the thyristors. In practice, some inductance is inserted in series with the parallel capacitor compensated load. For understanding the basic principle of the time-sharing, this practical refinement may be neglected since it does not detract the basic principle of operation. A detailed analysis of the inverter circuit [24] reveals that the following criteria be satisfied for operation of the inverter at the full load.

I. The phase displacement between the output voltage and current of the parallel compensated load circuit is about 30° with the current lagging the voltage.

II. The time-sharing capacitors (C_{1A}, C_{3A}, C_{5A}, C_{1B}, C_{3B} and C_{5B}) are selected such that the peak capacitor voltage is equal to the peak load voltage when the current is at the full load value with the inverter frequency set at the nominal value.

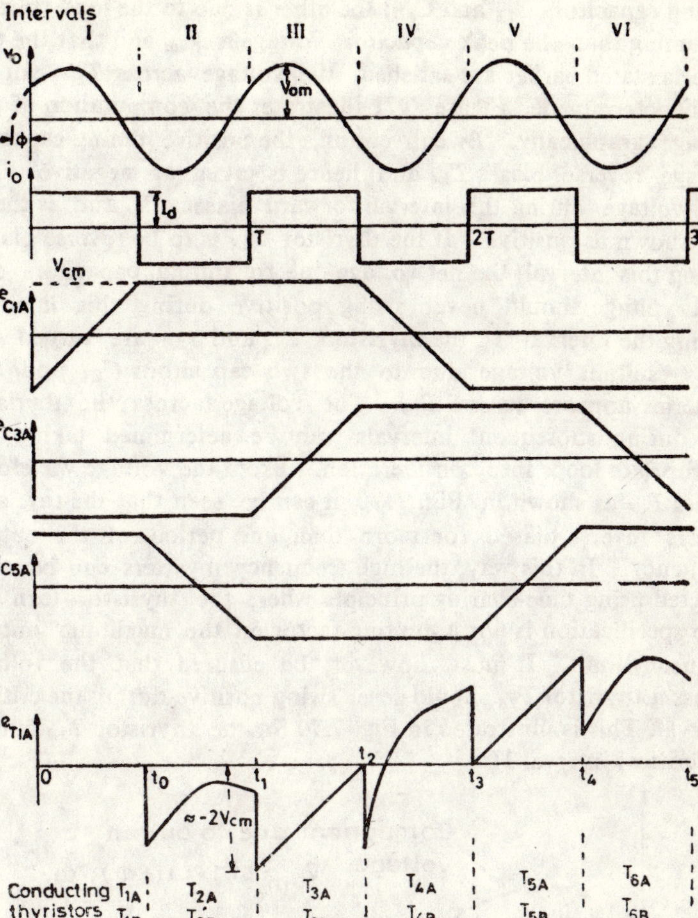

Fig. 7.70. Time variations of different circuit variables indicated in Fig. 7 69.

The waveform depicted in Fig. 7.70 incorporates these features. The voltage across the thyristor T_{1A} is illustrated to explain the principle of time sharing. During the interval I, T_{1A} and T_{1B} conduct. In the interval II, the thyristors T_{2A} and T_{2B} are turned on. The voltage across the thyristor T_{1A} during this interval can be determined by considering the loop formed by T_{1A}, C_{1A}, parallel capacitor compensated load, C_{5B}, T_{2B} and back to T_{1A}. Evidently, the voltage across T_{1A} is due to two components: one is due to two series timing capacitors C_{1A} and C_{5B}; the other is due to the load voltage. Assuming that the peak capacitor voltage is V_{cm} and that the two criteria stated earlier are satisfied, the voltage across T_{1A} can be easily determined. Figure 7.71 illustrates the computation of this voltage graphically. By convention, the positive timing capacitor voltage reverse biases T_{1A} and hence is taken as negative. The load voltage during this interval forward biases T_{1A} and is therefore shown as positive. If the thyristor T_{1A} is to be reverse biased during this interval, the net voltage due to timing capacitors and load voltage should never swing positive during this interval. During the interval III, the thyristors T_{3A} and T_{3B} are turned on. The resultant voltage due to the two capacitors C_{1A} and C_{3A} in series appears across T_{1A}. The voltage across the thyristor T_{1A} during subsequent intervals can be determined taking the appropriate loops into consideration. From the voltage waveform across T_{1A} as shown in Fig. 7.70, it can be seen that the thyristor T_{1A} is reverse biased for more than one period of the output frequency. In this way, the high frequency inverters can be constructed using time-sharing principle where the thyristor turn off time specification is not a limiting factor on the maximum output frequency limit. It must, however, be ensured that the voltage across a thyristor T_{1A} should never swing positive during the critical interval. This is illustrated in Fig. 7.70 for the thyristor T_{1A} during the critical interval II.

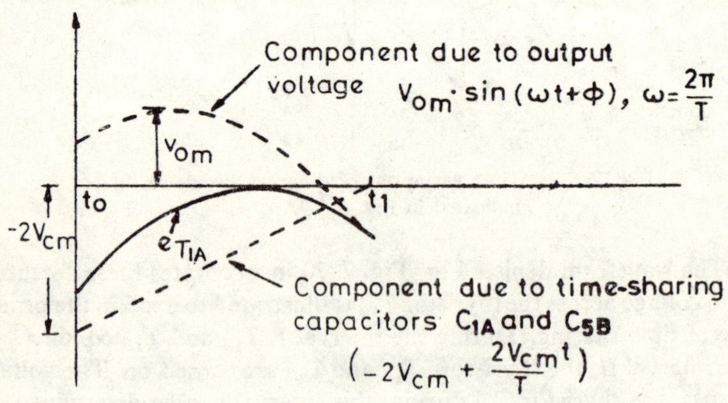

Fig. 7.71 Voltage across T_{1A} during the interval II from t_0—t_1.

The thyristor voltage (e_{T1A}) may swing positive during the critical period II if the output power decreases keeping the load phase angle, ϕ constant. Since the output power is proportional to the product of the inverter output current and cos ϕ for a given output voltage, the output current decreases in proportion to the decrease in the load power. The timing capacitor voltages also decrease accordingly. As a result, these voltages may not cancel out the load voltage completely and the voltage across the outgoing thyristor T_{1A} may swing positive during the critical period t_0-t_1. This reduces the turn-off time drastically. To overcome this problem, the phase angle is increased with a decrease in the output power without altering the load current in a significant way. The outgoing thyristor is thus reverse biased during the critical period t_0-t_1 since the load current is maintained almost constant. The input voltage to the inverter can be made continuously controllable by using a controlled rectifier before the reactor, L_d.

7.5 INVERTER GATING CIRCUITS

The gating requirements of different types of inverters have been described in sections 7.1 to 7.4. In this section the implementation of the gating circuits is given. Some representative circuits have been described for the voltage and the current source inverters.

The gating requirements of the inverter thyristors fall in two general categories. The first consists of 'main' thyristors which form the power switches. These thyristors usually conduct for 120 degrees or 180 degrees. Due to the inductive nature of most loads the main thyristors are gated continuously and with high frequency modulation. The second category consists of 'auxiliary' or 'commutation' thyristors. These thyristors are turned on to activate a commutation circuit to turn off main thyristor. The auxiliary thyristors themselves are self commutated usually due to a L–C oscillation. Therefore, the auxiliary thyristors are turned on by a short, e.g. 100 μs gate pulse. Some inverters employ complementary or auto sequential commutation in which turning on the next main thyristor in a sequence turns off the previously conducting thyristor as well. In such cases the auxiliary thyristors are not required.

In addition to providing required gate pulses to the thyristors the gating circuit must also provide control over the frequency of the inverter. This is usually done by a voltage controlled oscillator (V.C.O.) which adjusts the frequency of the gating sequence.

Another important feature of the inverter gating circuits is the provision for current limits. The commutation circuits used in the inverters fail if the load current exceeds a maximum specified value. Thus, there is a need to shut off the inverter if the instantaneous current exceeds the limit of commutation. Further, limits are incorporated to prevent continuous over loading to protect the load and the thyristors.

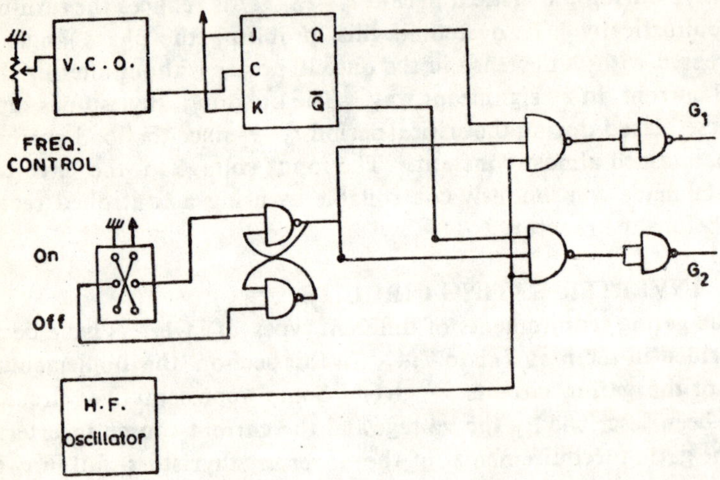

Fig. 7.72. Gating circuit for single phase complementary commutated inverter.

Figure 7.72 shows the circuit diagram of the gating circuit for a single phase complementary commutated voltage source inverter. The V.C.O. produces a clock of adjustable frequency which triggers the J–K Flip-Flop. The outputs Q and $\bar{Q}$ of the Flip-Flop are the envelopes of complementary gating signals. A high frequency ($\simeq$ 30KHz) oscillator is used to modulate the output Q and $\bar{Q}$. A debounced switch provides the on/off signal to start and stop the gating signals. The detailed circuit diagram of a high frequency oscillator and a V.C.O. are shown in Figs. 7.73 and 7.74(a) respectively.

Figure 7.73 shows the circuit diagram of a gating circuit for a three phase current source inverter. The V.C.O. provides the clock to drive a 3 stage ring counter. The six states of the ring counter are decoded by 3 input Nand gates and inverted to obtain the required gating signals without the high frequency carrier

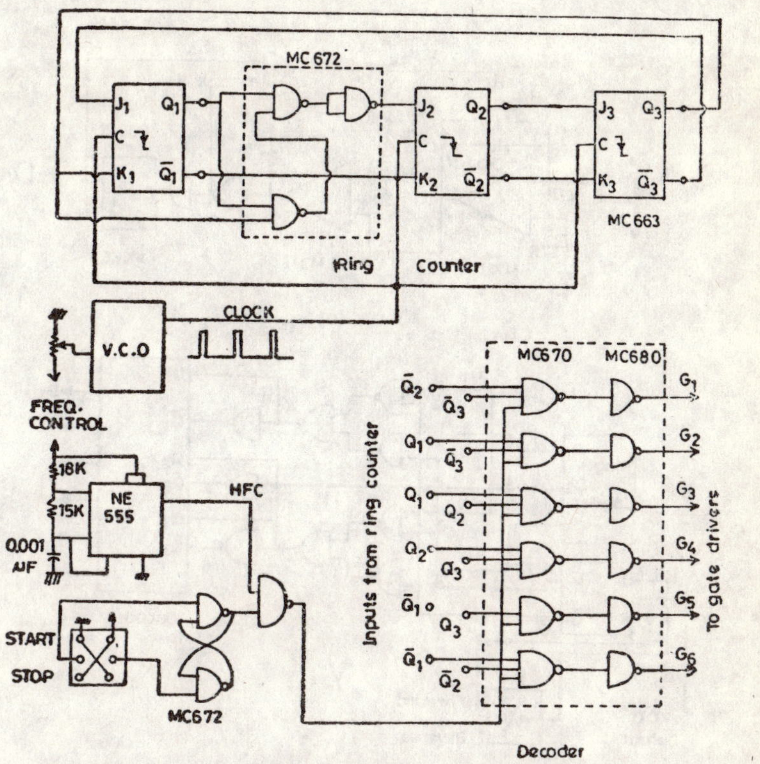

Fig. 7.73. Logic circuit for three-phase CS inverter.

(HFC) are shown in Fig. 7.75 which also shows the states of the ring counter Q_1, Q_2, Q_3. The three stage ring counter can have any of the eight possible initial states at the time instant when power is turned on. The required counter has only six useful states, shown in Fig. 7.75. Thus, there is a possibility of latching up in the unused states. The latch up in unused states is prevented by modifying the logic of J_2 input as shown in Fig. 7.73. Normally J_2 is equal to Q_1. But if the initial state of the counter is 010 or 101 then it automatically changes to one of the regular states of Fig. 7.75 at the next clock pulse.

In some applications the phase sequence of gating signals is also controlled. In Fig. 7.75 signal G_i lags signal G_{i-1} by 60° The reversed phase sequence calls for G_i leading G_{i-1} by 60°. The phase reversal can be carried out by interchanging signals Q_3 and Q_1 at the input to the decoder in Fig. 7.73. Figure 7.74(b) show

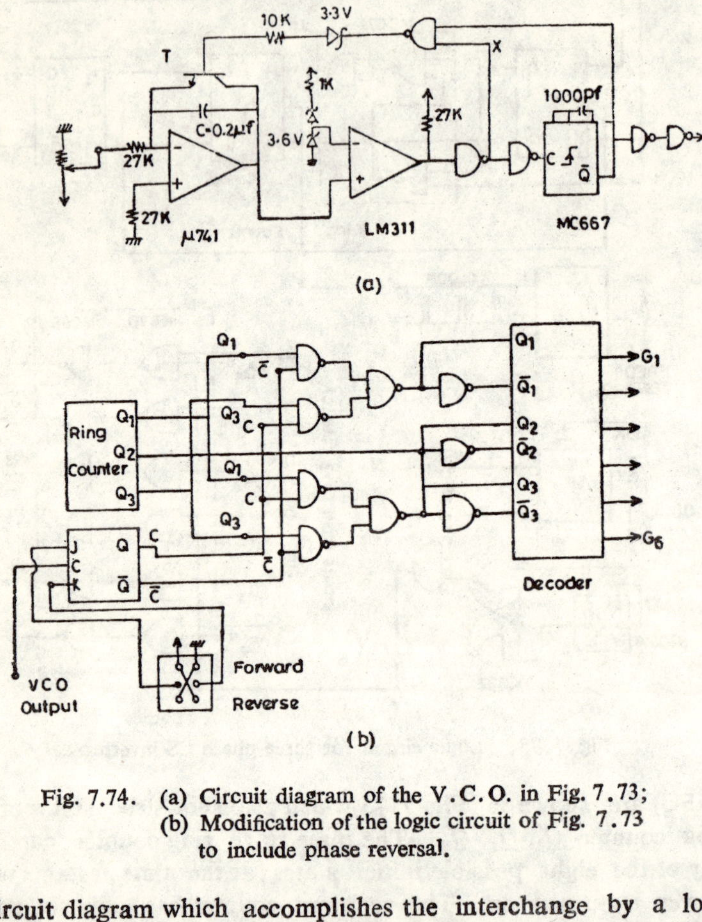

Fig. 7.74. (a) Circuit diagram of the V.C.O. in Fig. 7.73; (b) Modification of the logic circuit of Fig. 7.73 to include phase reversal.

a circuit diagram which accomplishes the interchange by a logic signal C. The signal C is set by a manual switch and synchronised to the V.C.O. output by a J.K. Flip-Flop.

The voltage and the current source type Pulse Width Modulated (PWM) inverters require complex gating circuitry. The sinusoidal PWM inverter further has many different schemes for harmonic reduction. A detailed description of some PWM gating circuits is given in [26], [27]. With the development of LSI circuits the PWM gating circuits have been placed in one monolithic IC e.g. HEF4752 [28]. In future it may be common to obtain standard/custom LSI circuits for complex inverter systems.

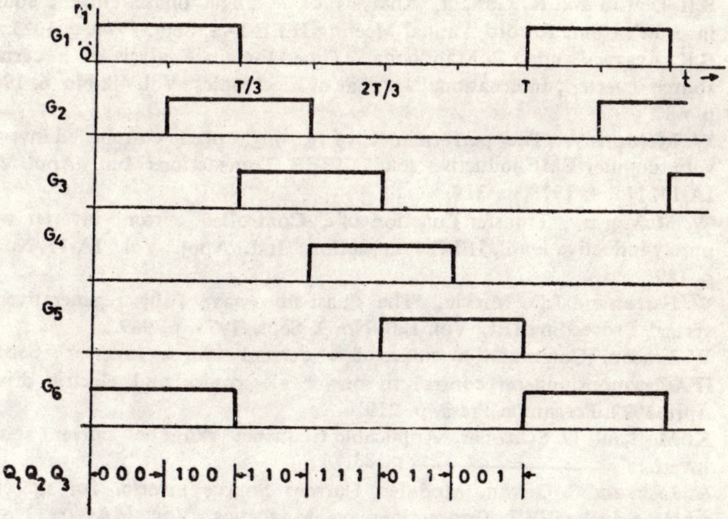

Fig. 7.75. Output waveform for the logic circuit of Fig. 7.73.

REFERENCES

1. B.D. Bedford and R.G. Hoft, 'Principles of Inverter Circuits', John Wiley & Sons, 1964, pp. 165–182 (contributed by W. McMurray).
2. S.B. Dewan, 'Power Semiconductor Circuits', John Wiley & Sons, 1975.
3. J.M.D. Murphy, 'Thyristor Control of AC Motors', Pergamon Press Ltd., 1973.
4. S.B. Dewan and D.L. Duff, 'Optimum Design of an Input-Commutated Inverter for ac Motor Control', IEEE Trans. Industry and General Applications, Vol. IGA-5, Nov./Dec. 1969, pp. 699–705.
5. H.S. Patel and R.G. Hoft, Generalised techniques of harmonics elimination, and voltage control in thyristor inverters—Part I harmonic elimination, IEEE Trans., IA-9, pp. 310–317, 1973.
6. K.A. Krishnamurthy, S.B. Mahajan, G.N. Revankar and G.K. Dubey, 'Selective reduction of harmonics in inverters', Int. J. of Electronics, 1979, Vol. 46, pp. 321–330.
7. G.N. Revankar and S.M. Sriraghavan, 'Inverters for uninterruptible power supplies', Jr. IETE, Vol. 28, No. 8, 1982, pp. 414–423.
8. S.M. Sriraghavan, B.D. Pradhan and G.N. Revankar, 'Multistage pulse-width-modulated inverter system for generating stepped-voltage waveforms', Proc. IEE Vol. 125, No. 6, June 1978, 529–530.
9. K.S. Rajashekharan and J. Vithayathil, 'Harmonics in voltage source PWM inverters', Int. J. Electronics, Vol. 50, No. 5, 1981, pp. 325–337.

10. S.B. Dewan and K. Gallant, 'Analysis of a single phase current source inverter', Conf. Record Annual Meeting IEEE-IAS, Sept. 1974, p. 1025.
11. S.K. Agarwal and O.P. Mandhana, 'Time Domain Analysis of a current source inverter', International Journal of Electronics Vol. 48, No. 6, 1980, p. 487.
12. W. McMurray, 'The performance of a single phase current fed inverter with counter EMF-inductive load', IEEE Transactions Ind. Appl. Vol. IA-14, No. 4, 1978, p. 319.
13. W. McMurray, 'Transfer Function of a Controlled current inverter with purely inductive load', IEEE Transactions Ind. Appl. Vol. IA-15, No. 2, p. 189.
14. W. Farrer and J.D. Miskin, 'The quasi-sine wave fully regenerative inverter', Proceeding IEE, Vol. 120, No. 9, Sept. 1973, p. 969.
15. W. Lineau, 'Commutation modes of a current source inverter', Second IFAC symposium on control in power electronics and electric drives, April 1977, Pergamon Press, p. 219.
16. K. Moll and D. Schroder, 'Applicable frequency range of current source inverters',————————, p. 231.
17. A. Joshi and S. Dewan, 'Modified Current Source inverter for sq. cage motor drive', IEEE Transactions on Magnetics, Vol. MAG-14, No. 5, Sept. 1978.
18. M.K. Parasuram and B. Ramaswami, 'Analysis and design of current fed inverter', Second IFAC symposium on power electronics and electric drives, April 77, Pergamon Press, p. 235.
19. G. Revankar, Reduction in Commutation time in a current source inverter', Journal Institution of Engineers (India) Vol. 62, Pt EL4 pp. 158–64, Feb. 1982.
20. R.L. Steigerwald, 'Characteristics of a current fed inverter with commutation applied through load neutral point'. Conf. Rec. IEEE/IAS Annual Meeting 1978, pp. 502–16.
21. T.H. Chin, 'A new controlled current type inverter with improved performance', Conf. Rec. IEEE/IAS International Semiconductor Power Converter Conf. Oct. 1977, pp. 185–92.
22. N. Nicholson, et al., 'Chopper commutated current source inverters', Conf. Rec. IEEE/IAS Annual meeting 1976, p. 1101.
23. B.R. Pelly, 'Latest developments in static high frequency power source for induction heating,' IEEE Trans. Ind. Electron. Contr. Instrum., Vol. IECI-17, No. 4, pp. 297–312, June 1970.
24. G.N. Revankar and Vijay A. Patwardhan, 'Time-sharing high frequency inverter circuit', IEEE Trans. Ind. Electron. Contr. Instrum., Vol. IECI-21, No. 3, pp. 199–203, August 1974.
25. R. Thompson, 'High Frequency SCR sinusoidal inverter,' Proc. IEE (Part B), April 1963.
26. T.L. Grant and T.H. Barton, 'A highly flexible controller for a pulse width modulation inverter,' Conf. Record IEEE-IAS Annual Meeting, Oct. 1978, p. 486.

27. A. Abbondanti, 'A digital modulator for PWM inverter', pp. 493.
28. Phillips Integrated Circuit Data Handbook, Part 4, Oct. 1980, p. 689.
29. H. K. Anand, A. Joshi and G.K. Dubey 'Digital simulation of single phase inverter with a counter EMF inductive load', International Journal of Electronics, 1983, Vol. 54, No. 4, pp. 485–504.

8. POWER FACTOR CONTROL

Presently thyristors are used to switch in capacitors and/or inductors across the supply lines to compensate the load reactive power. Continuous and very fast control of reactive power has been made possible using thyristors. Section 8.1 deals with the improvement of the supply power factor when the loads are concentrated and supplied from one bus terminal. This is usually called 'load compensation'. In the interconnected AC transmission systems, the loads are neither concentrated nor a meaningful measurement of the power factor is possible at a bus. In this case the power factor is not directly controlled but rather the voltage at a bus is regulated by switching on or off appropriate elements through thyristor switches. This aspect is briefly discussed. Further, the improvement of the supply power factor externally, specially when the thyristor converters form a major portion of the load on the AC supply system, has also been explained.

It is also possible to improve the supply power factor inherently by modifying the converter circuit configuration itself and/or control techniques. These aspects are covered in section 8.2.

8.1 STATIC REACTIVE POWER COMPENSATION

8.1.1 Introduction

The industrial loads such as induction motors and induction furnaces account for most of the reactive power drawn from the AC supply. The transformers used in power transmission and distribution and others which may form part of the load or the supply consume reactive power owing to the magnetizing current drawn by them. Besides, the household loads such as fluorescent lights, fans and other appliances using fractional horsepower AC motors draw reactive power from the AC supply system. The loads are usually not balanced in the three phases of the three-phase supply network. The negative sequence component which arises due to the unbalanced loads is also considered as a sort of the reactive current from the AC supply system. Thus, the demand for the reactive power from the supply network terminal increases to a large amount. This increased reactive power causes the system to operate at low power factor. The supply power factor is critical for an economical design, and efficient and reliable operation of power systems.

Some of the disadvantages of low supply power factor are the following.

1. Because of the large current, the losses (proportional to I^2) in the generators and transmission lines increase causing overheating of the system components. The cross-sectional area of the bus bars and the switch gears have to be increased since they have to be designed for the maximum current which would correspond to the minimum power factor for the same KW to be supplied from the AC supply system.

2. The *KVA* loading of source generators increases as the power factor decreases. This results in an overloaded condition and reduces the capacity for additional load growth. That is, the increased *KVA* load on the generators prevents any possibility of additional KW loading which may be placed on the generators even though the prime-mover capacity is available.

3. It causes poor voltage regulation at the load. Further, in *HVAC* systems with long transmission lines, the overvoltages due to load rejection affect insulation co-ordination.

4. As the load voltage decreases due to poor voltage regulation, the power transfer capability is adversely affected.

5. The investment in system facilities per KW of load supplied increases with a decrease in the supply power factor.

These disadvantages can be overcome to a large extent if the power factor is improved externally. In fact, it is a practice since the inception of AC power transmission to install reactive power compensators for power factor improvement. The reactive power compensators are external devices which supply and compensate the lagging reactive power consumed by the load thereby releiving the burden on the AC supply. These compensators are also known as power factor correcting devices. The reactive power compensators are connected across the supply terminals to relieve the transmission lines from the excess current. Hence, they are called the shunt compensators. Thus the function of the shunt compensators is to minimise the voltage fluctuation at a given terminal and to improve the supply power factor by compensating the load reactive power. In general, the problem of compensation by reactive power compensators is viewed from two aspects, load compensation and voltage support.

8.1.1.1 *Load Compensation*

The objective is to reduce or cancel the reactive power (*VAr*) demand of large, and fluctuating industrial loads, such as electric arc furnaces, rolling mills, etc., and to balance the real power drawn from the AC supply lines. These types of heavy industrial loads are normally concentrated in one plant and supplied from one network terminal, and therefore can be handled best by a local compensator connected to the same terminal.

8.1.1.2 *Voltage Support*

This is generally related to the voltage at a given terminal of a transmission line. The objective is to balance the voltages of the three phases, regulate or control the voltage at a given terminal whenever it deviates from a reference value due to the disturbances of both loads and generation. Because of the complex interconnected AC network, it may not be possible or practical to measure the quantities that would meaningfully characterize the load. Instead, the transmission line voltage is regulated by the shunt compensators because it determines the transmittable power. The transmission network compensation is achieved in a manner different from that of the load compensation although both compensation schemes are fundamentally the same in respect of compensation of the load reactive power. Here the compensation currents are expressed in terms of the terminal voltages, and the impedance of the AC system plays a significant role in the transmission network compensation.

With the passage of time, the shunt compensators have gone through several modifications with the rapid developments in the power system as regards the increase in the operating voltage level and complex interconnections of large networks. In what follows, a review of such reactive shunt compensators has been presented chronologically.

8.1.2 Shunt Reactive Power Compensators

8.1.2.1 *Fixed Capacitor Banks*

Shunt capacitors were first employed for the power factor correction in the year 1914 [1]. The leading current drawn by the shunt capacitors compensates the lagging current drawn by the load. They are installed across the supply lines at industrial service inputs.

Initially, the use of the capacitors was limited due to high cost per *KVAr*, large size and weight. However, over the years, the capacitor usage has increased phenomenally due to reduction in selling price, improved design and manufacturing methods resulting in small size and weight, and above all, better understanding of the system benefits that result from their use.

The selection of shunt capacitors is dependent on many factors the most important of which is the amount of lagging *VAr* taken by the load. In the case of widely fluctuating loads, the *VAr* of the load also varies over wide limits. Thus a fixed capacitor bank may often lead to either over-compensation or under-compensation resulting in lower power factor in the AC power supply system.

8.1.2.2 *Switched Capacitors*

The wide range of variation in lagging *VAr* on the system has caused the necessity for dynamic *VAr* compensation, that is, controlled compensation of the reactive power to achieve desired power factor at all load conditions. This is achieved using switched capacitors. Depending on the total *VAr* requirement, a number of capacitors are used which can be switched into or switched out of the system individually. The control is accomplished by continuously sensing the load *VAr*. If more compensation is required, then the required number of capacitors are switched into the circuit so as to take the extra loading *VAr* and conversely, the required number of capacitors are switched out if the load *VAr* falls. The smoothness of control is solely dependent on the number of capacitor switching units used. A very fine control of the power factor can be achieved at the expense of economy by using a large number of such small units. The switching is conventionally accomplished using relays and circuit breakers. One such method uses a master control relay in addition to a time-delay relay for each unit [1]. For one-step automatic control the master relay energizes the closing element of the time delay relay, and if the master-relay contacts stay closed for the time required for the time delay relay contacts to make, then the operating circuit is energized and the capacitor breaker closes. A similar process in reverse trips the capacitor breaker. For more accuracy and reliability, multistep control [1] is preferred. However, these methods using mechanical switches and relays invariably suffer from the drawback of being sluggish, unreliable, introducing switching transients and requiring frequent maintenance. With the advent of high power solid state devices, thyristors have replaced the mechanical switches. The rugged electronic circuits have made the system highly reliable. Thyristor switching of a static capacitor has made it possible to achieve virtually continuous control of reactive power generation on a large scale [2]. Each small unit of capacitors is switched on and off individually using thyristors as switching elements as shown in Fig. 8.1(a). The supply current is also shown neglecting the commutation interval. The switching-on transients are avoided by selecting the switching-on instant at the time when the network voltage corresponds in magnitude and polarity to the capacitor voltage. Switching off transients are not a problem since turning-off of the thyristors occurs at the next current zero after the removal of the gate pulse. The capacitor then remains charged to either positive or negative peak value of the network voltage and is prepared for a new switching-on which will be free from transients. The time required for switching on or off is made up of the time to detect the magnitude of the desired change in the reactive power (one-half cycle), plus a variable time (maximum of one-half cycle) waiting

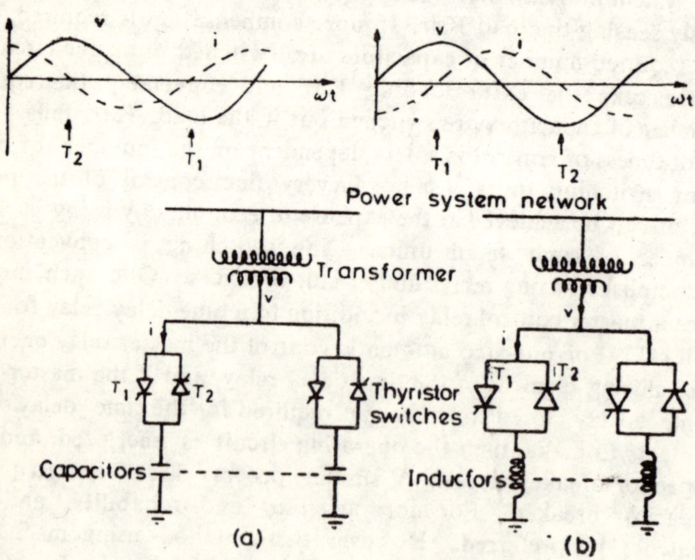

Fig. 8.1. Thyristor witched capacitors (a) and inductors (b).

for suitable switching conditions. Switching-on or off is therefore achieved within one cycle of the power frequency. Usually, a step-down transformer, as shown in Fig. 8.1 is necessary to suit the voltage ratings of thyristors and capacitors. Switched inductors working on the same principle are also extensively used in transmission systems for direct compensation of the charging capacitance of a transmission line [2]. Such a scheme is illustrated in Fig. 8.1(b) along with the supply voltage and current waveforms.

8.1.2.3 *Synchronous Condenser*

The synchronous machine when over-excited draws leading current from the supply system. This characteristic particularly makes it useful as a dynamic power factor correction device. The machine can provide continuous *VAr* balance when used with proper automatic exciter control system. The short time overload capacity of synchronous condenser is larger than that of capacitors. The synchronous condenser has also greater stabilizing effect upon the system voltage. However, it has also many drawbacks as compared to the shunt capacitors. The losses in the synchronous condensers are much greater than for the shunt capacitors. For synchronous condensers the full load losses vary from about $1\frac{1}{2}$ per cent to 3 per cent of the *KVA* rating whereas for capacitors the losses are about 1/3 per cent of the *KVA* rating [1]. The capacitors are more effective since they lend themselves to distribution at several locations closer to the load throughout the network. It will be very costly to distribute small synchronous condensers throughout the system. The *VAr* (reactive volt-amperes loading) rating of a capacitor installation can be easily changed with a variation in the load and system requirements day by day which is impractical with the synchronous condensers. The failure of a synchronous condenser compared to a single unit in a bank of capacitors is less likely to occur. However, any such failure in a synchronous condenser completely affects entire ability to produce *VAr*. This is not so in the capacitor banks since there are several other capacitors in a bank even if a particular unit fails. Synchronous condensers increase the short-circuit current of a system and thus the breakers of higher rating are required. The response of the synchronous condenser is slow due to the inherent mechanical inertia.

8.1.2.4 *Static Reactor Compensator*

The aim to achieve fine control over the entire *VAr* range as in the case of the synchronous condenser without sacrificing the advantages of static capacitors has been fulfilled by the development of the static reactor compensator. This essentially consists of a controllable reactor in parallel to a shunt capacitor as shown in Fig. 8.2.

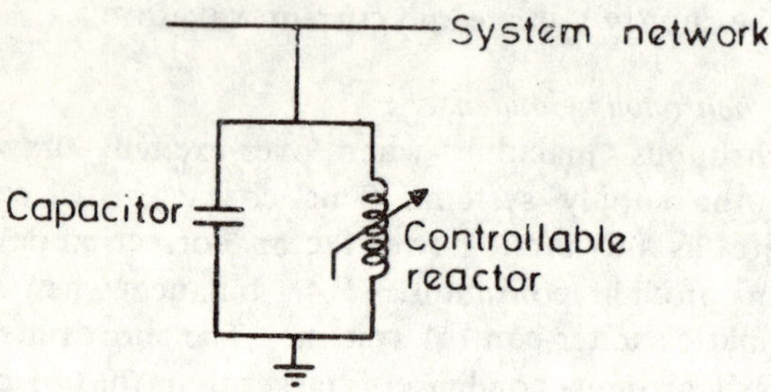

Fig. 8.2. Basic static reactor compensator.

By choosing proper values for the capacitor and the inductor, it is possible to achieve smooth and stepless control of *VAr* from lagging to leading over a wide range to compensate for the wide variations in the load *VAr*. This is accomplished by a continuous control of the effective fundamental reactance of the inductor. Since there is a greater requirement for the leading *VAr*, preferably a number of capacitors are switched in steps to provide a wide range of leading *VAr* [3]. The controllable reactor can be realized using different methods some of which are listed below:

(a) REACTORS WITH DIRECT CURRENT-CONTROLLED SATURATION [4]

The current in a reactor can be controlled by a direct current fed into a control winding mutually coupled to the main reactor winding. The flux-mmf (amp–turns) characteristic of such a reactor will be similar to that of a saturable-core reactor and is shown in Fig. 8.3 (a). The magnetization curve has two knee points *a* and *b* which determine the saturation limits. Between the points *a* and *b* the permeability is infinite, hence the core is unsaturated. Also the sum of the ampere-turns of all windings is zero. Again the permeability is finite over the regions *ac* and *bd* giving rise to a linear relationship between the sum of ampere-turns and the core flux. By controlling the direct current which sets the magnetic

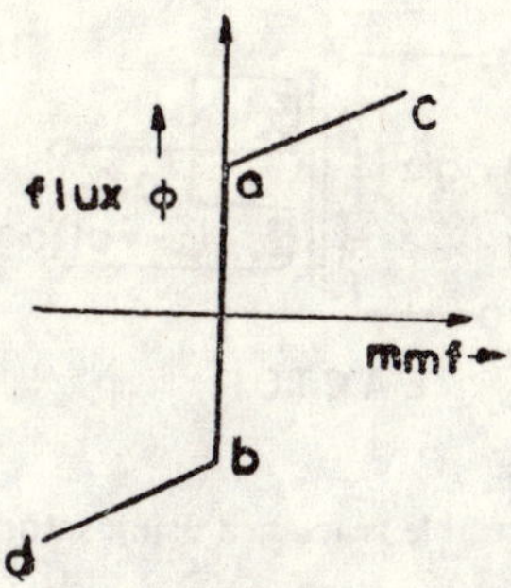

Fig. 8.3 (a) Flux versus *mmf* characteristic of a dc controlled shunt reactor ($I_{dc} = 0$).

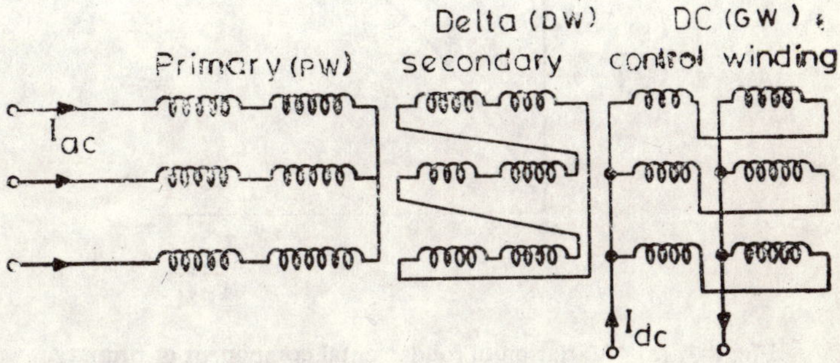

Fig. 8.3. (b) Reactor using dc current-controlled saturation.

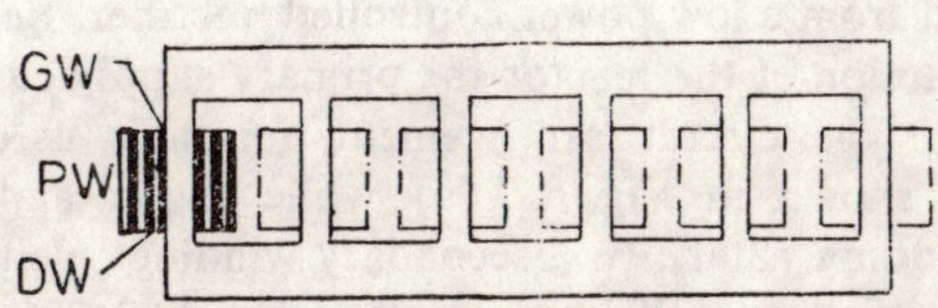

Fig. 8.3. (c) Six-limb core for a three-phase shunt reactor.

bias, the magnetic flux in the core can be varied. The fundamental component of the alternating current in the primary winding of the reactor can be varied from a very low value, say zero, when the magnetic flux in the core owing to direct current is low and to a maximum value when the magnetic flux in the core is high owing to a large component of the direct current. The control current can

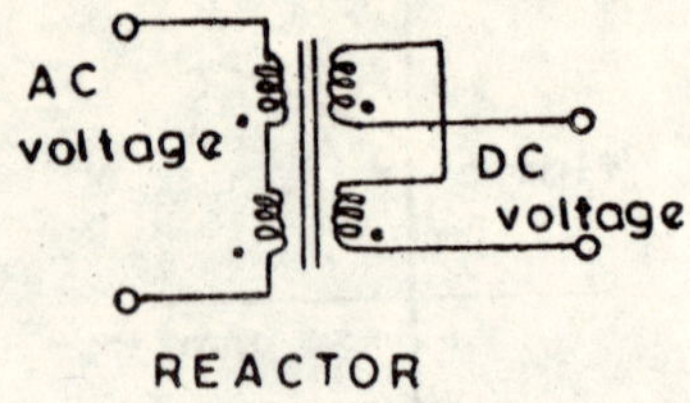

Fig. 8.3. (d) Variable reactance using saturable core reactor.

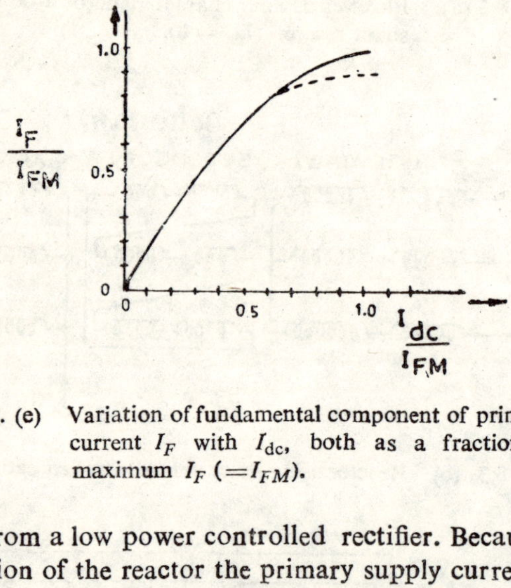

Fig. 8.3. (e) Variation of fundamental component of primary current I_F with I_{dc}, both as a fraction of maximum I_F ($=I_{FM}$).

be obtained from a low power controlled rectifier. Because of the switching action of the reactor the primary supply current is non-sinusoidal. The circuit arrangement normally used in HVAC systems is shown in Fig. 8.3 (b) while the core along with the primary winding (Star, PW), secondary winding (delta, DW) and the control winding (GW) are shown in Fig. 8.3 (c). The primary winding PW is designed for the harmonic currents of the order 1, 5, 7, 11, 13, ... The secondary delta connected winding DW is designed for the harmonic currents of the order 3, 9, 15, 21, ..., of which the third harmonic mostly influences the winding dimensions. It is rated for about 10 percent of the rated power. The control winding GW is designed for the harmonic currents of the order 0, 2, 4, 6, 8, ...

The DC control winding coupled to each phase is divided into two halves. One-half of the winding is reversed with respect to the other and connected in series with it as indicated in Fig. 8.3(d). The other DC windings coupled to the other two phases are also connected in a similar manner. By connecting the control winding in this way, the primary alternating current would be symmetrical thereby eliminating even harmonic currents in the primary winding.

The direct current thus changes the effective reactance posed to the power system. The variation of the fundamental component of the alternating supply current, as a function of the direct current I_{dc}, both being expressed as a fraction of the maximum value of the fundamental component of the supply current I_{FM} obtained when the core is fully saturated, is shown in Fig. 8.3 (e). The solid curve represents the case of negligible stray flux. The dotted curve shows the influence of the leakage between the windings whereby the maximum output is somewhat reduced. For an economical design of the DC winding, the ratio I_{dc}/I_{FM} is limited to about 0.85 since there is no significant increase in the output by increasing the ratio beyond this value. The higher harmonic currents are generally neglected while designing the windings and calculating the losses. There are eddy current losses in the inner delta and dc windings caused by the magnetic field of the outer primary winding.

This scheme requires a controllable DC source. As mentioned earlier, a thyristor controlled rectifier of a lower rating compared to that of AC power is essential. Though the operating principle is very simple, the transient behaviour of the reactor with DC controlled saturation is very complicated. It cannot be characterised by a simple time constant due to non-linear magnetisation characteristic. The response is sluggish owing to a large time constant of the DC control circuit.

(b) Thyristor-controlled Shunt Reactors (TCR) [7]

With the increase in the size and complexity of power system, fast reactive power compensation has become necessary in order to maintain the stability of the system. The thyristor-controlled shunt reactors have made it possible to reduce the response time to a few milliseconds. Thus, the reactive power compensators utilizing the thyristor-controlled shunt reactors are becoming increasingly popular. The basic operating principle of a thyristor-controlled reactor is illustrated in Fig. 8.4. A reactor operating in its linear region is connected to the AC system in series with the back-to-back connected thyristors as shown in Fig. 8.4 (a). The inductive reactive current is continuously controlled by the phase-control of the thyristors, one being fired in the positive half-cycle of the supply frequency and the other in the negative half-cycle. Thyristor circuits operating in this way are known as AC controllers. Figure 8.4 (b) shows the supply voltage and current waveforms for two different firing angles. For firing angle $\alpha = 90°$, maximum current is drawn from the supply. If the firing instants of the thyristors are delayed beyond 90°, the current becomes discontinuous. This is illustrated by the dotted line in Fig. 8.4 (b). The conduction interval decreases as the firing angle increases and finally no current flows for $\alpha = 180°$. The fundamental component of the current which is shown by the chain line in Fig. 8.4. (b) however, always lags the voltage by 90° and it is this component which accounts for the effective reactive power compensation. Due to the discontinuous conduction, the current has a certain finite harmonic content. A detailed treatment of static shunt reactor compensator utilizing thyristor-controlled reactor is presented in section 8.1.3.2. Normally for linear operation over a wide range, the gapped core type reactors are used [2]. Economics show that a step-down transformer is necessary to connect the reactor and the thyristors to the low voltage winding. The transient response time of thyristor-controlled reactor is within one cycle of the power frequency [2].

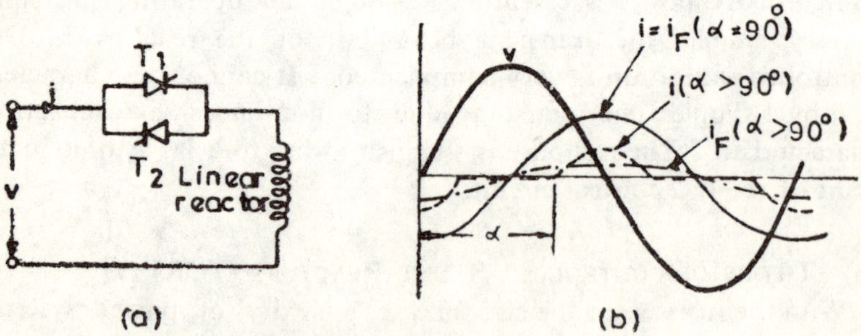

Fig. 8.4. Thyristor controlled reactor:
(a) Circuit diagram.
(b) Voltage and current waveforms.

(c) THYRISTOR-CONTROLLED HIGH-IMPEDANCE TRANSFORMER (TCT) [2, 3]

If a transformer is necessary for the connection of thyristor-

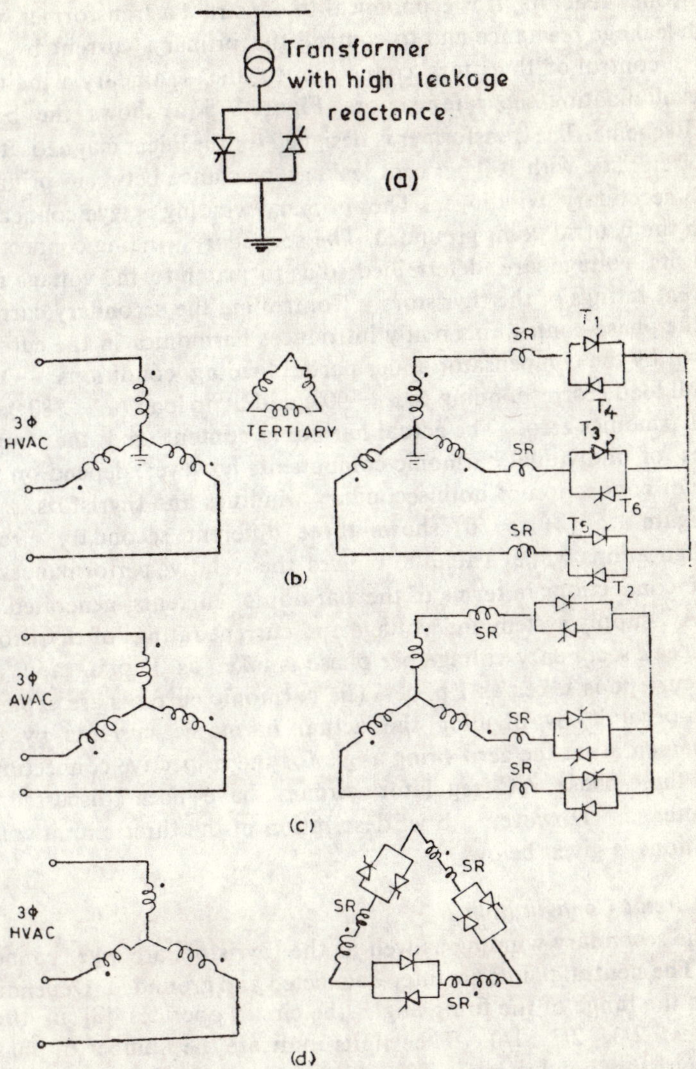

Fig. 8.5. Three different circuit configurations of secondary windings and thyristors:

(a) Variable reactance using thyristor controlled transformer (TCT).

(b) Star connected secondary with grounded neutral.

(c) ,, ,, ,, ,, isolated ,,

(d) Branched controlled delta.

SR: Additional reactor, if required.

controlled reactors, it is economical to construct a transformer with high leakage reactance and to control the primary current by the phase-control of thyristors connected to the secondary windings, thus eliminating separate reactors. Figure 8.5 (a) shows the basic TCT scheme. The transformer is designed for the linear magnetisation characteristic with 100 per cent leakage reactance between primary and secondary windings. The primary winding is wye connected with the neutral point grounded. The secondary winding connection and its voltage are determined so as to match to the voltage and current ratings of the thyristors. Controlling the secondary current by the phase-control inherently introduces harmonics in the current drawn by the compensator under partial loading conditions. Only at full load (corresponding to $\alpha = 90°$) and at no load ($\alpha = 180°$) are the harmonics zero. The actual harmonic content and the amplitudes of individual harmonic components however, depend on the type of connection of both secondary windings and thyristors.

Figure 8.5 (b) to (d) shows three different secondary circuit configurations while Table 8.1 gives the relative performances of these connections in terms of the harmonic currents generated in the AC supply system and voltage and current ratings of thyristors. The peak secondary voltage per phase is taken as 1 p. u. and the line current is taken as 1 p. u. The harmonic currents are expressed in per unit by dividing the actual harmonic currents by the fundamental at the zero firing angle for the respective connections. The three-phase AC regulator circuits have been discussed in Chapter 5. However, a brief description of the three circuit configurations is given below.

I. *Circuit Configuration 1*

The secondary winding as well as the thyristors are wye connected. The neutral points are inter-connected and grounded. Depending upon the range of the firing angle, the circuit operates [8] in three modes: 2/3, 2/1, 1/0. The digits indicate the number of phases or thyristors conducting. For example in mode 2/3, two thyristors and three thyristors alternately conduct. The other two modes are characterised accordingly. The maximum voltage appearing across a thyristor is given by the peak of the phase voltage.

II. *Circuit Configuration 2*

The secondary winding as well as the thyristors are connected in wye but there is no connection between the ground points. For the current to flow in the secondary, at least two thyristors are required to conduct. Depending upon the range of the firing angle, two different modes (3/2, and 2/0) of operation occur [8], [9]. The maximum possible voltage to which a thyristor is subjected can be shown to be $\pm 3/2$ times the peak of the phase voltage.

III. *Circuit Configuration 3*

The secondary winding as well as the thyristors are delta-connected. In this connection also, depending upon the firing angle range different modes of operation 2/3, 2/1, 1/0 take place [8]. The maximum voltage to which a thyristor is subjected will be the peak of the line-to-line voltage which is $\sqrt{3}$ times the peak of the phase voltage.

Table 8.1

V_{ph} :rms phase voltage
I:rms line current

Circuit configuration	Thyristor rating $\sqrt{2}\,V_{ph}$	Current I	Max. Harmonic Current in % 3	5	7	9	11
1	1.0	1.0	13.7	5.0	2.55	1.55	1.0
2	1.5	1.0	—	9.5	3.4	—	1.7
3	$\sqrt{3}$	$1/\sqrt{3}$	—	7.6	3.85	—	1.5

The triplen harmonics in the circuit configuration 1 can be eliminated by providing an additional delta-connected tertiary winding closely coupled to the secondary. Thus, the connection 1 is the most desirable one for minimum harmonic components and thyristor ratings although it requires some extra cost for the additional tertiary winding. The remaining objectionable higher order harmonic currents can be filtered out by connecting harmonic filters to the delta tertiary winding [3].

(d) Thyristor Controlled Reactor Compensator-Generalized University of Manitoba (UM) Concept

The thyristor controlled reactor (TCR) essentially consists of a fixed inductor in series with two thyristors connected in inverse

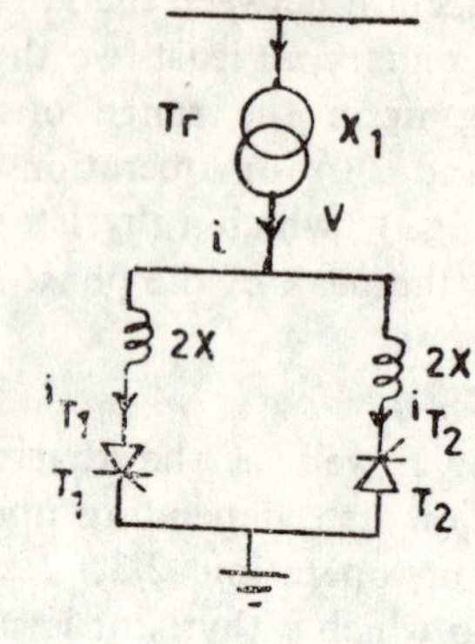

Fig. 8.6. TCR with an extended range for the firing angle, α.

parallel. By the phase control of the current through the thyristors, the effective value of the inductance is varied. The practical range of the firing angle α is from 90° to 180°. The range of the firing angle α is further extended to include 0 to 90° by a simple modification of the circuit by splitting the inductance into two equal values and connecting them as shown in Fig. 8.6. For $90° \leqslant \alpha \leqslant 180°$, the operation of this circuit is the same as that described for the TCR of Fig. 8.4. However, for $0 \leqslant \alpha \leqslant 90°$, there are periods when both thyristors conduct and gives rise to the circulating current in the parallel branches. The pertinent waveforms of the circuit of Fig. 8.6 are shown in Fig. 8.7.

Assuming the same supply voltage and equal maximum ratings for the TCRs shown in Figs. 8.4 and 8.6, the magnitude of the series inductor in the scheme with the extended control range is twice that with the conventional TCR of Fig. 8.4. This has a beneficial effect on the harmonic currents generated in the input supply. Since the inductance in series with each thyristor is doubled, the harmonic currents generated in the input supply are reduced to half of that for the TCR circuit of Fig. 8.4. The main drawback of the circuit of Fig. 8.6 is that the thyristors are not as such protected against over voltage stresses. In the event of an over voltage at the supply bus, if the thyristor say, T_1 is triggered into conduction, it may cause a large reverse voltage across the thyristor T_2 and eventually it may breakdown. This does not, however, occur with the circuit of Fig. 8.4. This is because, by turning the

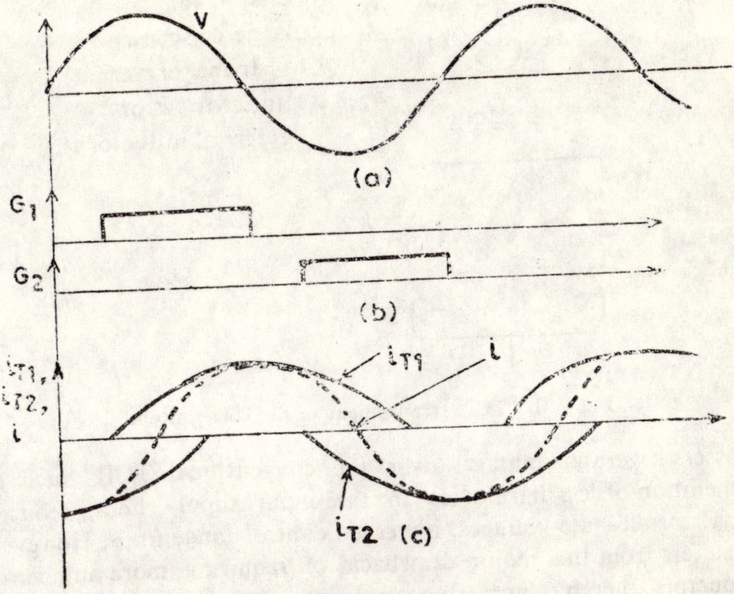

Fig. 8.7. Waveforms for the circuit of Fig. 8.6.
(a) Input voltage to the TCR.
(b) Gate pulses thyristors T_1 and T_2.
(c) Waveforms of thyristor and supply currents.

thyristor T_1 on, the voltage coming across the thyristor T_2 is limited to a small forward conduction drop of the thyristor T_1. The over-voltage with the circuit of Fig. 8.6 is reduced by connecting another inductor in series with the parallel branches of the circuit of Fig. 8.6. The resulting circuit arrangement of the generalized UM-concept [11] is shown in Fig. 8.8. It is possible to have a wide range of values for the inductors x_2 and x_3 for a given value of the maximum rating of the reactive compensator. Further the circuit arrangement of Fig. 8.8 represents a general case since the circuits of Figs. 8.4 and 8.6 are obtained by disconnecting the inductors x_2 and x_3 respectively. For three-phase systems, three similar units are connected in wye, wye with grounded neutral, or delta configuration as discussed in the preceding section for the conventional TCR compensator.

The TCR arising from the generalized UM-concept seems to

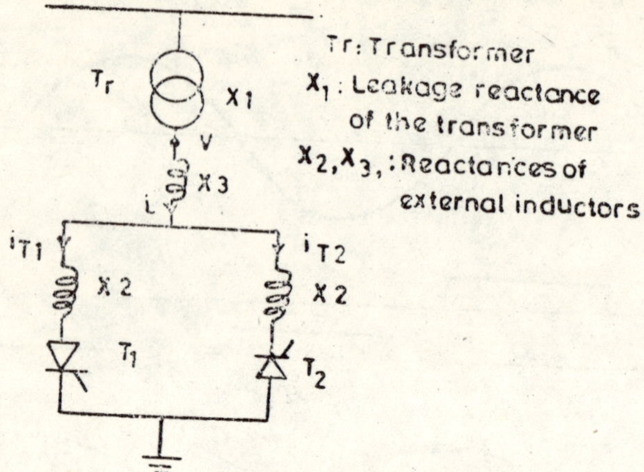

Fig. 8.8. TCR arrangement of the UM-concept.

possess several advantages over the conventional TCR such as generation of less harmonics in the input supply lines, reduced thyristor off-state voltages, increased control range for α. However, it suffers from the major drawback of requiring more number of inductors thereby increasing the cost. It is yet to be established whether the increase in the cost of the total installed rating of the reactors is justified by the improved performance.

8.1.3 Applications of Static Thyristor-Controlled Shunt Compensators for Load Compensation

8.1.3.1 *Fundamentals of Load Compensation*

Due to their fast response and very smooth *VAr* control, there has been a rapid increase in the number of static thyristor-controlled shunt compensators in industrial and utility systems for load compensation, transmission line voltage regulations and other purposes [3], [5] in industrial power systems. The load compensation means cancellation of the reactive power demand of large and fluctuating industrial loads such as electric arc furnaces, induction furnaces, rolling mill drives used in the steel industries etc., and to balance the real power drawn from the three phases of the AC supply system. In simpler terms, the load compensation involves two distinct acts; one being power factor correction and the other load balancing in the three phases. If both are carried out, they determine together the total reactive power needed for compensation.

Any 3-phase unbalanced linear steady state load can be compensated by three appropriate reactive impedances (capacitive and inductive) connected between the three phases [5]. The load on a power system varies with the time which can be represented by a series of "steady state" impedances at discrete time instants close to one another. The compensation, therefore, requires variable reactances that can be adjusted according to the load variation.

8.1.3.2 *Practical Realization of Variable Reactance* [7], [12].

Under steady-state conditions the compensating susceptances can be expressed in terms of the load impedance [5]. Since the load impedances are directly measurable, the compensating susceptances can always be determined by a direct use of appropriate analytical expressions. The different schemes for realizing variable reactance (or susceptance) have been outlined in section 8.1.2.4. The basic scheme for a single-phase using a fixed capacitor in parallel with a TCR is illustrated in Fig. 8.9.

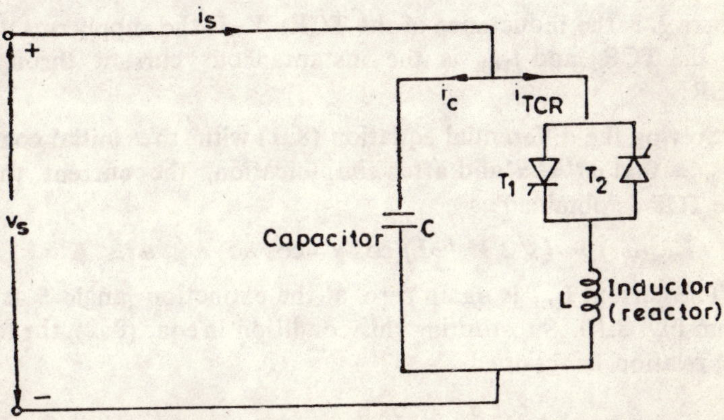

Fig. 8.9. Basic fixed capacitor, thyristor-controlled reactor type compensator.

Since the capacitor draws a fixed amount of current from the source, the resultant current drawn from the source is effectively controlled by controlling the conduction interval of thyristor switches using the phase-control. The supply voltage and the inductor current waveforms are shown in Fig. 8.10 for a particular value of the firing angle α. Theoretically the firing angle α can be varied from 90° to 180°.

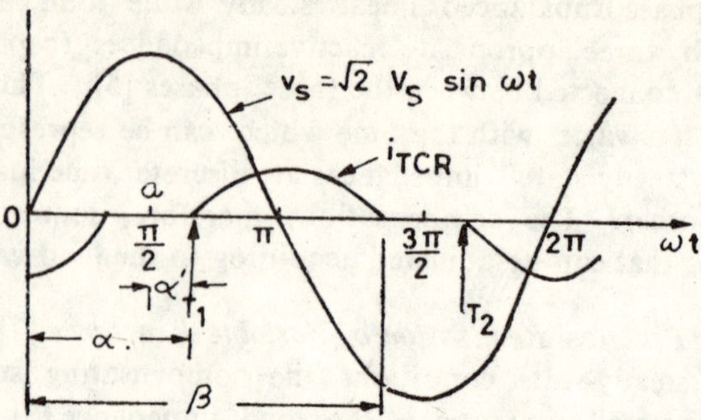

Fig. 8.10. Supply voltage and TCR current waveforms.

The current, i_{TCR} through the inductor when the thyristor T_1 is fired at a delay angle α is given by

$$L \frac{di_{TCR}(\omega t)}{dt} = \sqrt{2}\ V_S \sin \omega t \tag{8.1}$$

where L is the inductance of the TCR; V_S is the supply rms voltage to the TCR; and i_{TCR} is the instantaneous current through the TCR.

Solving the differential equation (8.1) with the initial condition $i_{TCR} = 0$ at $\omega t = \alpha$ and after simplification, the current through the TCR is obtained as

$$i_{TCR}(\omega t) = (\sqrt{2}\ V_S/\omega L)(\cos \alpha - \cos \omega t)\ \alpha \leqslant \omega t \leqslant \beta \tag{8.2}$$

The current i_{TCR} is again zero at the extinction angle β as seen from Fig. 8.10. Substituting this condition in eqn. (8.2), the following relation is obtained.

$$\cos \alpha = \cos \beta$$

Since the extinction angle is greater than the firing angle but less than 2π radians

$$\beta = 2\pi - \alpha \tag{8.3}$$

Similarly, considering the thyristor T_2 which is fired in the negative half-cycle of the supply frequency, the expression for the current is

$$i_{TCR}(\omega t') = (\sqrt{2} V_S/\omega L)(\cos \alpha - \cos \omega t');\ \alpha \leqslant \omega t' \leqslant \beta \tag{8.4}$$

where $\quad \omega t' = \omega t - \pi$

The current is almost sinusoidal and lags the voltage by 90° when $\alpha = 90°$ neglecting the commutation intervals of thyristors. The current becomes discontinuous for $\alpha > 90°$. And the discontinuous period increases with an increase in the firing angle α. The TCR current can be expressed by the Fourier Series

$$i_{TCR}(\omega t) = \frac{a_0}{2} + \sum_{n=1}^{\infty} (a_n \cos n\,\omega t + b_n \sin n\,\omega t) \tag{8.5}$$

The Fourier coefficients are obtained as follows:

$$\frac{a_0}{2} = \frac{1}{2\pi}\int_0^{2\pi} i_{TCR}(\omega t)\, d(\omega t) \tag{8.6}$$

$$a_n = \frac{1}{\pi}\int_0^{2\pi} i_{TCR}(\omega t) \cos n\omega t\, d(\omega t) \tag{8.7}$$

$$b_n = \frac{1}{\pi}\int_0^{2\pi} i_{TCR}(\omega t) \sin n\omega t\, d(\omega t) \tag{8.8}$$

The fundamental and harmonic coefficients using eqns. (8.7) and (8.8) can be shown to be

$$a_1 = \frac{\sqrt{2}\,V_S}{2\pi\omega L}[\sin 2\alpha - \sin 2\beta - (2\beta - 2\alpha) - 4\cos\alpha\{-\sin\beta + \sin\alpha\}] \tag{8.9}$$

$$b_1 = \frac{\sqrt{2}\,V_S}{2\pi\omega L}[-\cos 2\alpha + \cos 2\beta - 4\cos\alpha(\cos\beta - \cos\alpha)] \tag{8.10}$$

$$a_n = \frac{\sqrt{2}V_S}{2\pi\omega L}\left[\frac{2}{n+1}\{\sin(n+1)\alpha - \sin(n+1)\beta\} - \frac{2}{n-1}\{\sin(n-1)\alpha - \sin(n-1)\beta\} - \frac{4\cos\alpha}{n}\{-\sin n\beta + \sin n\alpha\}\right],\ n \triangleright 1 \tag{8.11}$$

$$b_n = \frac{\sqrt{2}V_S}{2\pi\omega L}\left[\frac{2}{n+1}\{-\cos(n+1)\alpha + \cos(n+1)\beta\} - \frac{2}{n-1}\{-\cos(n-1)\alpha + \cos(n-1)\beta\} - \frac{4\cos\alpha}{n}\{\cos n\beta - \cos n\alpha\}\right] n \triangleright 1 \tag{8.12}$$

The rms value of the nth order harmonic and its phase angle ψ_n are given by

$$I_n = (a_n^2 + b_n^2)^{1/2}/\sqrt{2}$$

$$\psi_n = \tan^{-1}\left(\frac{a_n}{b_n}\right)$$

The per-unit values of the fundamental and different harmonics are computed for different α's and are shown in Fig. 8.11. Substituting the value of β from eqn. (8.3) in eqns. (8.9) and (8.10) yields

$$a_1 = (-\sqrt{2}V_S/\pi\omega L)\,(2\pi - 2\alpha + \sin 2\alpha)$$

$$b_1 = 0$$

Hence, the fundamental component I_1 is,

$$I_1 = I_{TCRF}\,(\alpha) = (a_1^2 + b_1^2)^{1/2}/\sqrt{2}$$

$$= (V_S/\pi\omega L)\,(2\pi - 2\alpha + \sin 2\alpha)\cdot \pi/2 \leqslant \alpha \leqslant \pi \tag{8.13}$$

where I_{TCRF} is the fundamental rms current through the TCR. Phase angle, $\psi_1 = \tan^{-1}\,(a_1/b_1) = 90°$ (lagging).

The TCR (thyristor-controlled reactor) current changes from its maximum value to zero when the firing angle α is varied from 90° to 180°. It may be more convenient to fix the reference point or origin at 90° as indicated by the point '*a*' in Fig. 8.10. Let us recognize a new firing angle notation such that

$$\alpha' = \alpha - (\pi/2) \tag{8.14}$$

Substituting the relation (8.14) in the expression (8.13), the fundamental component of the TCR current is obtained as

$$I_{TCRF}(\alpha) = (V_S/\omega L)\,[1 - (2/\pi)\alpha' - (1/\pi)\sin 2\alpha']$$

$$0 \leqslant \alpha' \leqslant \pi/2 \tag{8.15}$$

Thus, the magnitude of the total fundamental current drawn by the static compensator is

$$I_{SF}(\alpha) = I_C - I_{TCRF}(\alpha) \tag{8.16}$$

$$= V_S\left[\omega C - \frac{1}{\omega L}\left(1 - \frac{2}{\pi}\,\alpha' - \frac{1}{\pi}\,\sin 2\alpha'\right)\right] \tag{8.17}$$

By controlling the fundamental component of *TCR*, it can be seen from eqn. (8.17) that the static reactor compensator draws either net capacitive or inductive current from the AC supply depending upon whether the current $I_{SF}(\alpha)$ is positive or negative respectively.

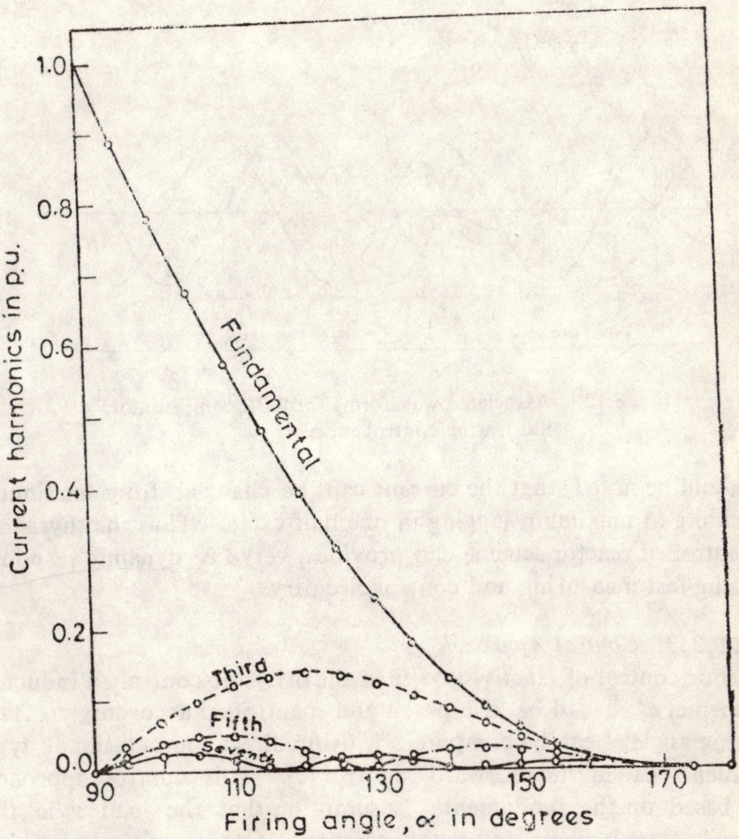

Fig. 8.11. Fundamental and harmonics values in TCS current.

The control process assuming that the compensator is under steady state operation before and after the initiation of the firing angle is illustrated in Fig. 8.12. The controlled inductor current $i_{TCR}(\alpha)$, the fixed capacitor current i_c, and the total current drawn from the AC supply $i_s(\alpha)$, with its fundamental component $i_{SF}(\alpha)$ are shown together with the applied voltage v_s as α' is increased from 0° to 90° gradually. The inductive reactance ωL has been assumed to be smaller than the capacitive reactance $1/\omega C$ in this illustration so as to achieve a range of control in both the inductive and capacitive domains. As is evident, the effective reactance and the compensating current can be adjusted at discrete instants of time, not more often than once in each half-cycle. However, it

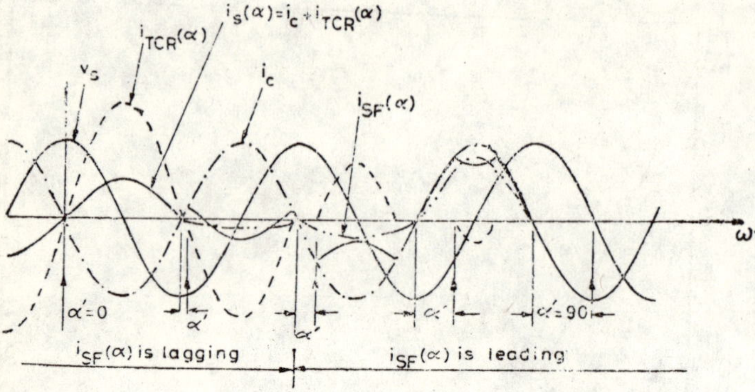

Fig. 8.12. Associated waveforms for TCR compensator illustrating control process.

should be noted that the current can be changed from maximum leading to maximum lagging in one half-cycle. Thus the thyristor-controlled reactor scheme can provide a very fast dynamic response using fast measuring and control circuitry.

8.1.3.3 *Control Approach*

For control of reactive power in the thyristor-controlled inductor scheme, α' has to be computed and controlled accordingly. The firing angle α' can be computed by using direct computational type which is called "feedforward" control [5]. This control approach is based on the fundamental assumption that the load is in the steady state between any two consecutive instants of time at which the current in the compensator is changed. Thus, between these time instants the relevant load currents (or powers, or impedances) can be sensed, and the required compensating currents can be computed using appropriate steady state relations. In 3-phase systems, the processing of each of the three steady state equations for the three phases expressing the required compensating current in terms of quantities characterising the load (currents, power or impedances) is carried out over the time intervals which are mutually displaced by one-third of the period of the AC system voltage. The three compensating currents are measured independent of each other, and the required firing angle α' is determined using the relation (8.17). Thus, the feedforward control scheme requires two main functional elements as illustrated by the block

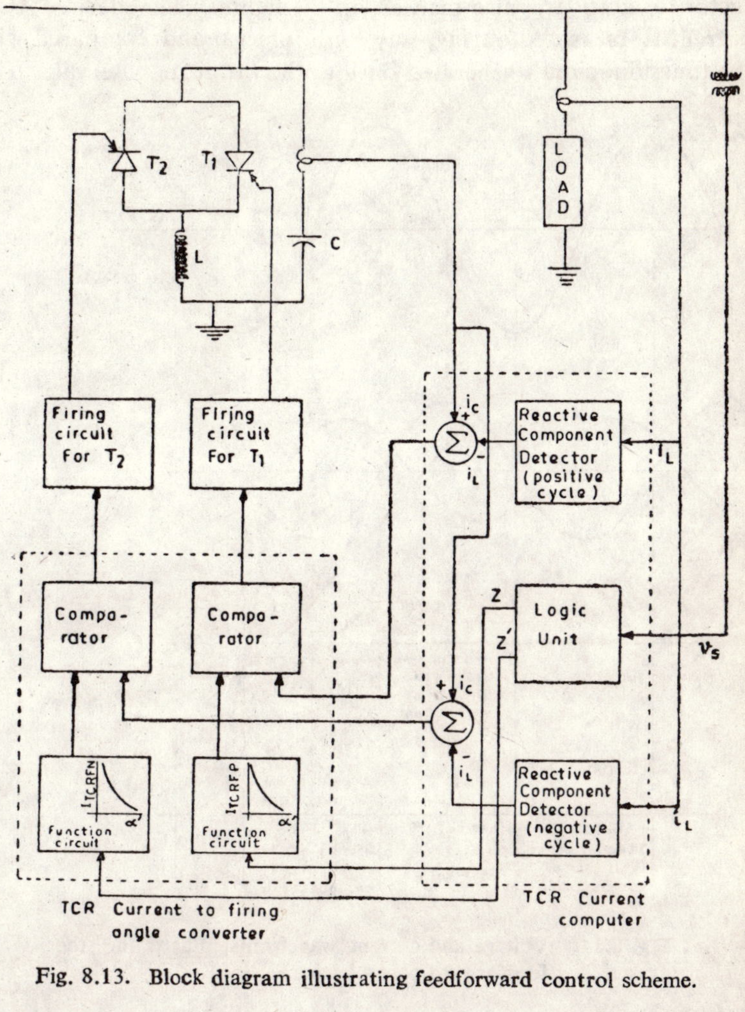

Fig. 8.13. Block diagram illustrating feedforward control scheme.

diagram of Fig. 8.13, one to compute the desired inductor current, and the other to determine the firing angle corresponding to the desired inductor current.

The waveforms illustrating the control process using feedforward control technique are depicted in Fig. 8.14. The capacitive and inductive reactances are assumed to be of equal magnitude. The time interval during which the reactive power is computed and the period during which the thyristors are fired to correct the power factor to unity are also marked in the figure. The desired TCR current is measured during, say, kth interval and compared with the function circuit generated during the kth firing interval. The

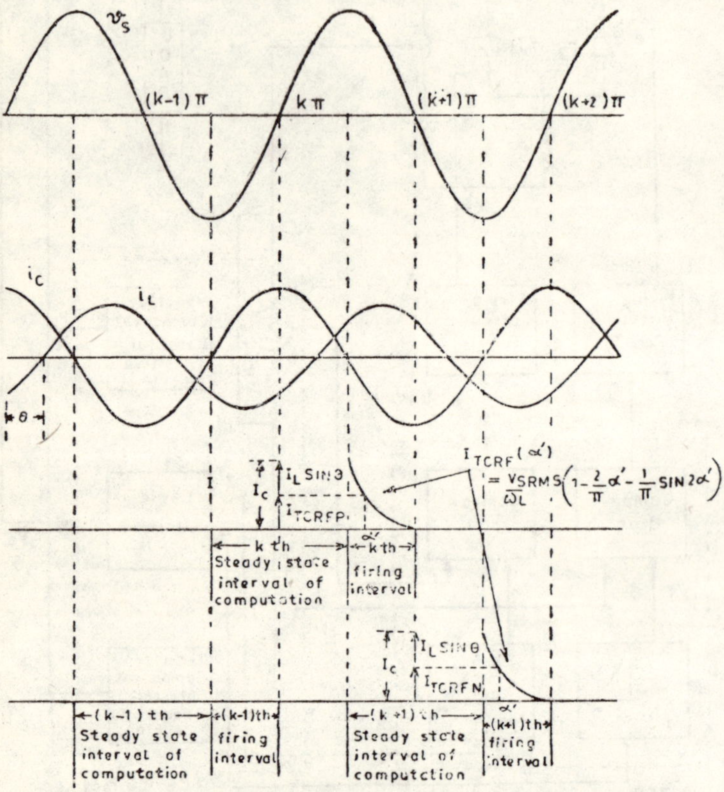

Fig. 8.14. Voltage and current waveforms illustrating the feedforward control scheme.

gate pulse corresponding to the firing angle α' is applied to the gate of the thyristor T_1. The gate pulse determined from the function circuit during the $(k + 1)$th firing interval as a result of measurement of the reactive power during the $(k + 1)$th interval is applied to the gate of the thyristor T_2. Thus, the TCR current is computed in every half-cycle and corrected by altering the firing angle of the appropriate thyristor during the subsequent firing interval if the load changes. Otherwise, the same firing angle is maintained. The oscillograms obtained from an experimental setup are illustrated in Figs. 8.15 and 8.16.

The experimental setup consists of a fixed capacitor of 31.90 micro-farad, a TCR of 319 mH, and a load of a fixed inductor of 319 mH and a variable resistance 0 to 220 ohms. The capacitor and TCR are selected such that the supply power-factor is held at unity even when the load resistance is zero, i.e., when the load current is totally reactive and is to be compensated fully by the fixed capacitor.

Fig. 8.15 shows the waveforms with the TCR regulator under steady state operating conditions while Fig. 8.16 shows the waveforms illustrating the dynamic response of the TCR regulator whenever the load resistance is changed. The load current and the TCR current with a large load resistance is shown in Fig. 8.15 (a) while the same waveforms with a small load resistance is shown in Fig. 8.15 (b). In the former, the load reactive power being small the TCR draws more current thus compensating the capacitor current

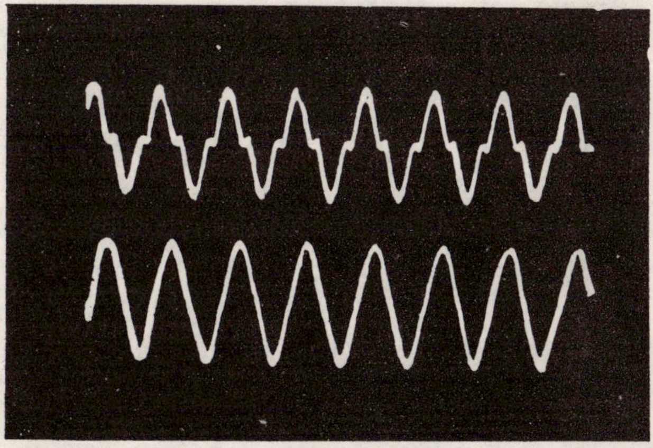

Fig. 8.15. (a) Large load resistance. Botrom trace: Load current. Top trace: TCR current.

while in the latter, the load reactive power being large, the TCR draws less current and the load current compensates the capacitor current thereby making the supply power factor unity in both cases. The large and small firing angles at which the TCR is operating can clearly be seen from the oscillograms of Figs. 8.15 (a) and (b).

Figure 8.16 (a) shows the dynamic response of the TCR when the

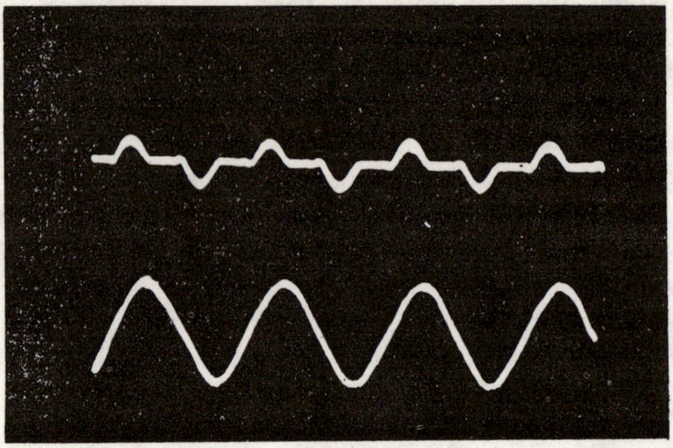

Fig. 8.15(b). Small load resistances. Top trace: Load current. Bottom trace: TCR current.
Oscillograms illnstrating waveforms under steady state operating conditions.

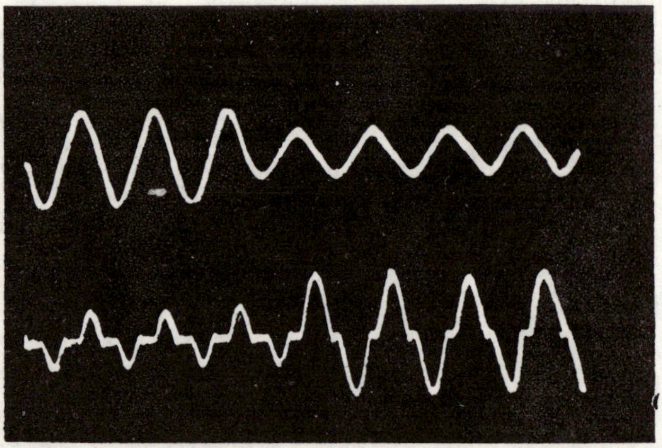

Fig. 8 16(a). Load resistance is changed from a low to high value.

load resistance is suddenly increased from a low value while Fig. 8.16 (b) shows the waveforms when the load resistance is suddenly changed from a high value to a low value. The oscillogram of Fig. 8.16(c) depicts the load current (top trace) and TCR current (bottom trace) waveforms with the load resistance changed from high to low and then to a high value again. In all these oscillograms of Fig. 8.16, the upper trace shows the load current while the lower trace shows the TCR current.

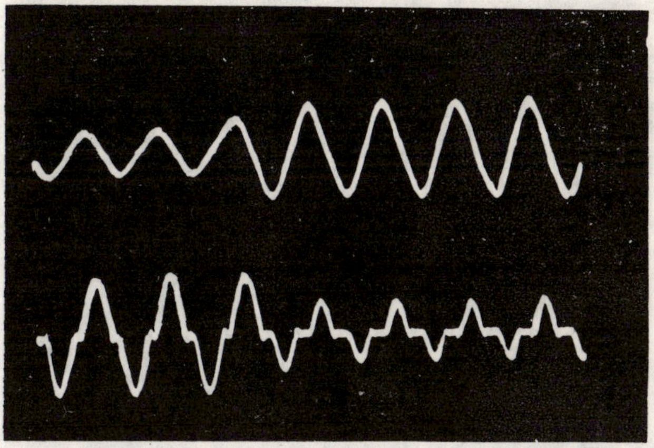

Fig. 8.16. (b) Load resistance changed from a high to a low value.

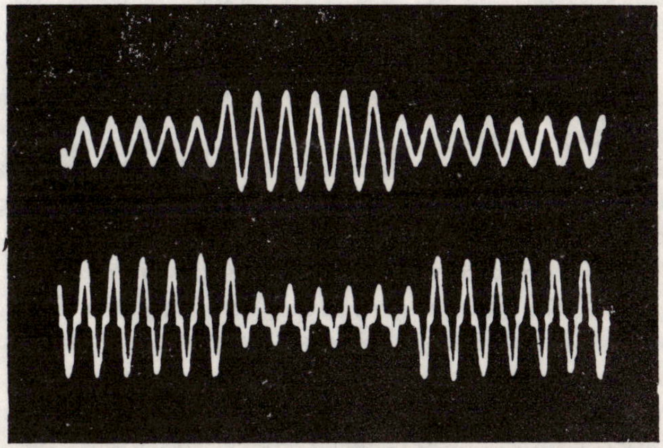

Fig. 8.16. (c) Load resistance is changed from low-high-low.

These oscillograms clearly indicate the fast response of 10 ms for compensating the load reactive power from the instant of the load change.

The oscillogram of Fig. 8.16 (d) indicates the response of the TCR regulator (bottom trace) whenever the load is continuously changed. The load current continuously changes while the TCR also responds continuously as can be seen clearly from Fig. 8.16 (d).

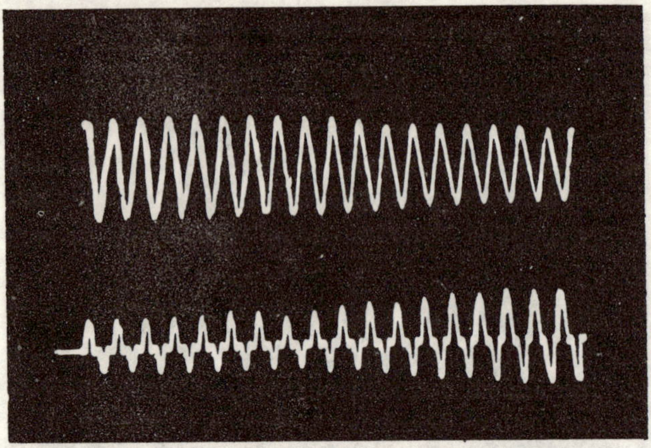

Fig. 8.16. (d) Load resistance is continuously changed
Oscillograms illustrating waveforms under dynamic resistance.
Top trace: Load current.
Bottom trace: TCR current.

8.1.4 Thyristor Controlled Reactor (TCR) for Transmission Line Voltage Support [5, 10]

The TCR is also utilized for the transmission network compensation. Unlike in the load compensation, the compensation currents are used to regulate the terminal voltage at a given bus. The TCR basically involves a variable reactance as noted from section 8.1.3.2. If the inductive reactance (or susceptance) is controlled continuously over a limited range of the operating voltage, the relationship between the fundamental frequency component of inductive current and the applied voltage are shown in Fig. 8.17 (a). The reactance (or susceptance) is effectively varied between the X_{L-MIN} (B_{L-MAX}) and the X_{L-MAX} (B_{L-MIN}) limit, over a voltage band $2e$. The I_L versus V characteristic shows a linear relationship within the $2e$ control band. By using a fixed capacitor in parallel with the controlled reactor, the fundamental characteristic relating V and I ($=I_L - I_C$) results as shown in Fig. 8.17 (b). The harmonic currents as in the case of the load compensation have been neglected.

Consider a transmission network shown in Fig. 8.18 with the thyristor controlled reactor (TCR) type static VAr system (SVS) installed at the bus 1 to regulate its voltage. From the economic considerations, the SVS is generally connected to the low voltage side of a transformer when applied for voltage control on HVAC

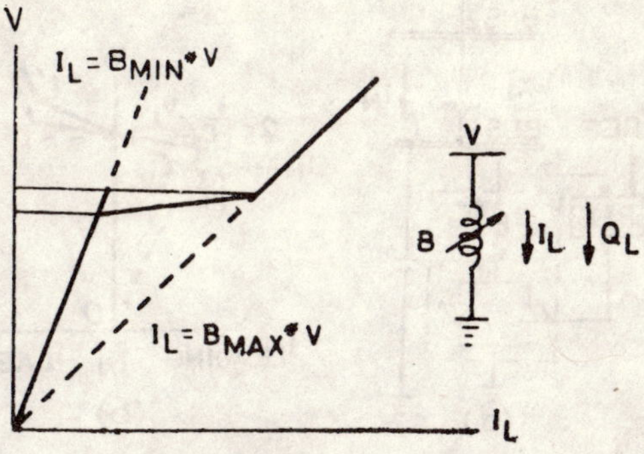

Fig. 8.17. (a) **Static characteristics of controlled reactor voltage versus current I_L.**

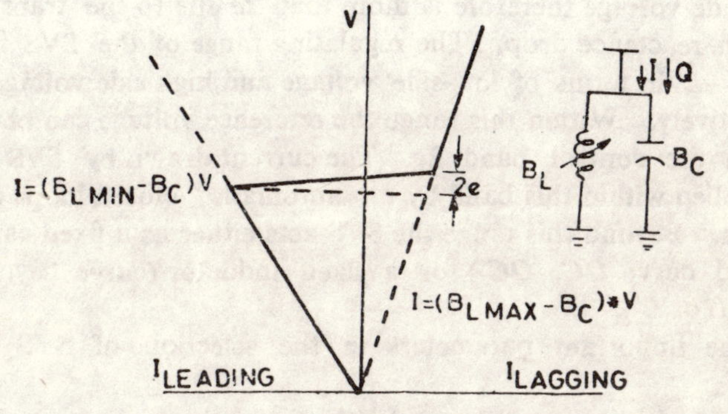

Fig. 8.17. (b) **Steady state SVS characteristics voltage versus net SVS current, I_L**

systems. However, the controlled voltage can be either from the low voltage side as in Fig. 8.18 (a) or from the HVAC bus derived through a suitable potential transformer. The $V-I$ characteristic for the configuration of Fig. 8.18 (a) is shown in Fig. 8.18 (b). Two curves are shown in this figure; one relating high side bus voltage (V_2) to I_N and the other relating low side voltage (V_1) to I_N. The difference between the end points of the two characteristics is the voltage drop across the transformer due to I_N flowing through it. The characteristics in Fig. 8.18 (b) have been drawn assuming that the SVS has been designed to control the low side voltage within the desired 2e voltage band. The overall regulation of the

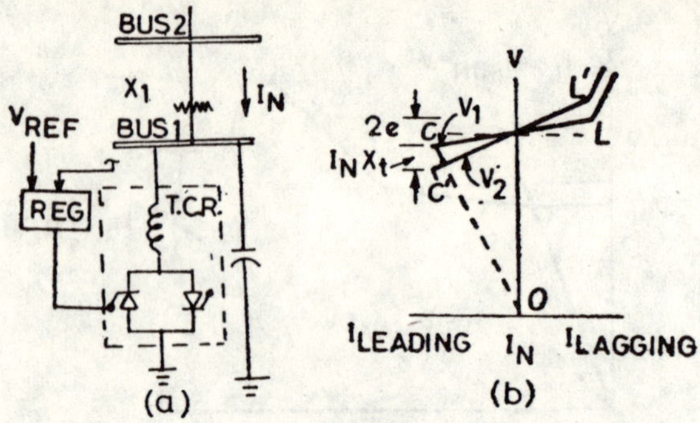

Fig. 8.18. SVS with Low-side (V_1) Voltage control;
(a) Schematic of SVS.
(b) Steady state characteristics.

high side voltage therefore is more than $2e$ due to the transformer leakage reactance drop. The regulating range of the SVS is C–L and C'–L' in terms of low-side voltage and high side voltage band respectively. Within this range the reference voltage can be varied within the control band $2e$. The current drawn by SVS can be controlled within this band by an automatic closed-loop control system. Beyond this range the SVS acts either as a fixed capacitor (dotted curve OC, OC') or a fixed inductor (curve beyond the point L or L').

Some important parameters in the selection of SVS are as follows:

1. Maximum capacitive and inductive ratings which decide the end points of characteristics shown in Fig. 8.18 (b).
2. Desired band of voltage control.
3. The zone within the control band wherein the SVS will stand by in the steady state.
4. Speed of response.

The SVS when applied to the voltage support [3] improves both transient and dynamic stability, reduces the insulation levels due to reduction of overvoltages and provides a fast control of reactive power.

8.1.5 Reduction of Reactive Power for Three-phase Converter Systems

Presently, thyristor converters are widely used in many industrial applications such as paper and textile mills, rolling mills in steel industries, electro-chemical, metallurgical industries etc. The reactive power requirement increases with an increasing number of thyristor controlled loads operating on the AC supply system. The increase in the reactive power is caused by two components of supply current: (1) Fundamental component owing to the displacement in phase with respect to the corresponding phase voltage. (2) Harmonic distortion current which is generated in the AC lines since the current drawn by thyristor converters is non-sinusoidal.

The power factor deteriorates and hence the reactive power consumption increases with line-commutated phase controlled thyristor converters at large phase angle delays. The power factor can be improved significantly by generating a source of reactive power externally and compensating most of the load reactive power [6]. It is possible to compensate even the reactive power corresponding to the harmonics drawn by thyristor converters. In this system, the instantaneous values of reactive component flowing into the converter circuit are measured during each half-cycle of line frequency. On the basis of the measured values, the compensation current for each phase is calculated and generated by a regulated reactive power source in a closed-loop control. Then, the generated compensation currents are injected into the input lines of the converter circuit.

Figure 8.19 shows the block diagram of a three-phase reactive power compensator which is connected in parallel to the converter but supplied from the same AC supply lines. The supply voltages may be expressed as

$$\left.\begin{aligned} v_a &= \sqrt{2}\,V_s \sin \omega t \\ v_b &= \sqrt{2}\,V_s \sin\left(\omega t - \frac{2\pi}{3}\right) \\ v_c &= \sqrt{2}\,V_s \sin\left(\omega t - \frac{4\pi}{3}\right) \end{aligned}\right\} \tag{8.18}$$

where V_s is the supply rms phase voltage. Let the three-phase load currents be $i_1 = (i_{1a}, i_{1b}, i_{1c})$, each phase current consisting of two components. That is,

$$\begin{bmatrix} i_{1a} \\ i_{1b} \\ i_{1c} \end{bmatrix} = \begin{bmatrix} i_{pa} \\ i_{pb} \\ i_{pc} \end{bmatrix} + \begin{bmatrix} i_{qa} \\ i_{qb} \\ i_{qc} \end{bmatrix} \tag{8.19}$$

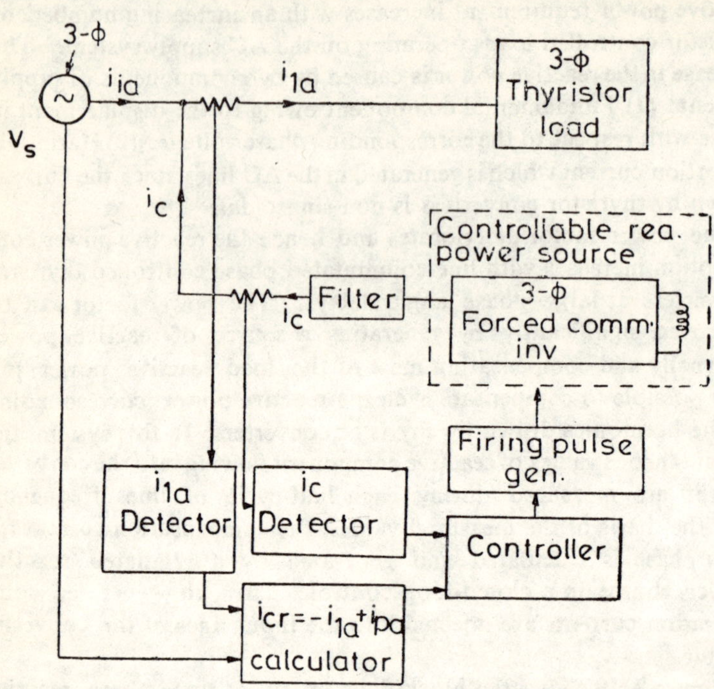

Fig. 8.19. Basic block diagram of a three-phase reactive power compensator (shown for phase, *a*).

The first vector on the right represents real component of currents which has the same frequency and phase as the supply voltage, i.e.,

$$\begin{bmatrix} i_{pa} \\ i_{pb} \\ i_{pc} \end{bmatrix} = \begin{bmatrix} I_p \cdot \sin \omega t \\ I_p \cdot \sin \left(\omega t - \frac{2\pi}{3}\right) \\ I_p \cdot \sin \left(\omega t - \frac{4\pi}{3}\right) \end{bmatrix} \tag{8.20}$$

where I_p is the peak of each line current, and a balanced three-phase load is assumed.

The second term on the right hand side of eqn. (8.19) represents the reactive components of load current consisting of the fundamental ($I_1 \sin \psi$ for phase a) and higher order frequency components owing to non-sinusoidal waveform.

From eqn. (8.19), the required compensation currents i_c which should be injected into the input supply lines (Fig. 8.19) for reducing the reactive power are

$$\begin{bmatrix} i_{ca} \\ i_{cb} \\ i_{cc} \end{bmatrix} = \begin{bmatrix} -i_{qa} \\ -i_{qb} \\ -i_{qc} \end{bmatrix} = \begin{bmatrix} -i_{1a} + i_{pa} \\ -i_{1b} + i_{pb} \\ -i_{1c} + i_{pc} \end{bmatrix} \tag{8.21}$$

When these compensation currents are added to the load currents, we obtain the supply currents $i_l = (i_{la}, i_{lb}, i_{lc})^t$ which contain only real components. Thus from eqns. (8.19) and (8.21),

$$\begin{bmatrix} i_{la} \\ i_{lb} \\ i_{lc} \end{bmatrix} = \begin{bmatrix} i_{1a} + i_{ca} \\ i_{1b} + i_{cb} \\ i_{1c} + i_{cc} \end{bmatrix} = \begin{bmatrix} i_{pa} \\ i_{pb} \\ i_{pc} \end{bmatrix} \tag{8.22}$$

Equation (8.22) shows that the power factor can be made unity. The block diagram of the compensation scheme is illustrated in Fig. 8.19. The required value of compensation current i_{cr} is computed using multipliers, adders, sample and hold functions. This is compared with the compensation current i_c. The controller then sets appropriate trigger pulses to the reactive power regulator which generates the required value of i_c and feeds to the ac line. The reactive power source is a three-phase inverter with a reactor at the output terminals. Pulsewidth modulation is employed in the control of the inverter. The function of the filter is to eliminate the modulation harmonics of the three-phase inverter. The compensator can respond very fast with a maximum correction time of one-half cycle of supply frequency (10 m sec.), and can compensate the load reactive power even during transient changes of the load current.

If the load is unbalanced, the load current consists of positive and negative sequence currents. The zero sequence currents are usually not present. It can be shown that the average power contributed by the negative sequence currents will be zero assuming that the supply voltages are balanced with positive sequence. Therefore the negative sequence currents can be considered as a kind of reactive components. By changing the compensation currents on the basis of positive sequence real components of the load current, the load currents can be balanced out at the supply terminals. As a result, the reactive power from the ac supply systems is either eliminated or reduced to a minimum possible level. The supply currents become almost sinusoidal making the supply power factor close to unity [5, 6].

8.1.6 Conclusions

It is possible to change the current flow through the static shunt compensator using thyristor-controlled inductor from maximum leading to maximum lagging in one-half cycle of the supply frequency. The thyristor-controlled compensator is finding widespread application not only in industrial power systems for the load compensation but also in the utility systems for the transmission line voltage regulation. The major problem is related to accurate and reliable measurement of signals such as voltage, current etc., and their processing for deriving power, reactive power etc. Considering the problems of measurements and control of these signals, the thyristor controlled reactor compensator can provide a response time within 20 to 30 m sec [3]. Such a fast time response for reactive power control is definitely not possible with other reactive power compensation schemes.

If the load consists of thyristor converters, the supply current becomes non-sinusoidal. It may not be possible to compensate the entire load reactive power by using TCR along with a fixed capacitor. However, an external reactive power source comprising of a forced commutated inverter and a fixed reactor may eliminate the reactive power completely. The forced commutated inverter which operates under pulsewidth modulation at a high frequency may generate higher order harmonics. But these can be filtered out using small-sized filter components.

8.2 POWER FACTOR IMPROVEMENT AND HARMONIC CONTROL OF CONVERTER FED SYSTEMS

Thyristor converters are now widely used because of their high efficiency, minimum maintenance, fast response, less space, light weight and flexibility in control. However, thyristor converters have poor power factor and draw nonsinusoidal current from ac lines. The harmonics that are generated interfere with other loads and the supply system. Hence attempts are being made to control the converter operation, so as to reduce the harmonics and to improve the power factor.

In ac/dc converters the supply voltage can be assumed to be sinusoidal and the line current is nonsinusoidal. In such a system p.f. is given by

$$\text{P.F.} = \frac{\text{Active power}}{\text{Apparent power}} = \frac{VI_1 \cos \phi_1}{VI} \tag{8.23}$$

where V = rms value of supply voltage; I = rms line current; I_1 = rms value of fundamental component of line current and ϕ_1 = phase difference between V and I_1

Now $$\text{P.F.} = \left(\frac{I_1}{I}\right) \cos \phi_1 = \mu \cos \phi_1 \tag{8.24}$$

where μ is called the distortion factor and $\cos \phi_1$ the displacement factor or fundamental p.f.

P.F. can be improved either by increasing μ or $\cos \phi$ or both of them. μ is also a measure of harmonic content in line current. Its higher value indicates lower magnitude of harmonics in the line current.

Single phase fully controlled and half controlled converters are shown in Fig. 8.20. The waveforms of the supply voltage v_S, load voltage v_d, line current i_s and the fundamental component of line current i_{s1} are shown in Fig. 8.21. The load current has been assummed to be a perfect dc. As seen in these figures, the displacement angle ϕ_1 is the same as α for a fully controlled converter and $\alpha/2$ for a half controlled converter. In both the converters as α is increased to reduce the output voltage, the displacement factor

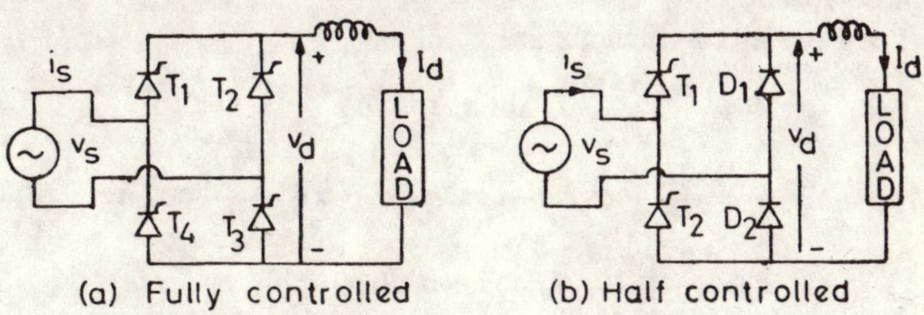

Fig. 8.20. Single-phase converters.

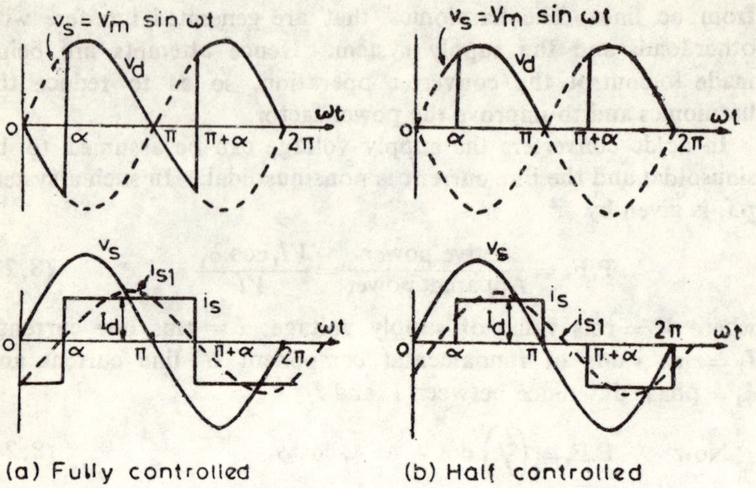

Fig. 8.21. **Source voltage and current, and load voltage, waveforms of 1-phase converters**

decreases, reaching a zero value when the output voltage is 0. Furthermore, due to non-sinusoidal waveform of line current the distortion factor, and therefore, the p.f. will always be less than unity.

Some important performance expressions for these converters are derived below.

(a) FULLY-CONTROLLED CONVERTER

The line current waveform can be described by the Fourier series of eqn. (7.70). Due to the symmetry of the line current waveform between positive and negative half cycles, even harmonics will be absent.

The Fourier co-efficients are

$$A_n = \frac{2}{\pi}\int_{\alpha}^{(\pi+\alpha)} I_d \sin n\omega t \; d(\omega t)$$

$$= \frac{2I_d}{n\pi}[\cos n\alpha - \cos n(\pi+\alpha)] = \frac{4I_d}{n\pi}\cos n\alpha \qquad (8.25)$$

$$B_n = \frac{2}{\pi}\int_{\alpha}^{(\pi+\alpha)} I_d \cos n\omega t \; d(\omega t)$$

$$= \frac{2I_d}{n\pi}[\sin n(\pi+\alpha) - \sin n\alpha] = -\frac{4I_d}{n\pi}\sin n\alpha \qquad (8.26)$$

The rms *n*th harmonic is

$$I_n = \left(\frac{A_n^2 + B_n^2}{2}\right)^{1/2} = \frac{2\sqrt{2}\, I_d}{n\pi} \tag{8.27}$$

The displacement angle of *n*th harmonic is given by

$$\phi_n = \tan^{-1}\frac{B_n}{A_n} = -n\alpha \tag{8.28}$$

Values of I_1 and ϕ_1 are obtained from eqns (8.27) and (8.28),

$$I_1 = \frac{2\sqrt{2}\, I_d}{\pi},\ \phi_1 = \alpha \tag{8.29}$$

$$\text{R.M.S. line current} = I_d \tag{8.30}$$

From eqns. (8.24), (8.29) and (8.30)

$$\text{P.F.} = \frac{2\sqrt{2}}{\pi} \cos\alpha \tag{8.31}$$

Equations (8.27) and (8.28) can be employed to obtain the magnitudes of the various harmonics present in the line current.

The average output voltage V_d is given by

$$V_d = \frac{1}{\pi}\int_{\alpha}^{(\pi+\alpha)} V_m \sin\omega t\, d(\omega t) = \frac{2V_m}{\pi}\cos\alpha$$

If the maximum value of V_d is denoted by V_{d0} then,

$$\text{the per unit output voltage} = \frac{V_d}{V_{d1}} = \cos\alpha \tag{8.32}$$

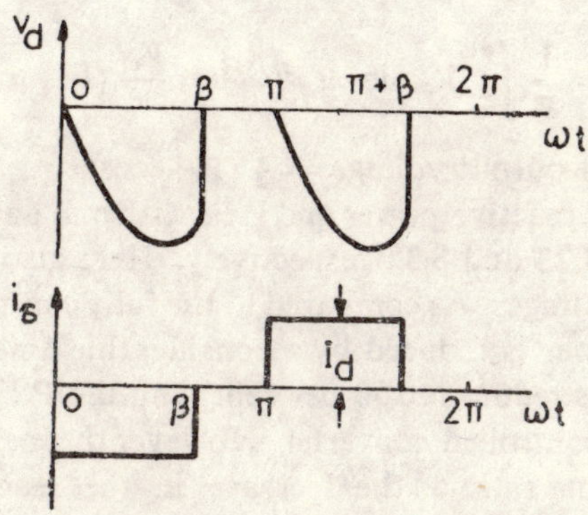

Fig. 8.22. Waveforms during inversion for a fully controlled converter with controlled flywheeling.

The variation of reactive power with per unit output voltage is shown in Fig. 8.23 (semi-circle *ABC*). The variation of P.F. with per unit output voltage is shown in Fig. 8.35. The reactive power is the highest at the zero output voltage and the P.F. is low for the low values of output voltage.

(b) Half Controlled Converter

The performance expressions can be obtained in the same way as for the fully controlled converter,

$$A_n = \frac{2}{\pi}\int_0^{\pi} I_d \sin n\omega t \, d(\omega t) = \frac{2I_d}{n\pi}(1+\cos n\alpha) \tag{8.33}$$

$$B_n = \frac{2}{\pi}\int_0^{\pi} I_d \cos n\omega t \, d(\omega t) = \frac{2I_d}{n\pi}\sin n\alpha \tag{8.34}$$

$$I_n = \frac{2\sqrt{2}}{n\pi} I_d \cos\frac{n\alpha}{2} \text{ and } \phi_n = -\frac{n\alpha}{2} \tag{8.35}$$

$$I_1 = \frac{2\sqrt{2}}{\pi} I_d \cos\frac{\alpha}{2} \text{ and } \phi_1 = -\frac{\alpha}{2} \tag{8.36}$$

$$I = \left[\frac{1}{\pi}\int_{\alpha}^{\pi} I_d^2 \, d(\omega t)\right]^{1/2} = I_d\left(\frac{\pi-\alpha}{\pi}\right)^{1/2} \tag{8.37}$$

From eqns. (8.24), (8.36) and (8.37),

$$\text{P.F.} = 2\left[\frac{2}{\pi(\pi-\alpha)}\right]^{1/2}\cos^2\alpha/2 \tag{8.38}$$

The average output voltage is

$$V_d = \frac{1}{\pi}\int_{\alpha}^{\pi} V_m \sin\omega t \, d(\omega t) = \frac{V_m}{\pi}(1+\cos\alpha)$$

and

The per unit output voltage $= \frac{1}{2}(1+\cos\alpha)$ (8.39)

Variations of reactive power and P.F. with p.u. output voltage are shown in Figs. 8.23 and 8.35 respectively. Here also the P.F. is low at low output voltages. As compared to the fully controlled converter, the reactive power is reduced by a considerable amount (maximum reactive power is reduced to 50 per cent) and the P.F. is also improved in the half controlled converter. However the increase in the P.F. is not in the same ratio as the decrease in the reactive power due to the simultaneous decrease in distortion factor. Due to discontinuous current, harmonic content in the source current increases

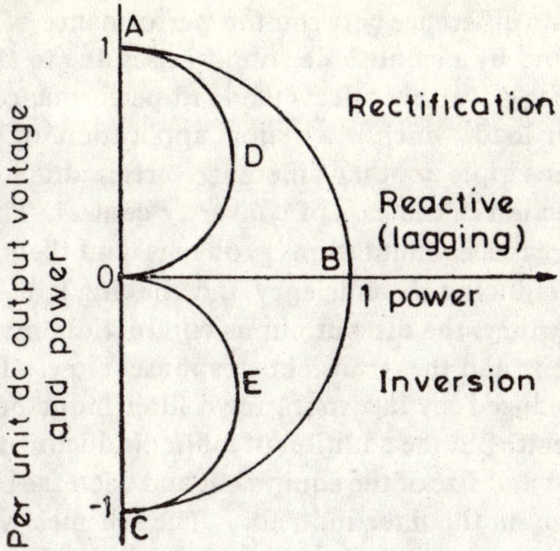

Fig. 8.23. Variation of reactive power with DC output voltage:
(a) Fully controlled converter: semicircle *ABC*
(b) Half controlled converter: semicircle *ADO*
(c) Fully controlled converter with controlled flywheeling } semicircle *ADO* and *OEC*.

causing a decrease in the distortion factor. If the dc motor performance is examined it is found that a considerable reduction in the armature current ripple and in the zone of discontinuous conduction is realised thus giving a substantial improvement in the motor performance. On the whole the performance of a half controlled converter is superior to that of a fully controlled converter. The main limitation of a half controlled converter, as compared to a fully controlled converter is that it cannot operate as a line commutated inverter and the regenerative braking of the motor is not possible.

A number of schemes have been developed for the improvement of P.F. and reduction of harmonics. Some use natural commutation and some employ forced commutation. Generally the natural commutation is preferred because of reliable operation and the lower equipment cost, though the forced commutation methods can give better P.F. and allow harmonics to be reduced to a greater extent.

In order to have a proper evaluation of the relative merits of the various schemes of power factor improvement and harmonic reduction, one should also consider their effect on the performance of load. The main difference between the performance of a load fed by a converter and by a smooth dc supply arises due to the presence of ac ripple which adversely affects the load performance. In the case of a dc motor load, which is a major application of converters, in addition to ac ripple another difference arises due to the discontinuous conduction in the case of converter control. The presence of ac ripple increases commutation problems and the heating of the motor, thus reducing the efficiency and making it necessary to the derate it. Further, the discontinuous conduction makes the speed regulation poor and the transient response slow. These adverse effects are reduced by incorporating a filter inductor in the motor armature circuit. But the addition of a filter inductor means increase in cost, weight and size of the equipment and decrease in the efficiency due to losses in the filter inductor. The schemes which keep the ripple within the prescribed limits and allow the elimination of discontinuous conduction with minimum inductance are perferred.

The distortion factor provides information about the harmonic content in the supply current. In order to have a correct evaluation of the effect of harmonics, one should also have information about the predominant harmonics present. If they are of high frequency, a simple line filter can be employed to filter them out. On the other hand, if their frequency is low, they cannot be filtered out without considerable reduction in the efficiency and unless a costly filter is used.

Important performance features of various methods of P. F. control and harmonic reduction will be described here. The detailed comparative study is described in [13].

8.2.1 Methods Employing Natural Commutation [13]

There are three methods of natural commutation.

(a) Use of controlled flywheeling [14–16].

(b) Asymmetrical control [17].

(c) Sequence control and simultaneous control [18–20].

8.2.1.1 *Use of Controlled Flywheeling*

The advantages of the half controlled converter over the fully controlled converter are described above. These advantages of the half controlled converter and the regeneration capability of the fully controlled converter are simultaneously realised by operating the fully controlled converter with the controlled flywheeling. The details of the controlled flywheeling are discussed in Chapter 3. The output voltage and the source current waveforms during rectification are the same as shown in Fig. 8.21(b) for the half controlled converter. The waveforms for inversion are shown in Fig. 8.22. During inversion the output voltage is controlled by varying the angle of disconnection, β. Various performance expressions are given in Table 8.2. The reactive *KVA* now varies along *ADO* during conversion and along *OEC* during inversion (Fig. 8.23). As compared to the fully controlled operation, the maximum reactive *KVA* is reduced to 50 per cent, both for conversion and inversion, and at the zero output voltage it is reduced from the maximum to zero. PF versus per unit output is shown in Fig. 8.35 by the same curve which is applicable to the half controlled operation. During inversion the PF curve will be the mirror image of the one shown for conversion.

8.2.1.2 *Asymmetrical Control*

In this scheme, the converter is operated with different firing angles during positive and negative half cycles of the supply voltage. The output voltage and source current waveforms of a half controlled converter with asymmetrical control are shown in Fig. 8.24. In one half cycle firing angle is α and in the next it is zero.

This scheme has a number of drawbacks. Due to asymmetrical waveform of the line current, dc components and even harmonics are introduced. The dc component is the most undesirable as it causes

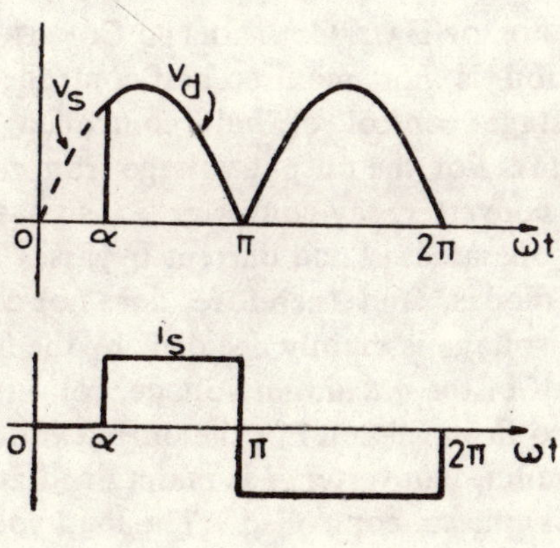

Fig. 8.24. Asymmetrical control.

the saturation of the supply transformers. The dc component can be eliminated if firing angles are altered in successive cycles. Another serious disadvantage is the large increase in the armature current ripple and the zone of discontinuous conduction. A very large value of filter inductance is required to keep the ripple within the permissible limit. Because of these limitations this scheme is not used in practice.

8.2.1.3 *Sequence Control*

These schemes employ more than one converter connected in series. Load voltage is the sum of the output voltages of the individual converters. A scheme using '*m*' converters connected in series is called *m*-stage control.

In *m* stage sequence control the variation of output voltage from 0 to V_{d0}/m, where V_{d0} is the maximum load voltage, is obtained by controlling one converter and bypassing other $(m - 1)$ converters. When the voltage reaches V_{d0}/m, the converter under control will be operating with the firing angle equal to zero. For variation of voltage from V_{d0}/m to $2\ V_{d0}/m$, second converter is also brought under control and the remaining $(m - 2)$ converters are bypassed. Thus, as the load voltage requirement increases more and more converters are activated and finally all the converters will be in operation when the output voltage is in the range of $\frac{(m-1)\ V_{d0}}{m}$ and V_{d0}. Due to the operation of converters in sequence, this is called ***sequence control*** of series converters.

(a) Sequence Control of Half Controlled Converters

When regeneration is not required half controlled bridges are employed. Two stage control of half controlled converters is shown in Fig. 8.25(a). For the output voltage from zero to half of the maximum, one converter, say converter *I*, is phase controlled and another is not operated. Load current bypasses converter *II* through the two diodes, and therefore, does not contribute any voltage. The load voltage is mainly decided by the firing angle of converter *I*. At half of the maximum voltage, converter *I* will be at full advance (zero firing angle). For the output voltage more than half of the maximum, converter *I* is maintained at full advance and converter *II* is phase controlled. The load voltage and line current waveforms for this case are shown in Fig. 8.25(b). From 0

to α only converter I is in operation, and therefore, the secondary of transformer connected to this converter is carrying the load current I_d. At α, converter II is activated by firing the appropriate thyristor. The load voltage jumps to the sum of voltages of the two converters. Since the load current now flows through the transformer secondary connected to II bridge also, total ampere turns of the secondaries double causing the source current to jump to twice the value.

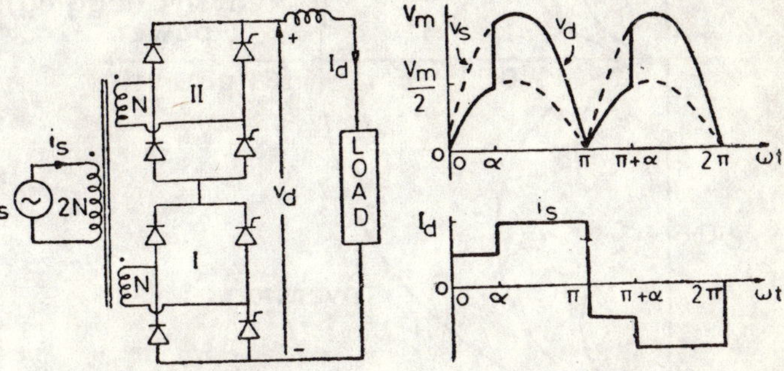

(a) (b)

Fig. 8.25. Two stage sequence control of half controlled converters:
(a) Two half controlled converters in series.
(b) Load voltage and source current waveforms.

Let the number of turns in the primary and secondary be $2N$ and N respectively. If the peak of the source voltage is V_m, then the peak of ac voltage input to each converters will be $V_m/2$. If the load current is I_d, then the source current will be $I_d/2$ when only one converter is conducting and I_d when both the converters are conducting.

The performance equations for the output voltage less than half (when only one converter is in operation) can be obtained from eqns. (8.33) to (8.39) by replacing V_m by $V_m/2$ and I_d by $I_d/2$. For output voltages greater than half they are derived from the waveforms of Fig. 8.25(b) and are listed in Table 8.2. Per unit output voltage and real power versus reactive power variation is shown in Fig. 8.26 by semicircles 4 and 5. For comparison reactive power for single stage half and fully controlled converters is also shown by semicircle 1 and 2 respectively. As compared to the single stage half controlled converter, the maximum reactive power is

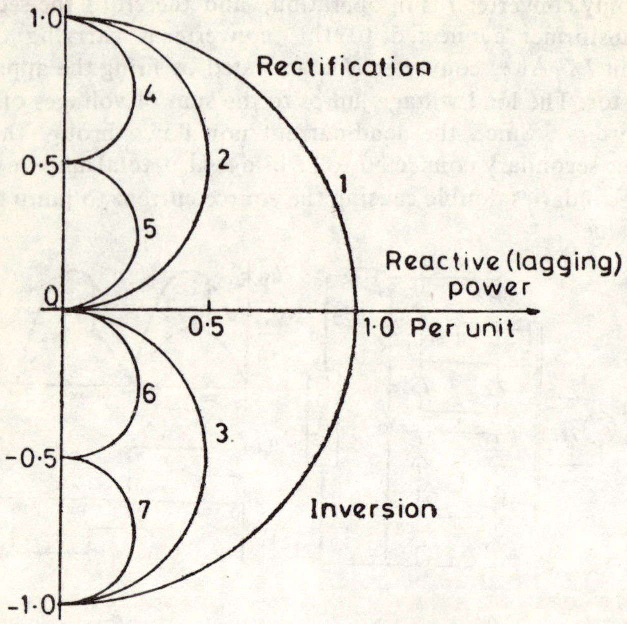

Fig. 8.26. Variation of reactive power with DC output voltage:

(a)	Two stage sequence control of half controlled converters.	Semicircles 4, 5
(b)	Two stage sequence control of fully controlled converters.	,, 2, 3
(c)	Two stage sequence control of fully controlled converters with controlled flywheeling.	,, 4, 5, 6, 7
(d)	Single stage half controlled converter.	,, 2
(e)	Single stage fully controlled converter.	,, 1

reduced to half and the reactive power at the zero output voltage is reduced from maximum to zero. P.F. versus per unit output voltage curve is shown in Fig. 8.35. Considerable improvement in P.F. is obtained compared to the single stage control.

Sequence control also permits considerable reduction in harmonics. Also the ripple in motor current is reduced and the value of inductance required to make current continuous is lower.

Further improvement in P. F. and reduction in source current harmonics is obtained by using more than two bridges in series. It is a common practice to use more than two bridges in series in traction application.

The harmonic content in source current is minimised by operating two series connected converters with simultaneous control [28]. In the simultaneous control both the bridges remain in operation in the whole range of the output voltage. Their firing angles α_1 and α_2 for a given output voltage are chosen so as to minimise the harmonic content. The control becomes complicated compared to the sequence control.

(b) SIMULTANEOUS CONTROL OF FULLY CONTROLLED CONVERTERS;

When regeneration is required, fully controlled bridges are employed. In simultaneous control, two or more series connected bridges are employed. Two stage simultaneous control circuit is shown in Fig. 8.27(a). In this case both the bridges remain in operation. The output voltage is controlled by keeping one converter fully advanced or fully retarded and controlling the firing angle of the other converter. For a positive output voltage let bridge *I* is chosen to be fully advanced as shown in Fig. 8.27(b). Now as the firing angle of bridge *II* is controlled from $\alpha = 0$ to 180°, the output voltage will change from 1 p.u. to 0. Now bridge *II* is fully retarded and the firing angle of bridge *I* is controlled. As the firing angle of bridge *I* is changed from 0 to 180° the output voltage will vary from 0 to -1 p.u.

The load voltage and source current waveforms for rectification and inversion are shown in Figs. 8.27(b) and (c). The output voltage and source current waveforms are identical to those with a single stage fully controlled converter with controlled flywheeling discussed in sec. 8.2.1.1. The reactive power variation is also the same and is shown in Fig. 8.26 by semicircles 2 and 3. Variation of P.F. with output voltage is also the same and is shown in Fig. 8.35. This comparison shows that the use of the controlled flywheeling in the single stage fully controlled converter gives the same performance as the two stage simultaneous control of fully controlled converter and also retains the regeneration capability.

(c) SEQUENCE CONTROL OF FULLY CONTROLLED CONVERTER WITH CONTROLLED FLYWHEELING [20]

In this case two or more fully controlled bridges connected in series are employed and they are operated with the controlled

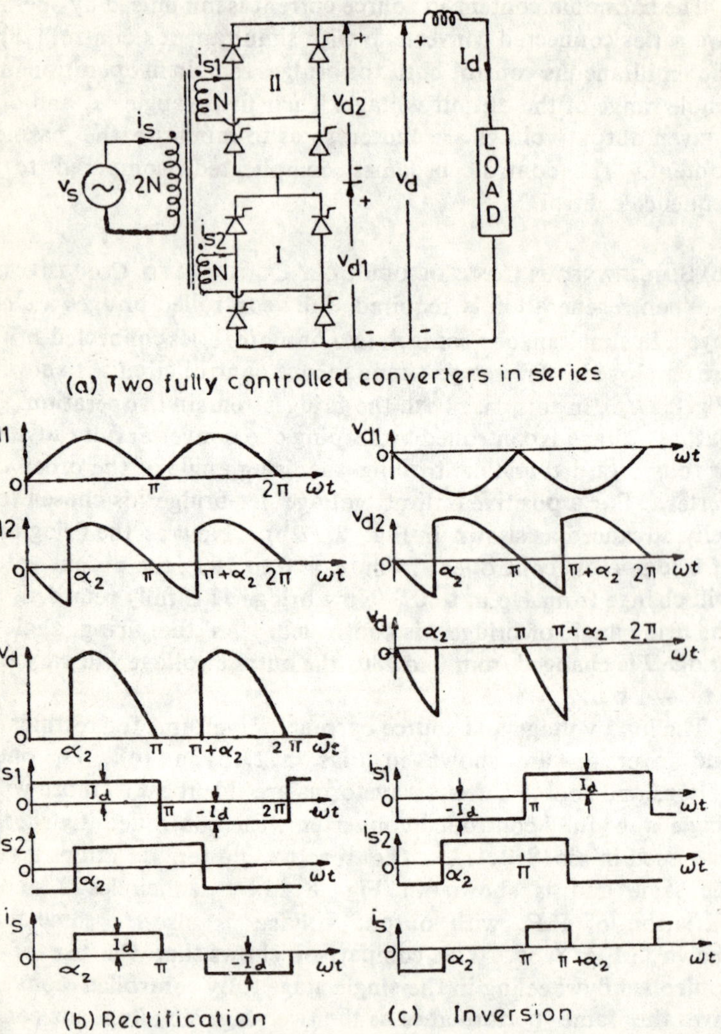

(a) Two fully controlled converters in series

(b) Rectification (c) Inversion

Fig. 8.27. Two stage sequence control of fully controlled converters.

flywheeling. Due to the use of fully controlled converters regeneration can be obtained.

Two stage control will have the circuit shown in Fig. 8.27(a). For output voltages between 0.5 p.u. and 1, one of the bridges, say *I*, is maintained at full advance and another is operated with the controlled flywheeling. The load voltage and source current waveforms are the same as shown in Fig. 8.25(b) for the two stage sequence control of half controlled converters. For output voltages between 0.5 p.u. and 0, converter *I* is operated in the rectification mode, and for output voltages between 0 and −0.5 p.u., it is operated in the inversion mode and converter *II* is bypassed by turning on the series connected thyristor pairs. The relevant waveforms are shown in Figs. 8.21(b) and 8.22 respectively. For the output voltage equal to −0.5 p u., converter *I* will be in operation with the angle of disconnection β (Fig. 8.22) equal to 180°. Now converter *II* is also operated in the inversion mode and the output voltage varies from −0.5 p.u. to −1 p.u. when its angle of disconnection β changes from 0° to 180°. The load voltage and the source

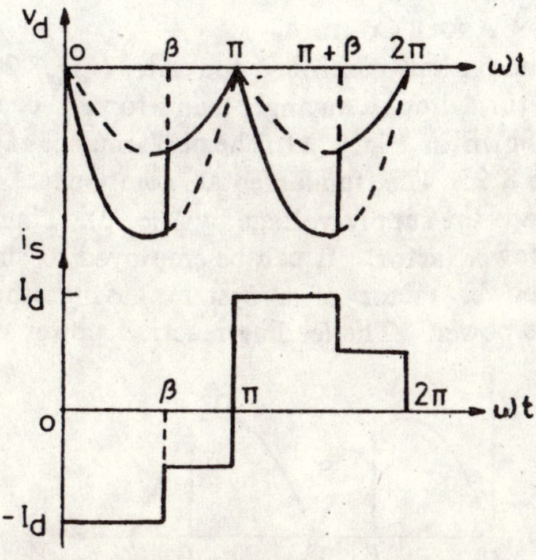

Fig. 8.28. Output voltage waveforms of sequence control of converter with controlled flywheeling for $-0.5 > V_d > -1$ p.u.

current waveforms for this case are shown in Fig. 8.28. Various performance expressions are given in Table 8.2.

The output voltage and real power versus reactive power curves are shown by semicircles 4, 5, 6 and 7 in Fig. 8.26. As compared to the two stage simultaneous control of fully controlled converters, the maximum reactive power is only half and at the output voltages of ±0.5 p.u., it is reduced to 0. The harmonic analysis of the source current shows that the total harmonic content is increased. However, this is not a serious disadvantage because the increase is mainly in the high frequency harmonics which can be filtered out using an inexpensive filter. P.F. versus output voltage curves are shown in Fig. 8.35. There is an appreciable improvement in the P.F. compared to the two stage simultaneous control of fully controlled converters.

8.2.2 Methods Employing Forced Commutation [1]

The following forced commutation methods are described here:

(a) Extinction angle control [21].
(b) Symmetrical pulse width modulation [21, 23-25]
(c) Selective harmonic elimination [25-28].
(d) Sinusoidal pulse-width modulation [28].
(e) Sequence control with forced commutation [29].

(a) Extinction Angle Control

In this scheme the half controlled converter (Fig. 8.20(b)) is used. Thyristor T_1 is turned on at an angle 0 and forced commutated at an angle β as shown in Fig. 8.29. The performance expressions are given in Table 8.2. The fundamental component of the source current now leads the supply voltage by $\frac{1}{2}(\pi-\beta)$. Thus, the scheme has a leading power factor. It can be employed to simultaneously improve the power factor of a system having other loads with lagging reactive power. The leading reactive power variation will

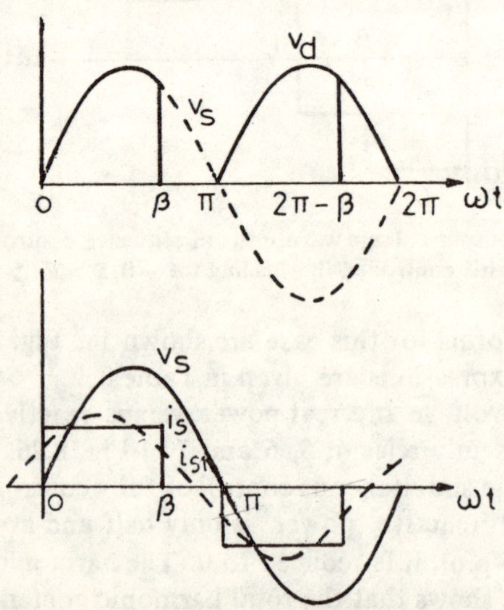

Fig. 8.29. Extinction angle control waveforms.

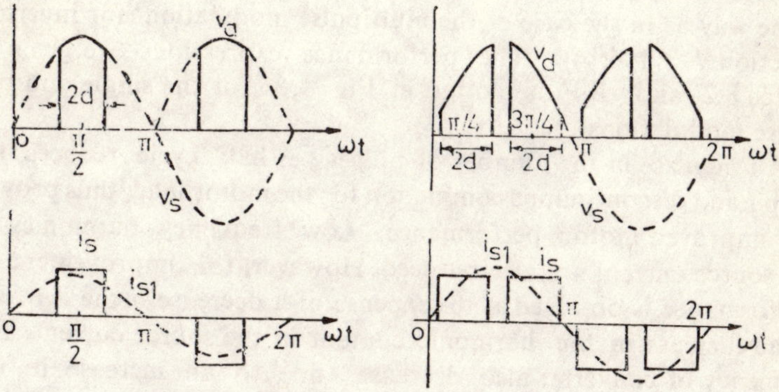

Fig. 8.30. Symmetrical pulse-width modulation.

be the mirror image of the semicircle ADO of Fig. 8.23 and the power factor variation is shown in Fig. 8.35. Thus, in all respect the behaviour of the extinction angle control is the same as phase control, except that it has a leading P.F. and the phase angle control has a lagging p.f.

(b) SYMMETRICAL PULSE WIDTH MODULATION (SYMMETRICAL PWM)

Symmetrical pulse width modulation consists of equidistant pulses of equal widths, consisting of one or more than one pulse per half cycle, symmetrically located with respect to angle $\pi/2$. Fig. 8.30 shows the load voltage and source current waveforms for single and two pulse modulations. A half controlled or a fully controlled converter with an arrangement of forced commutation for its thyristors is employed. In the single pulse modulation with the half controlled bridge, during the positive half cycle thyristor T_1 is turned on at $(\pi/2-d)$ and forced commutated at $(\pi/2+d)$, giving a pulse of width $2d$ centred at $\pi/2$. Thyristor T_2 is controlled in the same way during the negative half cycle. In the two pulse modulation, during the positive half cycle thyristor T_1 is turned on at $(\pi/4-d)$ and forced commutated at $(\pi/4+d)$; it is turned on again at $(3\pi/4-d)$ and forced commutated at $(3\pi/4+d)$. Thus the pulses of width $2d$ located at $\pi/4$ and $3\pi/4$ are obtained. Thyristor T_2 is controlled in the same way during the negative half cycle.

For all the output voltages, the fundamental component in source current remains in phase with the source voltage, giving a displacement factor equal to unity and zero reactive power.

Harmonic components in source current can be derived in the

same way as in the case of the multipulse modulation for inverters (Section 7.2.1.4(b)). The performance expressions are given in Table 8.2 and P.F. is plotted in Fig. 8.35 for the single and two pulse modulations.

An increase in the number of pulses per half cycle reduces the ripple and discontinuous conduction for the motor load, thus providing improved motor performance. Low frequency harmonics in the source current are also reduced. However, this improvement in performance is obtained at the expense of a decrease in the P.F. due to an increase in the harmonic content in the source current. The efficiency of converter also decreases due to an increase in the switching losses. Furthermore, because of the thyristor turn off time limitations, commutation is not reliable with a large number of pulses per half cycle.

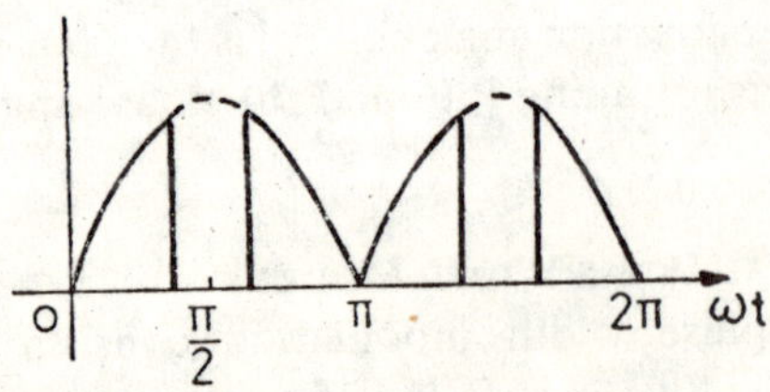

Fig. 8.31. Two pulse modulation with one forced commutation per half cycle.

A case of some interest is the modulation shown in Fig. 8.31. This can be named two pulse modulation with a single forced commutation per half cycle. A comparison of this modulation with the extinction angle control and the single pulse modulation of Fig: 8.30 gives some interesting information. The single pulse modulation gives the highest P.F. and the least amount of harmonics in the source crrrent, however the load current ripple is the highest and the largest value of filter inductance is required to eliminate discontinuous conduction in motor load. Thus, this modulation is the best as far as its effect on the supply is concerned and is the worst from the consideration of the motor performance. On the other hand, the two pulse modulation with one forced commutation per half cycle (Fig. 8.31) gives the minimum ripple in the load current and the inductance required to eliminate discontinuous conduction in motor load is also minimum. However, the P.F. is the lowest and the harmonic content is the highest. Thus, this modulation is the best from the consideration of the load performance and the worst from the consideration of the source performance.

(c) SELECTIVE HARMONIC ELIMINATION METHOD

In this scheme, certain unwanted harmonics are eliminated from the supply current. Basic principle is as follows:

For the elimination of any M harmonics, M pulses per quarter cycle are chosen. Let these pulses be located at positions $B_1, B_2, \ldots,$

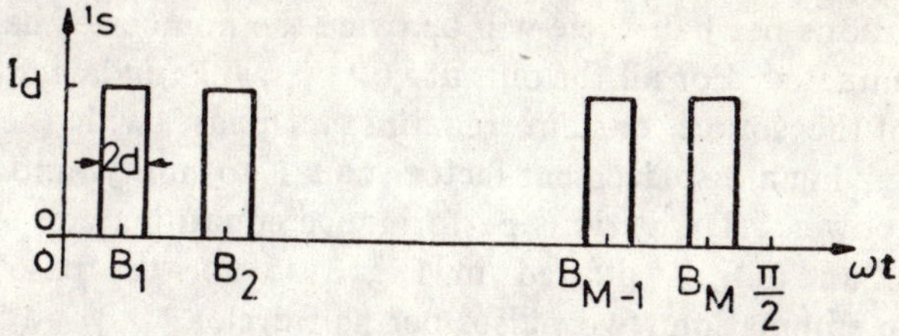

Fig. 8.32. Selective harmonic elimination.

B_M. Each pulse has a width $2d$. Waveforms of the source current during a quarter cycle is shown in Fig. 8.32. The amplitude of nth harmonic in the source current can be derived in the same way as eqn. 7.38(a) for the inverter case. Thus

$$I_n = \frac{8I_d}{n\pi} \sin nd \sum_{k=1}^{M} \sin nB_k \tag{8.40}$$

M harmonics can thus be eliminated by equating eqn. (8.40) to zero

$$\sum_{k=1}^{M} \sin nB_k = 0 \tag{8.41}$$

Eqn. 8.41 yields M nonlinear algebraic equations with M unknowns and they have to be solved for $B_1, B_2, \ldots, B_M$. The pulses are now located at $B_1, B_2, \ldots, B_M$ and symmetrically in the other quarter cycles. The output voltage is varied by adjusting the width $2d$ of the pulses.

Example: To illustrate the procedure, elimination of 5th, 7th and 11th harmonics is considered. As three harmonic components have to be eliminated, three pulses per quarter cycle have to be considered. Thus $M = 3$, $n = 5$, 7 and 11, and from eqn. (8.41)

$$\sin 5B_1 + \sin 5B_2 + \sin 5B_3 = 0$$
$$\sin 7B_1 + \sin 7B_2 + \sin 7B_3 = 0$$
$$\sin 11B_1 + \sin 11B_2 + \sin 11B_3 = 0 \tag{8.42}$$

These nonlinear algebraic equations can be solved by Newton Raphson numerical technique on a digital computer. The solution for the above set of equations is as follows:

$$B_1 = 17°,\ B_2 = 36° \text{ and } B_3 = 52°$$

In this approach, the number of pulses and the number of forced commutations per half cycle will be twice the number of harmonics to be eliminated. For all the output voltages, the fundamental component of the source current remains in phase with the source voltage, giving a displacement factor equal to unity and a zero reactive power. The various performance equations are shown in Table 8.2 and P.F. is plotted in Fig. 8.35 for the case of 3rd harmonic elimination (two pulses per half cycle).

The power factor is better than in the symmetrical pulse width modulation with two pulses per half cycle and in the two pulse modulation, though it is lower compared to the symmetrical single pulse modulation. The ripple in the load current and the inductance required to eliminate discontinuous conduction are lower in this scheme compared to the symmetrical single pulse modulation but higher compared to the symmetrical two pulse modulation. The main advantage of this scheme is that one could select the unwanted harmonics and eliminate them. For example, in main line traction one could eliminate those harmonics which have frequencies near to the track circuit frequency thus ensuring that the operation of the signals is not disturbed due to the harmonics generated by converters. Since low frequency harmonics are difficult to filter, this method can be employed to eliminate low frequency harmonics (say 3rd, 5th and 7th), then the filtering of the remaining harmonic can be done using a less expensive filter. The main limitation of this method is that the maximum voltage available with the eliminations of harmonics is limited. For example, for the third harmonic elimination two pulses located at 60° and 120° are obtained. When the pulse has a width $\pi/3$, these two pulses combine to give single pulse modulation. The maximum voltage available with the third harmonic elimination will therefore be 0.866 per unit.

(d) SINUSOIDAL PULSE WIDTH MODULATION (SINUSOIDAL PWM)

Another method of eliminating and reducing low frequency harmonics is the sinusoidal pulse width modulation. In this method, the width of a pulse is made a sinusoidal function of its location. The firing instants of the thyristors of the converters of Fig. 8.20 are decided by the interactions of a rectified sinusoid v of amplitude A_2 with a triangular wave of amplitude A_1 (Fig. 8.33(a)). Both these waveforms are synchronised with respect to mains supply. Converter connects the source to the load for the duration for which the rectified sinusoid v exceeds the triangular wave e, otherwise it freewheels. The waveforms of the load voltage and source current are shown in Fig. 8.33(b) and (c) respectively.

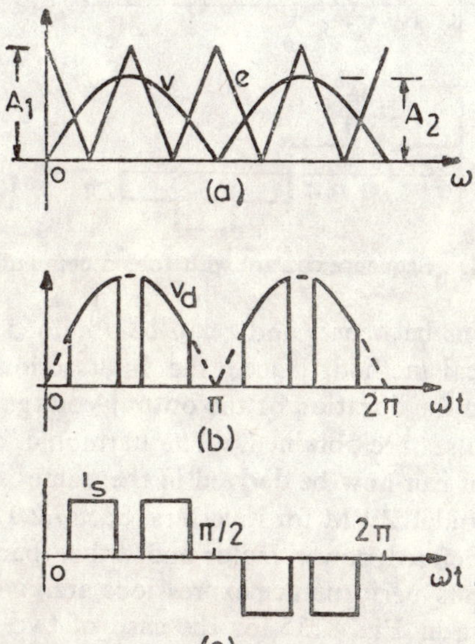

Fig. 8.33. Sinusoidal pulse width modulation.

The output voltage is varied by changing the amplitude of the rectified sinusoid v or the ratio A_2/A_1. The ratio A_2/A_1 is called the modulation index.

The triangular wave can be timed to have either its peak or zero with the zero of the sinusoid. In the former case, for m cycles of the triangular wave in a half cycle of the supply wave, there will be *m* pulses per half cycle. In the later case they will be $(m-1)$. For example in the figure shown there are two cycles of triangular wave in a half cycle of the supply wave giving 2 pulses per half cycle. When the zero of the triangular wave coincides with that of the sinusoid, three cycles of triangle wave, per half cycle of the supply wave, will be required to produce two pulses per half cycle. Both the approaches, for the same number of pulses, provide nearly the same performance, particularly for *m* equal to 5 and above.

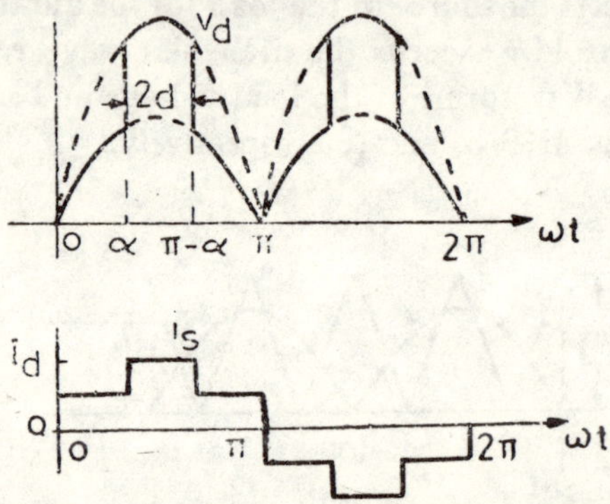

Fig. 8.34. **Sequence control with forced commutation.**

The intersections between *v* and *e* can be obtained using Newton-Raphson numerical method. Once the intersections are known, the location and the duration of the output voltage pulses and the source current pulses are obtained. The harmonic components in the source current can now be derived in the same way as in the case of the sinusoidal PWM for inverters (Sec. 7.2.1.4 (d)). As the pulses are well defined, power factor and other parameters can be evaluated. Various performance expressions are given in Table 8.2 and p.f. is shown in Fig. 8.35 for the case of two pulses per half cycle.

For all the output voltages, the fundamental of the source current remains in phase with the supply voltage, giving a displacement factor equal to unity and zero reactive power. The sinusoidal PWM (SM) with two pulses per half cycle when compared to the two pulse symmetrical PWM (SPM) and the third harmonic elimination method (THE), it is found that the power factor and the ripple in the load current are the maximum in the THE and minimum in SPM. Thus the scheme which gives the best performance at the source terminals gives the worst performance at the load terminals

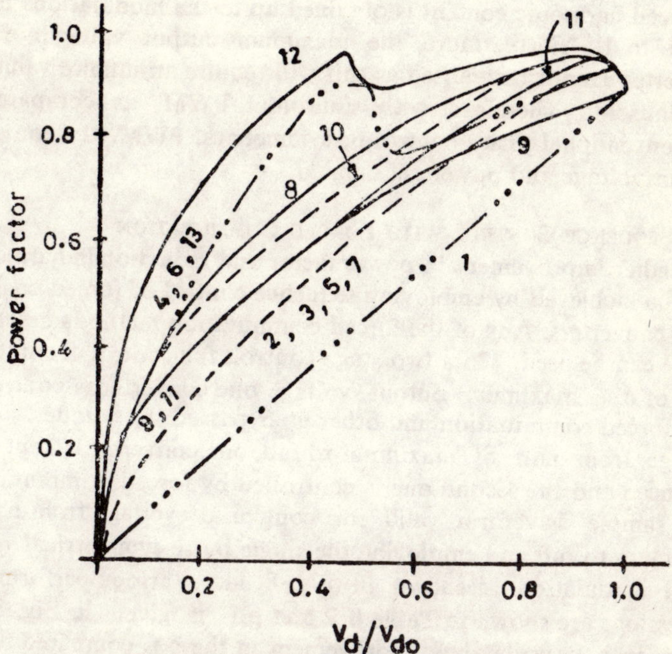

Fig. 8.35. Power factor vs per unit output voltage curves for various converter control techniques.

1. Conventional fully controlled operation.
2. Conventional half controlled operation.
3. Fully controlled converter with controlled flywheeling.
4. Two stage sequence control of conventional half controlled operation.
5. Two stage simultaneous control of fully controlled converters.
6. Sequence control of two stage fully controlled converters with controlled flywheeling.
7. Extinction angle control.
8. Symmetrical PWM with one pulse per half cycle.
9. Symmetrical PWM with two pulses per half cycle.
10. Selective harmonic elimination for third harmonic removal.
11. Sinusoidal PWM with two pulses per half cycle.
12. Sequence control of symmetrical single pulse modulation.
13. Sequence control of two stage converter with extinction angle control.

and vice-versa. SM offers a compromise of performances at both the terminals.

Important features related to harmonic contents are described in section 7.2.1.4 (c) for inverters. These are applicable here also.

Reduced harmonic content is obtained up to the modulation index equal to 1. Furthermore, the maximum output voltage of the converter cannot be realised as this will require an infinite value for the sinusoid. Therefore, with sinusoidal PWM, as compared to the conventional phase control or symmetrical PWM, the converter output voltage and power get derated.

(e) SEQUENCE CONTROL WITH FORCED COMMUTATION

Further improvement in power factor and reduction in harmonics can be achieved by employing sequence control of forced commutated converters. Any of the forced commutation methods described above can be used. In a two stage control, for control from 0 to half of the maximum output voltage one converter is controlled with forced commutation and other is bypassed by diodes. For voltage from half of maximum to full, one converter is kept fully advanced and the second one is controlled by forced commutation. One sample waveform valid for control of voltage from half of maximum to full and employing the single pulse symmetrical pulse width modulation is shown in Fig. 8.34. Various performance expressions are shown in Table 8.2 and p.f. is given in Fig. 8.35. There is a considerable improvement in the p.f. compared to the single stage control.

8.2.3 Methods for Implementation of Forced Commutation

This section describes two methods for the implementation of the pulse width modulation techniques described in the previous section.

Fig. 8.36 shows a single phase fully controlled bridge with forced commutation. It will be used when regeneration is required, otherwise half controlled bridge can be used. Capacitor C, and inductors

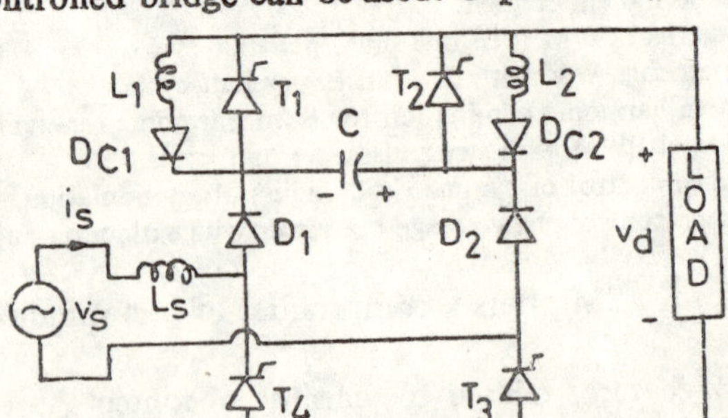

Fig. 8.36. Fully controlled converter for multi-pulse modulation.

L_1 and L_2, which have equal inductance, form a commutation circuit for forced commutation of thyristors T_1 and T_2. Additional diodes D_1 and D_2 in the power circuit and diodes D_{c1} and D_{c2} in the commutation circuit are provided to prevent the capacitor from loosing its charge. The pulse generation and the commutation during the positive half cycle is discussed below.

Let thyristors T_1 and T_3 be conducting initially, providing a positive pulse of source current and a positive pulse of voltage across the load. C is charged with the right hand plate positive during the previous commutation. When this positive pulse of the load voltage (and the source current) is to be terminated, thyristor T_2 is turned on. This connects the negatively charged capacitor across thyristor T_1 and turns it off. The load current, which is assumed constant at I_d, now flows through the path consisting of the source, diode D_1, capacitor C, thyristor T_2, the load and thyristor T_3. Diode D_2 does not conduct immediately as it is reverse biased by the source and the capacitor voltages. C reverses its charge partly due to the flow of load current I_d and partly through the resonance with L_1. When the capacitor voltage becomes greater than the source voltage with the left plate positive, diode D_2 gets forward biased and starts conducting. The current through the source cannot change abruptly due to the presence of the source inductance, therefore it decreases slowly. The load current I_d, which is constant, is shared between two paths; one consisting T_3, the source, D_1, C, T_2 and the load, and the other consisting T_3, D_2, T_2 and the load. When the source current falls to zero, entire load current flows through diode D_2 and this is the end of the commutation interval. The capacitor is now charged with the left plate positive with a voltage greater than the source voltage. Now the source current and the load voltage are zero.

When next pulse is to be produced in the present positive half cycle, thyristor T_1 is triggered. Thyristor T_2 is reverse biased by the capacitor and is turned off. The load current now flows through T_3, D_2, C, T_1 and the load. D_1 does not conduct immediately as it is reverse biased because the voltage across the capacitor is higher than the source voltage. When the capacitor voltage falls below the source voltage, D_1 also conducts and the load current is shared between the paths T_3, the source, D_1, T_1 and the load, and T_3, D_2, C, T_1 and the load. C reverses its charge due to the flow of the load current and the resonance through L_2. When the load current is fully transferred to the source, D_2 stops conducting and the commutation interval ends. Now C is charged with the right plate positive. This polarity is the same which was assumed during the previous current pulse.

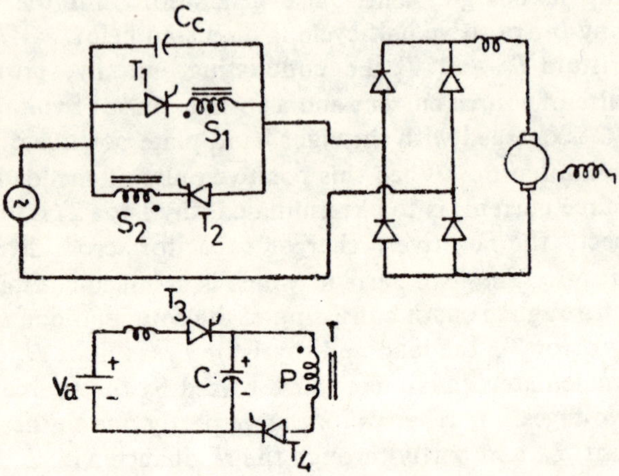

Fig. 8.37. Power circuit for multi-pulse modulation.

When regeneration is not required one can use the circuit shown in Fig. 8.37. It is suitable for single or multiple forced commutations per half cycle. It consists of an ac regulator with an arrangement of forced commutation for its thyristors followed by a diode bridge. During the positive half cycle, T_1 is employed and T_2 is used during the negative half cycle. The commutation circuit consists of thyristors T_3 and T_4, capacitor C, transformer T and the auxiliary source V_a. Secondaries S_1 and S_2 of transformer T are connected in series with thyristors T_1 and T_2 respectively with the polarities as shown in the figure.

To obtain a pulse of the source current and the load voltage, during a positive cycle, T_1 and T_3 are fired simultaneously. This connects the load to the source. At the same time C gets charged, with the polarity as shown in the figure, from the auxiliary source V_a. When the source current pulse is to be terminated, T_4 is fired. This connects C to the primary of transformer T. The secondary S_1 of transformer T reverse biases T_1 and turns it off. Capacitor C_c is provided to offer a path for reverse recovery current of T_1 during commutation. When T_1 turns off, the load current is diverted away from the source to the freewheeling path consisting of diodes.

To get the next pulse of the source current, during the same half cycle, T_1 and T_3 are fired again.

REFERENCES

1. 'Electrical Transmission and Distribution Reference Book', Westinghouse Electric Corporation, USA, Publication.
2. L.O. Barthold, H. Becker, J. Datzell, et al., 'Static Shunt Devices for Reactive Power Control,' CIGRE Report 31-08, 1974.
3. K. Reichert, J. Kauferle and H. Glavitsch, 'Controllable Reactor Compensator for More Extensive Utilization of High Voltage Transmission Systems,' CIGRE Report 31-04, 1974.
4. H. Becker, et al., 'Three-phase Shunt Reactors with Continuously Controlled Reactive Current,' CIGRE Report 31-13, 1972.
5. L. Gyugyi, R.A. Otto, T.H. Putman, 'Principles and Applications of Static Thyristor-controlled Shunt Compensators', IEEE Transactions on PAS, Vol. 97, No. 5, Sept/Oct. 1978, pp. 1935–1945.
6. Kumio Tsuboi, et al., 'Reduction of reactive power and balancing of supply currents for three-phase converter systems', ISPCC–1977 Conference, pp. 365–370.
7. S.R. Doradla and B.K. Patel, 'A thyristor reactive power compensator for fast-varying industrial loads,' Int. J. Electronics, 1981, Vol. 51, No. 6, pp. 763–777.
8. W. Shepherd, 'Thyristor control of AC circuits,' Crosby Lockwood Staples Ltd., England, 1975.
9. S.B. Dewan and A. Straughen, 'Power Semi-conductor 'Circuits,' Wiley-Interscience, New York, 1975.
10. R.L. Hauth and R.J. Maran, 'The performance of Thyristor-controlled static VAr Systems in HVAC Applications. Part I-Fundamental Relationships, 'Conference Record of IEEE/PES Summer Meeting in Los Angeles, California, July 19, 1978.
11. A.E. Hammad and R.M. Mathur, 'A new generalized concept for the design of thyristor phase-controlled VAr compensators. Part I Steady State performance, IEEE Trans. on PAS, Vol. PAS-98, No. 1 Jan./Feb. 1979, pp. 219–226.
12. S.R. Doradla, S. Bhattacharya, M. Krishnamoorty and V. Suryanarayana, "A thyristor controlled reactor (TCR) scheme for compensating fast-changing industrial loads", presented at the national seminar on Power Electronics, Poona, Dec. 1982.
13. H.K. Patel and G K. Dubey, 'Comparative study of single phase converter control schemes', Int. J. Electronics, Vol. 54, No. 1, 1983, pp. 63–76.
14. W. Farrer and D.F. Andrew, 'Fully controlled regenerative bridge with

half controlled characteristics', Proc. IEE, Vol. 125, No. 2, pp. 109–112, 1978.

15. P.N. Mipanic, 'The through pass inverter and its application to induction motor speed control', IEEE Trans. on PAS, Vol. PAS–87, p. 234, 1968.

16. W. Durry, W. Farrer and B.L. Jones, 'Performance of thyristor bridge converters employing flywheeling', Proc. IEE, Vol. 127, pp. 268–276.

17. P. Mehta and S. Mukhopadhyay, 'Improvement in dc motor performance by asymmetrical triggering, Part I—one quadrant drive,' IEEE Trans. Vol IA–2, 1975, p. 172.

18. T. Matsuhashi, S. Nagai and Y. Amemiya, 'A method of reducing harmonic content in the ac side of multisplit thyristor-controlled rectifier circuits', Electrical Engg. in Japan, Vol. 98, No. 2, 1978, pp. 70–75.

19. S. Mukhopadhyay, 'A new concept for improving the performance of phase controlled converters', IEEE Trans. Ind. Appl., Vol. IA–14, pp. 584–603, 1978.

20. H.K. Patel and G.K. Dubey, 'Modified sequence control technique for improving the performance of regenerative bridge converters', IEEE Trans. Ind. App vol. IA–19, No. 5, pp. 682–689. 1983.

21. P.C. Sen and S.R. Doradla, 'Symmetrical and extinction angle control of solid state series motor drive', IEEE Trans. IECI, Vol. IECI–23, pp 31–38, 1976.

22. P.C. Sen and S.R. Doradla, 'Evaluation of control schemes for thyristor controlled dc motors,' IEEE Trans. IECI, Vol. IECI-25, pp. 247–255, Aug. 1978.

23. P. Mehta, S. Mukhopadhyay and E. Orhun, 'Forced commutated ac-dc converter-controlled dc drives', IEE conference on Power Electronics—Power Semiconductors and Their Applications, Conference publication 123, p. 146, 1974.

24. H.K. Patel and G K. Dubey, 'Evaluation of time ratio control schemes for thyristor controlled dc drives', Jr Intn. of Engrs., Electrical Engg. Div., 1983.

25. K.A. Krishnamoorthy, G.K. Dubey and G.N. Revankar, 'Converter control with selective reduction of line harmonics', Proc. IEE, Vol. 125, No. 2, pp. 141–145, 1978.

26. K.A. Krishnamoorty, G.K. Dubey and G.N. Revankar, 'General method for selective reduction of line harmonics', Proc. IEE, Vol. 125, No. 11, pp. 1269–70, 1978.

27. G.N. Revankar, G.K. Dubey and K.A. Krishnamurthy, 'Present trends in line-current harmonic reduction in single phase thyristor converters', Jr. IETE, Vol. 25. No. 8, pp. 349–355, 1979.

28. T. Kataoka, K. Mizumachi and J. Miyairi, 'A pulse-width controlled ac to dc converter to improve power factor and waveforms of ac line current', IEEE Trans. on Ind. Appl., Vol. 15, 670–675, 1979.

29. K.A. Krishnamurthy, G.K. Dubey and G.N. Revankar, 'Sequence control of converters with less tripplen harmonics', Jr. Intn. of Engrs., Electrical Engg. Div., Vol. 59, pp. 100–105, 1978.

9. MEASUREMENT AND SENSING IN THYRISTORISED SYSTEMS

The electrical systems can be divided into ac and dc systems. Conventionally, ac system voltage and current are either sinusoidal or can at least be approximated to be sinusoidal. Unlike conventional ac systems, the thyristorised ac systems either have non-sinusoidal voltage or non-sinusoidal current or both the current and voltage are non-sinusoidal. If a measuring instrument is not chosen taking care of this aspect of thyristorised ac circuits, erroneous results are obtained. Similarly, conventionally, dc circuit is supposed to have perfect dc voltage and current. But the thyristorised dc circuits have harmonics in addition to the dc component either in voltage, or in current or both in voltage and current. Here also erroneous readings are obtained if the appropriate measuring instruments are not used. This chapter discusses the choice of instruments for the measurement of electrical signals in thyristorised systems.

For monitoring system behaviour, abnormal and fault conditions, closed loop operation etc. one is required to sense various electrical and mechanical quantities, the prominent among these are current, voltage and speed. This chapter will also discuss various methods of current, voltage and speed sensing in thyristorised systems.

9.1 MEASUREMENT OF ELECTRICAL QUANTITIES

Before one is able to consider the choice of suitable instruments for thyristorised systems, it will be useful to examine expressions for various electrical quantities.

9.1.1 Expressions of Various Electrical Quantities

(a) Voltage and Current

Any periodic non-sinusoidal waveform $x(t)$ with a period T can be described by the following Fourier series

$$x(t) = A_0 + \sum_{n=1}^{\infty} (A_n \sin n\omega t + B_n \cos n\omega t) \qquad (9.1)$$

where

$$\omega = 2\pi/T \text{ and}$$

$$A_0 = \frac{1}{2\pi}\int_0^{2\pi} x(\omega t)\, d(\omega t) \tag{9.2}$$

$$A_n = \frac{1}{\pi}\int_0^{2\pi} x(\omega t) \sin n\omega t\, d(\omega t) \tag{9.3}$$

$$B_n = \frac{1}{\pi}\int_0^{2\pi} x(\omega t) \cos n\omega t\, d(\omega t) \tag{9.4}$$

A_0 above is the dc component, which is also the average value of $x(t)$, X_{av}. Thus,

$$X_{av} = A_0 \tag{9.5}$$

Eqn. (9.1) can be rewritten in the form

$$x(t) = X_{av} + \sum_{n=1}^{\infty} \sqrt{2}\, X_n \sin (n\omega t + \varphi_n) \tag{9.6}$$

where X_n and φ_n are the rms value of nth harmonic and its phase angle respectively.

$$X_n = \sqrt{(A_n^2 + B_n^2)/2}$$

and

$$\varphi_n = \tan^{-1} (B_n/A_n) \tag{9.7}$$

The rms value of the quantity $x(t)$, X_{rms}, is given by

$$X_{rms}^2 = \frac{1}{2\pi}\int_0^{2\pi} x^2(\omega t)\, d(\omega t)$$

or

$$X_{rms}^2 = X_{av}^2 + \sum_{n=1}^{\infty} X_n^2 \tag{9.8}$$

When x is a pure dc quantity

$$X_{rms} = X_{av} \tag{9.9}$$

The above expressions derived for periodic variable $x(t)$ are applicable to current and voltage in an electrical circuit. Thus, if V_{av} and V_n are respectively the average value and rms value of nth harmonic then,

$$V_{rms}^2 = V_{av}^2 + \sum_{n=1}^{\infty} V_n^2 \tag{9.10}$$

Similarly

$$I_{rms}^2 = I_{av}^2 + \sum_{n=1}^{\infty} I_n^2 \tag{9.11}$$

When current and voltage have only dc components

$$V_{rms} = V_{av} \text{ and } I_{rms} = I_{av} \tag{9.12}$$

(b) Power

Power, consumed or delivered, in an electric circuit is given by

$$P = \frac{1}{T}\int_0^T v(t)\,i(t)\,dt \tag{9.13}$$

Eqn. (9.13) is a general expression valid for any periodic waveform, sinusoidal or non-sinusoidal.

Substituting the expressions of voltage and current in terms of their Fourier components and then solving eqn. (9.13) gives

$$P = V_{av}I_{av} + V_1I_1 \cos \varphi_1 + V_2I_2 \cos \varphi_2 + \dots V_nI_n \cos \varphi_n \dots \tag{9.14}$$

It may be pointed out here that the interaction of voltage and current of the same frequency only produces steady-state average power. Interaction between the voltage of one frequency and the current of another frequency produces only oscillating power, the average value of which is zero.

Let us now consider a particular case involving sinusoidal supply voltage and non-sinusoidal current. This situation will arise when a thyristor system draws non-sinusoidal current from a source whose capacity is very large. Though the current is non-sinusoidal source voltage remains sinusoidal. Under this situation

$$P = V_1\, I_1 \cos \varphi_1 \tag{9.15}$$

This expression will also hold good for the case with sinusoidal current and non-sinusoidal voltage.

Similarly, in dc circuits, when either voltage or current is perfect dc and another quantity contains ac ripple, power will be given by the expression

$$P = V_{av}\, I_{av} \tag{9.16}$$

(c) Power Factor

Basic definition of power factor is as follows:

$$\text{P. F.} = \frac{\text{Real power}}{\text{Apparent power}} \tag{9.17}$$

For the general case of non-sinusoidal voltage and current, real power is given by eqn. (9.14). Now,

Apparent power $= V_{rms} \times I_{rms}$ (9.18)

where V_{rms} and I_{rms} are given by eqns. (9.10) and (9.11).

Substituting these expressions in (9.17) gives

$$\text{P. F.} = \frac{V_{av}\, I_{av} + V_1 I_1 \cos \varphi_1 \ldots + V_n I_n \cos \varphi_n + \ldots}{\left(\sqrt{V_{av}^2 + \sum_{n=1}^{\infty} V_n^2}\right)\left(\sqrt{I_{av}^2 + \sum_{n=1}^{\infty} I_n^2}\right)} \quad (9.19)$$

A particular case arises when the voltage is sinusoidal but the current is non-sinusoidal. Here,

$$\text{P. F.} = \frac{V_1 I_1 \cos \varphi_1}{V_1 I_{rms}} \quad (9.20)$$

9.1.2 Measurement Techniques for Thyristorised AC Circuits

9.1.2.1 *AC Voltage and Current Measurements*

Expressions (9.10) and (9.11) describe the rms values of non-sinusoidal voltage and current. For the accurate measurement of rms values of non-sinusoidal voltage and current an instrument with wide frequency range is required. When the Fourier series converges very fast, the contribution of the harmonics to the rms value is negligible and therefore, an instrument with a limited frequency range can be employed.

For measurement of rms values of non-sinusoidal signals at power frequencies dynamometer instruments can be employed. Such a instrument contains a fixed and a moving coil arranged so that the torque acting on the moving coil is proportional to the product of the currents flowing in each coil. When its fixed and moving coils are connected in series torque and therefore deflection, will be proportional to the average value of current or voltage squared. It can, therefore, be used as a high quality ammeter or voltmeter for the measurement of rms values. Inaccuracies do arise from the inadequate response to the high frequency components due to attenuation by the inductance of coils and spurious losses by eddy currents. Nevertheless accuracy is adequate for most of the applications. The frequency range may be from 150 Hz to 2000 Hz. For better accuracy a meter with a frequency range on higher side is preferred. Such instruments are costly.

The rms value of non-sinusoidal signals can also be measured using instruments based on thermal effect. Thermocouple instruments use a thermocouple to measure the temperature of a resistance shunt. The heating effect, which depends on the rms value of the current flowing, is converted into a small dc voltage. This voltage is used to drive a moving coil dc meter movement. These instruments measure true rms values independent of waveforms and do not suffer from the high frequency limitations of dynamometer instruments. These instruments can be calibrated either on ac or dc and with careful design, frequency error should be negligible up to 10 KHz. These instruments are economical and well suited to application in thyristorised systems. The main reason that such instruments are not popular, especially in an industrial environment, is the difficulty in providing protection for thermocouples which may be fused rather easily on over load. Other instruments using thermal effect are the hot wire instruments which use the expansion of a taut wire due to heating to provide the deflection.

Due to the advance in large scale integrated circuits, it is now possible to make electronic instruments which can accurately measure true rms value. Such electronic instruments are economical and have excellent frequency range. The readings of such meters are obtained either using moving coil meter for deflection or a digital read out. Such instruments will find wide application in future. These instruments should be distinguished from other electronic instruments which rectify ac measurements and display a mean value of the result, possibly multiplied by a form factor of 1.11. Later type of instruments are totally unsuitable for measurement of rms values of non-sinusoidal signals. It is, therefore, necessary to carefully read the specifications of the electronic instruments before employing for the measurement of non-sinusoidal signals.

Other electrical instruments commonly used for ac measurement of sinusoidal signals are moving iron instruments and rectifier type moving coil instruments. These instruments are however not suitable for the measurement of rms value of a non-sinusoidal signal. Though the moving iron instruments measure the average of the square of current or voltage, they have two severe limitations. Their frequency range is limited, in some cases the limit being only slightly greater than 100 Hz. Secondly, the presence of ferromagnetic material produces significant error in reading due to hysteresis and varying eddy current effects caused by the harmonic components. The rectifier type moving coil instruments indicate the mean value of rectified signal multiplied by the assumed form factor of 1.11. Since the value of form factor will change in the presence of harmonics these instruments are not suitable for the measurement of rms value of a non-sinusoidal signal.

When the values of current and voltage are in excess of the range of available instruments, it is customary to use a current transformer or shunt with an ammeter and a potential transformer with a voltmeter. A current transformer introduces the ratio error and phase shift. With the increase in frequency, ratio error increases and phase shift decreases. In a well designed current transformer these errors are small (nearly 0.2 per cent and 0.5 degree) and can be compensated. In a potential transformer these errors are even less than the current transformer. Non-inductive shunts are available which can be employed for the accurate measurements.

It may be occasionally necessary to obtain the fundamental component of a non-sinusoidal signal. One may use tuned filters to suppress dc and harmonic components and use any ac meter to measure fundamental component.

9.1.2.9 *Measurement of Power in AC Circuits*

(a) CASE WITH NON-SINUSOIDAL VOLTAGE AND NON-SINUSOIDAL CURRENT

When both voltage and current are non-sinusoidal, the power is given by eqn. (9.14). The accurate measurement of power is only possible by using wattmeters with wide frequency range. The dynamometer wattmeters can be used for non-sinusoidal signals at power frequencies. The dynamometer wattmeters also have the same limitations as the dynamometer ammeters or voltmeters.

One can also use a Hall-effect wattmeter. It makes use of Hall-effect multiplier. Due to high speed of response it is possible to measure instantaneous power at power frequencies. For the measurement of average power, the output of the Hall-effect multiplier is filtered by a lowpass filter and fed into a moving coil instrument. If the field structure is designed with low inductance (or is suitably compensated) then the frequency errors are negligible and the power can be measured fairly accurately, independent of the waveforms of voltage and current.

Electronic wattmeters can be constructed employing IC multiplier chip, the output of which is filtered by a low pass filter to give a voltage proportional to average power. The readings of such meters are obtained either using a moving coil meter for deflection or a digital read out. These meters have wide frequency range, and therefore, permit fairly accurate measurement of power, independent of the waveforms of voltage and current.

Cases involving non-sinusoidal voltage and non-sinusoidal current occur in the ac side of voltage source and current source inverters, the load side of ac regulators and cycloconverters and the ac side of converters, ac regulators and cycloconverters fed by a weak ac source.

(b) CASE WITH NON-SINUSOIDAL CURRENT AND SINUSOIDAL VOLTAGE

When converters, ac regulators, cyclo-converters and inverters are supplied from a stiff ac source, while the current is non-sinusoidal, voltage is either sinusoidal or can at least be approximated to be sinusoidal. In such cases only the fundamental component will be responsible for producing power. One can employ wattmeters with low frequency range. Thus all the wattmeters in use for the measurement of power in sinusoidal power frequency ac systems can be used for this case without a significant error.

(c) CASE WITH SINUSOIDAL CURRENT AND NON-SINUSOIDAL VOLTAGE

In ac systems involving highly inductive load supplied by non-sinusoidal voltages, one can approximate the current to be sinusoidal. Due to the reasons mentioned above one can use any wattmeter designed for the measurement of power in sinusoidal systems.

9.1.2.3 *Measurement of Power Factor*

(a) CASE WITH NON-SINUSOIDAL VOLTAGE AND NON-SINUSOIDAL CURRENT

In this case, power factor is given by eqn. (9.19). It can be obtained by measuring real power, V_{rms} and I_{rms}. Real power can be measured using dynamometer wattmeter or Hall effect wattmeter or electronic wattmeter as described in the previous section. V_{rms} and I_{rms} can be obtained using dynamometer instruments, or thermal instruments or electronic rms meters.

(b) CASES WITH NON-SINUSOIDAL VOLTAGE AND SINUSOIDAL CURRENT, AND SINUSOIDAL VOLTAGE AND NON-SINUSOIDAL CURRENT

When voltage is non-sinusoidal and current is sinusoidal power is produced only by the fundamental component. This power can be measured with satisfactory accuracy using a wattmeter designed to measure power in sinusoidal circuits. V_{rms} can be obtained using a dynamometer or thermal or electronic rms meter. I_1 can be obtained by any ac ammeter. When voltage is sinusoidal and current is non-sinusoidal similar procedure can be employed.

9.1.3 Measurement Techniques for Thyristorised DC Circuits

Here the term dc circuits will include, in addition to dc systems, the branches of ac circuits carrying unidirectional current.

9.1.3.1 *Voltage and Current Measurements*

In dc circuits one requires the measurement of dc component (average value) to calculate torque of dc machines, ratings of semiconductor components, ratings of batteries etc. One also requires the measurement of rms values of current to assess the copper losses and the thermal condition of resistive elements in the load, busbars, cables, fuses etc.

The rms values of current and voltage can be obtained using dynamometer instruments, or thermal instruments or electronic rms meters.

The dc component (average value) of voltage and current can be obtained by using permanent magnet moving coil instruments. In these instruments, a current carrying coil is suspended in the field produced by a permanent magnet. The instantaneous torque is proportional to instantaneous value of current, but the deflection depends on the average value of torque which in turn is proportional to the average value of current. Presence of harmonics does introduce error, which is however, very small and can be ignored. When the waveform is very much distorted and precise measurements are desired, ac components of the waveform should be prevented from reaching the instrument.

When the quantity to be measured has a value larger than the ratings of available instruments, one can employ a shunt in parallel with the instrument.

9.1.3.2 *Power Measurements*

When both the voltage and current waveforms have ac components in addition to dc components, power is given by eqn. (9.14). One can measure power employing dynamometer, Hall-effect or electronic wattmeters. This type of situation exists on the dc side of converters, the load side of choppers, and the dc side of voltage source inverters fed by a weak dc source.

In the case of dc circuits fed by a stiff dc supply, the current may have ripple in addition to dc component, but voltage can be assumed to be dc without ac components. Such a case arises on the source side of choppers and voltage source inverters fed by a stiff dc source. Here power is given simply as a product of the source voltage and the average value of current, which can be measured using permanent magnet moving coil instruments.

In current source inverters, if sufficiently large choke is employed the input current to the inverter can be approximated to be a perfect dc. The input voltage, however, will have an ac ripple. Here also the power is obtained simply as a product of current and average value of voltage, which can be measured using a permanent magnet moving coil instruments.

In the case of dc motors fed by thyristorised power supplies, though the torque has an ac component due to the presence of ripple in the armature current, the fluctuation in speed is negligible and the back emf E can be assumed to be a pure dc quantity. The power developed by the machine can be calculated as a product of the back emf and the dc component of the armature current.

9.1.3.3 *Summary of Suitable Instruments for Thyristor Systems*

The requirements of instruments for various thyristorised systems is summarised in Table 9.1 to Table 9.5. Suffix S refers to source side and suffix L refers to load side. In these tables electronic meters referred are electronic rms meters and wattmeters, specially designed for the measurements in non-sinusoidal systems. In the case of current source inverters V_s, I_s and P_s are the quantities measured after the dc inductor and before the inverter circuit. These tables mainly consider the measurements at power frequencies. For high frequency inverters, the frequency ranges of various instruments should be carefully examined for their suitability.

9.1.4 Other Measurements and Recording of Waveforms

Amplitudes of various harmonics can be measured using harmonic analyser of appropriate frequency range.

Peak value of any signal, its rate of rise or its complete waveform can be seen on an oscilloscope. In thyristorised systems,

Table 9.1. Measurements in converters

Quantity	Value	Suitable Instruments	Meter reading
V_S, I_S	rms	Weak source: dynamometer, thermal, electronic	V_S (rms), I_S (rms)
		With stiff power supply: any ac meter for V_S	V_{S1}
P_S		Weak power source: dynamometer, Hall-effect, electronic	P_{S1}
		Stiff power supply: any ac watt-meter	P_{S2}
$P.F.$		Weak power supply, $P.F. = \dfrac{P_{S1}}{V_S\,(\text{rms}).\,I_S\,(\text{rms})}$ Stiff power supply, $P.F. = \dfrac{P_{S2}}{V_{S1}.\,I_S\,(\text{rms})}$	
V_L, I_L	rms	Dynamometer, thermal, electronic	
	average	Permanent magnet moving coil	
P_L		Dynamometer, Hall-effect, electronic	

Table 9.2. Measurements in choppers

Quantity	Value	Suitable Instruments	Meter reading
V_S, I_S,	rms	Weak source: dynamometer, thermal, electronic	V_S (rms), I_S (rms)
		Stiff source: permanent magnet moving coil for V_S	V_S (av)
	average	permanent magnet moving coil	V_S (av), I_S (av)
P_S		Weak source: dynamometer, Hall-effect, electronic wattmeters	
		Stiff source: V_S (av). I_S (av)	
V_L, I_L	rms	dynamometer, thermal, electronic	
	average	permanent magnet moving coil	
P_L		dynamometer, Hall-effect, electronic wattmeters	

Table 9.3. Measurements in voltage source inverters

Quantity	Value	Suitable instruments	Reading
V_S, I_S	rms	Weak source: dynamometer, thermal, electronic	
		Stiff source: permanent magnet moving coil for V_S	V_S (av)
	average	permanent magnet moving coil	V_S (av), I_S (av)
P_S		Weak source: dynamometer, Hall-effect, electronic wattmeters	
		Stiff source: $P_S = V_S \cdot I_S$ (av),	
V_L, I_L	rms	dynamometer, thermal, electronic	V_L (rms), I_L (rms)
P_L		dynamometer, Hall-effect, electronic wattmeters	P_L
P. F.		$P. F. = P_L/(V_L \text{ (rms)}. I_L \text{ (rms)})$	

Table 9.4. Measurements in current source inverters

Quantity	Value	Suitable instruments	Reading
V_S^*, I_S^*	rms	DC current with ripple: dynamometer, thermal, electronic	
		DC current with negligible ripple: I_S by permanent magnet moving coil	I_S (av)
	average	permanent magnet moving coil	V_S (av), I_S (av)
P_S^*		DC current with ripple: dynamometer, Hall-effect, electronic wattmeters	
		DC current without ripple: V_S (av). I_S (av)	
V_L, I_L	rms	dynamometer, Hall-effect, electronic	V_L (rms) I_L (rms)
P_L		dynamometer, Hall-effect, electronic wattmeters	P_L
P. F.		$P. F. = P_L/(V_L \text{ (rms)}. I_L \text{ (rms)})$	

*These quantities are measured after the dc inductor and before the inverter.

Table 9.5. Measurements in AC regulators and cyclo-converters

Quantity	Value	Suitable Instruments	Readings
V_S, I_s	rms	Weak source: dynamometer, thermal, electronic	V_S(rms), I_S(rms)
		Stiff source: any ac meter for V_S	V_{S1}
P_S		Weak source: dynamometer, Hall-effect, electronic wattmeters	P_{S1}
		Stiff source; any ac wattmeter	P_{S2}
P.F.		Weak source: $P.F. = P_{S1}/(V_S(\text{rms}) . I_S\ (\text{rms}))$	
		Stiff source: $P.F. = P_{S2}/(V_{S1} . I_S\ (\text{rms}))$	
V_L, I_L	rms	dynamometer, thermal, electronic	V_L(rms), I_L(rms)
P_L		dynamometer, Hall-effect, electronic	P_L
P.F.		$P.F. = P_L/(V_L(\text{rms}) . I_L(\text{rms}))$	

one needs to examine the waveforms between any two points of the power circuit (not just between any point and the earth). For this, one can either employ a differential oscilloscope or an oscilloscope isolated from the mains through an *individual* transformer. While working with an oscilloscope with isolating transformer, it is advisable to use a rubber pad insultation on the floor for the safety of the operator. Furthermore, one is required to observe more than one waveform simultaneously. It is, therefore, necessary to employ a multi-channel oscilloscope. Voltages in power circuits are higher then those permitted in a normal oscilloscope. Either the oscilloscope should be capable of dealing with power circuits voltages directly or a suitable probe for step down arrangement should be employed. A frequency range of 10 MHz is adequate for most of the measurements.

Often the presence of spurious signals (noise) in firing circuits, results into the mal-operation of thyristorised circuits. If these pulses are very narrow and sharply rising, a oscilloscope with a frequency range of the order of 50 to 60 MHz may be required.

9.2 SENSING IN THYRISTORISED SYSTEMS

Sensing of various quantities in thyristorised systems is required for the closed loop control, protection, detection of abnormal and fault conditions etc. This section discusses the methods of sensing of current, voltage, power and speed.

9.2.1 Current and Voltage Sensing

Thyristor controllers do not have any capacity to withstand overcurrents. Their transient and steady-state current ratings are the same. In order to limit the currents to safe values all closed loop (for control of speed, output voltage, etc.) systems are operated either with inner current control loop (also known as subordinate current control loop or cascade current loop) or current limit control loop. For the safety of the controller and convenient operation, even the open-loop systems are provided with current limit control loop. In order to implement these control loops one requires an arrangement for sensing current. Since the current need not be precisely controlled, approximate sensing of current signal is required. Similarly, fault conditions involving short circuit, overload, thyristor failure, blowing of fuse and failure of a thyristor to fire etc. can be detected by sensing the current [3]. Here also only approximate sensing of the current signal is required. The closed-loop current controlled systems, designed to supply regulated current may however, require precise sensing of current. In many dc drives speed is sensed by sensing the armature induced voltage. The armature induced voltage is sensed by obtaining a difference between the motor terminal voltage and *IR* drop. For obtaining *IR* drop, sensing of current *I* is required.

In order to avoid interaction between the control circuit, consisting of low voltage and current, and the power circuit, involving high voltage, high current and substantial amount of harmonics (noise), isolation must be provided between the two.

9.2.1.1 *DC Current and Voltage Sensing*

The following techniques are employed for sensing dc currents:

(i) A dc transductor;

(ii) A Hall-effect probe;

(iii) A resistance shunt with;

(a) modulator and demodulator for isolation;

(b) with opto-isolator for isolation;

(iv) Indirect method.

(i) *DC transductor*: A dc transductor is a magnetic amplifier. Current to be sensed is fed to a low resistance control winding to get a signal proportional to the current at the output. The output is filtered by a lowpass filter to remove ac components. The device in general is not able to provide information about the direction of current as the output is always of the same polarity. A schematic of the dc transductor and its characteristics are shown in Fig. 9.1. By additional winding for the biasing, its output signal can be made zero for the zero value of dc current I_d.

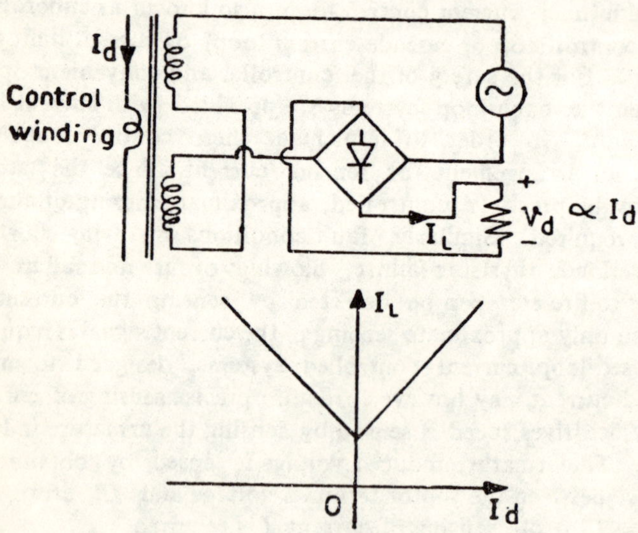

Fig. 9.1. DC transductor for current sensing.

(ii) *Hall-effect probe*: Hall-effect transducer is finding very wide application recently. It gives an output, which is the product of two input signals. One signal can be chosen to be a constant current and another the current to be sensed, to get a signal proportional to the current. A low pass filter is employed to remove ac components.

(iii) *Resistance shunt*: Specially designed non-inductive shunts can be used for current sensing. The main limitation of the shunt is that it provides only small output voltages. For example, a standard shunt develops from 7.5 mV to 75 mV at the rated current. The use of higher resistance shunt results in increased power dissipation and drift of the shunt resistance with the temperature. When accurate sensing is not required, resistive part of the circuit itself is employed, instead of shunts. For example, in the case of dc motor control, often the interpole winding is employed for current sensing. In low power and low cost systems, isolation may not be provided between the power and control circuits. In that case drop across the shunt is directly sensed without incorporating isolation circuits. When isolation is required, one of the following arrangements may be used.

(a) ISOLATION USING MODULATORS AND DEMODULATORS

One isolation arrangement is shown in Fig. 9.2. The dc signal v_d proportional to current I_d is inverted to ac (modulated) by

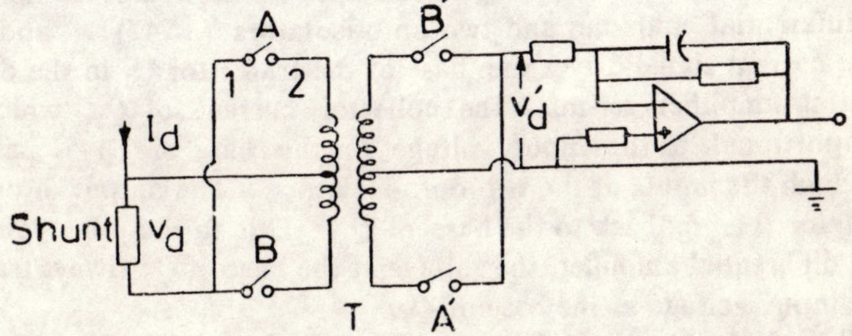

Fig. 9.2. Current sensing circuit using transistor chopper.

means of transistor choppers A and B, which close and open alternately at a frequency of 1 to 10 KHz. The ac signal is passed through a isolation transformer T and then rectified, by means of another pair of choppers A' and B'' so as to give a dc voltage v_d'

proportional to v_d. A and A', and B and B' close and open simultaneously, giving phase sensitive rectification (demodulation). The signal v_d' is filtered using an operational amplifier filter.

Second isolation arrangement employs opto-isolators.

(b) Isolation Using Opto-isolators

Recently opto-isolators are finding wide applications in thyristor controllers. Hence this arrangement will be described relatively in greater detail.

Figure 9.3 gives a scheme (Scheme *I*) of isolation using opto-isolators (also known as opto-couplers). The output of the shunt

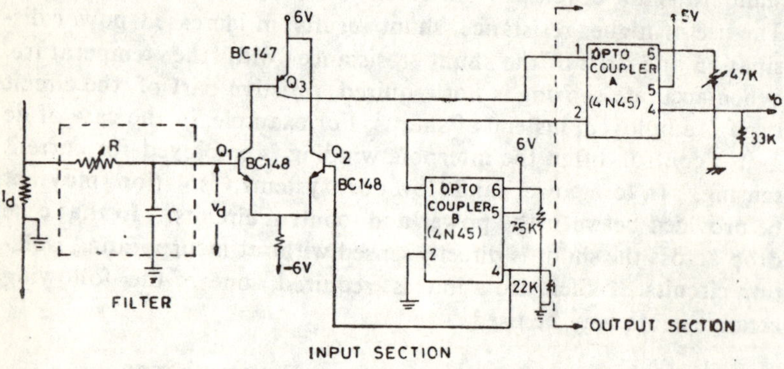

Fig. 9.3. Current sensing using opto-isolator (Scheme *I*).

is filtered by a low pass *RC* filter to remove ac components and given to an opto-isolator stage. The opto-isolator stage comprises a differential amplifier and two opto-isolators (4*N*45) *A* and *B*. The current signal drives the base of the transistor Q_1 in the differential amplifier set-up. The collector current of Q_3, which is proportional to the input voltage (at the base of Q_1) is passed through the inputs of the two opto-isolators. The output of opto-isolator *B* is fedback to the base of Q_2. Due to the high gain of the differential amplifier, the voltage at the base of Q_2 always tracks the input voltage at the base of Q_1.

Opto-isolators *A* and *B* are chosen to be identical. The output of *A* is therefore proportional to the output of *B* which in turn is proportional to the input current signal. A floating power supply, which is isolated from the power supply to *A*, is used for the differential amplifier and *B*.

The gain of the opto-isolator is a function of temperature. If a single opto-isolator is employed its gain will be temperature dependent and non-linear. By employing two indentical opto-isolators in the feedback loop temperature dependence of the gain and the coupling non-linearity are compensated. An alternative circuit (Scheme *II*) incorporating these features is shown in Fig. 9.4.

One LED couples the input section of the circuit to the output section, and the other LED provides the feedback for closed loop operation. The preset voltage level at the input of op-amp A_1 sets the LED's in their most linear operating range. When an input signal is applied to the inverting terminal, opamp A_1 drives LEDs to a level at which the collector current from Q_1 makes $I_{R2} = I_{R1}$ so that the inverting input is at virtual ground (the diode D_1 protects LEDs against negative overvoltages and it is not part of the feedback loop). Because LEDs are in series and matched,

$$I_{R3} = I_{R5}$$

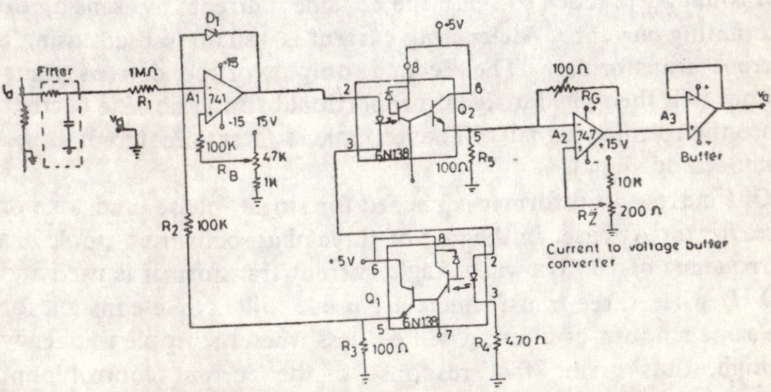

Fig. 9.4. AC current sensing using opto-isolators.

In the output section, the collector to base capacitance of Q_2 tends to decrease the frequency response of the circuit. This tendency is overcome by operating op-amp A_2 in the current to voltage converter mode, which maintains the signal voltage at the inverting input and across Q_2 at virtual ground. Amplifier A_3 acts as a buffer. The output voltage is given by

$$\begin{aligned} v_0 &= I_{R5}R_G = I_{R3}R_G \\ &\simeq (R_2 + R_3)\, I_{R2}R_G/R_2 \\ &\simeq (R_2 + R_3)\, v_d R_G/R_1 R_3 \end{aligned}$$

Therefore,

$$v_0/v_d \simeq (R_2 + R_3)\, R_G/R_1R_3$$

For the initial adjustment of potentiometers R_B and R_Z, a sine wave is fed into the input. The output is displayed on the scope, and R_B is adjusted for symmetrical clipping as the signal amplitude is raised to $3V$. Then the input terminals are short circuited, and potentiometer R_Z is adjusted so that the output voltage is zero. Finally, a 1–V signal is applied and R_G is adjusted to get the desired output level. The input signal can have any value from 0 to 3V. The circuit shown operates from dc to 50 KHz. The degree of isolation depends upon the isolation resistance of power supplies used for the input and output sections.

(iv) *Indirect method*: The currents in ac lines of fully controlled converters, carry information about dc side currents. Therefore, it is a common practice to sense the dc side current by sensing the alternating current. Alternating current is usually sensed using a current transformer. The rectified output of the current transformer will then yield a signal proportional to the dc side current. Since the rectified signal will have ripple, a filter is required to get a smooth dc signal.

One current transformer is needed for single phase and one or three for three phase. In the case of three phase converter, ripple has a frequency of 100 Hz when single current transformer is used and 300 Hz when three transformers are used. Filter time constant for the same amount of filtering will be less when the ripple frequency is high, thus giving fast response of the current control loop. Therefore, three-phase transformer connection is preferred in a three phase system. A typical circuit for three-phase converters is shown in Fig. 9.5. The major limitation of this scheme is that it is not able to sense the direction of current. Presently, for fully controlled converters, this method is most commonly used due to lower cost, simplicity and reliability.

Above description gives the methods of sensing dc current. DC voltages are sensed using a potentiometer connection formed by a good quality resistors. If the ripple is present, output is filtered. When isolation is not necessary this signal can be directly used. If isolation is required, it can be obtained using the same methods as described for current sensing.

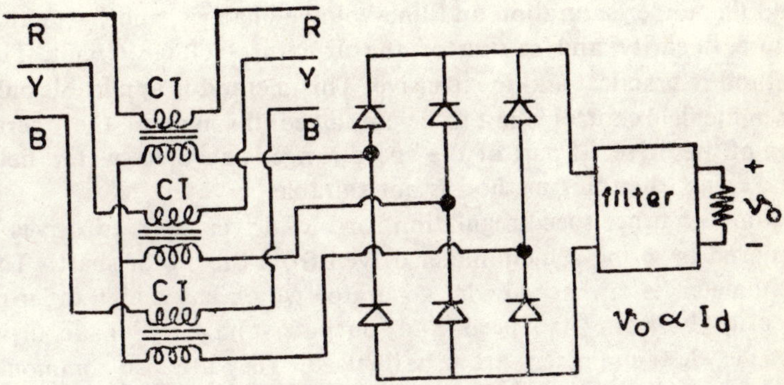

Fig. 9.5. Current sensing in three phase fully controlled converter and three phase ac systems.

9.2.1.2 *Sensing of AC Currents and Voltages*

AC current is sensed using a current transformer. When used in low frequency circuits care should be taken to avoid saturation. When the output is to be in the form of dc, the output of the current transformer is rectified and filtered. For sensing current in a three phase ac system, the circuit shown in Fig. 9.5 is employed.

AC voltages are sensed using potential transformers. When operated at low frequencies care should be taken to avoid saturation.

9.2.2 Sensing of Power

The instantaneous and average powers at power frequencies can be sensed by using Hall-effect probe, or electronic multiplier. For the average power the output of these devices is filtered using a low-pass filter to get a smooth dc signal.

9.2.3 Sensing of Speed

Speed sensing is required for closed-loop speed control of dc and ac drives. The accuracy of speed regulation very much depends on the accuracy of speed sensing device.

In dc drives employing separately excited motor, speed is sensed by sensing the induced voltage, which in turn is sensed by sensing the motor terminal voltage and the armature resistance drop and obtaining the difference of the two. The regulating accuracy of the armature voltage speed sensing is of the order of ± 2 per cent of the base speed. The signal is directly affected by the deviations of the field flux under saturation and line-voltage changes, and by temperature, linearity, and setting of the armature-voltage circuit. The method is practical and inexpensive. This method is applicable only when the field control is not to be employed throughout the operation of the drive. If part of the speed range is covered by the field weakening, then this method is not suitable.

More accurate speed regulation for dc drives and ac drives is achieved by using a tachometer driven from the motor shaft. The tachometer is an ac or dc generator which has a high order of linearity between its speed and output voltage. For dc drive systems, dc tachometers are usually used. They are also commonly used for ac drives. The dc tachometer is built with a permanent magnet field and sometimes with silver brushes to reduce the brush-to-commutator voltage drop. It is normally wound with a relatively high internal impedance so that it may require external temperature compensating resistors and a high impedance load. Typical voltage is 10 V per 1000 rpm.

Tachometers are available to secure regulating accuracies of ±0.1 per cent of the base speed. To achieve such regulation the speed regulating control loop must have very high gain and may be difficult to stabilize over the operating range of the motor. The commutator of the tachometer produces a ripple on the output voltage. At high speeds, the ripple frequency can be easily filtered; however, at low speeds, the ripple may require a filter with larger time constant which will be large enough to affect the dynamics of the system. Special large diameter tachometers, with a large number of commutator segments, are built to operate without large ripple at low speeds.

If very high speed accuracies are required as in computer peripherals, paper mills etc. digital tachometers are used. This method employs a shaft encoder which gives output pulses having a frequency proportional to the angular velocity of the motor. The shaft encoder consists of an aluminium or plastic disc mechanically coupled to the motor shaft. In the case of aluminium disc a number of holes or slots are uniformly made around the periphery of the disc. In the case of transparent plastic disc it is alternately painted black to provide alternately transparent and non-transparent parts. An opto-coupler unit consisting of a light source and a light sensing device is so mounted that the disc will run between the light source and the sensor. The sensor senses the light source whenever a slot/hole/transparent part crosses the opto coupler device and an output pulse is produced. The number of output pulses per revolution is equal to the number of slots/holes/transparent parts on the disc, and the frequency of the output pulse train is proportional to the speed of the shaft. These pulses are counted over a specific period T to obtain a number proportional to speed. One can also obtain information relating to the speed by measuring period T for a fixed number of pulses.

REFERENCES

1. K.N. Stanton, 'Instrumentation for thyristor control', IEEE Trans. on Industry and General Applications, Vol. IGA-4, Nov./Dec., 1968, pp. 638–643.
2. J.F. Lindsay, 'Measurement problems in determining the efficiency of thyristor supplied motor drives', IEEE Trans. on Industry Applications, Vol. IA–15, Jan./Feb., 1979, pp 8–13.
3. Y.V.V.S. Murty, G.K. Dubey and R.M.K. Sinha, 'Fault diagnosis in three phase thyristor converters using microprocessors', IEEE Trans. on Industry Applications, vol. IA–20 No. 6, Nov./Dec. 1984, pp. 1490–97.
4. M.V. Kotappa, 'Chopper control of battery operated vehicle', M. Tech. Thesis, I.I.T. Kanpur, 1983.

10. IC LOGIC AND CONTROL CIRCUITS

The power controllers described in chapters three to eight require low power control circuits for driving the gates of the thyristors. These circuits are made up of SSI and MSI integrated circuits and perform the required logic and control functions. The logic and control circuits can take various forms and use many different devices. The specific circuits for each type of power controller have been described in earlier chapters. The objective of this chapter is to describe in detail the basic principles for the design of the logic and control circuits. This has been achieved by discussing the various standard IC packages and their applications as described below.

The basic digital circuits using gates, flip-flops, counters etc. are built around standard IC logic families such as DTL, TTL, HTL and CMOS. The characteristics of these logic families is described in Section 10.1. The principles of Boolean algebra, the concept of a basic gate and interfacing between different logic families are discussed in this section.

Section 10.2 describes the design of combinatorial circuits. The description of such circuits using truth table and K-maps is given. The realisation of combinatorial circuits using logic gates, multiplexers and ROMs is discussed. The design of a BCD to seven segment decoder has been given to illustrate the design procedure.

Section 10.3 describes the design of sequential circuits. The truth tables of various kinds of flip-flops are given. The design of counters is described with examples. A brief description of TTL IC counters is given.

Section 10.4 to 10.7 deal with analog circuits. These circuits use the integrated operational amplifier as the basic element. The characteristics of the operational amplifier are given in Section 10.4. Some linear circuits, e.g. inverting, non-inverting and differential amplifiers, integrators, filters etc. are discussed in Section 10.5. The operational amplifiers can also be used in some non-linear circuits such as precision rectifiers, diode function generators, comparators etc. These circuits are discussed in Section 10.6.

Section 10.7 describes timing circuits such as monostables and astables. The popular IC timer NE 555 and its applications are given.

Section 10.8 discusses the Phase Locked Loop (PLL) which is widely used in d.c. servo systems. The linearised model of the PLL and its applications are discussed along with specifications of the NE 565 IC.

Section 10.9 describes the principles of analog to digital (A/D) and digital to analog (D/A) conversion. These circuits are important due to increasing use of digital hardware in signal processing and control.

Section 10.10 describes some practical considerations in the implementation of logic and control circuits for thyristorised power controllers. The problems of isolation, grounding and noise are discussed in this section.

10.1 DIGITAL LOGIC CIRCUITS

The term 'Digital Logic Circuits' in general denotes some special kinds of circuits. These circuits use devices which have two distinct states. These states are termed as 'Logical one' or High and 'Logical zero' or Low. A variety of devices, e.g. Pneumatic valves, electromagnetic relays and transistors exhibit the two state behaviour. In electrical engineering the term digital logic or digital circuit or logic circuit refers to the low power electronic circuits using transistors as basic switches. In this section these circuits are described in detail. After reviewing the basic prepositions of Boolean Algebra the concept of a basic gate and its specifications are discussed. This is followed by detailed description of the present day integrated circuit logic families like the DTL, HTL, TTL and the CMOS. Basic operation of the inverter in each family is described and the salient electrical specifications of the gate are brought out. Finally the special problems when ICs of different logic family are interconnected are discussed.

10.1.1 Boolean Algebra

The analysis of digital circuits is based on the postulates of 'Boolean Algebra'. This algebra gives the rules for the manipulation of binary variables. The binary variable, by definition, has only two states denoted by '1' and '0'. The states may represent two voltage levels, e.g. five volts and zero volts. The Boolean algebra defines three basic mathematical operations on the binary variables called AND (.), OR (+) and NOT (—). Table 10.1 gives the definitions of these operations. Fig. 10.1 shows the symbols for various

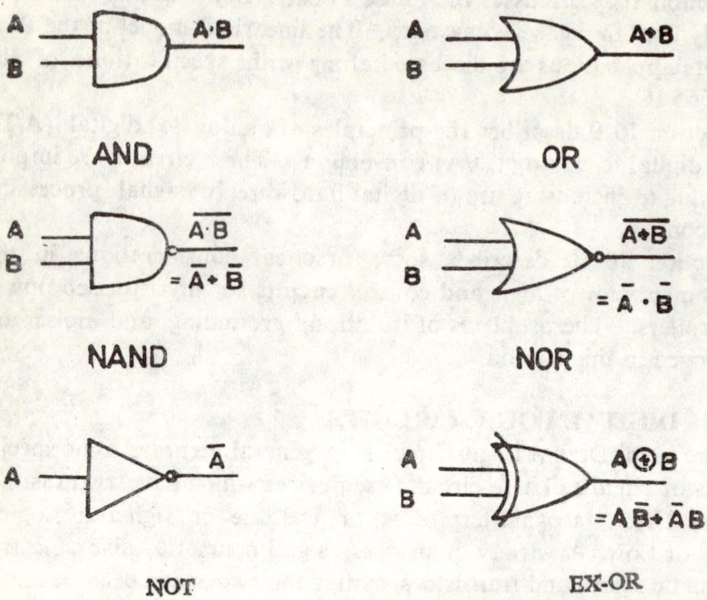

Fig. 10.1 Logic gates and their symbols.

Table 10.1. Basic operations on the Boolean variables x and y

Variables		AND	OR	NOT	NAND	NOR	EX-OR
x	y	$x.y$	$x+y$	$\overline{x}/\overline{y}$	$\overline{x.y}$	$\overline{x+y}$	$x+y$
0	0	0	0	1/1	1	1	0
0	1	0	1	1/0	1	0	1
1	0	0	1	0/1	1	0	1
1	1	1	1	0/0	0	0	0

electronic circuits called 'gates' which implement these operations.

Table 10.1 also shows the definitions of the operations NAND and NOR. These are derived from the AND and OR respectively by complementation. However it can be shown that the three basic operations, AND, OR and NOT, can be written in terms of either the NAND or the NOR operation alone. Therefore, entire electronic digital circuits based exclusively on the NAND operation or the NOR operation can be designed. The function Exclusive-OR can also be expressed in terms of AND, OR and NOT. The Ex-OR function is widely used in circuits as a basic operation. Table 10.2 lists some basic identities which are useful in simplifying expressions involving logical variables.

Table 10.2. Some basic identities in Boolean variables

$x + 0 = x$	$x + y = y + x$	$x + y + z = (x + y) + z = x + (y + z)$
$x.1 = x$	$x.y = y.x$	$x.yz = (xy).z = x.(yz)$
$x + 1 = 1$	$x + xy = x$	$xy + xz = x(y + z)$
$x.0 = 0$	$x(x + y) = x$	$(x + y)(x + z) = x + yz$
$x + x = x$	$x + \bar{x}y = x + y$	$(x + y)(y + z)(z + \bar{x}) = (x + y)(z + \bar{x})$
$x.x = x$		$xy + yz + z\bar{x} = xy + z\bar{x}$
$x + \bar{x} = 1$		$(x + y)(\bar{x} + z) = xz + \bar{x}y$
$x\bar{x} = 0$		
$\overline{(\bar{x})} = x$		

A further useful result from Boolean Algebra is called De Morgan's theorem. This theorem states that,

$$\overline{x + y + z + \ldots} = \bar{x}.\bar{y}.\bar{z}\ldots,$$

$$\overline{x.y.z\ldots} = \bar{x} + \bar{y} + \bar{z} + \ldots$$

i.e. the complement of a sum (product) of variables is equal to the product (sum) of its complements. This can be further generalised into,

$$\overline{f(x_1, x_2, x_3\ x_n, +, .)} = f(\bar{x}_1, \bar{x}_2, \bar{x}_3\ \bar{x}_n, ., 4,)$$

i.e. the complement of a Boolean function can be obtained by interchanging the AND (.) and the OR (+) operations and complementing the variables. For example if

$$F(x, y, z, w) = \bar{X}.(Z + \bar{W}.Y) + Z.Y \text{ then,}$$

$$\bar{F}(x, y, z, w) = [X + \bar{Z}(W + \bar{Y})].(\bar{Z} + \bar{Y})$$

A more detailed discussion of Boolean Algebra is given in [1].

10.1.2 Concept of a Basic Gate

The binary variables are realised in electronic digital circuits by voltages which have two distinct levels. The voltage levels act upon some electronic circuits called 'gates' to implement the operations given in Table 10.1. An important characteristic of the digital circuits is that they are highly modular in nature. The gates form the basic building block from which other components, e.g. Flip-flops, Shift Registers and Memories are built. These components in turn are interconnected to form more complex sub-systems, e.g. Multipliers and Accumulators.

The complex digital systems presently use Integrated Circuits (IC) technology to construct a large number of circuits on a single silicon chip. The ICs are normally classified according to the number of components per chip. The Small Scale Integration (SSI), Medium Scale Integration (MSI) and Large Scale Integration (LSI) circuits accommodate respectively about 10–100, 100–1000 and 1000–10,000 components per chip. The VLSI technology is further increasing the density to greater than 100,000 components/chip. All digital ICs in principle can be thought of as an inter-connected set of basic gates. For every logical family the manufacturer chooses one or two basic gates as the unit cells for all IC designs. As a result the manufacturer needs to specify the electric characteristic of the basic gate only. The functions performed by each individual IC are then specified only as the overall input output relationship. Thus there are the following kinds of specifications applicable to a digital IC.

(a) Functional Specifications in terms of a truth table.

(b) Parametric Specifications in terms of electrical characteristics of the basic gate.

(c) Dynamic Specifications in terms of transient response parameters and maximum operating frequency.

These specifications are discussed in detail in the following:

Functional specifications: These specifications usually take the form of a Truth Table. Table 10.3 is a truth table for the basic gates. The table lists the logical levels at the output for all logical combinations of the inputs. In this case A and B are input signals which may be outputs of other gates It is seen that Table 10.3 is essentially same as Table 10.1. The important difference however is that A and B in Table 10.3 and Figure 10.1 represent voltages which take binary values but are subject to the limitations of the Parametric and the Dynamic specifications. On the other hand in Table 10.1, x and y are mathematical variables which assume binary values without any other restrictions.

The functional specifications of more complex logical systems, for example counters, are elaborate as there are numerous input states. In general for a system with n binary inputs there are 2^n input states. Some inputs for example Clocks and Strobes periodically change between the logical zero and one states and are treated separately. The truth table is supplemented in complex systems by waveforms obtained at various nodes during a cyclic operation. The functional specifications of a complex system may also include a brief description of the actions of different inputs.

Table 10.3. Truth table for the basic gates of Fig. 10.1

Input state		Output of the Gate					
A	B	AND	NAND	OR	NOR	Ex-OR	NOT
0	0	0	1	0	1	0	1
0	1	0	1	1	0	1	1
1	0	0	1	1	0	1	0
1	1	1	0	1	0	0	0

Parametric specifications: These specifications state the electrical characteristics of the digital IC. As mentioned earlier, each family of ICs uses a basic gate as a building block. Therefore it is customary and convenient to state the parametric specifications for a given logic family in terms of the 'Basic Inverter' of that family.

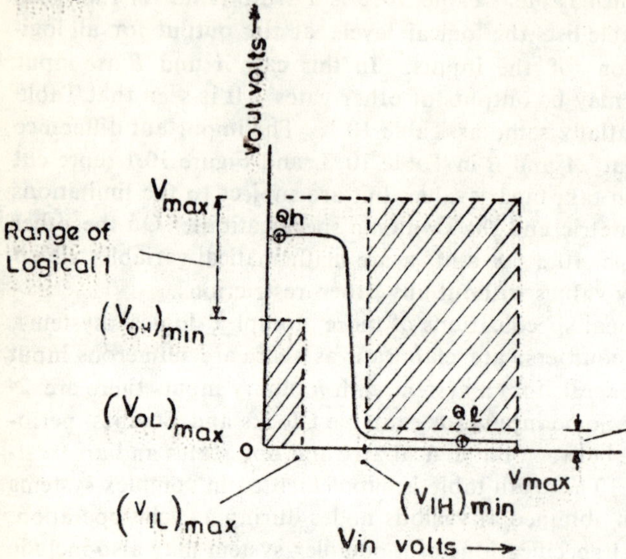

Fig. 10.2. Transfer characteristic of a typical IC inverter. Hatched areas show prohibited operating regions.

Figure 10.2 shows the input-output or transfer characteristic of a typical basic inverter. The characteristic is *continuous* and consists of two flat portions separated by a high gain region with large slope. In most inverters the high gain region is monotonic and there are only two points with slopes equal to unity. The characteristic shows that for low values of input voltage the output voltage is high, e.g. point Q_h in Fig. 10.2 and vice-versa. The input and output voltages are lower than an upper limit, V_{max}, which is determined by the power supply to the inverter. If the input signal is binary, stable operation is possible at some points, e.g. Q_h and Q_l in Fig. 10.2. An important requirement of the logic family is that output voltages of any gate corresponding to logical zero and logical one, be recognised as such, without ambiguity, at the input terminals of any other gate of the same family. A further complication arises from the variations in the transfer characteristics due to manufacturing tolerances, temperature variations, loading and power supply fluctuations. The IC manufacturers therefore specify the set of parameters which all members of the family must satisfy. These parameters are the worst case values of the input and the output voltages. Figure 10.2 shows the maximum value of the output voltage at logical zero, $(V_{0L})_{max}$, and the maximum input voltage, $(V_{IL})_{max}$, which is recognised as logical zero. The basic inverter is designed and tested to keep a margin between the above mentioned values. This is called 'Noise Margin' and provides unambiguous logic levels even in the presence of limited d.c. noise. Thus the d.c. noise margin at logical zero is

$$NM_0 = (V_{IL})_{max} - (V_{0L})_{max} \tag{10.1}$$

Similarly, the worst case values at logical one are $(V_{0H})_{min}$ and $(V_{IH})_{min}$ and the noise margin is given by

$$NM_1 = (V_{0H})_{min} - (V_{IH})_{min} \tag{10.2}$$

In terms of Fig. 10.2 these specifications mean that all transfer characteristics lie outside the hatched area. The permissible ranges of the output voltages are $(V_{0H})_{min} \leqslant V_{0H} \leqslant V_{max}$ and $0 \leqslant V_{0L} \leqslant (V_{0L})_{max}$. Similarly, the permissible ranges for the input voltages are $0 \leqslant V_{IL} \leqslant (V_{IL})_{max}$ and $(V_{IH})_{min} \leqslant V_{IH} \leqslant V_{max}$. Figure 10.3 shows the relative values of various voltage levels of the inverter characteristics.

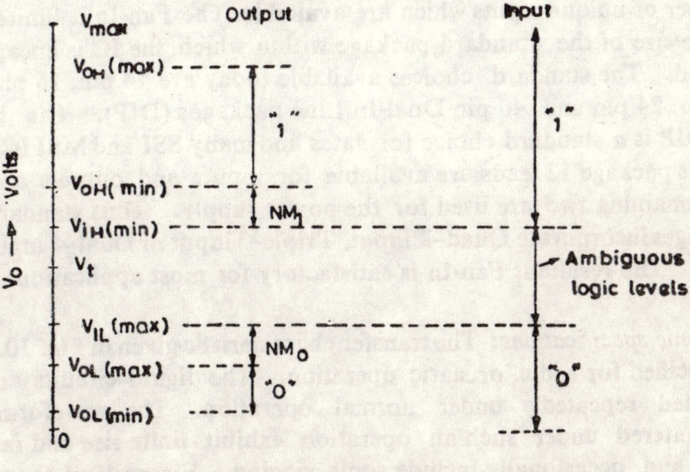

Fig. 10.3. Relative values of voltage levels for the inverter characteristics of Fig. 10.2.

Loading rules: Each gate requires a certain amount of input current. The drive requirements depend upon the logical level and are different for each logic family. Similarly the output drive capability depends on the logical level and the type of logic family. Since most systems are built from basic inverters of the same family by interconnections, it is convenient to specify the drive capabilities within one family by simple loading rules. For this purpose the load due to each input of the basic inverter is defined as the Unit Load.

For each member of the particular logic family the manufacturers specify the following:

(i) A number indicating the number of unit loads each input terminal presents to the output of the preceding stage.

(ii) A number indicating the number of unit loads each output can drive. This number is known as the 'Fan-Out' of the particular output.

The numbers mentioned above permit easy calculation for interconnecting different ICs and also different stages within an IC. The manufacturer guarantees that if the Fan-Out of the gate is not exceeded the characteristics will remain in the permissible region of Fig. 10.2.

The 'Fan-In' of a particular type of logic gate is simply the number of unique inputs which are available. The Fan-In is limited by the size of the standard package within which the IC is encapsulated. The standard choices available today are 14 pin, 16 pin, 20 pin, 24 pin and 40 pin Dual-In-Line packages (DIP). The 14 pin DIP is a standard choice for gates and many SSI and MSI ICs. In this package 12 leads are available for inputs and outputs and the remaining two are used for the power supply. Thus standard packages incorporate Quad–2 input, Triple–3 input or Dual–4 input gates. The resultant Fan-In is satisfactory for most applications.

Dynamic specifications: The transfer characteristic given in Fig. 10.2 is specified for a d.c. or static operation. The digital circuits are switched repeatedly under normal operation. The waveforms encountered under such an operation exhibit finite rise and fall times and occasionally include some ringing. Figure 10.4 shows the output voltage waveform of the inverter corresponding to a voltage pulse at the input. The rise and fall times at the output are specified as the time required to traverse from the 10% to the

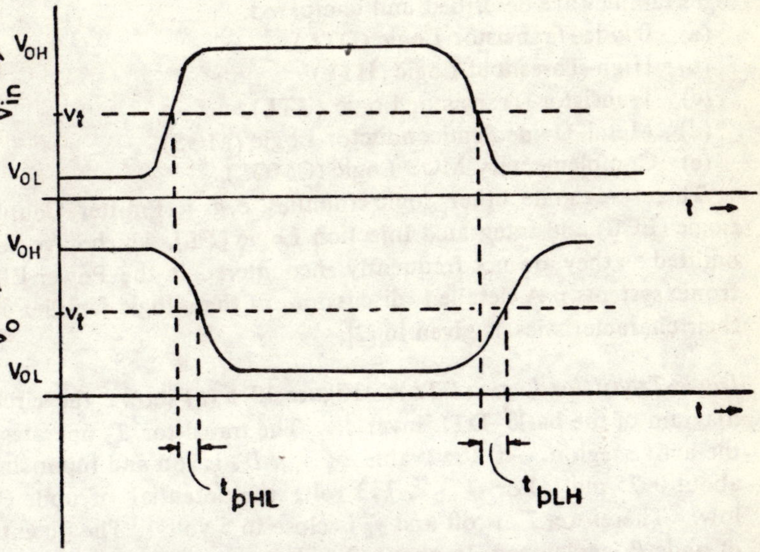

Fig. 10.4. Transient response of the IC inverter.

90% of the voltage swing. The delay time is measured between points where the waveforms pass through the switching threshold voltage v_t. The threshold voltage v_t may be defined as the point on the d.c. transfer characteristic of Fig. 10.2 where the input and the output voltages are equal. The delay time between a rising input and a falling output is termed t_{pHL} while that between a falling input and a rising output is termed t_{pLH}. Sometimes a single delay parameter t_{pd} is specified which is the average of t_{pHL} and t_{pLH}. Another common specification is the pair delay of the gate which is equal to the sum of t_{pHL} and t_{pLH}. The dynamic specifications determine the delay introduced in propagation of the digital signal through the gate. This delay places a basic limit on the operating frequency of the digital circuits.

10.1.3 Integrated Circuit Logic Families

The mass manufacture of ICs has enabled rapid design of the logic systems. A basic design decision is the choice of the specific family of ICs which are to be interconnected to form the particular logic system. Therefore, it is important to study and compare the characteristics of the available IC logic families. In this section the internal structure and the characteristics of the following logic families are described and compared.

(a) Diode–Transistor Logic (DTL),
(b) High–Threshold Logic (HTL),
(c) Transistor–Transistor Logic (TTL),
(d) Metal–Oxide Semiconductor Logic (MOS),
(e) Complementary MOS Logic (CMOS).

There are some other logic fomilies e.g. Emitter–Coupled Logic (ECL) and Integrated Injection Logic (I^2L), which have been omitted as they are not frequently encountered in the Power Electronic systems. A detailed discussion of these logic families and their characteristics is given in [2].

Diode-Transistor Logic (*DTL*): Figure 10.5 (a) shows the circuit diagram of the basic DTL inverter. The transistor T_1 operates in the active region. At low value of v_{iA}, D_A is on and input sinks about 0.75 ma. For $V_{iA} < 1.3$ volts the potential of node P is low. Therefore, T_2 is off and v_0 is close to 5 volts. The potential of node P is clamped to about $2.15V$ as v_{iA} is increased above $1.3V$ and D_A turns off. Transistor T_2 turns on and finally saturates when v_{iA} exceeds about 1.4 volts. The resulting transfer characteristic is shown in Fig. 10.5 (b). The use of transistor T_1 gives a current gain and allows a reduction in R_b compared to the earlier discreet DTL circuits. A reduced value of R_b means faster

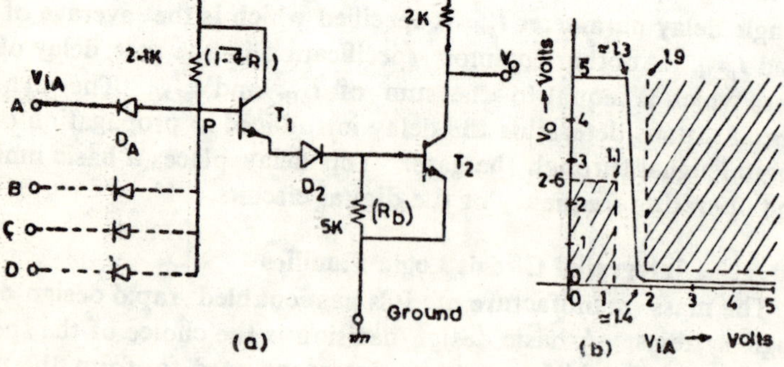

Fig. 10.5. (a) Circuit diagram. (b) Transfer characteristics of a DTL inverter.

turn-off of T_2. When driving other DTL gates T_2 sinks current when v_0 is at logical zero. Thus, the voltage level at logical zero increases with load. When v_0 is at logical one, the current requirement is small as the input diodes of the driven gate are off. However, the charging of the output capacitance takes place through R_C. Therefore, t_{pLH} is about twice of t_{pHL} being about 60–80 nS and 20–30 nS respectively.

The inverter using a single input diode D_A can be converted to a NAND gate by connecting extra diodes shown dotted in Fig. 10.5 (a). When all the inputs A to D are at logical one the current is diverted to the base of T_2 and turns it on. Hence, the NAND logic function. The circuit diagram shown in Fig. 10.5(a) shows an output stage with a passive pull up resistor of 2K. In some DTL ICs a pull up resistor of 6K is used with consequent increase in t_{pLH}. The use of output stage with passive pull up permits a connection of two ICs in parallel. This results in the 'Wired–OR' function. In order to increase the fan-out of the gate, the output stage can be changed to an active pull up. This is normally done in DTL buffers. However, the active pull up output stage can not be used in the Wired–OR configuration.

The specifications of a typical DTL inverter are given in Table 10.4.

High Threshold Logic (*HTL*): The high threshold logic developed by Motorola Company, is a modification of the DTL logic. Fig. 10.6(a) shows the circuit diagram of the HTL inverter. A

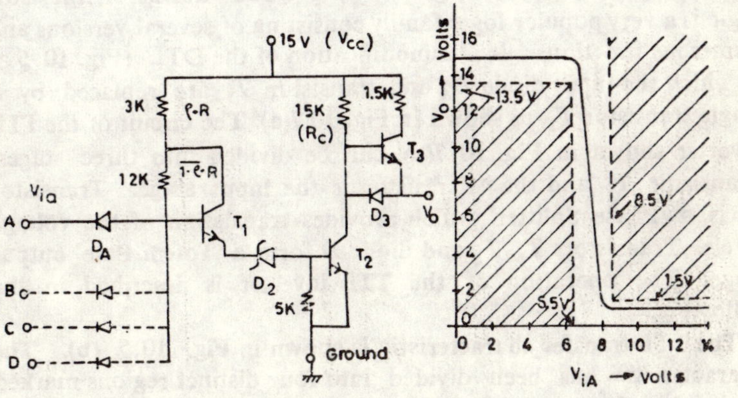

Fig. 10.6. (a) Circuit diagram. (b) Transfer characteristics of a HTL inverter.

comparison of Figs. 10.5 (a) and 10.6 (a) shows that in the HTL logic diode D_2 is a zener diode rather than a forward biased diode. The use of a zener diode with a breakdown voltage of about 6.7 volts changes the threshold voltage to a high value of 7.5V. The supply voltage V_{cc} has been raised to 15V. To limit the power dissipation, resistors R and R_c are of higher value compared to DTL. The HTL gate in Fig. 10.6 (a) has an active pull up output stage comprising of transistors T_3 and diode D_3. A passive pull up stage without these devices is also used. The passive pull up stage has larger rise time but allows the use of Wired–OR function. The HTL inverter can be modified into a NAND gate by the addition of the inputs B, C, D and the diode as shown by dotted lines in Fig. 10.6 (a).

Figure 10.6 (b) shows the transfer characteristics of HTL inverter. It is seen that the threshold voltage, i.e. the point where v_0 equals V_{tA}, is approximately 7.5 volts. The various input and output voltage limits are also shown. The noise margin of the HTL family is 5 volts which is an order of magnitude higher than any other logic family. Therefore, the HTL logic family is superior to other logic families in the high noise environments such as those present in Power Electronic applications. However, the HTL family suffers from a limited choice of available functions, higher cost and lower speed.

The specifications of a typical HTL inverter are shown in Table 10.4.

Transistor-Transistor Logic (TTL): The transistor-transistor logic is a very popular logic family consisting of several versions and numerous functions. It is a modification of the DTL (Fig. 10.5 a) in which the input diode D_A and transistor T_1 are replaced by a single transistor T_1, as shown in Fig. 10.7(a). The circuit of the TTL inverter shown in Fig. 10.7(a) can be divided into three stages. Transistor T_1 and the 4K resistor is the input stage. Transistor T_2 is a 'Phase Splitter' which provides translation of the voltage levels. Transistors T_3, T_4 and diode D form a 'Totem Pole' output stage. The operation of the TTL inverter is described in the following.

The TTL transfer characteristic is shown in Fig. 10.7 (b). The characteristic has been divided into four distinct regions marked in the above mentioned figure. In *region I* at low values of V_{tA}, transistors T_2 and T_4 are off and T_3 is on. The output voltage

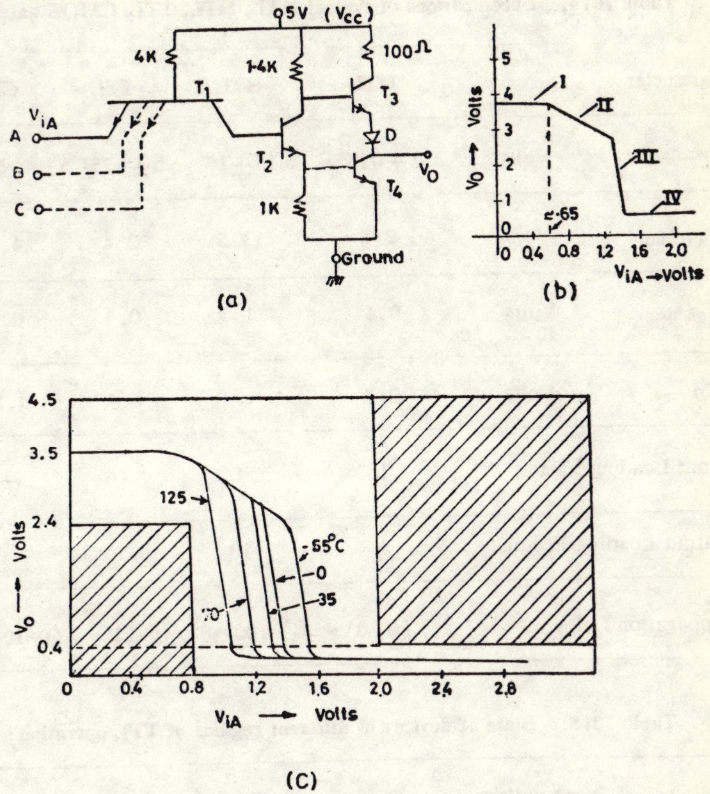

Fig. 10.7. (a) Circuit diagram of a medium speed TTL circuit.
(b) Transfer characteristics of a TTL inverter.
(c) Transfer characteristics of a TTL inverter at different temperatures.

V_0 is two diode drops below V_{CC} and is about 3.7V. The *B–E* junction of T_1 acts similar to the diode D_A in DTL (Fig. 10.5 a). The input sinks approximately 1mA of current. Since T_2 is off the collector-emitter voltage of T_1 is about 0.1V and collector current is the small leakage current of the *C–B* junction of T_2. When V_{iA} exceeds 0.65V, *region II*, the base voltage of T_2 rises above the cut-in value and T_2 begins to conduct. In region II T_2 is in the active region while T_4 is still off. T_2 acts like an amplifier with a gain of approximately 1.4. As the input voltage increases the base voltage of T_2 increases. Consequently the collector voltage

Table 10.4. Specifications of typical DTL, HTL, TTL, CMOS gates

Parameter		DTL	HTL	TTL	CMOS
V_{CC}	Volts	5 ± 0.5	15 ± 1	$5 \pm .25/.5$	5
$(V_{0H})_{min}$	Volts	2.5	13.5	2.4	4.99
$(V_{0L})_{max}$	Volts	0.4	1.5	0.4	0.01
NM	Volts	0.7	5	0.4	1.5
Input Loading Factor		1	1	1	—
Output Loading Factor		8	10	8–10	—
Propagation Delay nS		20-30	100	6–12	50–100

Table 10.5. State of devices in different regions of TTL operation

Region	Input voltage	T_1	T_2	T_3	T_4
I	$0-0.65$	on	off	on/sat	off
II	$0.65-1.35$	on	on	on	off
III	$1.35-1.45$	on	on	on	on
IV	>1.45	Inverse Active	sat	off	sat

of T_2 and the output voltage V_0 falls. The slope of the V_0-V_{iA} characteristic in region II is about 1.4. Throughout region II the base voltage of T_4 is increasing while that of T_3 is reducing. When V_{iA} exceeds 1.35 volts transistor T_4 turns on and shunts the 1K emitter resistance of T_2. Consequently, the gain of T_2 increases causing a rapid fall in the output voltage in *region III*. The gain in region III can be shown to be approximately six. When the input voltage reaches 1.45 volts the transistor T_3 shuts off, T_2 and T_4 reach saturation. In the *region IV* the input voltage is high, the output voltage is low and T_4 sinks the output current. The output voltage V_0 is the collector-emitter voltage of the saturated transistor T_4 and varies between 0.1 to 0.3 volts. Transistor T_1 operates in the 'inverse active' region as the base current is almost completely diverted to the collector. The inverse gain of the transistor T_1 is very low (0.2). Therefore, the input current at terminal A is small being approximately 14μA. Table 10.5 summarizes the state of the transistors in the four regions of operation. A detailed analysis of the TTL circuit is presented in [2].

The characteristic drawn in Fig. 10.7 (b) is general but the various break points are dependent on the transistor parameters and the temperature. Figure 10.7 (c) shows the temperature dependence and voltage limits. The manufacturer guarantees that over the entire temp. range the transfer characteristic lies outside the hatched area. The noise margin of the TTL logic family as seen from Fig. 10.7 (c) is only 0.4V and is thus smaller than DTL and HTL (Table 10.4). For this reason considerable care must be exercised in layout of this logic family. A detailed discussion of the noise considerations is given in [4] and chapter 5 of [5].

The TTL inverter can be extended to a NAND gate by adding the inputs B and C shown dotted in Fig. 10.7 (a). The use of active pull up does not permit Wired-OR connection. Some TTL gates provide 'Open Collector' output stage. In this arrangement the collector of T_4 is brought out and must be tied to V_{cc} by an external pull up resistor. The value of pull up resistor can be chosen to suit the desired rise time for the specific load. The circuit shown in Fig. 10.7 (a) is a 'medium speed' TTL inverter. In a high speed version of the basic circuit the emitter resistance of T_2 is reduced from 1 K to 600 Ω. Alternately it is replaced by an active pull down stage consisting of another transistor and resistors. The attempt in both cases is to provide for a fast removal of the charge stored in the base-emitter junction of T_4. The transfer characteristic with active pull down is different from Fig. 10.7 (b) in that the region II is missing and region I extends till about 1.35 volts. The inverter then directly enters region III with a higher slope. Another important version of TTL family is the 'Schottky-clamped' inverter. In this version each transistor has a built-in diode across the C-B junction which prevents its being forward biased. Therefore, all the transistors operate in the active i.e. non-saturated region. Consequently, the stored charges are smaller and the propagation delay is reduced. A detailed discussion on high speed and Schottky TTL inverters is given in [4].

Specifications of a typical medium speed TTL inverter are given in Table 10.4. The power supply voltage of TTL family is nominally 5V. The 7400 series calls for regulation within $\pm$ 0.25V while the costlier military grade 5400 series can tolerate $\pm$ 0.5V fluctuation. The use of the totem-pole output stage provides a sourcing current of approximately 10mA. The output transistor T_3 (Fig. 10.7(a)) remains in the active region for loads up to about 5mA and then saturates thus increasing the output impedance from 70Ω to 100Ω. When the output is at logical zero however ample base drive is provided to T_3 and it can sink 10-15mA of current keeping V_0 lower than the limit of 0.4V. The fan out of medium speed TTL driving itself is specified as 10. However, the fan out varies if different versions of the family are mixed. A detailed fan out tabulation is given in [5].

Metal-oxide semiconductor (MOS) logic: The logic families discussed so far use the bipolar transistor as the basic switching element. In a similar manner it is possible to use MOS transistors. The use of MOS transistors provides the advantages of low power dissipation and high packing density. The high packing density is achieved by eliminating the use of resistors and instead using transistors exclusively.

Figure 10.8(a) shows a MOS inverter using p-channel transistors. In this circuit the top transistor acts like a nonlinear load resistor for the driver or the inverter transistor. Figure 10.8(b) shows the transfer characteristic of such an inverter. The input and the output voltages have been normalised in this figure to the supply voltage V_{DD}. V_T is the threshold voltage of the transistors. m is a parameter depending on the supplies and is given by

$$m = V_{DD}/2(V_{GG}-V_T)-V_{DD} \tag{10.3}$$

It is interesting to note that from Fig. 10.8(b) that the load and driver transistors must be dissimilar in order to obtain a sharp fall in the output voltage. This is expressed by the ratio K_D/K_L. The constants K_D and K_L depend upon the width to length ratio of the channels of the driver and the load transistor respectively.

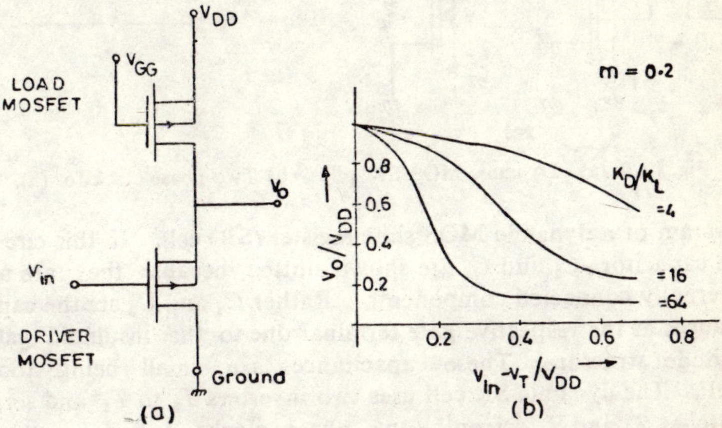

Fig. 10.8. (a) Basic *p*-MOS inverter. (b) Transfer characteristics of (a).

Many different versions of the basic MOS inverter have been fabricated. The different combinations are designated by letters indicating channels (*p* or *n*), mode (*E*nhancement or *D*epletion) and region (*T*riode or *S*aturated). While fabrication process for *p*-channel devices is simpler, the *n*-channel devices provide higher speed because of inherent mobility advantage of electrons over holes. The enhancement mode devices are preferred due to reduction of one step in fabrication over the depletion mode devices. Thus in MOS inverters the most popular combinations are NELT and NELS. A detailed analysis and design procedure for MOS inverters is given in [6]. Due to their complex processing but high packing density the MOS logic family is economical only when used in LSI circuits. Most practical LSI devices use NMOS technology.

The MOS devices also provide a class of logic circuits which have no counterpart in the bipolar family. Such circuits are termed '*Dynamic*' MOS circuits. Figure 10.9(a) shows the circuit

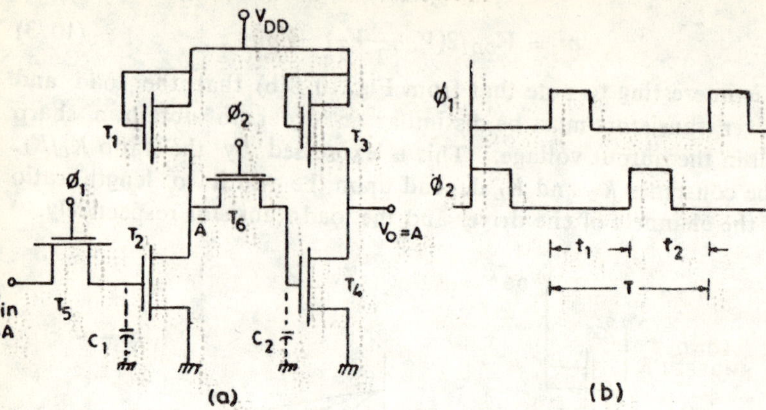

Fig. 10.9. (a) Dynamic MOS SR cell. (b) Two-phase clock for (a).

diagram of a dynamic MOS shift register (SR) cell. In this circuit the capacitors C_1 and C_2 are shown dotted because they are not externally connected components. Rather C_1 and C_2 are the capacitances at the respective gate terminal due to the insulated gate-cathode structure. These capacitances are small being about 5 pF. The dynamic SR cell uses two inverters T_1 to T_4 and series switches T_5 and T_6 driven by two phase clocks ϕ_1 and ϕ_2 respectively. During time t_1 (Fig. 10.9 (b)) T_5 is on and charges C_1 to the level of V_{in}. The output of the inverter T_1-T_2 is therefore $\bar{A}$ being complement of the input signal A. During time t_2, T_5 is off while T_6 is turned on. The capacitor C_2 is then charged to the level of the output voltage of inverter T_1-T_2. Therefore, the final output voltage at the end of t_2 is equal to the input A. However, since the charges are stored in the capacitors C_1 and C_2 they tend to leak off unless replenished periodically. Therefore, ϕ_1 and ϕ_2 are applied continuously to preserve the output voltage equal to A. This process is called 'refreshing' the dynamic cell. Many other versions of the dynamic MOS SR cell exist to obtain more efficient operation [6]. The dynamic MOS circuits are mainly used in LSI memories and shift registers where the refresh cycle is provided by special in-built circuits.

Complementary MOS (CMOS) logic: The MOS inverter shown in Fig. 10.8(a) uses two transistors of same polarity. A modified inverter is shown in Fig. 10.10(a). Here the driver is a *n*-channel MOS transistor while the load is a *p*-channel MOS transistor.

Therefore, it is called the complementary MOS (CMOS) inverter. The transfer characteristics of the inverter is shown in Fig. 10.10 (b). In region *I* where the input voltage V_{in} is low the *p*-channel transistor is on while the *n*-channel transistor is off. In region *V* when the input voltage is high the *n*-channel transistor is on while the *p*-channel transistor is off. Thus, under quiescent conditions the CMOS inverter draws extremely low power from the supply voltage V_{DD} as one of the two transistors is off. This reduces the overall power consumption of the CMOS family to a value much lower than other logic families and gives it a great advantage.

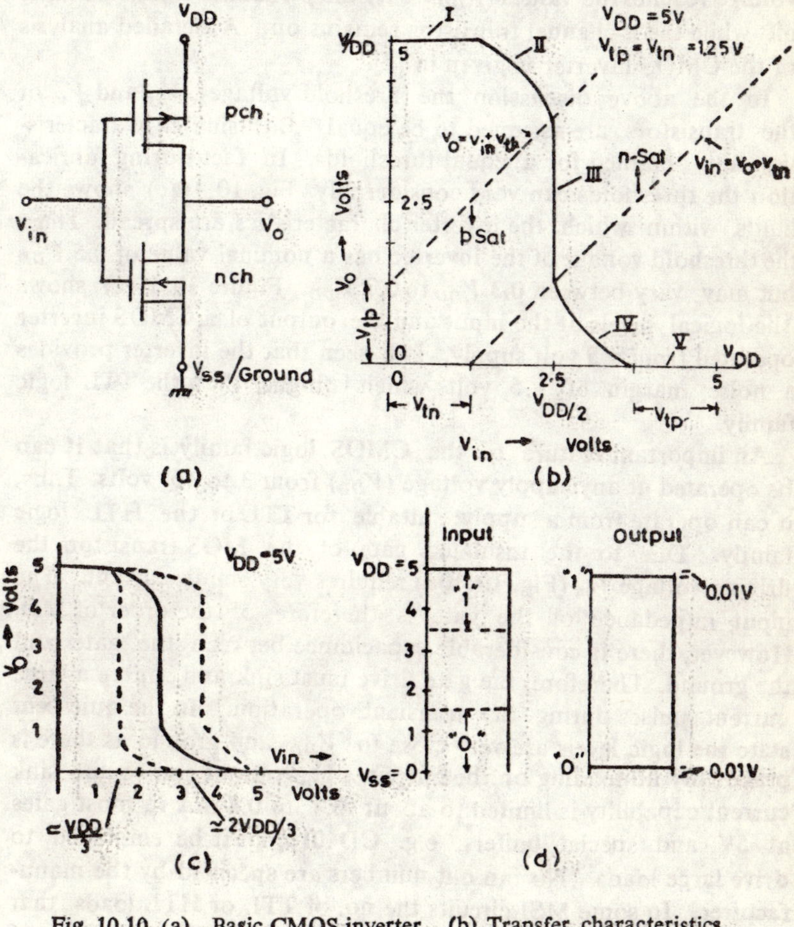

Fig. 10.10. (a) Basic CMOS inverter. (b) Transfer characteristics. (c) Variation in characteristics. (d) Logic levels for (a).

The dotted lines in Fig. 10.10(b) show the boundary between saturation and on regions. Thus in region *II* *p*-channel transistor is on while the *n*-channel transistor is saturated. The output voltage of the inverter falls slowly. When the input voltage reaches $V_{DD}/2$ and if both the transistors are identical, the output voltage falls abruptly to a low value shown as region *III*. In region *III* both transistors are saturated. If the transistors are not identical the fall is not abrupt but is still very steep. In region *IV* the *p*-channel transistor is saturated while the *n*-channel transistor is on. The output voltage falls slowly towards zero. When the input voltage reaches the values $(V_{DD}-V_{tp})$ the *p*-channel transistor turns off while the *n*-channel transistor remains on. A detailed analysis of the CMOS inverter is given in [2].

In the above discussion the threshold voltages, V_{tp} and V_{tn} of the transistors are assumed to be equal. But similar characteristics are obtained for unequal thresholds. In fact during fabrication the thresholds can vary considerably. Fig. 10.10(c) shows the limits within which the inverter characteristics are spread. Thus, the threshold voltage of the inverter has a nominal value of 0.5 V_{DD} but may vary between 0.3 V_{DD} to 0.7 V_{DD}. Figure 10.10(c) shows the logical levels at the input and the output of a CMOS inverter operated from a 5 volt supply. It is seen that the inverter provides a noise margin of 1.5 volts which is higher than the TTL logic family.

An important feature of the CMOS logic family is that it can be operated at any supply voltage (V_{DD}) from 3 to 15 volts. Thus, it can operate from a supply suitable for TTL or the HTL logic family. Due to the insulated gate of the MOS transistors the driving voltage V_{in} (Fig. 10.10a) requires very small current. The input impedance of the gate is therefore of the order of MΩ. However, there is considerable capacitance between the gate and the ground. Therefore, the gate drive must sink and source a large current pulse during the transient operation. In the quiescent state the logic levels are very close to V_{DD} and ground as there is practically no loading on the output. The output source and sink current capability is limited to about 0.4 to 0.8 mA in most gates at 5V and special buffers, e.g. CD4010, must be employed to drive large loads. No fan out numbers are specified by the manufacturer. In some MSI circuits the no. of TTL or HTL loads that can be driven, typically one or two, is specified. Specifications of a typical CMOS gate operating from 5V supply are given in Table 10.4.

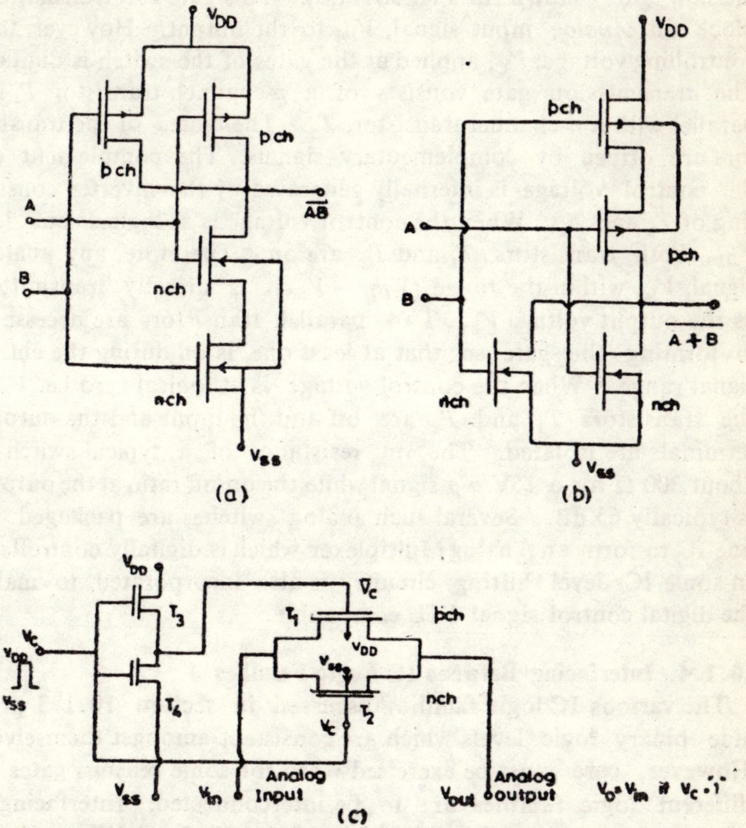

Fig. 10.11. CMOS gate structure. (a) NAND gate. (b) NOR gate. (c) Transmission gate.

The CMOS logic family employs series/parallel structures to obtain NAND and NOR functions as shown in Fig. 10.11 (a) and (b). The basic operation of these gates is similar to that of the inverter. For example in the NAND gate when both *A* and *B* are at logical one both the *n*-channel transistors are turned on while the *p*-channel transistors are turned off. This reduces the output to logical zero. If any of the inputs goes to zero the corresponding *n*-channel transistor shuts off while the *p*-channel transistor turns-on and the output goes to logical one. The NOR gate of Fig. 10.11(b) can be analysed similarly.

The CMOS logic family also provides a very unique transmission gate shown in Fig. 10.11(c). This gate can transmit or block an *Analog* input signal, V_{in}, to the output. However, the controlling voltage, V_C, applied at the gates of the switch is digital. The transmission gate consists of a *p*-channel transistor T_1 in parallel with a *n*-channel transistor, T_2. The gates of the transistors are driven by complementary signals. The complement of the control voltage is internally generated by the inverter consisting of T_3 and T_4. When the control voltage is at logical one i.e. V_{DD}, both transistors T_1 and T_2 are on. Therefore, any analog signal, V_{in}, within the range $(V_{DD} - V_{SS})$, is directly transmitted as the output voltage V_0. Two parallel transistors are necessary in forming the gate so that at least one is on during the entire signal range. When the control voltage is at logical zero i.e. V_{SS}, the transistors T_1 and T_2 are off and the input and the output terminals are isolated. The on resistance of a typical switch is about 300 Ω for a 15V *p-p* signal while the on/off ratio at the output is typically 65 dB. Several such analog switches are packaged in one IC to form an Analog Multiplexer which is digitally controlled. In some ICs level shifting circuitry is also incorporated to make the digital control signal TTL compatible.

10.1.4 Interfacing Between IC Logic Families

The various IC logic families discussed in section 10.1.3 provide binary logic levels which are consistent amongst themselves. However, care must be exercised when for some reasons gates of different logic families are to be interconnected. Interfacing is usually done to combine the best features from different logic families e.g. to obtain high speed of TTL with complex functions of CMOS or noise immunity of HTL. In many cases manufacturers have provided special ICs to convert from one logic family to another.

The basic problem of interfacing between the logic families is the compatibility of the voltage levels and the drive current requirements. As there are many versions within each family the manufacturer's date sheets must be consulted for each interfacing problem. Some common interfaces are discussed below.

TTL driving CMOS: The output current capability of TTL gates is sufficient to drive CMOS inputs. However, when both are operated from 5V supply, the TTL output voltage of 2.4 to 3.7V during logical one is not compatible with input range of CMOS (Fig. 10.10c). A simple solution is to use an external 2 to 6K Ω pull up resistor to increase the voltage to nearly 5 volts. A low value of the pull up resistor increases the power loss during the logical zero state while a high value reduces the speed during low to high transition. As the current required by CMOS gates is almost negligible a large fan out is possible. Level shifting buffers are required if CMOS is operated at voltages higher than 5 volts.

CMOS driving TTL: The output voltages are compatible in this case but the limited current capabilities of CMOS pose a problem. In order to sink the 1.6 mA current from a medium speed TTL, about four CMOS gates must be in parallel to obtain the required sinking current capability. Alternately CMOS buffer gates which can sink up to 6 mA must be used. A more elegant solution is to use low power TTL which sinks only 0.18 mA. In typical cases a fan-out of four is easily attainable.

CMOS driving HTL: CMOS can drive about 5 HTL loads satisfactorily if buffers are used with 15 V supply.

HTL driving CMOS: External pull up resistor is not necessary in this interface but the noise immunity with active pull up output state is reduced to 3.5V. With passive pull up, usually 15K Ω, or open collector output stage the noise immunity of 5 V can be obtained. Sufficient output current capability is available in HTL gates but the speed is the main limitation if higher pull up resistors are used to reduce the power loss. A more detailed analysis of interfaces between CMOS and other logic families is given in [7].

10.2 DESIGN OF COMBINATORIAL CIRCUITS

The basics of digital logic and logic families have been discussed in the last section. In this section the design of a particular class of digital circuits called 'combinatorial circuits' is described. General structure of a combinatorial circuit is shown in Fig. 10.12. The circuit has n binary inputs $X_1, X_2, \ldots, X_n$ and m binary outputs $Z_1, Z_2, \ldots, Z_m$. The combinatorial circuit itself may consist of interconnection of several basic gates discussed in section 10.1. In fact the gates shown in Fig. 10.1 are examples of the simplest form of combinatorial circuit with two inputs and one output. Another example is shown in Fig. 10.13 which has three inputs and one output.

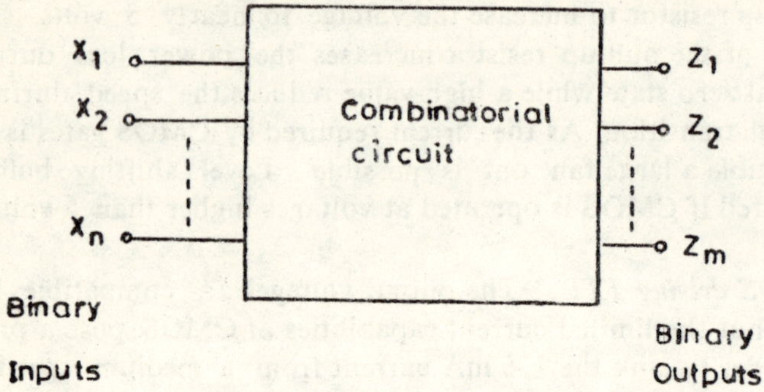

Fig. 10.12. General structure of a combinatorial circuit.

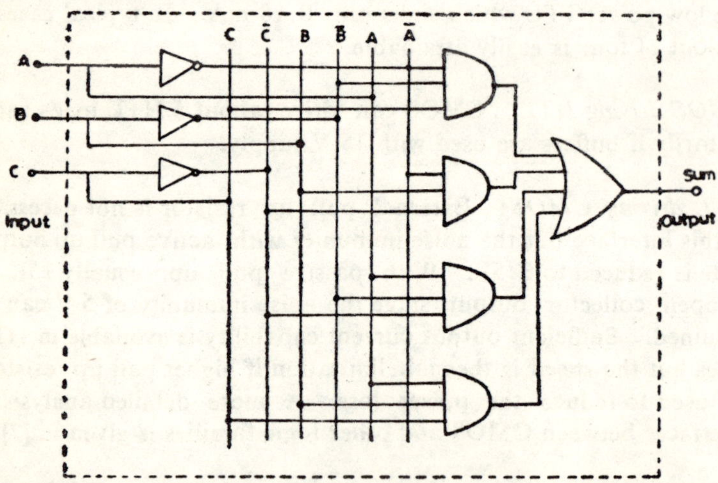

Fig. 10.13. A combinatorial circuit implementing the function 'sum' of table 10.6 and eqn. (10.5).

Each of the binary inputs $X_1, X_2, \ldots, X_n$ in Fig 10.12 can assume a value of either '0' (i.e. Logical Zero, Low or False) or '1' (i.e. Logical one, High or True). When each input variable is assigned a value, either zero or one, a set of 0's and 1's is obtained. This set is called the state of the input variables or simply the input state. Similarly the set of 0's and 1's corresponding to the output variables $Z_1 Z_2 ... Z_m$ is called the output state. The defining characteristic of the combinatorial circuits is that the output state depends solely on the present input state. Since the output state is independent of the past input states the combinatorial circuit is called a memoryless system. On the other hand, the sequential circuits discussed in the next section, have a memory and the output state depends on the present as well as the past input states.

The first step in the design of the combinatorial circuit is the assignment of variables. This step involves translation of the particular problem from a common language e.g. English to a precise mathematical description in terms of logical variables. Some examples of this exercise are given in [1]. This description takes the form of a truth table or a K-map which may be reduced to a Boolean expression. The next step is the minimisation of the Boolean expression. Finally the minimised or in some cases the original Boolean expression is realised using Gates, Multiplexers or Read only Memories (ROMs). These design steps are discussed in detail in the following and illustrated with examples.

10.2.1 Truth Table and K-Map

No general rules can be given for constructing the truth table except to state that for n input variables, there are exactly 2^n input states. The truth table shows the value, 0 or 1, of each of the output variables for each input state. For example, Table 10.6 shows the truth table of a One Bit Full Adder. The circuit has three inputs A, B and C and two outputs Sum and Carry. There are eight input states corresponding to the three input variables. The last column in Table 10.6 shows the input states in an algebraic form. The given binary expression equals one only in the particular state. The Boolean expression for the outputs can be directly written by combining the input state expressions corresponding to which the output equals one by OR operations.

Thus,

$$\text{Carry} = \bar{A}BC + A\bar{B}C + AB\bar{C} + ABC \tag{10.4}$$

and

$$\text{Sum} = \bar{A}\bar{B}C + \bar{A}B\bar{C} + A\bar{B}\bar{C} + ABC \tag{10.5}$$

The R.H.S. of eqn. (10.4) is a Boolean expression containing three variables A, B and C. Therefore 'Carry' can be termed as a Function of the variables A, B and C. Similarly 'Sum' is also a

Table 10.6 Truth table of one bit full adder

Inputs			Outputs		Input State in Algebraic Form
A	B	C	Carry	Sum	
0	0	0	0	0	$\overline{A}\,\overline{B}\,\overline{C}$
0	0	1	0	1	$\overline{A}\,\overline{B}\,C$
0	1	0	0	1	$\overline{A}\,B\,\overline{C}$
0	1	1	1	0	$\overline{A}\,B\,C$
1	0	0	0	1	$A\,\overline{B}\,\overline{C}$
1	0	1	1	0	$A\,\overline{B}\,C$
1	1	0	1	0	$A\,B\,\overline{C}$
1	1	1	1	1	$A\,B\,C$

Function of variables *A*, *B* and *C*. It is understood that all variables are Boolean, Binary or Digital.

Karnaugh Map (*K-Map*): An alternative representation of the input output relationship is the Karnaugh Map. A K-map is a two-dimensional representation of the truth table of any combinatorial circuit. It can be constructed for any number of input variables, though it is useful as a design tool only up to four variables. In general, an *n*-variable K-map is constructed by listing the various combinations of *m* input variables along one co-ordinate axis and those of the remaining $n-m$ input variables along the other co-ordinate axis. If *n* is even one chooses $m = \frac{1}{2}n$ for the sake of symmetry and convenience, while for odd values of *n*, *m* and $n-m$ are chosen to be consecutive integers. Along either axis, the input combinations are so ordered that only one bit has different values in two neighbouring input combinations. The configurations of 2-variable, 3-variable and 4-variable K-maps are given in Fig. 10.14.

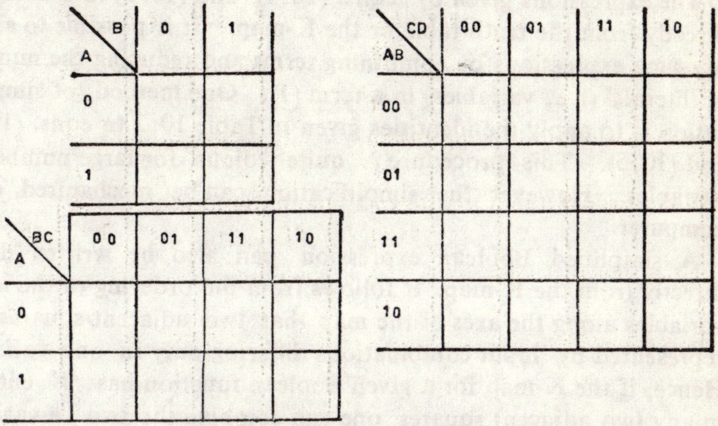

Fig. 10.14. Configurations of 2-variable, 3-variable and 4-variable K-maps.

The K-map for a given Boolean function of variables is constructed by entering '1's and '0's in the squares of an *n*-variable K-map, where the given Boolean function has the value '1' and '0' respectively. This is shown in Fig. 10.15 by the K-maps for the Sum and the Carry outputs of the full adder. Each of these is a three variable K-map as the full adder has three inputs. Note that each term, called minterm [1], in eqns. (10.4) and (10.5) corresponds to a square having an entry '1' in the K-maps for Sum

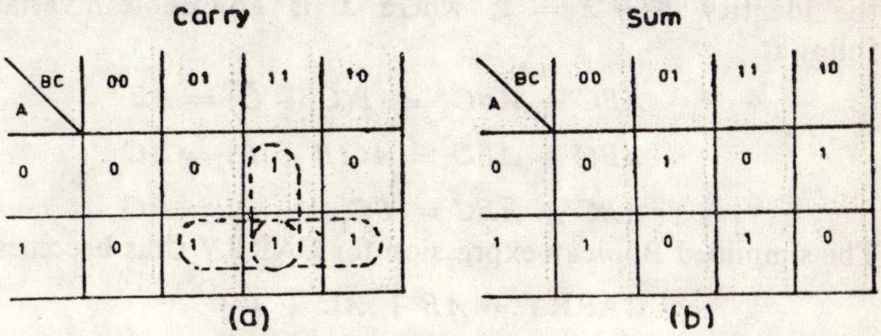

Fig. 10.15. 3-variable K-maps for (a) carry and (b) sum of full-adder (Table 10.6).

and Carry respectively. It is thus possible to write equations (10.4) and (10.5) directly from K-maps.

10.2.2 Minimisation of Boolean Expressions

The expressions given by eqns. (10.4) and (10.5) are obtained directly from the truth table or the K-map. It is possible to simplify such expressions by combining terms and reducing the number of 'literals' (i.e. variables) in a term [1]. One method of simplification is to apply the identities given in Table 10.2 to eqns. (10.4) and (10.5). This procedure is quite tedious for large number of variables. However the simplification can be mechanised on a computer.

A simplified Boolean expression can also be written down directly from the K-map. It follows from the ordering of the input variables along the axes of the map that two adjacent squares are represented by input combinations differing only in one variable. Hence, if the K-map for a given Boolean function has '1' entered in any two adjacent squares, one can combine the two n-variable minterms representing these two squares into a single $(n-1)$-variable term. This would clearly lead to a reduction in the number of literals appearing in the final Boolean expression.

Let us consider the CARRY output of a full-adder to illustrate this process. The 3-variable K-map is shown in Fig. 10.15 (a) with the 3 sets of adjacent pairs of squares indicated by loops. Clearly, a single 2-variable term AB represents the pair of two squares corresponding to the minterms ABC and $AB\bar{C}$, similarly, the 2-variable term AC can replace the pair of minterms ABC and $A\bar{B}C$, and the term BC can replace the pair of minterms ABC and $\bar{A}BC$. In fact, these pairings are equivalent to algebraic simplifications based on the identity $X + \bar{X} = 1$, where X is any Boolean variable, as follows:

$$ABC + AB\bar{C} = AB(C + \bar{C}) = AB$$

$$ABC + A\bar{B}C = AC(B + \bar{B}) = AC$$

$$ABC + \bar{A}BC = BC(A + \bar{A}) = BC \qquad (10.6)$$

The simplified Boolean expression for CARRY thus becomes

$$\text{CARRY} = AB + AC + BC \qquad (10.7)$$

It is obvious from the foregoing algebraic considerations that the two extreme columns in a 3-variable K-map have to be treated as 'adjacent'. Indeed, it is as though the rectangle is wrapped around a cylinder so that the two vertical edges coincide.

The combination of two adjacent squares to simplify the Boolean expression can obviously be extended to the combination of two such 'adjacent' pairs of squares, which would then be represented by an $(n-2)$-variable term. Two 'adjacent' pairs of pairs can again be combined, to yield a $(n-3)$-variable term, and so on. The number of squares which can thus be grouped together so as to be represented by a single term is, therefore, a power of 2, e.g., 2, 4, 8,... Such groups are referred to as *binary groups*; a binary group of 2^m adjacent squares is represented by an $(n-m)$-variable term. As practical applications of the K-map seldom extend for values of n beyond 4, the various binary groups that can be formed in a 4-variable K-map consist of groups of two, four or eight squares.

Based on the foregoing considerations, the rules for obtaining a Boolean expression for a Boolean function of n variables in the minimal 'sum of products' form can be stated as follows.

1. Construct the K-map for the output variable of interest from the truth table or statement of the function to be performed, entering '1's in the squares where the output has the value '1', '0's in the squares where the output has the value '0'. Enter ϕ (superimposition of '1' and '0') in the squares where the value of the output is of no concern for the normal functioning of the circuit being designed. These are called 'Do Not Care' states.

2. Write down the minterms for the '1's which do not have any neighbouring '1'.

3. Form binary groups of 2 with those of the remaining '1's which cannot be combined to form larger binary groups, and write down the $(n-1)$-variable terms required to include all such '1's.

4. If $n > 2$, repeat step 3 for binary groups of 4, and obtain the necessary $(n-2)$-variable terms.

5. If $n > 3$, repeat step 3 for binary groups of 8, and obtain the necessary $(n-3)$-variable terms.

6. Continue the process with larger binary groups, if necessary. Such a situation will generally not arise as the K-map becomes unwieldy for $n > 4$.

7. In the formation of binary groups in any of the foregoing steps, any of the '1's can be included in as many groups as necessary for forming the largest possible groups.

8. The squares having entry ϕ may be included into or left out of the binary groups being formed to maximise the size of these groups.

10.2.3 Realisation of Boolean Expressions

The minimisation techniques given in section 10.2.2 are useful as they reduce the hardware necessary for realisation of logic. The combinatorial ciruit as represented by a Boolean expression can be realised using any of the following integrated circuits and their combinations.

(i) Logic Gates

(ii) Multiplexers (MUX)

(iii) Read Only Memories (ROM)

The three realisations given above use respectively the SSI, the MSI and the LSI technologies. The above mentioned realisation techniques are discussed in detail in this section.

10.2.3.1 *Realisation Using Logic Gates*

This technique can be used with simple functions where the number of variables is small, say up to four. For example,the function Sum given by eqn. (10.5) cannot be simplified further as seen from the K-map of Fig. 10.15(b). Figure 10.13 shows a straight forward realisation of the function using AND OR and NOT gates. The four terms in equation (10.5) are realised by the four AND gates and the OR gate provides the final output.

The realisation using gates is not unique. For example, the Sum function of eqn. (10 5) can also be realised using NAND gates or NOR gates. Under these conditions the realisation which minimises the number of gates seems to be the best choice. However, this is not so because the gates are packaged in certain combinations, i.e. dual 4 input, triple 3 input and quad 2 input (section 10.1.2) to obtain best Fan-In from a standard 14 pin dual-in-line package. Therefore, the best realisation is the one which uses the minimum number of IC packages rather than the minimum number of gates. A list of some available TTL SSI IC packages is given in Appendix A–1, (p. 669–674).

10.2.3.2 *Realisation Using Multiplexers*

A digital multiplexer, often called MUX is an IC using MSI

technology. It is used to mechanise the realisation of the Boolean expressions. It can also be used in combination with other gates to obtain a realisation with minimum IC packages.

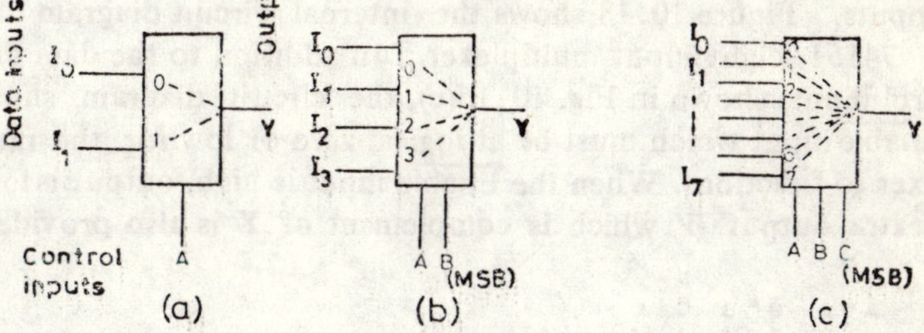

Fig. 10.16. (a) A 2-input MUX. (b) A 4-input MUX. (c) A 8-input MUX.

Figure 10.16 shows the schematic diagrams of multiplexers with two, four and eight inputs. The output Y is connected to a particular input data line depending upon the state of the control inputs. For example, in the eight input multiplexer the output Y is connected to the input I_3 when the control input 'CBA' equals 011 i.e. three. This is also sometimes referred to as 'selecting' I_3 to appear at the output. The number of input lines in a MUX can be increased by forming a tree structure. Figure 10.17 shows

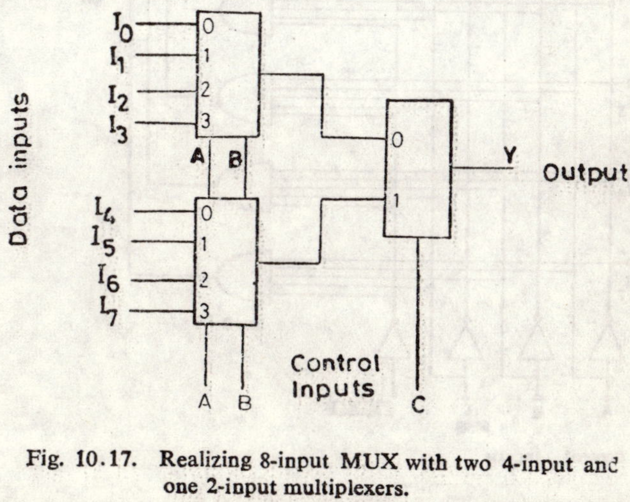

Fig. 10.17. Realizing 8-input MUX with two 4-input and one 2-input multiplexers.

how the two four input multiplexers are combined to form an eight input multiplexer. In this case the two lower order bits '*BA*' select input lines from both the multiplexers. These are further selected in the two input multiplexer by the MSB '*C*' of the control inputs. Figure 10.18 shows the internal circuit diagram of a TTL 74151 eight input multiplexer. In addition to the data and control inputs shown in Fig. 10.16(c), the circuit diagram shows an $\overline{\text{Enable}}$ input which must be at logical zero or low for the multiplexer to function. When the $\overline{\text{Enable}}$ input is high, output is low. An extra output *W* which is complement of *Y* is also provided.

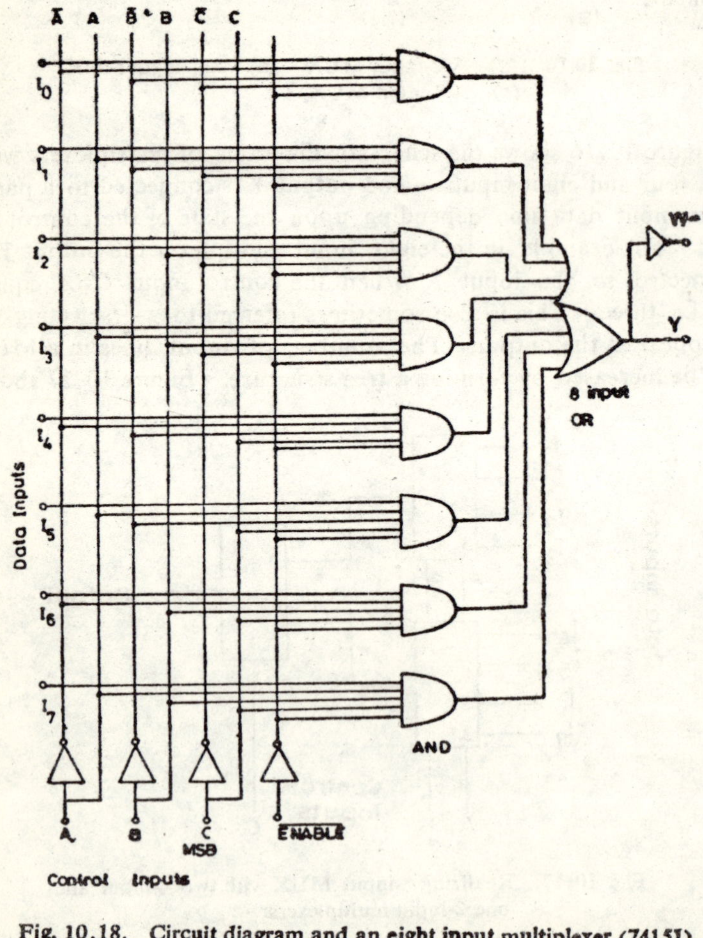

Fig. 10.18. Circuit diagram and an eight input multiplexer (74151),

The eight data inputs, three control inputs, enable, two outputs and two power supply lines complete the sixteen pin package. The multiplexers are available as Quad two input, Dual four input and Single eight input in 16 pin DIP. The multiple multiplexers in an IC package have common control inputs and enable line. The 16 input multiplexer requires a larger 24 pin package.

From the circuit diagram of Fig. 10.18 it is clear that the multiplexers of Fig. 10.16 are combinational circuits which implement the following Boolean expressions in a single MSI IC.

Two input Mux $$Y = I_0\bar{A} + I_1A \quad (10.8)$$

Four input Mux $$Y = I_0\bar{A}\bar{B} + I_1A\bar{B} + I_2\bar{A}B + I_3AB \quad (10.9)$$

Eight input Mux $$\begin{aligned} Y = {} & I_0\bar{A}\bar{B}\bar{C} + I_1A\bar{B}\bar{C} + I_2\bar{A}B\bar{C} \\ & + I_3AB\bar{C} + I_4\bar{A}\bar{B}C + I_5A\bar{B}C \\ & + I_6\bar{A}BC + I_7ABC \end{aligned} \quad (10.10)$$

Direct mechanisation: The operation of the multiplexer as a selector shows that it is possible to implement the Truth Table of the Boolean expression directly by treating the control inputs as variables and by providing the corresponding output states as the data inputs. In this realisation the number of control inputs is equal to n, the number of variables and the 2^n data inputs are either at logical one i.e. V_{cc} or logical zero i.e. ground.

Reduction of variables: A more economical method of realisation is suggested by eqn. (10.9). A comparison of equation (10.9) of the multiplexer and (10.4) of function 'carry' shows that the inputs I_0 to I_3 can be obtained directly in terms of variable C. The result is as follows.

$$I_0 = 0, I_1 = C, I_2 = C, I_3 = C + \bar{C} = 1 \quad (10.11)$$

Similarly, the data inputs for the function 'Sum' eqn. (10.5), can be obtained. The same result can be obtained directly by comparison of the K-maps as shown in Fig. 10.19 (a). Fig. 10.19 (b) shows the realisation of the functions Sum and Carry using a two input multiplexer. The diagram shows that the Boolean expressions of three variables have been reduced to one variable by using the four input multiplexer. In general using a four input multiplexer two variables can be factored out from a function of n variables.

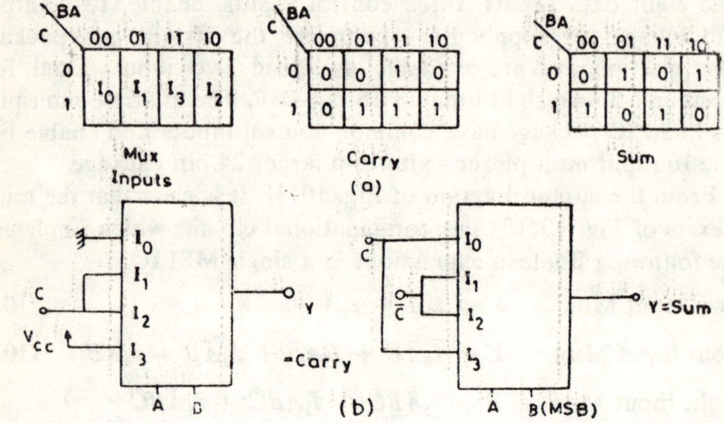

Fig. 10.19. (a) K-maps. (b) Realisation using mux of sum and carry.

The remainder are four separate 'residue functions' of $(n-2)$ variables which can be connected to the four data inputs of the multiplexer. Similarly an eight input and a sixteen input multiplexer can be used to factor out three and four variables respectively. Figure 10.20 shows that the reduction of variables is not unique when n is large, four in this example. The above figure shows six possible ways in which the two variables out of four can be chosen to act as the control inputs of the multiplexer.

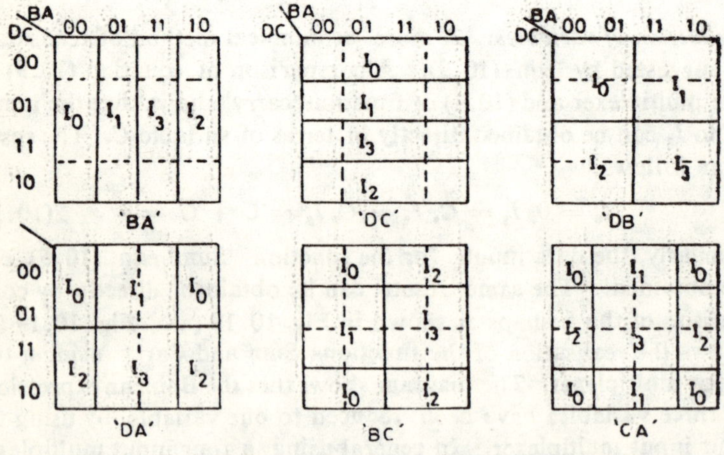

Fig. 10.20. Six possible choices in reducing a function of four variable by two '*XX*' shows chosen control inputs.

The reduction of variables using multiplexers may not always result in the minimum IC packages. Therefore, a detailed comparison of realisation using gates and multiplexers should be carried out for the final choice. Table 10.7 shows the functions which can be realised with only one IC, using reduction of one variable. This can be compared with the number of gate packages which may be used in alternate realisations of the functions.

Table 10.7. Realisation of functions using one multiplexer IC

Function	Type of Multiplexer
Any four functions of same two variables	Quad two-input
Any two functions of same three variables	Dual four-input
Any function of four variables	Eight input
Any function of five variables	Sixteen input

The reduction of variables using multiplexers reduces the problem to the realisation of the residue functions. The residue functions are connected to the inputs, I_0, I_1 etc., of the multiplexers. Sometimes, the form of residue functions is greatly simplified by the appropriate choice of control inputs of the multiplexer. In realising two or more functions together it is advantageous to have common residue functions. In realising particularly complicated functions it may be necessary to use multiplexers to generate the residue functions themselves. Thus, a complex function can be realised as a cascade connection of multiplexers with one level generating residue functions for the next level.

An absolute maximum: The residue functions of two variables can be generated with very few gates. Therefore, we can calculate the maximum number of packages required in the worst case as given below.

There are sixteen functions of two variables say A and B. Six of these are direct being A, $\bar{A}$, B, $\bar{B}$ 1 and 0. The others are $\bar{A}\bar{B}$, $A\bar{B}$, $\bar{A}B$, AB, $(A\bar{B} + \bar{A}B)$ and their complements. All the later ten functions can be generated using five two input NAND gates and five inverters. This gives a total of 3 packages. The absolute maximum package count for realisation using multiplexers is shown in Table 10.8. It is assumed that the residue functions of two variables are generated using gates, i.e. a maximum of 3 packages. Further as 16 input multiplexers require twice the space of an eight input multiplexer they are counted as two. Thirty-two and sixty-four input multiplexers are assumed to be constructed by cascading the required number of sixteen input multiplexers. It is assumed that the variables and its complements are available for use as inputs. The absolute maximum shown in Table 10.7 can be compared with the number of gates required in a particular problem to determine the minimum IC packages required for realisation.

10.2.3.3 *Realisation Using Read-Only Memory*

The truth table of a Boolean expression can be realised by using a Read-Only Memory (ROM). General block diagram of a memory system is shown in Figure 10.21. The memory is accessed by an 'address' consisting of b bits. Since each bit can have a value of zero or one, 2^b unique addresses can be obtained.

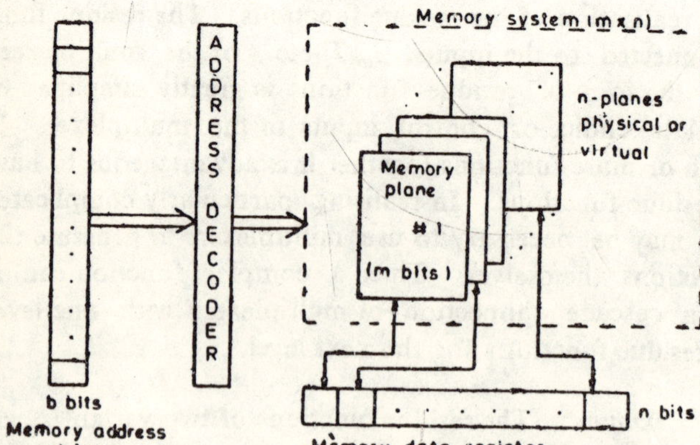

Fig. 10.21. A Typical memory system.

Table 10.8. Maximum package count in realisation of Boolean functions using multiplexers

S. No.	Number of variables	Type of multiplexer	Max no. of ICs
1	2	Quad-2 input	1
2	3	Dual-4 input	1
3	4	Eight input	1
4	5	Sixteen input	2
5	6	Sixteen input + Gates	5
6	7	Thirty-two input + Gates	8
7	8	Sixty-four input + Gates	12

For each address a *n* bit word is obtained as the output data. In a typical LSI IC the address varies from 10 to 12 bits. The output data consists of 8 bits.

In order to implement the Boolean expression the address lines are treated as the input variables. Each bit of the output then represents the value of a Boolean function. For example in Table 10.6 the input 'ABC' can be thought of as the address while the outputs Carry and Sum can be treated as the corresponding data which is stored in the memory. The process of writing the data from the truth table of a given function into the read only memory is called 'Programming' or 'Burning' the ROM. The data written into the ROM at the time of programming is final and cannot be changed. A ROM of size 1K words which has ten address lines and the output has eight bits. Thus it can realise eight different functions, each of ten variables. The variables are common to all functions and all the eight outputs are available simultaneously. It is clear that the ROM realisation of Boolean functions is ideal when a large number of functions of a large number of common variables is to be realised.

10.2.4 Example: BCD to Seven Segment Decoder

In this section the procedure for designing a combinatorial circuit using K-maps is illustrated with an example of a circuit called the Seven Segment Decoder. The circuit accepts a four bit input '*ABCD*', *A* being the MSB. This four bit input represents a Binary Coded Decimal (BCD) number between zero and nine. The seven outputs of the circuit are to drive the seven bars of a display. The bars are arranged and selectively lighted in such a manner so

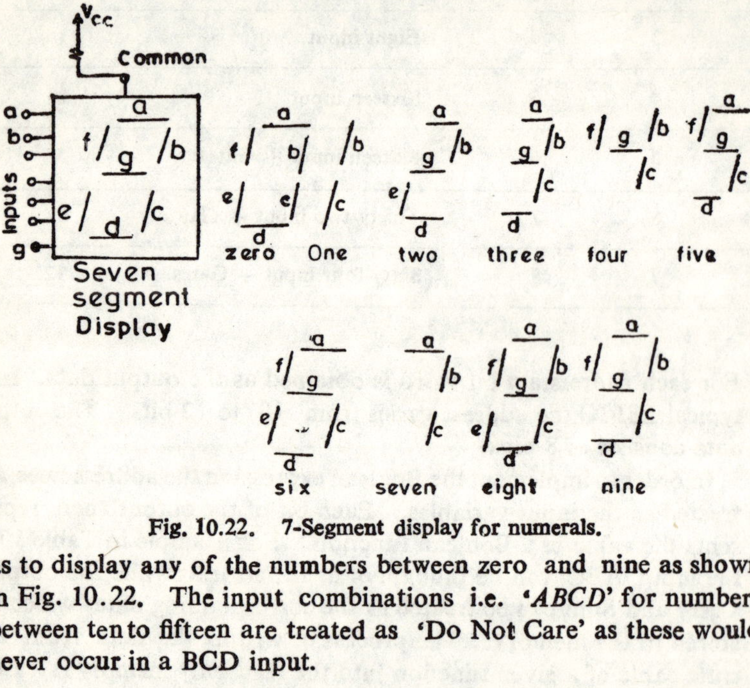

Fig. 10.22. 7-Segment display for numerals.

as to display any of the numbers between zero and nine as shown in Fig. 10.22. The input combinations i.e. '*ABCD*' for numbers between ten to fifteen are treated as 'Do Not Care' as these would never occur in a BCD input.

K-map: The seven segments are light emitting devices which are constructed with a common cathode. Thus, a logical zero must

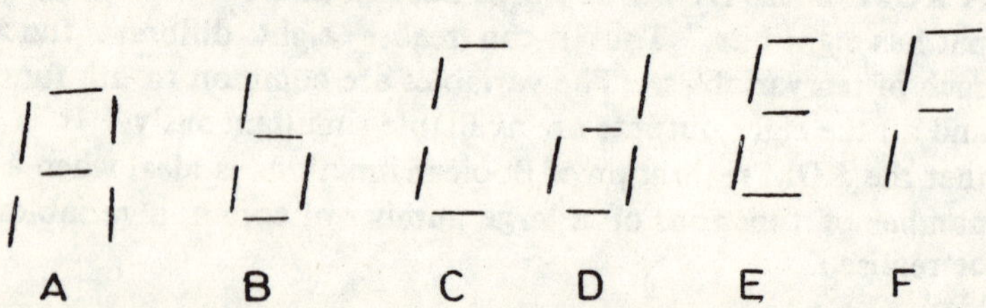

Fig. 10.23. Additional displays for HEX display.

be applied at an input of the display to light up a particular segment. From Fig. 10.22 it can be easily seen as to which segments are to be lighted or kept dark for displaying a particular decimal number. Each number corresponds to a *BCD* input. This information is presented in Table 10.9. From this table the state of each segment i.e. lighted or dark corresponding to each of the nine combinations of input '*ABCD*' can be determined. These states are entered in K-maps for each of the segments in Fig. 10.24. The position of each number is shown in a separate map in the bottom left hand corner of Fig. 10.24. The entries in the K-map are 0 corresponding to those combinations of the input state '*ABCD*' where the particular segment is lighted or on. For example, in Table 10.9 segment *a* is on when '*ABCD*' equals 0, 2, 3, 5, 6, 7,

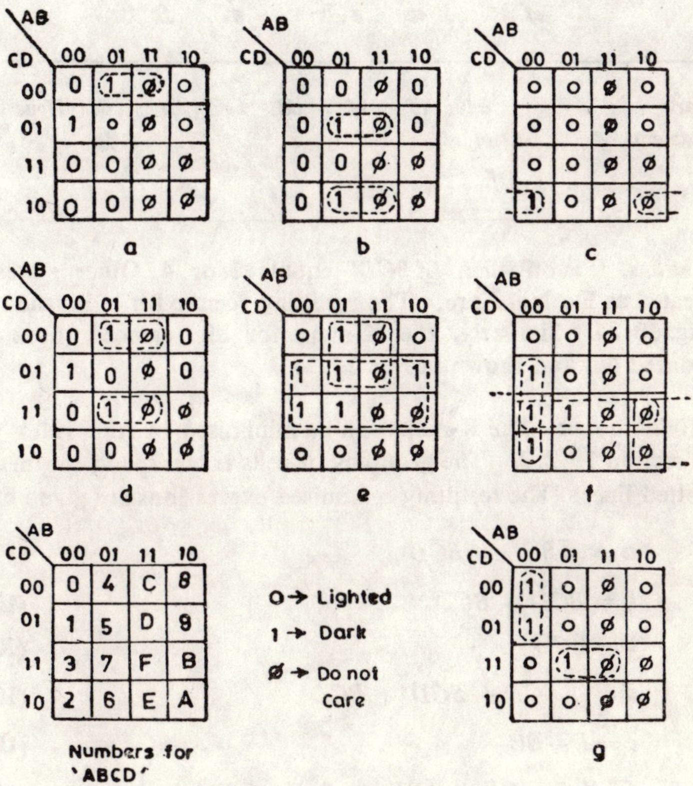

Fig. 10.24. K-map for the seven segment display of Fig. 10.16 with '*ABCD*' input.

Table 10.9. Lighted segments for seven segment display

Decimal Digit	0	1	2	3	4	5	6	7	8	9
BCD input										
'AB	00	00	00	00	01	01	01	01	10	10
CD'	00	01	10	11	00	01	10	11	00	01
Lighted Segments	*ab cd ef*	*bc*	*ab d eg*	*ab cd g*	*bc fg*	*ac df g*	*ac de fg*	*ab c*	*All*	*ab cd fg*
Dark Segments	*g*	*ad ef g*	*cf*	*ef*	*ad e*	*be*	*b*	*de fg*	*None*	*e*

8 and 9. It is off when '*ABCD*' equals 1 or 4. Other states are treated as Do Not Care. The resulting K-map for *a* is shown in Fig. 10.24. Similarly, the K-maps for all segments have been worked out and shown in Fig. 10.24.

Minimisation: The K-maps can be minimised by the rules given in section 10.2.2. The grouping of cells is marked in the map by dotted lines. The resulting minimised expressions are give below.

$$a = (\bar{A}\bar{B}\bar{C}D + B\bar{C}\bar{D}) \tag{10.12}$$

$$b = B\bar{C}D + BC\bar{D} \tag{10.13}$$

$$c = \bar{B}C\bar{D} \tag{10.14}$$

$$d = (\bar{A}\bar{B}\bar{C}D + B\bar{C}\bar{D}) + BCD \tag{10.15}$$

$$e = D + B\bar{C} \tag{10.16}$$

$$f = \bar{B}C + CD + \bar{A}\bar{B}D \tag{10.17}$$

$$g = \bar{A}\bar{B}\bar{C} + BCD \tag{10.18}$$

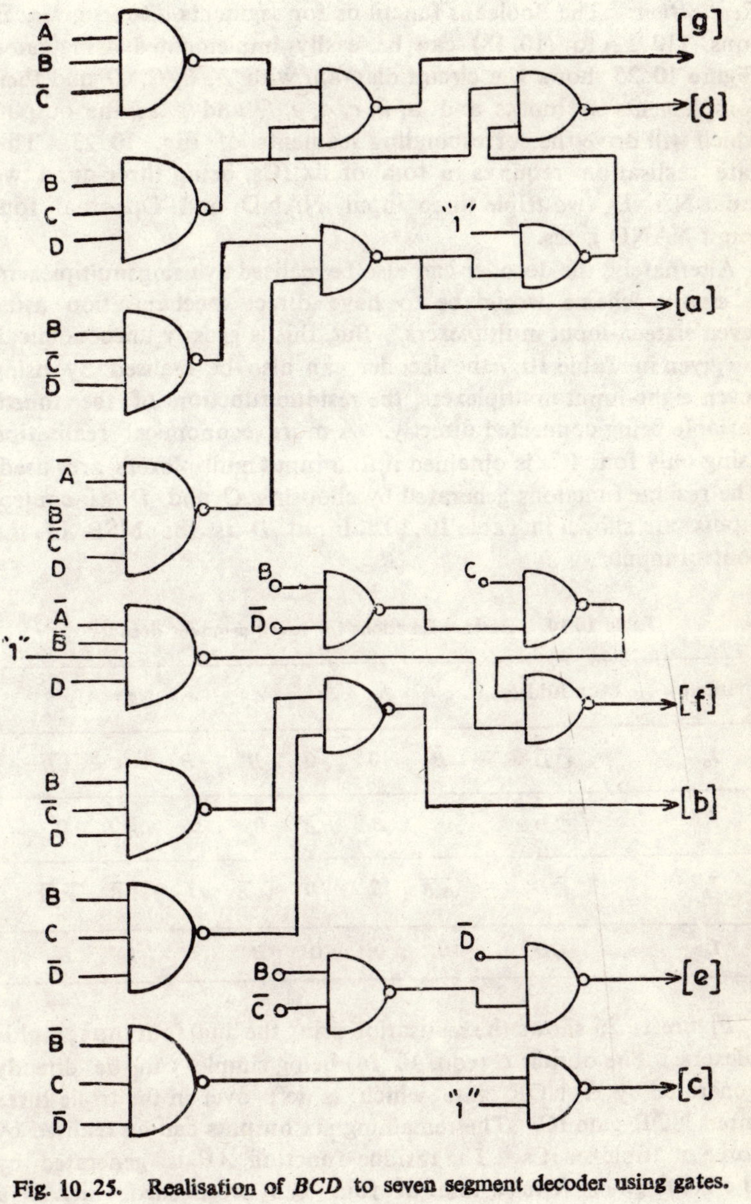

Fig. 10.25. Realisation of *BCD* to seven segment decoder using gates.

Realisation: The Booleans functions for segments a to g given in eqns. (10.12) to (10.18) can be easily implemented using gates. Figure 10.25 shows the circuit diagram with A, B, C, D and their complements as inputs and a, b, c, d, e, f, and g as the outputs which will drive the corresponding segments of Fig. 10.23. This gate realisation requires a total of six ICs, being three-quad two input NAND, two-triple three input NAND and One-dual four input NAND gates.

Alternately, the decoder can also be realised by using multiplexers. A simple scheme would be to have direct mechanisation using seven sixteen-input multiplexers. But, this is grossly uneconomical. As given in Table 10.7 the decoder can also be realised by using seven eight-input multiplexers, the residue function of the fourth variable being connected directly. A more economical realisation using only four ICs is obtained if four input multiplexers are used. The residue functions generated by choosing C and D as control inputs are shown in Table 10.10. Input D is the MSB of the control inputs.

Table 10.10. Residue functions for seven segment decoder

Input	Mux Address	a	b	c	d	e	f	g
I_0	$\bar{C}\bar{D}$	B	0	0	B	B	0	$\bar{A}\bar{B}$
I_1	$C\bar{D}$	0	B	$\bar{B}$	0	0	$\bar{A}\bar{B}$	0
I_2	$\bar{C}D$	$\bar{A}\bar{B}$	B	0	$\bar{A}\bar{B}$	1	$\bar{A}\bar{B}$	$\bar{A}\bar{B}$
I_3	CD	0	0	0	B	1	1	B

Figure 10.26 shows the realisation using the dual four input multiplexers. The output C (eqn. 10.14) being simple can be directly generated by a NOR gate which is left over in the triple three input NOR gate IC. The remaining six outputs can be realised by three multiplexer ICs. The residue function $\bar{A}\bar{B}$ is generated by the NOR gate. Thus a total of four ICs are needed which is a considerable saving over the gate realisation of Fig. 10.25 in which six ICs are required.

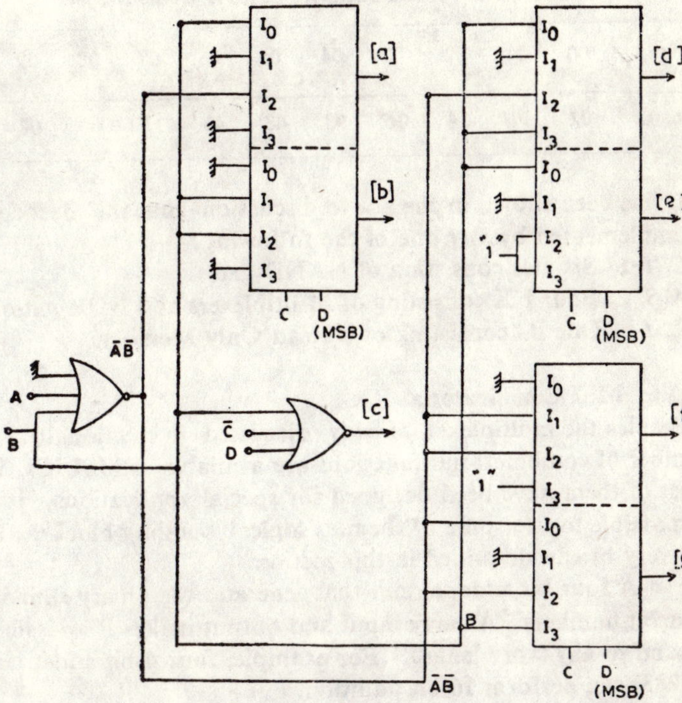

Fig. 10.26. Realisation of *BCD* to seven segment decoder using multiplexers.

The decoder can also be implemented using a small Read-Only Memory (ROM). For this purpose a table showing address and data is prepared for programming the ROM. The address is the input '*ABCD*' while the data can be arranged in an 8 bit word '*abcdefgho*'. As there are only seven segments the eighth bit has been made zero. The data can be read directly from the K-maps of Fig. 10.24. For example, for the address i.e. '*ABCD*' equal to '0001' the data '*abcd efgo*' equals '1001 1110'. Similarly, all other words can be obtained. The result is given in Table 10.11. The address and data are written in Hex format. Thus, in the example given above address is 01 and the data is *gE*. The ROM ICs usually are larger than the sixteen words required here. For example *TTL* 7488A ROM is of size 32 words × 8 bit. sixteen words of this ROM i.e. one half of the available capacity is sufficient to implement the decoder.

Table 10.11. ROM Table for Seven Segment Decoder

Address	00	01	02	03	04	05	06	07	08	09	A–F
Data	02	9E	24	0C	98	48	40	1E	00	08	$\phi\phi$

It has been shown in the above discussion that the decoder can be implemented by any one of the following:

SSI : Six ICs consisting of NAND gates.
MSI: Four ICs consisting of Multiplexers and NOR gates.
LSI : One IC consisting of a Read Only Memory.

10.2.6 MSI Combinatorial ICs

Besides the multiplexer already discussed in section 10.2.3, a number of combinatorial functions are available as MSI ICs. While most of them have been designed for special applications, it may be possible to use some of them as logical building blocks. These are very briefly described in this section.

1. A four bit adder circuit that generates the binary sum of two four bit numbers. A carry input and output makes it possible to expand to any word length. For example, four 4-bit adder circuits (74283) can perform 16-bit addition.

2. A look-ahead carry generator, which can be used with above adder circuits for faster operation on a 16-bit addition. For longer words, an additional circuit is used for each 16 bits (e.g. 74182).

3. A four bit Binary Magnitude comparison circuit, which compares two binary numbers A and B and produces three outputs; $A > B$, $A < B$ and $A = B$. Expansion inputs allow expansion to any word length in a-tree structure. For example up to 24 bits can be compared using six circuits in a two level tree structure. A three level tree can compare up to 120 bits or can be partially implemented to compare 32 bits with 10 packages (e.g. 7485).

4. A multiple-function arithmetic logic unit (ALU), which can perform four-bit addition, subtraction, Exclusive OR, OR, NOR, NAND, AND or comparison as specified by a control code, e.g. 74181. It also stores the result and shifts it to left or right with or without the carry. The word length can be expanded to any size with just one circuit required for each four bits. For very high

speed operation, the carry look ahead circuit described above can be used.

5. A multiplier which multiplies two 4-bit binary numbers and produces a 4-bit output. Two circuits together can produce an eight bit product. Expansion is not easy with these circuits. For example, an 8×8 bit multiplier with 16 bit output requires 16 packages—only eight of which are multipliers (e.g. 74284).

10.3 DESIGN OF SEQUENTIAL CIRCUITS

The sequential circuits as the name implies can be made to operate in a certain sequence of states. The general structure of a sequential circuit is shown in Fig. 10.27. It consists of a combinatorial circuit and a memory. The memory block acts a feedback element for the combinatorial circuit. Thus a set of inputs to the combinatorial circuit are the external inputs, $X_1X_2...X_n$. Another set of inputs are internal and are the outputs $Q_1Q_2...Q_n$ of the memory. Similarly the outputs of the combinatorial circuit can be divided into a set of outputs $Z_1, Z_2...Z_n$ of the sequential

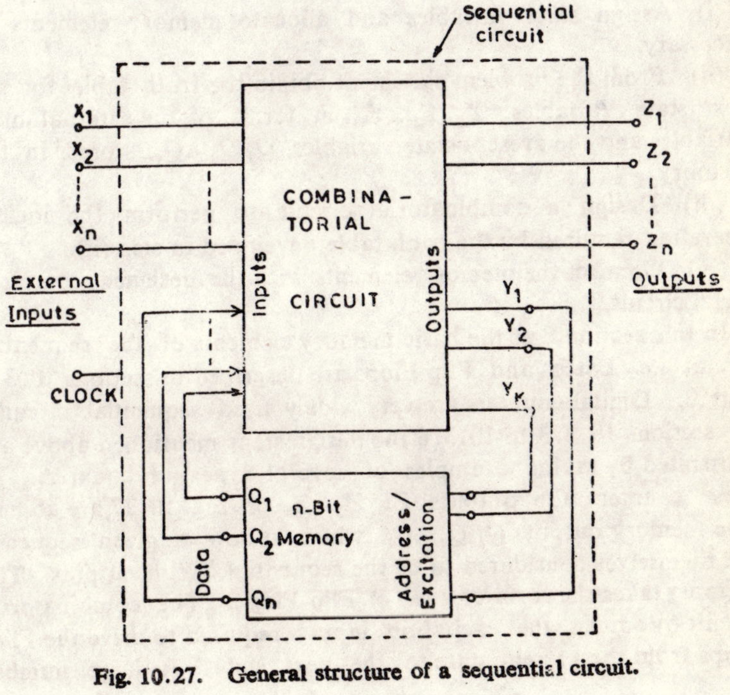

Fig. 10.27. General structure of a sequential circuit.

circuit and a set of addresses or excitations Y_1, Y_2...Y_n for the memory. A periodic input called Clock is usually applied to the sequential circuits so as to synchronise its operation. Such systems are called Synchronous Sequential Circuits or Sequential Machines. It is not absolutely necessary to have the inputs X_1, X_2...X_n in Fig. 10.27 if the inputs Q_1, Q_2...Q_n are adequate for the particular application. Similarly the outputs Z_1, Z_2...Z_n may be same as excitations Y_1, Y_2...Y_n or derived from them. The outputs 'Q_1 Q_2...Q_n' are called the present State of the sequential circuit. The present state along with the external inputs, if any, defines the outputs and the excitations Y_1Y_2...Y_n. Often the memory is simply a delay element (e.g. a set of D Flip-Flops). In such cases the outputs Y_1Y_2...Y_n can be called the next-state of the sequential circuit. The clock is usually a short duration pulse which advances the sequential circuit from one state to the next.

A formal mathematical procedure for the design of sequential circuits is given in Chapters 9 and 12 of [1]. Briefly the design follows the following steps.

(i) Assign state variables and allocate memory elements as necessary.

(ii) From the problem statement obtain the truth table for the next state Variables, Y_1 Y_2...Y_n, in terms of the external input variables and the present state variables, Q_1 Q_2...Q_n, stored in the memory.

(iii) Design a combinatorial circuit to perform the logical operation required by the truth table developed in step (ii).

(iv) Connect the memory elements with the designed combinatorial circuit.

In this section first the basic memory elements of the sequential circuit i.e. Latch and Flip Flops are described in sections 10.3.1 and .2. Digital counters are very widely used sequential circuits. In sections 10.3.3 to 10.3.6 the design steps mentioned above are illustrated by taking examples of several types of counters. In these counters the outputs Z_1 Z_2...Z_n of Fig. 10.27 are absent. The memory outputs Q_1 Q_2...Q_n which follow a given sequence are themselves considered to be the sequential circuit outputs. The memory takes the form of a set of Flip-Flops. The combinatorial circuit produces the excitation inputs required to drive the Flip-Flops from the present state to the next state. A large number of counters are presently available as MSI ICs. The IC counters have been briefly described in section 10.3.7.

10.3.1 Asynchronous Memory Element-Latch

The simplest memory element consists of two NOR gates or two NAND gates connected in a loop, as shown in Fig. 10.28. The inputs of the NOR latch are denoted by S (Set) and R (Clear) while those of the NAND latch is denoted by $\bar{S}$ and $\bar{R}$, so that both the circuits have the same truth table, also given in Fig. 10.28. The truth table is easily verified from an inspection of the circuits. Note that $S=1, R=1$ is the only input combination for which $Q=\bar{Q}$, and the truth table is different for the two circuits of Figs. 10.28 (a) and (b). As a result, return to the HOLD mode either from the SET mode or from the CLEAR mode presents no ambiguity in the value of Q, but a return to the HOLD mode from the combination $S=1, R=1$ makes the values of Q and $\bar{Q}$ totally unpredictable. Latches are therefore never used with $S=1, R=1$. One normally

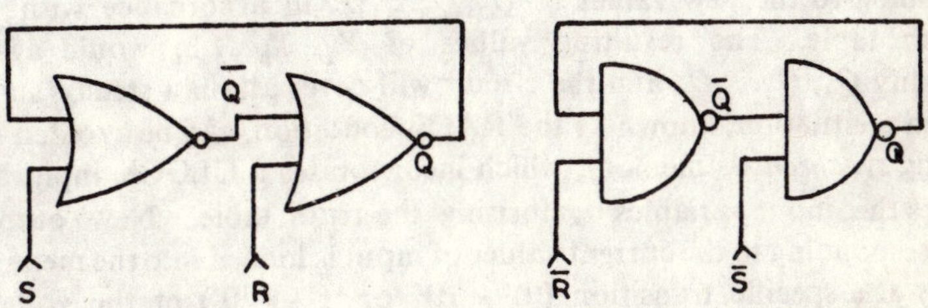

(a) NOR realisation (b) NAND realisation

S	R	Q	$\bar{Q}$	MODE
0	0	0 1	1 0	HOLD
1	0	1	0	SET
0	1	0	1	CLEAR
1	1	0 1	0 1	(NOR) (NAND)

Fig. 10.28. Latch realisations and truth table.

maintains the HOLD mode with $S=0$, $R=0$. Momentary 'activation' of the S input by making $S=1$ and then back to $S=0$, with $R=0$, sets the latch. The state $Q=1$ is maintained indefinitely as long as $S=0$, $R=0$. Similarly, momentary 'activation' of the R input, with $S=0$, clears the latch, the state $Q=0$ being maintained thereafter as long as $S=0$, $R=0$.

A latch is an *asynchronous* memory element, as the information is stored as soon as the input is applied. This creates problems in the design of sequential circuits, which have the general structure shown in Fig. 10.27. The variables X_1, X_2...X_n are the external inputs to the whole circuit. These, along with the variables Q_1, Q_2...Q_n stored in the n memory elements, form the inputs to the combinatorial circuit. Some of the outputs of the combinatorial circuits Y_1, Y_2...Y_k act as the inputs to the memory elements, while the rest, Z_1, Z_2...Z_n form outputs. If the memory is asynchronous, i.e. Q_1, Q_2...Q_n instantaneously respond to any change in Y_1, Y_2...Y_k. The combinatorial circuit would immediately produce outputs corresponding to the new values of Q_1, Q_2...Q_n in accordance with its truth table. The resulting values of Y_1, Y_2...Y_k would again modify Q_1, Q_2...Q_n and the circuit will never attain a steady state. Such a situation, known as the RACE condition, can be avoided by using *synchronous* memory, which incorporates a CLOCK input besides the input variables comprising the truth table. New output corresponding to the current values of input is loaded into the memory only at a specific transition ('0' → '1' or '1' → '0') of the voltage applied to the CLOCK terminal.

10.3.2 Synchronous Memory Element-FLIP-FLOP

The basic 1-bit synchronous memory element is the FLIP-FLOP, which is simply a latch with the clocking provision added. Asynchronous SET and CLEAR terminals may also be provided in a flip-flop, thereby enabling the user to use a flip-flop synchronously or asynchronously depending on the circuit requirements. Fig. 10.29 shows the circuit symbols and the truth tables of the four important types of flip-flops. Q_n denotes the value of Q after the nth clock edge, and Q_{n+1}, after the $(n+1)$th. The R-S flip-flop is just an extension of the latch, with same truth table, requiring the input combination $R=1$, $S=1$ to be carefully excluded from the possible values. The J-K flip-flop incorporates some more logic to make the input combination $J=1$, $K=1$ useable, the other three combinations having the same effect as the corresponding combinations of the *S* and *R* inputs in a latch. The *D* flip-flop utilises half of the capability of a *J-K* flip-flop, operating only either in the SET mode or in the CLEAR mode. The other two modes—STORE and TOGGLE (making $Q_{n+1} = \bar{Q}_n$) available in a *J-K* flip-flop are the only modes of operation of a *T* flip-flop. In each of these flip-flops, either the positive or the negative

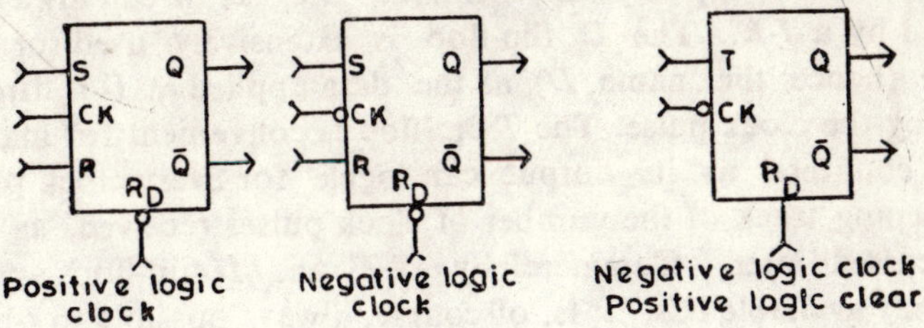

S	R	Q_{n+1}
0	0	Q_n
1	0	1
0	1	0
1	1	UNDEFINED

(a) R S Flip flop

T	Q_{n+1}
0	Q_n
1	$\bar{Q}_n$

(b) T Flip flop

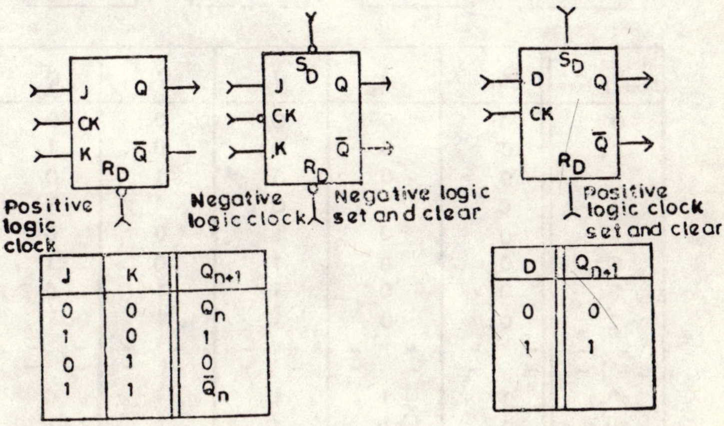

J	K	Q_{n+1}
0	0	Q_n
1	0	1
0	1	0
1	1	$\bar{Q}_n$

(c) J-K Flip flop

D	Q_{n+1}
0	0
1	1

(d) D Flip flop

Fig. 10.29. R-S, T and J-K flip-flops.

transition of the clock input may be considered to be the *active edge* causing the output to be updated, the flip-flop is said to use a *positive-logic clock* or a *negative-logic clock* accordingly. The choice of the active edge is indicated by adding a small circle at the clock (*CK*) terminal if a negative-logic clock is used.

Of the four kinds of flip-flops, the *J-K* flip is obviously the most versatile and is as such most widely used in a variety of applications. The *R-S* flip-flop finds limited use, as it can always be replaced by a *J-K*. The *D* flip-flop is extensively used for data storage (hence the name *D*), as the data applied at *D* is directly stored by the clock pulse. The *T* flip-flop is convenient for making simple counters, as its output can toggle for every clock pulse, thus keeping track of the number of clock pulses received, as will be discussed later. Commercially, *J-K* and *D* flip-flops are the commonly available ICs. It is, of course, always possible to obtain any of the four kinds of flip-flops by using one given flip-flop and gates, if necessary.

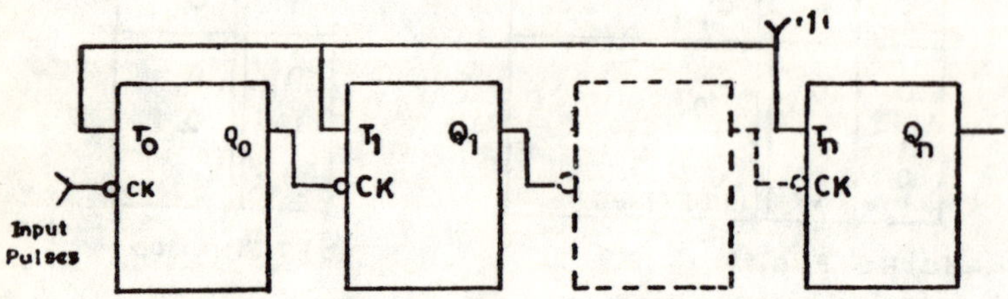

No. of Input pulses	Q_n	Q_{n-1}	Q_2	Q_1	Q_0
0	0	0	0	0	0
1	0	0	0	0	1
2	0	0	0	1	0
3	0	0	0	1	1
4	0	0	1	0	0
5	0	0	1	0	1
6	0	0	1	1	0
7	0	0	1	1	1
– –	– –	– –	– –	– –	– –
– –	– –	– –	– –	– –	– –
2^n-2	1	1	1	1	0
2^n-1	1	1	1	1	1
2^n	0	0	0	0	0

Fig. 10.30. *n*-bit ripple counter.

10.3.3 Ripple Counters Using T Flip-Flops

The simplest way of obtaining a counter is to cascade T flip-flops in the manner shown in Fig. 10.30, with the output (Q) of each flip-flop acting as the clock of the next. Such a counter is called a RIPPLE COUNTER, and counts the clock pulses applied to the first flip-flop in the binary code as indicated by the truth table. As the flip-flops have all been assumed to use a negative-logic clock, and since all of them are in the TOGGLE mode ($T = 1$), every $1 \rightarrow 0$ transition of Q_0 causes Q_1 to toggle, every $1 \rightarrow 0$ transition of Q_1 causes Q_2 to toggle and so on. The result is a straight binary counter, whose output $Q_n, Q_{n-1} \ldots Q_3\ Q_2\ Q_1\ Q_0$ represents, in the binary code, the number of input pulses received. The counter recycles after 2^n pulses have been received. The number of different states that any counter has in one complete cycle is called its CYCLE LENGTH. Thus, for an n-bit ripple counter, the cycle length is given by

$$N = 2^n.$$

Let us next explore the possibility of obtaining a cycle length other than a power of 2. It is obvious that if the required cycle length N is such that

$$2^n > N > 2^{n-1},$$

then the counter has to consist of n flip-flops. An n-bit counter can at most have 2^n different states representing as many different counts. The simplest way to achieve any specified cycle length N is to allow the counter to count from 0 to N in the usual fashion

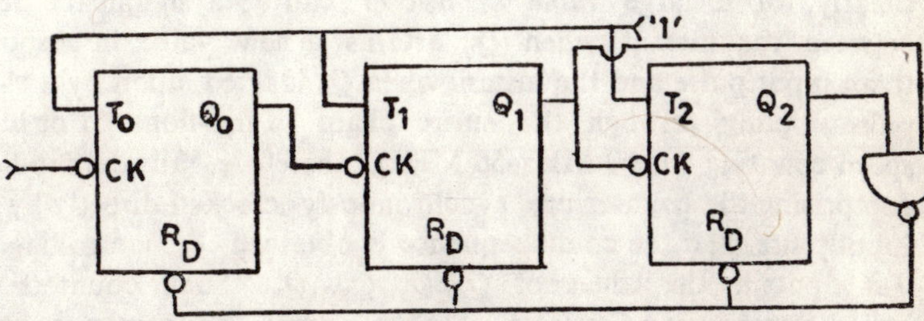

Fig. 10.31. Divide-by-6 ripple counter.

using the ripple counter configuration, and force the counter to go back to the all-zero state corresponding to zero count as soon as the $(N + 1)$th input pulse is received. This is illustrated in Fig. 10.31 where a 3-bit ripple counter has been modified by this technique to give $N = 6$. The output of the NAND gate remains '1' for all the counts from 0 to 5, and as such the CLEAR inputs of the flip-flops, taken to have a negative-logic clear (i.e., $Q = 0$ irrespective of other inputs when $R_D = 0$, and Q controlled by the T and CK inputs when $R_D = 1$) remain disabled. The counter, thus, counts from 0 to 5 in the binary code regardless of the additional NAND gate. As soon as the sixth pulse arrives, the NAND gate output goes to '0', thereby forcing Q_0, Q_1 and Q_2 immediately to '0' and thus returning the counter to the initial all-zero state. The counter therefore has only six states, the state corresponding to count 6 being just momentary, and a cycle length $N = 6$ has been obtained.

10.3.4 Synchronous Counters using J-K Flip-Flops

Ripple counters suffer from a speed limitation due to the fact that the new value of output is established with a progressive delay as one goes from the first flip-flop to the second, from the second to the third and so on. Thus, when the eighth pulse arrives, the count is changed by Q_0, Q_1, Q_2 and Q_3 changing in the following sequence.

1. Q_0 makes the transition '1' → '0'
2. Q_1 makes the transition '1' → '0'
3. Q_2 makes the transition '1' → '0'
4. Q_3 makes the transition '1' → '0'

Clearly, for a large value of n, there can be a significant delay between the instant when Q_0 attains a new value in response to an input pulse and the instant when Q_n is acted upon by a clock pulse rippling through the entire chain of flip-flops. For high-speed counting (at 10 MHz–50 MHz), therefore, all the flip-flops comprising the counter are synchronously clocked directly by the input pulse, and the count sequence is obtained by controlling the J-K inputs by the values of $Q_0, Q_1, Q_2 \ldots Q_n$. Such counters are called *Synchronous Counters*. Unlike a ripple counter, a synchronous counter does not have any natural sequence, and can therefore be designed for any prescribed sequence. One merely has to obtain the required values of the J-K inputs of each flip-flop at each state, and design a combinatorial circuit for generating the J-K inputs from the flip-flop outputs according to the resulting truth table.

It is convenient, at this stage, to express the truth table of a *J-K* flip-flop in the form of an EXCITATION table as given below.

Table 10.12. Excitation table of J-K flip-flop

Existing value of Q	Next value of Q	Required J	Required K
0	0	0	ϕ
0	1	1	ϕ
1	0	ϕ	1
1	1	ϕ	0

Let us illustrate the procedure for the design of synchronous counters by two examples. The first example is a 3-bit counter with

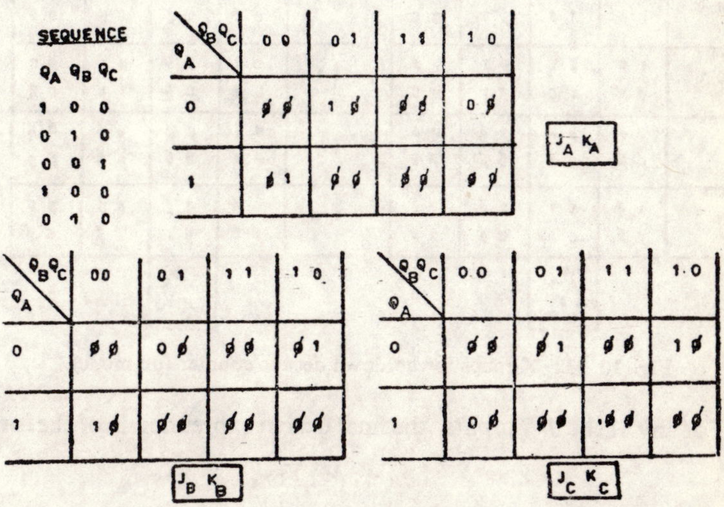

Fig. 10.32. K-maps for a 3 bit counter with a given sequence.

the sequence 100, 010, 001, 100,...at its output $Q_A\,Q_B\,Q_C$. The required values of J_A, K_A, J_B, K_B, J_C, K_C can be obtained by using Table 10.12 and can be directly written in the form of K-maps, as given in Fig. 10.32. Note that all states not included in the desired sequence are DON'T CARE states, as the counter will never enter one of these states after being properly initialised to one of the states in the desired sequence. Exploiting these DON'T CARE states the following simple expressions for the J-K inputs are obtained by the usual K-map technique for minimisation (Section 10.2.2).

$$J_A = Q_C, \quad J_B = Q_A, \quad J_C = Q_B, \quad K_A = K_B = K_C = 1.$$

The design of an up/down decade counter is taken up as the other example. The K-maps for the up-counting sequence 0000, 0001, 0010, 0011, 0100, 0101, 0110, 0111, 1000, 1001, 0000...and the down-counting sequence 0000, 1001, 1000, 0111, 0110, 0101, 0100, 0011, 0010, 0001, 0000...are separately shown in Fig. 10.33 and 10.34. The states 1010, 1011, 1100, 1101, 1110 and 1111 are DON'T CARE states. The design expressions for the J-K inputs are given in Table 10.13.

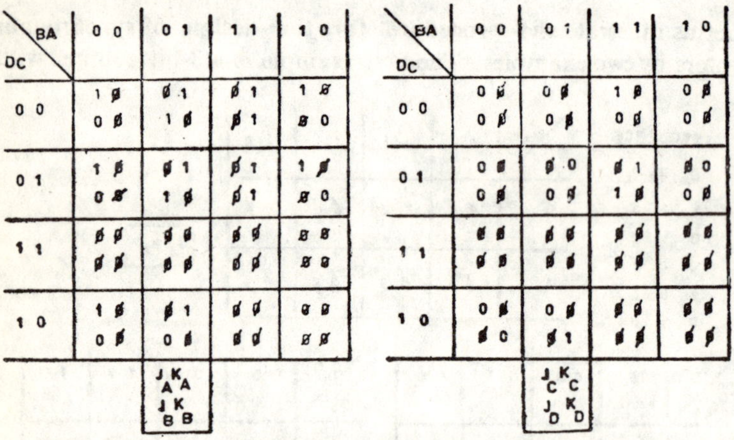

Fig. 10.33. K-maps for up/down decade counter (up mode).

For any of the J-K inputs, the final design expression is of the form

$$Y = \overline{M} \cdot Y_{UP} + M \cdot Y_{DN},$$

Table 10.13. Design Expressions for Up/Down Decade Counter

	J_A	K_A	J_B	K_B	J_C	K_C	J_D	K_D
Up	1	1	$\overline{D}\,A$	A	$B\,A$	$B\,A$	$C\,B\,A$	A
Down	1	1	$(D+C)\overline{A}$	$\overline{A}$	$D\,\overline{A}$	$\overline{B}\,\overline{A}$	$\overline{C}\,\overline{B}\,\overline{A}$	$\overline{A}$

BA \ DC	00	01	011	10
00	1 Ø 0 Ø	Ø 1 0 Ø	Ø 1 Ø 0	1 Ø Ø 1
01	1 Ø 1 Ø	1 Ø 0 Ø	Ø 1 Ø 0	1 Ø Ø 1
11	Ø Ø Ø Ø	Ø Ø Ø Ø	Ø Ø Ø Ø	Ø Ø Ø Ø
10	1 Ø 1 Ø	Ø 1 0 Ø	Ø Ø Ø Ø	Ø Ø Ø Ø

J_A K_A
J_B K_B

BA \ DC	00	01	11	10
00	0 Ø 1 Ø	0 Ø 0 Ø	0 Ø 0 Ø	0 Ø 0 Ø
01	Ø 1 0 Ø	Ø 0 0 Ø	Ø 0 0 Ø	Ø 0 0 Ø
11	Ø Ø Ø Ø	Ø Ø Ø Ø	Ø Ø Ø Ø	Ø Ø Ø Ø
10	1 Ø Ø 1	0 Ø Ø 0	Ø Ø Ø Ø	Ø Ø Ø Ø

J_C K_C
J_D K_D

Fig. 10.34. K-maps of up/down decade counter (down mode).

where Y_{UP} is the expression for *UP* counting and Y_{DN}, expression for DOWN counting, M being the mode control; $M = 0$ for up counting and $M = 1$ for down counting.

10.3.5 Shift Register Counters

A synchronous counter can also be built with *D*-flip-flops. Such counters are known as shift register counters. A shift register consists of a number of *D* flip-flops connected in tandem, as shown in Fig. 10.35. Shift registers of various lengths (no. of flip-flops) are commercially available as ICs. Long shift registers generally have only the *D*-input to the first flip-flop and the *Q*-output of the last flip-flop available at accessible terminals, besides the clock and power supply terminals. Shift register counters will therefore be considered only with configurations using a single serial input and a single serial output. The two choices for generating the input D_1 from the output Q_n are: (i) $D_1 = Q_n$ and (ii) $D_1 = \bar{Q}_n$. These two configurations, shown in Fig. 10.36 are called the RING COUNTER and the JOHNSON (or MOBIUS) COUNTER respectively. It is easy to see that an *n*-bit ring counter can have a maximum cycle length of *n*, while an *n*-bit Johnson counter has a maximum cycle length of 2*n*. Let us illustrate this by considering

3-bit and 4-bit ring as well as Johnson counters. Depending on the initial state, the counters go into different sequences as given in Tables 10.14 and 10.15.

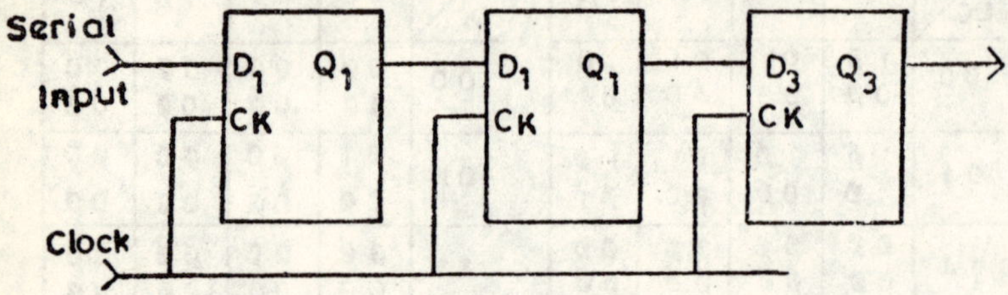

Fig. 10.35. Structure of a shift register.

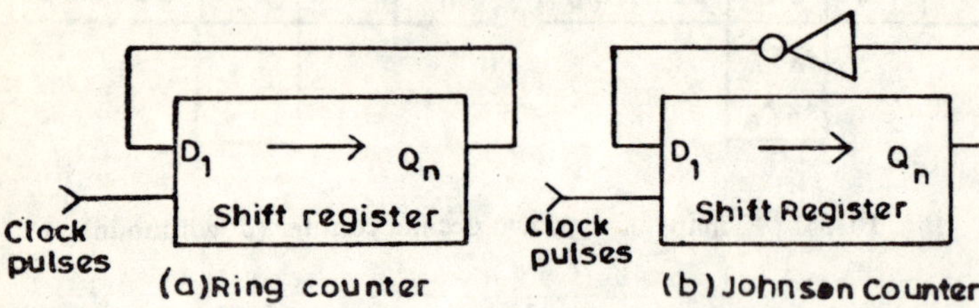

Fig. 10.36. Shift register counter.

Table 10.14 Sequences of 3-bit and 4-bit ring counters

001	011	111	1000	1100	0111	1010	1111
100	101		0100	0110	1011	0101	
010	110		0010	0011	1101		
			0001	1001	1110		

Table 10.15. Sequences of 3-bit and 4-bit and Johnson counters

000	010	0000	0101
100	101	1000	0010
110		1100	1001
111		1110	0100
011		1111	1010
001		0111	1101
		0011	0110
		0001	1011

10.3.6 Synchronous Counter Using ROM

Figure 10.37 shows a counter which follows closely the sequential machine structure shown in Fig. 10.27. The ROM implements the combinatorial circuit while the four *D*-Flops form the memory with a common clock. No external inputs are applied and the state outputs are the outputs of the sequential circuit. The 16×4 bit ROM can be programmed to obtain the required sequence of states. For example if the ROM contains the data shown in Table 10.16 the counter outputs $Q_3Q_2Q_1Q_0$ will follow the following

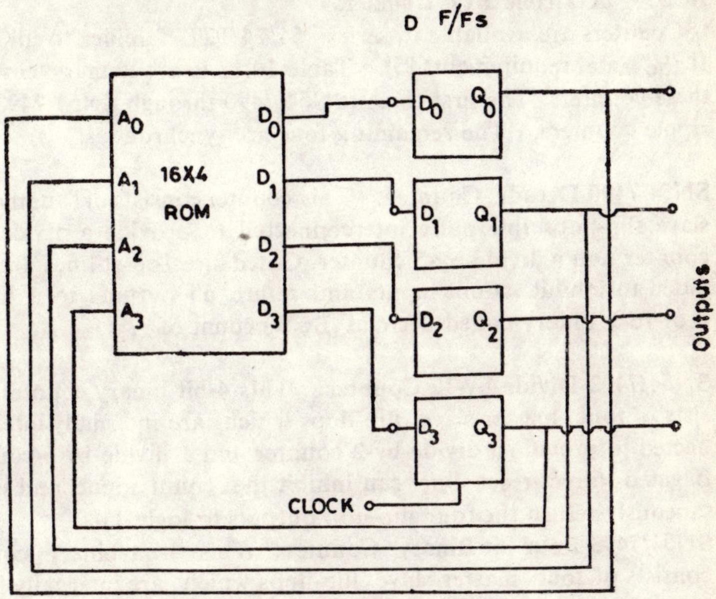

Fig. 10.37. Programmable counter using ROM.

Table 10.16. Data for counter ROM

Address $(A_3A_2A_1A_0)H$	0	1	2	3	4	5	6	7	8	9	*A*	*B*	*C*	*D*	*E*	*F*
Data $(D_3D_2D_1D_0)H$	1	8	9	1	2	*A*	*B*	3	4	*C*	*D*	5	6	*E*	*F*	7

sequence when clocked. *E*, *F*, 7, 3, 1, 8, 4, 2, 9, *C*, 6, *B*, 5, *A*, *D*, *E*, ... Figure 10.37 shows that the output state of the *DF/Fs* is also an address to the ROM. Therefore, for example, if the present state is *E*, the ROM output is *F*. At the next Clock pulse the *F*/*F*s store *F* and also present it to ROM as an address. The arrangement of Fig. 10.37, therefore, acts like a 15 state counter, whose sequence is decided by the data in ROM. If the *F*/*F*s are in the unused state 0 the ROM directs it to state 1 and the sequence automatically begins at 8. Any other sequence can be easily obtained by using a ROM with another set of data.

10.3.7 Available TTL Counters

Counters are available in series 54/74 TTL families to fill many of the usual requirements [5]. Table 10.17 is a summary covering these counters. The first three (SN54/7490 through SN54/7493) are ripple counters. The remaining four are synchronous.

SN54/7490 Decade Counter: This counter consists of four master-slave flip-flops internally interconnected to provide a divide-by-2 counter and a divide-by-5 counter. Gated direct-reset lines are provided to inhibit count inputs and return all outputs to a logical 0 or to a binary-coded-decimal (BCD) count of 9.

SN54/7492 Divide-by-12 Counter: This 4-bit binary counter consists of four master-slave flip-flops which are internally interconnected to provide a divide-by-2 counter and a divide-by-6-counter. A gated direct-reset line can inhibit the count inputs and simultaneously return the four flip-flop outputs to logical 0.

SN54/7493 Four-bit Binary Counter: This 4-bit binary counter consists of four master-slave flip-flops which are internally interconnected to provide a divide-by-2 counter and a divide-by-8 counter. A gated direct-reset line returns the four flip-flop outputs to logical 0.

SN54/74190 and SN54/74191 Synchronous Up/Down Counters: These are synchronous 4-bit binary up/down counters with parallel carry. The former has modified steering logic for an 8–4–2–1 BCD decade count whereas the latter is connected for normal binary counting. Four master-slave flip-flops change state with a logical 0 to logical 1 transition of the count input. Direction of counting is determined by the state of the up/down control line. Logical 0 is count up, logical 1 is count down. The enable input allows the counter to be

Table 10.17. Summary of series 54/74 counters

Type	Mode	Fully programmable	Clear	Bidirectional	Over/underflow output	Typical clock frequency max.	Typical power dissipation, mW
SN54/7490	÷2, ÷5, ÷10, Decade	No, preset to 1001 only	Gated	No	No	18 MHz	160
SN54/7492	÷2, ÷3, ÷6, ÷12	No	Gated	No	No	18 MHz	155
SN54/7493	÷2, ÷4, ÷8, ÷16	No	Gated	No	No	18 MHz	160
SN54/74190	Synchronous Decade	Direct	No	Yes	Yes	30 MHz	325
SN54/74191	Synchronous Binary	Direct	No	Yes	Yes	30 MHz	325
SN54/74192	Synchronous Decade	Direct	Yes	Yes	Yes	32 MHz	325
SN54/74193	Synchronous Binary	Direct	Yes	Yes	Yes	32 MHz	325

inhibited although clock pulses are present. A logical 0 enables the counter. Both counters can be parallel-loaded at data inputs when the load is at 0. The load and clear operations are independent of clock and counter state. The SN54/74190 and SN54/74191 have been designed so that external logic is minimized when cascading packages. Two outputs are provided for this purpose. Maximum or minimum output is 1 if the counter either contains 15 and is counting up or contains 0 and is counting down. In the case of SN54/74190, maximum count is 9. Ripple count enable is 0 when enable input is 0, max or min output is a 1, and count input is 0.

SN54/74192 and SN54/74193 Synchronous Up/Down Counters: These are synchronous 4-bit binary counters with parallel carry. The SN54/74193 is connected for normal binary counting, whereas the SN54/74192 has modified steering logic for an 8–4–2–1 BCD decade count. Four masterslave flip-flops change state with a 0 to 1 transition of count input. Direction of counting is a function of which count input is used. The unused clock input must be at a logical 1.

Both counters can be parallel-loaded at data inputs. When the load is at a logical 0, the load and clear operation are independent of clock and counter state. High on the clear input resets at flip-flops to 0. The SN54/74192 and SN54/74193 have been designed so that external logic is minimized when cascading packages. Carry is 1 when the maximum count has been reached and count-up input is 0. Borrow-out is 1 when 0 has been reached and count-down input is 0.

10.4 OPERATIONAL AMPLIFIER CHARACTERISTICS

The operational amplifier (op amp) is a basic analog building block for electronic systems. It is commonly used in systems required for instrumentation, computation, communication and control. It is a high gain direct coupled amplifier that is used in conjunction with an external feedback network to obtain desired gain and bandwidth.

The ideal op amp discussed in section 10.4.1 is a differential input, infinite gain amplifier with infinite bandwidth, infinite input impedance and zero output impedance. The analysis of circuits using ideal op amps can be easily carried out using circuit laws.

The practical op amp falls short of the ideal behaviour in many respects. Section 10.4.2 discusses the effect of the finite gain and bandwidth of the op amp. The effect of finite gain is to reduce the overall gain of the closed loop system. The finite bandwidth of the op amp introduces conditions in which the closed loop system may be unstable.

The imperfections of the practical op amp are characterised by the manufacturers in terms of various parameters. The definitions of these parameters are given in section 10.4.3.

10.4.1 The Ideal Operational Amplifier

In analysing circuits containing op amps it is convenient to start with the assumption that the op amp is an 'ideal' element. The op amp is represented by the symbol shown in Fig. 10.38. The ideal op amp has the following properties.

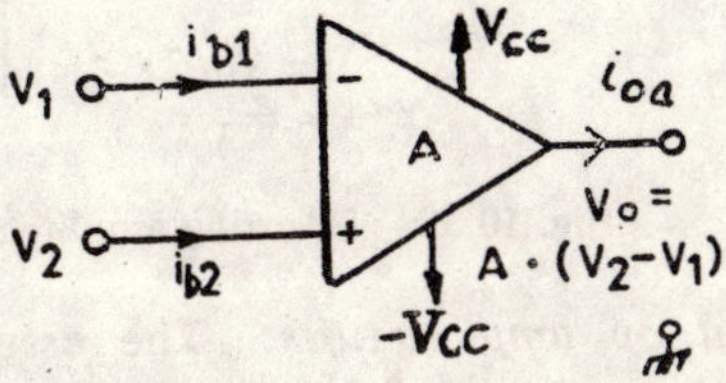

Fig. 10.38. Ideal op amp.

(i) The differential gain A (Fig. 10.38) is extremely large, $A \to \infty$ at all signal frequencies.

(ii) The input currents i_{b1} and i_{b2} (Fig. 10.38) are zero for all values of input voltages V_1 and V_2, i.e. the input impedance Z_{in} is infinite.

(iii) The op amp can source or sink any output current i_{0a} up to a limit, i_{0m}, i.e. the output impedance Z_0 is zero.

(iv) The output voltage v_0 can assume any value between $+V_{cc}$ and $-V_{cc}$, where V_{cc} is the d.c. power supply voltage.

The ideal operational amplifier of Fig. 10.38 is thus a difference amplifier of infinite gain, zero output impedance and infinite input impedance. Such an amplifier can be used with advantage in a 'feedback' connection. In a majority of circuits 'negative feedback' is used, which provides the following advantages.

(a) The gain of the closed loop system is insensitive to the exact value of A, as long as A is large. The closed loop gain is dependent only on the parameters of the feedback network.

(b) The effect of non-linearities in the forward path on the closed loop gain is considerably reduced.

(c) Non-linear input-output characteristics can be easily synthesised by introducing non-linear elements (i.e. diodes and transistors) in the feedback path.

(d) Effects of disturbances within the forward path are reduced.

(e) The input and output impedances of the overall system can be controlled.

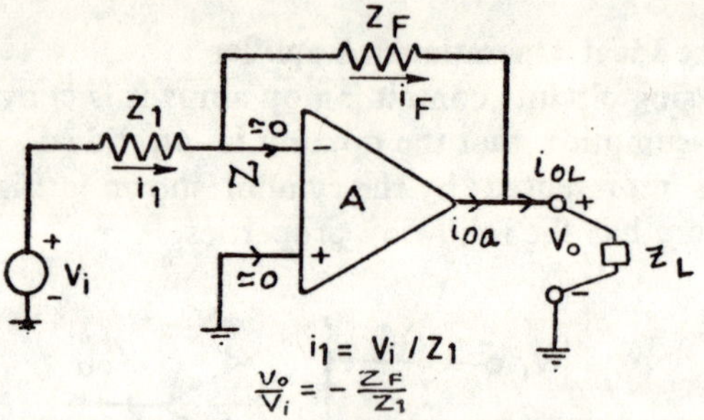

Fig. 10.39. Inverting amplifier.

Analysis of ideal op amp circuits: The assumption of ideal op amp simplifies the analysis of the circuits using feedback around the op amp. Figure 10.39 shows an 'INVERTING AMPLIFIER' using op amp, *A*. If the op amp is assumed to be ideal then the voltage between its input (+ and −) terminals is zero ($\underset{A\to\infty}{\text{Lt}}\ V_0/A$). The node *N* is therefore a *Virtual* ground. The currents flowing into the op amp terminals are zero as shown in Fig. 10.39. Applying Kirchoff's current law (KCL) at node *N* (Fig. 10.39).

$$i_1 = i_F \tag{10.19}$$

substituting (V_1/Z_1) for i_1 and $(-v_0/Z_F)$ for i_F in eqn. (10.19), the gain of the amplifier is given by,

$$(V_0/V_i) = -(Z_F/Z_1) \tag{10.20}$$

If the impedances Z_F and Z_1 are resistances then the circuit of Fig. 10.39 acts as an amplifier with a 180° phase shift and a constant gain for all frequencies of the input signal. The current i_F is related to the load current i_{0L} by the following equation.

$$i_F = (i_{0L} - i_{0a}) \tag{10.21}$$

The current i_F is dependent upon the input signal and the current

i_{0L} is dependent upon the load. The amplifier adjusts its output current i_{0a} to satisfy equation (10.21).

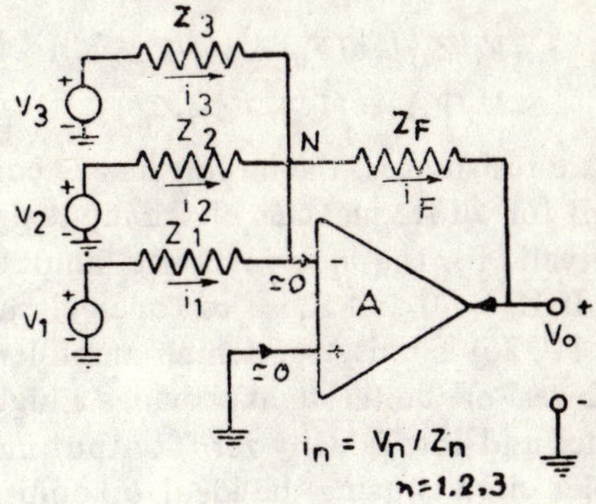

Fig. 10.40. Summing amplifier.

Figure 10.40 shows a modification of the inverting amplifier which can be used for summing a number of inputs. The node N is at virtual ground and KCL gives,

$$i_1 + i_2 + i_3 = i_F \tag{10.22}$$

Substituting (V_1/Z_1), (V_2/Z_2), (V_3/Z_3), $(-V_0/Z_F)$ for i_1, i_2, i_3 and i_F respectively in (10.22) gives,

$$V_0 = -V_3(Z_F/Z_3) - V_2(Z_F/Z_2) - V_1(Z_F/Z_1) \tag{10.23}$$

Thus the amplifier sums up signals V_1, V_2 and V_3 with specified gains. Equation (10.21) is valid for the summing amplifier with a load at the output.

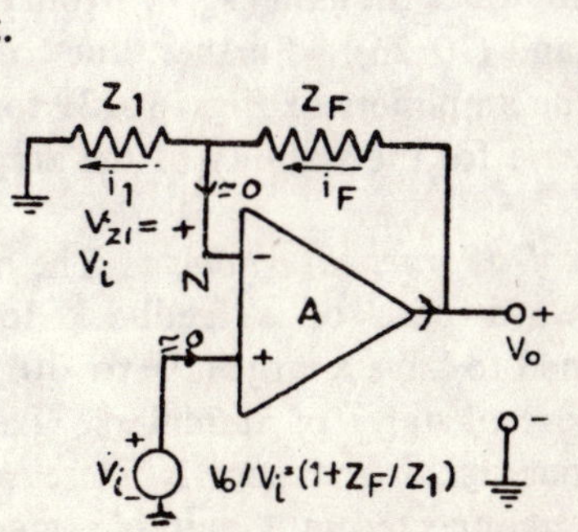

Fig. 10.41. Non-inverting amplifier.

Figure 10.41 shows a 'NON-INVERTING' amplifier circuit. Here the node N is at the potential V_i, so that the voltage between the input terminals (+ and −) is zero. By KCL at node N (Fig. 10.41),

$$i_1 = i_F \tag{10.24}$$

or,
$$(V_i/Z_1) = (V_0 - V_i)/Z_F \tag{10.25}$$

or,
$$(V_0/V_i) = \{1 + (Z_F/Z_1)\} \tag{10.26}$$

If Z_F and Z_1 are resistances, the circuit has a constant gain and zero phase shift for all frequencies of the input signal V_i. Equation (10.21) is also valid for the non-inverting amplifier with a load at the output. If $Z_F = 0$ and $Z_1 = \infty$ (open circuit) then the ratio (V_0/V_i) in eqn. (10.26) is unity. Such an amplifier circuit is called a 'Voltage follower' or 'buffer' as it provides a high (ideally infinite) input impedance and low (ideally zero) output impedance.

More complex circuits using the ideal op amp can be analysed similarly by applying circuit laws at the nodes. It should be noted however that the output voltage V_0 of the op amp is limited to $\pm V_{cc}$ and the output current, i_{0a} is limited to a specified value

10.4.2 The Non-ideal Op Amp, Effects of Finite Gain

The ideal op amp characteristics given in section 10.4.1 cannot be obtained in an integrated circuit (IC) implementation. There are various imperfections e.g. finite gain, finite bandwidth, finite input and output impedances, slew rate, common mode gain etc. In this section the effects of finite open loop gain A (Fig. 10.38) and bandwidth of the op amp upon performance of circuits of Figs. 10.39 and 10.41 are discussed. It has been shown that the finite value of A modifies the gain (V_0/V_i) from the expressions given by eqns. (10.20) and (10.26). Further, due to the variation of A with frequency, the amplifiers of Figs. 10.39 to 10.41 may become unstable. A criteria for the stability of the amplifiers is discussed.

Open loop gain of a practical op amp: The op amp is invariably used in the forward path of a feedback loop. Therefore, the amplifier is designed to have a large (10^5 to 10^6) d.c. gain. The op amp consists of several stages of amplifiers. The gain of each stage reduces as the input signal frequency is increased. The stages of amplifiers are interconnected in a suitable manner to produce the desired open loop gain characteristic. Figure 10.45 shows the gain vs. frequency (Bode) plot of a type 741 op amp. This type of characteristic is the usual one for the "internally compensated" op amps.

In such op amps no changes in the gain-freq. curve can be made from the outside. The characteristic of Fig. 10.45 can be represented by the general transfer function.

$$A(s) = \frac{a_0}{1 + s/\omega_0} \tag{10.27}$$

In eqn. (10.27) a_0 is the d.c. gain (105db in Fig. 10.45) and ω_0 is the corner frequency (7 Hz in Fig. 10.45).

In some op amps (e.g. LM 309) there are special terminals to provide an external 'compensation' network. In such op amps it is possible to alter the open loop corner freq. ω_0 (eqn. 10.27) or obtain a Bode plot different from Fig. 10.45. In the following the use of a general purpose internally compensated op amp with $A(s)$ of the form given by eqn. 10.27 has been assumed.

Closed loop systems using op amp: The linear circuits using op amps generally employ negative feedback from the output of the op amp to the input. Figure 10.42(a) shows the general block diagram of a negative feedback system. For such a system the closed loop gain is given by,

$$A_{cL} = \frac{V_0}{V_i} = \frac{\alpha A}{1 + A\beta} = \frac{\alpha/\beta}{1 + (1/A\beta)} \tag{10.28}$$

Figure 10.42(b) shows the non-inverting amplifier, for this circuit $\alpha = 1$ as V_i is directly applied to the positive terminal of the op amp. The feedback factor $\beta = R_1/(R_1 + R_2)$, is frequency independent as the network is resistive. Figure 10.42(c) shows the inverting amplifier. For this circuit, the input voltage at the negative input terminal of the op amp is given by

$$V_- = V_i \frac{R_2}{R_1 + R_2} + V_0 \frac{R_1}{R_1 + R_2} \tag{10.29}$$

Therefore $\alpha = -R_2/(R_1 + R_2)$ and $\beta = R_1/(R_1 + R_2)$. By substituting $A(s)$ from eqn. (10.27) and appropriate values of α and β in eqn. (10.28), expression of A_{cL} can be obtained. The quantity $A\beta$ is called the 'loop gain' of the system. If A is large such that $(1/A\beta) \ll 1$, then eqn. (10.28) shows that A_{cL} approaches the ideal value of α/β. The effect of finite gain, therefore, is the reduction of the ideal gain (α/β) by a factor $A\beta/(1 + A\beta)$. A more important effect is that if the loop gain AB approaches the value (-1), the denominator of A_{cL} (eqn. 10.28) tends to zero. Then the system becomes unstable and ceases to function as an amplifier.

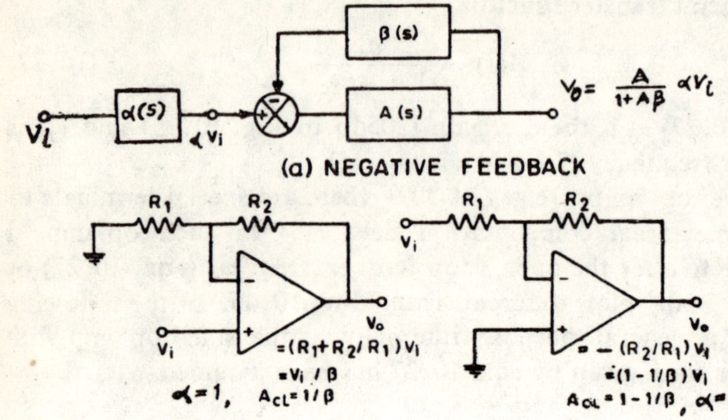

(b) Non-inverting gain (c) Inverting gain

Fig. 10.42.

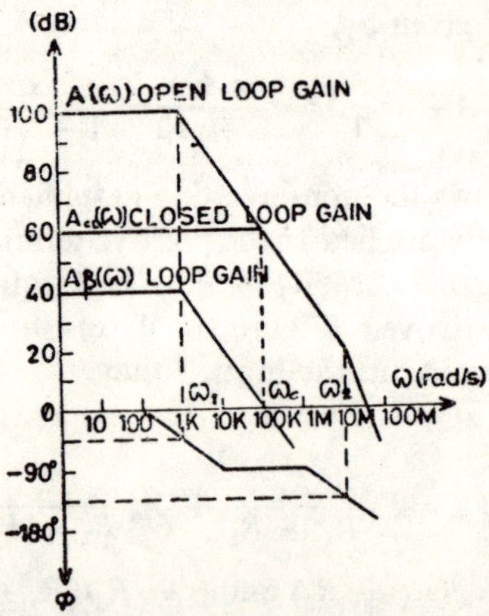

Fig. 10.43. **Bode plot showing relationship of the open loop gain $A(\omega)$. The loop gain $A\beta(\omega)$ and the closed loop gain $A_{\alpha}(\omega)$.**

Frequency response and stability: The block diagram of Fig. 10.42(a) is a linear system and its stability can be determined by the application of the standard methods e.g. Nyquist criterion, Bode plot etc. For a circuit with resistive feedback network such as Fig. 10.42(b), the feedback factor β is independent of frequency. Figure 10.43 shows the Bode Plots of the open loop gain A (eqn. 10.27), the closed loop A_{cL} (eqn. 10.28) and the loop gain $A\beta$. Since β is the constant Bode plot of $A\beta$ differs from A by a constant amount. For low values of frequency $\omega \ll \omega_0$, A_{cL} is given by

$$A_{cL} = \frac{A}{1 + A\beta} \simeq \frac{A}{A\beta} \tag{10.30}$$

At $\omega = \omega_c$, the loop gain is reduced to zero dB. This point is also the corner frequency for the closed loop gain A_{cL}. At higher frequencies ($\omega \gg \omega_c$), A_{cL} is given by,

$$A_{cL} = \frac{A}{1 + A\beta} \simeq A \tag{10.31}$$

and the Bode plot of A and A_{cL} merge together.

Figure 10.44 shows a typical Bode plot, similar to Fig. 10.43

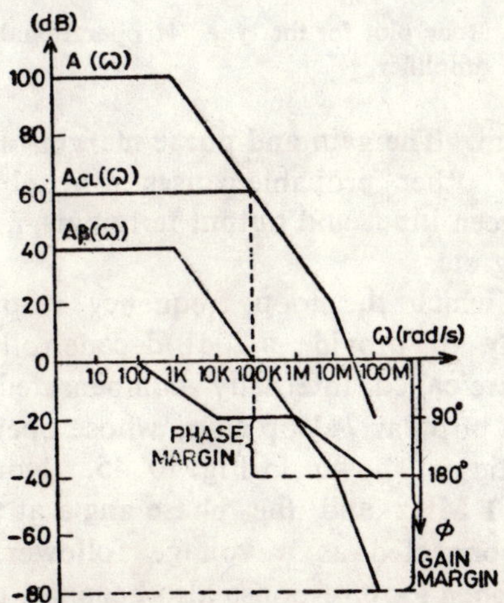

Fig. 10.44. Bode plot illustrating a phase margin of 90° and a gain margin of 80dB.

which illustrates the concepts of gain and phase margins. The loop gain $A\beta$ should have magnitude of zero dB and a phase shift of 180° ($A\beta = -1$) for system to be unstable. In Fig. 10.44 when $A\beta$ equals zero dB the phase shift is only −90°. Therefore, there is a 'phase margin' of 90°. Similarly when phase angle of $A\beta$ is 180°, its magnitude is −80 dB. Therefore, there is a 'gain margin' of

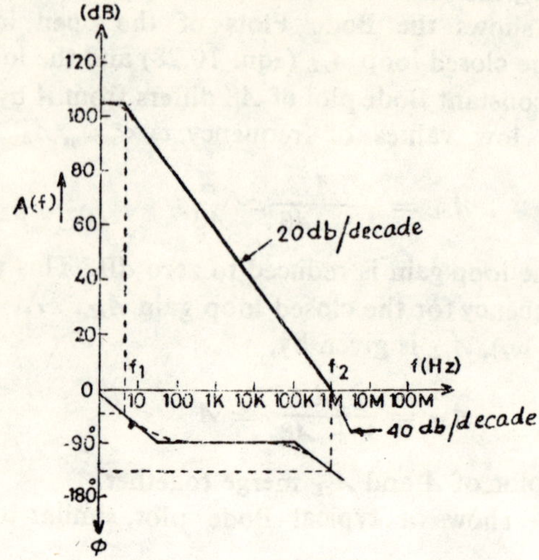

Fig. 10.45. Bode plot for the type 741 operational amplifier.

80dB in this system. The gain and phase margins are safety factors to guard against other probable causes of instabilities, e.g. stray capacitances between input and output terminals, inadequate power supply decoupling etc.

Op amps for which the loop frequency response is already tailored internally to provide a 20dB/decade roll off through the unity gain point are called internally compensated. An example of this type is the popular 741 op amp, whose open loop gain and phase characteristic is shown in Fig. 10.45. Note that the unity gain frequency is 1 MHz and the phase angle at this frequency is −135°. When connected as a voltage follower, the unity gain frequency of the open loop response also becomes the frequency at which the loop gain is unity. Thus, we have a phase margin of 45° or more for all applications of 741 where the closed loop voltage gain is equal or greater than unity.

For op amps which are not internally compensated, the manufacturer provides an external compensation terminal where a capacitor or an RC network may be connected for adjusting the frequency response. The advantage of external compensation is that the dynamic response of the overall circuit using the op amp may be improved by choosing proper compensation network. The procedure for designing such compensators is discussed in detail in [9].

10.4.3 The Non-ideal Op Amp: Definitions of Parameters

As mentioned in section 10.4.2, the performance of practical op amps falls short of the 'ideal' behaviour and op amp manufacturers specify various parameters which provide a measure of this nonideality of the op amp. A discussion of some of these important parameters has been given in this section.

(a) INPUT OFFSET VOLTAGE V_{0s}

The ideal op amp will show zero output voltage when both inputs are at zero potential. However, in all practical op amps a small voltage at the output is observed under this input condition. The input offset voltage, V_{0s}, is the differential d.c. input voltage required to provide zero output voltage with no input signal or source resistance. Model for the input offset voltage is shown in Fig. 10.46 with an inverting feedback circuit comprising of R_f and R_1.

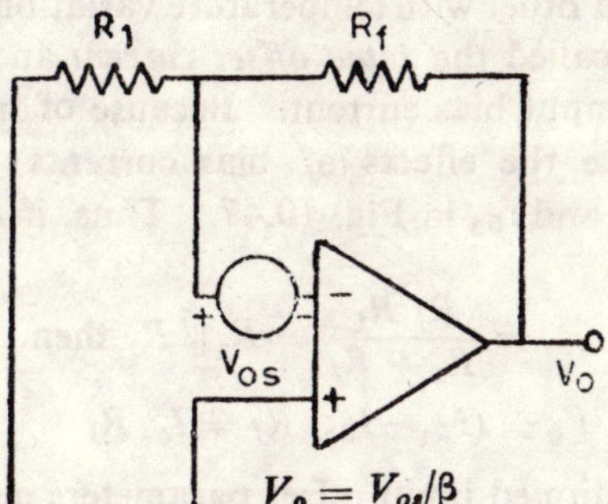

$$V_0 = V_{0s}/\beta$$
$$\beta = R_f + R_1/R_1$$

Fig. 10.46. Model of input offset voltage.

(b) INPUT BIAS CURRENT I_B

The ideal op amp does not draw any input current. However, all practical op amps need a bias current to flow in each input terminal which again may lead to an appearance of nonzero output voltage with zero input signal or offset voltage.

The bias current I_B is the d.c. current required at either input to provide zero output voltage with no input signal or offset voltage. Model for the input bias currents is shown in Fig. 10.47. The offset voltage and bias currents and their dependence and sensitivity to various parameters like temperature, time etc. represent

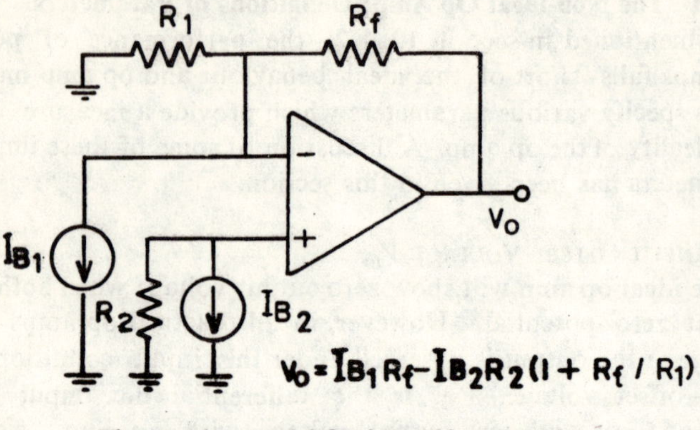

Fig. 10.47. Model for input bias current.

major sources of error in op amp circuits. In most op amps the bias currents I_{B1} and I_{B2} are of the same order of magnitude and tend to track each other with temperature variations. The difference $I_{0s} = I_{B1} - I_{B2}$ is called the *input offset current* and is usually much smaller than the input bias current. Because of tracking tendency, one can minimize the effects of bias currents by equating the coefficients of I_{B1} and I_{B2} in Fig. 10.47. Thus, if R_2 is chosen such that,

$$R_2 = \frac{R_f R_1}{R_1 + R_f} = R_1 \| R_f \text{ then,}$$

$$V_0 = (I_{B1} - I_{B2})\, R_f = I_{0s}\, R_f \qquad (10.32)$$

The above mentioned input offset parameters may be measured by setting up the circuit of Fig. 10.48 which also represents the d.c. equivalent of most practical op amp circuits. The output voltage V_0 is related to these parameters as follows.

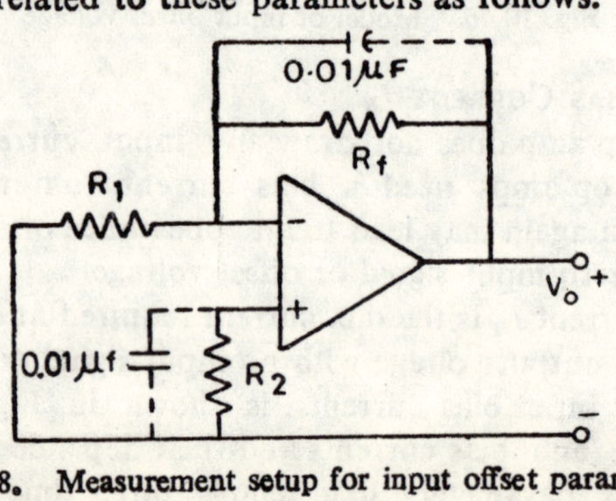

Fig. 10.48. Measurement setup for input offset parameters.

$$V_0 = \left(1 + \frac{R_f}{R_1}\right) V_{0s} + R_f I_{B1} - \left(1 + \frac{R_f}{R_1}\right) R_2 I_{B2} \quad (10.33)$$

The parameters V_{0s}, I_{B1}, and I_{B2} may be obtained by choosing R_1, R_f and R_2 appropriately so that only one or other of these parameters is dominant in Eqn. (10.33) and measuring the output voltage V_0. This choice may be as per Table 10.18.

(c) INPUT IMPEDANCE

An ideal op amp has infinite input and zero output impedance. In practical op amps, the output impedance is non-zero but small, and input impedance is high and has two components. The impedance between the two input terminals is the *differential input impedance*. Z_{id}; whereas impedance from either input to ground is called the *common mode* input impedance Z_{icm}. The models for

Table 10.18.

Dominant parameter	R_1	R_f	R_2	Approx. V_0
V_{0s}	Low (10-100)	Medium (10K-100K)	0	$\frac{R_f}{R_1} V_{0s}$
I_{B1}	Open	High (1M-10M)	0	$R_f I_{B1}$
I_{B2}	Open	0	High (1M-10M)	$-R_2 I_{B2}$

the input and output impedance are shown in Figs. 10.49 and 10.50. For most of the applications of the op amp over the useful

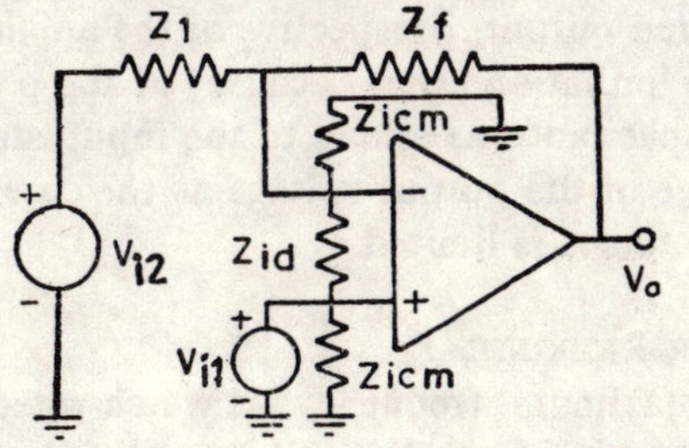

Fig. 10.49. Model for open l oop input impedance.

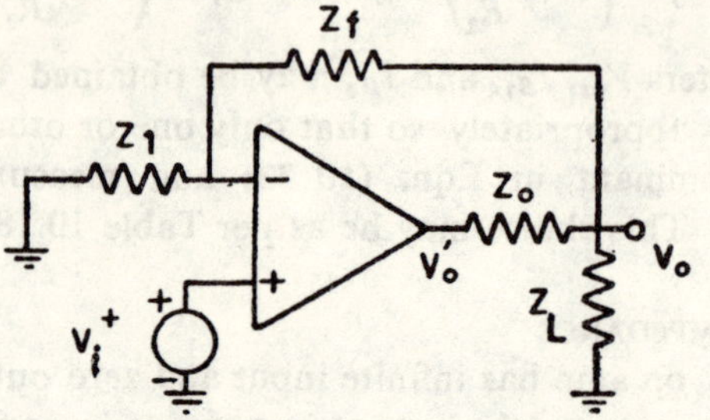

Fig. 10.50. Model for open loop output impedance.

frequency range, these impedances may be taken as resistive. Moreover Z_{icm} is usually ignored.

(d) COMMON MODE-REJECTION RATIO

This is the ratio of differential voltage gain to the common mode voltage gain. It is a measure of the asymmetry of signal responses of + and − input terminals of the op amps. The CMRR is discussed further in section 10.5 where the differential amplifier has been analysed.

(e) RATED OUTPUT

This is the peak value of the output voltage and current which can be simultaneously supplied. These are the safe limits which define the amplifier output capabilities. Since the output current limiting can limit the output voltage swing under load the two output signal ratings are interdependent and specified together.

(f) SLEW RATE, S_r

This is the maximum rate of change of output voltage when supplying the rated output, irrespective of the applied input signal. The slewing rate limitation arises because of the presence of phase compensation capacitance as a load to the input stage. This slows the rate of change of the output voltage as the current available to charge this capacitance is limited.

(g) FULL POWER RESPONSE, f_p

This is the maximum frequency at which rated output can be supplied without significant distortion in the sinusoidal output waveform. This response limit is often the result of the finite slewing rate. In this case, rate-limiting affects sinusoidal signals by limiting their slopes. By equating the slope of a sine wave to the slewing rate, full power response is related to S_r by $f_p = \dfrac{S_r}{2\pi E_{0r}}$ where E_{0r} is the peak rated output voltage.

The characteristics of some of the representative set of op amps are listed in Table 10.19 for reference in design applications.

10.5 LINEAR APPLICATIONS OF OP AMPS

In this section some basic linear circuits using op amps have been discussed. These circuits represent standard configurations which are widely in use. They can further be used as building blocks for more complex system, while still retaining the basic simplicity of designing with op amps.

Section 10.6.1 describes the basic differential amplifier. The circuit has been analysed in detail and the effects of amplifier parameters, such as finite gain, bias currents and offset voltage are discussed.

Section 10.5.2 describes the instrumentation amplifier using two stage amplification. The expressions for differential gain and *CMRR* are given.

Section 10.5.3 describes the integrator and differentiator circuits using op amps. The Bode plots of various circuits are used to obtain the operating frequency ranges. Integrator error due to op amp parameters is also given.

Section 10.5.4 describes the voltage to current and current to voltage converters. Circuits with both floating and grounded load are discussed.

Section 10.5.5 describes voltage references and voltage regulators using op amps and zener diodes.

Section 10.5.6 describes the design of active filters using op amps. Realization techniques using infinite gain multiple feedback, controlled source and state variable feedback are discussed.

10.5.1 Differential Amplifiers

The differential amplifier is a basic circuit widely used in the signal acquisition technology. It is a direct coupled amplifier which amplifies the difference between two signals, in the presence of a large common mode voltage. The differential amplifier also provides attenuation of common mode voltage and converts a differential

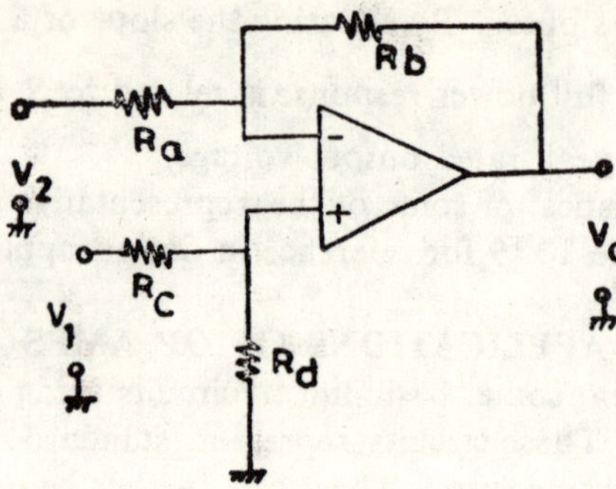

Fig. 10.51. Differential amplifier.

voltage into one that is referenced to a common ground. The common mode voltage may represent a d.c. level, 50 cycle 'Hum', or noise pickup. Such signals are usually obtained from transducers such as thermocouple and a strain-gauge bridges.

Figure 10.51 shows a simple differential amplifier using four matched resistors and an op amp. This circuit has been analysed in detail in this section.

Analysis assuming ideal op amp: The output voltage V_0 (Fig. 10.51) can be obtained as the sum of components due to inputs V_1 and V_2. Let the op amp be ideal i.e. $A = \infty$, $Z_i = \infty$, $Z_0 = 0$. If V_1 is applied with $V_2 = 0$, the circuit acts as a non-inverting amplifier (Fig. 10.41) with a gain of $(R_b/R_a) + 1$. If V_2 is applied with $V_1 = 0$, the circuit acts as an inverting amplifier (Fig. 10.39) with a gain of $(-R_b/R_a)$. Therefore, if V_1 and V_2 are applied simultaneously, the output voltage V_0 is given by,

$$V_0 = V_{01} + V_{02} \tag{10.34}$$

where

$$V_{01} = V_1 \cdot (R_d/R_c + R_d) \cdot (1 + R_b/R_a) \tag{10.35}$$

$$V_{02} = -(R_b/R_a) \cdot V_2 \tag{10.36}$$

Substitution from Eqns. (10.35) and (10.36) into (10.34) gives,

$$V_0 = \frac{R_b}{R_a}\left[\left\{V_1 \cdot \left(1 + \frac{R_a}{R_b}\right) \Big/ \left(1 + \frac{R_c}{R_d}\right)\right\} - V_2\right] \tag{10.37}$$

Thus, to obtain a voltage proportional to the difference signal $(V_1 - V_2)$, the resistors must be chosen such that,

$$R_a/R_b = R_c/R_d \tag{10.38}$$

Substitutions of (10.38) in (10.37) gives,

$$V_0 = R_b/R_a\,(V_1 - V_2) = A_d(V_1 - V_2) \tag{10.39}$$

The quantity (R_b/R_a) is called the 'differential gain' of the amplifier and is represented by symbol A_d.

Effect of mismatch in resistors: When the differential amplifier of Fig. 10.51 is constructed, the resistances R_a, R_b, R_c, R_d have tolerances. Thus, the actual value of each resistor may be different from its nominal value by a specific fraction, say δ. In such a case Eqn. (10.38) is not exactly satisfied. Therefore, the value of output voltage V_0 is not zero when V_1 equals V_2. It is convenient to calculate gain due to a common mode voltage by expressing V_1 and V_2 as,

$$V_1 = V_{cm} + (V_d)/2,\ V_2 = V_{cm} - (V_d)/2 \tag{10.40}$$

where $V_{cm} = (V_1 + V_2)/2$ is the common mode signal and V_d is the differential signal $(V_1 - V_2)$. Assuming worst case combination of mismatch between the resistors, it can be shown that

$$A_{cm} = V_{ocm} \,/\, V_{cm} = \frac{R_b}{R_a} \cdot \frac{4\delta}{1 + R_b/R_a} \tag{10.41}$$

Thus, the diff. amp has a common mode rejection ratio, A_d/A_{cm}, which is proportional to $1/\delta$ and is given by

$$(\text{CMRR})_d = \frac{1 + A_d}{4\delta} \tag{10.42}$$

Effect of amplifier common mode gain: In addition to the mismatch in the resistors, the non-ideal op amp itself has a gain for common mode signals. It is specified by the common mode rejection ratio, $(\text{CMRR})_a$, being the ratio between the gains of the op amp for the differential and the common mode signals. The finite circuit CMRR and op amp CMRR combine to provide the following expression for the output voltage V_0 of the differential amplifier.

$$V_0 = (V_2 - V_1)A_d + \frac{V_2 + V_1}{2} A_d \left\{ \frac{1}{(\text{CMRR})_d} + \frac{1}{(\text{CMRR})_a} \right\} \tag{10.43}$$

Effect of finite amplifier gain: If the op amp has a finite gain A then the + and − terminals of the op amp are not at the same potential. From Fig. 10.51

$$V_- = V_+ - V_i = \frac{R_d}{R_c+R_d}\cdot V_1 - \frac{V_0}{A} \qquad (10.44)$$

Assuming input currents of the op amp to be zero,

$$V_0 = V_- - (V_2 - V_-)\, R_b/R_a$$
$$= V_-(1 + R_b/R_a) - V_2 \cdot R_b/R_a \qquad (10.45)$$

Substituting for V_- from (10.44) in (10.45),

$$V_0 = \frac{1 + R_b/R_a}{1 + R_c/R_d}\cdot V_1 - V_2\cdot\frac{R_b}{R_a} - \frac{V_0}{A}\left(1 + \frac{R_b}{R_a}\right)$$
$$= \frac{R_b}{R_a}\left\{\frac{1 + R_a/R_b}{1 + R_c/R_d}(V_1) - (V_2)\right\} - \frac{V_0}{A}\left(1 + \frac{R_b}{R_a}\right) \qquad (10.46)$$

If $R_a/R_b = R_c/R_d$ then,

$$V_0 = \frac{R_b}{R_a}(V_1 - V_2) - \frac{V_0}{A}\left(1 + \frac{R_b}{R_a}\right)$$

or

$$A_d = \frac{V_0}{V_1 - V_2} = \frac{R_b/R_a}{1 + \left(1 + \frac{R_b}{R_a}\right)\cdot\frac{1}{A}} \qquad (10.47)$$

If the open loop gain of the amplifier is given by,

$$A = \frac{a_0}{1 + j f/f_c} \qquad (10.48)$$

Then,

$$A_d = \frac{R_b}{R_a}\,\frac{1}{1 + \left(1 + \frac{R_b}{R_a}\right)(1 + j f/f_c)\cdot\frac{1}{a_0}}$$
$$= \frac{R_b}{R_a}\,\frac{1}{1 + \frac{1}{a_0}\left(1 + \frac{R_b}{R_a}\right) + j\frac{f}{a_0 f_c}\left(1 + \frac{R_b}{R_a}\right)} \qquad (10.49)$$

If $a_0 \gg \left(1 + \frac{R_b}{R_a}\right)$,

Then, $$A_d = \frac{R_b}{R_a}\cdot\frac{1}{1 + j f/f_A}$$

where $$f_A = \frac{a_0 f_c}{\left(1 + \frac{R_b}{R_a}\right)} \qquad (10.50)$$

Thus, if inequality given above is satisfied, the difference amplifier has a gain of (R_b/R_a) at frequencies lower than the cut off frequency f_A.

Effect of bias currents and offset voltage; The amplifier offset voltage, v_{os} appears between the input terminals and is amplified by a factor $(1 + R_b/R_a)$. The bias current I_{B2} at + terminal (Fig. 10.47) flows through $(R_d \parallel {}^1R_c)$ and creates an effective d.c. input which is amplified. The bias current I_{B1} at—terminal flows through R_b and directly causes an output offset. Therefore, total d.c. offset error is given by,

$$V_0 = v_{os}\left(1 + \frac{R_b}{R_a}\right) - I_{B2}\frac{R_d\,R_c}{R_d + R_c}\left(1 + \frac{R_b}{R_a}\right) + I_{B1}\,R_b \qquad (10.51)$$

From equation (10.51) it is seen that errors due to bias currents are of opposite signs. Further if $R_c = R_a$ and $R_d = R_b$, the error due to bias currents reduces to $R_b\,(I_{B1} - I_{B2})$.

Limitations of differential amplifier: Circuit shown in Fig. 10.51 has the limitations of low input impedance and the difficulty of varying gain. Increasing the value of resistors increases the input impedance. However, the upper limit is decided by the errors given by Eqn. (10.51). The variation of gain requires that resistors be changed in a matched fashion and usually only step variation is possible.

10.5.2 Instrumentation Amplifier

Instrumentation amplifier is the workhorse of the signal acquisition technology and is, often, the answer when high input impedance is desired and the source impedance is unbalanced or changing. It is useful if variable gain is required and one would not like to trim many resistances for maximum CMRR and gain accuracy.

The configuration most often used is shown in Fig. 10.52. Note that if the inputs are tied together and driven the first stage acts as a unity follower and the common mode gain is unity. Note also that the feedback network need not be balanced.

The second stage is the usual differential amplifier of section 10.5.1. The CMRR of this stage is

$$(\text{CMRR})_d = \frac{1 + K}{4\delta} \qquad (10.52)$$

assuming that $(\text{CMRR})_a \gg (\text{CMRR})_d$

The overall CMRR of the instrumentation amplifier described here, therefore, is the product $A_d\,(\text{CMRR})_d$ where A_d is defined in Fig. 10.52. Thus,

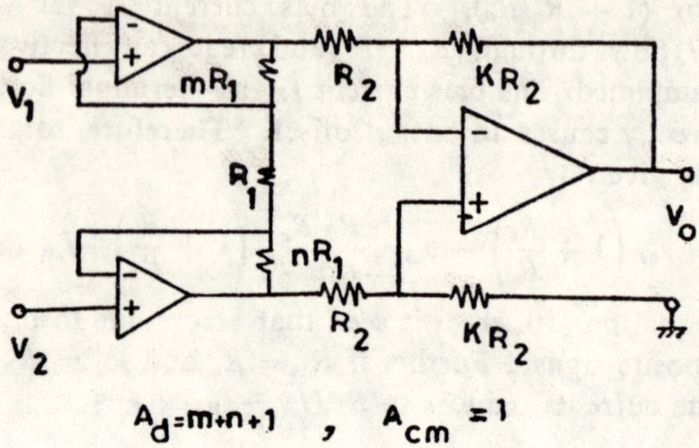

Fig. 10.52. Instrumentation amplifier.

$$(\text{CMRR})_{\text{Inst. amp}} = A_d \cdot \frac{1+K}{4\delta} = \frac{(m+n+1)(1+K)}{4\delta} \tag{10.53}$$

For $K = 1$ and $m = n \gg 1$

$$(\text{CMRR})_{\text{Inst. amp}} = \frac{m}{\delta} \tag{10.53}$$

10.5.3 Integrators and Differentiators

A. *Integrator*

The analog integrator is used extensively in signal processing and signal generating amplifications. It uses an op amp in the inverting configuration as shown in Fig. 10.53. Assuming ideal op amp with gain A.

$$\frac{v_{in} - v_{-}}{R} = i$$

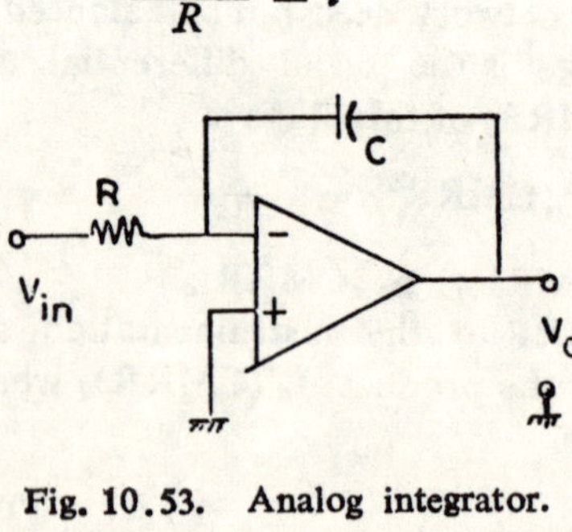

Fig. 10.53. Analog integrator.

$$v_- - v_0 = \frac{1}{C}\int_0^t i\,dt = \frac{1}{RC}\int_0^t (v_{in} - v_-)\,dt,$$

where $$v_- = -\frac{v_0}{A}.$$

If $$A \to \infty, v_- \to 0 \text{ and } v_0 = \to \frac{1}{RC}\int_0^t v_{in}\,dt \qquad (10.55)$$

To be realistic, one must take into account the effect of op amp offset voltage V_{0s} and the bias current I_B as shown in Fig. 10.54. Applying Kirchoff current law at node a and computing the output voltage, we get,

$$i_1 = i_2 + I_B$$

$$v_0 = -\frac{1}{RC}\int_0^t v_{in}\,dt + \frac{1}{RC}\int_0^t V_{0s}\,dt + \frac{1}{C}\int_0^t I_B\,dt + V_{0s} \qquad (10.56)$$

The first term is the desired result whereas remaining terms contribute to the errors. The integral error terms are the most troublesome as they keep increasing continually with time. These two ramp voltage errors continue to increase and drive the amplifier to saturation. Therefore, these error components set the upper limit on permissible integration time. The bias current error can be reduced by increasing C. This implies that R be reduced accordingly to achieve a specified value of RC time constant. However there is lower limit on R as well, because of the input source drive capability.

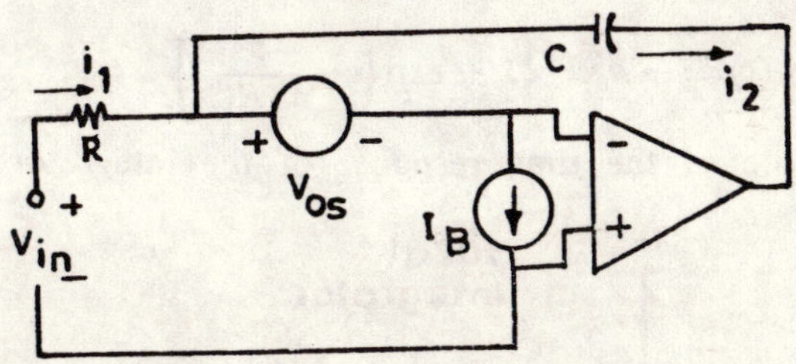

Fig. 10.54. Integrator with non-ideal op amp.

Converting the bias current to offset current by inserting a resistance R between the non-inverting input and ground will also reduce the error. The error ramp due to voltage offset is fixed by chosen value of RC time constant.

When the finite gain and bandwidth are taken into account the resulting integrator response function is given by

$$\frac{V_0(s)}{V_{in}(s)} = \frac{-a_0/RC\,\tau_0}{\left(s+\frac{1}{a_0RC}\right)\left(s+\frac{a_0}{\tau_0}\right)} \tag{10.57}$$

if $a_0 \gg 1$ and $(a_0.\ RC) > \tau_0$. Here the open loop frequency response of the op amp is approximated by $A = \frac{a_0}{1+s\tau_0}$. An ideal integrator has a single pole at the origin whereas this function has two poles on the negative real axis. At low frequency, the real integrator deviates from the ideal because of finite gain a_0. At high frequencies, the deviation is due to the finite bandwidth of the op amp given by amplifier pole at $1/\tau_0$.

The time domain response of the integrator to a step input $v_i(t) = Vu(t)$ is

$$v_0(t) = -a_0 V\left\{1 - \frac{\exp(-t/a_0 RC)}{1-(\tau_0/a_0^2 RC)} + \frac{\exp(-ta_0/\tau_0)}{(a_0^2 RC/\tau_0)-1}\right\} u(t) \tag{10.58}$$

For small t,

$$v_0(t) \simeq \frac{-V}{RC}\left[t+\frac{\tau_0}{a_0}\left\{-1+\exp\left(-\frac{ta_0}{\tau_0}\right)\right\}\right] u(t) \tag{10.59}$$

and the error is a time delay term due to finite bandwidth. For large t,

$$v_0(t) \simeq -a_0 V\left\{1-\exp\left(-\frac{t}{a_0 RC}\right)\right\} u(t) \tag{10.60}$$

The magnitude of the time response of the integrator given by

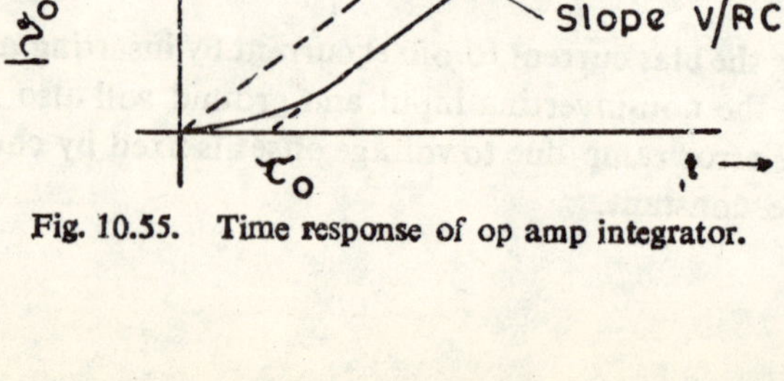

Fig. 10.55. Time response of op amp integrator.

eqn. (10.58) is shown in Fig. 10.55. For accurate integration, the process should be terminated at a time much less than a_0 RC and at an amplitude much less than a_0 V and 'Saturation Voltage' of op amp.

B. *Differentiator*

Interchanging R and C in Fig. 10.53 provides us with an analog differentiator shown in Fig. 10.56. However, this circuit is not practical as can be seen from the study of its transfer function given below.

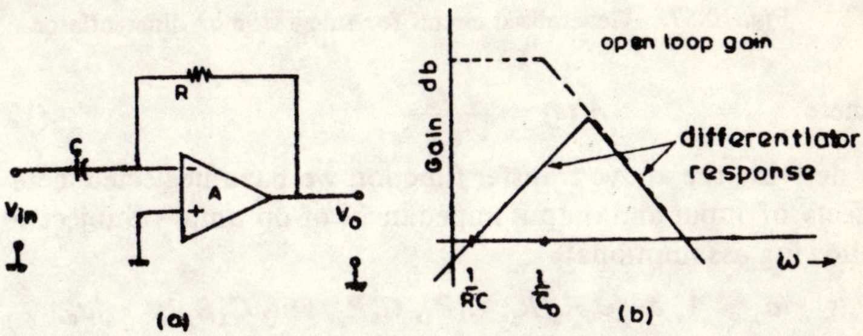

Fig. 10.56. (a) Analog differentiator. (b) Bode plot of (a).

$$\frac{V_0(s)}{V_i(s)} = \frac{-RCs\,\omega_n^2}{s^2 + 2\xi\,\omega_n\,s + \omega_n^2} \tag{10.61}$$

where

$$\omega_n^2 = \frac{a_0}{RC\,\tau_0} \tag{10.62}$$

$$\xi = \tfrac{1}{2}\sqrt{\frac{(\tau_0 + RC)^2}{RC\,\tau_0}\cdot\frac{1}{a_0}} \tag{10.63}$$

Thus, the damping factor ξ is very small indicating a very lightly damped circuit with complex poles, near the $j\omega$ axes. Thus the circuit has strong tendency towards instability. Another problem is its high gain at high frequencies. This contributes to the accentuation of high frequency components of amplifier noise, which may obscure the differentiated signal in some cases. These problems may be eliminated in the general circuit shown in Fig. 10.57 which may be used either as an integrator or differentiator by appropriate choice of components. The transfer function is given by

$$H(s) = \frac{V_0(s)}{V_i(s)} = -\frac{A(s)}{1 + \left(R_1 + \frac{1}{sC_1}\right)\left(sC_2 + \frac{1}{R_2}\right)\{A(s) + 1\}} \tag{10.64}$$

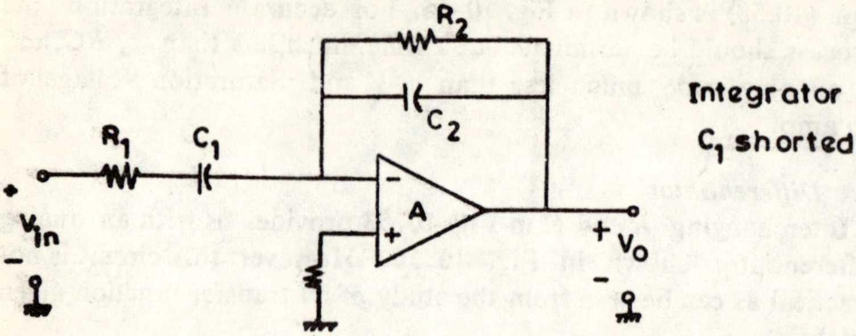

Fig. 10.57. Generalised circuit for integrator or differentiator.

where
$$A(s) = \frac{a_0}{1 + s\tau_0} \tag{10.65}$$

In deriving the above transfer function we have neglected here the effects of input and output impedances of op amp. Subject to the following assumptions:

$$a_0 \gg 1,\ a_0 \gg R_2/R_1,\ C_1R_1,\ C_2R_2, \text{ and } C_1R_2 \gg \tau_0/a_0,$$

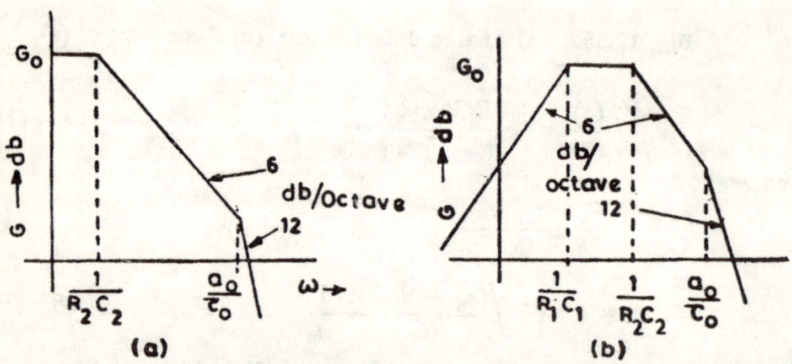

Fig. 10.58. Frequency response of (a) Practical integrator (b) Practical differentiator.

the approximate expression for $H(s)$ is obtained as

$$H(s) = -\frac{sC_1R_2}{s^3C_1R_1C_2R_2\tau_0/a_0 + s^2C_1R_1R_2C_2 + s(C_1R_1 + C_2R_2) + 1} \tag{10.66}$$

Replacing C_1 by a short circuit, we get a practical integrator.

$$H(s) = -\frac{1}{s^2C_2R_1\tau_0/a_0 + sC_2R_1 + R_1/R_2} \tag{10.67}$$

The resulting frequency response is shown in Fig. 10.58(a) where

$G = 20 \log_{10}/H(\omega)/$is the gain in db. The circuit works as an integrator over the range

$$\frac{1}{R_2C_2} \leqslant \omega \leqslant \frac{a_0}{\tau_0}$$

Over the frequency range for which $\omega C_1R_1 \ll 1$ as well as $\omega C_2R_2 \ll 1$,

$$H(s) = -\, sC_1R_2 \tag{10.68}$$

and the circuit acts as a differentiator. The approximate frequency response under the condition $C_1R_1 \gg C_2R_2 \gg \tau_0/a_0$ is shown in Fig. 10.58(b).

10.5.4 Voltage to Current Converters and Current to Voltage Converters

As implied by its name, the voltage to current converter (VIC) or the voltage controlled current source (VCCS) converts an input voltage variation to an output current variation independent of the load. It is used as a building block in active filter realizations and also in many communication and instrumentation circuits. Some of the commonly used a circuits for a floating load Z_L are shown in Figs. 10.59 and 10.60. The current i_1 is independent of the value of Z_L and equals V_1/R_1. In Fig. 10.59, both the signal

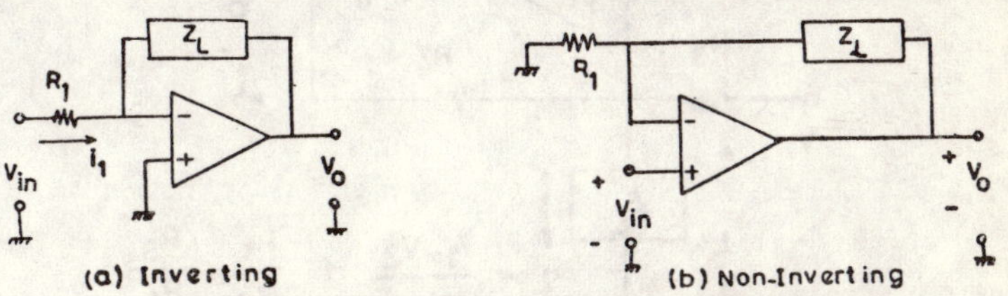

Fig. 10.59. *V* to *I* converters, floating loads.

source and the operational amplifier must be capable of supplying the desired amount of load current. The source current requirement is however much less for circuit (b) of Fig. 10.59 because of the noninverting mode of op amp operation. In the circuit of Fig. 10.60, most of the current is supplied by the op amp and only a small amount by the signal source. The load current is

$$i_L = \frac{v_1}{R_1}\left(1 + \frac{R_2}{R_3}\right) \tag{10.69}$$

Here the amplifier must be capable of providing most of the

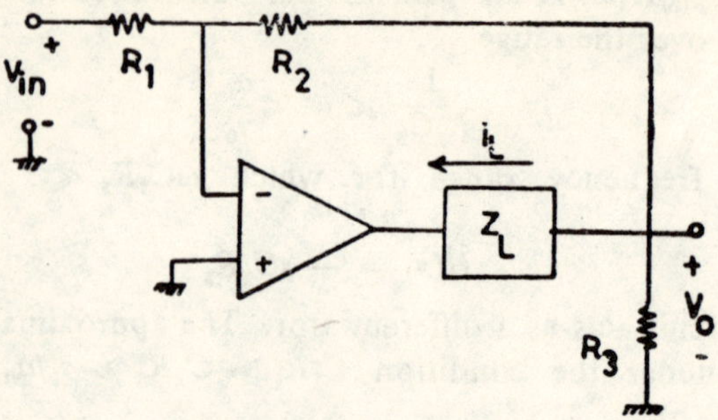

Fig. 10.60. *V* to *I* converter.

current to load and must also be capable of providing an output voltage equal to $v_{0\,max} \simeq i_{L\,max} \cdot (Z_L + R_3)$.

A voltage controlled current source for loads which are grounded on one side, is shown in Fig. 10.61.

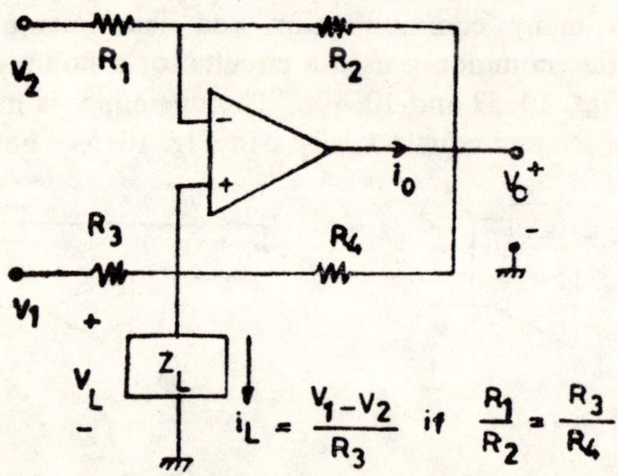

Fig. 10.61. A VCCS.

The output current i_L, in general is related to v_1 and v_2 by the following expression

$$i_L = \frac{v_1 - v_2}{R_3} + \frac{v_2}{R_0} - \frac{v_L}{R_0}$$

The output resistance as seen by the load v_L is obtained as

$$R_0 = \frac{R_1 R_3 R_4}{R_1 R_4 - R_2 R_3} \tag{10.71}$$

The load current i_L will, therefore, be independent of v_L if R_0 is infinite and this requires that

$$R_1R_4 = R_2R_3 \quad \text{or} \quad \frac{R_1}{R_2} = \frac{R_3}{R_4} \tag{10.72}$$

Under this condition,

$$i_L = \frac{v_1 - v_2}{R_3} \tag{10.73}$$

and the circuit behaves as a *voltage controlled current source* with the *differential transconductance*

$$G_m = \frac{i_L}{v_1 - v_2} = \frac{1}{R_3} \tag{10.74}$$

The output voltage and current supplied by the op amp under this condition with resistive load are,

$$v_0 = \frac{R_4}{R_3}(i_L R_L - v_1) + i_L R_L + i_L R_4 \tag{10.75}$$

and

$$i_0 = \left(\frac{1}{R_1} + \frac{1}{R_3}\right)(i_L R_L - v_1 + i_L R_3) \tag{10.76}$$

Subject to the restrictions on $v_{0\,max}$ and $i_{0\,max}$ imposed by the op amp, it may be verified that VCCS with single input is capable of delivering more i_L if $v_2 = 0$ and the input is applied as v_1. In this case,

$$v_0 = 1\left(+\frac{R_4}{R_3}\right) i_L R_L \tag{10.77}$$

$$i_0 = \left(\frac{1}{R_1} + \frac{1}{R_3}\right) i_L R_L \tag{10.78}$$

From (10.77) and (10.78) it is seen that for a prescribed $|v_0|_{max}$ and $|i_0|_{max}$, and for a given range of R_L variation largest I_L is obtained if $R_4 \ll R_3$ and $R_1 \gg R_3$. Therefore, for a given op amp and given R_L range, one can design a VCCS for a desired value of G_m by using equations (10.72), (10.74), (10.77) and (10.78).

A current to voltage converter is easily realized as shown in Fig. 10.62. Such a circuit is useful for acquiring signals from photocells and photomultiplier tubes which are basically current sources. The circuit presents almost zero load impedance because of the virtual ground property of the inverting input terminal. The lower limit on measurement of current input is determined by the bias current of the inverting input. For high resolution measurement, FET input op amps may be used. Capacitor C_F across the feedback resistance R_F is used to reduce high frequency current noise. Output impedance of the circuit is very low.

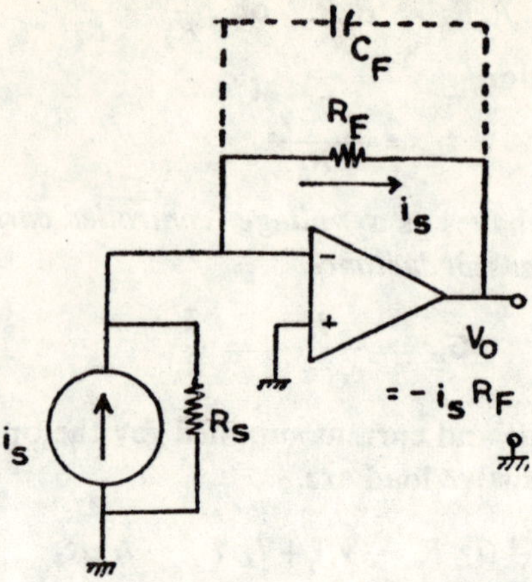

Fig. 10.62. *I* to *V* inverter.

10.5.5 Reference Voltage Sources and Voltage Regulators

Op amp can very conveniently be used as *reference voltage source*

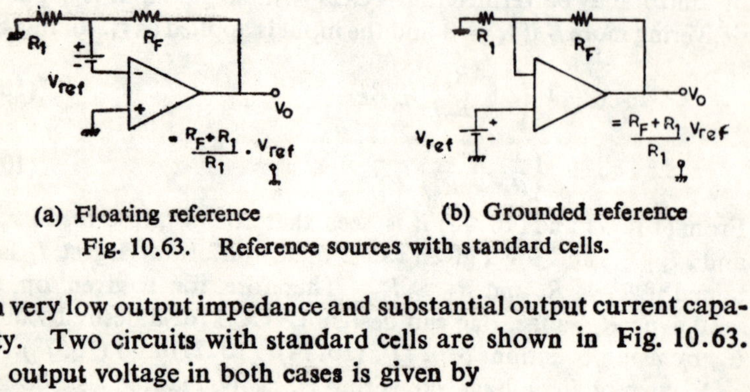

(a) Floating reference (b) Grounded reference

Fig. 10.63. Reference sources with standard cells.

with very low output impedance and substantial output current capability. Two circuits with standard cells are shown in Fig. 10.63. The output voltage in both cases is given by

$$V_0 = V_{ref}\left(1 + \frac{R_F}{R_1}\right) \tag{10.79}$$

The effective output impedance R_{0e} is given by

$$R_{0e} = \frac{R_0}{A\beta} = \frac{R_0(R_1 + R_F)}{AR_1} \tag{10.80}$$

where R_0 is open-loop output impedance and A is the open loop gain. The load regulation is given by

$$\text{Regulation } \% = \frac{R_0}{A R_L} \times 100 \tag{10.81}$$

where R_L is the minimum load impedance.

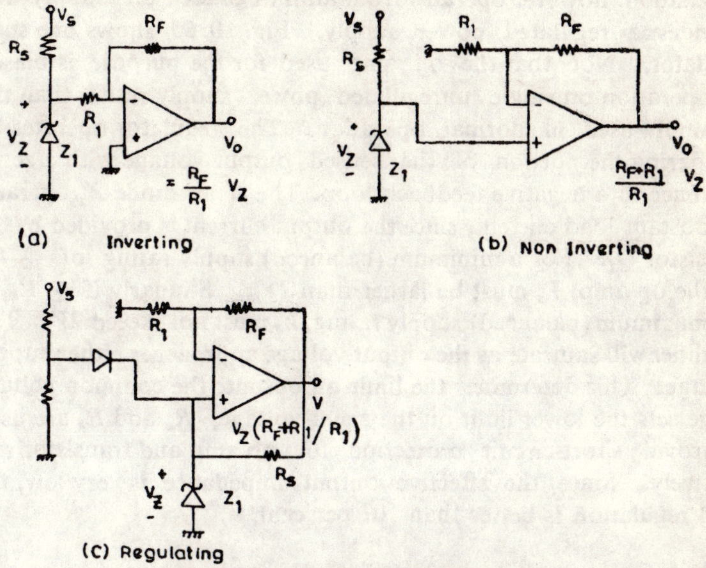

Fig. 10.64. Zener reference sources.

Op amp reference voltage sources using zener diodes as reference are shown in Fig. 10.64. Load regulation is the same as before and given by Eqn. (10.81). Regulation with respect to input voltage V_s will also depend on the dynamic resistance of the reference zener diode Z_1. Improved regulation with respect to input voltage is obtained in circuit of Fig. 10.64(c) which provides the output reference voltage as the source for zener diode current.

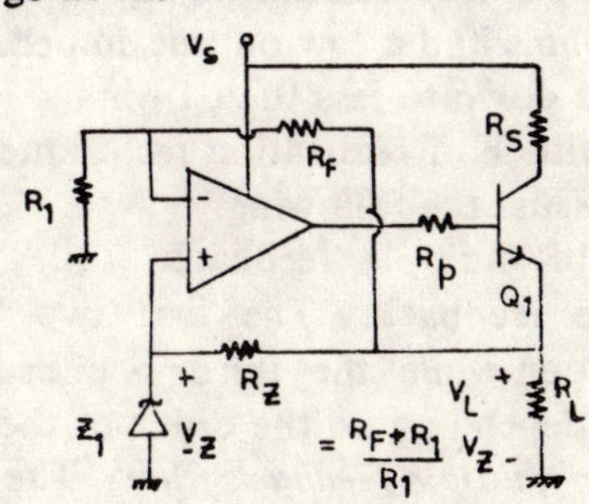

Fig. 10.65. A positive voltage regulator.

The voltage references described above may also be regarded as voltage regulators. For higher output currents, a power booster may be added inside the loop. A voltage regulator, as commonly understood, however operates from an unregulated d.c. supply and provides a regulated power supply. Fig. 10.65 shows one such regulator. Note that the op amp used for the purpose, is biased for operation on single unregulated power supply rather than the $\pm$ supply used in normal operation. The regulator operates by comparing the portion of the sensed output voltage with a zener reference in a negative feedback loop. The zener diode Z_1 operates at constant load current, since the output current is provided by the transistor Q_1. For a minimum (balanced) supply rating of $\pm V_m$ for the op amp, V_s must be larger than $2V_m$. Similarly if $\pm V_M$ is the maximum (balanced) supply rating, V_s must not exceed $2V_M$. The amplifier will saturate as the output voltage approaches either supply voltage. This determines the limit on output; the common voltage range sets the lower limit on the zener voltage, R_p and R_s are used to provide short-circuit protection for op amp and transistor respectively. Since the effective output impedance is very low, the load regulation is better than .01 per cent.

10.5.6 Active Filters

One of the most important applications of op amps is in the realization of active filters in the lower frequency range. These filters achieve the desired bandwidth and frequency selective characteristics without the use of bulky inductors which would otherwise be needed in this frequency range. Other advantages are their small size and weight, economy, ease of tuning and multifunctional capability. They also provide excellent isolation capabilities with input impedance ranging from few kilo ohms to thousand of megohms and a low output impedance ranging from few hundred ohms down to less than 1 ohm.

Out of the multitude of realization techniques for active filters, we shall briefly discuss the following.

(i) Infinite gain multiple feedback, (ii) controlled source, (iii) state variable feedback. The first two uses one op amp in their basic realization while the third one uses more than one op amp, the number depending on the order of the transfer function.

(i) *Infinite-gain multiple feedback filter*. The basic circuit of this type for realizing a voltage transfer function with a pair of complex conjugate S-plane poles with zeros restricted to origin or infinity is shown in Fig. 10.66. The op amp is used here in inverting mode.

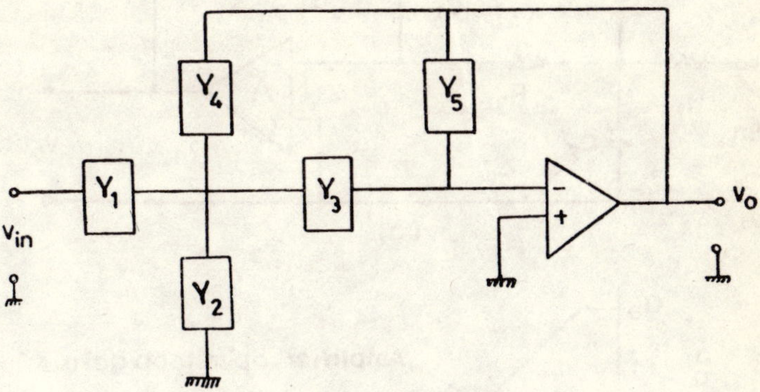

Fig. 10.66. Infinite gain feedback circuit.

Each element Y_i represents a single resistor or capacitor. In terms of the admittances Y_i, the voltage transfer function is

$$\frac{V_0(s)}{V_{in}(s)} = \frac{-Y_1Y_3}{Y_5(Y_1+Y_2+Y_3+Y_4)+Y_3Y_4+\frac{1}{A}\{(Y_3+Y_5)(Y_1+Y_2+Y_4)+Y_3Y_5\}} \tag{10.82}$$

In the limiting case of the open loop gain $A \to \infty$, we obtain

$$\frac{V_0(s)}{V_{in}(s)} = \frac{-Y_1Y_3}{Y_5(Y_1+Y_2+Y_3+Y_4)+Y_3Y_4} \tag{10.83}$$

A realisation of multiple feedback circuit is shown in Fig. 10.67 (a). The transfer function of the filter assuming the op amp to be ideal is given by the following equation

$$\frac{V_0(s)}{V_{in}(s)} = \frac{-1/R_3R_1}{S^2C_2C_5+SC_5(1/R_1+1/R_3+1/R_4)+1/R_3R_4} \tag{10.84}$$

This equation can be compared with the standard second order transfer function.

$$\frac{V_0(s)}{V_{in}(s)} = \frac{-H_0\omega_0^2}{S^2+\alpha\omega_0S+\omega_0^2} \tag{10.85}$$

The values of the d.c. gain H_0, damping factor α and the corner frequency ω_0 can be calculated by comparing (10.84) and (10.85). The values of components in Fig. 10.67(a), given H_0 and ω_0, can be calculated as follows.

1. Choose C and calculate k,

$$k = \omega_0 C \tag{10.86}$$

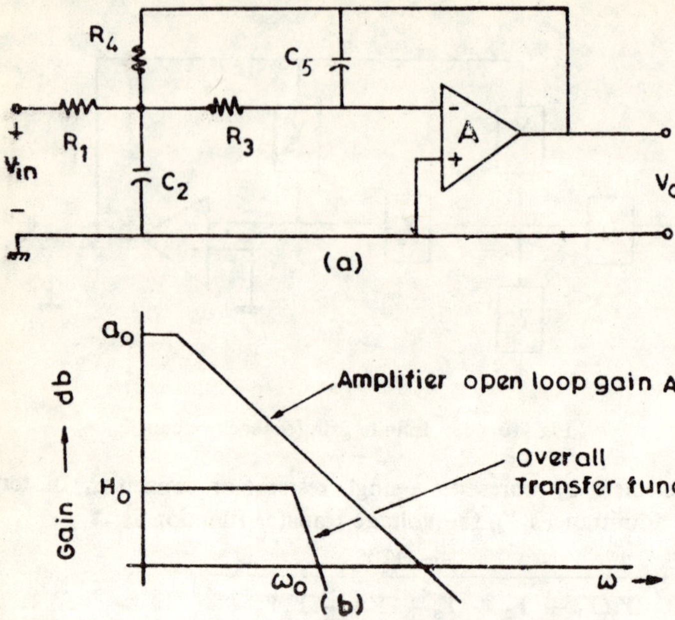

Fig. 10.67. (a) Multiple loop LB filter.
(b) Frequency response.

2. Choose α equal to $\sqrt{2}$ for "maximally flat" response and 40dB/decade roll off.

3. Calculate the component values C_5, C_2, R_1 and R_3

$$C_5 = C \tag{10.87}$$

$$C_2 = 4(H_0 + 1)\, k/\omega_0\alpha^2 \tag{10.88}$$

$$R_1 = \alpha/2H_0k \tag{10.89}$$

$$R_3 = \alpha/2(H_0 + 1)k \tag{10.90}$$

$$R_4 = H_0 \cdot R_1 = \alpha/2k \tag{10.91}$$

For a Low Pass filter with $f_0 = 100$ Hz and $H_0 = 20$dB the following values are obtained from above equations.

$$C = 0.1\mu\text{F},$$
$$k = 6.28 \times 10^{-5},$$
$$C_5 = 0.1\mu\text{F},$$
$$C_2 = 2.2\mu\text{F},$$
$$R_1 = 1125\Omega,$$
$$R_3 = 1020\Omega,$$
$$R_4 = 11.25K\Omega.$$

(ii) *Controlled-source filter.* In general, a controlled source is an active network element which has an output voltage or current which is a function of some single input voltage or current but is unaffected by any of the other voltages or currents in the network. This gives rise to four kinds of controlled sources. In our design, we shall confine ourselves only to the voltage controlled voltage source (VCVS). This is easily implemented with an op amp as shown in Fig. 10.68. The input impedance is very large (tens to hundreds of

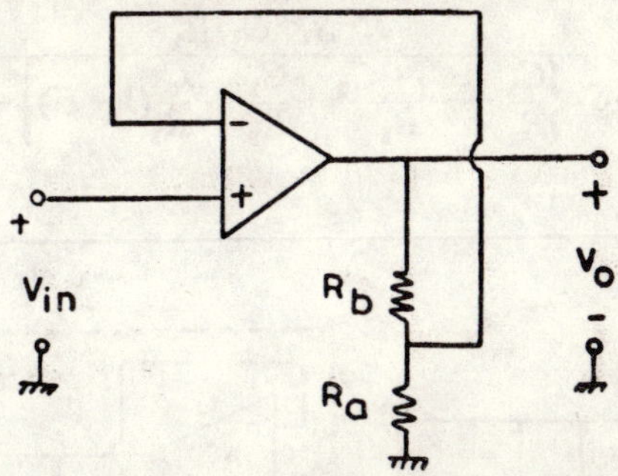

Fig. 10.68. A voltage controlled voltage source (VCCS).

megohms) while the output impedance is very low, usually less than 100 Ω for k between 1 and 10. The voltage transfer function is

$$\frac{V_0(s)}{V_{in}(s)} = 1 + \frac{R_b}{R_a} = K \tag{10.92}$$

Figure 10.69 shows the use of VCVS as an ideal gain block in conjunction with other circuit elements, to realize voltage transfer

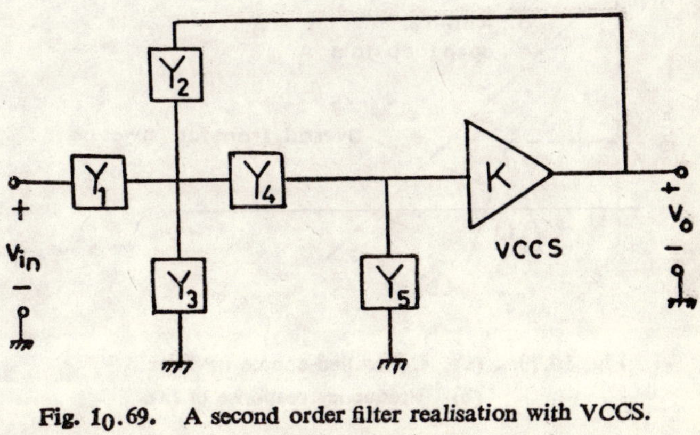

Fig. 10.69. A second order filter realisation with VCCS.

functions of the same type as in infinite gain multiple feedback configuration discussed earlier. The voltage transfer function is given by

$$\frac{V_0(s)}{V_{in}(s)}=\frac{KY_1Y_4}{Y_5(Y_1+Y_2+Y_3+Y_4)+Y_4\{Y_2(1-K)+Y_3+Y_1\}} \quad (10.93)$$

Figure 10.70 shows a Band Pass filter which uses the controlled source configuration. The transfer function of the filter assuming the op amp to be ideal is given by the following equation,

$$\frac{V_0(s)}{V_i(s)}=\frac{SK\,C_1/R_1}{S^2C_1C_2+S\cdot\left\{\frac{C_1}{R_3}+\frac{C_1+C_2}{R_1}+\frac{C_2}{R_2}+\frac{C_1}{R_2}(1-K)\right\}+\frac{1}{R_3}\left(\frac{R_1+R_2}{R_1R_2}\right)} \quad (10.94)$$

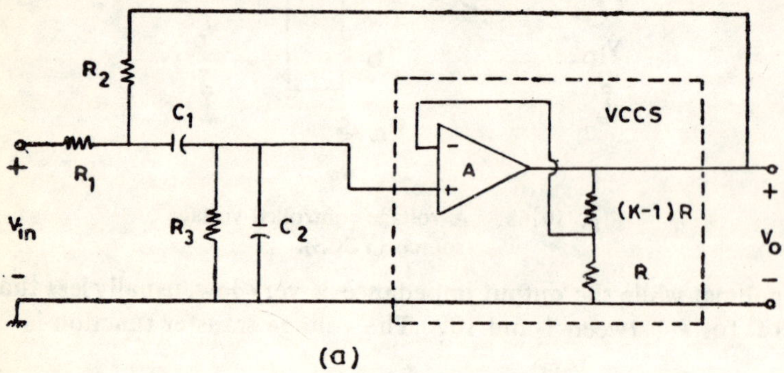

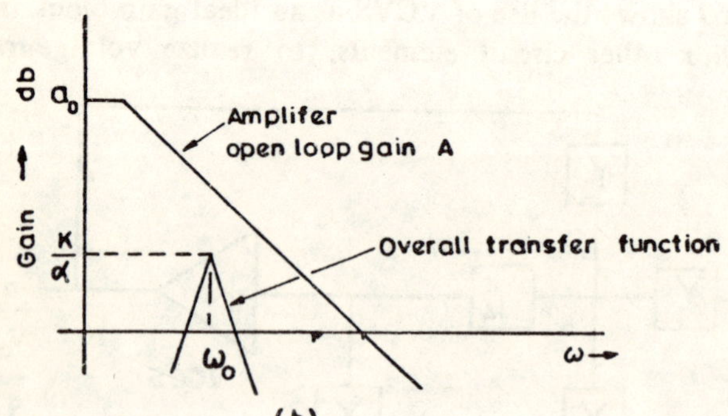

Fig. 10.70. (a) Controlled source BP filter.
(b) Frequency rosponse of (a).

This equation is of the same form as the following standard transfer function of the Band Pass filter

$$\frac{V_0(s)}{V_i(s)} = \frac{H\omega_0 s\alpha}{s^2 + \alpha\omega_0 s + \omega_0^2} \tag{10.95}$$

The quantity $1/\alpha$ is called the Quality Factor of the filter and is denoted by Q. It is a measure of the 'sharpness' of the frequency response of the filter. The values of H, ω_0 and for the filter of Fig 10.70(a) can be obtained by comparing eqns. (10.94) and (10.95). The values of the components in Fig. 10.70(a) for given values of ω_0 and Q can be obtained from the following design procedure.

1. Choose C_1 and calculate k

$$k = \omega_0 C_1 \tag{10.96}$$

2. Calculate the value of K

$$K = \frac{1}{3}\left(6.5 - \frac{1}{Q}\right) \tag{10.97}$$

3. Calculate the values of C_1, R_1, R_2 and R_3 from the following equations.

$$C_2 = 0.5\, C_1 \tag{10.98}$$

$$R_1 = 2/k \tag{10.99}$$

$$R_2 = R_1/3 \tag{10.100}$$

$$R_3 = 2R_1 \tag{10.101}$$

For a Band Pass filter with centre frequency of 300 Hz and a Q of 10, the components values obtained from above equations are as follows.

$C_1 = 0.02\mu\text{F}$, $k = 37.8\times 10^{-6}$, $K = 2.133$,

$C_2 = 0{\cdot}01\mu\text{F}$, $R_1 = 53\text{K}\Omega$, $R_2 = 17.7\text{K}\Omega$, $R_3 = 106\text{K}\Omega$.

(iii) *State-variable filter*: A general state-variable configuration which realises a general n-th order transfer function of the form

$$\frac{V_0(s)}{V_1(s)} = \frac{a_0 + a_1 s + \dots + a_{n-1} s^{n-1} + a_n s^n}{b_0 + b_1 s + \dots + b_{n-1} s^{n-1} + b_n s^n} \tag{10.102}$$

is given in Fig. 10.71(a). $V_a(s)$ is $V_1(s)$ divided by the denominator of eqn. (10.102). This configuration makes use of op amps in the same way as they would be used in an analog computer realization of transfer functions using integrators and summers. A particularly useful topology in this connection uses 3 to 4 op amps for realization of a general transfer function of 2nd order. It is called *Biquad* is also available as an IC. One such realization is shown in Fig. 10.71(b). A straight forward analysis of the circuit assuming ideal op amps and neglecting R^1 yields,

$$V_2 = -\frac{V_1}{SCR};\ V_3 = \frac{-V_2}{SCR} = \frac{V_1}{S^2C^2R^2}$$

$$V_+ = V_{in}\frac{m}{m+1} + V_2\frac{1}{m+1}$$

$$V_1 = V_+(n+1) - n\,V_3$$

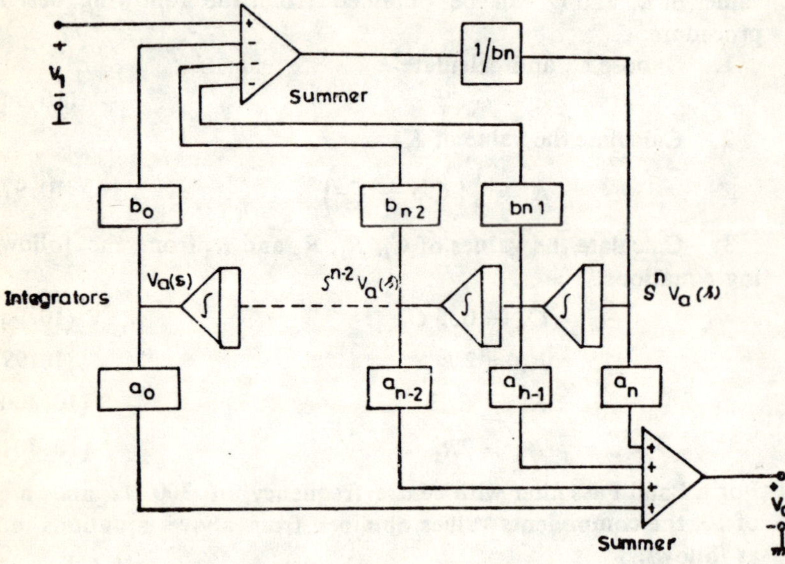

(a)

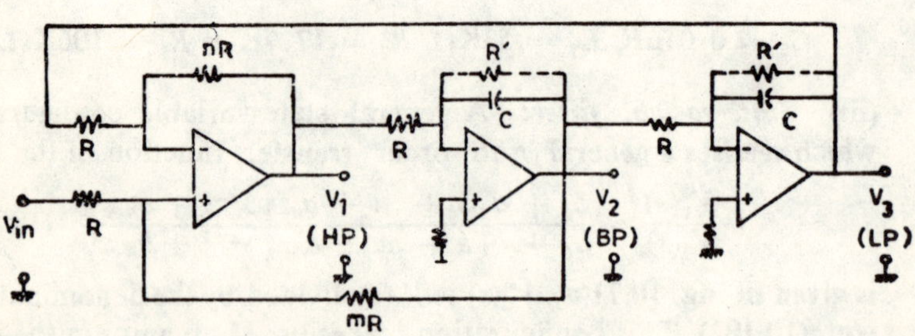

(b)

Fig. 10.71. (a) A general realisation of state
(b) A biquad state variable filter .

$$= \frac{m(n+1)}{n+1} V_{in} - \frac{n+1}{m+1} \cdot \frac{V_1}{SCR} - \frac{nV_1}{S^2C^2R^2} \quad (10.103)$$

The circuit provides low pass, high pass and band pass response characteristics at the indicated points. Thus

$$H_{HP}(s) = \frac{V_1}{V_i} = \frac{H_0 s^2}{s^2 + \alpha\omega_0 s + \omega_0^2} \quad (10.104)$$

$$H_{BP}(s) = \frac{V_2}{V_i} = \frac{\alpha H_0 \omega_0 s}{s^2 + \alpha\omega_0 s + \omega_0^2} \quad (10.105)$$

$$H_{LP}(s) = \frac{V_3}{V_i} = \frac{H_0 \omega_0^2}{s^2 + \alpha\omega_0 s + \omega_0^2} \quad (10.106)$$

The gain constant H_0, the damping factor and natural frequency ω_0 are related to the circuit element values as shown in Table 10.20.

Table 10.20. Parameters of Biquad Filter

Function	ω_0	α	H_0
High Pass (*HP*)	$\sqrt{n}/RC$	$\frac{1}{\sqrt{n}} \frac{n+1}{m+1}$	$\frac{m(n+1)}{(m+1)}$
Band Pass (*BP*)	—do—	—do—	$-m$
Low Pass (*LP*)	—do—	—do—	$\frac{m(n+1)}{n(m+1)}$

From Table 10.20, we observe that we can control natural frequency ω_0 by varying n and damping factor α, by changing m for a particular choice of ω_0.

To obtain variable phase shift with constant gain magnitude characteristic, one may employ circuits of Fig. 10.72 where phase shifts are controlled by potentiometers. The circuits realize the transfer function,

$$\frac{V_0(s)}{V_i(s)} = \pm \frac{1 - sRC}{1 + sRC} \quad (10.107)$$

The phase shift developed by two circuits are

$$\phi_a = -2\tan^{-1}\omega RC \text{ and } \phi_b = -180^\circ + \phi_a$$

The appropriate upper frequency limit of operation for this purpose is determined by the open loop gain and phase shift characteristics of the op amp employed.

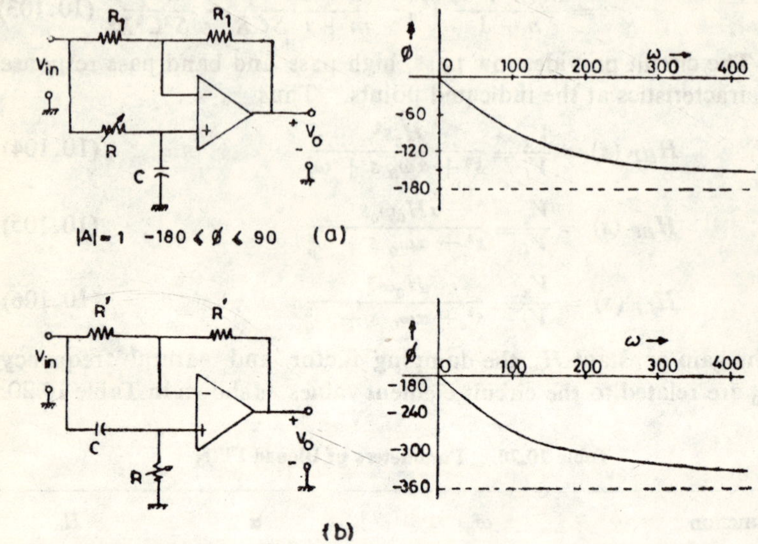

[A]=1, $-360 \leqslant \phi \leqslant -180$

Fig. 10.72. Op amp phase shifters and frequency response for $RC=10^{-2}$.

10.6 NON-LINEAR APPLICATIONS OF THE OP AMP

There are a class of applications of the op amp which have non-linear input-output characteristics. A classification of these circuits based on the source of non-linearity is as follows.

(*A*) *Circuits with non-linear feedback*: In these circuits the op amp is in the linear region but the feedback factor β (Fig. 10.42a) is varied. Non-linear devices e.g. diodes and transistors are employed to change the value of β for some preset values of the inputs.

Examples of such circuits are the Precision Rectifier, the Diode Function Generator and the Log Amplifier.

(*B*) *Circuits with zero/positive feedback*: In these circuits, under steady state, the op amp is either in the positive or the negative saturation. The op amp is used as a high gain d.c. amplifier either without feedback or with positive feedback to switch between the saturation levels.

Examples of such circuits are the comparator and the Schmitt Trigger.

(*C*) *Circuits with switches*: In some circuits the op amp is combined with a switch to perform some non-linear functions. The switches are opened and closed periodically to obtain the required input-output relationship.

Examples of such circuits are the Peak Detector and the *S/H*.

(*D*) *Circuits with stable oscillations*: In these circuits op amps are combined with passive (*R-L-C*) networks to obtain stable oscillations. The output is usually a square or sinusoidal wave whose amplitude and the time period are adjustable. The non-linear elements are employed to limit the amplitude of the output.

Examples of such circuits are the Square Wave Generator and the Sinusoidal Oscillator.

In the following sections few basic non-linear circuits using the op amp are described. These circuits are representative of the types commonly used and several variations depending on the application requirements are possible.

10.6.1 Circuits with Non-linear Feedback

The introduction of non-linear elements in the op amp circuits gives rise to a variety of functions. The non-linear elements (NLE) can be placed at the input, in the forward path or the feedback path of the amplifier using the op amp.

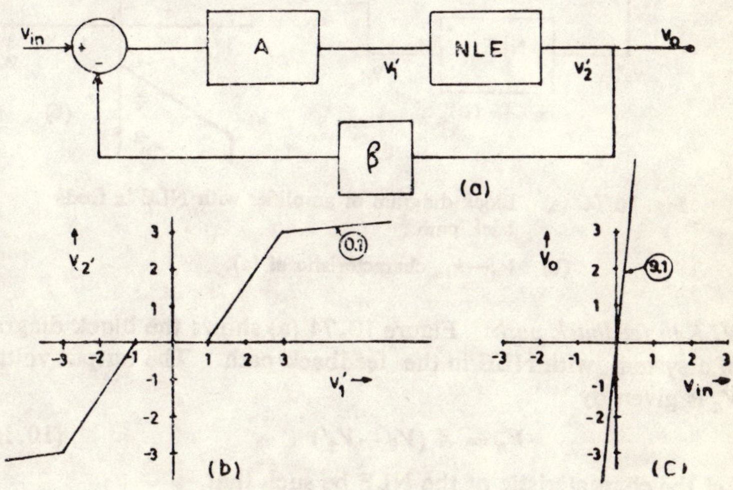

Fig. 10.73. (a) Block diagram of amplifier with NLE in forward path Input-output characteristics of (b) NLB (c) Amplifier.

NLE in the forward path: Figure 10.73(a) shows the block diagram (c.f. Fig. 10.42a) of a $-ve$ feedback system with NLE in the forward path. The non-linearity may represent some inherent characteristics of the amplifier itself e.g. saturation of output stage. Figure 10.73(b) shows the input-output characteristic of the NLE. If gain A is 1000 (60 db) the V_0 vs. V_{in} characteristic of the system is shown in Fig. 10.73(c). The high forward gain of 1000, effectively reduces the effect of non-linearity on the output. In particular, when $\beta = 0.1$,

(a) The dead band is reduced to $-10^{-3} \leqslant V_{in} \leqslant 10^{-3}$.

(b) The closed loop gain is (1500/151) for $-3 \leqslant V_{in} \leqslant +3$ except for the small dead band.

(c) The closed loop gain is (100/11) for

$$V_{in} \leqslant -3 \text{ and } V_{in} \geqslant +3$$

Therefore, if the forward gain is high the effect of introducing a NLE on the closed loop gain is negligible. However, if the incremental slope of the NLE is zero, i.e. for ideal saturation the output will also saturate.

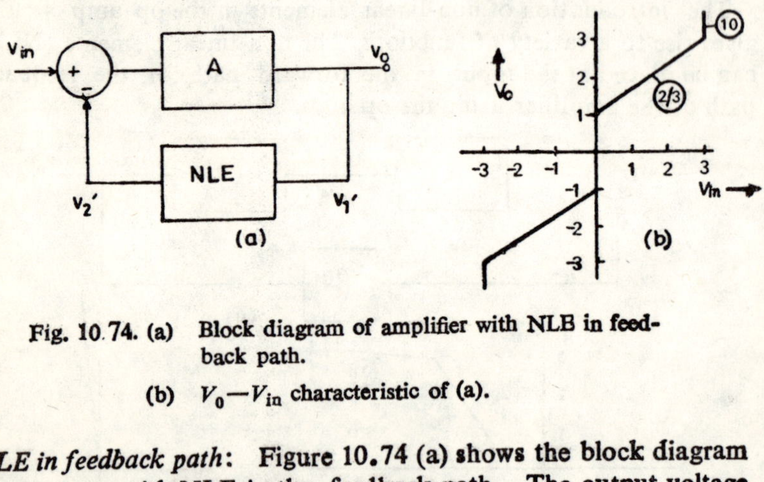

Fig. 10.74. (a) Block diagram of amplifier with NLB in feedback path.

(b) $V_0 - V_{in}$ characteristic of (a).

NLE in feedback path: Figure 10.74 (a) shows the block diagram of a system with NLE in the feedback path. The output voltage V_0 is given by

$$V_0 = A\,(V_{in} - V_2') \tag{10.108}$$

Let the characteristic of the NLE be such that,

$$V_2' = f\,(V_1') \tag{10.109}$$

where the function f is known either graphically or analytically substituting in (10.108) gives,

$$V_0 = f^{-1}\left(V_{\text{in}} - \frac{V_0}{A}\right) \tag{10.110}$$

If $A \to \infty$

$$V_0 \simeq f^{-1}(V_{\text{in}}) \tag{10.111}$$

where f^{-1} is the inverse of the function f. For example, let NLE having the characteristic shown in Fig. 10.73 (b) be used in Fig. 10.74 (a). The V_0-V_{in} characteristic of the system can be derived from that of Fig. 10.73 (b) and equation (10.111). For a given value of V_{in} $(= V_2')$ the corresponding output v_0 $(= V_1')$ can be obtained from Fig. 10.73 (b). The resulting characteristic is shown in Fig. 10.74 (b).

Thus, the inverse function can be obtained by connecting a NLE in the feedback path of the op amp.

Figure 10.75 shows the circuit implementations of the feedback systems using NLE. The op amps in Fig. 10.75 provide the high gain amplifier "A" shown in Fig. 10.73 and 10.74.

Limiters: The concept of a limiter is widely used in non-linear circuit application of the op amps. Usually a limiter element limits the "Voltage" at a node in the circuit while the currents are allowed to vary to suit the inputs.

The ideal limiter: Fig. 10.76 shows an ideal limiter. It consists of an ideal diode D in series with an ideal voltage source V_B. The voltage across the limiter cannot exceed V_B due to the conduction of the diode D.

Figure 10.77 (a) shows an ideal limiter connected in series with the load. The V_0-V_{in} characteristic is shown in (b). The output voltage is zero till the input voltage reaches a threshold V_B. The slope S is $(R_L/R_L + R_S)$.

Figure 10.78 (a) shows an ideal limiter connected in shunt with the load. The V_0-V_{in} characteristic is shown in (b). The slope S is $(R_L/R_L + R_S)$.

Practical forms of limiter circuits use some elements to obtain the source V_B of Figs. 10.77 and 10.78. A zener diode and diodes with resistive dividers are commonly used. The limiter circuits are combined with op amp circuits to obtain non-linear input-output characteristics. Such circuits are described in the following.

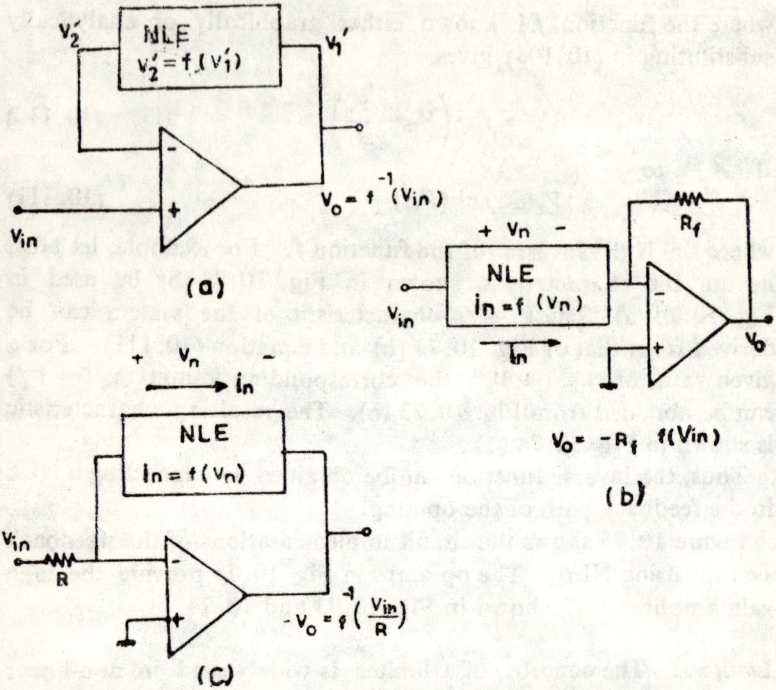

Fig. 10.75. Op amp circuits with non-linear element (NLE). (a) Inverse function generation using non-linear gain. Use of non-linear trans-conductance to generate (b) Function f (c) Inverse f^{-1}.

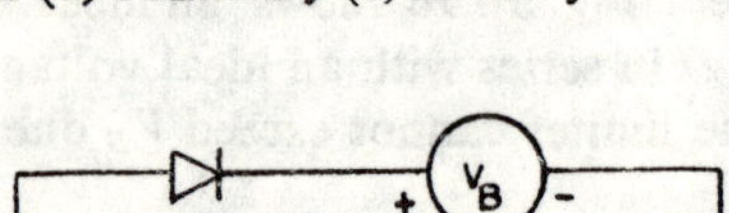

Fig. 10.76. Ideal limiter element.

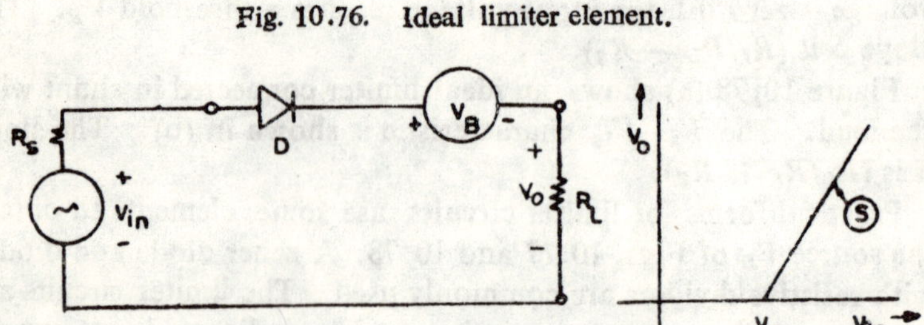

Fig. 10.77. (a) Series limiter. (b) $V_0 - V_{in}$ characteristic of (a).

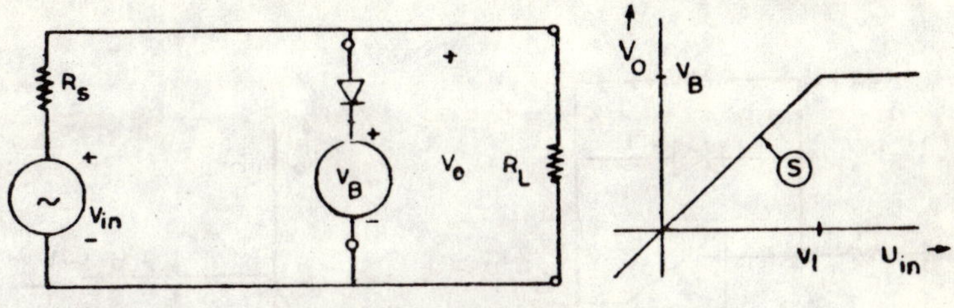

Fig. 10.78. (a) Shunt limiter.
(b) $V_0 - V_{in}$ characteristic of (a).

Precision diode rectifiers: The op amp amplifiers (Figs. 10.39, 10.41) can be modified to amplify in one direction only. Figure 10.79 (a) shows an inverting amplifier with a diode D connected across R_f. The −ve input of the op amp is a virtual ground. Thus, if $V_0 < 0$, the diode D begins to conduct and the voltage V_0 is limited to the diode drop, about −0.7V. Figure 10.79 (b) shows the V_0 vs V_i characteristic of the circuit. If the input is a sinusoidal voltage, then,

$$V_0 = -V_{in} \cdot \frac{R_f}{R_1} \text{ if } V_{in} < 0$$

$$= -0.7\text{V} \text{ if } V_{in} > 0 \tag{10.112}$$

Thus, the circuit acts like a half wave rectifier.

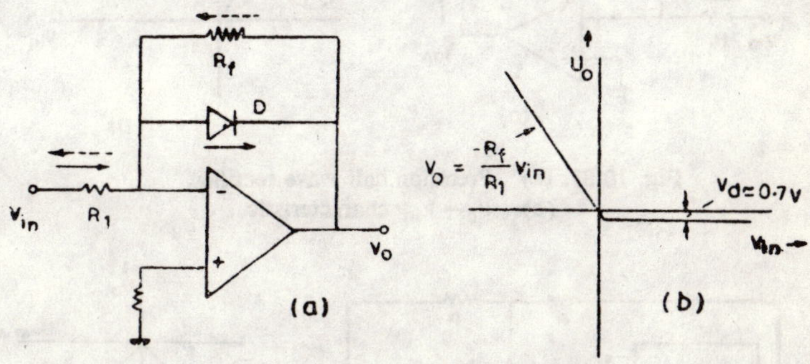

Fig. 10.79 (a) Inverting ampfifier with diode limiter, $V_0 > 0$.
(b) $V_0 - V_{in}$ characteristic of (a).

Fig. 10.80 (a) shows a similar half wave rectifier obtained by limiting the output of a non-inverting amplifier. Figure 10.80 (b) shows the corresponding $V_0 - V_i$ characteristic.

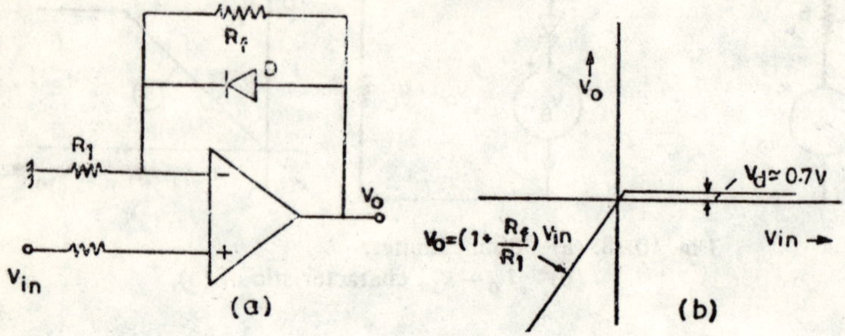

Fig. 10.80. (a) Non-inverting amplifier with diode limiter, $V_o < 0$.
(b) $V_0 - V_{in}$ characteristic of (a).

The circuits shown in Figs. 10.79 and 10.80 limit the output voltage to 0.7V in one direction. Therefore, if the magnitude of the input signal is comparable to 0.7V, the rectification is not perfect. The 'Precision Rectifier, of Fig. 10.81 overcomes this limitation by using two diodes. The output voltage V_0 is precisely zero

Fig. 10.81. (a) Precision half wave rectifier.
(b) $V_0 - V_{in}$ characteristic.

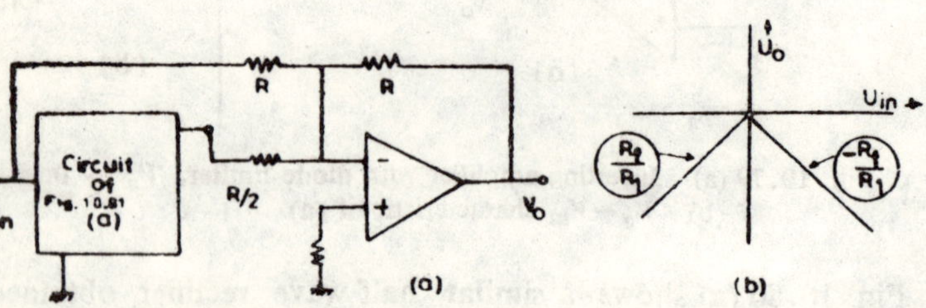

Fig. 10.82. (a) Precision full wave rectifier.
(b) $V_0 - V_{in}$ characteristic.

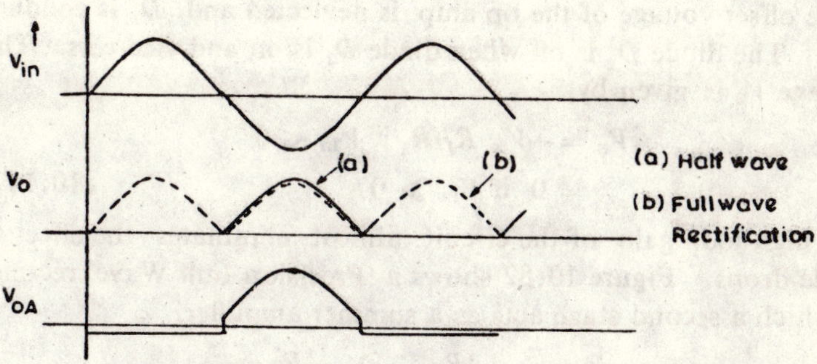

Fig. 10.83. Voltage waveform in precision rectifiers.

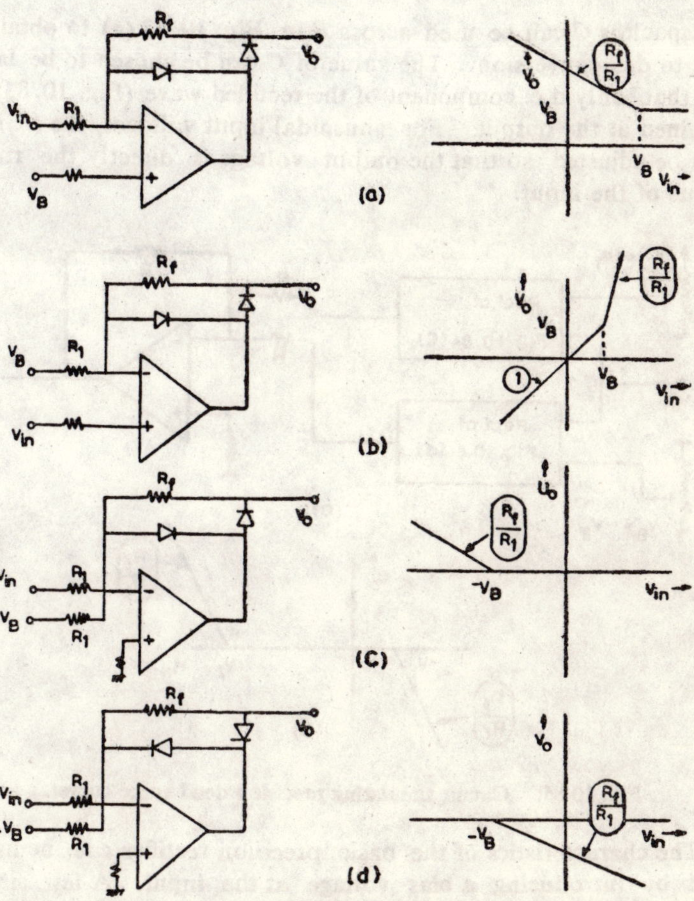

Fig. 10.84. Precision rectifiers with bias voltage V_B at input.

if the offset voltage of the op amp is neglected and D_1 is conducting. The diode D_2 is off when diode D_1 is on and vice-versa. The voltage V_0 is given by,

$$V_0 = -V_{in} R_f/R_1 \text{ if } V_{in} < 0$$
$$= 0 \text{ if } V_{in} > 0 \qquad (10.113)$$

The high loop gain of the circuit almost eliminates the effect of diode drops. Figure 10.82 shows a 'Precision Full Wave' rectifier in which a second stage acts as a summer amplifier.

$$V_0 = -\frac{R_f}{R_1} V_{in} + \frac{2R_f}{R_1} V_{01} = -\frac{R_f}{R_1} V_{in} \qquad (10.114)$$

A capacitor C can be used across R in Fig. 10.82 (a) to obtain a a.c. to d.c. conversion. The value of C can be chosen to be large so that only d.c. component of the rectified wave (Fig. 10.83) is retained at the output. For sinusoidal input voltages, the (R_f/R_1) can be adjusted so that the output voltage is directly the r.m.s. value of the input.

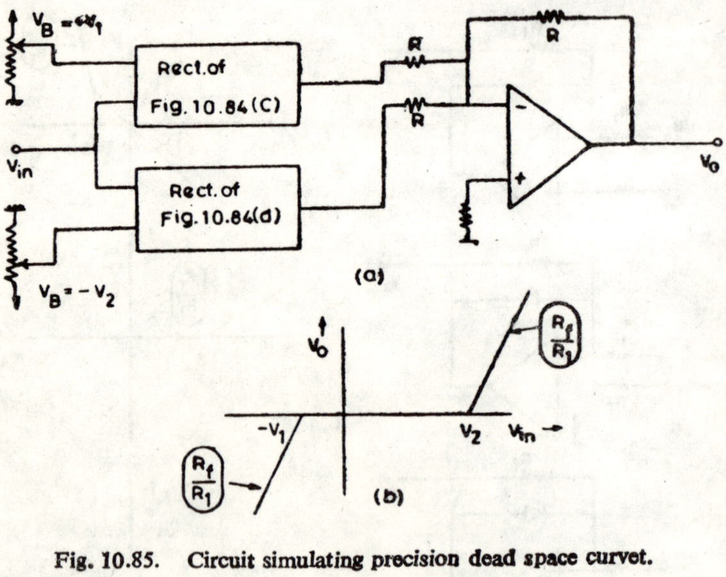

Fig. 10.85. Circuit simulating precision dead space curve.

The characteristics of the basic precision rectifier can be modified by introducing a bias voltage at the input. A few simple circuits of this type and their input-output characteristics are shown in Fig. 10.84. The circuits of Fig. 10.84 can be combined to obtain a combination of characteristics. Figs. 108.5 and 10.86 show some examples.

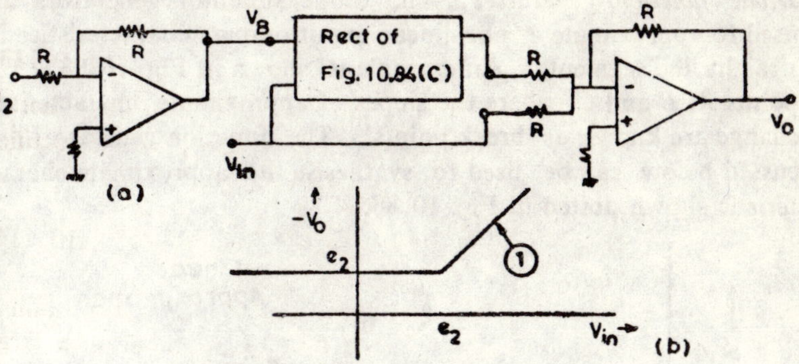

Fig. 10.86. Precision maximum selector.

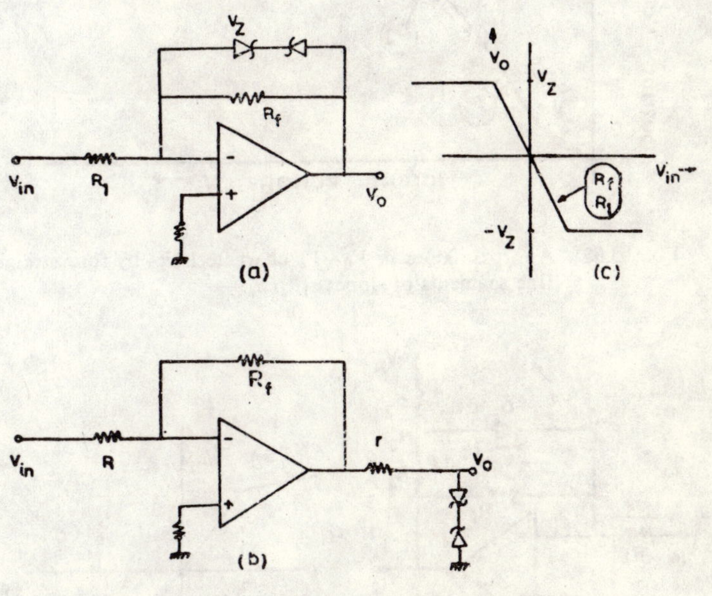

Fig. 10.87. (a), (b) Inverting amplifiers with zener limiters.
(c) $V_0 - V_{in}$ characteristic.

The output of the inverting or non-inverting amplifier can also be limited by using a zener diode as a shunt limiter. The zener diode can be directly connected in the feedback path as shown in

Fig. 10.87 (a). The zener diode can also be connected at the output in series with a isolation resistor r as shown in Fig. 10.87 (b).

Diode function generators: The diode function generators are used to approximate a non-linear input-output characteristics by straight line segments. An example is shown in Fig. 10.88. The points A, B and C where the slopes of approximate characteristic change are known as 'break points'. The function generators discussed below can be used to synthesise the approximate characteristic shown dotted in Fig. 10.88.

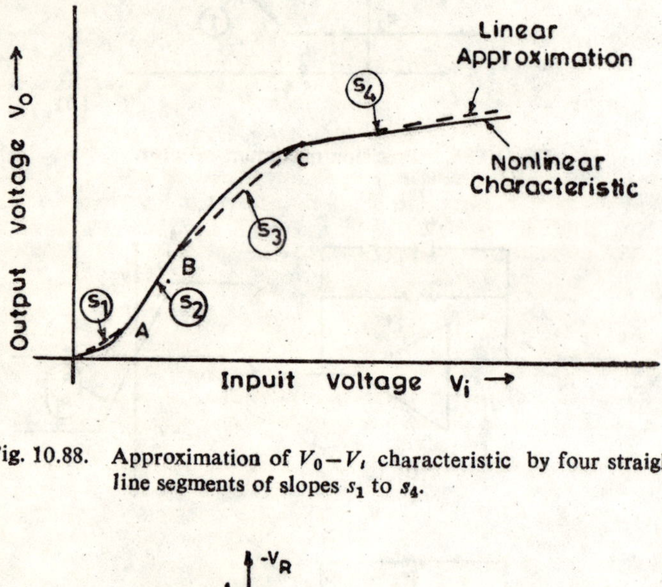

Fig. 10.88. Approximation of V_0-V_i characteristic by four straight line segments of slopes s_1 to s_4.

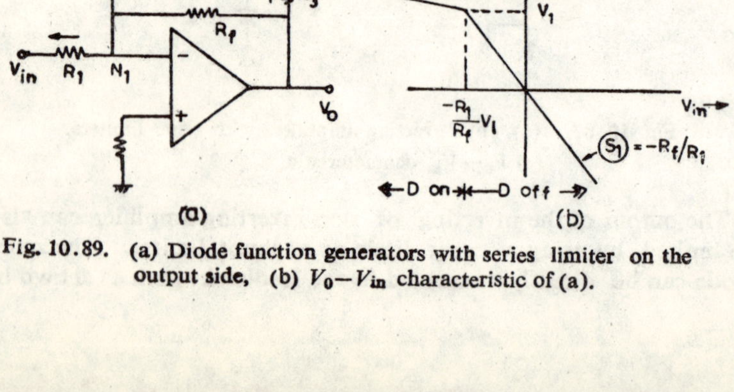

Fig. 10.89. (a) Diode function generators with series limiter on the output side, (b) V_0-V_{in} characteristic of (a).

Figure 10.89 shows a diode function generator with series Limiter on the output side. If the diode D is assumed to be off, the output voltage is given by,

$$V_o = -(R_f/R_1)\,V_{in} \tag{10.115}$$

The cathode of D is at virtual ground. The voltage across diode V_d is given by,

$$V_d = (V_0)(R_2/R_3 + R_2) + (-V_R)(R_3/R_2 + R_3) \tag{10.116}$$

The diode D remains off if V_d is less than a threshold voltage V_r Applying the condition $v_d \leqslant v_r$ to eqn. (10.116) gives,

$$v_0 \leqslant v_1 \tag{10.117}$$

where

$$v_1 = v_r(R_3 + R_2/R_2) + V_R(R_3/R_2) \tag{10.118}$$

From eqn. (10.115) (10.117), the condition for D to be off is,

$$v_{in} > (-R_1/R_f)\cdot v_1 \tag{10.119}$$

When the input voltage is reduced below the value given by (10.119) the diode D starts conducting. The voltage V_d is then equal to the forward drop of the diode which can be assumed to be approximately 0.7V. When the diode D is conducting, by KCL at nodes N_1 and N_2 (Fig. 10.89a).

$$(v_0/R_f) + i_d + (v_{in}/R_1) = 0 \tag{10.120}$$

$$\left(\frac{v_0 - v_d}{R_3}\right) - \left(\frac{v_d + v_R}{R_2}\right) = i_d \tag{10.121}$$

Eliminating i_d between above equations gives,

$$v_0 = -(R_f'/R_1)\,v_{in} + (R_f'/R_3)\cdot V_{l1} \tag{10.122}$$

where,

$$V_{l1} = v_d\cdot(R_3 + R_2/R_2) + v_R(R_3/R_2) \tag{10.123}$$

$$R'_f = R_f \parallel {}^{l}R_3 \tag{10.124}$$

A comparision of eqn. (10.118) and (10.123) shows that the v_{l1} and v_1 are equal if v_d is assumed to be equal to v_r. The $v_0 - v_l$ characteristic with this assumption in shown in Fig. 10.89(b). The non-ideal nature of the diode results in a rounding of the practical characteristic near the break point. The function generator can be extended by adding more limiters like the $R_2 - R_3 - D$ in Fig. 10.89(a) between the output and node N_1 with different values of the bias voltage V_R.

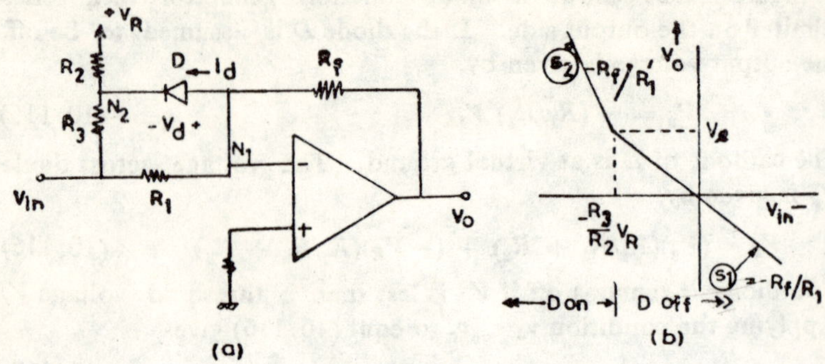

Fig. 10.90. (a) Diode function generator with series limiter on the input side.
(b) $V_0 - V_{in}$ characteristic of (a).

Figure 10.90 shows a diode function generator with the series limiter on the input side. The node N_1 is at virtual ground. When diode D is off the circuit functions like an inverting amplifier with gain of (R_f/R_1). The potential of node N_2 when the diode D is off is positive and v_d is given by,

$$v_d = -\{(v_{in} R_2/R_2 + R_3) + (V_R \cdot R_3/R_2 + R_3)\} \quad (10.125)$$

The diode D remains off till v_d is less than a threshold value V_r. Applying the condition $v_d \leqslant V_r$ to equation (10.125) gives the condition for D to be off.

$$V_{in} \geqslant -V_r \cdot (R_3 + R_3/R_2) - V_R(R_3/R_2) \quad (10.126)$$

When the diode D is conducting, KCL at nodes N_1 and N_2 gives,

$$v_0/R_f = -(v_{in}/R_1 - i_d) \quad (10.127)$$

$$\frac{V_{in} + V_d}{R_3} + \frac{V_R + V_d}{R_2} + i_d = 0 \quad (10.128)$$

Eliminating i_d between eqns. (10.127) and (10.128) gives,

$$V_0 = -V_{in} R_f/R'_1 + v_l \quad (10.129)$$

where $$v_l = -(V_d \cdot R_f/R_3 + V_R \cdot R_f/R_2 + V_d \cdot R_f/R_2) \quad (10.130)$$

$$R_1' = (R_1 \| R_3) \quad (10.131)$$

The input-output characteristic of the circuit is shown in Fig. 10.90 (b). The sharp break point is obtained on the assumption that V_r and V_d in eqns. (10.126) and (10.130) are equal. The function generator can be extended by adding more limiters like the

$R_2 - R_3 - D$ in Fig. 10.90 (a) between the input and the node N_1 with different values of the bias voltage V_R.

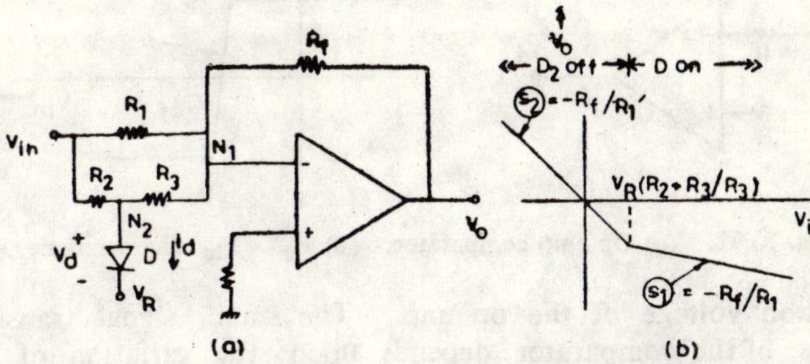

Fig. 10.91. (a) Diode function generator with shunt limiter on the input side.
(b) $V_0 - V_{in}$ characteristic of (a).

Figure 10.91 shows a diode function generator with a shunt limiter on the input side. The node N_1 is at virtual ground while node N_2 is at the potential $(V_{in}R_3/R_2 + R_3)$. D is off if V_{in} is less than $V_R\,(1 + R_3/R_2)$. The circuit acts like an inverting amplifier with gain R_f/R_1'. R_1' is the parallel combination of R_1 and $(R_2 + R_3)$. When the diode D is conducting voltage of node N_2 is clamped is $(V_R + V_d)$ which is nearly constant. The voltage v_0 then changes with the gain (R_f/R_1). Figure 10.91 (b) shows the input-output characteristic of the circuit assuming that the effect of V_r and V_d is negligible. The circuit can be extended by putting more stages like $R_2 - R_3 - D$ between the input and the node N_1 with different values of the bias voltage V_R.

10.6.2 Circuits with Zero or Positive Feedback

The high gain of the op amp drives it into positive or negative saturation for very small input voltages. Thus, when used without the output feedback it forms an excellent comparator. Figure 10.92 (a) shows an op amp comparator. The open loop gain A decides the d.c. $V_0 - V_{in}$ characteristic. The polarities of the output voltages, V_s, can be reversed if the input and the reference terminals are interchanged. In practical op amps the output saturation voltage V_s is not stable, being dependent on the power supply. Thus, any of the limiter circuits described earlier may be used to accurately limit the output to a value lower than the

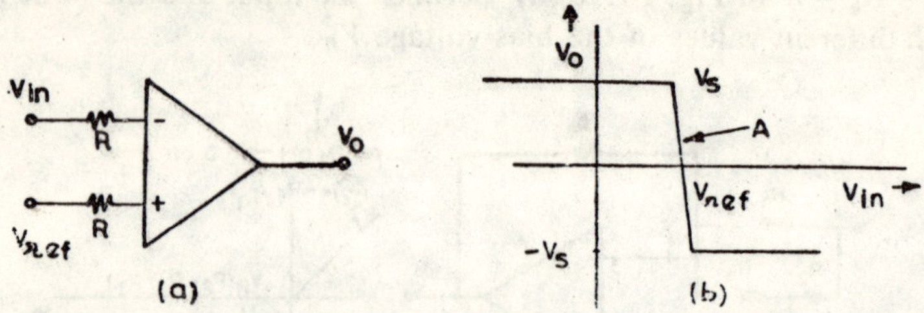

Fig. 10.92. (a) Op amp comparator. (b) $V_0 - V_{in}$ characteristic.

saturation voltage of the op amp. The small signal transient response of the comparator depends upon the variation of the

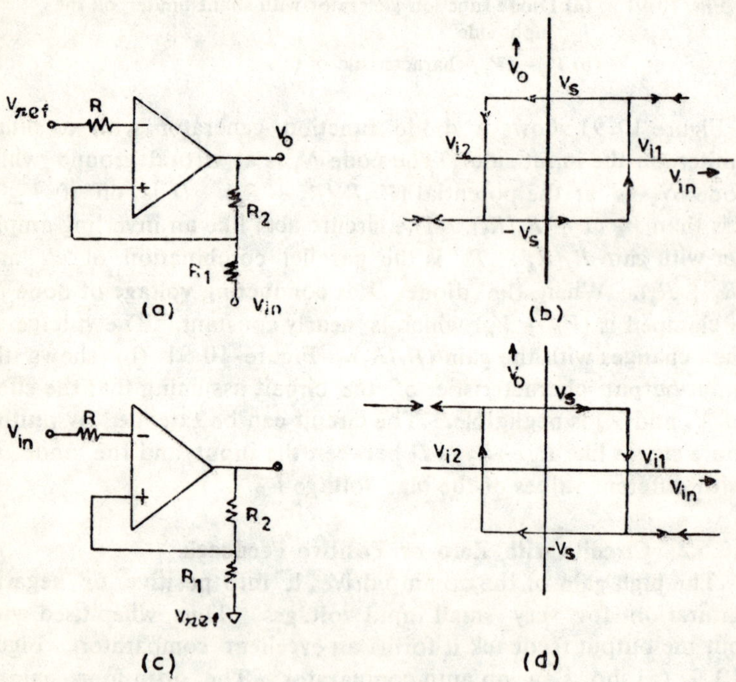

Fig. 10.93. Comparators with adjustable hyteresis: (a) Non-inverting comparator. (b) $V_0 - V_{in}$ characteristic of (a). (c) Inverting comparator. (d) $V_0 - V_{in}$ characteristic of (c).

op amp gain with frequency. The internally compensated op amps make poor comparators as their response is sluggish. The large signal response of the comparator is limited by the slew rate of the op amp. The accuracy of the trip point, where the input equals V_{ref}, is affected by the offset voltage and the bias currents of the op amp. Each input is applied through a resistor R to present equal resistance to bias currents. The threshold voltage in most IC comparators (e.g. 710, 311, 321) is less than few mV.

Figures 10.93 (a) and (c) show modified comparators in which positive feedback is used to obtain a hysteresis in the $V_0 - V_{in}$ characteristic. The transient response of these circuits is faster than the open loop comparator due to the positive feedback. The input voltages V_{i1} and V_{i2} at the trip points can be easily calculated by equating the expressions for the voltages at the positive and negative input terminals. The calculated values are as follows.

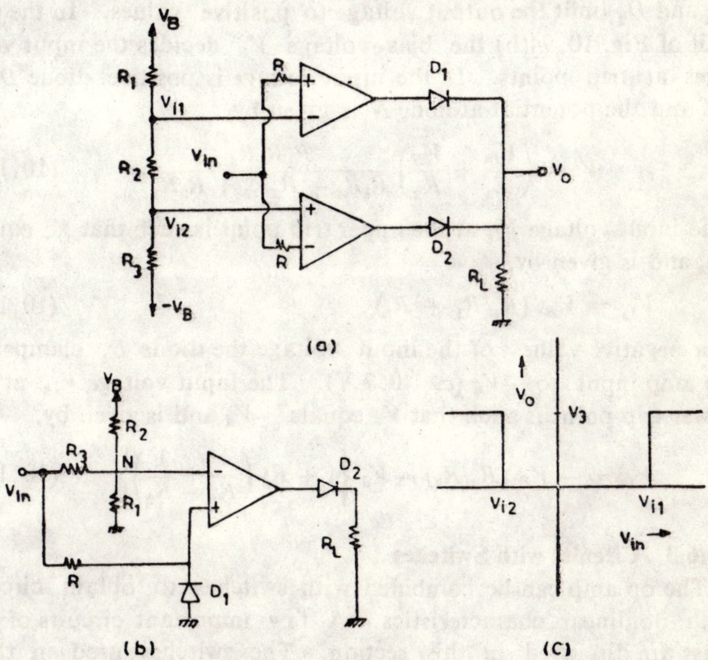

Fig. 10.94. (a) and (b) Window comparators. (c) $V_0 - V_{in}$ characteristic of (a) and (b).

For the non-inverting comparator of Fig. 10.93(a)

$$V_{i1} = V_{ref}\cdot(1 + R_1/R_2) + V_s\,(R_1/R_2) \tag{10.132}$$

$$V_{i2} = V_{ref}\cdot(1 + R_1/R_2) - V_s\,(R_1/R_2) \tag{10.133}$$

For the inverting comparator of Fig. 10.93 (c)

$$V_{t_1}=\{V_{ref}+V_s\,(R_1/R_2)\}\frac{R_2}{R_1+R_2} \qquad (10.134)$$

$$V_{t_2}=\{V_{ref}-V_s\,(R_1/R_2)\}\frac{R_2}{R_1+R_2} \qquad (10.135)$$

The circuits of Fig. 10.93 are also known as 'Schmitt Trigger'.

Figure 10.94 shows two possible configurations for the window comparator and its $V_0 - V_{in}$ characteristic. The circuit of Fig. 10.94(a) is a combination of two open loop comparators of the type shown in Fig. 10.92(a). The voltages V_{t_1} and V_{t_2} can be adjusted by suitable choice of resistors R_1, R_2 and R_3. The diodes D_1 and D_2 limit the output voltage to positive values. In the circuit of Fig. 10.94(b) the bias voltage V_B decides the input voltages at trip points. If the input voltage is positive, diode D_1 is off and the potential at node N is given by,

$$V_-=\left(\frac{V_{in}}{R_3}+\frac{V_B}{R_2}\right)\frac{R_1R_2R_3}{R_1R_2+R_2R_3+R_3R_1} \qquad (10.136)$$

The input voltage V_{t_1} at the upper trip point is such that V_- equals V_{in} and is given by,

$$V_{t_1}=V_B.\,(R_1/R_1+R_2) \qquad (10.137)$$

For negative values of the input voltage the diode D_1 clamps the op amp input to $-V_d$ ($\simeq -0.7$ V). The input voltage V_{t_2} at the lower trip point is such that V_- equals $-V_d$ and is given by,

$$V_{t_2}=-\,V_B\,(R_3/R_2)-V_d\left\{1+R_3\left(\frac{1}{R_1}+\frac{1}{R_2}\right)\right\} \qquad (10.138)$$

10.6.3 Circuits with Switches

The op amp can be combined with switches to obtain circuits with nonlinear characteristics. A few important circuits of this class are discussed in this section. The switches used in these circuits are usually electronic switches. Diode bridges and field effect transistors are commonly used as switches. The CMOS transmission gate (Fig. 10.11) can also be used to switch signals into op amp circuit.

Modulator: Figure 10.95 (a) shows a modulator in which the positive input of the op amp is connected to ground through a

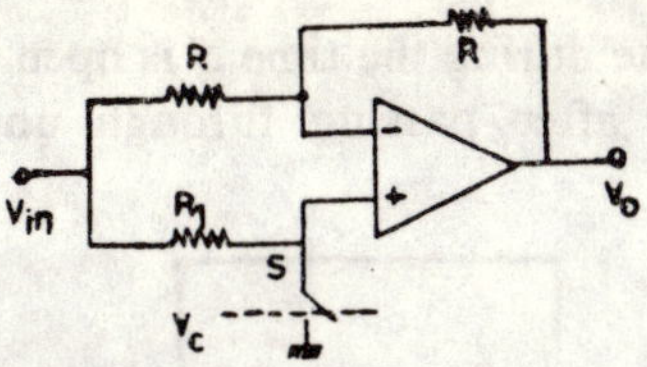

V_c	S	V_o
"0"	On	V_{in}
"1"	Off	V_{in}

(a)

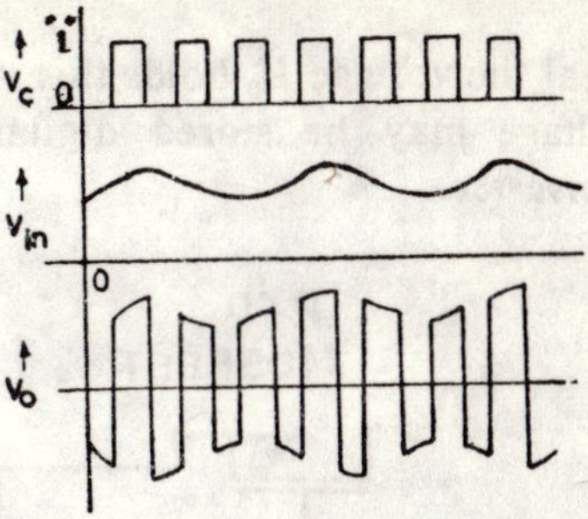

(b)

Fig. 10.95. (a) Op amp modulator. (b) Waveforms of (a).

switch S. When S is open there is no current through R_1. Due to feedback both the inputs of the op amp are at V_{in}. Therefore the output voltage is also equal to V_{in}. When S is closed the circuit works as an inverting amplifier and the output equals $-V_{in}$. When S is a fast electronic switch, e.g. an FET, the output can be rapidly switched between $\pm V_{in}$. Therefore, as shown in Fig. 10.95 (b) a slowly varying d.c. signal can be modulated to obtain a modulated a.c. signal. Such signals are used in amplification of slow varying d.c. signal to prevent the signal changes from being masked by drifts of the amplifiers. Special op amp modules called chopper stabilised op amps are used for this purpose [2].

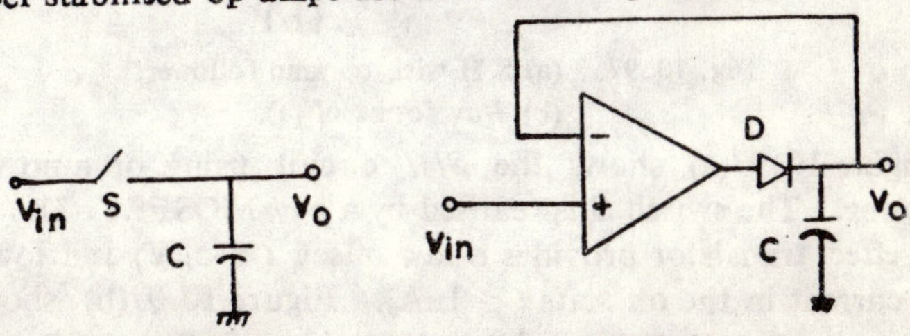

(a) Basic S/H. (b) Peak detector.

Fig. 10.96.

Sample and hold: This is an important application of the op amp in which the signal is sampled at a given instant and its value is stored in the form of capacitor voltage. The basic principle of Sample and Hold (S/H) circuit is shown in Fig. 10.96(a). Here when switch S is closed the output voltage V_0 tracks the input voltage. When S is opened the instantaneous value of V_{in} is equal to the output voltage V_0. If the switch and the capacitor are ideal the voltage V_0 holds this value during the time S is open. The voltage may be stored digitally after passing through an A/D converter.

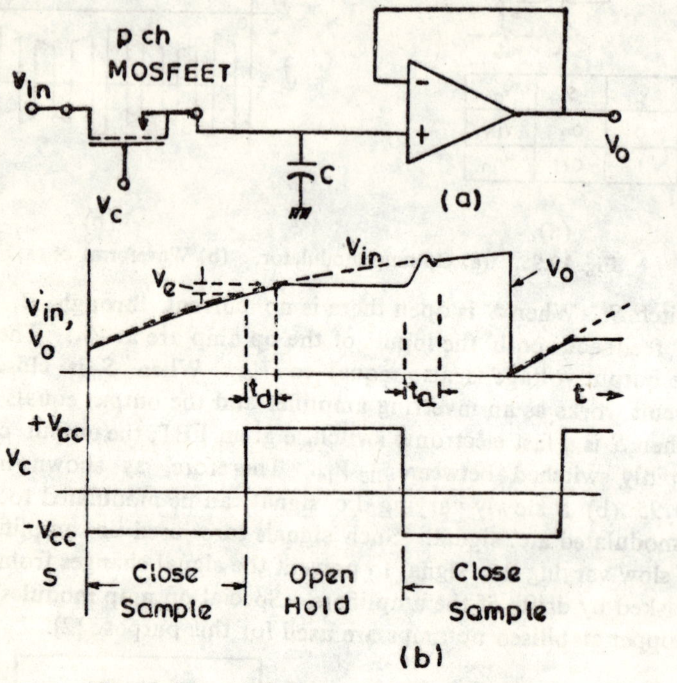

Fig. 10.97. (a) S H with op amp follower.
(b) Waveforms of (a).

Figure 10.97(a) shows the S/H circuit using op amp voltage follower. The switch S is realised by a *p-ch* MOSFET. The use of field effect transistor provides a low offset ($< 5\mu V$) and low leakage current in the off state ($< 1nA$). Figure 10.97(b) shows the input and output voltages of the S/H circuit. The FET switch S is closed when V_c is negative and is open when V_c is positive.

The S/H circuits are subject to some errors in their operation. The FET switch has a finite closing time. Therefore, a small time interval t_d elapses between the Hold command of the control voltage v_c and the actual closure of the switch. This causes the stored sample to be slightly different from the value required to be stored. The resulting error, V_e in Fig. 10.97(b), is called the 'aperture error'. The time delay t_d (Fig. 10.97b) is small, less than 100 ns, therefore, the error is significant only if the input is changing rapidly. When the switch is changed over from Hold to Sample there may be a considerable difference between the voltage of the capacitor and the instantaneous value of v_{in}. The capacitor C therefore charges or discharges through the on resistance of the switch and the input resistance of the signal source. Therefore, there is an acquisition time, t_a (Fig. 10.97b) during which the output V_0 is not equal to the input V_{in}.

The switch can be changed over to Hold only after the signal has been acquired, i.e. after interval t_a. The acquisition time depends upon the value of RC time constant of the charging circuit.

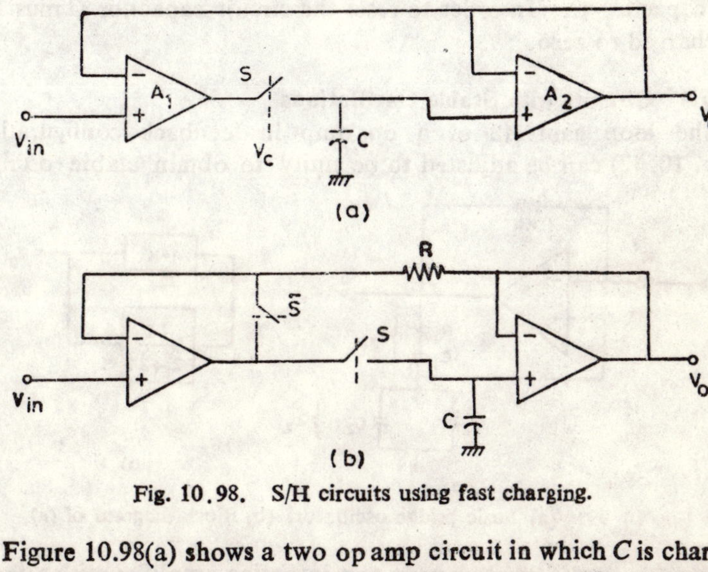

Fig. 10.98. S/H circuits using fast charging.

Figure 10.98(a) shows a two op amp circuit in which C is charged through the op amp A_1 during the acquisition time. The acquisition time is therefore reduced. Amplifier A_2 acts like a voltage follower to provide the output V_0 equal to the capacitor voltage V_c. The input is sampled when S is closed and held when S is open. When S is open the amplifier A_1 tends to saturate. Fig. 10.98(b) shows a modified circuit in which a complementary switch $\overline{S}$ is closed when switch S is open. Thus, saturation of A_1 is avoided. Resistor R limits the current due to difference between V_{in} and the sampled output V_0. This circuit is available as an IC, NE 5537.

The charge stored on the capacitor is also subject to error during the Hold period. The leakage in the dielectric of the capacitor can be reduced if the Polycarbonate, Polyethylene or Teflon is used. A further source of error is the offset voltage and the bias current of the op amp. Op amps with FET input stage are used to reduce the bias current error. Some other types of S/H circuits and their errors are discussed in [8].

Peak detector: Figure 10.96 (b) shows a simple circuit which can be used to detect the peak value of a waveform. The op amp acts like a voltage follower. The capacitor C is charged by the op amp output voltage through D if V_{in} is greater than V_0. If V_{in} falls below V_0 the op amp output saturates in the negative direction and D is reverse biased. Thus, V_0 is retained at the peak value by the capacitor C. In order to reset the circuit capacitor C must be discharged to zero.

10.6.4 Circuits with Stable Oscillations

The loop gain $A\beta$ of a op amp in feedback configuration (Fig. 10.42) can be adjusted to be unity to obtain stable oscilla-

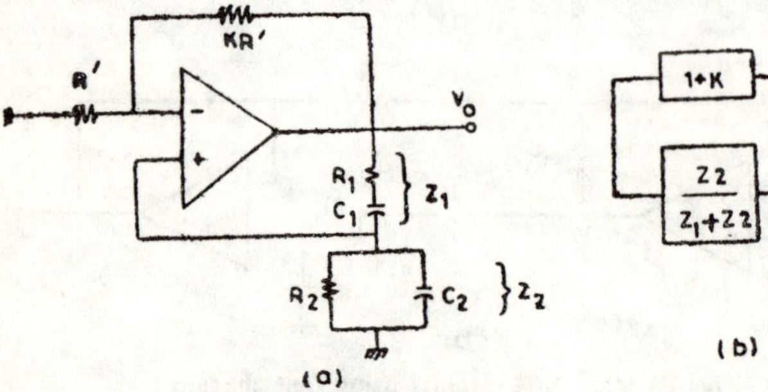

Fig. 10.99. (a) Basic bridge oscillator. (b) Block diagram of (a).

tions. Figure 10.99(a) shows a non-inverting amplifier with a feedback. The block diagram of the system is shown in Fig. 10.99(b). The stable oscillations can be obtained if,

$$(1+K)\frac{Z_2}{Z_1+Z_2}=1 \tag{10.139}$$

where

$$Z_1=\frac{1+j\omega R_1C_1}{j\omega C_1}, Z_2=\frac{R_2}{1+j\omega R_2C_2} \tag{10.140}$$

Substituting eqn. (10.140) in (10.139) and equating the real and the imaginary parts gives the following conditions.

$$\omega^2 R_1 R_2 C_1 C_2 = 1 \qquad (10.141)$$

$$\frac{R_1}{R_2} + \frac{C_2}{C_1} = K \qquad (10.142)$$

Further if,

$$R_1 = R_2,\ C_1 = C_2 \qquad (10.143)$$

Then,

$$\omega = \frac{1}{RC} \quad \text{and} \quad K = 2 \qquad (10.144)$$

The circuit, thus, seems to sustain oscillations if eqns. (10.141) and (10.142) or (10.143) and (10.144) are satisfied. However, this calculation does not specify the amplitude of the oscillations i.e. the value of voltage V_0. It can, however, be concluded that sinusoidal voltage of an unspecified magnitude may be obtained at V_0. The circuit shown in Fig. 10.99(a) if assembled as such, however, may not have any oscillations. The reason is the absence of any mechanism to precisely satisfy the conditions of eqns. (10.141) and (10.142). Any small deviation from the conditions due to component tolerances, power supply fluctuations etc. tends to suppress the oscillations. Thus, in order to obtain sustained oscillations additional feedback must be employed. Figure 10.100(a) shows a circuit in which the gain of the non-inverting amplifier is variable. The output voltage V_0 is rectified by diode D_1 and filtered. The d.c. output is connected to the gate of the JFET. The resistance of the FET is thus a function of the output voltage and increases with it.

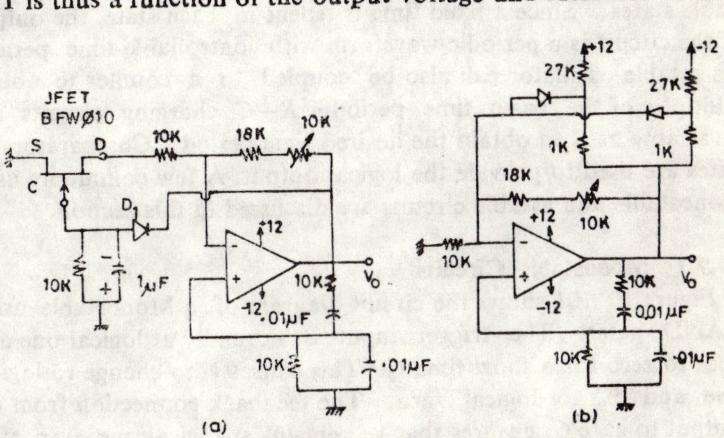

Fig. 10.100. Bridge oscillator with amplitude control by (a) Variable gain, (b) Diode limiters.

The gain K is therefore reduced if V_0 increases and vice-versa. The system, therefore, is able to sustain oscillations. Figure 10.100(b) shows a circuit in which the output voltage is limited by the diode limiters described in section 10.6.1. The clamping effectively reduces the gain and sustains oscillations. However, if the variable 10K feedback resistance is increased beyond a limit, the output voltage is not sinusoidal but shows clipping in both directions.

There are many other kinds of sinusoidal oscillators in which op amps can be used, e.g. Hartley and Colpitts oscillator. These are described in detail in [9]. Many non-linear oscillators producing square wave output of variable frequency can be devised using the op amp in open loop mode. These are described in the next section under Timing Circuits.

10.7 TIMING CIRCUITS

The timing circuits are commonly used in control applications for generating a logic signal which stays in a predetermined state i.e. one or zero for a fixed time interval. The timing is initiated by a trigger pulse which is a logic signal of very short duration. Such circuits are also known as One Shots or Monostables, because they change from a stable to a quasi-stable state at the start of the time interval. The return to the stable state marks the end of the time interval. Another class of timing circuits are called Astable oscillators because they are continuously shifting between two quasi-stable states. Since a fixed time is spent in each state, the output of the circuit is a periodic waveform with controllable time period. An astable oscillator can also be coupled to a counter to obtain multiples of the basic time period. $R-C$ charging circuits are invariably used to obtain the desired time period. Comparators or gates are used to provide the logical output. A few commonly used monostable and astable circuits are discussed in this section.

10.7.1 Monostable Circuits

Figure 10.101 shows the circuit diagram of a Monostable using NAND gates. The trigger input is normally at logical one and goes to zero for a short time t_p. This causes v_t to change to logical one and V_0 to logical zero. The feedback connection from the output to gate G_1 ensures that v_0' remains at logical one even after v_t has returned from zero to one. The output of gate G_1 charges the RC circuit. The voltage v_t starts from about 3.5 volts and falls

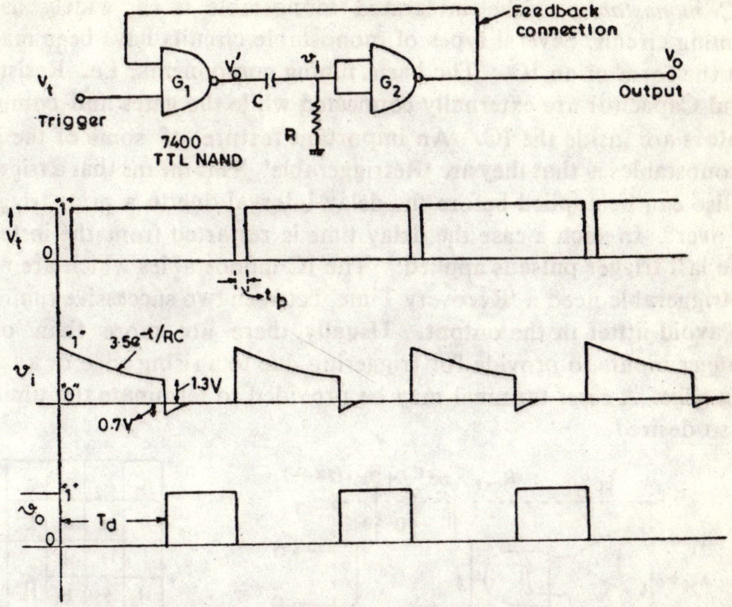

Fig. 10.101. Monostable circuit using NAND gates.

exponentially towards the threshold voltage of the gate G_2. When v_i falls to about 1.3 volts the output returns to logical one. The delay time T_d in Fig. 10.101 is given by

$$T_d = RC \ln \left[\frac{V_{IQ} + V_{HL}}{V_{IH}} \right] \tag{10.145}$$

In above equation for TTL gate V_{IQ} is nearly zero, V_{HL} is 3.7 volts and V_{IH} is nearly 1.3 volts. The gate G_2 is being used as a comparator and its characteristics are subject to variations due to temperature. Further the output transition time is also dependent on gate parameters. The built in diodes at the input of G_2 clamp v_i to about -0.7 V at the end of the delay interval T_d.

If the feedback connection in the circuit of Fig. 10.101 is removed the voltage v_0' will return to logical zero as soon as v_i returns to logical one. The monostable then can function only if t_p is larger than T_d. The feedback connection thus provides the 'regeneration' to make T_d independent of t_p. Similar monostables can be constructed using NOR and NAND gates of TTL, CMOS or HTL families.

IC monostables: The integrated monostable is a widely used timing circuit. Several types of monostable circuits have been made in the form of an IC. The basic timing components, i.e. Resistor and Capacitor are externally connected while the gates and comparators are inside the IC. An important feature of some of the IC monostables is that they are 'Retriggerable'. This means that a trigger pulse can be applied before the delay interval due to a prior trigger is over. In such a case the delay time is restarted from the instant the last trigger pulse is applied. The IC monostables which are not retriggerable need a 'Recovery Time' between two successive timings to avoid jittter in the output. Usually, there are more than one trigger inputs to provide for triggering due to a rising edge or a falling edge. A reset terminal may be provided to terminate the timing if so desired.

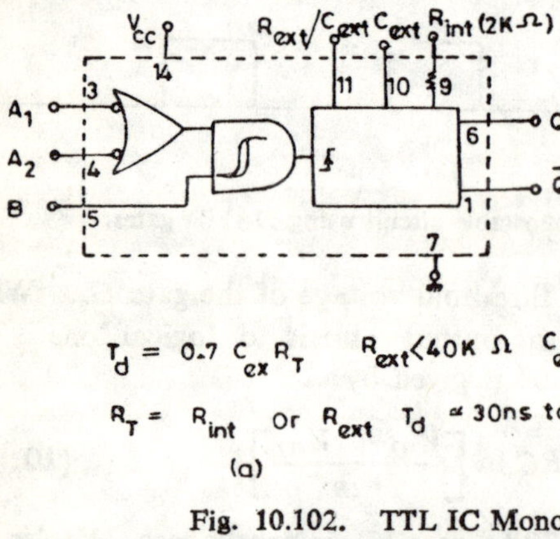

$T_d = 0.7\ C_{ex} R_T$ $R_{ext} < 40K\ \Omega$ $C_{ext} < 1000\ \mu F$

$R_T = R_{int}$ or R_{ext} $T_d \simeq 30$ns to 28 Sec

(a)

A_1	A_2	B	Q
H	H	X	L
↓	H	H	
H	↓	H	⎍
↓	↓	H	
X	X	L	L
L	X	↑	⎍
X	L	↑	

(b)

Fig. 10.102. TTL IC Monostable 74121.
(a) Block diagram,
(b) Function table.

Figure 10.102 shows the block diagram of a TTL monostable 74121 and its function table. The internal trigger of the monostable is sensitive to a rising edge as shown. The inputs A_1 and A_2 go through a NAND gate and input B is connected directly to a Schmitt AND gate. A negative i.e. falling edge at A_1 or A_2 or both while B is held High i.e. at logical one, starts the timing. Alternatively a positive i.e. rising edge at B starts timing if A_1 or A_2 or both are held Low i.e. at logical zero. An internal timing resistor of 2K ohm is provided by connecting pin 9 to supply. If an external resistor is used, it is connected between pin 11 and supply keeping pin 9 open. The timing capacitor C_{ext} is connected between pin 10 and 11. The limits on the component values and the formula for calculating the delay interval T_d are given in Fig. 10.102. The rising edges at inputs A_1 and A_2 must change faster than 1v/μs while input B can respond to pulse edges as slow as 1v/sec. The monostable is not retriggerable. For adequate recovery the duty cycle is limited to 67% if the internal timing resistor is used. With external timing resistors the duty cycle rises to 90% when R_{ext} equals 40K ohms.

Figure 10.103 shows a retriggerable TTL monostable 74122. In this monostable the duty cycle is unlimited. A clear input is provided to terminate the timing if required. Inputs A_1 and A_2 start the timing at falling edges. Inputs B_1 and B_2 start the timing at rising edges. The internal trigger point needs a rising edge and receives the signal $(\bar{A}_1 + \bar{A}_2) \cdot B_1 \cdot B_2 \cdot CLR$. The function table of Fig. 10.103(b) gives the details of various triggering conditions. For values of C_{ext} lower than 1000pF the delay varies below 2 to 10μs. The variation can be obtained from manufacturer's data book. For values of C greater than 1000pF the delay interval is given by equation given in Fig. 10.103(a).

Monostable ICs are also available in HTL (e.g. MC667) or CMOS

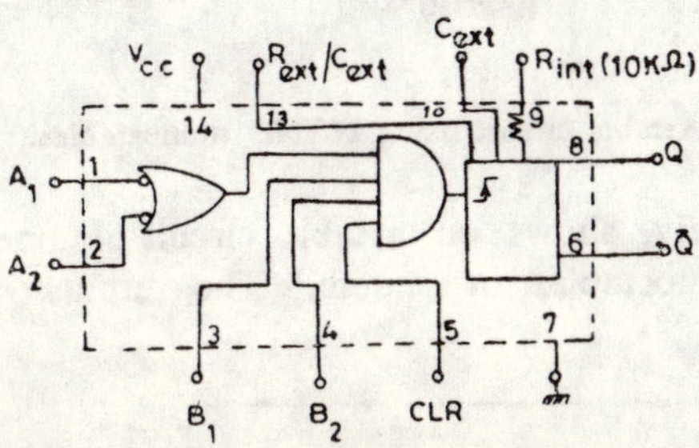

$$T_d = 0.32\, R_T\, C_{ext} \left[1 + \frac{0.7}{R_T} \right] \text{nS}$$

$R_T = R_{int}$ Or R_{ext} in KΩ , $5K \leq R_{ext} \leq 50K$

C_{ext} in pF & >1000pF

(a)

CLR	A_1	A_2	B_1	B_2	Q
H	L	X	↑	H	⎍
H	L	X	H	↑	⎍
H	X	L	↑	H	⎍
H	X	L	H	↑	⎍
H	H	↓	H	H	⎍
H	↓	H	H	H	⎍
H	↓	↓	H	H	⎍
↑	L	X	H	H	⎍
↑	X	L	H	H	⎍
L	X	X	X	X	L

(b)

Fig. 10.103. TTL IC retriggerable monostable 74122

(a) Block diagram.
(b) Function table.

(e.g. MC1428B) logic families. The manufacturers data book should be consulted for details of these ICs.

10.7.2 Astable Circuits

The astable circuits continuously oscillate between two quasi-stable states. The output is therefore a pulse train whose frequency and duty ratio can be controlled. This basic oscillator can be used as a timing circuit. In some cases the output of the oscillator is fed to a counter to obtain larger time periods.

Figure 10.104 shows two TTL monostable circuits which are connected in tandem to operate as an astable circuit. Each monostable output acts like a trigger to the next one and generates the two delay intervals in one period of oscillation. The resistor R must be lower than 800 ohms so as to keep the drop due to sinking current 1.6 mA below the threshold voltage. The time period of the astable can be varied by varying the value of time constant RC.

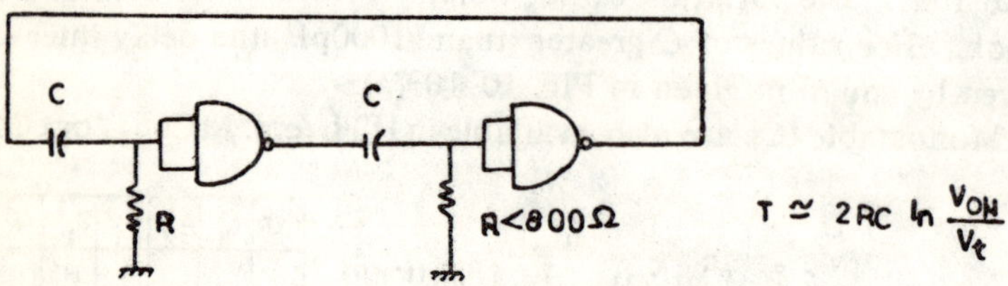

Fig. 10.104. TTL Astable circuit using NAND monostables.

Figure 10.105, similarly, shows an astable circuit obtained by connecting two IC monostables in tandem. The outputs Q_1 and

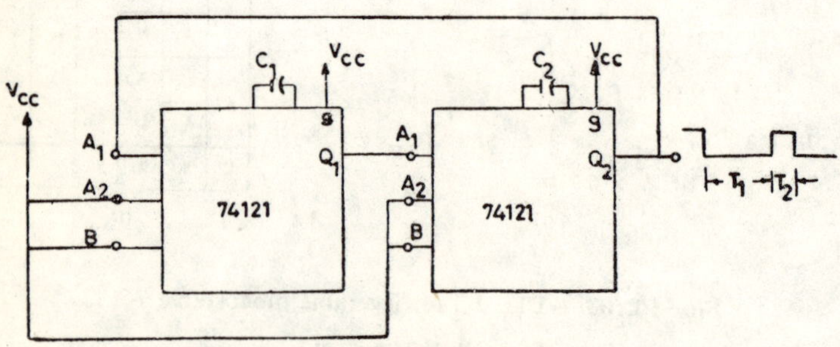

Fig. 10.105. Astable circuit using 74121 monostable.

Q_2 form two pulse trains. Two duty ratio of the pulse train can be adjused suitably choosing C_1 and C_2.

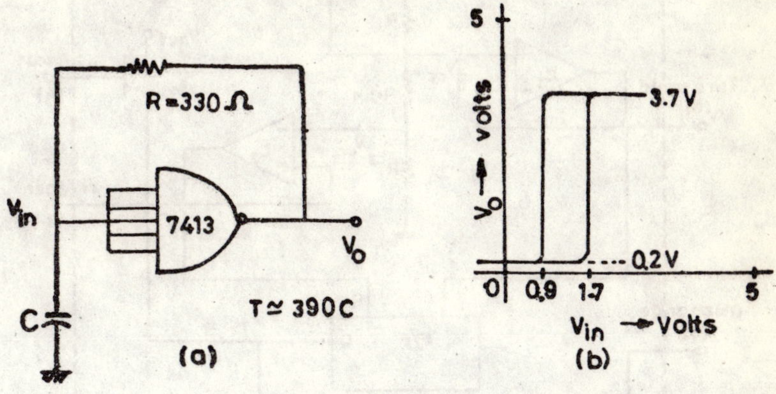

Fig. 10.106. (a) Astable circuit using 7413 Schmitt inverter.
(b) Transfer characteristics of 7413.

Figure 10.106 shows a popular astable circuit made from an RC and a 7413 Schmitt inverter whose characteristics are also shown. The voltage V_{in} oscillates between the trip points of the Schmitt inverter i.e. 1.7V and 0.9V. The frequency of the circuit with R equal to 330 ohms is approximately given by (1/390C). In the configuration of Fig. 10.106 the 7413 Schmitt inverter can be replaced by any of the op amp comparator circuits given in section 10.6.2. This circuit is sometimes called a 'Relaxation Oscillator'.

10.7.3 555 IC Timer

The IC timer (NE555) is a versatile device which can be connected in many different configurations. Figure 10.107 shows the block diagram of the IC. It consists of two high-gain comparators C_1 and C_2 and a Flip-Flop. The supply voltage V_{CC} is divided by a chain of three resistors. The comparator C_1 is biased at $2V_{CC}/3$. Whenever the voltage at Threshold input, pin (6), exceeds $2V_{CC}/3$, the output of the comparator C_1 goes low and sets the Flip-Flop. This causes the discharge transistor T_s to turn on. The comparator C_2 is biased at $V_{CC}/3$. Whenever the voltage at the Trigger input, pin (2), falls below $V_{CC}/3$, the output of the comparator C_2 goes low and resets the Flip-Flop. This causes the discharge transistor T_s to turn off. The output of the Flip-Flop is connected to an

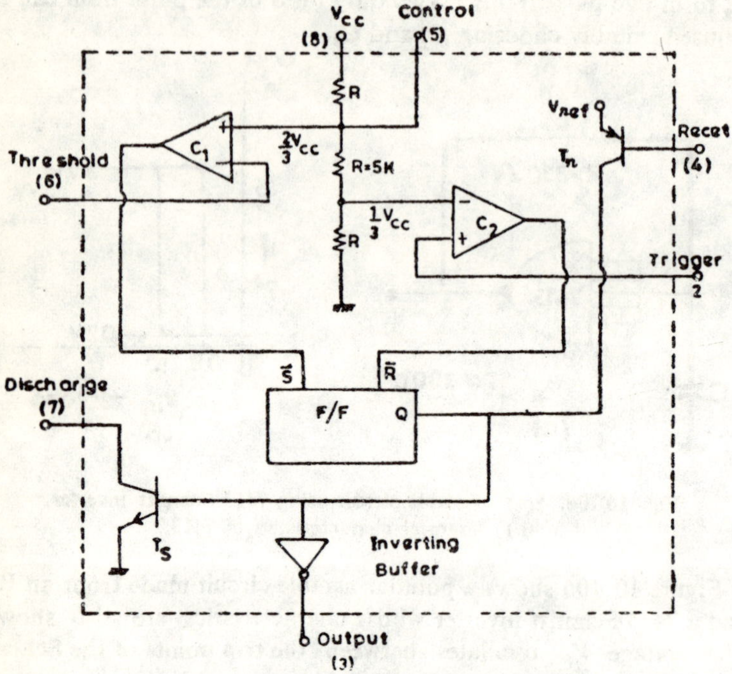

Fig. 10.107. Block diagram of 555 IC timer.

inverting buffer which provides the output. The discharge transistor T_s can also be independently turned on by applying a low voltage at the Reset input, Pin 4. The use of a resistive divider chain makes the time periods in monostable and astable operation, independent of the supply voltage V_{CC}. The IC timer can, therefore, operate from any supply voltage between 5 and 15 volts. The bias voltages to the comparators C_1 and C_2 can be varied by applying a dc control voltage, V_5, at pin (5). The bias voltages for C_1 and C_2 are then V_5 and $V_5/2$ respectively.

Figure 10.108(a) shows the connections for operation as an Astable. The timing components R_A, R_B and C are externally connected as shown. Figure 10.108(b) shows the waveforms during a period T. The capacitor voltage v_c in the steady state varies between $V_{CC}/3$ and $2V_{CC}/3$. During the time interval T_1 the transistor switch T_s is off. The capacitor C charges through R_A and R_B towards V_{CC}. The instant the capacitor voltage v_c rises above the value $2V_{CC}/3$, comparator C_1 sets the F/F and turns on T_s. The

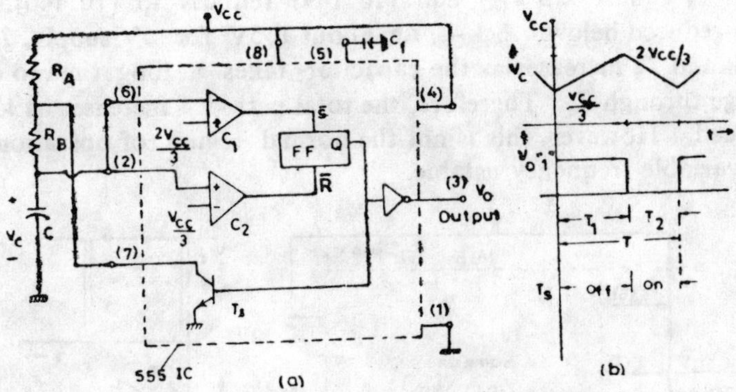

Fig. 10.108. (a) 555 IC timer connected astable.
(b) Waveforms of (a).

junction of R_A and R_B is now almost at ground potential and C discharges through R_A into T_s. As the capacitor voltage v_c crosses the value $V_{CC}/3$, comparator C_2 resets F/F and turns off T_s. The cycle then repeats itself. The exponential rise and fall of v_c between $V_{CC}/3$ and $2V_{CC}/3$ gives the following expressions for the time intervals T_1 and T_2,

$$T_1 = (R_A + R_B)C \ln 2 \quad (10.146)$$

$$T_2 = R_B C \ln 2 \quad (10.147)$$

The period of oscillation therefore is

$$T = T_1 + T_2 = (R_A + 2R_B)\, C \ln 2 \quad (10.148)$$

The frequency of oscillation $1/T$ can therefore be controlled by varying the values of R_A, R_B and C. The ratio T_2/T_1 can also be controlled by varying R_A and R_B.

In Fig. 10.108 pin 5 has been bypassed to ground by a small filter capacitor C_f, usually 0.01 μF. This is to avoid any stray pick up at pin 5. The frequency of oscillations can be controlled by a voltage at pin 5 between maximum of V_{CC} and a minimum of approximately 0.3 V_{CC}. The time interval T_2 remains constant at the value given by eqn. (10.147). The time interval T_1 (Fig. 10.108b) increases as the control voltage V_5 is increased and is given by the equation,

$$T_1 = (R_A + 2R_B)\, C \ln \left[\frac{V_{CC} - 0.5\, V_5}{V_{CC} - V_5}\right] \quad (10.149)$$

when V_5 equals 2/3 V_{CC} eqn. (10.149) reduces to (10.146). If V_5 is reduced below 0.3 V_{CC}, i.e. about 1.5V for 5V supply, T_1 is small and T_2 increases as the capacitor takes a long time to discharge through T_s. Therefore, the total period T increases as V_5 is reduced. However, this is not the normal region of operation of the variable frequency astable.

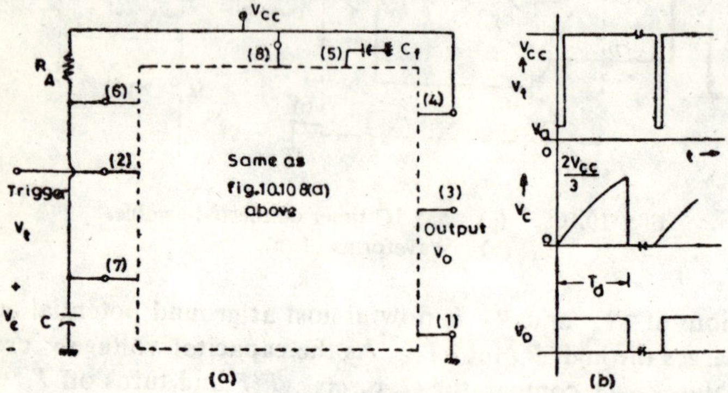

Fig. 10.109. (a) 555 IC Timer connected as monostable.
(b) Waveforms of (a).

Figure 10.109(a) shows the connection of the 555 IC timer as a Monostable circuit. The input of comparator C_2 is used as the trigger input of the monostable. When the trigger voltage V_t is pulled low the F/F is reset and T_s turns off. The capacitor C therefore charges through R_A towards V_{CC}. When the capacitor voltage V_C reaches 2/3 V_{CC} comparator C_1 sets the F/F and turns on T_s. The capacitor C is discharged very rapidly through T_s. Figure 10.109b shows the resulting waveforms. The lower level of the trigger voltage, V_a, should be below 1/3 V_{CC} to change the state of comparator C_2. In most applications V_a is zero. The duration of the trigger must also be smaller than the delay period T_d. The delay period T_d is given by

$$T_d \simeq 1.1\, R_A C \tag{10.150}$$

The monostable of Fig. 10.109 is not retriggerable. If a trigger pulse is applied at pin 2 during the delay period T_d it has no effect as the F/F is already reset. The monostable can however be reset by applying the negative going trigger voltage V_t at pin 4. This turns on the transistor switch T_s and discharges the capacitor. The timing then restarts from the positive going edge of the reset pulse. The monostable pulse width can be modulated by applying a low frequency control voltage at pin 5. Several monostables can also be connected in tandem to obtain several delay intervals.

10.8 PHASE LOCKED LOOP (PLL) AND ITS APPLICATIONS

In this section a system called 'Phase Locked Loop' has been described. Although known and used since early 1930's, it is now an important building block for signal processing tasks in control and instrumentation applications. In the following sections a brief review of its basic principles of operation, linearized analysis and design and some of its applications are given.

10.8.1 Basic Principle of Operation

The PLL is a feedback system comprising of a phase comparator, a low pass filter and an error amplifier in the forward signal path and a voltage controlled oscillator (VCO) in the feedback path (Fig. 10.110). With no signal input to the system, the error voltage $v_d(t)$ is equal to zero. The VCO operates at a set frequency

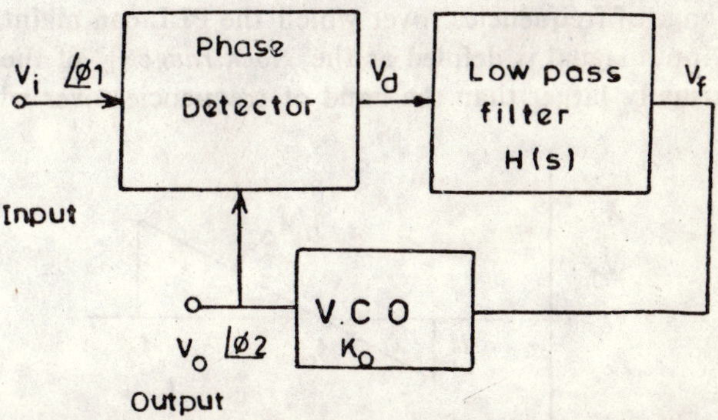

Fig. 10.110. Block diagram of a phase locked loop (PLL).

$\omega_0 = 2\pi f_0$, known as its free running frequency. If an input signal $v_i(t)$ is applied to the system, the phase comparator compares the phase and the frequency of the input with the VCO frequency and generates an error voltage $v_d(t)$ related to the frequency and the phase difference between the two signals. This error voltage is then filtered, amplified and applied to the control terminals of the VCO. In this manner the control voltage $v_f(t)$ forces the VCO frequency to vary in a direction which reduces the frequency difference between f_0 and the input signal. If the input signal freq. f_i

is sufficiently close to f_0, the feedback nature of PLL causes the VCO to synchronize or 'lock' with the incoming signal. To understand this locking phenomenon a little better, let us suppose that initially the loop is out of lock, but the frequencies of the input signal v_i and output v_0 are close to each other. v_d, the signal out of the phase detector will then be a beat note, whose frequency is equal to the frequency difference $(f_i - f_0)$. This signal is applied to the VCO as it falls well within the pass band of the low pass filter. The instantaneous frequency of the VCO is therefore changing and at some instant of time when VCO frequency equals the input frequency, lock will result. v_f assumes a d.c. level sufficient to hold the VCO frequency in lock with the input frequency. Any attempt to change the frequency of VCO by external means (like tuning) etc. will be nullified by the action of the feedback loop. Similarly, any change in the input frequency will result in an instantaneous phase change between v_i and v_0 and hence a d.c. level change in v_d. This shift will change the VCO frequency to maintain lock.

The range of frequencies over which the PLL can maintain lock with an input signal is defined as the *'lock range f_L'* of the system. This is usually larger than the band of frequencies over which the

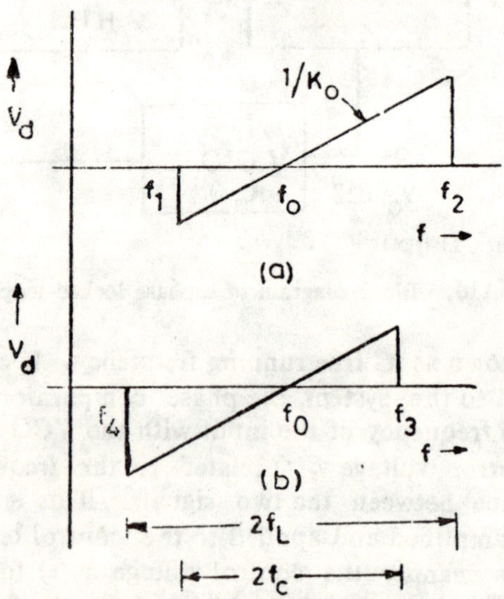

Fig. 10.111. V_d vs input frequency, f.

PLL can acquire lock with an incoming signal. This latter range of frequencies is knwon as the '*capture range*', f_C.

Figure 10.111 shows the typical frequency to voltage curve of PLL. The input is assumed to be a sine wave whose frequency is swept slowly over a broad frequency range and the vertical scale is the corresponding loop error voltage v_d. In Fig. 10.111 (a) the input frequency is being gradually increased. The loop does not respond to the signal until it reaches a frequency f_1, corresponding to the lower edge of the capture range. Then the loop suddenly locks on the input, causing a –ve jump of the loop error voltage. As the input frequency is increased further v_d varies with frequency with slope equal to the reciprocal of VCO gain and goes through zero at $f_i = f_0$. The loop tracks the input until the input frequency reaches f_2 corresponding to the upper edge of the lock range. Then the PLL losses lock, the error voltage drops to zero and the VCO frequency returns to f_0. If the input frequency is now slowly swept back, the cycle repeats itself as shown in Fig. 10.111 (b). The loop recaptures the signal at f_3 and tracks it down to f_4. The frequency spread between (f_1, f_3) and (f_2, f_4) correspond to the total capture range $2f_c$ and lock range $2f_L$ of the system, i.e.

$$f_3 - f_1 = 2f_c$$
$$f_2 - f_4 = 2f_L$$

10.8.2 Linearized Analysis of the PLL

Let the input signal v_i be a sinusoid of frequency ω_0 and phase $\varphi_1(t)$ while the VCO output signal v_0 is at the same frequency but whose phase is $\varphi_2(t)$ i.e.

$$v_i = 2V_i \sin\{\omega_0 t + \varphi_1(t)\} \tag{10.151}$$
$$v_0 = 2V_0 \cos\{\omega_0 t + \varphi_2(t)\} \tag{10.152}$$

An ideal multiplier can be used as a phase detector. The output of the phase detector is

$$v_d = v_i \cdot v_0 = K_D V_i V_0 [\sin\{\varphi_1(t) - \varphi_2(t)\} + \sin\{2\omega_0 t + \varphi_1(t) + \varphi_2(t)\}] \tag{10.153}$$

where K_D is the detector gain constant.

The low pass filter removes the high frequency component at $2\omega_0$ and hence the output of the filter may be written as

$$v_f = K_D V_i V_0 \sin\{\varphi_1(t)-\varphi_2(t)\} * h(t) \tag{10.154}$$

where * stands for convolution and $h(t)$ is the impulse response the filter corresponding to the transfer function $H(s)$. Assuming that the system is in lock and operating with a small error as any good control system should, we may write

$$v_f \simeq K_D V_i V_0 \varphi_e(t) * h(t). \tag{10.155}$$

where $\varphi_e(t) = \varphi_1(t)-\varphi_2(t)$

The transfer characteristic of the VCO is given by,

$$\varphi_2(t) = K_0 \int_0^t v_f \, dt, \quad \frac{d\varphi_2}{dt} = K_0 V_f \tag{10.156}$$

Taking Laplace transform of (10.155) and (10.156) we have

$$v_f(s) = K_D V_i V_0 \varphi_e(s) H(s)$$

$$\varphi_2(s) = \frac{K_0 V_f(s)}{s} = \frac{K_0 K_D V_i V_0 \varphi_e(s) H(s)}{s} \tag{10.157}$$

Also $\qquad \varphi_e(s) = \varphi_1(s)-\varphi_2(s) \qquad (10.158)$

From (10.157) and (10.158), the closed loop gain of the PLL is given by

$$C(s) = \frac{\varphi_2(s)}{\varphi_1(s)} = \frac{K_0 K_D V_i V_0 H(s)}{s + K_0 K_D V_i V_0 H(s)} = \frac{G(s)}{s + G(s)} \tag{10.159}$$

where $G(s) = K_0 K_D V_i V_0 H(s)$. $G(s)/s$ is the open loop gain of the PLL. Again from eqns. (10.157) and (10.158), we have

$$\frac{\varphi_e(s)}{\varphi_1(s)} = \frac{s}{s + K_0 K_D V_i V_0 H(s)} \tag{10.160}$$

Eqns. (10.159) and (10.160) represent the phase transfer functions of the loop.

Application of the final value theorem yields

$$\lim_{t\to\infty} \varphi_e(t) = \lim_{s\to 0} s\,\varphi_e(s) = \lim_{s\to 0} \frac{s^2 \varphi_1(s)}{s+K_0 K_D V_i H(s) V_0} \tag{10.161}$$

For a step change $\Delta\varphi$ in input phase, $\varphi_1(s) = \Delta\varphi/s$ and hence

$$\lim_{t\to\infty} \varphi_e(t) = \lim_{s\to 0} \frac{s}{s + K_0 K_D V_i V_0 H(s)} \Delta\varphi = 0 \tag{10.162}$$

This implies that loop eventually nullifies any change of input phase by tracking it and hence in steady state the phase error goes to zero. For a step change $\Delta\omega$ in input frequency,

$$\varphi_1(s) = \frac{\Delta\omega}{s^2} \text{ and hence.}$$

$$\lim_{t\to\infty} \varphi_e(t) = \lim_{s\to 0} \frac{\Delta\omega}{s + K_0 K_D V_i V_0 \, H(s)}$$

$$= \frac{\Delta\omega}{K_0 K_D V_i V_0 \, H(0)} \tag{10.163}$$

In this case the resulting steady state phase error is dependent on the magnitude of the frequency step and d.c. loop gain of the system $K_0 K_D V_i V_0 \, H(0)$. The error can be reduced to a small value with large value of loop gain.

10.8.3 Loop Filter

In working with PLL's, it is important to consider not only the steady state performance as described above but also the transient performance. The transient performance is predominantly governed by the loop filter. In fact, it is the loop filter that makes the PLL, a versatile system building block. Transfer functions for different types of filters are derived in the following.

First order filter: The simplest loop filter is an RC filter. Such a filter is used for wide bandwidth applications. The transfer function for loop filter is shown in Fig. 10.112 (a).

$$H(s) = \frac{V_f(s)}{V_d(s)} = \frac{1}{1 + s\,R_1 C_1} \tag{10.164}$$

Substituting in eqn. (10.159) the closed loop gain of the PLL is given by

$$C(s) = \frac{\varphi_2(s)}{\varphi_1(s)} = \frac{K_0 K_D V_i V_0/\tau_1}{s^2 + s/\tau_1 + K_0 K_D V_i V_0/\tau_1} \tag{10.165}$$

where $\tau_1 = R_1 \, C_1$

Figure 10.112 (b) shows the open loop gain $C(s)$ for the PLL with first order filter. Equation (10.165) represents a second order system whose natural frequency, ω_n, and the damping factor ξ, are respectively given by the following equations

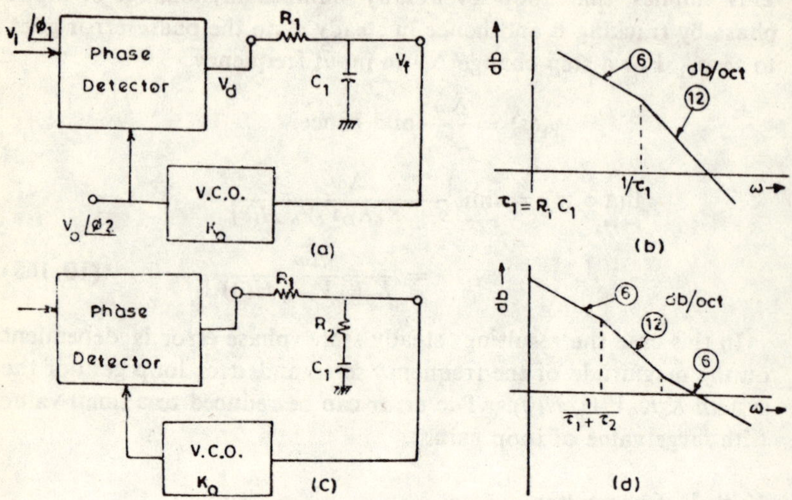

Fig. 10.112. (a) PLL with first order filter. (b) Open loop gain of (a). (c) PLL with second order filter. (d) Open loop gain of (b).

$$\omega_n = \left[\frac{V_i\, V_0\, K_0\, K_D}{R_1\, C_1}\right]^{1/2} \tag{10.166}$$

$$\xi = \tfrac{1}{2}\left[\frac{1}{R_1 C_1 K_0 K_D V_i V_0}\right]^{1/2} \tag{10.167}$$

This means that large τ_1 or high loop gain will reduce the damping factor and hence reduce stability. Therefore, if a narrow bandwidth is desired, the damping factor becomes very small and causes instability. With this filter, it is not possible to adjust bandwidth, and damping independently.

2nd order filter: The above problem can be overcome by using another filter of the type shown in Fig. 10.112 (c). Now the transfer function of the filter is

$$H(s) = \frac{V_d(s)}{V_f(s)} = \frac{s\,\tau_2 + 1}{s\,(\tau_1 + \tau_2) + 1} \tag{10.168}$$

where $\tau_2 = R_2\, C_1$.

The closed loop transfer function becomes

$$C(s) = \frac{\varphi_2(s)}{\varphi_1(s)} = \frac{K_0 K_D V_i V_0\,(s\,\tau_2 + 1)/(\tau_1 + \tau_2)}{s^2 + s\{(1+\tau_2 V_i V_0 K_0 K_D)/(\tau_1+\tau_2)\} + K_0 K_D V_i V_0/\tau_1+\tau_2} \tag{10.169}$$

and the loop natural frequency

$$\omega_n = \left(\frac{V_i V_0 K_0 K_D}{\tau_1 + \tau_2}\right)^{1/2} \tag{10.170}$$

while the damping factor becomes

$$\xi = \frac{1}{2}\left[\frac{K_0 K_D V_i V_0}{\tau_1 + \tau_2}\right]^{1/2}\left[\tau_2 + \frac{1}{K_0 K_D V_i V_0}\right] \tag{10.171}$$

If $1/K_0 K_D V_i V_0 \ll \tau_2$,

$$\xi \simeq \frac{\omega_n \tau_2}{2} \tag{10.172}$$

Figure 10.112 (d) shows the open loop gain of the PLL with second order passive filter. In practice, for a fixed loop gain $K_0 K_D V_i V_0$, the natural frequency of the loop may be chosen and will depend mainly on τ_1 since $\tau_2 \ll \tau_1$ in most cases. Then the damping factor may be determined by τ_2 and for all practical purposes will be an independent adjustment. The plots of above equation for design purposes are given in [10]. In these plots $V_i V_0$ is absorbed in $K_0 K_D$ as the input to a PLL in IC form is either externally or internally limited. V_0 is also fixed, VCO being inside the IC and operating at fixed level. The second order filter shown in Fig. 10.112 (c) is a passive filter. An active filter with R_1 at input and $R_2 - C$ in the feedback of the op amp also gives a second order PLL. The transfer function of a PLL using such an active filter is given in [10].

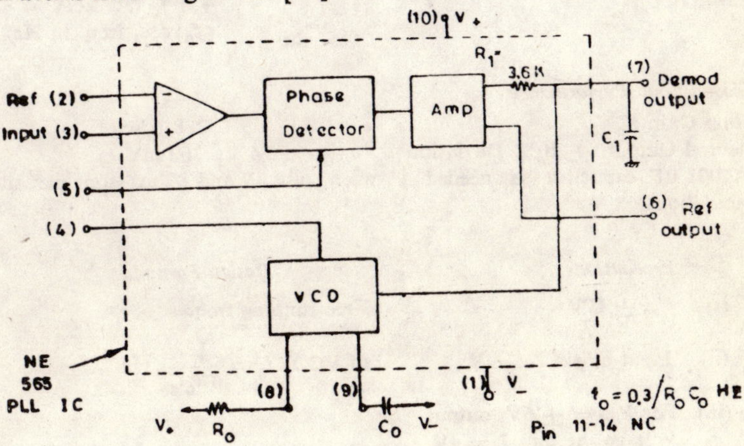

Fig. 10.113. NE 565 PLL IC R_0, C_0 and C_1 are external to IC.

10.8.4 A Phase Locked Loop in IC Form

Several forms of PLL are available in an IC chip [11]. One of these, NE 565, features a highly linear VCO producing a square wave output, and a double balanced phase detector. Some of the important characteristics of 565 are given in Table 10.21 for

$$V_{cc} = \pm 6\text{V},\ T_A = 25°\text{C}.$$

Pin connections as well as design formulae are given in Fig. 10.113, and Table 10.21.

Table 10.21. PLL operating characteristics

Phase Detector	
Input impedance	5k
Input Level for Limiting	10mV
Output Resistance	3.6k
Output Common Mode Voltage	4.5V
Offset Voltage (Between pins 6 and 7)	100mV
Sensitivity K_D	.68V/rad
Voltage Controlled Oscillator	
Stability, temperature	200 ppm/°C
Supply Voltage	200 ppm/V
Square Wave Output Pin 4	5.4 V_{pp}
Triangle Wave Output Pin 9	2.4 V_{pp}
Maximum Operating Frequency	500 KHz
Sensitivity K_0	4.1 f_0 rad/sec/V (f_0: osc. freq. in Hz)
Closed Loop Performance	
Loop Gain K_0K_D	2.8 f_0/sec.
Demod Output, $\pm$ 10% Deviation	300 mV

(A.001 μF capacitor is needed between pins 7 and 8 to stop parasitic oscillations)

Precautions	*Design Formula*
(i) $V_{cc} \pm 12$V	Free-running freq.
(ii) Input at pin 2 $\leqslant$ 4V	of VCO, $f_0 = \dfrac{1.2}{4R_0C_0}$ Hz
(iii) For $V_{cc} = \pm 6$V, output voltage at pin 7 = 4V, max. voltage swing at	

pin 7 $\leqslant$ 2V p-p.	Lock-range, $f_L = \pm \frac{8f_0}{V_{cc}}$ Hz
(iv) Both input terminals (pins 2 and 3) must receive identical dc bias, This bias may range from 0 to — 4V.	Capture-range, $f_c = \pm \frac{1}{2\pi}\sqrt{\frac{2\pi f_L}{\tau}}$ where $\tau = (3.6 \times 10^3) \times C_1$

10.8.5 Basic Applications

PLL has been used as a versatile building block in a variety of applications involving frequency or phase demodulation, signal conditioning and frequency synthesis. The operation of the PLL in some of these applications is briefly described in the following.

(i) *FM demodulation*: This application is obvious from the basic principle of operation of the PLL as a control system in frequency and phase. In this application, the linearity of the demodulated output is determined solely by the VCO voltage to frequency characteristic. The PLL can be used for detecting either wideband or narrowband FM signals with very high detection linearity. PLL demodulators form an integral part of sophisticated communication receivers with improved signal to noise ratio performance.

(ii) *Frequency Synchronization*: Using PLL, the frequency of a relatively poor oscillator can be phase locked with a low level but highly stable reference signal. The frequency stability of the reference is thus transferred to an oscillator from which one can obtain power at much higher level. In some applications, the synchronizing signal can be in the form of a low-duty-cycle burst at a specified frequency. The PLL can then be used to regenerate a coherent CW reference frequency by locking onto this short synchronizing pulse. In digital systems, PLL's are often used for synchronizing system clocks, for clock signal recovery in digital communication systems and for synchronizing disk or tape drive mechanisms in information and storage retrieval systems.

(iii) *Signal Conditioning*: PLL can be used as an extremely narrowband tracking filter to recover periodic signals buried in noise. Its advantages as compared to active RC filters are the following:

(a) High frequency capability: monolithic PLL's are available with frequencies of operation extending to tens of MHz.

(b) Independent control of selectivity and centre frequency: The centre frequency is set by the VCO free running frequency and the selectivity is determined by the low pass filter. Thus, there are no alignment problems.

(c) Few external components and ease of tuning.

(iv) *Frequency multiplication*: By incorporating a frequency divider (e.g. divide by N counter) into the feedback loop, as shown in Fig. 10.114 the PLL system functions as a frequency selective frequency multiplier. When the system is in lock, the two inputs to the phase detector are at the same frequency $f_0 = Nf_s$.

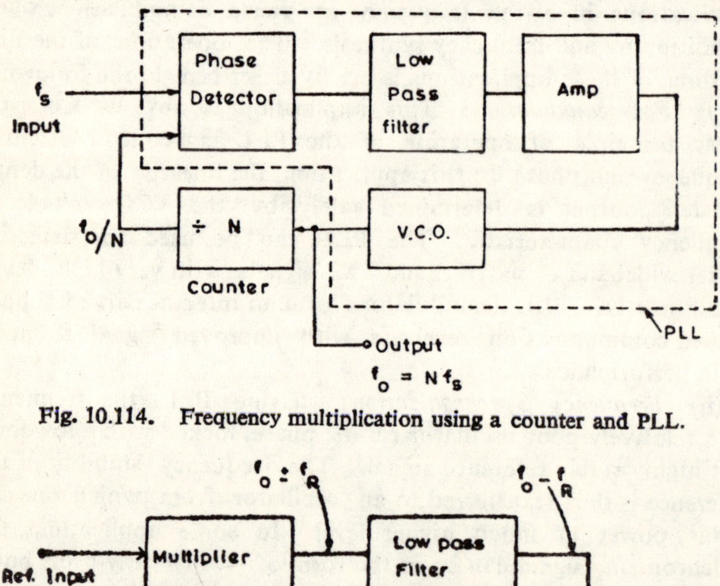

Fig. 10.114. Frequency multiplication using a counter and PLL.

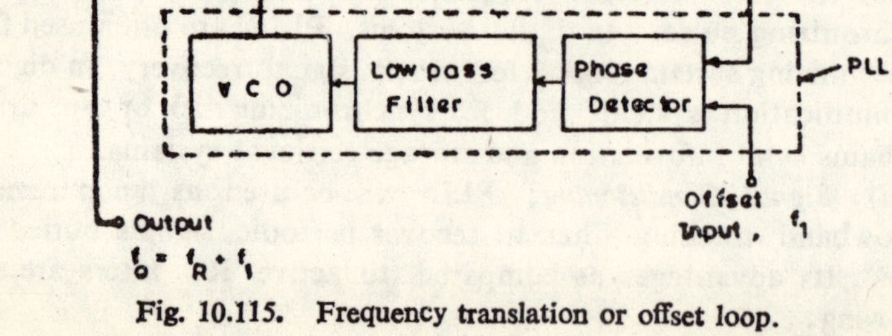

Fig. 10.115. Frequency translation or offset loop.

(v) *Frequency translation*: PLL can be used to translate the frequency of highly stable but fixed frequency oscillator by a small amount in frequency as shown in Fig. 10.115.

(vi) *Coherent Demodulator.* The carrier phase can be recovered from the A.M. Signal with the help of a PLL. The coherent reference signal thus recovered when multiplied with the AM received signal and filtered provides the demodulated output.

(vii) *A PLL application* of particular interest in Power Electronics is the speed control of motors. Such systems for speed control have scored over the conventional servo systems because they have fewer components, are inexpensive and are capable of much superior performance with speed accuracies of the order of .002%. Especially suitable for PLL control are systems in which motors must be synchronized to each other or to an existing clock. Examples of such applications are material handling conveyor belts, production lines, paper mill drives, computer tape and disk drives etc.

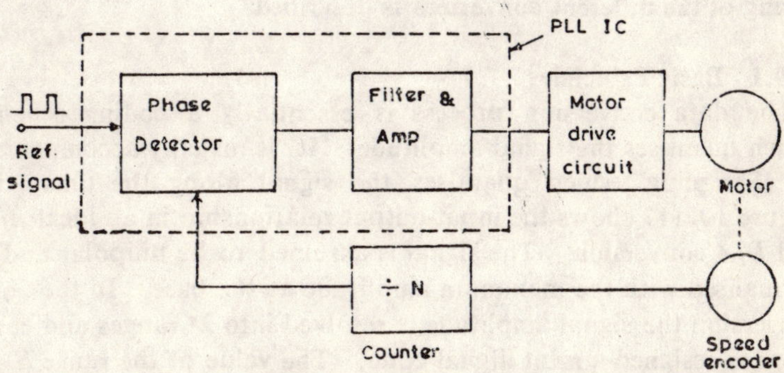

Fig. 10.116. Motor speed control system using PLL.

A PLL motor speed control system is shown in Fig. 10.116. The system operation is exactly the same as in PLL except that VCO of the PLL is replaced here by the motor and the speed encoder. In the proper steady-state operation of the loop, the motor always runs at a fixed speed as set by the reference signal frequency irrespective of the load conditions or other disturbances. By changing the reference signal frequency or by introducing a divide-by-N counter in the feedback path, the motor can be set to run at any desired speed within a certain range determined by the control dynamics of the system. The details of the design of this particular application of PLL, are given in [10]-[14].

10.9 A/D AND D/A CONVERSION

The data converter circuits are gaining importance as many control systems are being implemented wholly or partially in a computer. In order to couple the analog signals to the computerised controller the *A/D* conversion is necessary. Similarly to couple the controller outputs to indicators or actuators, it may be necessary to convert the digital signals to analog forms.

In this section the basic schemes for analog to digital (*A/D*) and digital to analog (*D/A*) conversion are explained. This is followed by discussion of the circuits for the *A/D* and the *D/A* conversion. As many circuits are available in the integrated form, only the basic principles and the block diagram of the circuits are discussed. A more detailed description of the particular integrated circuit can be obtained from the manufacturer's data book. Finally the interfacing of the different converters is described.

10.9.1 Basic Principles

The data conversion process is essentially a coding scheme which quantises the signal amplitude. It is usually accompanied by 'Sampling' which quantises the signal along the time axis. Figure 10.117 shows the input-output relationship in an ideal *A/D* and *D/A* conversion. The signal is assumed to be unipolar and is normalised with the maximum amplitude as the base. In the *A/D* conversion the signal amplitude is resolved into 2^n ranges and each range is assigned a n-bit digital code. The value of the range S is chosen to be $(2^n - 1)^{-1}$. The highest and the lowest range is reduced to $S/2$ so that the zero and the one p.u. values are at the centre of these ranges. This assignment results in an exact representation of the zero and the one p.u. values during the reverse, i.e. *D/A* conversion. For any finite number of bits, n, the *A/D* conversion is ambiguous within the range S. The quantity S is often called the 'Resolution' of the converter.

Figure 10.117 also shows the input-output relationship in the *D/A* conversion. The 2^n values of the n-bit digital input are assigned analog values corresponding to the centre of their respective ranges i.e. $0, S, 2S ... (2^n - 1)\,S$. The zero and the one p.u. points can be thought of as the mid-points of the highest and the lowest ranges,

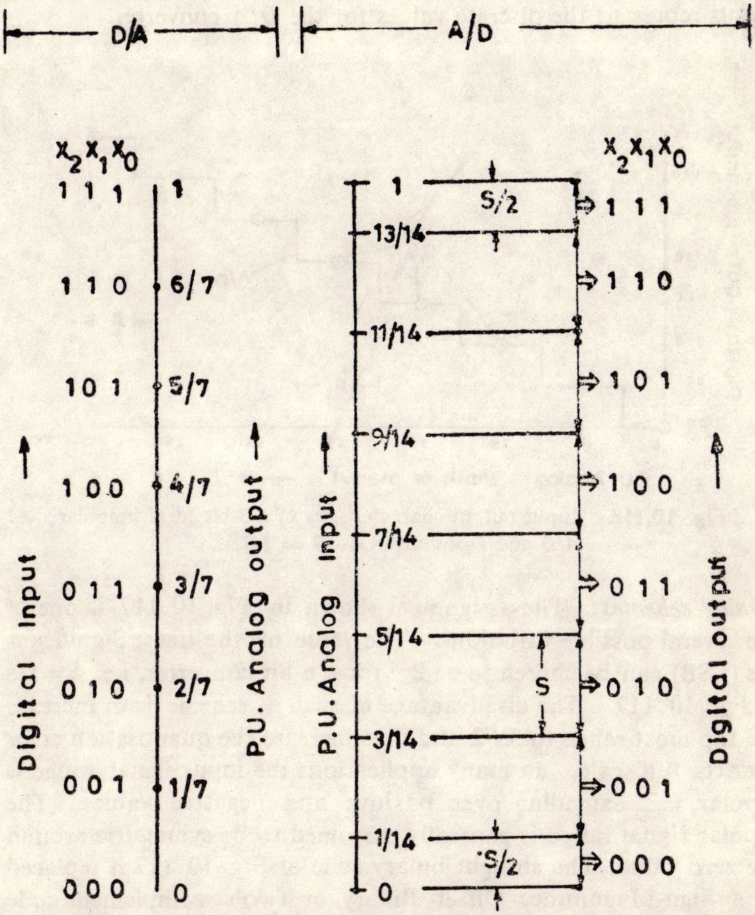

Fig. 10.117. Input-output assignments in unipolar 3-bit A/D and D/A conversion ($S = 1/7$).

halves of which extend beyond the signal range. It is clear from the Fig. 10.117 that the maximum error in a A/D conversion followed by a D/A conversion is $S/2$. The quantity $S/2$ is therefore called 'Quantization Error' and is due to the finite number of bits used in the digital representation. The quantization error can be reduced only by increasing n, the number of bits in the digital representation. Figure 10.118 shows the relationship between the analog and the digital quantities for the ideal 3-bit converters. The solid lines represents the characteristic of the A/D converter while the circled points represent the discrete values for the D/A converter.

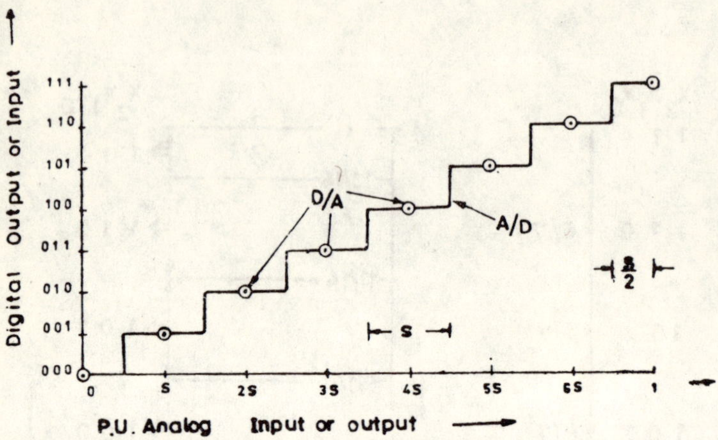

Fig. 10.118. Input output characteristics of a 3-bit ideal unipolar A/D and D/A converters $S = 1/7$.

Coding schemes: The assignment shown in Fig. 10.117 is one of the several possible variations. The value of the Least Significant Bit (LSB) can be chosen to be 2^{-n} for a n-bit converter, i.e. $S=1/8$ in Fig. 10.117. The disadvantage of such a scheme is to increase the top most range to $3S/2$ and thus increase the quantisation error near the full scale. In many applications the input signal range is bipolar, i.e. extending over positive and negative values. The bipolar signal range is generally assumed to be symmetric around the zero value. The straight binary code of Fig. 10.117 is replaced by a Sign-Magnitude, Offset Binary or Two's complement code for bipolar inputs. Figure 10.119 shows an example of such code assignments over the $\pm 1/2$ range. The offset binary code simply shifts the zero to the lowest range. In this code the Most Significant Bit (MSB) is one for the positive values and zero for the negative values. The Two's complement code represents the positive inputs as the straight binary code of Fig. 10.117. The negative inputs are represented by the two's complement of the code for the positive inputs of equal magnitudes. It is seen from Fig. 10.119 (b) that the full scale values, i.e. one half, are not exactly represented in the D/A conversion, unlike the zero. The exact conversion of the full scale and the zero values requires

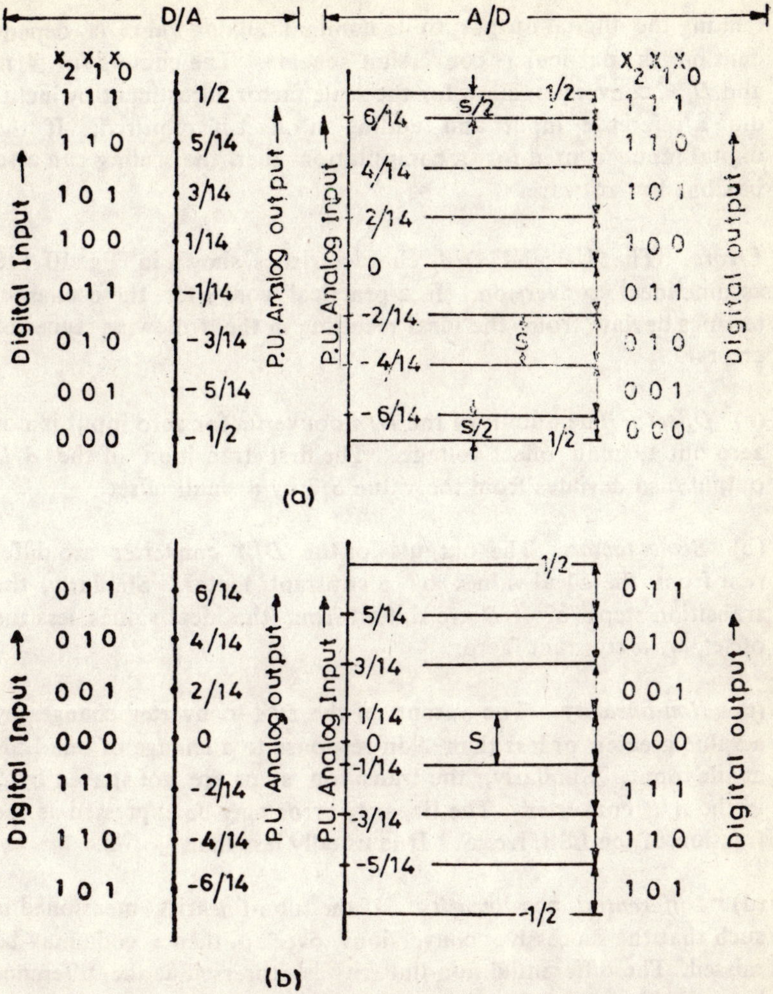

Fig. 10.119. Input output assignment in bipolar 3-bit A/D and D/A conversion on using (a) Offset binary, (b) Two complement codes $S = 1/7$.

increasing the value of LSB to $(2^n - 2)^{-1}$, i.e. 1/6. In the two's complement code the MSB is zero for positive values and one for negative values.

It is clear from the above that the digital output of an A/D conversion is a code. Therefore, the signal magnitude may not be equal to the decimal value of the binary output. The scale factor relating the digital output to its nominal analog value is dependent on the particular conversion scheme. The circuits for A/D and D/A, conversion allow for the scale factor adjustment by including a reference input and adding an offset if required. If the digital input is used for a computation then the scaling can also be done by software.

Errors: The A/D and D/A characteristics shown in Fg. 10.118 assume ideal conversion. In a practical converter these characteristics deviate from the ideal resulting in the following types of errors.

(a) *Offset*: The output of the D/A converter for zero input is not zero but a small offset voltage. The first transition of the A/D output also deviates from the value $S/2$ by a small offset.

(b) *Scale factor*: The outputs of the D/A converter are different from the ideal values by a constant factor. Similarly, the transition steps of A/D are shifted from the ideal values less the offset, by a constant factor.

(c) *Non-linearity*: The output of the D/A converter changes by a value greater or less than S in response to a change of one LSB at the input. Similarly, the transition steps are not spaced by S in the A/D converter. The linearity error may be expressed as the fraction of the LSB, i.e. S. It is usually less than $\pm S/2$.

(d) *Differential non-linearity*: If the non-linearity mentioned is such that the successive conversions overlap, then a code may be missed. The differential non-linearity is expressed as the difference between the successive non-linearity errors. If this value exceeds S then a code in the A/D output or a level in the D/A output may be absent. The output then is not monotonic, i.e. an increase in input may result in a decrease in the output.

Figure 10.120 gives the ideal and the practical input-output characteristics. Some types of errors are illustrated in this figure.

10.9.2 Scheme For A/D Conversion

The analog to digital conversion can be carried out in several ways. The various schemes for A/D conversion are discussed in

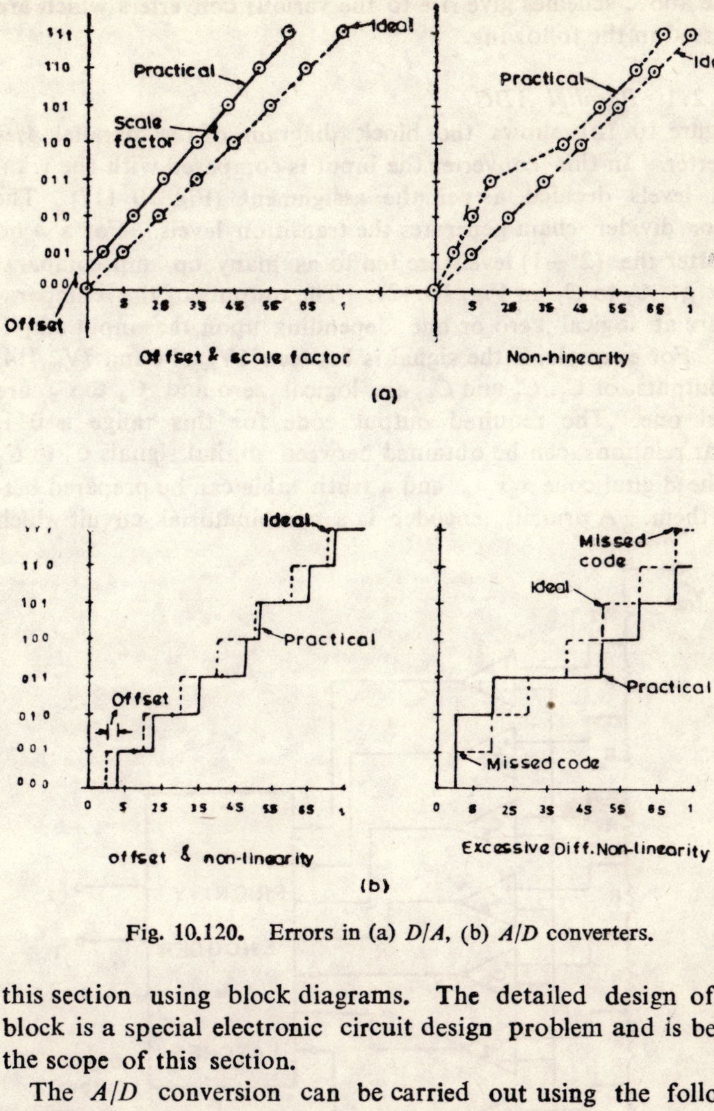

Fig. 10.120. Errors in (a) D/A, (b) A/D converters.

this section using block diagrams. The detailed design of each block is a special electronic circuit design problem and is beyond the scope of this section.

The A/D conversion can be carried out using the following schemes:

(a) Simultaneous conversion of all bits.

(b) Conversion of one level at a time.

(c) Conversion of one bit at a time.

(d) Indirection conversion.

The above schemes give rise to the various converters which are discussed in the following.

10.9.2.1 *Parallel ADC*

Figure 10.121 shows the block diagram of the parallel *A/D* converter. In this converter the input is compared with the transition levels decided as per the assignment (Fig. 10.117). The resistor divider chain generates the transition levels. For a *n* bit converter the $(2^n - 1)$ levels are fed to as many op amp comparators e.g. C_1 to C_7 in Fig. 10.121. The outputs of the comparators are at logical zero or one depending upon the input amplitude. For example, if the signal is between $5V_{ref}/14$ and $7V_{ref}/14$, the outputs of C_1, C_2 and C_3 are logical zero and C_4 to C_7 are logical one. The required output code for this range is 011. Similar relations can be obtained between digital signals C_1 to C_7 and the digital code $x_2x_1x_0$ and a truth table can be prepared between them. A priority encoder is a combinatorial circuit which

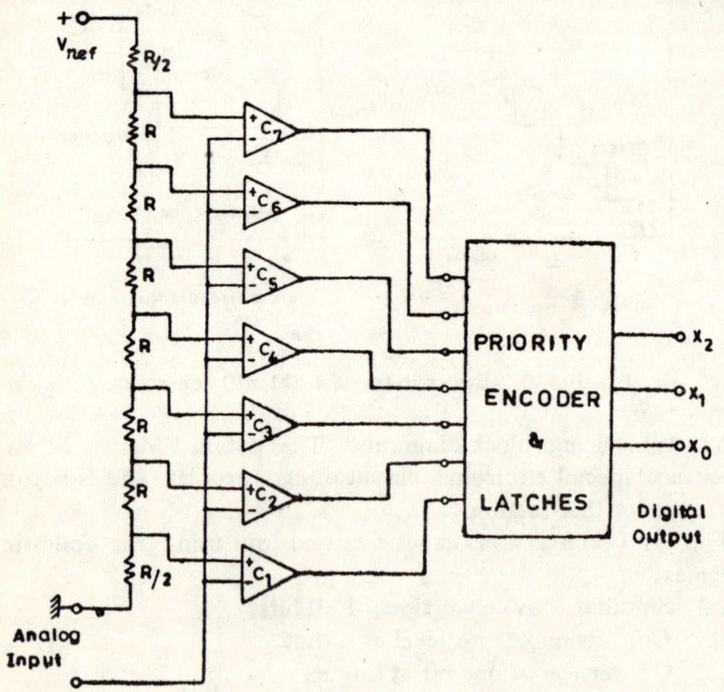

Fig. 10.121. Parallel *A/D* converter.

implements the above mentioned truth table and provides the digital output.

In this converter the comparators and the combinatorial logic provide the output almost instantaneously after the input signal is applied. Therefore, this converter is also known as the Flash *A/D* converter. All bits are available simultaneously at the output and the delay may be of the order of few nanoseconds if emitter-coupled logic is used. In the integrated circuit form the digital output is usually stored in a set of output latches at the end of conversion.

10.9.2.2 *Counting-Comparator or Ramp ADC*

In this *A/D* converter the conversion is done by comparing the input with the levels decided as per the assignment (Fig. 10.117). But, unlike the parallel comparators, the comparison is done using only one comparator and proceeding with one transition level at a time starting from the lowest. In order to present the levels to the comparator sequentially, a *D/A* Converter is used as shown in Fig. 10.122. The input to the *D/A* Converter is obtained from a binary counter driven by a gated clock. In Fig. 10.122 assume that the counter counts up and is cleared at the start of conversion. The comparator output is high at start and allows the clock pulses to increment the counter. Thus, at each clock pulse the counter and hence the output of the *D/A* Converter goes up by a LSB. The clock pulses are stopped when the output of the *D/A* converter exceeds the input and the output of the comparator goes

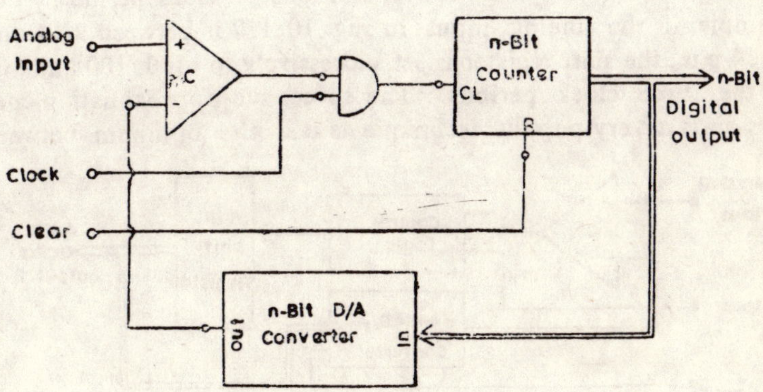

Fig. 10.122. Counting comparator *A/D* converter.

to zero. The output of the counter is the digital equivalent of the input in the straight binary code. The time of conversion in this scheme is proportional to the input magnitude and may take a maximum of (2^n-1) clock periods. The use of an up counter has the disadvantage that a reduction in the input is ignored by the converter until it is cleared. A modified converter which uses an Up–Down counter can overcome this difficulty. The output of the comparator can then be used to change the mode of the counter. Such an A/D converter is also called 'Tracking' or 'Servo' type converter. It has the advantage that it responds to the changes in the input.

10.9.2.3 *Successive Approximation ADC*

Figure 10.123 shows a block diagram of this type of A/D Converter. Here the conversion proceeds one bit at a time starting from the most significant bit (MSB). Let us assume that the shift register is initially cleared and the MSB is set to one. The output of the D/A converter now corresponds to the middle of the range. The comparator output changes to one (zero) if the input is in the upper (lower) half of the signal range. The control logic senses the output of the comparator and if it is zero, the MSB of the shift register is reset. Otherwise the MSB is left unchanged at one. The control logic sets the next bit and then resets it only if the comparator output is zero. The procedure is repeated till all bits have been tested. At the end of the n steps or clock cycles, the shift register contains the digital code for the analog input. In this procedure the n-bit A/D conversion takes n clock periods. For example if the analog input in Fig. 10.117 is between 9/14 and 11/14 p.u., the shift register is set successively to 100, 100 and 101 in the three clock periods. The successive approximation conversion is a very popular technique as it is the optimum between

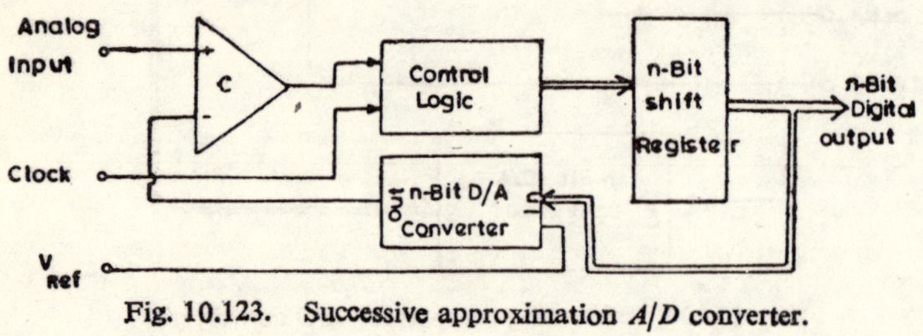

Fig. 10.123. Successive approximation A/D converter.

extensive hardware of parallel converter and the large time of conversion of the ramp converter.

10.9.2.4 *Dual Slope ADC*

This converter falls under the indirect conversion scheme and is popular in the instrumentation circuits. Figure 10.124 shows the

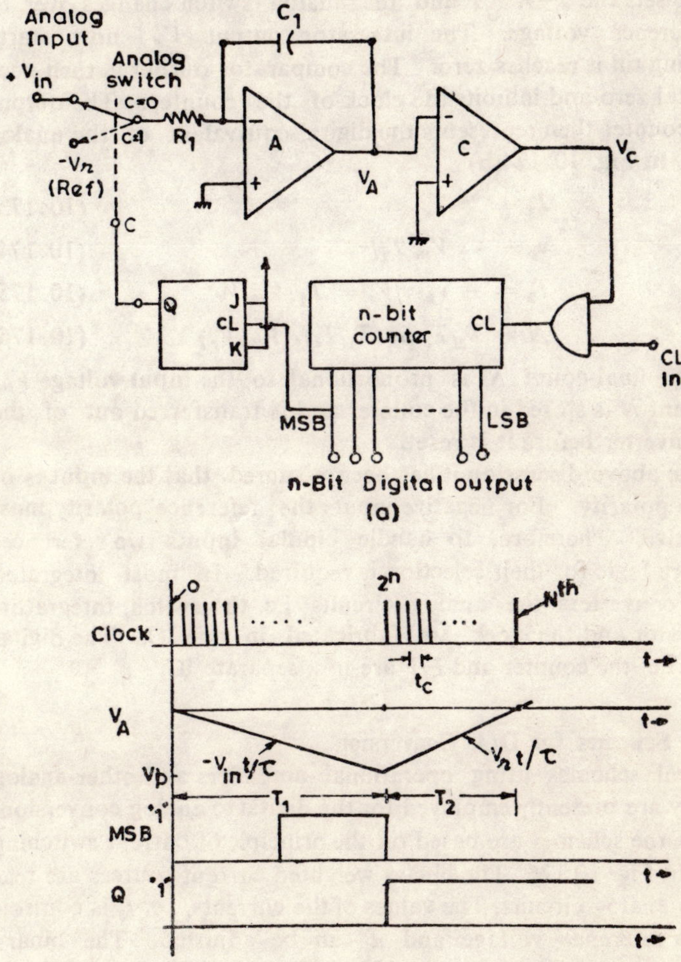

Fig. 10.124. (a) Block diagram.
(b) Waveforms during conversion for a dual-slope *A/D* converter. $\tau = R_1C_1$.

block diagram and waveforms for the converter. To start the conversion the capacitor C_1 is discharged by means not shown and the analog switch is connected to the input. The binary counter is also cleared at start. The input V_{in} is integrated for an interval equal to the 2^n clock periods. The output v_c is at logical one during this period. A the end of 2^n clock periods, the MSB of the counter sets the $J-K$ F/F and the analog switch changes over to the reference voltage. The integrator output V_A now starts increasing till it reaches zero. The comparator output v_c then goes to logical zero and inhibits the clock of the counter. The output of the counter then represents the digital equivalent of the analog input. In Fig. 10.124(b)

$$T_1 = 2^n \cdot T_c \tag{10.173}$$

$$v_p = -V_{in} T_1/\tau \tag{10.174}$$

$$T_2 = -v_p \cdot \tau/V_r = T_1 \cdot V_{in}/V_r \tag{10.175}$$

$$N = T_2/T_c = (T_1/T_c) \cdot (V_{in}/V_r) \tag{10.176}$$

Thus, the final count N is proportional to the input voltage V_{in}. The count N is stored in the counter and is transferred out of the A/D converter before it is reset.

In the above discussion it has been assumed that the input is of positive polarity. For negative inputs the reference polarity must be positive. Therefore, to handle bipolar inputs two references and extra logic for their selection is required. In most integrated circuit converters the analog circuits i.e. the switch, integrator, comparator and the clock are fabricated in one IC. The digital circuits i.e. the counter and F/F are in a separate IC.

10.9.3 Schemes for D/A Conversion

Several schemes using operational amplifiers and other analog circuitry are presently employed for the digital to analog conversion. Most of the schemes are based on the principle of current switching shown in Fig. 10.125. The binary weighted current sources are realised by analog circuits. The values of the currents, i.e. I_0 is controlled by a reference voltage and it can be adjusted. The binary variables x_3, x_2, x_1 and x_0 are either zero or one and accordingly connect the current source to either the ground or the negative input of the op amp. The output current I_l is the sum of the currents whose controlling bits are at logical one. The output

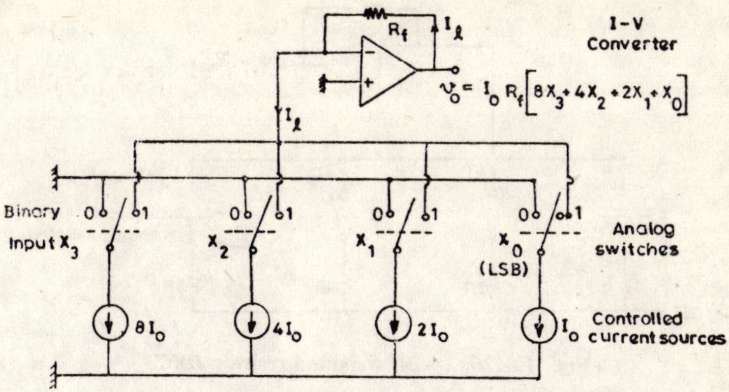

Fig. 10.125. Basic circuit configuration for *D*/*A* converter.

voltage v_0 is given by,

$$v_0 = I_0 R_f \, [8x_3 + 4x_2 + 2x_1 + x_0] \qquad (10.177)$$

The output voltage v_0 is thus proportional to the decimal value of the binary number '$x_3x_2x_1x_0$'. The weight of the digital bits can be changed from the conventional 8, 4, 2, 1 by changing the current sources. However, the binary weights are most commonly used. In many integrated *D*/*A* converters the output op amp is omitted and the current I_l itself is presented as the output.

Various circuits have been devised to implement the configuration of Fig. 10.125. Some widely used circuits are described in the following.

10.9.3.1 *Binary Weighted Resistor DAC*

Figure 10.126 shows the circuit diagram for this *DAC*. The binary weighted currents are generated in this circuit by using resistors and a reference voltage source. The SPST switch is closed whenever its controlling bit is one and is open whenever it is zero. The output voltage is given by eq. (10.177) where I_0 is $V_{ref}/8R$ and the on resistance of the switches is neglected.

The weighted resistor *DAC* is used for low resolution, e.g. 4-bit applications. The circuit is rather simple but has the following disadvantages.

(a) The circuit uses resistors of widely varying values. Therefore, the tolerance of the MSB resistor is very small in order to prevent the error from swamping the LSB current. It can be shown

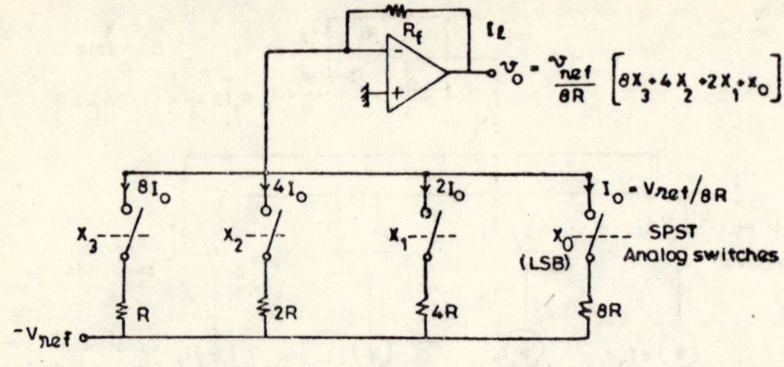

Fig. 10.126. 4-Bit weighted resistor *DAC*.

that the tolerance for the MSB should be better than $1/2^{n+1}$ for a *n*-bit converter. This prevents the use of more than 4 bits in practical converters.

(b) The load on the reference source V_{ref} is not constant. The load can be balanced by using a *DPDT* switch and diverting the current to ground when the bit is zero.

(c) The series resistance of the on switch affects the accuracy of the convertor.

10.9.3.2 DAC Using R—2R Ladder Network

The ladder network consists of identical sections connected in cascade. A section is shown in Fig. 10.127. The resistors R_1 and R_2 are chosen such that the input resistance R_i is equal to the load resistance R_L and the voltage attenuation (V_L/V_1) is a specified fraction, *a*. The values of R_1 and R_2 are given by,

$$R_2 = R_L\,(a/1 - a) \tag{10.178}$$

$$R_1 = R_L\,(1 - a) \tag{10.179}$$

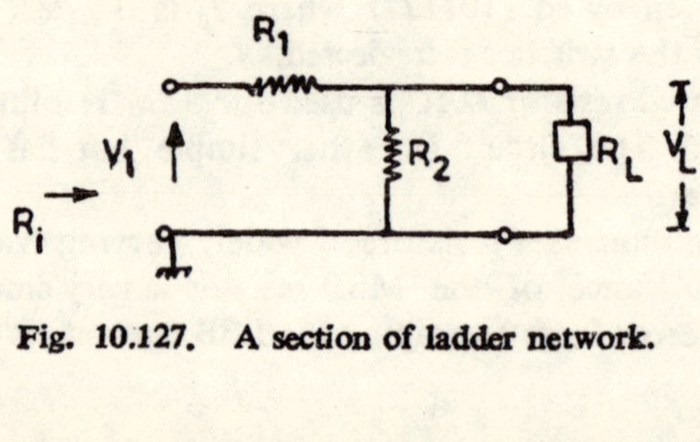

Fig. 10.127. A section of ladder network.

Thus, for an attenuation of one half ($a = 0.5$) and R_L equal to $2R$,

$$R_2 = R_L = 2R \tag{10.180}$$

$$R_1 = 0.5\,R_L = R \tag{10.181}$$

The sections of $R-2R$ resistors are connected in cascade to form the *DAC* as shown in Fig. 10.128. The input resistance to the right of each of the nodes N_0 to N_3 is $2R$. The ladder attenuates the reference source voltage V_{ref} by the factor a in each section. The lower ends of the resistors of value $2R$ are connected to the ground or the virtual ground of the op amp. The resistors of value $2R$ carry binary weighted currents as shown in Fig. 10.128. The output voltage v_0 is given by eq. (10.177) where I_0 is $(V_{ref}/16R)$. The current I_{ref} drawn from the reference source constant and is equal to $16I_0$. The $R-2R$ ladder can be easily extended to obtain the required number of bits as there are only two values of the resistors.

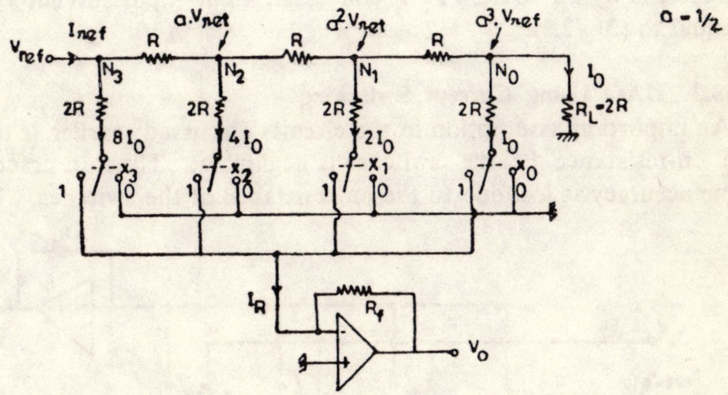

Fig. 10.128. *DAC* using $R-2R$ ladder network $I_0 = V_{ref}/16R$.

A variation of the *DAC* known as Inversely Operated $R-2R$ Ladder is shown in Fig. 10.129. Here the ladder network is loaded with $2R$ at both ends. The lower ends of the resistors of value $2R$ are connected either to the ground if the input bit is zero or to the voltage V_{ref} if it is one. The load current I_l can be calculated by applying each input bit one at a time and then using the superposition theorem. As the ladder is loaded symmetrically the input resistance to the right and left of each of the nodes N_0 to N_3 is $2R$. The output voltage v_0 is given by the following equation.

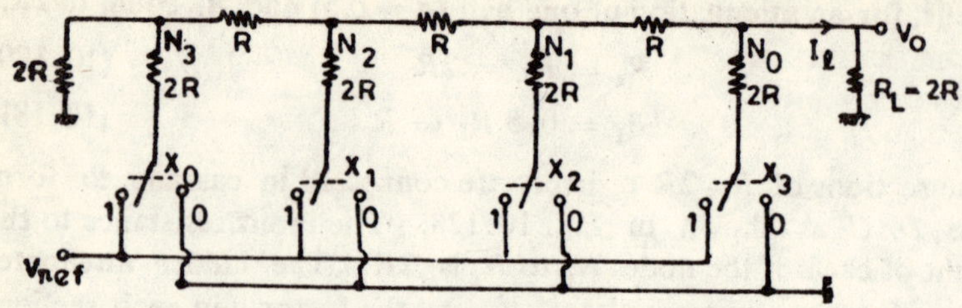

Fig. 10.129. *DAC* using inverse *R*—2*R* ladder network.

$$v_0 = \frac{V_{ref}}{24}(8x_3 + 4x_2 + 2x_1 + x_0) \tag{10.182}$$

The circuit of Fig. 10.129 can also be operated with the values of R_L other than $2R$. The network to the left of R_L is then equivalent to a source of voltage $3v_0/2$, where v_0 is given by eq. (10.182), with an internal source resistance of R. The load resistance can also be replaced by an op amp I—V converter. The input current then is equal to $(3v_0/2R)$.

10.3.3 DAC Using Current Switching

An important assumption in the circuits discussed earlier is that the on-resistance of the switches is negligible. Thus, in practice some accuracy is lost due to the on-resistance of the switches. The

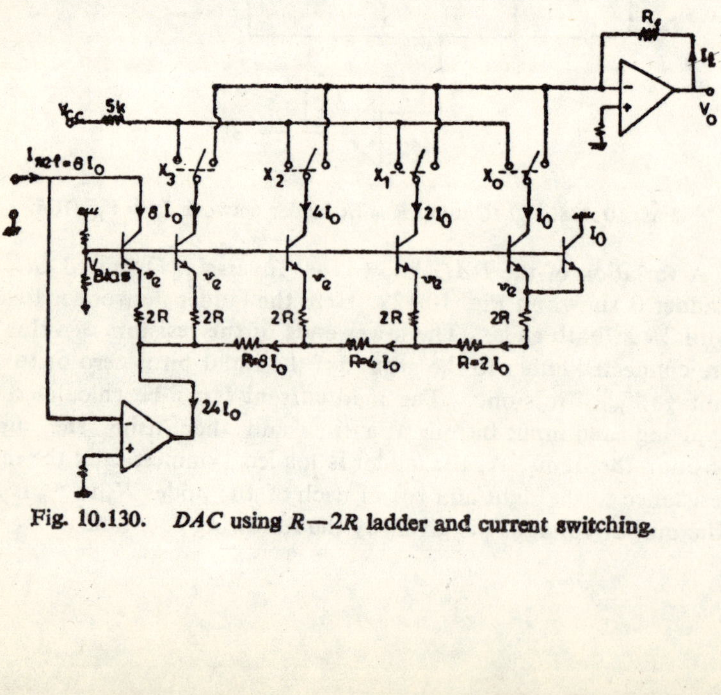

Fig. 10.130. *DAC* using *R*—2*R* ladder and current switching.

current switching technique overcomes this problem by using electronic current sources rather than voltage sources and resistances to realise the configuration of Figure 10.125. The $R-2R$ ladder network is commonly used as load in the electronic current source.

Figure 10.130 shows a circuit in which transistors have been used for realising the binary weighted current sources. The reference current decides the output voltage which is given by equation (10.177). An important problem in the design of this circuit is that the base-emitter drops of the transistors have to be made equal while their collector currents differ widely. This is achieved by scaling the emitter areas of the transistors to equalise the base-emitter drop. A mismatch in the base-emitter drops results in an error in the binary weights for the currents.

Figure 10.131 shows a circuit in which the $R-2R$ ladder is inversely operated by current sources of equal amplitude. The transistors now carry equal currents and the matching problem mentioned above is reduced. The op amp provides the ladder with an equivalent zero load resistance. The output voltage can be

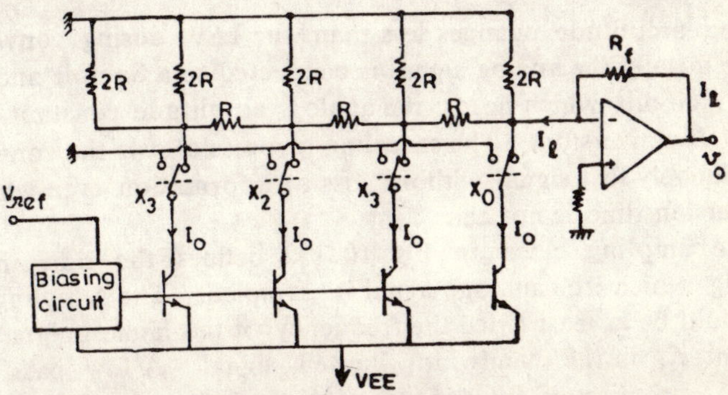

Fig. 10.131. *DAC* using inverse $R-2R$ ladder and current switching.

calculated by assuming the lower part of the circuit to be current sources of value I_0 and applying the super position principle. The output voltage is given by

$$v_0 = \frac{I_0 R_f}{8} [8x_0 + 4x_1 + 2x_2 + x_3] \qquad (10.183)$$

The reference current I_0 can be controlled by a reference input voltage through a biasing network.

10.9.4 Interfacing of Converters

The A/D and D/A converter circuits discussed in previous sections are usually connected to other system components e.g. Sample and Hold circuits and filters. The interfacing and timing considerations are discussed in this section.

Sample and hold: Figure 10.132 shows a general block diagram of a digital system using A/D and D/A converter. The circuits for A/D conversion discussed in section 10.9.2 assume that the input analog signal is held constant during the time of conversion. Only low frequency signals (< 100 Hz) change so slowly that the

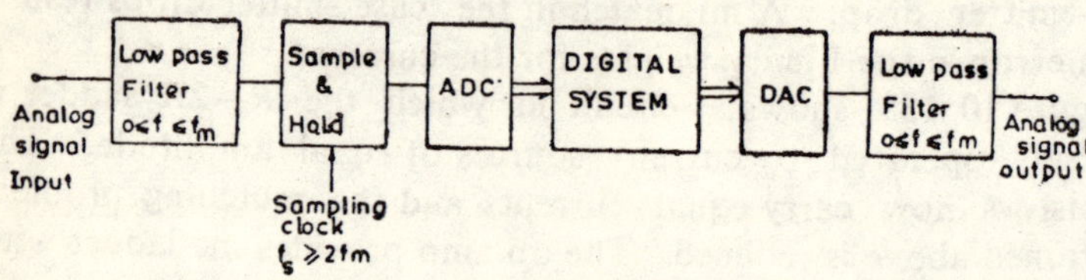

Fig. 10.132. Inter connections of converters and digital system.

analog amplitude changes less than the LSB during conversion. More usually the analog signal is connected to a Sample and Hold (S/H) circuit which holds the analog amplitude constant during the A/D conversion. The sampling process allows the conversion of relatively fast signal without loss of information even when the conversion time is not zero.

The sampling clock in Fig. 10.132 defines the time instants during which the analog signal is sampled. The sampling freq. f_s should be at least twice the frequency of the highest signal component f_m in the bandwidth limited signal. A low pass filter, called 'anti-aliasing' filter is often placed before the S/H to ensure a bandwidth limited signal. The sampling clock is also used to coordinate the ADC. The falling edge of the sampling pulse (Fig. 10 133) starts the conversion. At the end of the conversion the ADC generates a 'end of conversion' (EOC) signal which changes the S/H to sample mode and a new value of input is sampled for the next conversion.

Digital to analog conversion: The digital output of the ADC serves as an input to the digital system (Fig. 10.132). The output

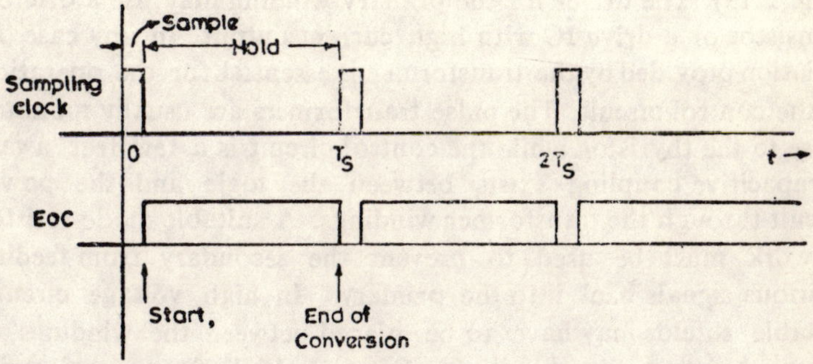

Fig. 10.133. Interfacing signals for *A/D* conversion.

of the digital system is connected to a DAC. Usually the digital output is stored in a buffer and remains unchanged for sufficient time for conversion to analog output. The DAC output is shaped by a low pass reconstruction filter so that components above the maximum frequency f_m are filtered out. This removes the glitches and other transient voltages generated during *D/A* conversion. In some systems a *S/H* block is introduced between the DAC and the low pass filter. This is done when digital storage is not possible due to shortage of hardware or multiplexing of output. The storage is then analog. The *S/H* and DAC are then coordinated by the end of conversion signals like those shown in Fig. 10.133.

10.10 IMPLEMENTATION OF GATING CIRCUITS

The building blocks described in sections 10.1 to 10.9 can be used to implement gating and control circuits for power electronic converters. Some examples of such circuits are given in the sections 3.5, 5.2, 6.7 and 7.5. Most of these circuits are waveform generators which use gates, flip-flop, counters, op amps etc. However, a few considerations specific to the power electronic applications are given below:

The thyristor firing methods have been considered in section 2.3. It is clear that the gate-cathode terminals of the thyristor are in the power circuit and are at a high potential with respect to the power ground. A popular method of coupling the logic circuit to the gate-cathode circuit is through a pulse transformer. The primary of the pulse transformer is connected to the logic circuit while the secondary winding is connected across the gate-cathode circuit

(Fig. 2.19). The driver for the primary winding may use a discrete transistor or a drive IC with high current output. In any case the isolation provided by the transformer is essential for the operation of the control circuit. The pulse transformers are usually mounted close to the thyristor while the control circuit is a few feet away. A capacitive coupling exists between the logic and the power circuit through the transformer windings. A suitable diode-resistor network must be used to prevent the secondary from feeding spurious signals back into the primary. In high voltage circuits suitable shields may have to be placed between the windings to reduce the capacitive coupling. Often a high frequency carrier gating is used to reduce the size of the pulse transformer. The duty cycle of the carrier gating must be about 1/3 to prevent the saturation of the transformer core. The voltage and current feedback signals obtained from the power circuit also need to be isolated. The current signal is often obtained from a shunt and the isolation is provided by an optical coupling which operates from an isolated power supply. A discussion of the current and voltage sensing is given in Chapter 9.

The logic and control circuits driving a power converter often work in an environment with substantial electrical noise. The switching of large currents in thyristors and transistor switches is a common source of radiated noise. There is conducted noise on the power supply and ground lines within the logic circuit. For these reasons power supply decoupling capacitors must be provided for each sub-group of ICs on a board. The ground lines must also run separately to a common ground point to prevent the noise from being injected into other circuits. A high d.c. noise immunity is obtained by using either HTL or CMOS logic at typically 15V power supply. Sometimes it may be necessary to use the slower HTL or CMOS at the peripheral interfaces and fast TTL for internal signal processing. Since the signals are usually not of very high frequency, the layout of the circuit is not very critical. However, it must be shielded against the radiated noise and designed to reduce the effects of conducted noise. A detailed discussion on the operation in noisy environment is given in [15] and [16].

The problem of isolation between the power and logic circuit is also present when the signals are being viewed during testing. The most important point to be kept in mind is that most of the oscilloscopes usually have only one ground, available at the probe as a clip or wire. Thus, in a multichannel scope signals are usually viewed with respect to a common ground.

When viewing power circuit waveforms a two wire supply can be used to power the oscilloscope. The chassis of the oscilloscope is then at the potential of the power circuit and should not be touched by the user of the scope. An alternative is to use double insulated power scopes where the outer body is insulated from the chasis. The oscilloscopes with differential inputs are also suggested but they are not suitable for viewing fast edges of the waveforms in the power circuit. The reason is the reduction in the CMRR of their input stage with frequency.

When viewing power circuit and logic circuit waveforms simultaneously on different channels of the scope, the grounds should be kept isolated. In a laboratory this can be achieved by supplying the logic circuit through a power supply whose ground is not connected to the ground of the power circuit. However, in the industrial systems a common ground between the logic and power circuits is often used. In such cases the logic and power circuit waveforms can be viewed with respect to the common ground only. Use of two different points as grounds will result in a direct short circuit between them through the chassis and should be avoided.

REFERENCES

1. Z. Kohavi—Switching and Finite Automata Theory (Book), McGraw Hill 1970., p. 41, Chapters 9 and 12.
2. H. Taub and D. Schilling—Digital Integrated Circuits (Book), McGraw-Hill, Kogakusha, 75.
3. Lane S. Garett. 'Integrated-Circuit digital logic families Part I—RTL, DTL, HTL Devices, IEEE Spectrum. Oct. 1970, p. 46.
4. Lane S. Garett, 'Integrated Circuit digital logic families Part II—TTL Devices, IEEE Spectrum, Nov. 1970, p. 63.
5. Robert L. Morris and John R. Miller., 'Designing with TTL integrated circuits' (Book), Mc-Graw Hill, 1971.
6. William M. Penny and Lillianhan (Editors)., 'MOS integrated Circuits' (Book), Van Nostrand Reinhold Co., 1972.
7. A. Havasy, 'Interfacing COS/MOS with other logic families', RCA application note, ICAN-6602, 1974.
8. J. Naylor, 'Digital and Analog signal applications of operational amplifiers', IEEE Spectrum, June 1971, p. 38.
9. J. Roberg, 'Operational Amplifiers' (Book), John Wiley & Sons, 1975, pp. 485-496.

10. F.M. Gardner, 'Phase Lock Techniques' (Book), Wiley Interscience, 1979, Chapters 2 and 4.
11. A. W. Moor, 'Phase Locked Loops for motor speed control', IEEE Spectrum, vol. 10, April, 1973, p. 61.
12. N. K. Sinha and N.H. Bailey, 'Speed Control of a DC Servo motor using Phase Locked Loops', IEEE Trans., Vol. IECI-23, No. 1, Feb. 1976.
13. K. Eapen, K. Venkatesan, S. C. Gupta, 'A Phase Locked Loop DC motor control system', Journal I. E. (India), EE Div., Vol. 59, June 1979.
14. D.M. Smithgall, 'A Phase Locked Loop Motor Control System', IEEE Trans., Vol. IECI-22, No. 41, Nov. 1975.
15. Verell Boaen, 'Designing logic circuits for high noise immunity', IEEE Spectrum, Jan., 1973.
16. Andrc Wavre, 'Application of integrated circuits to industrial control systems with High-Noise environments', IEEE Transactions, Vol. IGA-5, No. 3, May/June 1969, p. 278.

APPENDIX A-1

CATALOGUE OF POPULAR TTL AND MOS ICS

Table 1. 54/74 Series TTL/SSI Integrated Circuits

Gate	Function	Device
	Quad 2–input NAND	54/7400, 54/74L00, 54H/74H00
	Quad 2–input NAND, Open collector	54/7401, 54H/74H01
	Quad 2–input NOR	54/7402
	Quad 2–input NAND, open collector	54/7403
	Hex Inverters	54/7402, 54/74L04, 54H/74H04
	Hex Inverter, open collector	54H/74H05
	Hex inverter buffer/driver, open collector	54/7406
	Hex buffer/driver, open collector	54/7407
	Quad 2–input AND	54/7408
NAND/NOR	Quad 2–input AND, open collector	54/7409
	Triple 3–input NAND	54/7410, 54/74L10, 54H/74H10
	Triple 3–input AND	54H/74H11
	Hex inverter buffer/driver, open collector	54/7416
	Hex buffer/driver, open collector	54/7417
	Dual 4–input NAND	54/7420, 54/74L20, 54H/74H20
	Dual 4–input AND	54H/74H21
	Dual 4–input NAND, open collector	54H/74H22
	Dual 2–input NAND, open collector	54/7426
	8–input NAND	54/7430, 54/74L30, 54H/74H30
	Dual 4–input NAND buffer	54/7440, 54H/74H40

Gate	Function	Device
	Expandable dual 2–input AND-OR-INVERT	54/7450, 54H/74H50
	Dual 2-wide 2-input AND-OR-INVERT	54/7451, 54/74L51, 54H/74H51
AND-OR-INVERT	Expandable 4–wide 2–2–2–3–input AND-OR	54H/74H52
	Expandable 4–wide 2–2–2–3–input AND-OR-INVERT	54/7453, 54H/74H53
	2–2–2–3–input AND-OR-INVERT	54/7454, 54/74L54, 54H/74H54
	Expandable 2–wide 4–input AND-OR-INVERT	54/74/74L55, 54H/74H55
	Dual 4–input	54/7460, 54H/74H60
Expander	Triple 3–input	54H/74H61
	3–2–2–3 input AND-OR	54/74H62
	Edge-triggered J-K	54/7470
	J-K master-slave	54/7472
	Dual J-K master-slave	54/7473
	Dual D-type edge triggered	54/7474
	Dual J-K master-slave, present and clear	54/7476
	J-K master-slave	54/74104
	J-K master-slave	54/74105
	Dual J-K master-slave	54/74107
	Monostable nonretriggerable	54/74121
	R.S. Master-slave	54/74L71
	J-K Master-slave	54/74L72
	Dual J-K master-slave	54/74L73
Flip-Flop	Dual D-type edge-triggered	54/74L74
	Dual J-K master-slave	54/74L78
	J-K master-slave	54H/74H71
	J-K mater-slave	54H/74H72
	Dual J-K master-slave	54H/74H73
	Dual D-type edge-triggered	54H/74H74
	Dual J-K master-slave with preset and clear	54H/74H76
	Dual J-K master-slave	54H/74H78
	J-K negative edge-triggered	54H/74H101

Gate	Function	Device
	J-K negative edge-triggered	54H/74H102
	Dual J-K, negative edge-triggered	54H/74H103
	Dual J-K, negative edge-triggered	54H/74H106
	Dual J-K, negative edge-triggered	54H/74H108

Table 2. 54/74 Series Combination TTL/MSI Integrated Circuits

MST	Function	Device
Data selector	16–bit	54/74150
	8–bit, with	54/74151
	8–bit	54/74152
	Dual 4–line-to-1-line	54/74153
	Quad 2–input	54/74157
Encoder	Full BCD priority encoder	54/74147
	Cascadable octal priority encoder	54/74148
Code converter	6–line-BCD to 6–line Binary, or 4–line to 4–line BCD 9's/BCD 10's converters	54/74184
	6–bit-binary to 6–bit-BCD converter	54/74185A
	BCD-to-decimal (4 to 10 lines)	54/7442
	Excess-3-to-decimal (4 to 10 lines)	54/7443
	Excess-3-Gray-to-decimal (4 to 10 lines)	54/7444
	BCD-to-decimal decoder/driver (4 to 10 lines)	54/7445
	BCD-to-7-segment decoder/driver	54/7446
	BCD-to-7-segment decoder/driver	54/7447

MST	Function	Device
Decoder/ demultiplexer	BCD-to-7-segment decoder/driver	54/7448
	BCD-to-7-segment decoder/driver	54/7449
	BCD-to-decimal decoder driver	74141
	BCD-to-decimal decoder/driver	54/74145
	4–line-to-16-line decoder/demultiplexer	54/74154
	Dual 2–line-to-4-line decoder/demultiplexer	54/74155
	Dual 2–line-to-4-line decoder/demultiplexer	54/74156
	Gated full-adder	54/7480
	2–bit full-adder	54/7482
Arithmetic device	4–bit full-adder	54/7483
	Quad 2–input exclusive-OR gate	54/7486
	4-bit arithmetic logic unit	54/74181
	Look-ahead carry generator	54/74182
Comparator	4–bit magnitude comparator	54/7485 54/74L85, 54/74S85,

Table 3. 54/74 Series Sequential TTL/MSI Integrated Circuits

Function	Device
	Ripple (Asynchronous) counters
Decade	54/74L90, 54/74LS196, 54/7490, 54/7490A, 54/74176, 54/74196
4–bit binary	54/74L93, 54/74LS197, 55/74293, 54/7493A, 54/74177, 54/74197
Divide-by-12	54/7492A
	Synchronous counters
Decade	54/74160, 54/74162

Function	Device
Decade/up/down	54/74L192, 54/74190, 54/74LS190, 54/74192, 54/74LS192
4–bit binary	54/74L193, 54/74191, 54/74S191, 54/74193, 54/74LS193, 54/74161, 54/74163
	Shift registers
8–bit serial-in, serial-out	54/7491A, 54/74L91
4–bit parallel-in, serial-out	54/7494
8–bit parallel-in, serial-out	54/74165, 54/74166
8–bit serial-in, parallel-out	54/74164, 54/74L164
4–bit parallel-in	54/74L95, 54/74L99, 54/74LS295, 54/74LS95A, 54/74LS195
Parallel-out	54/74178, 54/74179, 54/7495A, 54/74195, 54/74S195
5–bit parallel-in, parallel-out	54/7496, 54/74L96
8–bit parallel-in parallel-out	54/74199
4–bit parallel-in parallel-out (bidirectional)	54/74194, 54/S194, 54/74LS194
8–bit parallel-in parallel-out (bidirectional)	54/74198
	Random-access memory (RAM)
16–bit register file	54/74170
16–bit multiple-port register file	74172
16–bit read/write memory	54/7481A, 54/7484A
64–bit read/write memory	7489
256–bit read/write memory	74200, 74S200, 74S206
	Read-only memory (ROM)
256–bit	54/7488A
256–bit programmable	74188A
512–bit programmable	54/74186
512–bit programmable	54/74186
1024–bit	54/74187

Table 4. MOS/LSI Integrated Circuits

Function	Device
Random-access memory (RAM)	
256–bit	TMS 1101, JC, NC
Fully decoded 1024-bit	TMS 1103 NC
High-speed content-addressable	TMS 4000 JC, NC
256–bit	TMS 4003 JR, NC
2048–bit dynamic	TMS 4020 NC
1024–bit	TMS 4023 NC
2048 dynamic	TMS 4025 NC
Read-only memory (ROM)	
2560–bit dynamic	TMS 2300 JC, NC
Row output character generator	TMS 2400 JC, NC
2560–bit static	TMS 2500 JC, NC
2048–bit static	TMS 2600 JC, NC
USASCII-to-selectric/Selectric-to-USASCII code converter	TMS 2602 JC, NC
EBCDIC-to-USASCII code converter	TMS 2602 JC, NC
EBCDIC-to-USASCII code converter	TMS 2603 JC, NC
USASCII-to-EBCDIC/Selectric-to-EBCDIC code generator	TMS 2604 JC, NC
USASCII, BAJDOT, Selectric EBCDIC code generator	TMS 2605 JC, NC
3071–bit static	TMS 2700 JC, JM, NC
1024–bit static	TMS 2800 JC, NC
1280–bit–static	TMS 2900 JC, NC
Series character generator	TMS 4100 JC, NC
4096–bit static	TMS 4400 JC, NC
Programmable logic array (PLA)	
17–bit, 18–output, 8–JKFFPLA	TMS 2000
13–b input, 10–output, 10–JKFFPLA	TMS 2200

11. POWER CONVERTER CONTROLS USING PLCs AND MICROCOMPUTERS

11.1 INTRODUCTION

The developments in the semiconductor technology made it feasible to put several thousands of transistors and resistors on a single chip. Such high density chips are referred to as large-scale-integration (LSI) or very-large-scale integration (VLSI) chips. As compared to SSI (small scale integration) and MSI (medium scale integration) in terms of logic gates, the LSI and VLSI are much more complex. Typically an LSI chip has complexity exceeding 100 gates and a VLSI chip, greater than 1000 gates. On the other hand, SSI has complexity of less than 50 gates and an MSI has a complexity between 50 to 100 gates. There are two obvious directions in which the capability of semiconductor industry in producing LSI or VLSI could be utilized. One could design an extremely complex and powerful chip that could perform specialised tasks; or else one could make a general purpose powerful chip which could be programmed for various specialised tasks. In either case the volume of production has to be very high to make the chip a low cost device. The tooling cost of a chip manufacture (a one-time cost) is very high and has to be shared by the large volume. Cost of raw material is comparatively negligible.

The general purpose powerful chip is referred to as microprocessor and is programmable in the same sense as a general purpose computer. This ability to change or expand the functions performed by the microprocessor chip without needing any rewiring or redesigning of the system, makes it attractive in a large number of applications and consequently a high volume demand making it cost-effective.

For the circuit engineers microprocessors filled in the gap where the combination of volume, performance and cost could not be effectively met by SSI, MSI or custom made chips. For any logic design for which the chip count goes beyond 100 or so, it is advisable to use a microprocessor provided deterioration in speed is acceptable (Microprocessors on a single chip is a MOS device and so it is inherently slower. Bit-slice microprocessors are bi-polar devices but they are slightly more involved for being put into application). On the other hand for computer engineers and scientists, it merely brought in another generation in technology. To begin with it was vacuum tubes, then came transistors, SSI, MSI and now LSI. To them, the microprocessor put along with memory and input-output makes a microcomputer which is hardly any different from the conventional computers.

Thus, there are two distinct directions in which microprocessors have been used:

(a) Computer engineers and scientists utilized this to penetrate in those areas where conventional computer applications were not cost-effective. Thus, emerged the customization of computer systems to suit process control applications, a variety of intelligent terminals, desk-top computers and distributed data processing.

(b) The circuit designers have viewed this as a highly complex logic chip. Besides replacing random logic by a microprocessor, they have brought out a variety of new instruments and incorporated intelligence in many of the conventional instruments such as CRTs, multimeters etc. Thus, for circuit designers microprocessor has been a new challenge for new innovations.

Today, microprocessor finds application in calculators, elevators, traffic lights, washing machines, weighing machines, tractors, video games, music synthesis, air line reservation terminals, automobiles and innumerable scientific, medical and business gadgets. Primarily these applications could be categorised into following four broad categories:

(i) Computing;
(ii) Processing of phyiscal parameters;
(iii) Process-control; and
(iv) Communications.

These applications emerge because of two essential properties of the microprocessors—firstly, it is almost a replica of central processing unit of a conventional computer on a single chip; and secondly, the chip is very inexpensive, consumes very little space and power. The first property endows it enormous flexibility in design and experimentation; while the second property makes its use economically feasible even in low priced gadgets.

In this chapter, we will discuss some of the applications of microprocessors and programmable logic controllers for controlling power converters. The programmable logic controllers, known as

PLCs have found widespread application in substituting the conventional relay logic used in control of power converters in our country. Gradually, microprocessors are replacing PLCs to offer greater flexibility and versatility. First of all, some of the basic terminologies and definitions relevant to our discussions are presented. This is followed by discussions on fundamentals of PLCs and then of microprocessors and microcomputers. A comparison of various types of controllers are presented thereafter. The section following this discusses aspects concerning development of a microprocessor based system for our application. The chapter concludes with giving detailed application design examples.

11.2 SOME DEFINITIONS AND TERMINOLOGIES

Bit: In all digital systems information is represented in binary form i.e. it can have two values 'true' or 'false' which may be represented by '1' and '0' respectively. '1' or '0' is called a binary digit and 'bit' is a short form of 'binary digit'.

Word size: Word size of a microprocessor refers to the number of bits which can be processed at a time by the microprocessor. A 8-bit word is known as a 'byte' and a 4-bit word a 'nibble'. Intel 4004 has a word length of a nibble and that of Intel 8085 and 8086 are one byte and two bytes respectively.

Word size of a memory refers to the number of bits which can be read and/or written simultaneously into the memory. It depends upon the manner in which memory is organised.

Register: When a number of bits are stored in a volatile storage and these bits can be entered or accessed simultaneously or sequentially, it is called a 'register'. A register is used to hold the digital information temporarily. The length of a register is equal to the number of bits it can hold.

Arithmetic Logic Unit (ALU): ALU is a digital circuit which performs arithmetic operations such as addition, subtraction, complementation, ANDing, ORing etc. on the contents of specified registers. The answer is stored in a specified register.

Normally the operation is performed on arguments stored in a special register called 'accumulator' and a temporary register. The result is stored back in the accumulator.

Central Processing Unit (CPU): CPU of a microcomputer consists of ALU and associated digital circuitry to control the functions of ALU and other peripheral devices such as memory and input/output units.

Microprocessor and Microcomputer: CPU on a chip is called 'microprocessor'. A microprocessor when connected with memory, input and output means, timing circuits results in a microcomputer. Microcomputers may also be implemented on a single chip with limited resources.

Address: An address is the digital information by which a word in the memory or a physical input/output device can be referred. A digital information can be held in a register and a register holding address is called 'address register'. An address register of n bit length can address $2n$ different memory locations or input/output devices. Normally microcomputers have 16 address lines and so 2^{16} different 16-bit patterns (addresses) can get generated.

Memory: A memory element such as flip-flop, magnetic core, fused semiconductor junction can be used to store one bit of digital information. The memory elements such as flip-flops are 'volatile' in nature in the sense that the stored digital information stays only as long as they are connected to power supply. Magnetic cores, permanently fused semiconductor junctions are of non-volatile nature.

The memory elements when arranged in an array or matrix form, make a digital memory system in which digital information is stored. The digital memory system can be viewed as a set of pigeon-holes, having pigeon-hole addresses and each pigeon-hole storing one or several bits of digital information. For storing several bits of information i.e. a word at a single address, the memory matrix is grown in the third dimension by stacking one above another. The number of such planes define the word length of a memory system. Thus, functionally a memory system consisting of m words of n bits of word length, when supplied with an address, an n-bit information could be read from or written into the memory system. The block-diagram of a memory system is shown in Fig. 10.21. In order to refer to m words in address, we require b bits of address register such that $2^b \geqslant m$.

In some memory systems information could be written as well as read. These are referred to as a RWM (read-write memory) or RAM (random-access memory). Access-time of a memory system is defined as the time taken to fetch a word. If the access time is same for any word in the system, it is called as random access memory (RAM), however, by convention the term RAM has been used also to mean read-write memory. The early RAMs have primarily been of magnetic core elements which are of non-volatile type. Semiconductor RAMs are volatile and they are available in different sizes with bipolar and MOS technologies. If the semiconductor memory elements are permanently fused, the information can only be read from the memory system and no information can be written into it. These are called read only memory (ROM) systems. ROMs are used for storing custom-designed tailored information. When ROMs are programmable i.e. where the user could fuse the bit patterns into the memory system, it is called PROM (programmable read only memory). Once PROM has been programmed, it cannot be altered. However, some PROMs are erasable with ultraviolet light. Such a memory system is known as EPROM (erasable programmable read only memory). Electrically alterable read only memory (EAROM) are also available now.

It is interesting to note that the memory system can be used for implementing any combinatorial function. A combinatorial circuit with b inputs and n outputs can be replaced by a memory system consisting of 2^b words of n bits each by storing the entire truth table in the memory. Design of programmable logic arrays is primarily based on this concept.

ROMs and RAMs are fabricated on LSI chips. Their typical densities are 4096 to 65,536 bits (4K to 64K bits, 2^{10} bits is referred to as 1K in computer terminology) per chip. For example a 2K eight-bit ROM will have 2048 words of 8 bit length of nonvolatile nature. To address 2K words we need 11 bits. Thus, for example, for an address 10111010111 (11 bits) the data stored in ROM may be 11010001 (8 bits). In case of RAMs the bits can be organised in several possible ways to yield varying word lengths as per our requirements. For example, a 16K bits RAM chip can be organised as 16,384 by 1-bit words or 8,192 by 2-bits words or 4,096 by 4-bits words or 2,048 by 8-bits words or 1,024 by 16-bits words.

RAMs being volatile memory are normally used as 'scratchpad' memory for storage of temporary data, programs or results, whereas ROMs are used to store those programs or data-tables which are required to be available in the microcomputer all the time.

RAMs and ROMs are also called 'main-memory' for the microcomputer.

Register-bank: A set of registers (general purpose) numbering between 4 to 16 are normally available in CPU of a microcomputer for storage of temporary information (data or instruction). These registers are addressed in a similar way as memory locations. For example, 3-bit address is required to address a register in a register-bank of 8 registers.

Secondary storage or back up storage: Large storage space on magnetic devices such as magnetic tapes, disks, drums, floppy etc. which are of sequential or semirandom access type are called secondary storage or back up storage. These are normally used to hold large system programs or large data files.

Instruction set: A microprocessor is designed to execute a set of commands such as add two numbers (say ADD), subtract (SUB), complement a number, compare two numbers, jump to an instruction conditionally or unconditionally etc. These set of commands form the instruction set for the microprocessor. The power and suitability of a microprocessor for an application depends upon this instruction set (repertoire) and the time that the device takes to execute them.

Instruction format: The form in which an instruction is presented to the CPU is called 'instruction format'. A microprocessor may have a fixed instruction format or variable format depending upon the nature of instruction. An instruction may be of 1 byte, 2 bytes, 3 bytes or any multiple of bytes length. Every instruction has two components: an operation code part and an address part (for the operands). The operation code part denotes the command for the microprocessor and the address part gives the locations of operands. For some operations no operand is required (examples being complement or rotate the contents of accumulator, the reference to accumulator is implicit in the op-code). A typical instruction format is shown below:

Op-code	Addressing mode	Address(es)
	operand(s)	address(es)

Here the 'addressing mode' defines the manner in which the given address is converted into the effective operand address. Depending upon the nature of op-code, there may be two operand addresses or one operand address (the other operand being implied in the special register, accumulator) or no address at all in cases where instruction does not need an address or accumulator is the implied operand address. As an example, if the op-code field length is of 3 bits, there are 2^3 (= 8) instructions available; if the addressing mode is of 2 bits, there are 2^2 (= 4) different addressing modes possible, and say 11 bits are used for the address field then 2^{11} (= 2048) locations can be directly addressed. In case the microprocessor uses single operands, the length of the instruction would be $3 + 2 + 11 = 16$ bits i.e. 2 bytes.

Program: Program is a set of instructions which when executed in sequence by the microprocessor, results in the desired answer(s). The program is normally stored in the main memory of the microcomputer (known as stored program concept). For execution, the microprocessor 'fetches' each instruction in sequence (the sequence may be altered by the program itself) from the main memory to CPU and 'executes' it i.e. performs the operation as dictated by the instruction. Thus, the microprocessor is in 'fetch' mode or in 'execute' mode and it alternately goes into the fetch mode and execute mode till it comes to a 'halt' instruction.

Bus: A microprocessor is required to communicate with peripheral devices and memory. This communication is normally provided through a set of conductors called a 'bus'. Normally, a microcomputer has an 'address bus', a 'data bus' and a 'control bus'. The address bus is used to send address bits to memory or to the external device which is being addressed. The data bus is used for transferring information bits (data or instruction) to or from memory or external devices. Control bus is used to communicate control signals between the microprocessor and the other subsystems comprising the microcomputer.

Normally a microprocessor uses its own internal bus to communicate between ALU, control unit and register bank.

Buses can be unidirectional or bidirectional depending upon whether they allow communication in single direction or both the directions. Normally address bus is unidirectional and data-bus and control bus are bidirectional. Sometimes a single bidirectional bus is used for transferring data as well as address.

Clock: The different subsystems of a microcomputer needs synchronization of their operations for any meaningful communication. This is provided by the 'clock' or 'timing circuit'. A clock may be 'single-phase' or 'multi-phase' as shown in the Fig. 11.1. The clock signals are periodic in nature and in case of multiphase clock the multiple periodic signals are synchronised with each other.

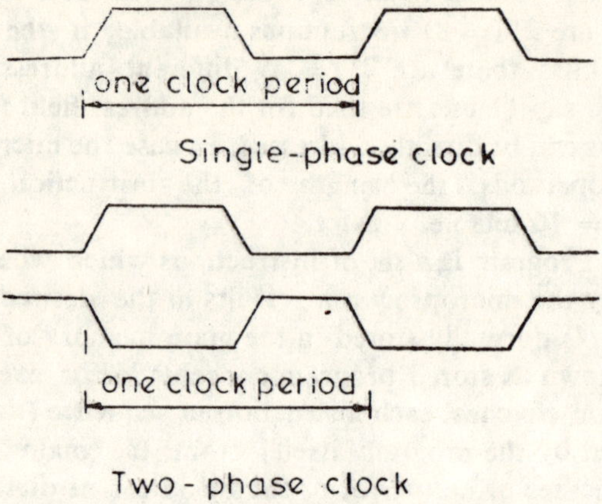

Fig. 11.1. Clock signal examples.

Interface: A circuit which matches the input/output requirements of two systems so that the two systems could communicate with each other meaningfully, is called an 'interface'. An interface may be a simple serial to parallel or parallel to serial bit converter, or may be programmable to act as an input/output ports (the place where the input/output devices are connected) of varying nature to suit application requirements. A wide variety of interface chips are commercially available. Intel 8251, 8273, 8255 or Motorola 6821 are some examples.

Number systems: This subsection presents basics of number system and their representation in microcomputer.

(a) REPRESENTATION OF NUMBERS

All digital systems process information which are represented in the binary form i.e. in the form of bits (binary digits). Bits can take values '0' or '1'. Such a form of representation is referred to as binary number system representation i.e. the number system with base as 2.

Consider decimal number system. The digits here can take values between 0 to 9 and the base of the number system is 10. For example a decimal number 142.25 can be written as

$$142.25 = 1\times10^2 + 4\times10^1 + 2\times10^0 + 2\times10^{-1} + 5\times10^{-2}$$

i.e. each place before the decimal point and also after the decimal point has a weight with which the digits are multiplied and then summed up to yield its value. The weights here are in powers of 10 and so the base is 10. In general a number N in a number system with a base (or radix) b is represented as follows:

$$N = (a_{i-1}, a_{i-2} \ldots a_0 a_{-1} \ldots a_{-j}) = a_{i-1} \times b^{i-1} + a_{i-2} \times b^{i-2} + \ldots + a_0 \times b^0 + a_{-1} \times b^{-1} + \ldots + a_{-j} \times b^{-j}$$

where b is the base or radix of the system, i is the number of integral digits (i.e. whole number) and j is the number of fractional digits.

A number may not have a fractional part or integral part. Such a representation system is called a 'weighted number system'.

Given below are some examples of numbers with different bases:

$$(1101001.101)_{\text{binary i.e. base}=2} = 1 \times 2^6 + 1 \times 2^5 + 0 \times 2^4 + 1 \times 2^3 + 0 \times 2^2 + 0 \times 2^1 + 1 \times 2^0 + 1 \times 2^{-1} + 0 \times 2^{-2} + 1 \times 2^{-3}$$

$$= \left(64 + 32 + 0 + 8 + 0 + 0 + 1 + \frac{1}{2} + \frac{1}{4} + \frac{1}{8}\right)_{\text{decimal}}$$

$$= (105.875)_{\text{decimal}}$$

$$(61.72)_{\text{octal i.e. base}=8} = 6 \times 8^1 + 1 \times 8^0 + 7 \times 8^{-1} + 2 \times 8^{-2}$$

$$= \left(48 + 1 + \frac{7}{8} + \frac{2}{64}\right)_{\text{decimal}}$$

$$= (49.3125)_{\text{decimal}}$$

$$(25.3)_{\text{base}=6} = 2 \times 6^1 + 5 \times 6^0 + 3 \times 6^{-1}$$

$$= \left(12 + 5 + \frac{3}{6}\right)_{\text{decimal}}$$

$$= (17.5)_{\text{decimal}}$$

$$(2F.C)_{\text{Hexadecimal i.e. base}=16} = 2 \times 16^1 + F \times 16^0 + C \times 16^{-1}$$

$$= \left(32 + 15 + \frac{12}{16}\right)_{\text{decimal}}$$

$$= (47.75)_{\text{decimal}}$$

In case of hexadecimal the digits can take values from 0 to 15, these are represented as 0 to 9 and then A to F in that order i.e. $A = 10$, $B = 11$, $C = 12$, $D = 13$, $E = 14$ and $F = 15$.

In computers numbers are always represented in binary form but octal, hexadecimal and decimal systems are used more conveniently by the users to represent information.

(b) Conversion of Numbers from Binary, Octal, Hexadecimal to Decimals

From the above examples it is obvious that such conversions are straight forward. Conversion from any number system to decimal is obtained by weighted sum of the digits where the weights depend upon the position of the digit in the number.

(c) Conversion of Decimal to Binary

The steps of the conversion are described below:

(i) separate out the integral part and fractional part of the decimal number;

(ii) divide the number by 2, record the remainder to the next position towards the most significant position (if it is the first remainder to be recorded then it is put at least significant position) in the resulting bit pattern;

(iii) in case the quotient obtained in step (ii) is zero, stop the division and the resulting bit pattern represents the conversion of the integral part of the number; now proceed to step (v) to convert the fractional part;

(iv) in case the quotient obtained in step (ii) is non-zero, make the number equal to quotient and go to step (ii);

(v) take the fractional part of the number and multiply it by 2;

(vi) store the integral part of the number in the next bit position towards the least significant bit position (in case it is the first bit being stored then it is stored at the most significant position) of the resulting bit pattern;

(vii) in case fractional part is zero, stop and the resulting bit pattern represents the conversion for the fractional part.

Else go to step (v).

This is explained in the following example:

decimal number to be converted to binary = 13.675

		remainder	
2	13	1	↑
2	6	0	
2	3	1	read along the direction of arrow
2	1	1	
	0		

$(13)_{decimal} \equiv (1\ 1\ 0\ 1)_{binary}$

most significant bit (the first 1); least significant bit (the last 1)

	Integral part	
$.675 \times 2 = 1.350$	1	
$.350 \times 2 = 0.750$	0	read along the direction of arrow
$.750 \times 2 = 1.500$	1	
$.500 \times 2 = 1.000$	1	↓

$(.675)_{decimal} \equiv (.\ 1\ 0\ 1\ 1)_{binary}$

binary point (the point); most significant bit (the first 1); least significant bit (the last 1)

$(13.675)_{decimal} \equiv (1101\ .\ 1011)_{binary}$

(d) Conversion from Decimal to Octal

It is similar to the above except that the operation of division and multiplication are done with respect to 8 instead of 2.

(e) Conversion from Decimal to Hexadecimal

It is also similar to the operation shown in (c) except that the operation of division and multiplication are done with respect to 16 instead of 2.

(f) CONVERSION FROM BINARY TO OCTAL:

The digits of the octal number system can take values between 0 to 7. These when represented in binary are form as shown in Table 11.1.

Table 11.1.

Octal digit	*Binary form*
0	000
1	001
2	010
3	011
4	100
5	101
6	110
7	111

Note that 3 bits are needed to represent digits of octal number system. Thus, the procedure for conversion of a binary number to octal is as follows:

(i) Make groups of 3 bits starting right to left for the integral part and left to right for the fractional part. Add as many 0s as required towards most significant bit position in case of integral part and towards least significant position in case of fractional part.

(ii) Convert each group of 3 bits to the corresponding octal digit using the above table which is direct decimal equivalent of the number.

Example:

1 101 001 001 . 100 1

001 101 001 001 . 100 100

$\equiv (1\ 5\ 1\ 1\ .\ 4\ 4)_{octal}$

(g) CONVERSION FROM BINARY TO HEXADECIMAL (HEX in short):

The digits of hexadecimal number system can take values between 0 -- 15 (or 0 to *F*).

Their corresponding binary equivalent are shown in Table 11.2 (4 bits are required to represent a hexadecimal digit).

Table 11.2.

Hex digit	*Binary equivalent*
0	0000
1	0001
2	0010
3	0011
4	0100
5	0101
6	0110
7	0111
8	1000
9	1001
A	1010
B	1011
C	1100
D	1101
E	1110
F	1111

The conversion procedure is similar to that of conversion from binary to octal except that a group of 4 bits is made instead of 3 in case of octal. This is illustrated below:

1 0   1 0 1 1   0 0 0 1   1 1 1

$$\underbrace{0\,0\,1\,0}\quad\underbrace{1\,0\,1\,1}\quad\underbrace{0\,0\,0\,1}\cdot\underbrace{1\,1\,1\,0}$$

$\equiv (2B1.E)_{Hex}$

(h) Conversion from Octal to Binary

It is done by straight forward substitution of the 3 bit binary representation for each octal digit.

Example:

$$(1511.44)_{octal} \equiv (001\ \ 101\ \ 001\ \ 001.100\ \ 100)_{binary}$$

(i) Conversion from Hex to Binary

This is also straight forward substitution of 4-bit binary equivalent for each hex digit.

Example:

$$(2B1.E)_{Hex} \equiv (0010\ \ 1011\ \ 0001.1110)_{binary}$$

It is obvious from the above illustrations that conversion of binary to octal and Hex and vice-versa are straightforward. Hex being the most compact form of representing long binary string, it is most popularly used for writing machine code programs or for display of output in case of microcomputers.

(j) Binary Coded Decimal (BCD or 8421 code) Representation

When every digit of a decimal number is given a binary representation, the resulting code is called 'Binary Coded Decimal' number. Since there are 10 digits (0–9) in decimal number 4 bits are required to represent them. The 4 bits give representation for numbers 0 to 15. The combinations for 10 to 15 are treated illegal in BCD. The Table 11.3 shows the BCD codes.

Table 11.3.

Decimal digit	*BCD representation*	
	weights	8421
0		0000
1		0001
2		0010
3		0011

Decimal digit	*BCD representation*
4	0100
5	0101
6	0110
7	0111
8	1000
9	1001
↑	1010
Illegal	1011
codes	1100
│	1101
│	1110
↓	1111

Given below are some examples of BCD and straight binary representation:

$$(973.625)_{decimal} \equiv (1001\ \ 0111\ \ 0011\ \ \cdot\ \ 0110\ \ 0010\ \ 0101)_{BCD}$$

$$\equiv (1111001101\ \ \cdot\ \ 101)_{binary}$$

$$(10)_{decimal} \equiv (0001\ \ 0000)_{BCD}$$

$$\equiv (1010)_{binary}$$

It is evident from the avove examples that BCD needs more number of bits to represent decimal information as compared to straight binary representation. However, conversion from decimal to BCD and vice-versa are extremely simple. Arithmetic in BCD can be performed the same way as in case of decimal numbers.

(k) Representation of Signed Numbers

A n-bit word in a microcomputer can hold positive numbers from 0 to 2^{n-1}. For example, for a 8-bit word (a byte) the minimum and maximum numbers would be as follows:

$$\text{minimum} : 00000000 \equiv (0)_{decimal}$$

$$\text{maximum} : 11111111 \equiv (255)_{decimal}$$

One of the ways to represent negative numbers is to use one of the bits (say left most bit) for the purposes of sign. For example, when this sign bit is 0, the number is treated as positive, if it is 1 then it is treated as negative. This method of representation is referred to as 'sign-magnitude' representation of signed numbers. Thus, for a 8-bit word the minimum and maximum in sign-magnitude form would be:

	S	magnitude	
minimum :	1	1111111	$\equiv (-127)_{decimal}$
maximum :	0	1111111	$\equiv (+127)_{decimal}$

For subtraction of two numbers, it is convenient to represent the negative number as 1's complement (i.e. bit by bit complementation) of the corresponding positive number. Thus, in one's complement for example, (– 116) will get represented as follows:

	S	magnitude (in 7 bits)
$(+116)_{decimal}$	$\equiv$ 0	1110100
$(-116)_{decimal}$	$\equiv$ 1	0001011 (in 1's complement form)

It is evident that in both sign-magnitude form and in one's complement form, the decimal number zero gets represented in two ways as + 0 or – 0. This double representation for 0 vanishes when 2's complement is used to represent negative numbers besides giving some further advantages in performing subtraction. 1's complement when added with 1 in least significant bit position (l.s.b.) results in the corresponding 2's complement form. Thus,

$$
\begin{array}{rll}
(-116)_{decimal} \equiv & 10001011 & \text{(in 1's complement form)} \\
& \quad\quad + 1 & \\
\hline
& 10001100 & \text{(in 2's complement form)} \\
\hline
\end{array}
$$

Table 11.4 shows sign-magnitude, one's complement and two's complement of numbers when the word size is assumed to be 4 bits and sign bit is at the left most position. It is evident that the range of numbers represented in 2's complement exceeds by 1 on the negative side.

When numbers are represented in two's complement form, addition of two signed numbers is performed by mere binary addition of the numbers and the carry bit is ignored. In case of subtraction, 2's complement of subtrahend is added to the minuend in a similar fashion. In case of sign manitude form, the sign bit, carry bit and the overflow bit from the magnitude part addition are used to form the sign and magnitude of result. For subtraction, 1's complement of subtrahend is used. Details of arithmetic operations may be found in any book on digital arithmetic.

Table 11.4.

Signed decimal number	*Sign-magnitude form*	*One's complement form*	*Two's complement form*
0	0000 (+ 0)	1111 (− 0)	0000
	1000 (− 0)	0000 (+ 0)	
+ 1	0001	0001	0001
+ 2	0010	0010	0010
+ 3	0011	0011	0011
+ 4	0100	0100	0100
+ 5	0101	0101	0101
+ 6	0110	0110	0110
+ 7	0111	0111	0111
− 1	1001	1110	1111
− 2	1010	1101	1110
− 3	1011	1100	1101
− 4	1100	1011	1100
− 5	1101	1010	1011
− 6	1110	1001	1010
− 7	1111	1000	1001
− 8	—	—	1000

The above definitions and explanations provide a background to the material that follows. The reader will find a number of additional terminologies explained in the text.

11.3 PROGRAMMABLE LOGIC CONTROLLERS (PLCs)

Traditionally, electromechanical relays were used for implementing control logic and interlocking systems. The coils and contacts of the relays are specifically configured as per the requirements.

This required large space and maintenance and alternations in the system logic have always been difficult to carry out. The relays have gradually been replaced by solid-state switches. This increased reliability and speed. However, any alteration in the system logic required efforts similar to that in relay implementation. The programmable logic controllers (PLCs) offer a solution to this. Here, the desired logic function or interlocking sequence is not hard-wired but can be programmed and stored in memory. The contents of the memory can be easily altered and that leads to the change in control logic or interlocking sequence.

11.3.1 Basic Configuration of a PLC

A programmable controller which handles binary inputs and binary outputs related by logic statements is termed programmable logic controller. Figure 11.2 shows the basic configuration of a typical PLC. It consists of input module, output module, logic processing unit (LPU or sometimes also called CPU) and memory.

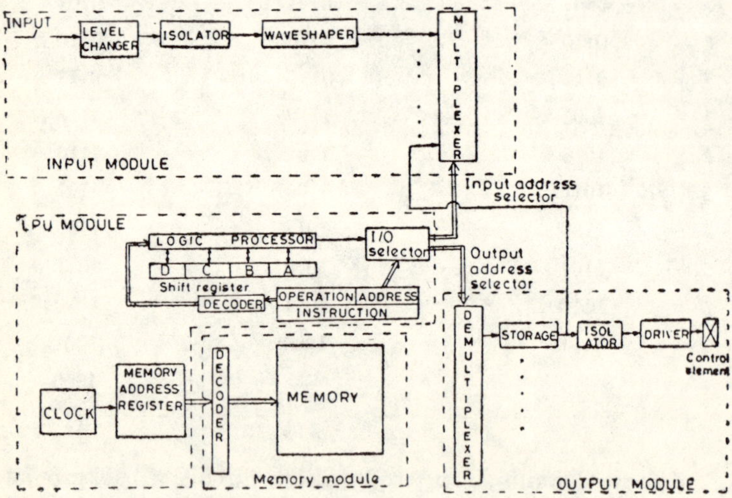

Fig. 11.2. Block schematic diagram of a PLC.

Most of the industrial control systems operate on inputs from binary elements such as limit switches, push buttons based on the settings of parameters such as temperature, pressure etc. The output of the system is again binary elements which cause opening or closing of valves, energisation or de-energisation of solenoids etc. The PLC checks the status of each input and performs logic as per the program stored in the memory and generates the required binary output. Scanning of input is done cyclically at a fast rate. The actual rate depends upon the number of input and output lines and the length of the program stored in the memory which constitute the control logic. The LPU fetches instruction from the memory and executes it in a sequential manner. An internal clock controls the sequential fetch of the instruction.

(a) Input Module

As shown in the Fig. 11.2 the input module consists of several inputs which are level changed (through zeners) and isolated (through optical isolators) to provide noise immunity. It is wave shaped (compatible to TTL or CMOS PLC) and then fed to one of the lines of the multiplexer. The line address of the multiplexer becomes the input address. When LPU feeds the address part of the instruction to the control input of the multiplexer, the output of the multiplexer show the status of the input line referred to in the instruction.

(b) Logic Processing Unit (LPU)

It consists of the following functional blocks:

(i) *Clock pulse generator and memory address register* (*or counter*)
It provides timing to the system. It increments the memory address register to enable sequential instruction fetch. At the end of the program, the memory address register is reset to its initial value and the process is repeated.

(ii) *Instruction fetch and decode logic*: It receives the content of the memory and decodes the operation to be performed. In case the operation is checking the input status or setting the output control element, the address part is fed to the control inputs of multiplexer or demultiplexer.

(iii) *Processing unit*: It consists of a shift register and logic circuits to perform logic operations on the bits of the shift register. To illustrate the process, let us consider a hypothetical PLC which has a 4-bit shift register (A, B, C and D bits) and supports instructions to perform the operations as logical AND, OR, NOT, READ input status and SET output control element. The manner in which the operations are performed are explained below:

READ XXX — read input status from the input line address XXX and push it into the shift register. It means that *A* bit of shift register will hold the input status read and the earlier content of *A* moves to *B*, that of *B* moves to *C* and of *C* moves to *D*.

AND — the logical AND of the bit *A* and bit *B* is stored back in bit *A* position. Content of *C* moves into *B* and that of *D* moves into *C*.

OR — It performs the logical OR and the rest is same as in case of AND instruction.

NOT — Store the complement of bit *A* back into *A*. The bits *B*, *C* and *D* are not disturbed.

SET XXX — Sets the output elements as per the output line address XXX to the value of the *A*-bit. Thus, if *A*-bit is '1' it may mean close the valve and a '0' as open the valve.

Some of the PLCs also provide monitoring facilities to test and debug.

Further details will be clear from the illustrative example given later.

(c) *Memory module*: It normally consists of a EPROM or EAPROM (electrically alterable PROM) to enable the user to reprogram the memory for additions or modifications. The memory capacity depends upon the specific application and may vary between 1 K to 16 K bytes.

(d) *Output module*: It consists of a demultiplexer which routes its input (from A–bit) to the appropriate output line control element. On receiving the SET command the LPU directs the address part to the control input of the demultiplexer. The output is to be stored so that its status could be checked. Optical isolator and driver is used to actuate the control. Sometimes a preset delay is also introduced in the output path.

11.3.2 Programming a PLC

Programming for the PLC is done either by using Boolean

equations representing the control logic or by using the conventional relay ladder diagrams. The detailed programming procedures are available in manufacture's manual.

As an illustration let us consider the relay diagram given in Fig. 11.3. The corresponding Boolean equations for the output are as given below:

$$\text{Out } 21 = I_n1 \ . \ I_n2$$

$$\text{Out } 22 = I_n1 \ . \ (\overline{I_n3} + I_n4)$$

$$\text{Out } 23 = ((\overline{I_n4} \ . \ I_n5) + \overline{I_n2}) \ . \ \overline{I_n6}$$

Using the hypothetical PLC instructions as outlined earlier, let us develop a program which performs the logic given in Fig. 11.3. Let us assume that the instructions are each of one byte length. For simplicity let us assume that three bits are used to represent

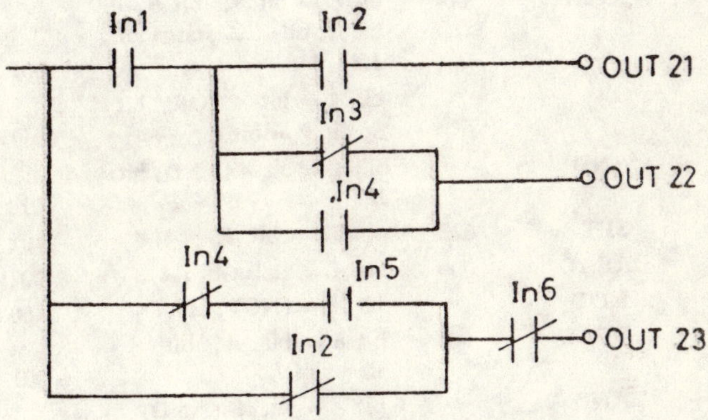

Fig. 11.3. An example relay diagram.

the operation of the instruction and the rest of five bits are used for input/output addressing. The bit assignment for the operations are as follows:

Operation	*Bit-configuration*
READ	001
AND	010
OR	011
NOT	100
SET	101

Memory location	*Instruction operation*	*Input/ output address*	*Comments*	*Actual memory map*
001	READ	001	Bit A = status of I_n1;	001 00001
002	READ	002	bit B = bit A; bit A = status of I_n2;	001 00010
003	AND		bit A = bit A (AND) bit B;	010 00000
004	SET	021	out 21 = bit A;	101 10101
005	READ	001	bit A = status of I_n1;	001 00001
006	READ	003	bit B = bit A; bit A = status of I_n3;	001 00011
007	NOT		bit A = (NOT) bit A;	100 00000
008	READ	004	bit C = bit B; bit B = bit A; bit A = status of I_n4;	001 00100
009	OR		bit A = bit A (OR) bit B; bit B = bit C;	011 00000
010	AND		bit A = bit A (AND) bit B;	010 00000
011	SET	022	out 22 = bit A;	101 10110
012	READ	004	bit A = status of I_n4	001 00100
013	NOT		bit A = (NOT) bit A	100 00000
014	READ	005	bit B = bit A; bit A = status of I_n5;	001 00101
015	AND		bit A = bit A (AND) bit B;	010 00000
016	READ	002	bit B = bit A; bit A = status of I_n2;	001 00010
017	NOT		bit A = (NOT) bit A;	100 00000
018	OR		bit A = bit A (OR) bit B;	011 00000
019	READ	006	bit B = bit A; bit A = status of I_n6;	001 00110
020	NOT		bit A = (NOT) bit A;	100 00000
021	AND		bit A = bit A (AND) bit B;	010 00000
022	SET	023	Out 23 = bit A	101 10111

Fig. 11.4. PLC program for control logic of Fig. 11.3.

The Fig. 11.4 illustrates the program, meaning of each of the instructions and the corresponding memory map.

To see how the program works, let us take a particular instance of input status and follow the execution trace of the program. This is illustrated below.

Initial status of inputs:

$$I_n1 = 1;\ I_n2 = 0;\ I_n3 = 1;\ I_n4 = 1;\ I_n5 = 0;\ I_n6 = 0$$

Execution of instruction at location	Status of shift register A, B, C, D	Comments
001	1000	
002	0100	
003	0000	
004	0000	Out 21 = 0
005	1000	
006	1100	
007	0100	
008	1010	
009	1100	
010	1000	
011	1000	Out 22 = 1
012	1000	
013	0000	
014	0000	
015	0000	
016	0000	
017	1000	
018	1000	
019	0100	
020	1100	
021	1000	
022	1000	Out 23 = 1

11.3.3 Program Modification

The major advantage of PLC over hard-wired logic is the flexibility that it offers in modifications and updating etc. Consider the example of Fig. 11.3 and let us say it is desired to modify the control logic to as shown in Fig. 11.5 (a).

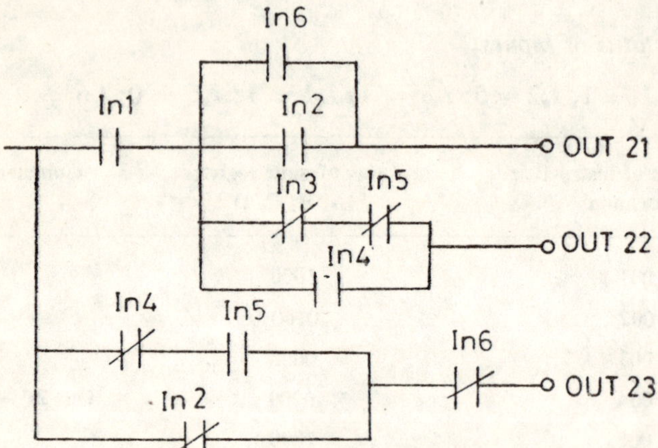

Fig. 11.5(a). Modified control logic.

The corresponding modified Boolean equations are as follows:

$$\text{Out } 21 = I_n1 \;.\; (I_n6 + I_n2)$$

$$\text{Out } 22 = I_n1 \;.\; (\overline{I_n3}.\overline{I_n5} + I_n4)$$

$$\text{Out } 23 = (\overline{I_n4} \;.\; I_n5 + \overline{I_n2}) \;.\; \overline{I_n6}$$

Fig. 11.5 (b) gives the corrresponding modified program. The steps required for program modification to achieve the altered logic are as follows:

1. Copy the contents of EPROM from locations 001 to 002 into a RAM.

2. Introduce the following instructions in locations 003 to 004 into the RAM

READ 006
OR

3. Copy the contents of EPROM from location 003 to 007 into the RAM locations 005 to 009.

4. Introduce the following instructions in locations 010 to 012 into the RAM

Memory location	*Instruction operation*	*Input/output address*
001	READ	001
002	READ	002
003	READ	006
004	OR	
005	AND	
006	SET	021
007	READ	001
008	READ	003
009	NOT	
010	READ	005
011	NOT	
012	AND	
013	READ	004
014	OR	
015	AND	
016	SET	022
017	READ	004
018	NOT	
019	READ	005
020	AND	
021	READ	002
022	NOT	
023	OR	
024	READ	006
025	NOT	
026	AND	
027	SET	023

Fig. 11.5 (b). Program corresponding to Fig. 11.5 (a).

READ 005

NOT

AND

5. Copy the rest of EPROM program into the RAM starting with the location 013.

6. Copy the contents of RAM onto the EPROM

7. The system starts running as per the modified control logic.

It should be apparent from the above discussion that an on-line reprogramming or modification is also feasible. In such a case the contents of EPROM are copied temporarily onto a RAM and the system program fetch and execute continues on the basis of the RAM contents. In the meantime EPROM contents are modified as desired. When the modification is over, the system is switched back to the EPROM. The programming panel allows such switchings.

The program entry consists of following steps:

(i) Plug the programming panel into the controller.

(ii) Load the program line by line through push button.

(iii) Check out and debug. The programming panel has a display of LED or CRT showing any individual circuit at a glance. The operator may make alterations and verify their effect.

(iv) Once the program is operational, transfer it to EPROM or EAPROM.

11.4 MICROPROCESSOR AND MICROCOMPUTER ARCHITECTURE

The actual layout and configuration of a microcomputer depends upon the specific application for which the system is being used and also upon the type of constituent microprocessor. A basic functional block diagram of a typical microcomputer is shown in Fig. 11.6. The microprocessor part is shown enclosed within dotted lines. The double lines indicate multiple-bit paths whereas single lines indicate control information or single bit data paths. A microprocessor when interfaced with appropriate input/output device and memory modules forms a microcomputer system. Sometimes the clock and timing circuit is also kept outside the microprocessor chip while some need only a crystal to be inserted to provide for the clock.

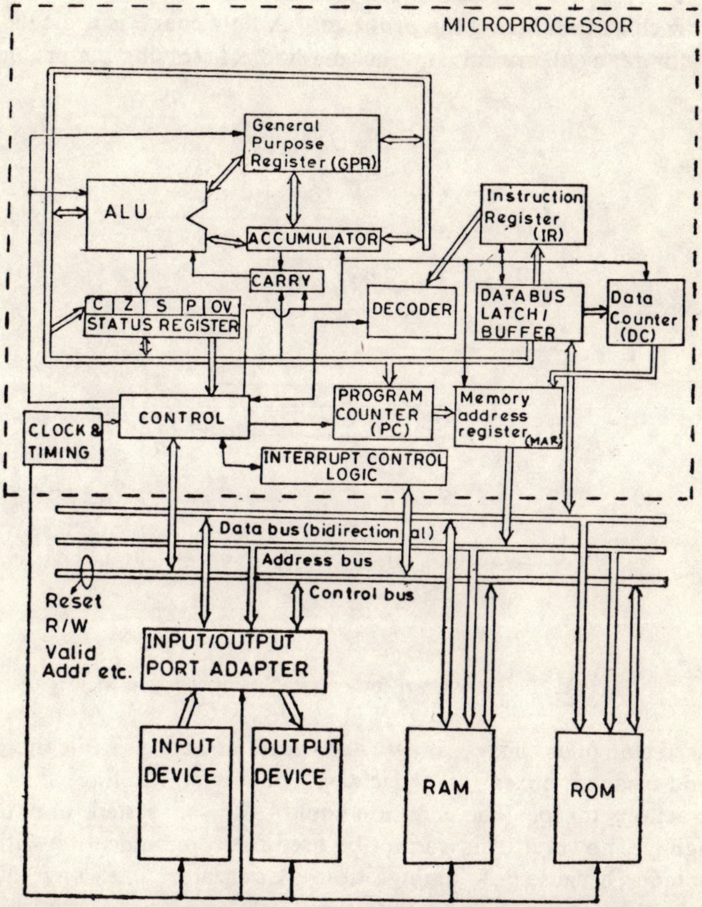

Fig. 11.6. Basic functional diagram of a typical microcomputer.

The program characterising the behaviour and role of microprocessor is stored in the main memory which also holds data. The microprocessor fetches instruction from the program memory and executes it. The instructions are fetched sequentially unless the sequence is altered by the program itself or it comes to a halt.

11.4.1 Working of a Typical Microcomputer

In order to examine working of a microcomputer, let us consider as an example a simple problem of counting up to ten. The counting restarts when the count reaches ten. Figure 11.7 gives a flow chart for solving the problem. A flow chart is a means of depicting an algorithm or the method of solving the problem.

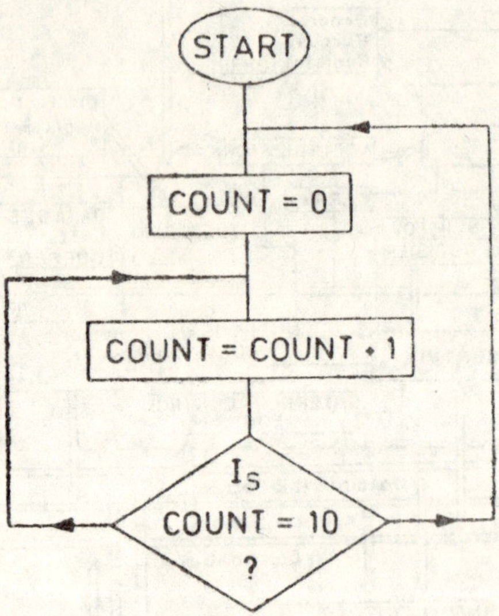

Fig. 11.7. Flow-chart for repeatedly counting up to ten.

The rectangular boxes show statements or actions while the diamond shaped boxes are decision boxes controlling flow of the actions taken. The solution could also be stated in natural language, however, this cannot be used for communicating with a machine because of ambiguities. Computer like any other machine is inanimate and needs precise statement and sequencing of commands which it can interpret and execute. A number of programming languages which are formal, precise and unambiguous have been developed to suit various application environments. For example for scientific computations FORTRAN (stands for FORmula TRANslation), PASCAL, ALGOL are used. For commercial applications COBOL (commercial business oriented language), for string processing SNOBOL, for list processing LISP etc. are popular. These programming languages are called higher level languages and are independent of the microprocessor or the computer. Converting a flow chart into a program in a higher level language is somewhat straightforward. The flow chart of Fig. 11.7 when written in FORTRAN is as given in Table 11.6.

Table 11.6. Fortran Program to Repeatedly Count up to Ten

```
102   ICOUNT = 0
101   ICOUNT = ICOUNT + 1
      IF (ICOUNT - 10) 101, 102, 102
```

However, the microprocessor or a computer cannot execute the statements in a higher level language directly. Normally a processor (a microprocessor or CPU of a computer) can execute only a set of basic instructions like clear a register, add to a register, complement contents of a register, subtract from a register, shift contents of a register, branch, stop etc. These instructions have to be presented in binary form. Such a language for programming is called machine language. Every microprocessor or CPU has its own machine language. Programming in machine language is difficult because one has to deal with numbers. Development, diagnosis or modification of a machine language program with say more than 100 statements is very cumbersome. Therefore, for convenience, a symbolic name is assigned to instructions and its components (operation code, address etc.). Normally the symbolic names for the operations are defined by the manufacturer. A programming language which uses symbolic names for operations and addresses along with some other structures to help programming at the machine level, is referred to as *assembly language.* Table 11.7 shows assembly language program for the counter example using Intel's 8085 microprocessor.

Table 11.7. 8085 Assembly Language Program for Counter Example

Label	*Instruction*	*Comments*
AGAIN	: MVI A, 0	; Move immediate 0 to register A
CONT	: INR A	; Increment register A
	CPI 10	; Compare immediate register
		; A with 10 and set the
		; flag accordingly
	: JC CONT	; Jump on carry flag to label CONT
		; (carry flag is set to 1 when contents A is less than 10)
	IMP AGAIN	; Jump unconditionally to label AGAIN

The three columns of an assembly language program are 'label', 'instruction' and 'comments'. Label is like a statement number in Fortran and is separated from instruction by a delimiter colon (:). An instruction needs to have a label only if there is another instruction in the program that refers to it. The comments help us in understanding the program and has no effect on the program. The comments are separated from the rest through a delimiter semicolon (;).

The first instruction corresponds to COUNT = 0 in Fortran. Instead of a variable COUNT here a register A (which is called accumulator) is being used. The next instruction corresponds to COUNT = COUNT + 1 in Fortran. The last three instructions correspond to the last statement of Fortran. The instruction CPI 10 causes the contents of register A to be compared with 10. In case contents of A is less than 10, the 'carry' flag is set to 1 (true). JC CONT causes jump to the instruction labelled as 'CONT' if the 'carry' flag is set. The program comes to the last instruction only when the 'carry' flag is not set and causes unconditional jump to a statement labelled 'AGAIN'. Thus, the labels CONT and AGAIN correspond to the statement numbers 102 and 101 respectively.

A program which converts an assembly language program into machine language program is called 'assembler'. Similarly the program which translates a higher level language program into machine language is called 'compiler'. Such programs are known as systems software.

Now, let us see how a machine language program looks like and how a microprocessor executes it. Table 11.8 shows 8085 machine language program corresponding to the assembly language program of Table 11.7 where starting address of the program is assumed to be $(0400)_{Hex}$. The numbers have been shown in the form of Hex codes (groupings of 4 bits) to make it compact. It should be noted that the word size of 8085 is 8 bits (i.e. 2 Hex digits) and there are 16 bits (i.e. 4 Hex digits) for addressing (i.e. 2^{16} directly addressable memory space).

It is evident from Table 11.8 that all instructions are not of equal length, some are single byte, some 2-byte, and some 3-bytes. This is typical of all microprocessors. Here the instruction INR A is of single byte, the instructions MVI A, 0 and CPI 10 are of two bytes where the first byte contains the opcode and the second byte

Table 11.8. 8085 Machine Language Program Corresponding to the Assembly Language Program of Table 11.7.

Assembly Language		*Memory location in HEX*	*Memory contents (in HEX)*	*(in Binary)*	*Nature of information*
AGAIN	: MVI A, 0	0400	3E	00111110	opcode
		0401	00	00000000	data
CONT	: INR A	0402	3C	00111100	opcode
	CPI 10	0403	FE	11111110	opcode
		0404	0A	00001010	data
	JC CONT	0405	DA	11011010	opcode
		0406	02	00000010	address
		0407	04	00000100	
	JMP AGAIN	0408	C3	11000011	opcode
		0409	00	00000000	address
		040A	04	00001000	

holds the data which is immediately required for executing the instruction, the instructions JC CONT and JMP AGAIN are of 3–bytes length, where the first byte is opcode and the second and third bytes give the lower order and higher order bytes of the referred address respectively. It should be noted that the machine language program is stored consecutively in the memory starting at some location. If the starting address is changed then some of the contents of the memory have also to be changed (e.g. contents of locations 0406, 0407, 0409 and 040A which hold addresses). A system program which does the job of storing a machine language into the memory appropriately is called 'loader'.

Now, let us consider that the program has been loaded into the memory and let us examine how a microprocessor (8085) executes it. Figure 11.8 gives a simplified block diagram of 8085. Figure 11.9 shows status of different registers during the fetch and execute cycles when the counter program is being executed. The program counter (PC) is used to keep track of the sequencing of instructions. To begin with PC is loaded with address of the first instruction which is 0400. Now the instruction is to be

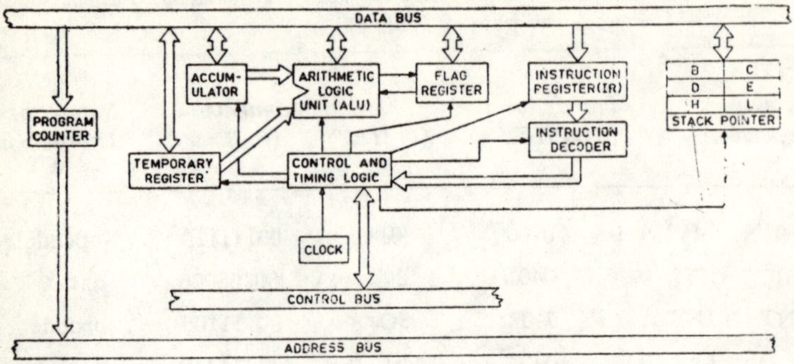

Fig. 11.8. 8085 Simplified block diagram.

fetched. The PC transfers its contents to address bus and the control unit issues a 'read' signal through the control bus. The memory module decodes the address of the address bus, retrieves its contents (which is 3E in this case), places it on the data bus and then signals to the microprocessor through the control bus that the information asked for is ready on the data bus. Figure 11.10 illustrates a typical timing diagram. The control unit of the microprocessor now transfers the contents of data bus to the instruction register because it knows that it is in fetch mode and in fetch mode the information brought in is an instruction. The instruction (which is 3E) is decoded and the control unit now looks for the information that is contained in the next byte as it knows now that the instruction is 'move immediate to accumulator' and the next byte contains the data. It increments the PC and a memory read operation is performed. However, this time the information on the data bus is not routed to instruction register but to the accumulator (A register) because it is in execute mode and the nature of the instruction being executed directs the next byte to be transferred to the accumulator. This completes the execution of first instruction. Now the control unit would fetch the next instruction. It increments PC and initiates memory read operation. The contents of data bus is transferred to IR which is 3C (Increment accumulator). The instruction is decoded and the control unit executes this instruction by incrementing the contents of the accumulator. Next, again the control unit increments PC to fetch next instruction. The memory read operation is initiated and the contents of the data bus is transferred

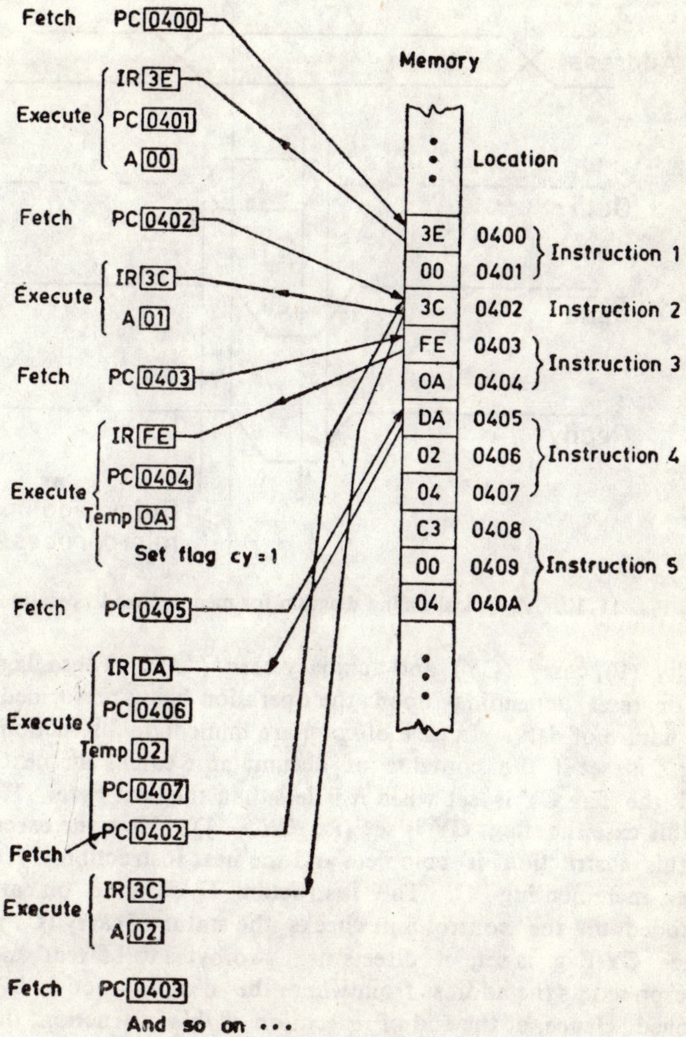

Fig. 11.9. Fetch and execute cycles and register status in execution of counter program.

to IR which is FE (compare immediate). The instruction is decoded, the next byte (0A) is read and transferred to the temporary register. Now the control unit issues a signal to ALU to compare the contents of accumulator and temporary register and set the flags accordingly. 8085 has five flags zero (Z), sign (S),

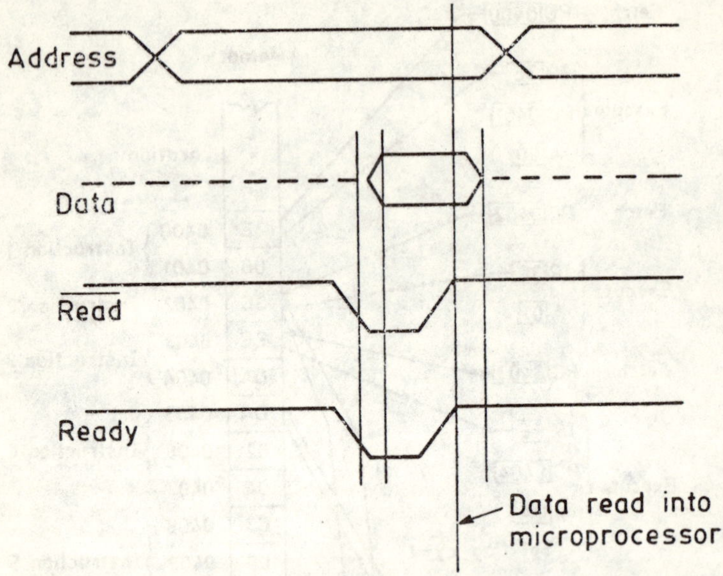

Fig. 11.10. A typical timing diagram for memory read operation.

parity (P), carry (CY) and auxiliary carry (AC). These flags are set or reset depending upon the operation being performed and the nature of data. In case of compare immediate instruction, the flag Z is set if the contents of accumulator equals the next byte and the flag CY is set when it is less than the next byte. Hence, in this case the flag CY is set (i.e. CY = 1). Now the execution of this instruction is complete and the next instruction is fetched after incrementing PC. The instruction DA (jump on carry) is decoded and the control unit checks the status of carry (CY) flag. Since CY flag is set, it directs next two bytes to be read and be interpreted as the address from where the next instruction is to be fetched. Hence, at the end of execution of this instruction, the PC gets the value as 0402. Now, the next instruction fetched is from location 0402 and process repeats. The instruction 5 of Fig. 11.9 will get executed only when CY flag is not set in instruction 3 i.e. when the contents of accumulator is not less than the second byte which is 10 in decimal.

It should be noted that the instructions and data are stored in the same memory and are intermixed. It is programmer's responsibility to ensure valid sequence of opcode and data. The microprocessor distinguishes them only by context i.e. whether it is in fetch mode or execute mode and knowing the nature of opcode. The system will go completely out of control in case data is misinterpreted for opcode and vice-versa.

The timing and control circuits unit is like a processor within the processor whose task is to sequence a set of basic actions (called micro-instructions) to realise execution of a machine language instruction. Typical microinstructions are transfer from one register to another, register to bus or bus to register, increment, complement, shift, rotate etc. A ROM in the microprocessor chip contains a program in terms of microinstructions (called microprogram) which contains what exactly is required to be done for executing each machine language instruction. For example, for the instruction MVI A(the opcode 3E), the following sequence of microinstructions may be required:

(a) Increment PC
(b) Transfer PC to address bus
(c) Set 'read' signal
(d) If 'ready' signal is there then transfer data bus contents to accumulator else wait for 'ready' signal
(e) Increment PC
(f) Branch to the microinstruction which corresponds to the beginning of fetch cycle.

The above six steps constitute the microprogram corresponding to the opcode 3E. Thus, when the instruction decoder decodes the opcode as 3E, the control unit branches to that ROM location (remember that this ROM is within the microprocessor chip) where the microprogram corresponding to opcode 3E is stored. Writing microprograms is very much distinct from writing programs to be executed by the microprocessor. The microprogramming is done by the manufacturers themselves and cannot be altered by the user. Such programs which are stored in ROM and are not visible to the user, are known as 'firmware' The 'hardware' refers to the physical devices and the circuits which supports the microinstructions. The firmware characterises the machine and defines a set of machine language to be used by the programmers. The 'software' refers to those programs which helps the user in programming and its execution, such as assemblers, loaders, monitors, compilers, operating systems etc.

Sometimes the manufacturer also gives facility to alter the

microprogram (redefine firmware): such microprocessors are called microprogrammable microprocessors. It is possible to define own opcodes in tune with the requirements of an application with the help of such microprocessors. Bitslice microprocessors are examples of this type.

In the foregoing discussions, we have explained the working of a typical microprocessor with the help of a simple example and simplified 8085 system. A few new terminologies have also been explained. All other microprocessors function in a similar fashion. The differences primarily lie in word size, addressing mechanisms, the way input/output is handled, interrupts etc. Details of these aspects are discussed later in this section. Aspects concerning development of a microprocessor based system and interfacing memory modules and peripherals are discussed in Section 11.6.

11.4.2 Input/Output in a Microcomputer

Depending upon the application a variety of input/output devices with varying speeds of data transfer are employed in practice. Switches, keyboard, analog signals, pulse trains, paper-tape, card, magnetic tape, disk, drum, floppies, cassettes etc. form some of the input devices. The output devices may be a teletype, printer, display, CRT, plotter, recorders, paper tape, card, magnetic tape, disk, drum, floppies, cassettes etc.

There are three broad categories of the nature of input/output in microcomputers. These are:

(a) Program controlled I/O
(b) Interrupt controlled I/O
(c) Direct Memory Access (DMA) controlled I/O

Under the program controlled input/output, the instructions for input/output are explicitly given in the program and the processor initiates, waits and supervises the transfer. Program controlled I/O is also of two types: memory mapped I/O and explicit I/O instructions. In memory mapped I/O, the I/O port is thought of as a memory location. The address decoder circuit distinguishes the memory from I/O ports. For example, in 8085 the total addressable space of 64K may be divided into two parts, 32K of memory (ROM and RAM) and 32K of I/O ports. This means that 32,768 I/O devices can be connected and addressed. However, the memory space has been reduced by the same amount. The greatest advantage of memory mapped I/O is that for all memory reference instructions such as load accumulator, store accumulator etc., the I/O as well as the operation is performed by the same instruction. For the programmer this acts just as another memory location with only difference in the speed of transfer.

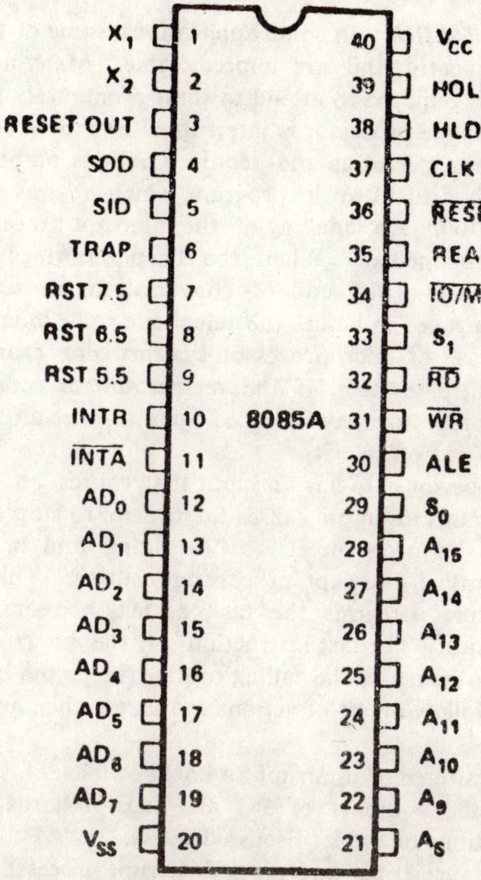

Fig. 11.11. 8085 Pin diagram.

A few microprocessors use a separate control line to specify that the address is for either I/O or memory. In 8085 this is called $IO/\overline{M}$ (pin 34 in Fig. 11.11). During memory transfer $IO/\overline{M}$ is low and during execution of I/O instructions IN and OUT, $IO/\overline{M}$ is high. This way the decoding circuit easily distinguishes memory and I/O and the entire addressable space is available for memory. It should be noted that the instructions IN and OUT are 2-byte instructions and 256 (2^8) devices can be addressed. However, memory mapped I/O instructions are 3-byte instructions and so require larger program memory space.

Interrupt Controlled I/O: In some applications, some of the events occur very frequently and are unpredictable. Many a time the microprocessor is required to attend to these events very promptly. In such situations the processor is interrupted by the event (at I/O), the microprocessor performs the required actions on being interrupted and finally returns to the program which it was executing before interruption. An analogy of the interrupt process may be given with a telephone call. When the telephone rings we attend to the telephone and at the end of conversation we resume the work we were engaged in before the telephone call. Similar situations arise in case of microprocessor systems. For example in a temperature control system, if the temperature exceeds a certain limit, the microprocessor may be called upon to raise an alarm and actuate relays to put off heaters.

The microprocessor chip has an input that causes an interrupt. A high signal on this input pin causes the control to stop the microprocessor from the program it was executing and branch to a special routine called interrupt processing routine. The interrupt processing routine performs the task as may be required by the interrupting device. The last instruction of the interrupt processing routine is to return to the calling routine (i.e. to the interrupted program). The following set of actions are taken when an interrupt occurs.

(i) Find the source of interrupt and acknowledge it.

(ii) Record the contents of PC and flags or status register in predefined locations or stack (discussed later).

(iii) Find the starting address of interrupt processing routine. Load PC with this address and enter in fetch mode.

At the end of interrupt processing routine the following actions are taken:

(i) Restore the flags or status register and PC from the prespecified locations or stack.

(ii) Enter into fetch mode and continue.

A processor may also be interrupted when it is executing an

interrupt processing routine by another device. Now priorities have to be assigned to different interrupts. The situation is analogous to a business executive having several telephone connections, some may be hot lines, some internal and each has assigned priority in terms of its importance. In such a case, the microprocessor checks if the incoming interrupt is of higher priority and if so, exactly similar steps are followed as in case of first interrupt. However, it is important that the status and return addresses be stored in stack like structure (discussed later).

In case of 8085 there are two groups of interrupt. One group of interrupt is caused by applying signals to individual pins, these are TRAP, RST5.5, 6.5 and 7.5 (pin numbers 6, 9, 8 and 7 respectively). The second group is controlled by INTR and INTA signals (pin numbers 10 and 11 respectively). For the first group the interrupt processing routine's address is defined by the manufacturer. These addresses are as given below:

Interrupt pin	Interrupt processing routine address in Hex
TRAP	0024
RST5.5	002C
RST6.5	0034
RST7.5	003C

The RST5.5 and 6.5 interrupts respond to a high level where RST7.5 responds only to a positive edge. The TRAP responds to a high level but does not respond a second time unless the level goes low and then to high again.

INTR is used as a general purpose interrupt. It is sampled only during the next to the last clock cycle of the instruction. If it is active, the PC will not be incremented and an interrupt acknowledge INTA will be issued. A RESTART or CALL instruction may be inserted during this cycle to branch to interrupt servicing routine. The INTR is enabled and disabled by software. It is disabled by reset and immediately after an interrupt is accepted.

TRAP has the highest priority followed by RST 7.5, 6.5 and 5.5 in that order. INTR has the lowest priority.

Direct Memory Access Controlled I/O: DMA is used for transferring data between memory and a peripheral device without help of the microprocessor. DMA data transfer rates are much higher than programmed transfer. DMA is often used by CRT and disk controllers to increase data transfer rate.

Microprocessor has a pin which when held high puts the processor in a 'hold' mode. The buses of the micro-computer are put under the control of DMA controller and data transfer takes place without involving microprocessor. 8085 has a HOLD input (pin number 39), a high on which indicates that DMA transfer is requested. The processor at the end of current machine cycle stops and places a high signal on hold acknowledge, HLDA output (pin number 38). The address, data and control buses of the microprocessor are put on high impedance and the peripheral device takes control of these buses to effect transfer. The buses must be tri-state buffers and a separate DMA controller chip is required to take control of the buses. At the end of the DMA transfer, the DMA controller pulls the HOLD line low and the microprocessor resumes its operation.

11.4.3 Pin Configurations and Buses

Typically a microprocessor has 40 pins (some recent ones have 64 pins). Figure 11.11 shows pin configuration for 8085. Sixteen pins are needed for addressing ($A_0 - A_{15}$) which forms the address bus. Address bus is unidirectional. All data leave and enter the processor via eight-bidirectional signal pins ($D_0 - D_7$). These consume 24 pins. However 8085 uses overlapped data and address pins. The lower order 8 address pins are bidirectional and are used for data as well as address (designated as $AD_0 - AD_7$). One additional line ALE (pin No. 30) for address latch enable is used to latch the lower order address bits. After the address has been latched the pins $AD_0 - AD_7$ act as data bus. Thus 8085 saves 7 pins. Two pins are used for read and write ($\overline{RD}$ and $\overline{WR}$ in 8085). 2 to 3 pins are used for power supply. 8085 uses +5 (V_{CC}) and ground (V_{SS}) for supply. One or two pins are used for clock purposes. 8085 uses X_1 and X_2 pins for connecting a crystal or R–C circuit. The rest of the pins are used for status and control of states to facilitate outside world communication with the microprocessor. These form the control bus. For 8085 these pins are HOLD, HOLDA, TRAP, RST 7.5, RST 6.5, RST 5.5, INTR,

$\overline{INTA}$, $IO/\overline{M}$, READY, CLK (OUT), RESET (OUT), $\overline{RESET\ IN}$, SOD, SID, S_0 and S_1. Many of these pins have been referred and discussed earlier. READY line is kept low during the read and write operation and the CPU waits for the READY to go high. CLK (OUT) gives twice the X_1, X_2 input period. RESET sets the PC to zero and resets the interrupt enable and HLDA flip flops. RESET (OUT) indicates that CPU is being reset and can be used as a system reset. SOD and SID are serial output and serial input data lines respectively. S_0 and S_1 are two special status outputs which provide additional information about the machine cycle in progress. It also indicates whether the processor is in HALT state. The states are as given under:

S_1	S_0	
0	0	HALT
0	1	WRITE
1	0	READ
1	1	FETCH

11.4.4 Status Flags

A CPU has a set of flip-flops which are automatically set or reset to reflect the results of ALU operations. These are called status flags. There are five status flags which are important:

(a) Carry status (CY), Auxiliary carry status (AC)
(b) Zero status (Z)
(c) Sign status (S)
(d) Overflow status (OV)
(e) Parity status (P)

Carry status: In order to perform multibyte arithmetic, any carry out of the high order bit of two data words must be recorded in the carry status, so that it may be propagated into higher order memory words. The carry status is also useful when performing multiword shift operations and comparison. 8085 has carry status flag and auxiliary carry status flag (AC). AC is set when an instruction causes a carry out of bit 3 into bit 4 of the result. AC is primarily used in DAA (Decimal Adjust Accumulator) instruction.

Zero status: This is set to 1 when a data manipulation operation generates a zero result; the flag will be reset to 0 otherwise. 8085 provides for this status.

Sign status: It is the contents of the sign bit (or its complement). The sign bit is always the highest order bit of any number, single word or multiword. This is also provided in 8085.

Overflow status: Overflow status when set to 1 indicates that the answer overflowed the answer space and is therefore wrong.

Parity status: This flag is set to 1 each time a data transfer operation detects a data byte with the wrong parity (odd or even). 8085 sets the parity status to module 2 sum of the bits of the result of the operation. Thus, it sets to 0 for even parity and to 1 for odd parity.

11.4.5 Addressing Schemes

The instructions that we considered earlier consisted of an opcode and one address, and the result of the operation was stored in register called accumulator. This is an example of single address instruction. However, there are a number of ways in which the operand addresses are specified. These are called addressing modes of an instruction. Normally the computer instruction format itself has the following three forms:

(i) 0 address instruction
(ii) 1 address instruction
(iii) 2 address instruction
(iv) 3 address instruction

In the 0 (zero) address instruction, the opcode is specified and the address of the operand is not specified but it is implied. In the one address instruction the opcode and the address of the one operand is specified. In the two address instruction, the opcode and addresses of the two operands are specified. The result is stored in the first operand place. In 3 address instruction the two addresses specify the operands and the third specifies the destination.

Addressing modes: Instructions can be classified according to the manner in which effective addresses are computed and operands obtained. The following are the various modes of addressing the operands.

Immediate addressing: In this mode of addressing the instruction actually contains the operand itself rather than an address of the operand (example: MVI in 8085).

Direct addressing: The operand is given by the address of the memory word where the operand is contained. This is called direct addressing (example: LDA, STA in 8085).

Register addressing: This mode of addressing is conceptually the same as direct addressing. In this mode of addressing, the address field contains the number of the register in which the operand is stored. Normally CPU has a number of general purpose registers (GPRs) which can be used for the purpose. 8085 has six eight-bit registers in addition to accumulator (B, C, D, E, H and L in Fig. 11.8). These registers can also be used in pairs to make 16 bits. These registers are useful to hold values of different program variables avoiding reference to memory which is more time consuming. Thus, the register bank acts as a scratch-pad. An example of register addressing is MOV r_1, r_2.

Indirect addressing: In direct addressing the address specifies which memory word or register contains the operand. Indirect addressing is a scheme in which the address specifies which memory word or register contains not the operand but the address of the operand. In 8085 the instruction specifies a register pair which contains the memory address where the operand is found. The higher order bits of the address are in first register and lower order bits in the second register (example: MOV r, M in 8085).

Indexed addressing: In this mode, the addresses consist of two parts: the number of an index register (may be one of the GPRs) and a constant. The address of the operand is the sum of the constant and contents of the index register. The need to increment or decrement an index register just before or after it is used is so common that many processors (example 6800) provide special instructions, or addressing modes or even special index registers, which automatically increment or decrement themselves. Automatic modification of an index register is called autoindexing.

This mode of addressing is useful for addressing arrays or for counting purposes.

Base register addressing: The address length can be reduced by having a number of base registers, each containing a pointer to some memory location. Address consists of two parts: a base register number (may be one of the GPRs) and an integer called the displacement. The addressing computation adds the displacement to the contents of the selected base register to produce the address of the operand.

The base register addressing is normally used to enhance the addressing capability of the machine or to move the chunk of information from one part of the memory to another.

It is seen that in both indexing and base register addressing, the contents of the selected register is added to a constant contained in the instruction itself in order to produce the address of the operand. Thus, there appears to be no difference between the two modes of addressing. However, traditional usage is that when the displacement has enough bits to address the entire memory itself, the scheme is called indexing. Otherwise it is called base register addressing. Furthermore, base register addressing schemes usually allow two registers to be specified, the operand address being the sum of both registers and the displacement.

6800 provides relative addressing which allows programmer to specify an address relative to the current instruction, hence, processor could jump ahead or backwards. 6800 also provides direct addressing mode in which only low order byte needs to be specified and the higher order byte is assumed to be zero. Thus, the first 256 memory locations are directly addressed with single byte address.

Stack addressing: To save CPU time and memory, it is desirable that instructions be as short as possible. The ultimate limit in reducing address lengths would be to have instructions with no address at all, just opcodes. This is accomplished by organizing the processor around a data structure called a *stack* or LIFO (Last in first out). A stack consists of data items (words, characters, bits etc.) stored in consecutive order in the memory. The first item pushed onto the stack is said to be at the bottom of the stack. The item most recently pushed onto the stack is said to be on the top of the stack. There is always a register that contains the address of the top of the stack and it is called the *stack pointer*.

Instructions with no addresses are used in conjunction with a stack. In this form of addressing the two operands are to be popped off the stack, one after another, the operation is performed and the result is pushed back onto the stack.

Stack is a useful data structure for recursive computation, subroutine linkages and interrupt handling.

11.4.6 Instruction Set

An instruction can be considered as a rule for transforming the old state vector into a new state vector (A state vector denotes all the changeable parts of a process i.e. a program under execution). Some instructions change the value of one or more memory words or register. Other instructions merely change the program counter. Normally instructions can be classified into the following classes.

Data transfer instructions: The operation of copying data from one place to another is one of the most fundamental of all operations. These instructions fall into four general groups: loading register from memory, storing registers into memory, register to register transfer, and memory to memory moves. Certain instructions not only move data but set the condition codes, depending on whether the data is negative, positive or zero (Examples: MVI: MOV r_1, r_2; MOV r, M etc. in 8085).

Dyadic operations: Dyadic operations are those that combine two operands to produce a result. One group of dyadic operations is arithmetic operations (add, subtract, multiply and divide). Another group of dyadic operations includes the logical instructions (OR, AND, EXCLUSIVE OR).

(Examples: ADD, SUB, ANA, DRA, XRA etc. in 8085).

Monadic operations: Monadic Operations have one operand and produe one result. Instructions which complement, increment, decrement shift or rotate the contents of a word or a byte, fall under this category. An important use of shifting is multiplication and division by powers of 2. Rotate operations are useful in conjunction with packing and unpacking bit sequences from words.

(Examples: CMA, RLC, RRC, RAL, RAR, INR, DCR etc. in 8085)

Comparison and conditional jumps: Nearly all programs need the ability to test their data and alter the flow of control, based on the results. A common method of doing so is to provide conditional jump instructions that test some condition and jump to a particular memory address if the condition is met. The CPUs have bits (flags) that are used to indicate specific conditions like occurrence of overflow, carry, zero, parity etc.

(Examples: JMP, JNZ, JZ, JC, JP, JM, JPO, CPI CMP etc. in 8085)

Procedure call instruction: A procedure (often called subroutine) is a group of instructions that perform some task and that can be invoked (called) from several places in a program. Development of large programs gets facilitated with the use of procedures besides improving its readability. Two types of instructions, one to invoke a subroutine and another to return from the subroutine are required. In 8085, these instructions are CALL and RET respectively.

When a CALL instruction is executed, the microprocessor saves the content of PC which is the 'return address' in the stack and branches to the address of subroutine. In the subroutine, when a RET is executed, the processor pops out (removes from top) the return address from the stack. The program then resumes execution at the instruction immediately following the CALL.

As pointed out earlier, stack is a special data structure which is like a bucket closed at one end and data can be inserted and taken out from only one end. Hence, the policy of retrieval of data is that of Last-in-first-out (LIFO). Such a data structure is very useful for subroutine linkages. Consider a situation where a subroutine calls another subroutine (nested subroutine call) as depicted in Fig. 11.12. Obviously, the return address cannot be saved in a register as at the time of SUB2, the return address from SUB1 will get lost. With stack, the return address L_m is pushed into the stack at the time of CALL SUB1 from the main routine and then L_{s1} is pushed when CALL SUB2 is executed in SUB1.

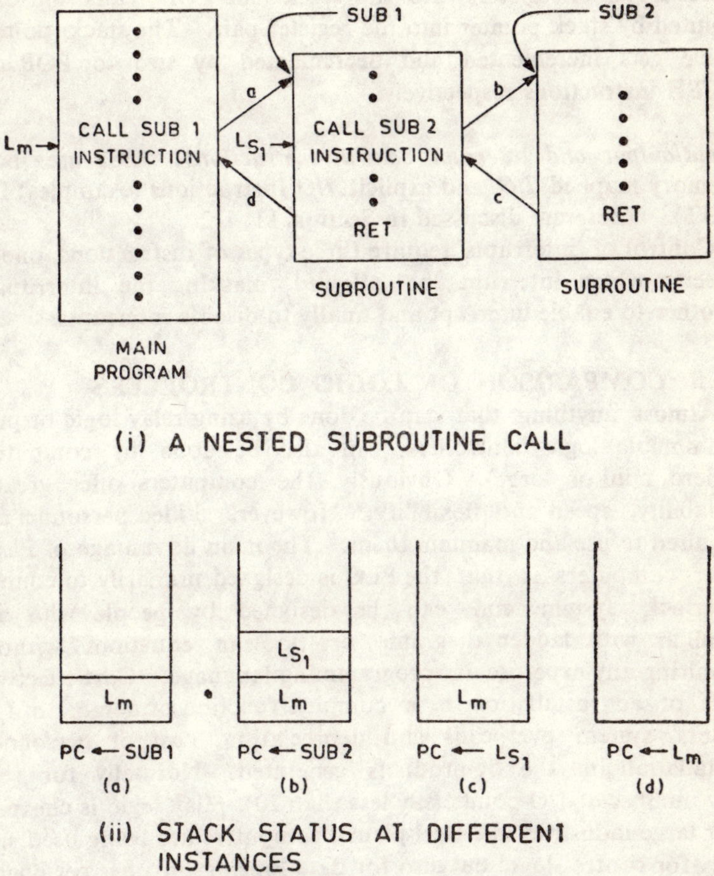

Fig. 11.12. Using stack for subroutine linkages.

When RET is executed in SUB2, the return address L_{s1} is popped off from the stack and the processor starts executing SUB1 from the location L_{s1} onwards. When a RET is executed in SUB1, the stack is again popped off and the processor starts executing main program from L_m onwards (See Fig. 11.12). Thus, stack controls the nested call to subroutines. Similar situations arise in nested interrupt handling and in recursive computation.

8085 uses the RAM for the stack. A special 16 bit register called stack-pointer contains the address of the stack top. CALL and RET manipulate the stack automatically. The PUSH instruction stores a *pair* of registers into the stack and POP reads the data pointed by stack pointer into the register pair. The stack pointer value gets incremented and decremented by two for POP and PUSH instructions respectively.

Input/output and interrupt control instructions: 8085 uses both memory mapped *I/O* and explicit *I/O* Instructions (examples: IN, OUT). These are discussed in Section 11.4.2.

Control of interrupts require three types of instructions, one to specify which interrupt be allowed (masking the interrupts), another to enable interrupt and finally to disable interrupts.

11.5 COMPARISON OF LOGIC CONTROLLERS

Almost anything that can be done by using relay logic or programmable logic controllers, can also be done by computers (micro, mini or large). Obviously, the computers offer greater reliability, speed and flexibility. However, skilled personnel are required to use and maintain them. The main advantage of PLCs over computers is that the PLC is designed primarily to control interlock system and can be designed by people who are familiar with ladder diagrams or Boolean equations without requiring any expertise in programming language. Cost-effectiveness of an installation is a complex function of number of I/O points, system overheads and depreciation, cost of personnel, installation and the by-products generated. Normally for very low number of I/O points (say less than 20), relay logic is cheaper. For large industries like steel plants, computers are being used not only for control-logic but also for data logging or other peripheral uses on time shared basis. Table 11.5 gives a gross comparison of different types of installations.

11.6 DEVELOPING A MICROPROCESSOR BASED SYSTEM

11.6.1 General Considerations

Normally the task of developing a microprocessor based system consists of the following four distinct stages:

(i) System analysis and design
(ii) Hardware realization of the system
(iii) Development of required software
(iv) Testing and Debugging

Table 11.5. A Comparison of Logic Controllers

S. No.	*Comparison Parameter*	*Relay Logic*	*PLC*	*Microcomputer*	*Minicomputer*
1.	Power supply fluctuations	Can withstand	Requires protection	Requires protection from transients	Requires protection from transients
2.	Temperature, Humidity and dust	No special provision required	No special provision required	A.C. room required. Dust control required when disks are used	A.C. room with dust control is required
3.	Reliability	Low	High	High	High
4.	Equipment size	Large	Small	Small	Small to medium
5.	Speed of operation	Low	High	High	High
6.	Sequencing of operations	Parallel	Sequential or cyclic	Time-multiplexed as desired	Time-multiplexed as desired
7.	Nature of stored logic	Hardwired	Memory with limited capacity	Large memory and floppy disks possible	Large memory and back-up storage
8.	Method of programming	Ladder-diagram	Ladder diagram or Boolean equations (simple instructions)	Assembly language or higher level languages	Higher level languages (normally)
9.	Ability to modify control program	Difficult	Reprogrammable (on-line)	Reprogrammable (on-line)	Reprogrammable (on-line)
10.	Relative cost	Low	Medium	Medium	High

(Contd.)

S. No.	*Comparison Parameter*	*Relay Logic*	*PLC*	*Microcomputer*	*Minicomputer*
11.	Fault-diagnosis	Difficult	Self-diagnosis feasible	Very simple	Very simple
12.	Users training	Knowledge of ladder diagram	Knowledge of simple instructions	Knowledge of programming on a microcomputer	Knowledge of higher level language
13.	Scope for expansion	Requires major modification and is costly	Low cost minor additions	Low cost minor additions	Low cost minor additions
14.	Scope for system being used for other purposes such as sharing other tasks or processes	Nil	Nil	Limited	Feasible
15.	Service and maintenance	Time-consuming, less skilled personnel	Simple, less time consuming, medium skilled personnel	Simple, less time consuming, digital circuits personnel or computer personnel	Less time consuming (mostly). computer or computer maintenance personnel

The system anaiysis and design stage normally consists of the following sub-stages:

(i) Obtain a block schematic of the system and identify the role of microprocessor.

(ii) Analyze the nature of input-output signals and identify the conversions required.

(iii) Estimate timing constraints for various blocks analytically or through simulation.

(iv) Choose a suitable microprocessor.

(v) Identify the tasks for hardware logic and the microprocessor.

(vi) Estimate the memory requirements for RAM and ROM.

(vii) Identify and evaluate interrupt requirements.

Unlike conventional computer, the microprocessors are always dedicated to the system. The peripheral equipment cost being several order higher than the microprocessor chip, no purpose is served by making it general purpose. The final program is stored into the PROM and is dedicated to the application. Development of hardware normally involves following stages:

(i) Select a single board microcomputer or custom made hardware depending upon the system complexity and the infrastructure available for hardware development. Develop and design layout of the entire system.

(ii) Select peripheral chips and develop hardware interface.

(iii) Interface required memory (RAM/ROM).

(iv) Develop interfaces for input/output.

(v) Implement interrupt structure.

(vi) Design power supply, protection environments, packages, chassis, wiring diagrams.

(vii) Develop diagnostic and monitoring routines, simulate input/output and test the hardware.

The software design and development may involve following stages:

(i) Choice of a suitable vehicle for software development and programming language for implementation. This very much depends upon the available resources like MDS, assembler, compilers, cross assemblers, cross-compilers etc. and also on the complexity of the required software and the constraints on memory space and time.

(ii) Functional partitioning of application software and their implementation.

(iii) Implementation of I/O and interrupt software.

(iv) Development of supervisory or executive software.

(v) Design and development of diagnostic software study.

(vi) Study of hardware and hardware trade offs if necessary.

(vii) Test the entire system software with simulated or real inputs.

Finally, the hardware-software integration requires field testing and debugging and then burning of PROMs.

It should be noted that many of these stages have to be iterated due to the difficulties faced in earlier stages.

There exists a number of ways in which the system development and integration is feasible. Some of the possible alternatives are as follows:

(a) A small and simple system, in the absence of other development aids, may be developed as follows:

(i) Breadboard the prototype hardware. Provide input through switches and output display through LED's.

(ii) Develop program in m/c language. Enter it through switches.

(iii) Test and debug the system through switches and displays.

(iv) Transfer the final developed program into PROM and implement the system.

(b) The program is developed using a simulator on a mini, midi or large computer. Simulator is a program normally written in a higher level language which accepts inputs in assembly language codes of the microprocessor and simulates the behaviour of the microprocessor. Once the program is finalised, it is transferred to the microcomputer board which has resident assembler (i.e. assembler in a ROM). The program is assembled. The object code is put into a PROM and the system is implemented. Program development using a simulator is comparatively easier as system resources and software aids of a conventional computer can be made use of. However, through simulation the real time behaviour of the program with actual input/output cannot be tested. In some of the applications, this may be an important consideration. The input-output system may also be simulated on the computer and a more realistic behaviour could be obtained.

The entire program development can also be done on the

microcomputer assembly which has resident assembler. In such a case, it would be desirable to have resident loader, monitor and editor programs as well. Editor is a program which allows modifications in the parts of the program. The actual input-output devices may be connected to the assembly and the behaviour of the program may be analysed.

(c) The program is developed using a cross-assembler or cross-compiler and tested through simulator on a mini, midi or large system. Cross-assembler is a program written in a higher level language which accepts assembly language code of the microprocessor and generates object-code (m/c language program) for the microprocessor on the mini, midi or large computer. The codes thus developed could be tested with the help of simulator if presented on the host computer. Alternatively the codes could be transported through paper tape, cassettes or PROM programmers depending upon the devices available on the host machine and the microcomputer assembly.

Cross-compilers accept programs in a higher level language and produce object code for the microprocessor.

It should be noted here that programs written in a higher level language are much more concise than assembly language programs for the same functions. Being machine independent, they are transportable. They are easier to read, understand and diagnose. However, there are two important considerations, the economics of storage space and running time of the program, that dictate the choice of programming language in microprocessors. In some situations, the demands of economy may preclude the use of a machine independent language altogether. Programs written in languages such as Fortran, Cobol, PL/I and others are known to be many times slower than handcoded assembly language programs using similar algorithms. Hence, choice of a programming language very much depends upon the application to which the microprocessor is being put to.

(d) In many problem situations, a complex integration of hardware and software requirements is desired. Moreover, the development cost of microprocessor based system is high primarily because of the high peripheral cost. Normally, the cost of the microprocessor is not even 10% of the total cost of the system. Therefore, an alternative approach is to interface the microprocessor with an existing minicomputer and utilize all peripheral resources and the software available with the minicomputer. A cross-assembler on the minicomputer and PROM burner with the system form excellent development aids. The entire system is developed and tested in an integrated hardware-software environment. It is also possible to simulate the other systems associated with the desired microprocessor based system and feed the simulated inputs. The closed-loop performance can also be studied.

Thus, we observe that depending upon the complexity of the desired system and the development tools available, one may follow any one of the above approaches or evolve a mixed approach.

To cope up with the problems of development, a large variety of microprocessor development systems (MDS) are commercially available today. Normally it consists of a microcomputer board with some RAMs, PROMs, EPROMs, peripheral interface cards and ROMs containing resident software such as assembler, monitor, loader, editor and diagnostic programs. Some MDSs also contain a PROM burner and Floopy disks. Usually an alphanumeric keyboard, a printer or typewriter and sometimes a paper tape reader/punch are provided for user communications. In more advanced MDS facilities for multi-microprocessors with multiuser terminals are provided around a disk-oriented time shared system. In-circuit-emulators (ICE) are provided to emulate other microprocessors. The user system under development may be directly interfaced with the ICE. The higher level programming languages

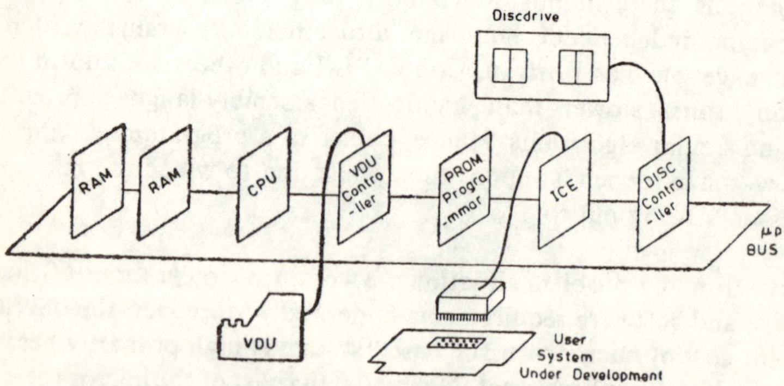

Fig. 11.13. A typical microprocessor development system.

such as PL/M, Pascal and Fortran are also made available. A typical MDS is shown in Fig. 11.13.

11.6.2 Microprocessors Based Drive Control System

The Fig. 11.14 shows a block diagram of a typical drive control system which is self explanatory. f_s is the feedback signal obtained

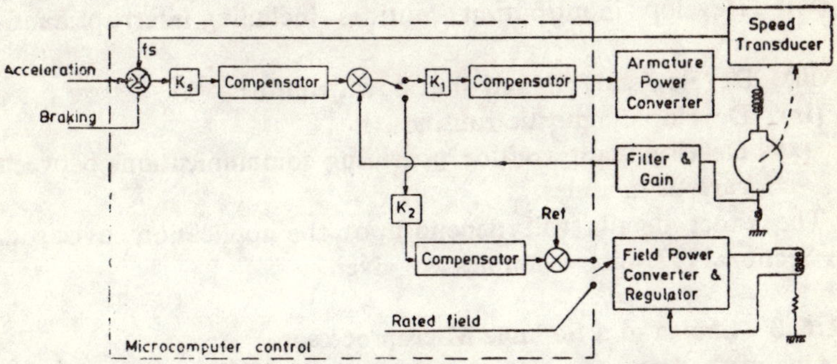

Fig. 11.14. Block diagram of a typical drive control system.

through a speed transducer. The tasks which are normally handled by the microcomputer is shown within dotted lines. Typically, the stages of developing a microcomputer based drive system are as follows:

(i) Based on the systems study, determine the control parameters and define control strategy.

(ii) Transform the control strategy into a digital control algorithm.

(iii) If required, the behaviour of the digital control algorithm be studied through simulation on a hybrid or digital computer system.

(iv) Develop the required hardware including protection and shielding from noise and surges. The input-output interfaces, wave shaping circuits, isolators should also be provided.

(v) Develop the required software (this is elaborated later).

(vi) Integrate hardware and software and test the system through simulated inputs.

(vii) Field-test the system.

The development of required software may involve the following steps:

(i) Partition the program into logical routines.
(ii) Look into the existing program library if they could be made use of.
(iii) Develop initialization routine.
(iv) Develop interrupt processing routines.
(v) Develop feedback control routines.
(vi) Develop drive sequencing routine.
(vii) Develop input/output routines including interrupt handling.
(viii) Develop start up and shut down routines.
(ix) Develop diagnostic routines.
(x) Develop main routine providing communications between various modules.

The exact details will depend upon the application envisaged. In Section 11.7 some examples are given.

11.6.3 Choice of a Suitable Microprocessor

One of the important aspects of comparison of microprocessors is their speed. The system time constant must be smaller than the computation time to ensure stability. Sometimes when this is not possible, a *separate* hardware is utilized to share some of the computational task. Other important aspects of comparison are maximum addressable memory space, code structure and instruction repertoire (e.g. their versatility, size and type of registers, indexing, availability of stacks, memory segmentation, etc.) and so on. One important aspect of code structure in design of drive control systems is the interrupt facility available.

An interrupt system must satisfy many requirements to succeed. The interrupt handling and servicing routines should be both efficient and as simple and straightforward as possible. Besides, the normal execution of instructions should not slow down because of the interrupt system even when interrupts are not occurring. Finally, where the input-output signals have widely varying degree of urgency and importance, a priority scheme and its programmability are important considerations. Several general purpose interrupt controller chips such as Intel 8259 are available to help implementation.

The reaction time, overhead and saturation are three important parameters which can be employed to compare different interrupt systems and to evaluate the suitability of an interrupt system for a given environment.

The reaction time can be defined as the length of time from the occurrence request for attention in the external environment to the execution of the first useful instruction related to processing the external signal. The overhead is the difference between the total time required to completely process an interrupting signal and the time spent in useful work related to the origin of the signal. When a computer cannot respond to incoming signals at the rate at which they arrive, and information conveyed by such signals is essentially lost, the system is said to have saturated. Saturation may occur due to any combination of long reaction times, high overhead, and high computational demands of the interrupt processing routines.

It should, however, be noted that the wise choice is seldom based on technical qualities of the processor chip. Availability and proper matching of auxiliary products such as memory and I/O chips, availability of development facilities, suitability of available software, familiarity of personnel, etc. often force the choice.

Now, having looked at the stages in the process of developing a microprocessor based system, there are three more aspects which a designer must be familiar with. These are:

(i) the methods of address decoding and selection;

(ii) how to augment RAM and ROM capacities of a microcomputer;

(iii) how to interface input-output units to a microcomputer.

The next three subsections deal with these aspects.

11.6.4 Address Decoding

The address decoder generates device select signals when a certain address or range of addresses is present on the address bus. As an example, consider that an address decoder is required to control eight input-output ports whose addresses are 7000 to 7007 (written in Hex) and the address bus consists of 16 lines. Figure 11.15 shows a possible configuration where an eight line decoder has been used. The lines A_2, A_1, A_0 provide the 3 bits for the eight lines while the input to the enable line is fed from a logic which fixes the rest of the bits of the address. A complete combinatorial circuit could also be used for the purpose. The combinatorial circuits are used when the number of devices to be addressed are small and they are not consecutive.

The address decoding for memories is similar to that used for a

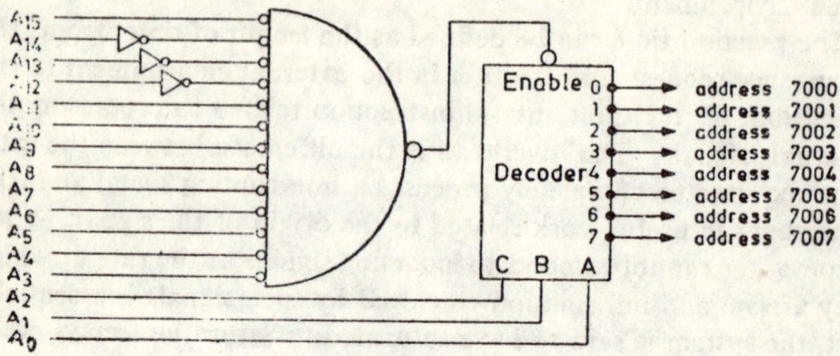

Fig. 11.15. Address decoder for eight I/O devices.

group of input-output ports. Consider, as an example, an address decoder which is required to address 1024 bytes locations of ROM where there are 4 ROM chips each of 256 bytes. Normally, each chip has its own address selector which can select 256 locations. One possible configuration of address decoder is shown in Fig. 11.16, where 1024 ROM locations are partitioned in 0-255, 256-511, 512-766 and 767-1023 address ranges to correspond to 256 locations on a chip. The chip select signal is generated through the 8 line decoder.

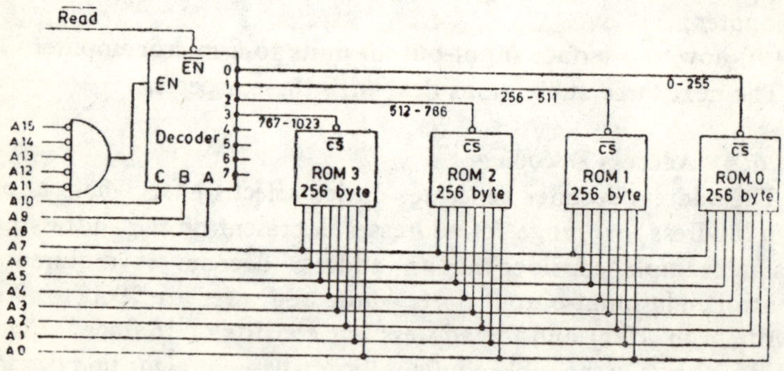

Fig. 11.16. ROM address decoding.

Addresses for RAMs are also decoded in a similar fashion with some extra control circuits to write into the RAM. RAM chips have a $\overline{\text{WRITE}}$ input in addition to the $\overline{\text{CS}}$ (chip select) input. For read/write operation the $\overline{\text{CS}}$ must be low. A low at $\overline{\text{WRITE}}$ enables the data bus contents to be transferred to the specified location whereas a high at $\overline{\text{WRITE}}$ enables the output of RAM to be made available to the data bus.

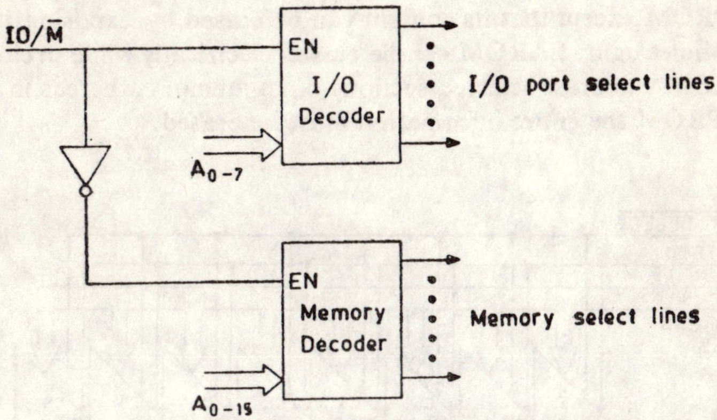

Fig. 11.17. I/O mapped address decoding.

Figure 11.17 shows an input/output mapped decoding circuit which increases the total addressable space by using an addition control line IO/$\overline{\text{M}}$. This extra control line is available on few microprocessors including 8085.

11.6.5 Connecting RAMs and ROMs

A large number of semiconductor memory devices are commercially available. Before we look at the problem of connecting them to a microcomputer, let us glance at their properties.

RAMs are the read/write memory devices which are of two types: static and dynamic. Static RAMs are made up of flip-flops and the state of flip-flop does not change unless a new data is stored or power is turned off. Dynamic RAMs use capacitors as storage elements. Since the storage cell is simple, the packing density is high. However, the charge on the capacitor leaks off after a few milliseconds and so they must be refreshed (i.e. data read and re-written). Since dynamic memories must be continuously refreshed, special circuits are needed to do this which tends to increase its complexity. Smaller systems use only static RAMs.

ROMs provide permanent storage of information. These are of four types: mask-programmed, programmable ROM (PROM), erasable PROM (EPROM) and electrically alterable ROM (EAROM). Mask-programmed ROMs are programmed by the IC manufacturers and are often used for high-volume products. The PROMs can be programmed through a PROM programmer. Once they are programmed, they cannot be changed. EPROM is similar to PROM except that its content can be erased by exposing it to ultraviolet light. EAROM can be erased electrically while in circuit. EAROM can erase selected sections of its memory whereas in case of EPROM the entire information must be erased.

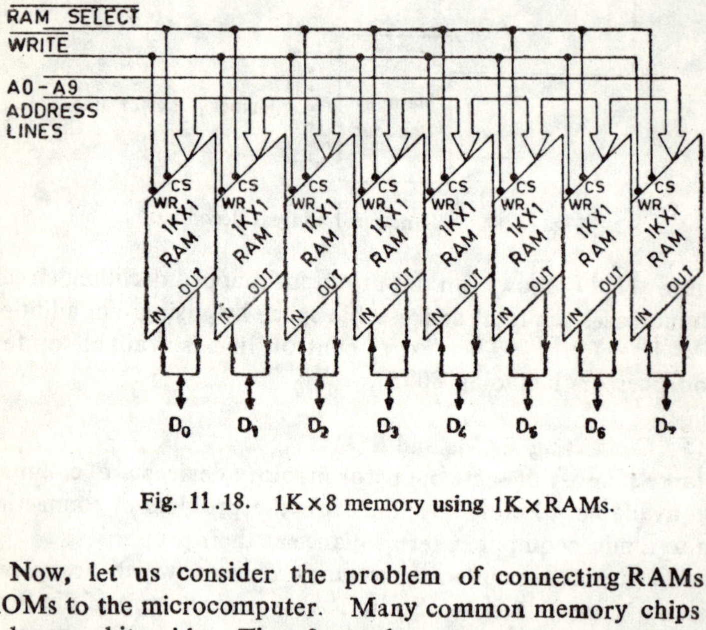

Fig. 11.18. 1K×8 memory using 1K×RAMs.

Now, let us consider the problem of connecting RAMs and ROMs to the microcomputer. Many common memory chips are only one bit wide. Therefore, there are two or more possible configurations of organising memory in terms of bytes. Figure 11.18 shows a 1K×8 RAM configuration using 1K×1 chips. Here, each chip is contributing a bit of the addressed byte. The Fig. 11.19 shows another configuration, where each RAM chip (1K×4) contributes 4 bits of the addressed byte. It also shows connections for a 2K ROM chip having addresses from 1024 to 3071.

The above examples provide background of some general interconnections. Other aspects of memory interfacing such as address decoding, control of I/O and data buses and their timing have been discussed earlier.

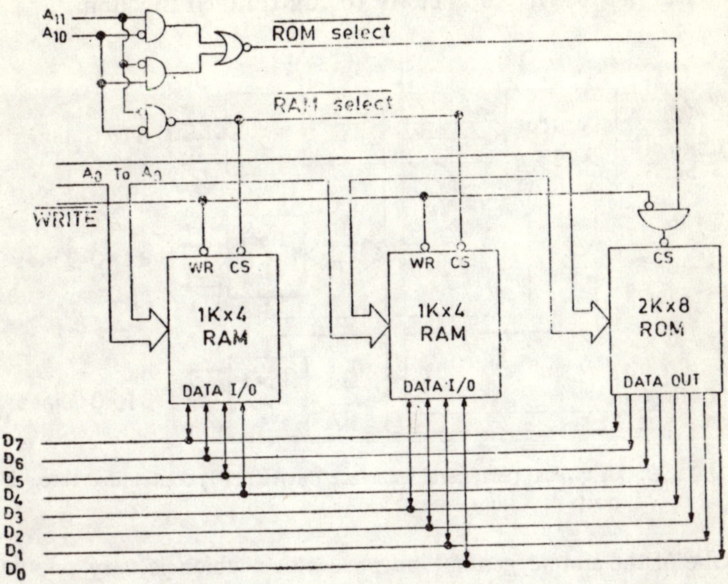

Fig. 11.19. A 2K ROM and 1K RAM (with 1K×4 RAMs) configuration.

11.6.6 Interfacing Input-Output Devices

The task of interfacing input-output devices to a microcomputer and its management takes a major share of designer's effort. The peripheral devices are of varying nature in terms of their speed and the control signals that are required to effect error free transfer of data. Normally, a peripheral controller circuit is used to generate specific signals required by the device and a general interface circuit is used to communicate between the microcomputer and the peripheral controller. For example, if the peripheral device is an impact printer, the peripheral controller would be required to produce signals to select character, drive print hammer, line feed, position the print etc. on the basis of the information received through the microcomputer interface circuit. Many a time the interface circuit is combined with the controller circuit in a single chip. A wide variety of specialized interface chips is commercially, available which include keyboard controllers, teletype controllers CRT display controllers, direct-memory access controllers, floppy disk controllers etc. Some of these LSI chips are even more complex than a microprocessor and some internally use a special purpose microprocessor to control their function.

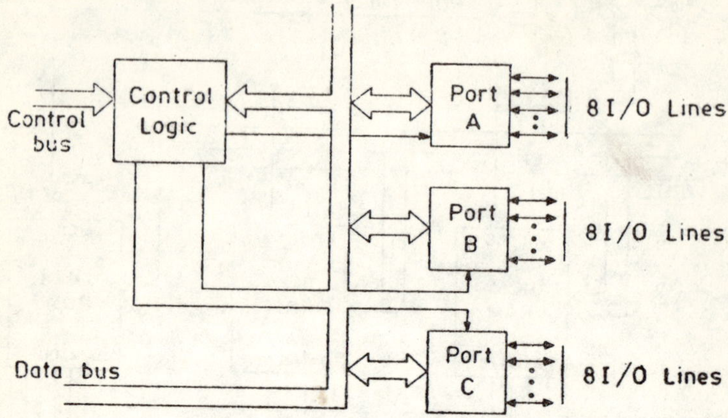

Fig. 11.20. A typical peripheral interface device (PPI) containing three I/O ports of 8 lines each.

One of the simple general purpose chips which is very popular is programmable peripheral interface (PPI) or parallel peripheral interface or parallel input-output (PIO) or peripheral interface adapter (PIA). Motorola 6821 and Intel 8255 are examples which have found widespread use. Figure 11.20 shows general configuration of these peripheral interface chips. Here, each port can be used either as an input port or output port or both. The nature of the port is controlled by the contents of a register on the chip. The contents of the register in turn is controlled by software (or firmware) and hence the functions of the chip becomes programmable. The chip also generates some control signals required for synchronizing communication and control interrupt under software control.

The 8255 PPI has three I/O ports that can be programmed in three main modes of operation under the control of a 8-bit control word buffer in the chip. In mode 0, each port performs simple input-output. The ports *A* and *B* must be inputs or outputs but not both and the port *C* is divided into two groups of 4 bits each which can be defined as input or output.

In mode 1, the functions of ports *A* and *B* remain the same as in mode 0, however, the 4 bits of port *C* are assigned for control signals for executing transfer (handshake) with port *A* and the other 4 bits of port *C* are used for similar purpose with port *B*.

In mode 2 of operation, the single port *A* is set to be a bidirectional data port and defines 5 bits of port *C* as control port.

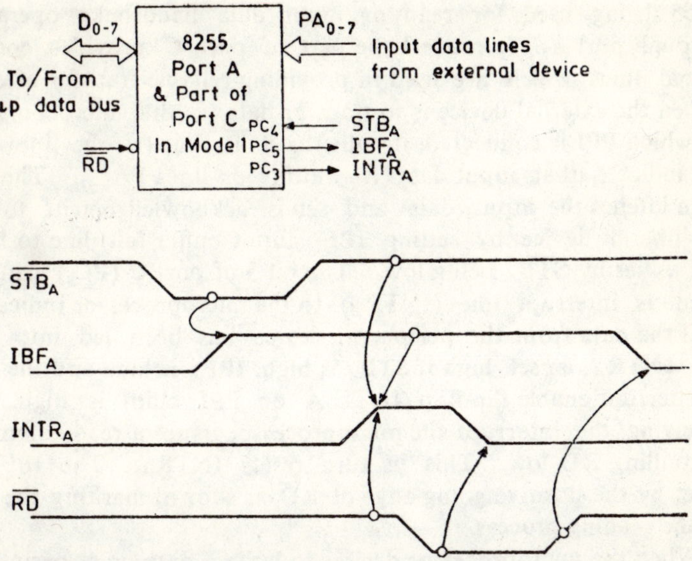

Fig. 11.21 (a). Read (input) data handshake operation through port *A* in mode 1.

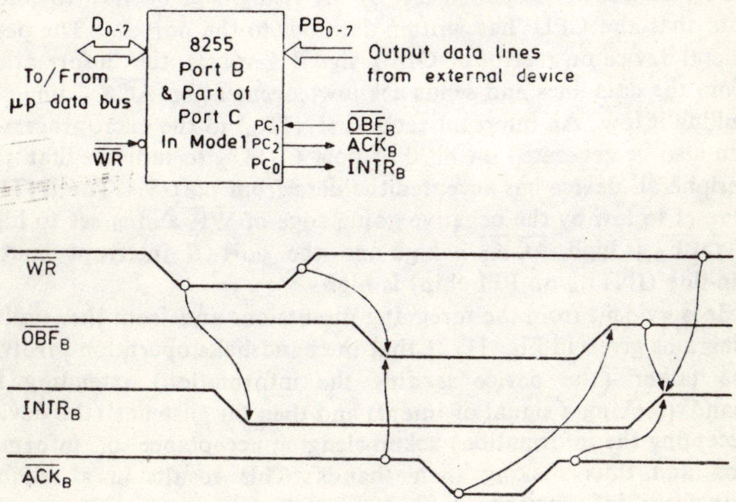

Fig. 11.21 (b). Write (output) data handshake operation through port *B* in mode 1.

As an example, let us consider the handshake operation for input and output in mode 1 of 8255 PPI. Figure 11.21 (a) shows 8255 being used for read or input data handshake operation through port *A*. In mode 1 the part of port *C* provides control signal lines which are used in providing reliable transfer of data. When the external device is to transfer data to the microcomputer to which PPI is connected, it pulls the $\overline{STB_A}$ line (strobe line) low to indicate that input data is waiting on lines PA_{0-7}. The PPI then latches the input data and sends acknowledgement to the peripheral device by setting IBF (input buffer full) line to high. IBF is set by $\overline{STB_A}$ being low. The bit 3 of port *C* (PC_3) can be used as interrupt line ($INTR_A$) to the microprocessor indicating that the data from the peripheral device has been led into port *A*. $INTR_A$ is set high if STB_A is high, IBF_A is high and the port *A* interrupt enable flip-flop (INTE A on PPI chip) is high. On receiving the interrupt the microprocessor issues a read command by pulling $\overline{RD}$ low. This in turn resets $INTR_A$. The IBF_A is reset by the positive going edge of $\overline{RD}$ as shown, marking the end of the reading process.

When the microprocessor desires to write a data on a peripheral device, it places the data on the data bus and sends a signal through $\overline{WR}$ line pulling it low as shown in Fig. 11.21 (b). The $\overline{OBF_B}$ output line is pulled low by the rising edge of $\overline{WR}$ to indicate that the CPU has written data out to the port *B*. The peripheral device on receipt of $\overline{OBF_B}$ signal receives the information from the data lines and sends acknowledgement on $\overline{ACK_B}$ line by pulling it low. An interrupt request ($INTR_B$) to the microprocessor can also be generated on bit 0 of port *C* (PC_0) to indicate that the peripheral device has accepted the data from port *B*. The $INTR_B$ is reset to low by the negative going edge of $\overline{WR}$ and is set to high if $\overline{OBF_B}$ is high, $\overline{ACK_B}$ is high and the port *B* interrupt enable flip-flop ($INTE_B$ on PPI chip) is high.

It is evident from the foregoing discussions and from the timing diagrams given in Fig. 11.21 that the handshake operation involves the 'talker' (the device sending the information) extending its 'hand' (seeking a signal of intent) and then the 'listener' (the device accepting the information) acknowledging acceptance of information and thus shaking their 'hands'. This results in a reliable transfer of information.

Another common type of interface chip which provides serial

inputs and outputs are commonly known as UARTs (Universal Asynchronous Receiver and Transmitter). The configuration of a typical UART is shown in Fig. 11.22. An UART accepts a byte of data from the microprocessor and then outputs it one bit at a time very much like a parallel-in serial-out shift register. It can also handle data in the other direction which converts a serial bit stream into a parallel from which can be directly used by the microprocessor. The function and format of information handling is controlled by a control register similar to that used in case of a PPI.

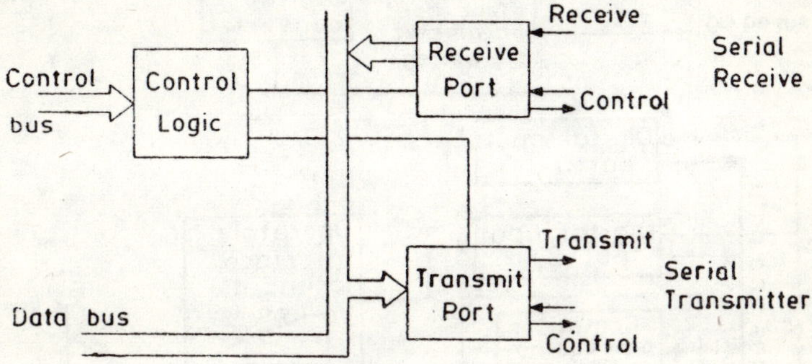

Fig. 11.22. A typical UART (universal asynchronous receiver and transmitter).

USART (Universal Synchronous/Asynchronous Receiver/Transmitter) is another general purpose interface chip which provides asynchronous as well as synchronous data communications. A typical USART is Intel 8251. In the UART mode of operation, the 8251 can accept from five to eight data bits from the microprocessor data bus and transmit them serially as part of a character on output. The transmitted character is composed of a start bit, the five to eight data bits, a parity bit (optional) and either $1\frac{1}{2}$ or two stop bits. If a character is not transmitted immediately following another character, any number of idle bits can be transmitted while a new character transmission is indicated by a start bit. The USART mode and data format are programmable using mode and command instructions.

11.7 APPLICATION EXAMPLES

11.7.1 A Typical Microprocessor Based Process Control System

Figure 11.23 shows a simple process control system. The process parameters are converted to electrical signals and are transformed to digital form using A/D converters. These are interfaced with the rest of the system through input ports which are connected to

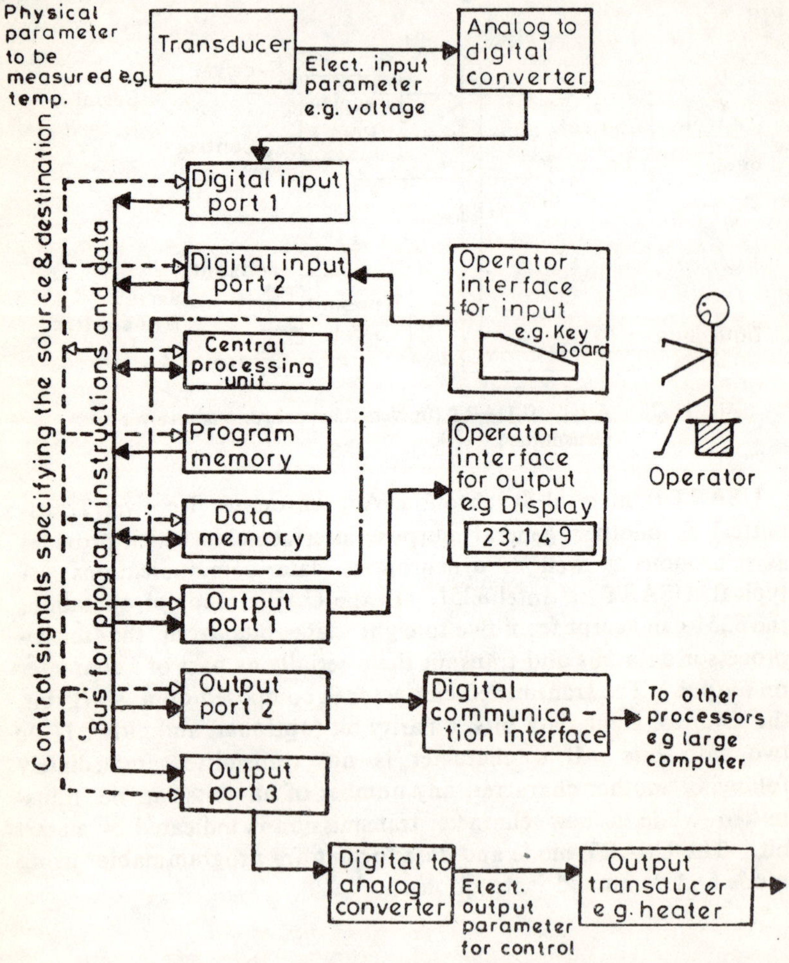

Fig. 11.23. **A microprocessor based system configuration for instrumentation and control.**

a system bus. The digital input port 1 as shown can be repeated for each process parameter. Another input port is also provided for keyboard entry. The microprocessor is interfaced to the system bus. One or more output ports are normally provided for displaying parameters for operator's feedback and datalogging. Another output port normally provides communication with a large computer. One or more output ports provide signals to control the physical parameters.

A real implementation is normally a variation of Fig. 11.23. The variations are normally in the manner in which communication network is built, in the number of microprocessors used for distributed computing and control and the manner in which the large computer is used to co-operate the activities of distributed processors.

It is readily seen from the above system configuration the role of microprocessors in 'processing', 'control' and 'communication'.

11.7.2 Speed Control of a DC Motor

A major demand of the present day industries is to have precise and wide range of speed control for drives with good speed resolution and reproducibility. In this example we discuss the design and implementation of a fully digital system using an Intel 8080 microprocessor and some random logic for speed control of a DC motor which is fed by a SCR chopper. The main advantages of such a digital technique are:

1. Elimination of measurement non-linearities (the bug-bear of analog systems).
2. Exact reproducibility of speed setting.
3. High accuracy.
4. Simplicity of implementation — with the motor as the only analog device in the entire system, no D/A converters etc. are needed.
5. Adaptability. By simply scaling up or down the chopper (putting components of a different rating) the system can be adapted to any larger or smaller DC motor with no changes in the rest of the hardware.
6. The biggest advantage of the use of a microprocessor is the resulting flexibility. The microprocessor is a general purpose digital electronic block whose function is determined by programming. Thus, by simply changing the program, any desired control

technique can be implemented. Besides, the basic approach to the design of any practical control system necessarily involves some trial and error procedures. While stability is the basic requirement in any control system design, it is necessary to make the most effective compromise between relative stability and steady-state accuracy. This in turn essentially involves some trial and error experimentation with different control equations. In a microprocessor based system this experimentation can be carried out by simply changing the program without any time consuming changes in the hardware. Thus, a microprocessor based system is highly suitable for the development of such a control system.

7. Programmed control is possible, i.e. speed can be controlled in any given sequence over a period of time. The design and implementation of the system is described in the subsequent sections.

11.7.2.1 *Overall Control Scheme*

(a) Speed Control of DC Motor

The generalized equation for DC Motors is

$$N = \frac{V_a - (T\, R_a / K_t\, Q_f)}{K_a\, Q_f}$$

where N = speed (rpm), V_a = DC voltage applied to the armature, T = torque, R_a = armature resistance, Q_f = field flux; K_a and K_f are constants.

One very important method of speed control of DC drives evident from the equation is to vary V_a. This is known as Armature Voltage Control. A high efficient method of varying the armature voltage when the available supply is DC, is the use of a Chopper. A chopper is basically an ON-OFF control which supplies pulses of source voltage whose average value is controlled by varying the duty cycle of the pulses.

(b) Overview of System Concept

Figure 11.24 illustrates the configuration of this control system. Motor speed is converted into pulses by an optical transducer. The microprocessor integrates these pulses over a period of time to compute the actual rpm of the motor. After measurement the feedback signal (actual rpm) is compared with the reference input

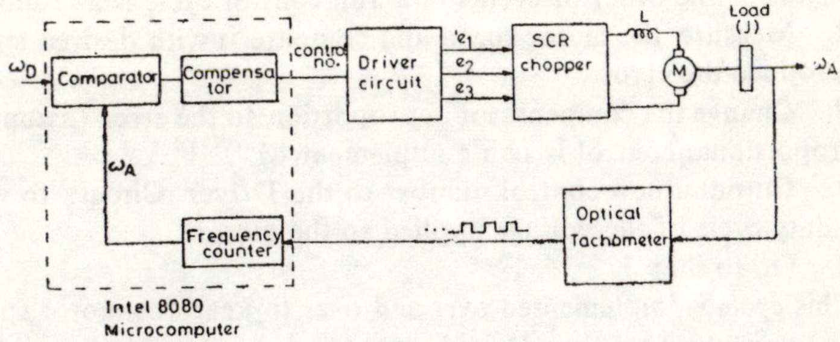

Fig. 11.24. Block diagram of the overall system.

(setting of desired rpm) and the error is then used to adjust the duty cycle of the voltage pulses applied by the chopper to the motor to correct the speed.

(c) Control Scheme Details

Depending on the particular compensation strategy being implemented, the microprocessor makes computations based on the sign and magnitude of the error and outputs a binary number (called control number) to the Driver circuit. The Driver circuit basically consists of programmable counters and it outputs three pulse trains (of pulses denoted by e_1, e_2 and e_3 respectively). The control number determines the phase shift between e_1 and e_2 which in turn controls the 'ON-time' of the SCR chopper. The third pulse e_3 is required only for certain types of choppers (which use three SCRs) and it has no relevance in the control action.

As an example, if a proportional control scheme is being implemented, the width of voltage pulses applied to the motor armature must be changed in proportion to the error. The constant of proportionality (or multiplying factor) will be the gain of the compensator. If the actual rpm (N_A) is less than the desired rpm (N_D), a number proportional to ($N_A - N_D$) must be added to the previous control number and vice-versa. Addition of this number to the control number will increase the phase shift between e_1 and e_2 thereby increasing the width of voltage pulses applied to the motor. This in turn will result in an increase in the speed of the motor.

The system acts as a sampled data feed-back control system as it is not continuous but measures and corrects the speed at discrete intervals. The order of events in a full control cycle is as follows:

1. Measure the motor speed and compare it with desired speed to produce the error.

2. Change the compensator in proportion to the error (assuming a proportional control is being implemented).

3. Output a new control number to the Driver Circuit to vary the duty cycle of the voltage applied to the motor.

4. Go to Step 1.

This cycle is implemented over and over to keep the motor speed at (or very close to) the desired speed. A wait period could be included between steps 3 and 4 to permit the motor to respond to any change in the duty cycle made at the end of step 3. For accurate measurement of speed the integration of pulses must be carried out over a fairly long interval of time. The speed of the motor is assumed to be essentially held constant during this sampling period. Although the duty cycle is kept constant during the entire cycle and changed only at the end of step 3 when a new control number is output to the Driver Circuit, the speed will not be constant during the sampling period. This is due to the fact that the motor and load assembly with its comparatively large time constants, responds slowly to the change in duty cycle. This is especially true in the absence of a wait period. Taking advantage of this desirability of a wait period, the above cycle is modified as follows (no change up to step 3).

5. Convert the measured rpm from binary to BCD and output to the Display Unit (consisting of nixie tubes).

6. Go to step 1.

(d) Microcomputer Configuration

Figure 11.25 gives the block diagram of the Intel 8080 Microcomputer System configuration realized after assembly from the SDK-80 kit (8080 system design kit). Two points of interest noticeable in the diagram are:

(1) Exclusion of clock logic (8224) from the CPU chip: This is characteristic of all early microprocessors. The use of a crystal controlled clock is highly desirable in the present application for accurate timing of the sampling interval during which the speed of the motor is measured.

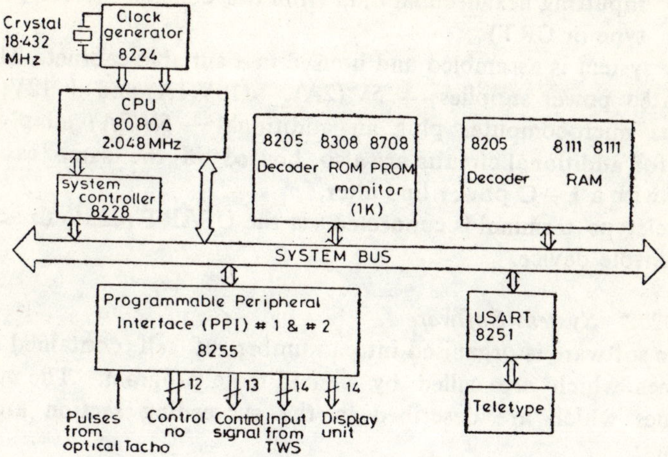

Fig. 11.25. Intel 8080 microcomputer system configuration.

(2) Exclusion of the bus interface logic (8228) from the CPU chip: This is due to the fact that 8080 CPU outputs an inadequate set of control signals. External bus interface logic is required to combine the control signals with instruction status signals (available on the data bus at the beginning of each machine cycle) to create an adequate control bus. The 8228 chip also performs the functions of bus driver to buffer the 8080 data bus from memory and I/O devices.

The system is provided with 512 bytes of RAM out of which the last 30 bytes are used as scratch pad memory by the ROM resident system monitor. The memory has an access time of 350 ns for the RAM (8111/2111A) and 450 ns for the EPROM (98708/2708) which means that the system can work at the maximum speed of CPU (2 MHz clock rate) without any delaying effect of the memory. A high speed system is desirable from considerations of response time in this real time application.

The system has an adequate monitor resident in 1K bytes of ROM. The monitor provides the following facilities which are very useful as software debugging aids:

(a) Displaying selected areas of memory
(b) Initiating execution of user programs
(c) Setting break points in user programs
(d) Modifyıng contents of memory and processor registers

(e) Inputting hexadecimal data from the console device (Teletype or CRT).

The system is assembled and housed in a sutiable cabinet. Three regulated power supplies; + 5V (2A), – 10V(1A) and + 12V(1A) for the microcomputer plus an additional + 5V(2A) supply to cater for additional circuits are also housed in the same cabinet along with a L–C power line filter.

A teletype terminal is connected via the USART (8251) to act as the console device.

11.7.2.2 *System Software*

The software is organised into a number of self contained subroutines which are called by a small main program. The major routines which are described in the succeeding section are as follows:

1. Frequency Counter Routine (CTR).
2. Routine for Starting the Motor (START).
3. Routine for Inputting Reference Speed (INPUT).
4. Routine for Displaying the RAM (DSPLY).
5. Routine for Control Compensation (CNTRL).

(a) FREQUENCY COUNTER ROUTINE (CTR)

Apart from the compensation program, the software frequency counter is the most critical part of the entire software. The MSB of port A (P_{A_7}) of PPI chip No. 1 (Port addr 14H, see. Fig. 11.25) is connected to the output of the optical encoder. Figure 11.26 gives the flow chart of the frequency counting routine in which the input bit P_{A_7} is checked every 36 microsecs for a change in voltage level; 36 micro secs being the execution time of each loop in the flow chart. The routine is separated into two loops. The upper loop is executed if the input is high and the microprocessor keeps on looping and checking the input. When the input goes low, the next check causes a jump to the lower loop. The frequency counter is now incremented (for a high to low transition has occurred) and the lower loop is initiated. The lower loop is identical to the upper loop except that it checks for a low to high transition. At the beginning of each loop, the timer, register pair BC, is incremented and checked for completion of the measurement period. In the actual program, the four possible loop paths (marked I, II, III and IV in the flow chart of Fig. 11.26) have the same execution time

viz. 36 microseconds. This is essential for maintaining the accuracy of the sampling interval.

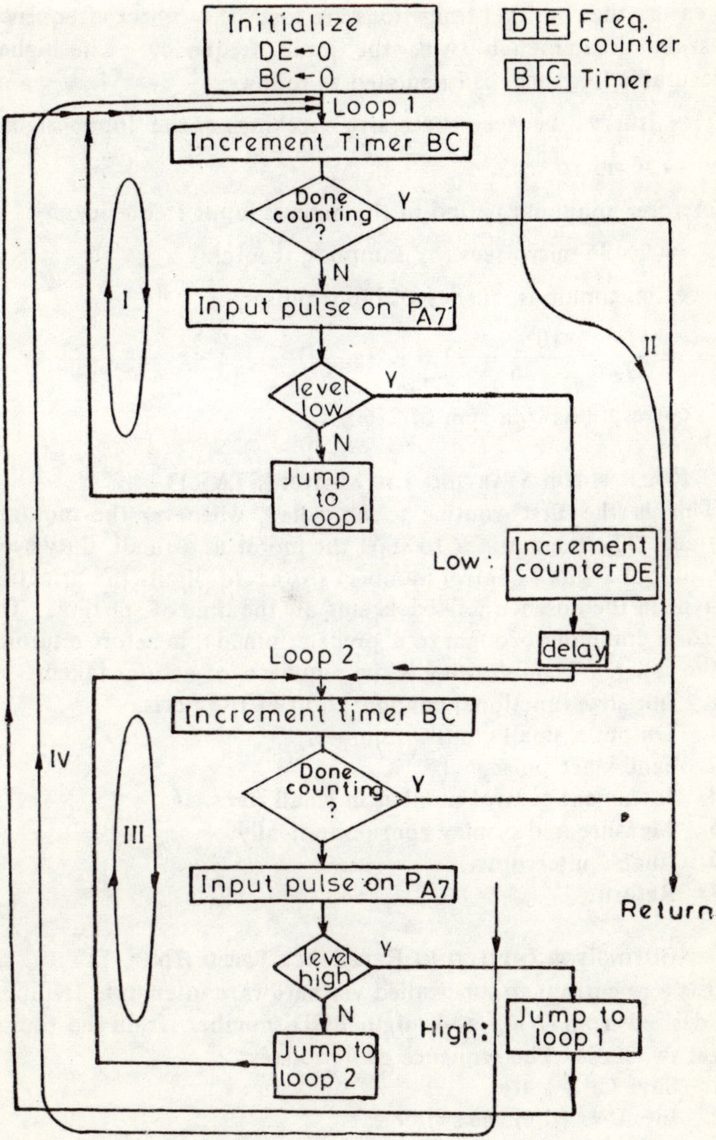

Fig. 11.26. Flow chart of frequency counter.

The coding of this routine is critical in terms of execution time, for the time between input checks (= loop execution time) dictates the maximum frequency that can be measured by this program. To ensure that no level transitions are lost, the 'check frequency' must be greater than twice the input frequency. The highest measurable frequency is calculated as follows:

T_r = Interval between successive checkings of the input signals

= 36 micro sec.

Therefore, minimum period of the highest input frequency

= 2×36 micro secs (by sampling theorem)

Hence, maximum frequency of input pulses

$$= \frac{1}{2T_r} = \frac{10^6}{2\times 36} = 13.8 \text{ KHz.}$$

This corresponds to a rpm of 2760.

(b) Routine for Starting the Motor (START)

This is the first routine to be called, whenever the motor is started. The main aim is to start the motor at a small duty cycle (i.e. output a small control number) so as to limit the armature current in the absence of back emf at the time of starting. The speed is gradually brought to a predetermined rpm before returning to the caller. The following is the sequence of actions taken:

1. Initialize functional configuration of I/O ports.
2. Output a small control number.
3. Send start pulse.
4. Increment control number in small steps.
5. Measure and display rpm periodically.
6. Enable interrupts.
7. Return.

(c) Routine for Inputting Reference Speed (INPUT)

It is a re-entrant routine called via hardware interrupt. It inputs the desired rpm (N_D), a 4 digit BCD number, from the thumb wheel switches. The sequence of actions is:

1. Save CPU state.
2. Input N_D from the switches.
3. Limit N_D to 2700 (max. measurable speed) if it is greater than 2700.

4. Convert the BCD value of N_D into binary.
5. Store the 2's complement of N_D in predetermined locations (2).
6. Restore CPU state.
7. Return.

(d) ROUTINE FOR DISPLAYING RPM (DSPLY)

It converts the actual rpm (N_A) stored in predetermined memory locations (by the frequency counter routine) into a BCD number and outputs it to the display unit. It uses a fast binary to BCD conversion subroutine using the DAA (Decimal Adjust Accumulator) instruction. The total execution time for the routine is 1.1 m sec.

(e) ROUTINE FOR CONTROL COMPENSATION (CNTRL)

This routine performs a simple type of compensation action viz. proportional control with gain of the compensator as 1. The sequence of actions is:

1. Subtract the reference rpm (N_D) stored in predetermined locations from the actual rpm (N_A) available in register pair DE at the end of frequency counter routine.
2. If $N_A - N_D = 0$, return (no control action needed).
3. If $|N_A - N_D| > 255$, limit the error to 255. This is to prevent too abrupt a correction being applied.
4. If the error $(N_A - N_D)$ is positive, subtract it from the previous control number.
5. If the error is negative, add it to the previous control number.
6. Output the new control number to the deriver circuit and also store it in predetermined locations.
7. Return (Total time: 98 μ sec).

If a value of gain other than 1 is to be used, the error can be multiplied by the gain factor before the correction is applied to the control number. By experimentation the proper value of the gain to optimize speed of response or steady state accuracy can be worked out.

(f) MAIN PROGRAM

The main program consists of the following few statements:

```
       Call  START
Loop:  Call  CTR
       Call  CNTRL
```

```
Call  DSPLY
JMP  LOOP
```

11.7.3 Fault Diagnosis in Three Phase Thyristor Converters using Microprocessor

Three phase thyristor converters are widely used in medium and high power applications such as large d.c. motor drives, generator excitation systems and h.v.d.c. transmission. In such applications it is of prime importance to build highly reliable systems incorporating fault diagnostic facility which would aid in locating the fault and isolating the faulty elements immediately. Further, corrective steps can be initiated immediately after system breakdown, thereby enhancing the reliability and maintainability of the system. With the availability of low cost microprocessors (μP), fault diagnostic schemes can be incorporated along with the control schemes with the addition of little hardware.

The system described here is built around a μP and therefore, it has the advantage of software control which drastically reduces the required hardware. The processor herein caters to mainly three needs.

(i) Fault diagnosis.
(ii) Thyristor firing (equidistant firing scheme).
(iii) Closed loop operation.

A CRT terminal is interfaced with the microcomputer system to display the nature and location of the fault.

11.7.3.1 *Types of Faults*

Various faults that occur in thyristor converter circuits are:

(i) Phase failure (PF).
(ii) Gate pulse failure (GPF).
(iii) Turn on failure of thyristor (TFT).
(iv) Short circuit across the thyristor (SCT).
(v) Short circuit across the d.c. terminals (SCD).

The above faults are illustrated in Fig. 11.27.

A PF fault occurs when one of the incoming phases is faulty. GPF may occur due to a fault in the pulse amplification or pulse isolation stages, which results in failure of corresponding thyristor to conduct. TFT may occur due to a fault internal to the device or insufficient pulse amplification. SCT may occur mainly due to the failure of the device or occasionally an accidental short between

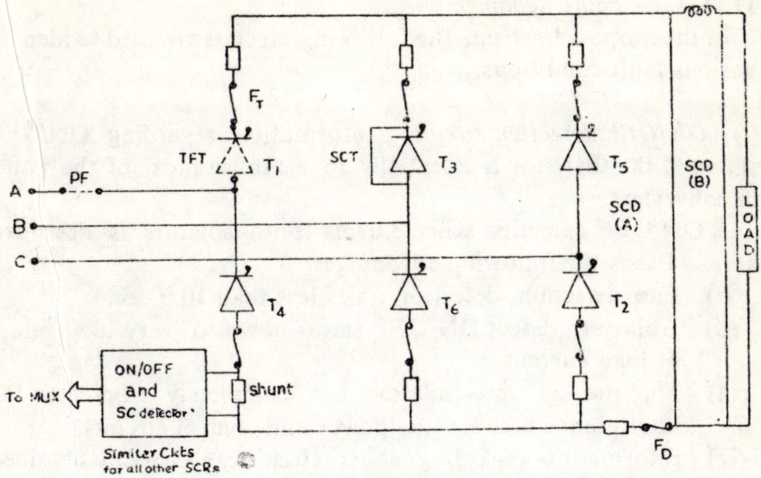

Fig. 11.27. Various faults in thyristor converter.

the anode and cathode terminals of the thyristor may lead to this fault. SCD may occur at two locations, (i) before the filter inductor SCD (A); (ii) after the filter inductor SCD (B).

(i) *SCD before the filter inductor*: Due to this fault the current shoots up instantaneously and eventually blows the fuses in series with the conducting thyristors. If the gating pulses to the incoming thyristors are not disabled, fuses of all the other thyristors also blow out. It is pertinent to note the fact that for converters of large rating, semiconductor fuses, which are connected in series with the thyristors, are very costly. The cost of the fuse being 10 per cent of the thyristor cost.

(ii) *SCD after the filter inductor*: A large value of filter inductance does not allow the current to rise abruptly even in a short circuited condition. If the converter is operated with a current loop, a properly designed controller can phase back the converter to a safe firing delay, thus protecting the converter.

All these faults cause a reduction in the converter output voltage and power, and also lead to unbalanced load condition resulting in increased input current harmonics.

11.7.3.2 *Fault Sending*

In the proposed scheme the following circuits are used to identify various fault conditions:

(i) *ON/OFF detection circuit*: Information regarding ON/OFF status of the thyristor is essentially to establish most of the faults in converters.

A ON/OFF detection scheme using opto-isolators is described here. It has the following advantages:

(i) Fast detection, detection times less than 10 μ sec.
(ii) Ability to detect ON/OFF status down to very low values of load current.
(iii) The use of opto-isolators has completely eliminated the interference between the power and control circuits.
(iv) Information regarding short circuit can be easily obtained from the same circuit.

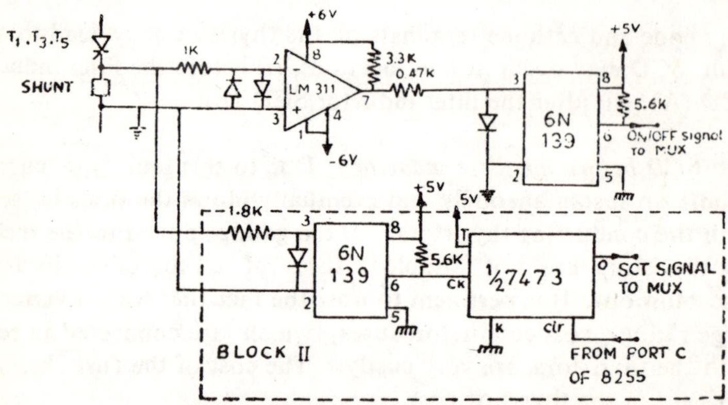

Fig. 11.28. ON/OFF and short circuit detection.

The circuit diagram of the scheme is given in Fig. 11.28. A low value shunt is connected in series with each thyristor. The voltage across which represents the current signal through the thyristor. This voltage is fed to a comparator. In the absence of current through the thyristor, the comparator is adjusted to be in negative saturation by using off set adjustment. This gives a low state at the output of opto-isolator indicating that the thyristor is OFF. Similarly, the presence of current through the thyristor as low as 2 per cent of the rated current, gives a high state at the output of opto-isolator indicating that the thyristor is ON. Two floating power supplies are used to power the detection circuits one each for the positive (T_1, T_3, T_5) and negative (T_2, T_4, T_6) group of thyristors.

(ii) *Short circuit detection*: The circuit diagram of this scheme is also given in Fig. 11.28. Under normal operating conditions the current through the thyristor is either zero or positive. This develops a positive or zero voltage across the shunt, driving the output of opto-isolator in block II high. Suppose one thyristor (say T_3 in Fig. 11.27) is shorted and another thyristor of the same group (T_5) is fired. A large reverse current flows through T_3. This reverse current develops a large negative voltage across the shunt, which drives the output of the corresponding opto-isolator low. A latch (7473) is used to hold this information.

The output of opto-isolator circuit in SCD fault detection is high under normal condition. The input bias current is so adjusted

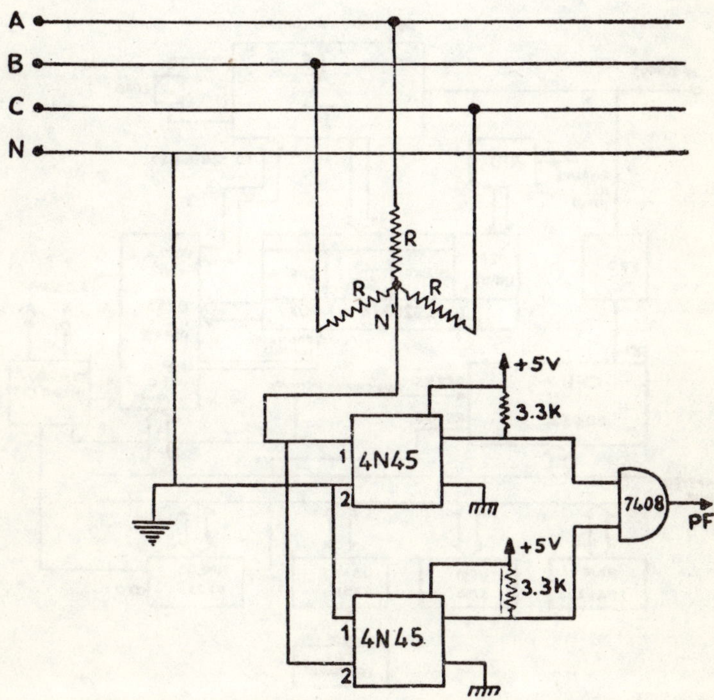

Fig 11.29. Circuit diagram of phase failure detection.

that the output goes low when an SCD fault occurs.

(iii) *Phase failure detection*; The circuit diagram of phase failure fault detection is given in Fig. 11.29. In the presence of all the three phases the potential between N and N' is zero (assuming balanced power supply). This produces high state at the output of both the opto-isolators, which are ANDed to produce a PFD signal which in this case, is high. Absence of any one of the phases causes an a.c. voltage between N and N'. The positive half cycle drives the output of one of the opto-isolators low and the negative half cycle drives the output of other opto-isolator low. Hence, PFD signal is always low indicating phase failure fault (PF).

11.7.3.3 *Hardware and Software Realisation*

The block diagram of the system is given in Fig. 11.30. The microcomputer system consists of one CPU (8085A), 2K EPROM

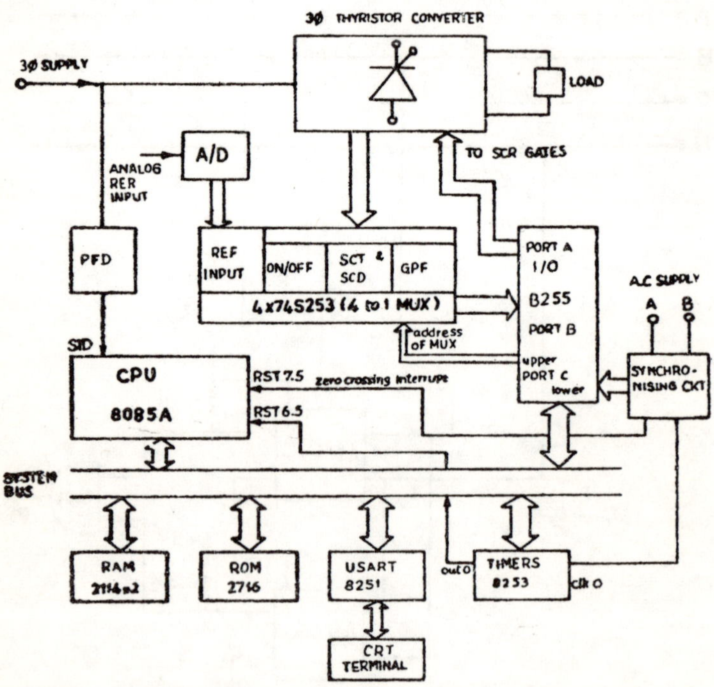

Fig. 11.30. Hardware realisation of fault diagnosis scheme.

(2716), 1K RAM (2×2114), one serial I/O port (8251A), three parallel I/O ports (8255A) and three programmable 16 bit timers (8253) all connected to a standard 8085A system bus.

(*i*) *Firing scheme*: The synchronising circuit (Fig. 11.30) senses the a.c. supply and uses a PLL to simulate the three phases (V_{ab}, V_{bc}, V_{ca}). It also generates a high frequency clock for the firing angle timer and six zero crossing pulses per supply cycle which are tied to RST 7.5 interrupt. The simulated *A*, *B* and *C* phases are connected to lower half of port *C* of 8255.

At each zero crossing the processor interrupted through RST 7.5 and the firing delay is loaded into the timer (8253). After the

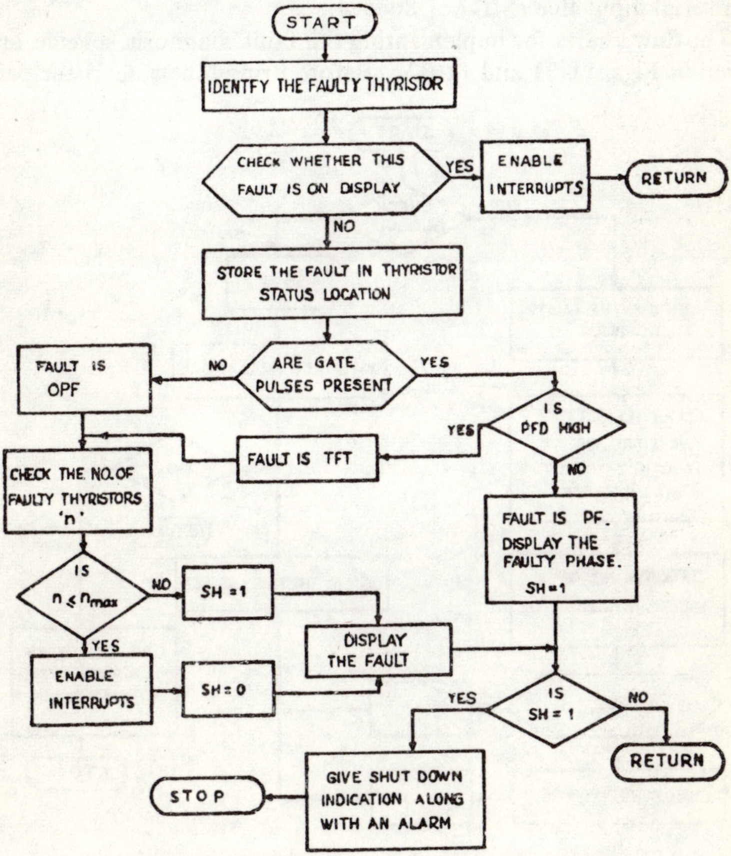

Fig. 11.31. Flow chart for short circuit routine.

delay, the timer interrupts the processor through RST 6.5. Upon receiving this interrupt the processor generates the firing pulses for the appropriate thyristors depending upon *A*, *B*, *C* status and firing angle range. The firing pulses from port *A* of 8255 are fed to the gates of the thyristors after optical isolation and pulse amplification.

(*ii*) *Fault diagnosis*: The outputs of the fault sensing circuits are connected to port *B* (8255) through an 8 bit 4 to 1 multiplexer (MUX). Any one of the inputs (8 bits) to the MUX may be selected under program control through 2 bits of port *C*. A CRT terminal is interfaced through an USART (8251) to display the nature and location of the fault. The PFD signal is connected to the serial input line (SID) of 8085 *A*.

The flow charts for implementing the fault diagnosis scheme are given in Figs. 11.31 and 11.32. Before firing a new thyristor pair

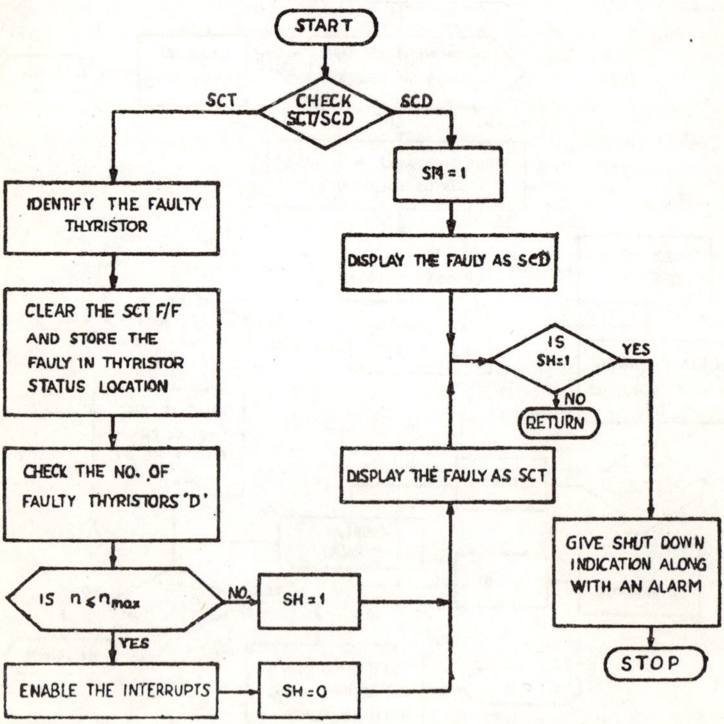

Fig. 11.32. Flow chart for on/off failure routine.

SCT/SCD is scanned. If a short circuit is detected processor enters into the short circuit routine. In this routine the corresponding fault is identified and displayed on the terminal. Immediately after firing a new thyristor pair, processor scans the ON/OFF status. On detection of a firing failure the processor enters into the ON/OFF failure routine. In this routine the corresponding fault is determined and displayed on the CRT terminal.

An SCD fault or a PF fault is serious enough to shut down the converter. In case of failure of only one thyristor shut down is not a must as it does not result in considerable reduction in power conversion. But, if more than one thyristor fails the power conversion ratio drastically falls, calling for a shut down of the converter. The converter is shut down by blocking the gate pulses to all the thyristors.

(iii) *Digital current feed back loop*: Most of the thyristor converters are operated with a current feedback loop which provides inherent over load and short circuit protection. The implementation of digital current feedback loop (DCFL) with proportional plus integral (PI) controller and some experimental results are presented in the following sections.

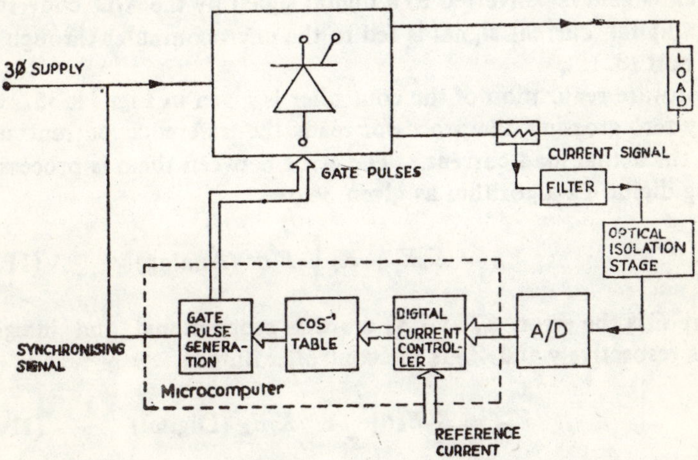

Fig. 11.33. Hardware realisation of digital current loop.

The block diagram of the DCFL is shown in Fig. 11.33. The current is sensed using a small resistor in series with the load.

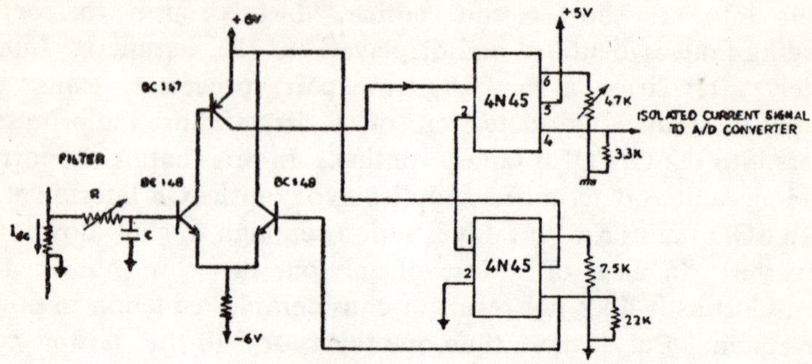

Fig. 11.34. Optical isolation of DC current signal.

The filter is a low pass RC network, whose time constant (T_F) may be adjusted. The filtered signal is then optically isolated. The optical isolation stage, shown in Fig. 11.34, comprises of a differential amplifier and two opto-isolators (4N45). A floating power supply which is isolated from the power supply of opto-isolator *A* is used for the differential amplifier and opto-isolator *B*. Assuming the opto-isolators to be identical, the output of opto-isolator *B* is proportional to the current signal. This analog current signal is converted to a digital signal by the A/D converter. The digital current signal is fed to the microcomputer through an I/O port (8212).

Software realization of the controller is given in Fig. 11.35. At every zero crossing, the processor reads the reference current and also the actual load current. The error between them is processed using digital PI algorithm as given below.

$$V_C = K_p E + K_I \int E\,dt \text{ (Analog)} \tag{11.1}$$

where E is the error, K_p and K_I are the proportional and integral gains respectively and V_C is the controller output.

$$V_{cn} = K_p E_n + \sum_{K=-\infty}^{n} K_I E_K \text{ (Digital)} \tag{11.2}$$

Equation (11.2) can be rewritten for ease of computation

$$\sum_{K=-\infty}^{n} K_I E_K = \sum_{K=-\infty}^{n-1} K_I E_K + K_I E_n \tag{11.3}$$

$$\sum_{K=-\infty}^{n-1} K_I E_K = V_{C(n-1)} - K_p E_{n-1} \tag{11.4}$$

$$V_{Cn} = V_{C(n-1)} + K_p[E_n - E_{n-1}] + K_I E_n \tag{11.5}$$

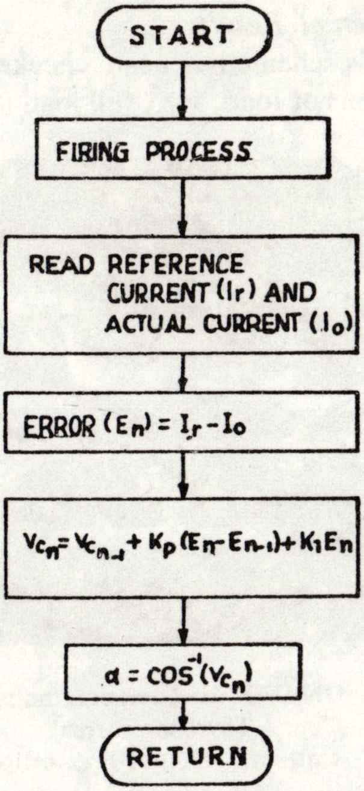

Fig. 11.35. Flow chart of digital current loop.

With the control voltage calculated from eqn. (11.5) the new firing angle 'α' corresponding to this voltage is obtained from arc cosine loop-up table. The entire computation time is very small compared to 3.33m sec (1/6th of the time period for 50 Hz supply).

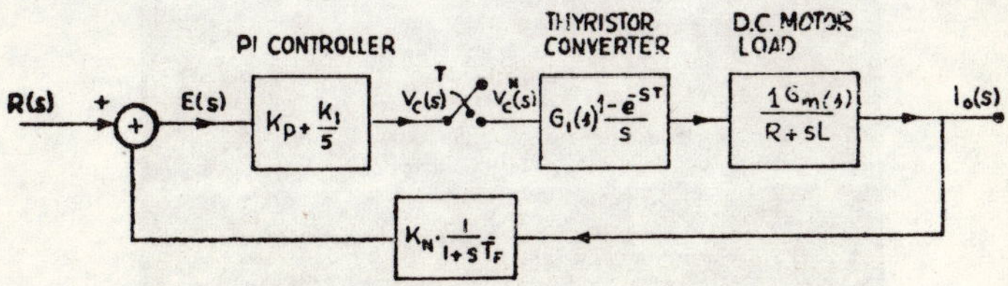

Fig. 11.36. Block diagram of digital current loop.

Block diagram of digital current loop is shown in Fig. 11.36 in

S-domain. *Z*-domain transfer function of the loop was obtained and PI controller parameters K_p and I_I were calculated so as to minimize the mean square error.

11.7.3.4 *Experimental Results*

The fault diagnosis scheme has been checked for various faults for different conditions of load, viz., full load to 2 per cent full load.

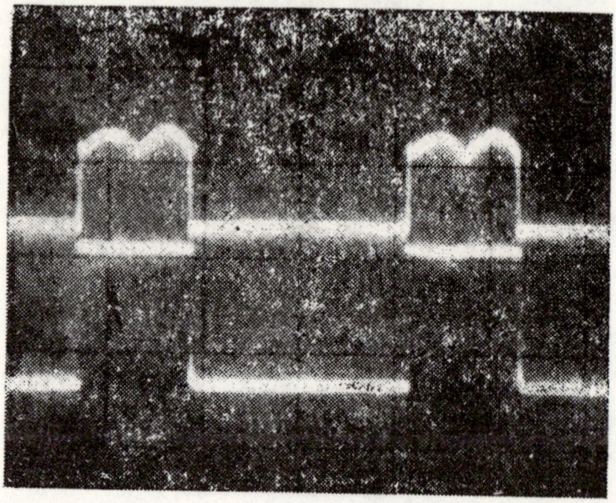

Fig. 11.37(a). ON/OFF detection-continuous conduction
Top: Thyristor current.
Bottom: ON/OFF detection signal.

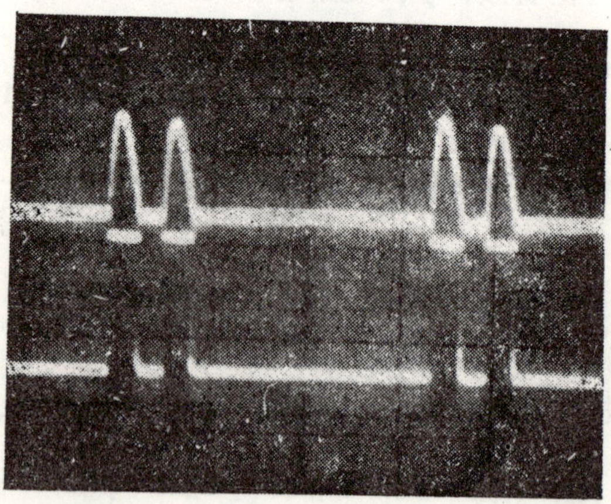

Fig. 11.37(b). ON/OFF detection discontinuous conduction.

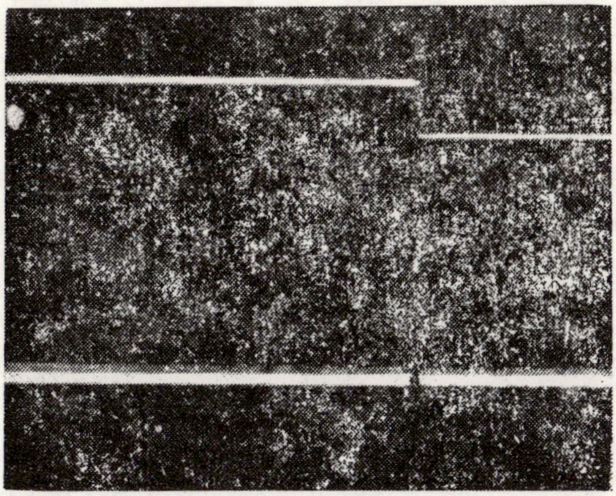

Fig. 11.37(c). Short circuit detection.
Top: Short circuit detection signal.
Bottom: Reverse current through the thyristor.

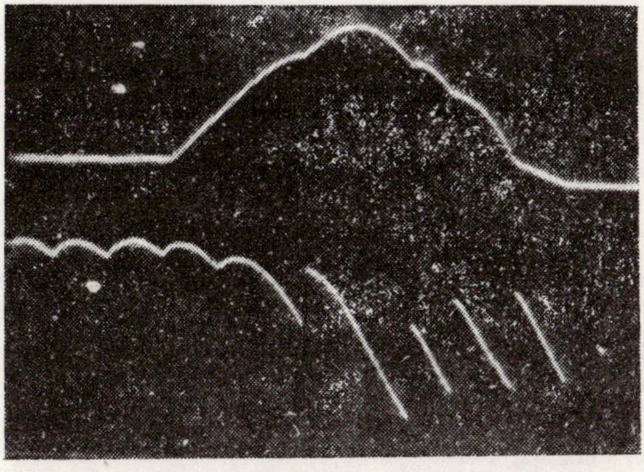

Fig. 11.37(d). Short circuit across the d.c. terminal.
Top: Converter output current.
Bottom: Converter output voltage.

The waveforms relating to ON/OFF detection with d.c. motor load with continuous and discontinuous conduction modes are given in Fig. 11.37(a) and Fig. 11.37(b). With a short circuit across the thyristor the SCD signal is shown in Fig. 11.37(c), which shows the opto-isolator output going low corresponding to a step increase of reverse current through thyristor. It can be seen that the reverse current goes to zero as the fuse blows out.

Figure 11.37 (d) shows the converter output current being adjusted to a set value after a short circuit occurs across the d.c. terminals after the filter inductor.

The appendix B gives a short listing of the programs developed for this application.

REFERENCES

1. T. R. Blakeslee, Digital Design with Standard MSI and LSI, 2nd Ed., Wiley-Interscience, Somerset, N. J., 1979.
2. J. L. Hilburn and P. M. Julich, Microcomputers/Microprocessors: Hardware, Software and Applications, Prentice-Hall, Englewood Cliffs, N. J., 1976.
3. R. L. Krutz, Microprocessors and Logic Design, John Wiley & Sons, N. Y., 1980.
4. L. A. Leventhal, 8080A/8085 Assembly Language Programming, Osborne and Associates Inc., Berkeley, 1978.
5. A. Osborne, An Introduction to Microcomputers, vol. I—Basic Concepts, Osborne and Associates Inc., Berkeley, 1976.
6. A. Osborne, An Introduction to Microcomputer, vol. II—Some Real Products, Osborne and Associates Inc., Berkeley, 1978.
7. B. A. Chetty, Role of Programmable Logic Controllers in India, Electronics-Information and Planning, vol. 8, No. 9, June 1981, pp. 651-664.
8. 8080/8085 Assembly Language Programming Manual, Intel Corporation, Santa Clara, California, 1978.
9. M6800 Microprocessor Applications Manual, Motorola Semiconductor Inc., Phoenix, 1975.
10. Y.V.V.S. Murty, G.K. Dubey and R.M.K. Sinha, 'Fault diagonosis in three phase thyristor converters using microprocessors', IEEE Trans. in Ind. Appl., vol. IA-20, Nov.-Dec. 1984, pp. 1490-1497.

APPENDIX A: INSTRUCTION SET FOR 8080/85

DATA TRANSFER GROUP

Move			*Move (contd.)*			*Move immediate*		
		Hex			Hex			Hex
MOV	A, A	7F	MOV	E, A	5F	MVI	A, byte	3E
	A, B	7B		E, B	58		B, byte	06
	A, C	79		E, C	59		C, byte	0E
	A, D	7A		E, D	5A		D, byte	16
	A, E	7B		E, E	5B		E, byte	1E
	A, H	7C		E, H	5C		H, byte	26
	A, L	7D		E, L	5D		L, byte	2E
	A, M	7E		E, M	5E		M, byte	36
	B, A	47		H, A	67			
	B, B	40		H, B	60		Load	
	B, C	41		H, C	61		Immediate	
	B, D	42		H, D	62			
	B, E	43		H, E	63	LXI	B, dble	01
	B, H	44		H, H	64		D, dble	11
	B, L	45		H, L	65		H, dble	21
	B, M	46		H, M	66		SP, dble	31
	C, A	4F		L, A	6F			
	C, B	48		L, B	68		Load/Store	
	C, C	49		L, C	69		LDAX B	0A
	C, D	4A		L, D	6A		LDAX D	1A
	C, E	4B		L, E	6B		LHLD adr	2A
	C, H	4C		L, H	6C		LDA adr	3A
	C, L	4D		L, L	6D		STAX B	02
	C, M	4E		L, M	6E		STAX D	12
	D, A	57		M, A	77		SHLD adr	22
	D, B	50		M, B	70		STA adr	32
	D, C	51		M, C	71			

(*Contd.*)

Move		*Move (contd.)*		*Move Immediate*	
	Hex		Hex		Hex
D, D	52	M, D	72	KCHG	EB
D, E	53	M, E	73		
D, H	54	M, H	74		
D, L	55	M, L	75		
D, M	56				

byte = constant, or logical/arithmetic expression that evaluates to an 8-bit data (second byte of 2-byte instructions)
dble = constant, or logical/arithmetic expression that evaluates to a 16-bit data (second and third byte of 3-byte instructions)
adr = 16-bit address (second and third bytes of 3-byte instructions)

Arithmetic and Logical Group

*Add**			*Increment**			*Logical**		
		Hex			Hex			Hex
ADD	A	87	INR	A	3C	ANA	A	A7
	B	80		B	04		B	A0
	C	81		C	0C		C	A1
	D	82		D	14		D	A2
	E	83		E	1C		E	A3
	H	84		H	24		H	A4
	L	85		L	2C		L	A5
	M	86		M	34		M	A6
ADC	A	8F	INX	B	03	XRA	A	AF
	B	88		D	13		B	AB
	C	89		H	23		C	A9
	D	8A		SP	33		D	AA
	E	8B					E	AB
	H	8C					H	AC
	L	8D					L	AD
	M	8E	Decrement**				M	AE
			DCR	A	3D	ORA	A	B7

(*Con d.*)

*Add**

Subtract*

		Hex
SUB	A	97
	B	90
	C	91
	D	92
	E	93
	H	94
	L	95
	M	96
SBB	A	9F
	B	98
	C	99
	D	9A
	E	9B
	H	9C
	L	9D
	M	9E

Double Add***

DAD	B	09
	D	19
	H	29
	SP	39

*Increment***

		Hex
	B	05
	C	0D
	D	15
	E	1D
	H	25
	L	2D
	M	35
DCX	B	0B
	D	1B
	H	2B
	SP	3B

Specials

DDA*	27
CMA	2F
STC***	37
CMC***	3F

Rotate***

RLC	07
RRC	0F
RAL	17
RAR	1F

*Logical**

		Hex
	B	B0
	C	B1
	D	B2
	E	B3
	H	B4
	L	B5
	M	B6
CMP	A	BF
	B	B8
	C	B9
	D	BA
	E	BB
	H	BC
	L	BD
	M	BE

Arith and Logical Immediate

ADI	byte	C6
ACI	byte	CE
SUI	byte	D6
SBI	byte	DE
ANI	byte	E6
XRI	byte	EE
ORI	byte	F6
CPI	byte	FE

* = all flags (C, Z, S, P, AC) affected

** = all flags except CARRY affected (exception INX and DCX affect no flags)

*** = only CARRY affected

BRANCH CONTROL GROUP						I/O AND MACHINE CONTROL	
Jump						Stack Ops	
		Hex					Hex
JMP	adr	C3				PUSH B	C5
JNZ	adr	C2				D	D5
JZ	adr	CA				H	E5
JNC	adr	D2				PSW	F5
JC	adr	DA					
JPO	adr	E2				POPB	C1
JPE	adr	EA				D	D1
JP	adr	F2				H	E1
JM	adr	FA				PSW*	F1
PCHL		E9					
						XTHL	F3
						SPHL	F9
Call							
CALL	adr	CD				Input/Output	
CNZ	adr	C4				OUT byte	D3
CZ	adr	CC				IN byte	D8
CNC	adr	D4					
CC	adr	DC					
CPO	adr	E4				Control	
CPE	adr	EC					
CP	adr	E4				DI	F3
CM	adr	FC				EI	FB
						NOP	00
Return			Restart			HLT	76
RET	C9		RST	0	C7		
RNZ	C0			1	CF	(8085 only)	
RZ	C8			2	D7		
RNC	D0			3	DF	RIM	20
RC	D8			4	E7	SIM	20
RPO	E0			5	EF		
RPE	E8			6	F7		
RP	F0			7	FF		
RM	F8						

APPENDIX B TO CHAPTER 11

```
; The program given in this appendix has been reproduced from
; The M. Tech. dissertation of Mr. Y. V. V. S. Murthy titled
: 'Control and Fault Diagnosis of Three Phase Thyristor
; 'Converters Using Microprocessor', EE, IIT, Kanpur.
; Feb., 1983.   See Text Section 11. 7. 3.
```

PART I

```
; THE FOLLOWING PROGRAM (IN INTEL 8085 ASSEMBLY LANG-
; UAGE) IS DEVELOPED TO FIRE THE 3-PHASE THYRISTOR CON-
; VERTER THE PROGRAM READS THE INPUT REFERENCE FROM
; PORT-B AND A, B, C, STATUS FROM PORT-C (8255)
; AND IT TRANSFERS THE GATE PULSES THROUGH
; PORT-A (8255)
; _______________________    ________________________
; INITIALISATION ROUTINE
; _________________________________________________
; 8255 IS PROGRAMMED IN MODE           O
; ___________________________________________________________
; PORT                  I/O              ADDRESS
; ______            _______           ___________
: PORT A-----------OUTPUT                40H
; PORT B------------INPUT                41H
; PORT C (UPPER)-------OUTPUT            42H
; PORT C (LOWER)-----INPUT               42H
; ___________________________________________________________
; 8253 IS PROGRAMMED IN THE FOLLOWING MODES
; ___________________________________________________________
; COUNTER NO.               MODE              ADDRESS
; ____________            _______            _ _ _ _ _
;      0                     0                  60H
;      1                     —               NOT USED
;      2                     —               NOT UEED
; ___________________________________________________________
;
```

```
   CMD55     EQU    83H;       Command word for 8255
   CM311     EQU    10H;       Command word for 8253 counter-0
   FST       EQU    0180H
   ACOS      EQU    0000H
   ORG       0000H
INIT:
   MVI A, 18H
   SIM
   MVI A, CMD55
   OUT   43H
   MVI   A, CM311
   OUT   63H
   ORG   0034H
   JMP   RST65
   NOP
   NOP
   NOP
   NOP
   NOP
   ORG   003CH
   JMP   RST75
   LXI   SP, 2040H
   EI
   LOOP: JMP LOOP
;  FIRING ROUTINE
;  ------------------------------------------
;  ZERO CROSSING INTERRUPT ROUTINE [RST 7.5]
;  ------------------------------------------
   RST75:
   IN 41H; Read input reference voltage
   CPI 40H; 1
   JC X1; Set the firing angle regions
   CPI 0C0H; 0 to 59, 60 to 119, 120 to 180
   JC X2;
   MVI B, 00H
   JMP X3
   X1: MVI B, 10H
   JMP X3
   X2: MVI B, 08H
   X3: MVI H, 01H
```

```
    RLC
    JC X4
    RRC
    MOV L, A
    MOV A, M; Get the ARC COSINE of the control voltage
    CPI OFFH
    JNZ X5
    MVI A, 224
    JMP X10
  X4: RRC
    CMA
    MOV L, A
    MOV A, M
    CMA
  X5: SUI 09H
    JZ X12
    JNC X10
  X12: MVI A, 01H
  X10: OUT 60H; Load the firing angle
    EI          ; in to the external counter
    RET
;  ________________________________________
;  TIMER INTERRUPT ROUTINE [RST 6.5]
;  ________________________________________
RST65:
  IN 42;  Read states of A, B, C phases
  RRC
  ANI 07H
  ORA B
  LXI H, FST; Get the starting address of the firing
  ADD L; sequence look up table in to the H, L
  MOV L, A; resistors
  MOV A, M
  OUT 40H; Fire the SCRs
  MVI A, CM311
  OUT 63H
  EI
  RET
;  ________________________________________
;  FIRING SEQUENCE LOOK UP TABLE
;  ________________________________________
```

```
; ORG FST
  DB 00H, 06H, 18H, 0CH, 21H, 03H, 00H, 00H, 0CH, 30H, 18H
  DB 00H, 00H, 18H, 21H, 30H, 0CH, 03H, 00H, 00H, 00H, 00H, 0
  DB 00H, 00H
; ____________________________
; ARC COSINE LOOK UP TABLE
; ____________________________
; ORG ACOS
  DB 255, 224, 212, 202, 194, 187, 180, 174, 168, 163, 158, 153
  DB 149, 144, 140, 136, 132, 128, 124, 120, 117, 113, 110, 106
  DB 103, 100, 97, 94, 91, 87, 84, 82, 79, 76, 73, 70
  DB 67, 65, 62, 59, 57, 54, 51, 49, 46, 44, 41, 39
  DB 36, 34, 32, 29, 27, 24, 22, 20, 17, 15, 13, 11
  DB 8, 6, 4, 2, 254, 252, 250, 248, 246, 244, 241, 239
  DB 237, 235, 233, 231, 229, 227, 225, 223, 220, 218, 216, 214
  DB 212, 210, 208, 206, 204, 202, 200, 198, 196, 194, 192, 190
  DB 188, 186, 184, 182, 180, 178, 177, 175, 173, 171, 169, 167
  DB 165, 163, 161, 159, 157, 155, 153, 151, 149, 148, 146, 144
  DB 142, 140, 138, 136, 134, 132, 130, 128
  END
```

PART II

```
; ________________________________________________
; THE FOLLOWING PROGRAM IN INTEL 8085A ASSEMBLY
; LANGUAGE IS DEVELOPED TO DETECT VARIOUS FAULTS
; IN 3-PHASE THYRISTOR CONVERTER
; THE FAULTS ARE ABBREVIATED AS FOLLOWS
; ________________________________________________
; PF: PHASE FAILURE FAULT
; GPF: GATE PULSE FAILURE
; SCT: SHORT CIRCUIT ACROSS THE THYRISTOR
; TFT: TURN-ON FAILURE OF THYRISTOR
; SCD: SHORT CIRCUIT ACROSS THE D. C. TERMINALS
; ________________________________________________
; 8251 IS PROGRAMMED IN THE FOLLOWING MODE
; 7 BITS (ASCII), NO PARITY, ASYNC*16, ONE STOP BIT
; ________________________________________________
; ADDRESS OF A/D CONVERTER IS OCOH
  ________________________________________________
CMD55   EQU  83H;    COMMAND WORD FOR 8255 PORT
CM311   EQU  10H;    COMMAND WORD FOR 8253 COUNTER-
CM312   EQU  76H;    COMMAND WORD FOR 8253 COUNTER-
MD51    EQU  4AH;    Mode Word for 8251
CMD51   EQU  27H;    Command word for 8251
CTIL    EQU  14H;    4800HZ Baud rate generator
CTIH    EQU  00H
FST     EQU  0380H
ACOS    EQU  0300H
LABEL   EQU  0400H
TFT     EQU  0400H
GPFL    EQU  040CH
SPFL    EQU  0418H
SCL     EQU  0424H
ERROR   EQU  0430H
CF      EQU  0439H
SHD     EQU  0533H
STATUS  EQU  2001H
ORG     0000H
```

```
INT:
    MVI A, 18H
    DB   30H
    MVI A, CMD55
    OUT   43H
    MVI   A, CM311
    OUT   63H
    MVI   A, CM311
    OUT   63H
    MVI   A,   CM312
    OUT   63H
    MVI   A,   CTIL
    OUT   61H
    MVI   A,   CTIH
    OUT   61H
    MVI   A,   CMD51
    OUT   81H
    IN    80H
    IN    80H
    LXI   SP,   21FFH
    LXI   H,   START
    CALL  LBL
    JMP   INIT
    ORG   0034H
    JMP   RST65
    NOP
    NOP
    NOP
    NOP
    NOP
    ORG   003CH
    JMP   RST75
INIT: LXI   H,   FIRE
    CALL   LBL
X11:  IN   81H
    RRC
    RRC
    JNC   X11
    IN   80H
    CPI   'F'
```

```
  JZ  X22
  LXI  H,  ERROR
  CALL  LBL
  JMP  X11
X22: LXI  H,  TABLE
  CALL  LBL
  MVI  A,  00H
  OUT  42H
  LXI  H,  STATUS
  MOV  M, A
  CMA
  OUT  42H
  EI
LOOP:  JMP  LOOP
; ______________________________________________
; ZERO CROSSING INTERRUPT ROUTINE RST 7.5
; ______________________________________________
RST75:
  PUSH  PSW
  PUSH  H
  IN OCOH; READ INPUT REFERENCE VOLTAGE
  CPI  40H;
  JC  X1; ! SET THE FIRING ANGLE
  CPI OCOH; ! REGIONS 0 to 60, 60 to 120 and 120
  JC  X2; !
  JMP  X3
X1: MVI  B,  10H
  JMP  X3
X2: MVI  B, 08H
X3: LXI  H,  ACOS
  RLC
  JC  X4
  RRC
  MOV  L, A
  MOV  A, M; GET THE ARC COSINE OF THE CONTROL
         VOLTAGE
  CPI  OFFH
  JNZ  X5
  MVI  A, 224
  JMP  X10
```

```
X4:  RRC
  CMA
  MOV  L, A
  MOV  A, M
  CMA
X5:  SUI  09H
  JZ  X12
  JNC  X10
X12:  MVI A, 01H
X10:  OUT 60H; LOAD THE FIRING ANGLE IN THE EXTERNAL
  POP  H
  POP  PSW
  EI
  RET
;  ________________________
;  RST 6.5 ROUTINE
;  ________________________
RST65:  PUSH PSW
  PUSH  H
  PUSH  D
  IN  42H; Read A, B, C status
  RRC
  ANI  07H
  ORA  B
  LXI  H, FST
  ADD  L
  MOV  L, A
  MOV  C, M
  MVI  A, CM311
  OUT  63H
  MVI  A, 60H
  OUT  42H
  CALL DELY
  IN  41H
  CPI  00H
  JNZ  SCFT; JMP SC routine
  MOV  A, C
  OUT  40H
  CALL DELY
  CALL DELY
```

```
  CALL DELY
  CALL DELY
  MVI  A, 20H
  OUT  42H
  CALL DELY
  IN  41H
  ANA C
  CMP C
  JNZ ONOF;  JMP ON/OFF routine
RT: POP  D
  POP  H
  POP  PSW
  EI
  RET
; ————————————————————
; SHORT CIRCUIT ROUTINE
; ————————————————————
;
SCFT:
  MOV  D, A
  ANI  40H
  JNZ  DCSC
  LXI  H, STATUS
  MOV  A, D
  ORA  M
  JPF  XX1
  MVI  A, OCOH
  OUT  42H; Clear SC flip-flop
  RRC
  OUT  42H
  EI
XX: LXI H, SCL
  CALL LBL
  MOV  A, D
  CALL FAULT
  DB  20H
  ANI  04H
  RNZ
  CALL SHDN
DCSC: MVI A, 80H
```

```
    OUT   40H;
    LXI   H, DCS
    CALL  LBL
    CALL  SHDN
XX1: MVI A, 80H
    OUT   40H;
    JMP   XX
; ___________________________
;  ON/OFF FAILURE ROUTINE
; ___________________________
;
ONOF:
    MOV H, A
    DB   20H
    RLC
    JC PF; Phase failure fault
    MOV A, H
    CMA
    ANC C
    JPE  SHDN
    MOV D, A
    LXI  H, STATUS
    MOV A, M
    CMP D
    JZ RT; Display over
    MOV A, D
    ORA M
    JPE  YY
XY3:  MOV M, A
    MVI A, OAOH
OUT  42H
    IN  41H
    ANA  D
    JZ GPF; Gate pulse failure
    EI
    LXI H, TFT; Turn on failure of thyristor
    CALL  LBL
    MOV A, D
    CALL FAULT
    JMP  XY4
```

```
YY: MVI A, OFFH
   DB  3OH
   MVI A, 80H
   OUT 40H
   JMP XY3
GPF: MOV A, D
   EI
   LXI H, GPFL
   CALL LBL
   MOV A, D
   CALL FAULT
   JMP XY4
PF: MOV A, H
   CMA
   ANA C
   MOV D, A
   LXI H, SPFL
   DI
   MVI A, 80H
   OUT 40H
   CALL LBL
   MVI A, 24H
   ANA D
   JNZ PHAA
   MVI A, 09H
   ANA D
   JNZ PHAB
   MVI E, 'C'
   CALL FOUND
   CALL SHDN
PHAA: MVI E, 'A'
   CALL FOUND
   CALL SHDN
PHAB: MVI E, 'B'
   CALL FOUND
   CALL SHDN
XY4: DB 20H
   ANI 07H
   JZ XXX
   CALL SHDN
```

```
XXX: EI
  RET
; ________________________________________________
; LBL:
; SUBROUTINE TO SEND MESSAGES ON TO CRT TERMINAL
; ________________________________________________
; FAULT:
; SUBROUTINE TO FIND THE FAULTY THYRISTOR
; ________________________________________________
; SHDN
; SHUT DOWN INDICATION SUBROUTINE
; ________________________________________________
; CRLF
; SUBROUTINE  TO  SEND  CARRIAGE  RETURN AND LINE FEED
; ________________________________________________
```

PART III

```
; ————————————————————————————————————————————
; THE FOLLOWING PROGRAM COMPUTES THE CURRENT LOOP
; WITH DIGITAL PI CONTROLLER
; REG. 'C' CONTAINS THE CONTROL VOLTAGE (Vc)
; REG. 'D' CONTAINS THE PRESENT ERROR (En)
; REG. 'E' CONTAINS THE PREVIOUS ERROR (En—1)
; CONTROL EQUATION IS Vcn=Vcn−1+KP*(En−En—1)+KI*En
; ————————————————————————————————————————————
;
;
   IN OCOH; Read actual current
   MOV D, A
   IN 41H; Read reference current
   CMP D
   JC NEG
   SUB D
   MOV D, A
   JMP YY3
NEG: MOV H, A
   MOV A, D,
   SUB H
   ORI 80H
   MOV D, A; Store the error in REG. D
YY3: MOV A, E
   RLC
   JC CH
   RRC
   ORI
   JMP YY1
CH: RRC
   ANI 7FH
YY1: MOV L, A
   MOV H, D
   CALL ADDN
   MOV H, A
   MOV A, D
```

```
   RLC
   JC XNEG
   RRC
   RRC
   RRC
   ANI 1FH
   JMP KID
XNEG: RRC
   RRC
   RRC
   ANI 1FH
   ORI 80H
KID: MOV L, A
   CALL ADDN
   MOV H, A
   MOV L, C
   CALL ADDN
   RLC
   JNC VCP
   MVL C, OOH
   JMP YY2
   VCP: RRC
YY2: MOV C, A
   MOV E, D
   RET
; ______________________________________________
; ADDN:
; THIS SUBROUTINE ADDS TWO NUMBERS IN SIGN
; MAGNITUDE FORMAT
; Z=X+Y
; 'Y' SHOULD BE IN REG. 'H'
; 'Y' SHOULD BE IN REG. 'L'
; THE RESULT 'Z' WILL BE STORED IN THE ACCUMULATOR
; ______________________________________________
```

PROBLEMS

Chapter 2

2.1 Can you apply a forward voltage across a thyristor soon after the reverse recovery current drops to zero?

2.2 A thyristor is triggered by a pulse train of 5 kHz. The duty ratio of the pulse train is 0.5. If allowable average gate power is 100 watts calculate the maximum allowable gate drive power.

2.3 (a) For a given SCR the gate-cathode characteristics has a spread given by the following equations:

$$I_G = mV^2 \quad I_G = mV^{3/2}$$

where I_G is in mA, V in volts, $m = 2$.
The gate source voltage is 15 V, and the load line has a slope of -112.5 V/amp. Calculate:

(i) Series resistance to be connected in the gate circuit.

(ii) Trigger voltage V_{GT} and trigger current I_{GT} given that the average gate power dissipation is 0.5 W.

(b) What are the other protective elements generally connected in the gate circuit.

2.4 (a) Mention different ratings of an SCR.

(b) Why does the average current rating decrease with a decrease in the conduction angle?

(c) A thyristor is used in a single-phase have-wave circuit supplying resistive load. The specification sheet shows that the maximum direct current rating of the device is 230 A. Calculate the average current rating of the thyristor if the firing angle $\alpha = 90°$.

(d) What is the effect of inserting an inductance in the load circuit on the average current rating of the thyristor.

2.5 For the circuit shown in Fig. P 2.1 answer the following.

(a) Calculate the maximum values of di/dt, dv/dt for the SCR.

(b) What are the rms and average current ratings of the SCR for $\alpha = 60°, 90°, 120°$. Neglect the effect of L (for this part) on the waveform of i. Draw the waveforms of i for these firing angles.

(c) What is the voltage rating of the SCR?

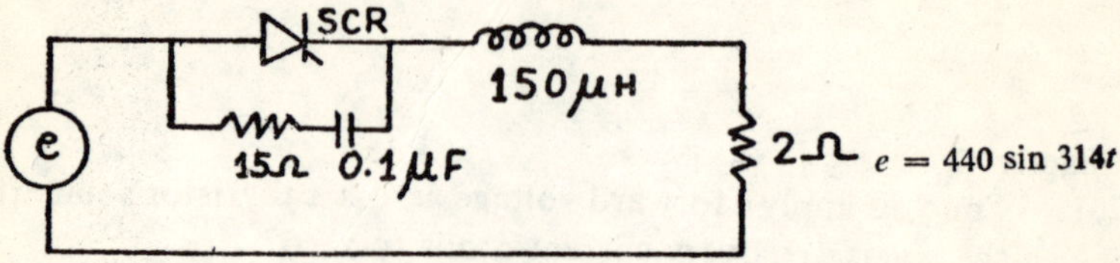

Fig. P 2.1

2.6 A thyristor is rated to carry an rms current of 200 A with an allowable case temperature of 80° C. Determine the following assuming the maximum allowable junction temperature $T_j = 125°$ C.

(a) Maximum allowable current if the thyristor is to carry the direct current continuously.

(b) If the thyristor is used in a single-phase half-wave circuit supplying a purely resistive load, what are the maximum allowable average currents for $\alpha = 0$ and 90°?

(c) Determine the case temperatures for the part (b) if the specification sheet shows that the average powers dissipated are 210 W and 165 W for $\alpha = 0$ and 90°, and the thermal impedances are 0.17° C/W and 0.19° C/W respectively.

2.7 Enumerate the differences between an RC firing circuit and UJT pulse firing circuit.

2.8 (a) What are different methods of commutation?

(b) Sketch the time variations of i_T, i_D, v_L, v_C if the thyristor is turned on with the initial capacitor voltage $v_c(0) = -100$ V in the circuit of Fig. P 2.2. Assume the elements and devices to be ideal.

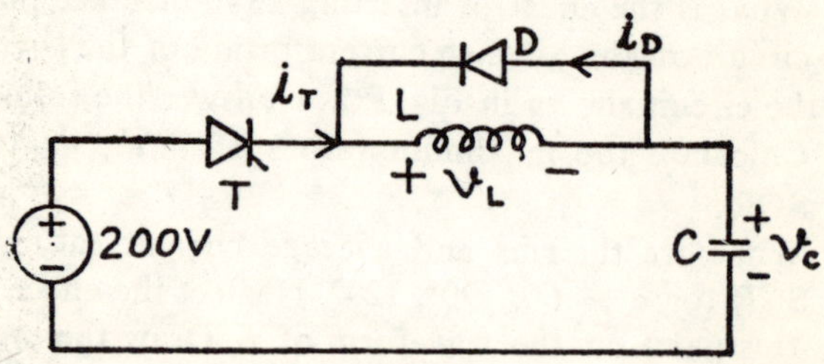

Fig. P 2.2

2.9 Figure P 2.3 shows a circuit illustrating line commutation. Assume the firing angle $\alpha = 70°$ and the load inductor to be ideal.

(a) Determine the turn off times offered to thyristors T_1 and T_2 if T_1 is turned on first.

(b) Same as (a) but the thyristor T_2 is turned on first.

(c) Sketch the voltage waveform across the thyristor T_1 in (a) and (b).

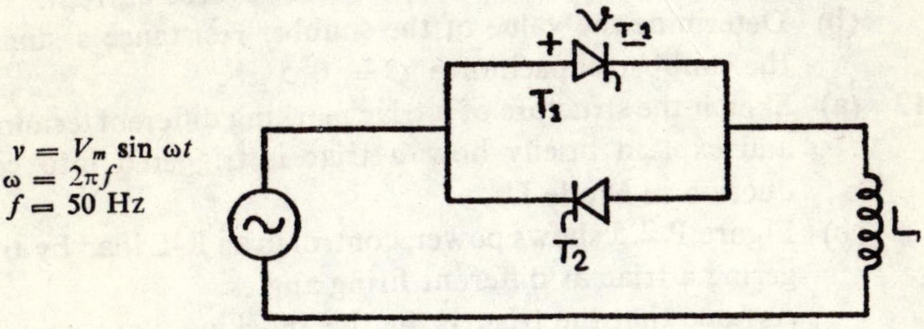

Fig. P 2.3

2.10 (a) Determine the expressions for L_1 and C in Fig. P 2.4 if the device turn off time is t_{off} and the maximum load current is I_0 (Assume constant load current).

(b) What is the effect of adding another branch shown in the dotted lines on the turn off time provided to T_1. Explain qualitatively whether the capacitance as determined in part (a) is sufficient for commutation at the full load current. If not, do you choose higher or lower value?

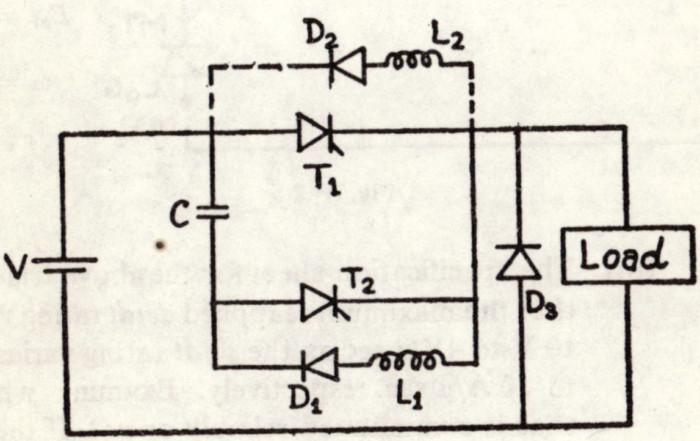

Fig. P 2.4

2.11 A thyristor is commutated by applying a reverse voltage across the thyristor. The maximum reverse recovery current is 20 A which is interrupted in a time interval of 5μ sec. Neglect load resistance and assume an inductance of 10 mH in series with the thyristor. Take source voltage which is applied across the series connection of the thyristor and inductance as −400 V.

(a) Calculate the peak voltage appearing across the thyristor at the time of interrupting the reverse current.

(b) Determine the value of the snubber resistance assuming the snubber capacitance $C = 0.5\ \mu F$.

2.12 (a) Sketch the structure of a triac marking different terminals and explain briefly how a triac is triggered into conduction in Mode I.

(b) Figure P 2.5 shows power control in an R-L load by triggering a triac at different firing angles. Assume that the triac is in the blocking state prior to the application of a triggering pulse.

(i) Derive an expression for the current i, in terms of α, e, R, L, and Load phase angle ϕ.

(ii) Compute $\frac{di}{dt}$, $\frac{de}{dt}$ at the end of conduction angle r with the following data.

$\alpha = 60°$, $\phi = 30°$, $R = 10\ \Omega$, $L = 20$ mH, $r = 150°$.

Triac blocking resistance $= 50\ k\Omega$.

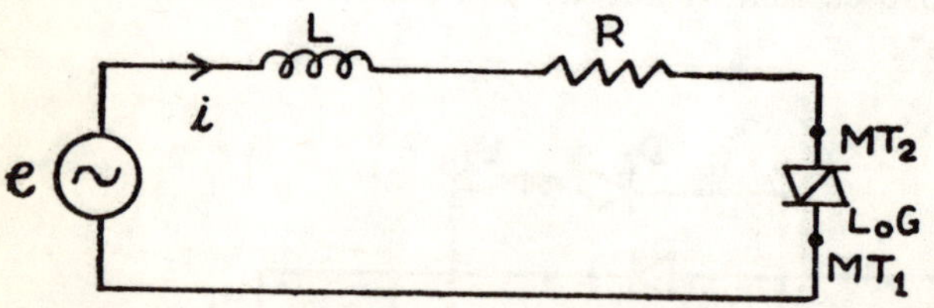

$e = E_m \sin \omega t$.

$E_m = \sqrt{2} \times 220$

$\omega = 2\pi f$

$f = 50$ Hz

Fig. P 2.5

(iii) The specification sheet for the above triac indicates that the maximum reapplied de/dt rating varies from 10 V to 4V/μ sec as the di/dt rating varies from 5 A to 10 A/μ sec respectively. Examine whether the triac is commutated reliably or not. If the commutation failure is likely, suggest a suitable method for limiting the de/dt rating so as to operate the triac with power control by varying the firing angle α.

2.13 Explain the essential difference with regard to commutation of (i) triac (ii) two thyristors connected in antiparallel.

2.14 A thyristor string consists of five thyristors in series. The string voltage is 5000 volts. Maximum permissible blocking voltage of each thyristor is 1200 volts. If the maximum blocking current is 15 mA, calculate the value of the equalising resistance to be connected in parallel with each thyristor.

2.15 A thyristor string is formed by the series and parallel connection of thyristors. The voltage and current ratings of the string are 6000 volts and 4000 amperes respectively. Available thyristors have the voltage and current ratings of 1200 volts and 1000 amps respectively. The string efficiency is 90% for both the series and parallel connections. Calculate the number of thyristor to be connected in series and parallel. If the maximum blocking current is 15 mA and $\Delta Q_{max} = 25$ microcoulombs, calculate the values of R and C in Fig. 2.43.

Chapter 3

3.1 A single-phase fully controlled thyristor bridge converter supplies a load consisting of R, L and V_C. The inductance L in the output circuit is so large that the output current may be considered to be virtually constant. Assume the thyristors to be ideal and take the following data:

RMS supply voltage = 220 V, Load resistance = 0.5 ohm

Output current, $I_0 = 10$ A.

(a) Calculate the firing angle, α if $V_C = 135$ V
(b) Calculate the firing angle, α if $V_C = -145$ V
(c) Which source (ac or dc) is supplying power to the load in (a) and (b)?
(d) Sketch the time variations of output voltage, v_0 in (a) and (b).

3.2 A single-phase fully controlled thyristor bridge converter supplies an inductive load. Assuming that the output current is virtually constant, and is equal to I_0, determine the following if the supply voltage = 220 V and the firing angle, $\alpha = 60°$.

(a) Average output voltage of the converter
(b) Supply rms current
(c) Supply fundamental rms current
(d) Displacement factor or fundamental power factor
(e) Supply power factor
(f) Supply harmonic factor.

3.3 (a) Derive an expression for the average output voltage of a three-phase bridge type thyristor converter as a function of the firing angle α, neglecting the source inductance.

(b) Suppose that the converter in part (a) is supplied from a non-ideal source. Determine the expression for the output voltage if the supply is represented by an ideal source in series with an inductance, L_S. Assume that there is no overlap in successive commutations.

(c) A three-phase fully controlled thyristor bridge converter supplies a dc voltage source of 400 V having an internal resistance of 1.8 ohm. The output current may be assumed to be constant at 20 A since there is a large inductance in the output circuit. The supply rms voltage per phase is 230 V and the source inductance in each phase is 5 mH. The source resistance may be neglected. Calculate the following:

(i) Firing angle α for an output voltage of 436 V.
(ii) Overlap angle in degrees.
(iii) Draw the time variation of the waveform of the voltage across the thyristor conducting the positive current in the phase, A for one complete cycle.

3.4 Consider any multipulse fully controlled converter operating at a triggering angle α, and supplying a load current which may be considered virtually constant. Show that, assuming the thyristors and the supply voltage as ideal, the mean ac power input from the supply system is equal to the mean output power.

3.5 A three-phase fully-controlled converter is employed to charge a battery with an emf of 95 V and an internal resistance of 0.1 ohm. The supply rms voltage is 110 V line-to-line and sufficient inductance is included in the output circuit to maintain the current virtually constant at 20 A. Determine:

(i) The delay angle, α required.

(ii) The power factor at the ac source terminals.

3.6 Consider a single-phase fully controlled thyristor bridge converter. Each thyristor has a snubber circuit, R-C connected across it. The source inductance is L_S and the load inductance is infinitely large. For the purpose of designing the snubber circuit, draw the appropriate equivalent circuit during the commutation of one pair of diagonally connected thyristors, and determine the equivalent snubber circuit resistance and capacitance.

3.7 A three pulse converter which eliminates dc magnetization of the supply transformer is shown in Fig. P 3.1. The transformer secondary is connected in the zig-zag configuration. The phase A of the primary winding of the transformer is coupled to the two secondaries a_1 and a_2. Similarly B and C are coupled to b_1, b_2 and c_1, c_2 respectively. Assume that the load current is constant.

(i) With respect to the voltage v_1, what is the earliest instant (zero firing angle) at which the thyristor T_1 is triggered into conduction and what is the range of the firing angle.

(ii) Derive an expression for the average output voltage.

(iii) Sketch v_A, v_1, i_{T_1}, i_{T_2}, i_A.

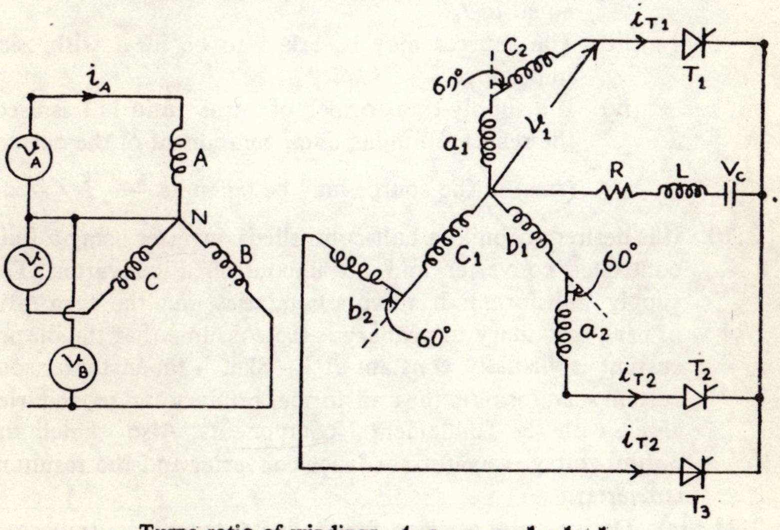

Turns ratio of windings, A, a_1, a_2 = 1 : 1 : 1

Fig. P 3.1

(iv) Derive expressions for the supply power factor, displacement factor, distortion factor.

3.8 The supply power factor of an ac to dc converter fed dc load is generally defined in two ways as follows:

(i) Ratio of the output power ($V_0 I_0$) to the total apparent power.

(ii) Product of distortion factor and displacement factor. Which definition in your opinion is more accurate, and why is it so? Under what conditions, both definitions lead to the same result?

3.9 Two three-phase line commutated fully controlled bridge converters are connected in series to produce a high direct voltage.

(a) Suggest 3 possible connections and/or control techniques.

(b) Sketch the source current waveform in each case for the output voltage of 0.5 p.u.

(c) Calculate the displacement factor in each case.

(d) Calculate the supply power factor in each case.

The following assumptions may be taken for granted.

(i) The maximum output voltage of each converter is the same.

(ii) The output current is continuous and constant being equal to I_0.

(iii) The sources may be taken to be ideal with zero impedance.

(iv) If a supply transformer of turns ratio 1:1 is used, the reflected fundamental component of the current (rms) in the source may be taken as $\frac{\sqrt{6}}{\pi} I_0 \angle -\alpha$.

3.10 It is desired to build a half-controlled converter using a fully controlled converter and an uncontrolled converter. The supply transformer has two secondaries and the turn ratio of each secondary to primary is 0.5. Assume that the output current is virtually constant at I_0. Sketch the instantaneous current waveforms in the transformer primary and secondaries along with the fundamental components. Also sketch the output voltage waveforms of each converter and the resultant converter.

3.11 (a) Derive an expression for the instantaneous output voltage of a three-phase *symmetrical half-controlled* converter.

The output voltage of the three-pulse fully controlled converter is given by the following expression.

$$v_0 = \frac{3\sqrt{6}V\cos\alpha}{2\pi} + \sum_{N=1}^{\alpha} \frac{3\sqrt{6}V}{2\pi}\left[\frac{1}{(3N-1)^2} + \right.$$

$$\left. + \frac{1}{(3N+1)^2} - \frac{2\cos\alpha}{(3N-1)(3N+1)}\right]^{1/2} \cdot \sin(3N.\omega t + \phi_{3N})$$

where V is the supply rms voltage per phase

α is the firing angle

ω is supply frequency in rad/sec.

(b) Plot the variation of the amplitudes of 3rd and 6th harmonics (in p.u. of maximum direct voltage) as a function of the firing angle, α.

3.12 (a) An asymmetrical single-phase half controlled bridge circuit with finite source and load inductances is shown in Fig. P 3.2. The source inductance may be considered to be small compared to the load inductance but cannot be neglected. The input voltage and output current waveforms are shown in Fig. P 3.3 along with the subintervals. There are four subintervals in each half cycle of the supply voltage.

(i) Sketch i_{T_1}, i_{D_1}, i_{T_2}, i_{D_2}, i_L and v_0.

(ii) Draw the equivalent circuit of the converter valid for each subinterval for one half-period of the supply voltage.

(b) Explain briefly what you have understood by 'Half-waving effect' in a single-phase symmetrical half-controlled converter.

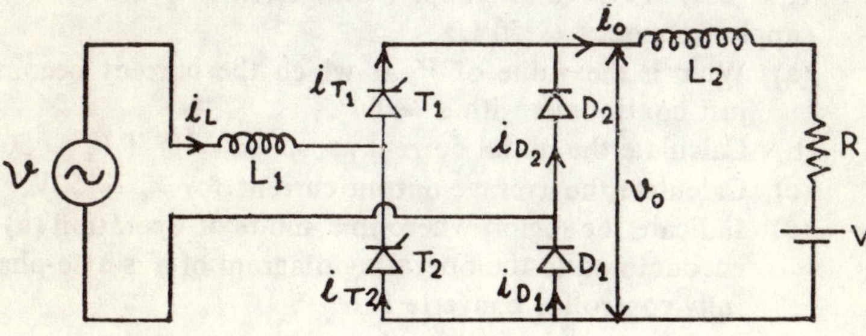

Fig. P 3.2

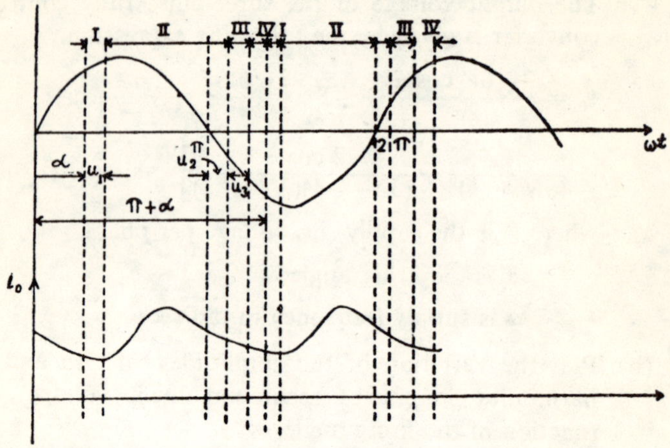

Fig. P 3.3

3.13 Consider a single-phase fully controlled converter with optional freewheeling having firing angle α_1 (angle of connection of the supply voltage to the load) and another firing angle α_2 (angle of disconnection of the supply voltage from the load). Assume that the converter output current is virtually constant at I_0. Derive expressions for:

(i) Average output voltage of the converter, V_0

(ii) Supply rms current, I_{SRMS}

(iii) Supply fundamental rms current, I_{1RMS}

(iv) Displacement factor or fundamental power factor, FPF

(v) Power factor, PF

3.14 A fully controlled converter is used to supply a load consisting of R, L, and V_c. The data given is as follows:
$R = 2\ \Omega$, $L = 25$ mH, supply voltage (rms) $= 250$ V
supply frequency $= 50$ Hz

(a) What is the value of V_c at which the current becomes just continuous with $\alpha = 60°$.

(b) Calculate the initial current at $\omega t = \alpha = 60°$ if $V_c = 50$V.

(c) Calculate the average output current for $V_c = 50$V.

(d) Indicate the region where this mode of operation (a) is encountered in the operating diagram of a single-phase fully controlled converter.

3.15 (a) Derive an expression for the average output voltage of a symmetrical three-phase half-controlled thyristor converter. Assume continuous output current. Consider both cases when $\alpha < 60°$ and $\alpha \geqslant 60°$.

(b) A symmetrical three-phase half-controlled thyristor converter is used to charge a battery. If the supply rms voltage (line to line) is 440 V, effective load resistance is 0.46 ohm and the battery voltage is 400 V, answer the following. Neglect source impedance.

(i) Calculate the firing angle to limit the load current to 100 *A*. Assume a large filter inductance in the load circuit.

(ii) Supply power factor.

(iii) Sketch the voltage waveform across any one thyristor and determine the turn off time for the conditions given in the problem.

3.16 (a) Show the operating diagram of a single-phase fully controlled converter and mark qualitatively the regions of discontinuous conduction for the load phase angle,

$$\phi = \tan^{-1} \frac{\omega L}{R} = 45°.$$

(b) For certain load parameters, it has been observed from the operating diagram of the single-phase fully controlled converter, the current is discontinuous. Whenever the incoming thyristors are triggered, they turn on at α. Sketch the output voltage and output current waveform if $\alpha < \theta$, $\theta = \sin^{-1} \frac{V_C}{\sqrt{2} V}$ where V is the rms supply voltage.

(c) It is desired to make the current continuous for the condition given in (b). Derive an expression for determining the boundary between discontinuous and continuous conduction regions.

3.17 A single-phase fully controlled converter is operated in such a way that it provides half controlled characteristics in the rectification mode.

(a) Making use of the operating diagram for rectifier operation shown in Fig. 3.35, determine whether the load current is continuous or discontinuous if $\alpha_1 = 20°$, $\phi = 60°$, $m = 0.4$, $R = 2.25\ \Omega$.

(b) Calculate the average current in (a) if $V_S = 220$ V.

(c) The load source is increased to 0.55 p.u. with the same firing angle. Will the current be still continuous? If not, determine the mode of discontinuous conduction.

(d) Calculate the additional inductance required in the output circuit for the output current to be just continuous using the diagram of Fig. 3.39 (a).

3.18 A single-phase fully controlled converter with half controlled characteristics is operated in the inversion mode.

(a) Referring the operating diagram of Fig. 3.37, determine whether the current is discontinuous or continuous if $\alpha_2 = 60°$, $\phi = 60°$, $R = 2.25\ \Omega$, $m = 0.35$. Sketch output voltage and output current waveforms using this data.

(b) Suppose the converter operates in the discontinuous mode with $\alpha_2 = 90°$ and $m = -0.4$. Calculate the additional inductance to be inserted in the output circuit for the current to be just continuous in this operating mode. You may refer the diagram shown in Fig. 3.39 (b). Supply rms voltage to the converter = 220 V, $f = 50$ Hz.

3.19 A three-phase dual converter consisting of two three-phase fully controlled bridges operates under circulating current operation. Calculate the peak value of the circulating current given the following data.

Supply rms voltage per phase = 230 V

$$f = 50 \text{ Hz}$$

Circulating current limiting reacter, $L = 20$ mH

Resistance of the reacter is neglected.

$$\alpha_1 = 60°$$

$$\alpha_2 = 120°$$

3.20 (a) Why is each thyristor in a 6-pulse bridge converter supplied with a gate pulse of 120° width ?

(b) Explain briefly with a block diagram 'cosine-wave crossing control' scheme for the three-phase bridge converter and enumerate the advantages and disadvantages of this method of triggering.

Chapter 4

4.1 The ideal cycloconverter of Fig. 4.1 provides output voltage, v_0 and output current, i_0 which are assumed to be sinusoidal. Sketch these waveforms indicating the intervals during which *P*-converter and *N*-converter conduct. Also, state clearly whether the average power flow is from the source to the load or vice versa.

(a) $\phi_0 = 0$
(b) $\phi_0 = 45°$ lagging
(c) $\phi_0 = 120°$ lagging
(d) $\phi_0 = 45°$ leading
(e) $\phi_0 = 120°$ leading.

4.2 The single-phase to single-phase cycloconverter of Fig. 4.3 (a) is supplying a resistive load. The phase delay angle α has the same value for the four thyristors. The supply transformer has the turns ratio 1 : 1 : 1. The frequency ratio, $f_0/f_s = 5$. Sketch the time variations of the following waveforms for $\alpha = 0$ and 60°.

(a) Gate pulses for the four thyristors.
(b) Output current.
(c) Supply current.

4.3 The single-phase to single-phase cycloconverter of Fig. 4.3 (a) is supplying an *R-L* load. The output to input frequency ratio is 1/5. Sketch to scale the time variations of output current and output voltage for one period at the output frequency for

(a) Load $pf = 0.866$ lagging and the firing angle $\alpha = 60°$.
(b) Load $pf = 0.5$ lagging and $\alpha = 30°$. The turns ratio of the input transformer is 1 : 1 : 1 and the supply voltage may be taken as 220 V rms at the frequency 50 Hz. Load circuit resistance = 2 Ω.

4.4 What precautions must be built into the control circuit if the circuit in problem 4.3 is to operate satisfactorily at the frequency ratios (f_s/f_0) that are not necessarily integer multiples.

4.5 (a) Derive an expression for the rms output voltage of a circulating current mode cycloconverter constructed using *m*-pulse converters.

(b) Calculate the maximum possible rms output voltage of a three-phase to single-phase converter operating in the circulating current mode if the cycloconverter is constructed employing 6-pulse *P*- and *N*-converters, each converter being supplied from a transformer having 2 : 1 turns ratio. The converter transformers (*Y-Y*) are supplied from a balanced three-phase supply source with an rms line to line voltage of 440 V. From the considerations of commutation, the firing angle in the inversion mode should not exceed 165°.

4.6 A single-phase to single-phase cycloconverter of Fig. P 4.1 is required to provide an output frequency which is 1/3 of the input frequency.

(a) Sketch the gating pulses and output voltage waveform for the maximum output voltage. Neglect turn off time intervals of thyristors.

(b) What should be turns ratio *n* if the peak output voltage is twice the peak supply votage?

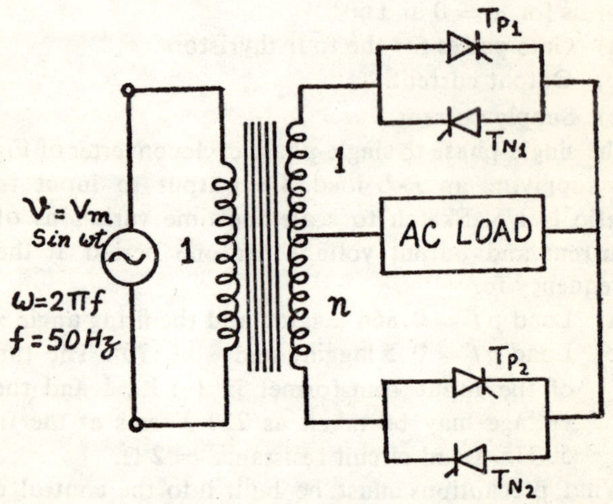

Fig. P 4.1

4.7 A three-phase to single-phase cycloconverter is constructed using three-pulse converters. It is operated in the circulating current mode with the following data:

$$\text{rms supply voltage} = \frac{440}{\sqrt{3}} \text{ V per phase,}$$

Reference signal : Sinusoidal

Modulation signal; Co-sinusoidal

Input to output frequency ratio = 5.
Supply frequency = 50 Hz.

(a) Draw to scale the time variations of the following waveforms for one period at the output frequency for the maximum output voltage ($r = 1$).

(i) output voltage of the positive converter
(ii) output voltage of the negative converter
(iii) output voltage at the mid-point of the circulating current reactor.

4.8 A three-phase to single-phase cycloconverter is constructed using three-pulse converters. It is operated in the non-circulating current mode with the following data.
rms supply voltage = $440/\sqrt{3}$
Input to output frequency ratio = 6
Supply frequency = 50 Hz
Load: $R = 2.5\ \Omega$, $\omega_0 L = 1.443\ \Omega$ at 5 Hz
Reference signals : sinusoidal
modulating signals: co-sinusoidal

(a) Draw to scale the time variation of the output voltage of the cycloconverter for one period of the frequency for the maximum output voltage ($r = 1$).
(b) repeat part (a) for $r = 0.5$.

4.9 Comment on the ac source current of a cycloconverter and explain why the overall supply power factor is poorer for a cycloconverter than for a converter.

4.10 A three-phase to single-phase non-circulating cycloconverter constituted by 6-pulse bridge converters is to supply an *R-L* load. The output frequency is 2 Hz and the inductive load reactance at this frequency is $\omega_0 L = 1.6\ \Omega$. The load circuit resistance is 2 ohms. The cycloconverter is supplied directly from a three-phase 440 V, (rms line to line) 50 Hz source. The cycloconverter uses cosinusoidal modulating signals and sinusoidal reference voltage. Calculate:

(a) RMS output voltage if $r = 0.5$. Neglect commutation intervals of thyristors.
(b) output and input powers neglecting losses within the converter
(c) RMS output current

(d) peak thyristor voltage

(e) supply power factor if the reactive component in the source due to load power factor and phase control process is 0.321 p.u. of maximum output current.

Chapter 5

5.1 A single phase ac regulator of Fig. 5.2 with a resistance load has the following details:

source voltage = 230 V, source frequency = 50 Hz and $R = 5$ ohms

Calculate:

(i) The greatest rms and average thyristor currents for any firing angle α.

(ii) The minimum circuit turn-off time for any firing angle α.

(iii) The range of the firing angle α for which the greatest forward or reverse voltage is applied to either of the thyristors and the magnitudes of these voltages.

(iv) The maximum value of the di/dt occurring in the thyristors.

(v) The ratio of the amplitudes of third harmonic and fundamental for $\alpha = \pi/4$.

5.2 A single phase ac regulator with an L load has the following details:

source voltage = 230 volts with a frequency by 50 Hz and $\omega L = 5$ ohms

Determine

(i) The control range of the firing angle.

(ii) The maximum possible value of the load current.

(iii) The maximum possible values of rms and average thyristor currents.

(iv) The range of firing angle α for which the greatest forward or reverse voltage is applied to either of the thyristors and the magnitudes of these voltages.

(v) The maximum value of di/dt occurring in the thyristors.

(vi) Calculate the conduction angles and show the nature of the waveforms of the load current, load voltage and thyristor voltage for $\alpha = 0$ and $\alpha = 135°$ assuming a gate pulse of duration π.

5.3 Repeat problem 5.2 for $R = 4$ ohms and $\omega L = 3$ ohms.

5.4 Calculate the average current, rms current and maximum

repetitive voltage ratings of thyristors of a 1-phase ac regulator of Fig. 5.2 supplying a 100 kW load with a source voltage of 230 V, and efficiency and power factors of 0.9 and 0.8 respectively.

5.5 A 3-phase Y-connected controller with a resistance load has the following details:
The line to line source voltage = 440 V, frequency = 50 Hz and R = 5 ohms. Calculate,
(i) The maximum average and rms thyristor currents for any value of the firing angle α.
(ii) The maximum forward and reverse thyristor voltages for any value of the firing angle α.
(iii) The control range of the firing angle.

5.6 Repeat 5.5 for R = 4 ohm and X_L = 3 ohms.

5.7 For the three phase ac regulator of Fig. 5.17(a) draw the waveforms of the gate pulses, the line current i_A, load line voltage v_{ab} and thyristor voltage v_{Aa} for $\alpha = \pi/2$.

Chapter 6

6.1 Consider the circuit of figure 6.5(a). It is desired to use it in the following configurations.
(a) Type D chopper in the second and the third quadrants, i.e. negative load current and bipolar load voltage.
(b) Type C chopper in the third and the fourth quadrants, i.e. negative load voltage and bipolar load current.
Make a table showing how the switches S_1 to S_4 must be opened, closed or modulated to obtain these configurations.

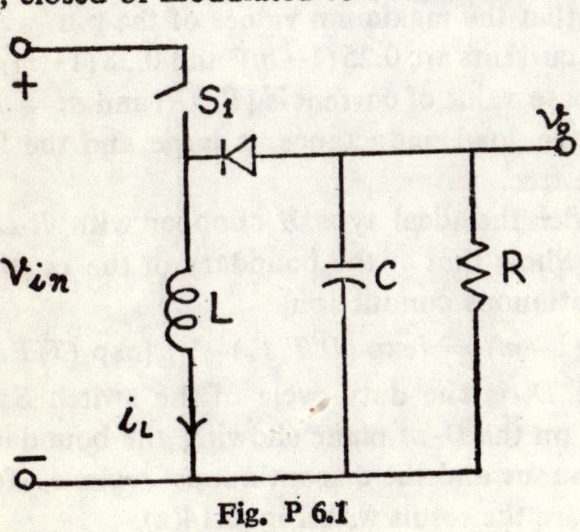

Fig. P 6.1

6.2 An inverting chopper is shown in Fig. P 6.1. Switch S_1 is operated under the PWM strategy (Sec. 6.2.1).

(a) Show that if $RC \gg T$, in the steady-state,

$$v_0 = -V_{in}D/(1-D),$$

where $D = T_{on}/T$

(b) Estimate the peak to peak ripple in i_L and v_0 for the following parameters. $V_{in}=12$ V, $L=8$ mH, $C=100\,\mu F$, $R=100$ ohm, $T=1$ mS, $T_{on}=0.25$ mS.

6.3 For the ideal type A chopper of Figure 6.12(a) show that the average input current is given by,

$$I_{in} = \frac{D(V_{in}-E_b)}{R} - \frac{L}{RT}(I_1 - I_2)$$

where $D = T_{on}/T$.

6.4 Consider the type *A* chopper using current commutation shown in Fig. 6.17(a). Neglect R_f and the components shown dotted. The inductance of the source has also been neglected. Let V_{in} = 230V, C_1 = 60 μF, L_1 = 30 μH. Assume i_o to be constant at 200 Amp. during commutation. Calculate the turn-off time provided to the main thyristor T_1 and the total time required for commutation.

6.5 For the component values given in problem 6.4 above, calculate the turn-off time provided by the voltage commutated chopper shown in Fig. 6.19(a). Are the values of L_1 and C_1 reasonable? If not, suggest alternative values using the design equations of Sec. 6.6.2.

6.6 For the ideal type *A* chopper with *R-L-*E_b load (Fig. 6.12a) show that the maximum values of the p.u. average and rms diode currents are $0.25(1-m)^2$ and $0.38(1-m)^{3/2}$ respectively. The base value of current is (V_{in}/R) and $m = E_b/V_{in}$. Assume that the load inductance is large and the load current is ripple free.

6.7 Consider the ideal type *B* chopper with *R-L-*E_b load (Fig. 6.1c). Show that at the boundary of the continuous and the discontinuous conduction,

$$(1-m) = \{\exp(D'T/T_a)-1\} / \{\exp(T/T_a)-1\},$$

where D' is the duty cycle of the switch S_2. Hence plot a curve on the *D–m* plane showing the boundary between the continuous and the discontinuous regions, for $T/T_a = 0.5$, compare the result with Fig. 6.14(a).

6.8 Consider the ideal type *A* chopper (Fig. 6.12a) with the following operating conditions.
$V_{in} = 220V, f = 400$ Hz, $D = 0.4$, $R = 0.5$ ohm, $L = 2$ mH
$E_b = 22V$, Calculate the following quantities.
(a) The average value of the load current
(b) The rms values of the first, second and the third harmonic of the load current.
(c) Approximate rms value of the load current using results of (a) and (b).
(d) The maximum and the minimum load currents I_1 and I_2.
(e) The average value of the supply current using the result of problem (6.3) above.
(f) The input power, the power absorbed by the source E_b and the power loss in the resistor *R*.
(g) Using result of (*f*) calculate the rms value of the output current. Compare with the result obtained in part (c), above.

6.9 Extend the circuit diagram of Fig. 6.15(a) to a four phase chopper. Assume that the output current is ripple free. Draw the waveforms of the input current for the four ranges of T_{on}. Show the conducting switches in various parts of a period, *T*.

6.10 (a) Figure 6.27 shows a voltage controlled oscillator (VCO) operating at 400 Hz and producing a clock and a ramp output. Suggest a circuit for realising this VCO.
(b) A 'current feedback' signal is assumed to be present Fig. 6.27 as an input. Describe a method for obtaining such a signal. The signal must be isolated from the power circuit.

Chapter 7

7.1 The bridge inverter of Fig. 7.2(a) supplies a single phase load with an output voltage of 230V and rms load current of 100A. If the load power factor varies widely calculate
(a) The rms thyristor current and
(b) repetitive peak voltage across the thyristors

7.2 The half bridge inverter of Fig. 7.1(a) supplies a load consisting of $R = 1$ ohm, $\omega L = 5$ ohm and $1/\omega C = 4$ ohm in series. The dc source voltage ($V/2$) is 115V. Calculate the fundamental, third harmonic and fifth harmonic components in the source current [Hint: Use Fourier series expansion for the output voltage

$$v_0 = \sum_{n=1}^{\infty} \frac{2V}{n\pi} \sin n\,\omega t,\ n = 1,3,5.]$$

7.3 In the bridge circuit of Fig. 7.2(a) the positive cycle of the load current is from

(a) $T/8$ to $5T/8$

(b) $7T/8$ to $11T/8$

Draw the nature of the load current waveform showing the devices under conduction during different intervals for one cycle of these two cases.

7.4 The bridge inverter of Fig. 7.2(a) supplies a RLC load. The source voltage is 230V, $R = 1$ ohm, $\omega L = 5$ ohm, and $T = 0.1$ m sec. All harmonics in the load current can be neglected. What should be the value of C for the load commutation to be possible. The thyristor turn off time is 10 μsec. Choose $t_c = 1.5\ t_q$

[Hint: use Fourier expansion for the load voltage

$$v_0 = \sum_{n=1}^{\infty} \frac{4V}{n\pi} \sin n\,\omega t,\ n = 1,3,5...]$$

7.5 McMurray inverter of Fig. 7.10(a) uses a commutation circuit consisting of C=25 μF and L=25 μH. The source voltage (V) is 230V dc. The load current varies from 50 to 150 Amps at the instant of commutation. Would this circuit provide reliable commutation if the thyristor turn off time is 25 μsec?

7.6 The modified McMurray inverter of Fig. 10(a) is fed by a dc source of 230 V. The dc source voltage may fluctuate by ± 20%. The current during commutation may vary from 10 to 100 Amps. If the turn off time of thyristors is 25 μsec calculate the values of C and L. Use a factor of safety of 2, i.e. $t_c = 2\ t_q$.

7.7 Modified McMurray-Bedford inverter of Fig. 7.16 is fed by a dc source of V=230 V. The dc source may fluctuate by ±15 percent. The current during commutation may vary from 15 to 110 Amps. If the thyristor turn-off time is 15 μsec, calculate the values of commutation circuit parameters. Use a factor of safety of 2 i.e. $t_c = 2\ t_q$. Do you get a unique solution?

7.8 Three phase inverter circuit of Fig. 7.20 is fed by a dc source of 230V. The dc source voltage may fluctuate by ± 30 volts.

The maximum inductor current during commutation is 50 Amps. If the thyristor turn off time is 20 μsec, calculate the values of C and L of the commutation circuit. Choose $t_c = 2t_q$.

7.9 The output voltage of the single phase bridge inverter of Fig. 7.2(a) is controlled by pulse-width modulation. It has two pulses situated at 45° and 135°. When the pulse width $2d = 45°$, the positive half cycle of the current is from 22.5° to $\pi + 22.5°$. The load current can be assumed to be sinusoidal.

Draw the gate pulses for all the thyristors. Draw the load voltage and current waveforms and show various devices under conduction during various intervals of a cycle of the load voltage.

7.10 A single phase C. S. I. (Fig. 7.44) has the following parameters:

$$C = 50\ \mu\text{F},\ I = 20\ \text{A},\ f = 1/T = 500\ \text{Hz}.$$

(a) Determine the range of R over which the inverter can be operated if the third harmonic in output current is to be kept lower than 15% of the fundamental.

(b) What is the lowest value of % third harmonic over the range of R in (a).

(c) Can this inverter be operated with thyristors having turn-off time of 250 μ S.

(d) Plot the power output as a function of R over the range in (a).

(e) If R is opened (i. e. $R = \infty$) calculate the peak value of the reverse voltage across the thyristors.

7.11 The harmonic content of the load current in the above problem, 7.10, can be improved if an inductor is used in series with the load. The inverter configuration of Fig. 7.45(a) is then employed with the R-L load. The circuit parameters are $L = 0.8$ mH, $C = 50\ \mu$F, I = 20 A, $f = 1/T = 500$ Hz.

(a) Calculate the range of R over which the inverter is in the overlap mode.

(b) For $R = 8$ ohm, calculate the third harmonic content in the load current as % of the fundamental.

(c) For $R = 8$ ohm, calculate the power dissipated in the load.

(d) How does the power dissipated in the load vary as R is varied from 8 ohm to 800 ohm.

(e) Calculate the turn-off time and the peak inverse voltage of the thyristors when R is zero ohms.

7.12 A three phase C.S.I. (Fig. 7.54) is operated at 10 Hz with a purely inductive load. $C = 30\ \mu F$, inductive load $= 20$ mH in each phase, $I_{dc} = 4A$.

(a) Calculate the peak capacitor voltage and the turn-off time of the thyristors.

(b) Calculate the maximum frequency of the inverter at which commutations (mode (i) and (ii)) occur in distinct time intervals.

(c) Calculate the maximum frequency of the inverter at which no more than three diodes are conducting simultaneously.

(d) If each of the thyristors have a series inductance of $5\ \mu H$ due to layout and leads, calculate the maximum rate of rise of anode current. Is this value within acceptable limits ?

(e) In Figure 7.54 the capacitors are connected in delta. What should be the value of the capacitors if they are to be connected in star ?

7.13 At the operating conditions of the above problem 7.12, plot the variables shown in Fig. 7.56. In addition plot the following:

(a) The capacitor voltage $V_{C_1}, V_{C_2} \cdots V_{C_6}$ taking reference directions shown in Fig. 7.54.

(b) The voltage across the input terminals of the inverter.

(c) The voltage across diode D_1 in Fig. 7.54.

7.14 A practical inverter has a current source consisting of thyristor converter operating from 3ϕ, 50 Hz supply and a series inductor L as shown in Fig. 7.41. Assume that the inverter is operating at 10 Hz and the inductor L is finite.

What are the frequencies present in the spectra of the dc link current, i_{dc}, and the input line current from the ac supply.

7.15 A series inverter has the following data:

$V = 230$ V, $L = 40\ \mu H$, $C = 10\ \mu F$, $R \leqslant 1$ ohm. Assume that the inverter is ideal. Calculate:

(a) RMS value of the fundamental voltage across L.

(b) RMS value of the fundamental voltage across C.
(c) RMS value of the fundamental load current.
(d) Average and RMS source currents.
(e) output power.

7.16 In the series inverter of Fig. 7.62,
$V = 230$ V, $L = 40\mu$H, $C = 5\mu$F, $R = 4$ ohms,
turn of time of thyristors used $= 10\mu$ sec. Determine:
(a) The maximum output frequency.
(b) The peak thyristor voltage.
(c) The peak thyristor current.
(d) The power output.
(e) The rms thyristor current.
(f) Sketch to scale the time variations of i and V_c.

7.17 In the series inverter of Fig. 7.62,
$V = 200$ V, $R = 2$ ohms, output frequency $= 5$ KHz, thyristor turn off time $= 20\ \mu$ sec.
(a) Calculate the values of L and C required if the peak thyristor voltage is to be 250 V.
(b) Calculate the output power for the part (a).

Chapter 8

8.1 (a) A reactive load is supplied from a single-phase 220 V ac power supply. The load current varies between the two extreme limits (3-j 0) A to (5-j 10) A. It is required to compensate the load reactive power completely and improve the supply power factor to unity using a fixed capacitor and a thyristor controlled reactor of linear characteristic. Determine the required values of capacitor and inductor.

(b) If the load current has changed to (5-j5), determine the firing angle at which the gate pulses are to be applied to the thyristors of the TCR. The fundamental component of the current drawn by the TCR is given by $\frac{-V}{\omega L}\left(1-\frac{2}{\pi}\alpha' - \frac{1}{\pi}\sin 2\alpha'\right)$ where $\alpha' = \alpha - \frac{\pi}{2}$ and α is the firing angle of a thyristor in $1-\phi$ ac regulator $\left(\frac{\pi}{2} < \alpha < \pi\right)$.

(c) Assuming that the current after load change is in steady state, what is the minimum time taken by the controller to correct the power factor to unity?

(d) Suppose that the TCR has acted immediately after the load has changed. What are the factors that would decide the final time for the controller to settle to a steady firing angle?

8.2 A wye-connected inductive load is supplied from a three-phase supply system having a line to neutral voltage of 10 kV with a short-circuit level of 250 MVA per phase. The impedance angle under short-circuit conditions is 78.69°. The mean power of the load is 25 MW per phase and the reactive power of the load per phase varies from 0 to 50 MVAr. Determine:

(a) Supply power factor and regulation under full load conditions.

(b) It is desired to maintain the load voltage constant at 10 kV under full load conditions using a reactive power compensator.

(i) What should be the reactive power rating of the compensator?

(ii) What is the supply power factor and its phase angle with the compensator?

(c) Suppose the compensator is employed to improve the power factor to unity under full load conditions.

(i) What should be the reactive power rating of the compensator?

(ii) What is the value of voltage regulation under full load conditions in this case?

8.3 Calculate the displacement factor, distortion factor and power factor for a fully-controlled converter for $\alpha = 60°$, assuming load current to be perfect dc.

8.4 A single phase half controlled converter employs forced commutation with a single pulse per half cycle. If the pulse is located from 30° to 120° during a half cycle, calculate the displacement factor, real and reactive fundamental power, and power factor. The source voltage is $230\sqrt{2} \sin 100 \pi t$ and the load current is 10 amps ripple-less dc.

8.5 Four stage sequence control of 1-phase fully controlled converter is often used in the main line traction requiring regenerative braking.

Draw the load voltage and source current waveforms and derive expressions for the per unit output voltage, fundamental and rms source current and power factor for the following ranges of the per unit output voltage E_{on}.

(a) $0.5 < E_{on} < 0.75$ and

(b) $0.75 < E_{on} < 1.0$.

Assume load current to be ripples dc and equal number of turns in all the transformer secondaries.

8.6 Repeat problem 8.5 if the fully controlled converters are operated with controlled flywheeling.

Chapter 10

10.1 A Binary Coded Decimal (BCD) signal appears on the four input lines of a logic circuit. It is required that the output of the circuit goes to 1 whenever the input is any one of the following decimal numbers

0, 2, 3, 5, 8

(a) Draw a *K*-map of the circuit.

(b) Design the above circuit using TTL NAND gates and inverters.

(c) Design the above circuit using 4 input multiplexers (Fig. 10.16b).

(d) Compare the number of chips required in (b) and (c).

10.2 Consider the circuit of Figure 10.7(a). Let a load resistance of 1K be connected between the output and the ground. For the following inputs determine the states (i.e. off/active/on) for the transistors T_2, T_3 and T_4. $ABC = 000, 001, 111$. Assume $\beta = 50$, $V_{CB} = -0.5$ in saturation and $V_{BE} = V_d = 0.5$ V at cutin.

10.3 The circuit shown in Fig. P 10.1 is claimed to be a F/F. Label the terminals and determine its truth table.

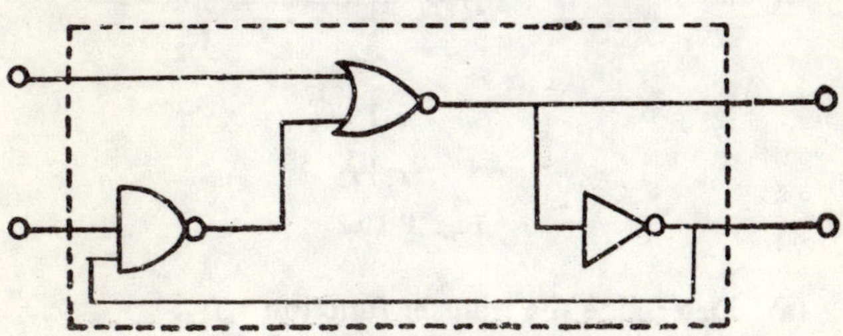

Fig. P 10.1

10.4 Design a synchronous three-state Up Down counter using *J—K F/Fs*. The counter has two inputs which control its operation as follows.

Up	Down	Function
0	0	Holds count
0	1	Counts down
1	0	Counts up
1	1	Reset to zero

The counter should not lose count while changing modes and should reset itself if found in an unused state.

10.5 You are given a *T—F/F*. It has the characteristic that $Q_{n+1} = Q_n$ if $T = 0$ and $Q_{n+1} = \bar{Q}_n$ if $T = 1$ on falling edge of the clock. Show how this *F/F* and a combinatorial circuit can be used to obtain a *S—R F/F*.

10.6 A ripple counter is to operate at a maximum frequency of 10 MHz. The propagation delay of each *F/F* is 10 ns. What is the maximum count the counter can provide?

10.7 The low pass Sallen and Key circuit is shown in Fig. P 10.2. Assume the op-amps to be ideal.

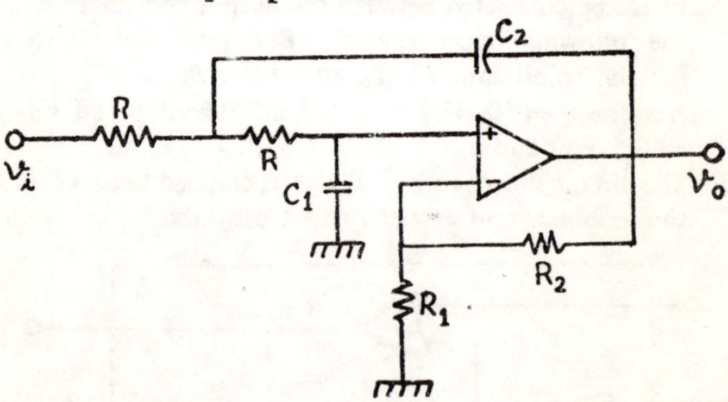

Fig. P 10.2

(a) Determine the transfer function v_0/v_1.

(b) Calculate the values of components to obtain,

(i) D.C. gain of 11,
(ii) Critically damped response,
(iii) Corner frequency of 100 kHz.

(c) Sketch the Bode Plot of (v_0/v_i) for the values determined in (b).

10.8 The All Pass Networks shown in Fig. 10.72 have the following values of the components.
$R = R_1 = 10$ K, $C = 1$ μF for (a),
$R' = 33$ K, $C = 0.22$ μF, $R = 10$ K for (b).
Plot the phase shift between v_0 and v_i as a function of input frequency.

10.9 The Howland current source is shown in Fig. P 10.3.

(a) Determine I_L as a function of v_1 and v_2. Assume the op-amp to be ideal.

(b) Now assume the op-amp has an offset voltage of 5 mV but is otherwise ideal. What is the error introduced in I_L due to this imperfection, as a function of α.

(c) If the error determined in (b) is to be $\triangleleft$ 5 μA, what is the minimum acceptable value of α.

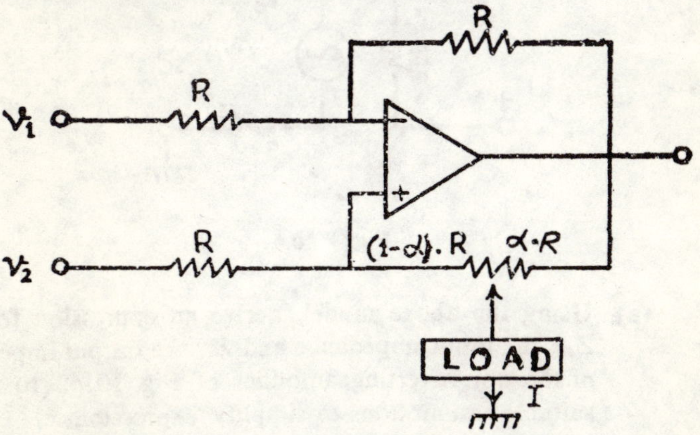

Fig. P 10.3

10.10 The open loop gain of an op-amp is given by the following equation

$$A(s) = \frac{7 \times 10^5}{(1 + S)(1 + 0.7 \times 10^{-6} S)}$$

The op-amp is connected as a 'voltage follower', ($Z_F \rightarrow 0$,

$Z_1 \to \infty$ in Fig. 10.41).

(a) Determine the expression for the gain v_0/v_i of the follower.

(b) Calculate the damping factor and the natural frequency of the system.

(c) Draw Bode Plot of the gain (v_0/v_i).

10.11 (a) Determine the gain (v_0/v_i) of the circuit shown in Fig. 10.57 over the frequency range of 2 Hz to 20 MHz. Let $C_1 = 2\ \mu F$, $R_1 = 2.7$ K, $C_2 = 0.02\ \mu F$, $R_2 = 2.7$ K, $A(s) = 10^6/(1 + 0.05s)$.

(b) Sketch the Bode plot of the gain calculated in (a).

(c) Identify the function this circuit performs in different frequency ranges.

10.12 A commonly used model for the op-amp is the VCCS with finite input and output resistances as shown in Fig. P 10.4.

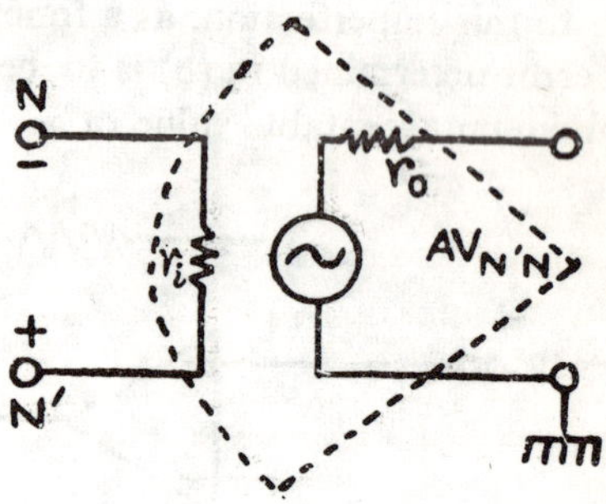

Fig. P 10.4

(a) Using the above model, derive an expression for the Z_{in}, the input impedance and Z_o, the output impedance of the non-inverting amplifier of Fig. 10.42 (b). Make suitable assumptions to simplify expressions.

(b) Let the amplifier be constructed using 741C op-amp having the following specifications.

$$r_i = 1M\Omega,\ r_o = 75\Omega,\ R_1 = R_2 = 10K\Omega$$

$$A(s) = a_0\,(1 + s/\omega_0)^{-1},\ a_0 = 100\ \text{db},\ f_0 = \omega_0/2\pi = 10\ \text{Hz}.$$

Evalute Z_{in} and Z_0 at 1 KHz using results of (a).

10.13 (a) For the circuit shown in Fig. P 10.5, calculate the gain v_o/v_i assuming the op-amp to be ideal.

(b) Calculate the natural frequency and the damping factor of the system for the following values.

$R_1 = 11\,\text{K}$, $R_2 = 25\,\text{K}$, $C_1 = 0.001\ \mu\text{F}$, $C_2 = 0.1\ \mu\text{F}$.

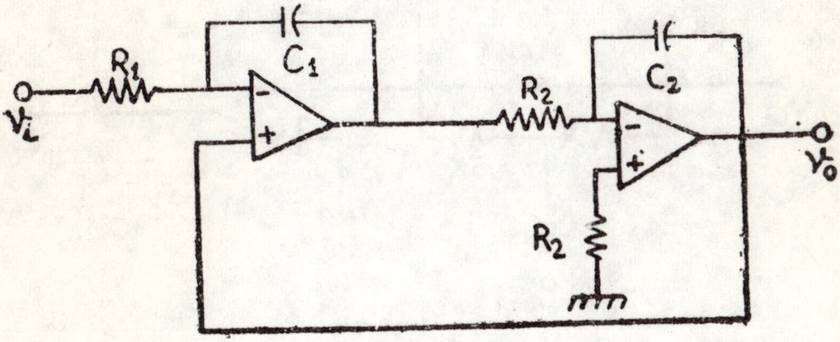

Fig. P 10.5

10.14 A certain feedback amplifier is modelled as shown in Fig. P 10.6. Assume the op-amp to be ideal.

(a) Calculate the gain v_o/v_i.

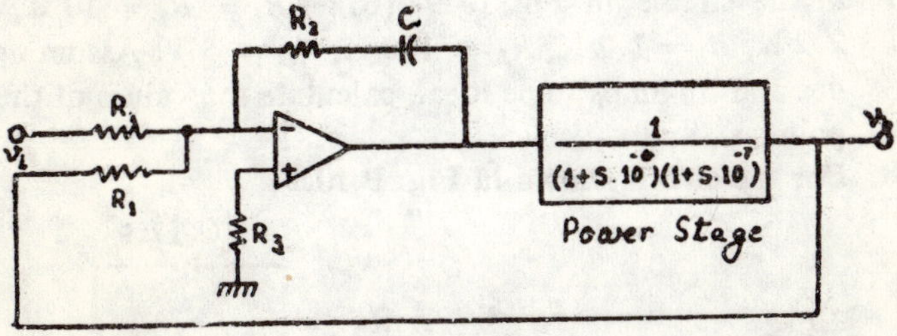

Fig. P 10.6

(b) Let $R_1 = 10\text{K}$. Calculate values of R_2 and C so that the response to a step input is critically damped.

10.15 Draw the v_o vs. v_{in} characteristic of the diode function generator shown in Fig. P 10.7 Assume the op-amps and the diodes to be ideal.

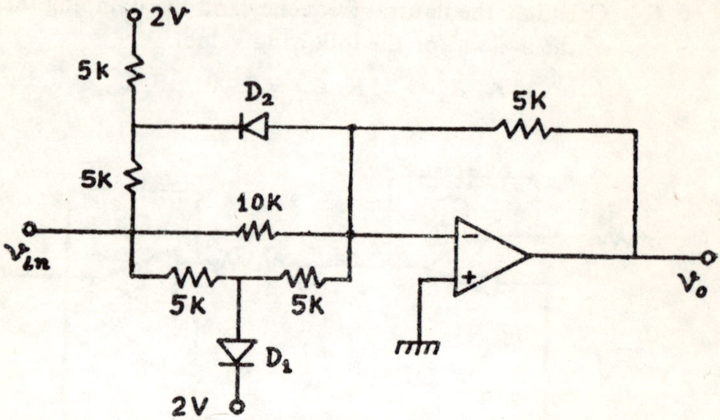

Fig. P 10.7

10.16 Consider the inverting comparator of Fig. 10.93(c). Let $V_{ref} = 0$, $R_1 = R_2 = 10K$. Assume that the op-amp saturates at $\pm$ 10 V. In the linear region the op-amp has a gain A given by,

$A = a_0/(1+S/\omega_0)$ where $a_0 = 100$ db, $\omega_0 = 10$ rad/sec.

The op-amp is in the state $V_{in} = 0$, $V_0 = +$ 10 V at $t = 0$. A step input of + 10 V is applied at $t = 0$. Calcalate the response of output voltage v_0. Ignore slew rate limitations.

10.17 In the circuit of Fig. 10.94 (b) let $R_1 = R_2 = 10$ K, $R_3 = 2.2$ K, $R = 2.2$ K, $R_L = 100$ K, $V_B = 5$ V. Assuming diodes and op-amps to be ideal, calculate the values of the trip-points v_{t_1} and v_{t_2}.

10.18 For the circuit shown in Fig. P 10.8.

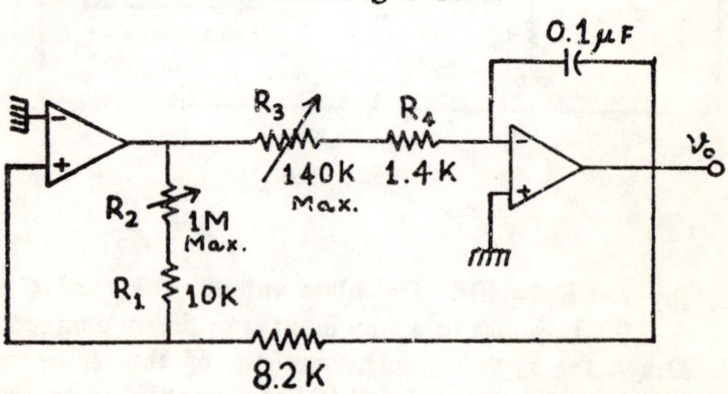

Fig. P 10.8

(a) Draw the waveform of voltage v_0 for $R_3 = 100$ K, $R_2 = 500$ K.

(b) Determine the effects of varying R_3 and R_2.

10.19 In the circuit of Fig. 10.129, $R = 7.5$ K, $V_{ref} = 1.6$ V. Calculate the output voltage v_0 and the current supplied by the source V_{ref} when $x_3\, x_2\, x_1\, x_0 = 0101$.

10.20 A multiplying DAC is connected as shown in Fig. 10.9. If the digital input X is the binary equivalent of decimal no. N, calculate the ratio (v_0/v_{in}).

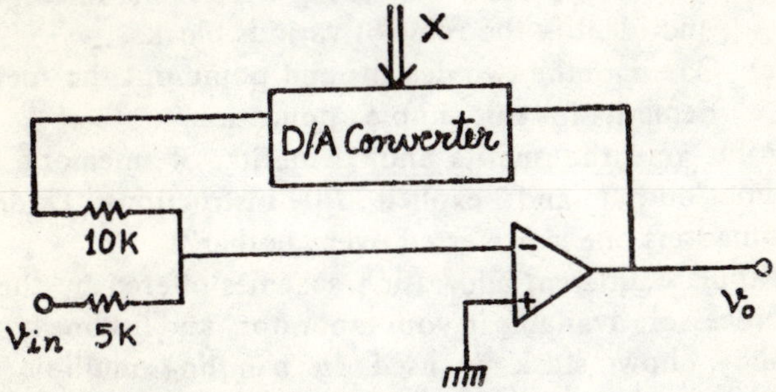

Fig. P. 10.9

Chapter 11

11.1 Convert the following decimal numbers to binary, octal and hexadecimal. Check your answer by reconverting them to decimal.

(i) 925.5625 (ii) 79.05 (iii) 128.625

11.2 For a 16 bit microcomputer, list the range of signed integers that a word can hold while using

(a) sign-magnitude form,

(b) one's complement form and

(c) two's complement form.

11.3 Think of the ways in which real numbers can be represented on a microcomputer and discuss their relative merits or demerits.

11.4 Consider an electrically heated furnace for which an electronic control mechanism is to be designed. The furnace has six doors, four of which are used for maintenance purposes and must be closed while in operation. Out of the other two doors only one should be open during operation. The operator enters the temperature through 'thumb-wheel' switches with decimal settings. The thermocouple reads the current temperature. It is desired that the control should raise alarm when any of the maintenance doors is open and raise another alarm if both the operation doors are open. It should generate a desired heater current by multiplying a constant K to the error in temperature.

(a) Design the control system using a PLC. Identify additional hardware blocks that may be required and list their roles.

(b) Design the controller using a suitable microcomputer and identify the roles of various blocks.

(c) Compare the two designs and point out the merits and demerits for this simple application.

11.5 Point out the merits and demerits of memory mapped input/output and explicit $I/0$ instructions. Under what situations one is preferred over another?

11.6 Tabulate different addressing schemes offered by the microprocessors available in your laboratory and list one typical use.

11.7 Show how stack is used in handling multiple priority interrupts.

11.8 Redesign the speed controller of Section 11.7.2 using another microprocessor available in your laboratory. List the major difficulties encountered and their possible solutions.

11.9 For the example given in Section 11.7.3, explore other possible ways in which interrupt generation and processing could be organized. Do you see any advantage in using a 16 bit processor (8086 or 68000) for this application?

11.10 Write programs for generating following waveforms using a microcomputer.

(a) triangular waveform

(b) trapezoidal waveform

(c) saw-tooth waveform

11.11 Reading exercises:

(a) B.W. Williams, Microprocessor control of DC 3 phase thyristor inverter circuit, IEEE Trans. on Ind. Electr. Control and Instr., vol. 27, No. 3, Aug. 1980.

(b) Y.T. Chan, Adam J. Chmiel, A Microprocessor based current controller for SCR-DC motor drives, IEEE Trans. on Ind. Electr. Control Instr., vol. 27, No. 3, August, 1980.

(c) S.K. Tso and P.T. Ho, Dedicated microprocessor scheme for thyristor phase control of multiphase converters, IEE Proc., vol. 128, pt. B, No. 2, March 1981.

(d) G. Olivier, V.R. Stefanovic and G.E. April, Microprocessor Controller for a thyristor converter with an improved power factor, IEEE Trans. on Ind. Electr. and Control Instr., Aug. 1981.

(e) N. Matsui and Y. Ohashi, A microcomputer based commutation failure free control system for the cascaded rectifier circuit, IEEE Trans. on Ind. Electr. and control Instr., vol 28, No. 4 Nov. 1981.

(f) Pei-Chong Tang, Shui-Shong Lu and Yung-Chun Wu, Microprocessor based design of a firing circuit for three phase full wave thyristor dual converter, IEEE Trans. on Ind. Electr., vol. IE–29, No. 1, Feb. 1982.

6.5 69 μs, values reasonable.
6.6 Hint: Integrate waveforms.
6.7 Hint: Write differential eqns.
6.8 (a) 132 A, (b) 18.65, 2.9, 1.29 A, (c) 133.35 A, (d) 165.4 and 99.94 A, (e) 53.63 A, (f) 11.798, 2.904, 8.894 kW, (g) 133.37 A.

Chapter 7

7.1 (i) 70.7 A (ii) 230 V in the forward direction.
7.2 Fundamental: 73.21 A rms, third harmonic: 2.52 A rms, fifth harmonic 0.855 A rms.
7.4 2.496 μf.
7.5 Yes.
7.6 $C = 24.266\,\mu$F, $L = 36.524\,\mu$H.
7.7 $L = 25\,\mu$H, $C = 50\,\mu$F, no unique solution.
7.8 $L = 40\,\mu$H, $C = 50\,\mu$F, no unique choice.
7.10 (a) $6 \leqslant R \leqslant \infty\,\Omega$, (b) 11.11% for $R \geqslant 63\,\Omega$, (c) Yes, for $8 \leqslant R \leqslant \infty\,\Omega$, (d) Power falls as R increases (e) 200 V.
7.11 (a) $7.2 < R < \infty\,\Omega$ (b) 10% (c) 1347 W (d) Power reduces monotonically as R is increased (e) 200 μS, 160 V.
7.12 (a) 119.2 V, 1.34 mS (b) 48.3 Hz (c) 79 Hz (d) 11.92 A/μs, Yes (e) 90 μF
7.14 The components of i_{dc} are of type $60n$ and the components of i_{ac} are of type 50, $(6\beta \pm 1)50$, $(6\beta \pm 1)50 \pm 60n$, Hz, where $n = 0, 1, 2, 3, 4, 5, \ldots$ $\beta = 1, 2, 3, 4, 5, \ldots$.
7.15 (a) 198.85 V (b) 212.1 V (c) $102.68 \angle -7.2°$ (d) 46.22 A, 72.61 A (e) 10.46 kW.
7.16 (a) 6.865 KHz (b) 240.4 V (c) 38.75 A (d) 1.98 kW (e) 15.73 A.
7.17 (a) $L = 49.7\,\mu$H, $C = 10.3\,\mu$F (b) 3.1 kW.

Chapter 8

8.1 (a) $L = 70.06$ mH, $C = 144.75\,\mu$F (b) $\alpha' = 23.5°$ (c) 10 m sec. (d) Time delay in measurement and control circuitry.
8.2 (a) 0.447 lagging, 47.45% (b) (i) — 56.35 MVAr (ii) 0.969 leading, 14.3° (c) (i) — 50 MVAr (ii) 2.5%
8.3 0.5, 0.9, 0.45.
8.4 0.96, 1.2665 kW, 0.732 kVAr (leading), 0.86.

Chapter 10

10.1 (d) Gate: 2 chips, $MUX = 2\frac{1}{2}$ chips.

10.2 For $ABC = \phi\phi\phi, \phi\phi 1, 111$, $T_2T_3T_4$ = off/active/off, off/active/off, sat/off/sat. respectively.

10.3 Input TOP = S, Bottom = $\bar{R}$, Output TOP = $\bar{Q}$, Bottom = Q, Set (S) has priority over Reset ($\bar{R}$).

10.4 $J_A = \bar{Q}_B\bar{U}D + Q_BU\bar{D}$, $J_B = \bar{Q}_AU\bar{D} + Q_A\bar{U}D$, $\bar{K}_A = \bar{U}\bar{D}Q_A\bar{Q}_B$, $\bar{K}_B = \bar{U}\bar{D}\bar{Q}_AQ_B$.

10.6 512.

10.7 (b) $R_1 = 1$ K, $R_2 = 10$ K, $C_1 = 4.44$ nF, $C_2 = 0.57$ nF.

10.9 (a) $I_L = (v_2 - v_1)/\alpha R$ Amp (b) $I_L = 1/\alpha$ μA (c) 0.2.

10.12 (b) $Z_{in} = 500$ MΩ $Z_0 = 150$ mΩ.

10.13 (b) 960 Hz, 0.066.

10.14 (b) $R_1 = 10$ KΩ, $R_2 = 25$ KΩ, $C = 40$ pf.

10.16 $V_0 = 20 - 10 \exp(.5 \times 10^6 t)$, $0 \leqslant t \leqslant 2.2$ μs.

10.17 $v_{i1} = 2.5$ V, $v_{i2} = -1.1$ V.

10.18 Increasing R_2 reduces amplitude, Increasing R_3 increases time period.

10.19 0.333 V, 0.124 mA.

10.20 $(-2/NK)$ K is DAC gain constant.

INDEX